A BRIEF TABLE OF INTEGRALS* (continued)

$$\int u \sin u \, du = \sin u - u \cos u. \qquad \int u^n \sin u \, du = -u^n \cos u + n \int u$$

$$\int u \cos u \, du = \cos u + u \sin u. \qquad \int u^n \cos u \, du = u^n \sin u - n \int u^{n-1} s.$$

$$\int e^{au} \sin nu \, du = \frac{e^{au}(a \sin nu - n \cos nu)}{a^2 + n^2}. \qquad \int e^{au} \cos nu \, du = \frac{e^{au}(a \cos nu + n \sin nu)}{a^2 + n^2}.$$

$$\int \sin au \sin bu \, du = -\frac{\sin(a + b)u}{2(a + b)} + \frac{\sin(a - b)u}{2(a - b)}, \qquad a^2 \ne b^2.$$

$$\int \cos au \cos bu \, du = \frac{\sin(a + b)u}{2(a + b)} + \frac{\sin(a - b)u}{2(a - b)}, \qquad a^2 \ne b^2.$$

$$\int \sin au \cos bu \, du = -\frac{\cos(a + b)u}{2(a + b)} - \frac{\cos(a - b)u}{2(a - b)}, \qquad a^2 \ne b^2.$$

$$\int \sinh u \, du = \cosh u. \qquad \int \cosh u \, du = \sinh u.$$

$$\Gamma(t) = \int_0^\infty e^{-u} u^{t-1} \, du, \quad t > 0; \qquad \Gamma(\tfrac{1}{2}) = \sqrt{\pi}; \quad \text{and} \quad \Gamma(n + 1) = n!, \text{ if } n \text{ is a positive integer.}$$

SOME POWER SERIES EXPANSIONS

$$f(x) = f(a) + (x - a)f'(a) + \frac{(x - a)^2}{2!} f''(a) + \cdots + \frac{(x - a)^n}{n!} f^{(n)}(a) + \cdots \qquad \text{(Taylor series)}$$

$$e^x = \sum_{n=0}^\infty \frac{x^n}{n!} \qquad \sin x = \sum_{n=0}^\infty \frac{(-1)^n x^{2n+1}}{(2n + 1)!} \qquad \cos x = \sum_{n=0}^\infty \frac{(-1)^n x^{2n}}{(2n)!}$$

$$(1 - x)^{-1} = \sum_{n=0}^\infty x^n \qquad (1 - x)^{-2} = \sum_{n=0}^\infty (n + 1)x^n \qquad \ln(1 - x) = -\sum_{n=1}^\infty \frac{x^n}{n}$$

$$\tan x = x + \tfrac{1}{3}x^3 + \tfrac{2}{15}x^5 + \tfrac{17}{315}x^7 + \tfrac{62}{2835}x^9 + \cdots$$

$$\arcsin x = x + \frac{x^3}{2 \cdot 3} + \frac{1 \cdot 3}{2 \cdot 4 \cdot 5} x^5 + \frac{1 \cdot 3 \cdot 5}{2 \cdot 4 \cdot 6 \cdot 7} x^7 + \cdots \qquad \arctan x = \sum_{n=0}^\infty (-1)^n \frac{x^{2n+1}}{2n + 1}$$

$$J_0(x) = \sum_{k=0}^\infty \frac{(-1)^k x^{2k}}{(k!)^2 2^{2k}} \qquad J_1(x) = \sum_{k=0}^\infty \frac{(-1)^k x^{2k+1}}{k!(k + 1)! 2^{2k+1}} \qquad J_n(x) = \sum_{k=0}^\infty \frac{(-1)^k x^{2k+n}}{k! \Gamma(n + k + 1) 2^{2k+n}}$$

* *Note:* An arbitrary constant is to be added to each formula.

D0220665

THIRD EDITION

FUNDAM
OF DIFF
EQUA

- **THIRD EDITION**

FUNDAMENTALS
OF DIFFERENTIAL
EQUATIONS

R. KENT NAGLE
EDWARD B. SAFF

University of South Florida

Addison-Wesley Publishing Company

READING, MASSACHUSETTS • MENLO PARK, CALIFORNIA
NEW YORK • DON MILLS, ONTARIO • WOKINGHAM,
ENGLAND • AMSTERDAM • BONN • SYDNEY • SINGAPORE
TOKYO • MADRID • SAN JUAN • MILAN • PARIS

Sponsoring Editor: Jerome Grant
Production Supervisor: Peggy McMahon
Marketing Manager: Andrew Fisher
Manufacturing Supervisor: Roy Logan
Cover Designer: Hannus Design Associates
Editorial/Production Services: The Wheetley Company, Inc.

Reprinted with corrections, May 1994.

Library of Congress Cataloging-in-Publication Data

Nagle, R. Kent.
    Fundamentals of differential equations/by R. Kent
    Nagle, Edward B. Saff.—3rd ed.
       p. cm.
    Includes index.
    ISBN 0-8053-5056-X
    1. Differential equations.   I. Saff, E. B., 1944—
II. Title.
QA371.N24 1993   515'.35-dc20   93-24033
ISBN 0-8053-5056-X

4 5 6 7 8 9 10-DO-9594

To our families,
   who endured our late nights,
   soothed our anxieties,
   and shared our enthusiasm.

Sandy   **Loretta**
Kevin, Jeffrey   Lisa, Tracy, Alison

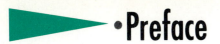 **·Preface**

OUR GOAL

While an introductory course in differential equations is a mainstay of undergraduate curricula in the sciences and engineering, the flavor of such a course varies considerably because of departmental requirements, instructors' tastes, and students' mathematical maturities and backgrounds. Our goal has been to write a flexible *one-semester* text that spans a variety of topics in the basic theory as well as applications of differential equations. At the same time we have striven to make the text "user friendly" through various design features.

PREREQUISITES

While some universities make *linear algebra* a prerequisite for differential equations, many schools (especially engineering) only require calculus. With this in mind, we have designed the text so that only Chapter 6 (Higher Order Linear Differential Equations) and Chapter 10 (Matrix Methods for Linear Systems) require more than high school linear algebra. Moreover, Chapter 10 contains a review section on matrices and vectors as well as specific references for the deeper results used from the theory of linear algebra. We have also written Chapter 9 (Systems of Differential Equations and Their Applications) so as to give an introduction to systems of differential equations—including methods of solving, applications, numerical procedures, phase plane analysis, and Poincaré maps—that does not require a background in linear algebra.

IMPROVEMENTS IN THIS THIRD EDITION

● **New Sections**   Three new sections have been added to the text; namely,

§4.9 Undetermined Coefficients Using Complex Arithmetic
§7.1 Introduction: A Simple Electrical Circuit
§9.7 Dynamical Systems, Poincaré Maps, Strange Attractors, and Chaos.

Engineering oriented students will find in the new optional Section 4.9 a streamlined approach to solving undetermined coefficients problems—one that appeals to the notions of complex impedance and phase angles. The new introduction, Section 7.1, to the Laplace transforms chapter emphasizes the advantages of the transform method in handling discontinuous forcing functions. The added Section 9.7 describes the Poincaré method for nonautonomous systems and provides a glimpse of several topics of current research activity. There we explain how chaotic systems arise in physical phenomena and emphasize the role of numerical computations in analyzing dynamical systems.

- **Technical Writing Exercises**   Communications skills are, of course, an essential aspect of professional activities. Yet few texts provide opportunities for the reader to develop these skills. Thus we have added at the end of most chapters a set of clearly marked technical writing exercises which invite students to make documented responses to questions dealing with the concepts in the chapter. In so doing, students are encouraged to make comparisons between various methods and to present examples that support their analysis.

- **New Group Projects**   Since they were well received in the first two editions, we have added several new Group Projects at the ends of the chapters. They are

    Chapter 1:   *Magnetic "Dipole"*
    Chapter 2:   *Designing a Solar Collector*
    Chapter 4:   *Convolution Method, Nonlinear Equations Solvable by First Order Techniques*
    Chapter 6:   *Transverse Vibrations of a Beam*
    Chapter 8:   *Airy's Equation, Buckling of a Tower*
    Chapter 9:   *Strange Behavior of Competing Species-Part I*
    Chapter 10:   *Strange Behavior of Competing Species-Part II*
    Chapter 11:   *Green's Function, Numerical Methods for $\Delta u = f$ on a Rectangle.*

- **Use of Computer Software**   The availability of computer packages such as MATHEMATICA®, DERIVE®, and MAPLE® provide an opportunity for the student to conduct numerical experiments and tackle realistic applications that give additional insights into the subject. Consequently, we have added several exercises and projects throughout the text that are designed for the student to employ available software in phase plane analysis, eigenvalue computations, and the numerical solutions of various equations.

- **New Problems**   A variety of new problems have been added to the exercises. Most of the additions are related to applications or the use of commercially available computer software.

- **Other Changes**   The instructor familiar with the second edition will notice that the discussion in Sections 7.7 and 7.8 concerning the impulse response function has been significantly expanded in order to emphasize its relationship to the general solution of an initial value problem using convolution. In addition, the discussion of Cauchy-Euler equations has been dispersed throughout Chapter 4 and not confined to a separate section. Furthermore, nonlinear second order equations solvable by first order techniques is now discussed in a group project at the end of Chapter 4 and no longer is a separate section.

FEATURES FOR THE INSTRUCTOR:

- **Choice of Applications**   Because of syllabus constraints, some courses will have little or no time for sections (such as those in Chapters 3, 5, and 9) that exclusively deal with applications. Therefore, we have made the sections in these chapters almost completely independent of each other. To afford the instructor even greater flexibility, we have built in a variety of applications in the exercises for the theoretical sections. In addition, we have included several projects that deal with such applications as aquaculture, designing a solar collector, and cleaning up the Great Lakes.

- **Linear Theory**   We have developed the theory of linear differential equations in a gradual manner. In Chapter 4 (Linear Second Order Equations) we present the basic theory for linear second order equations and discuss various techniques for solving these equations. Higher order equations are briefly mentioned in this chapter. A more detailed discussion of linear higher order differential equations is given in Chapter 6 (Higher Order Linear Differential Equations). For a beginning course emphasizing methods of solution, the presentation in Chapter 4 is sufficient and Chapter 6 can be skipped.

- **Proofs**   While more pragmatic students may balk at proofs, most instructors regard these justifications as an essential ingredient in a textbook on differential equations. As with any text at this level, certain details in the proofs must be omitted. When this occurs, we flag the instance and refer readers to either a problem in the exercises or to another text. For convenience, the end of a proof is marked by the symbol ◄◄◄ .

- **Exercises**   An abundance of exercises is graduated in difficulty from straightforward, routine problems to more challenging ones. Deeper theoretical questions, along with applications, usually occur toward the end of exercise sets. Throughout the text we have included problems and projects that require the use of a microcomputer. These exercises are denoted by the symbol

- **Group Projects**   At the end of each chapter are group projects relating to the material covered in the chapter. A project might involve a more challenging application, delve deeper into the theory, or introduce more advanced topics in differential equations. Although these projects can be tackled by an individual student, classroom testing has shown that working in groups lends a valuable added dimension to the learning experience. Indeed, it simulates the interactions that take place in the professional arena.

- **Optional Sections**   Several sections of the text are labeled (Optional). These sections can be omitted without affecting the logical development of the material. As mentioned earlier, the sections in Chapters 3, 5, and 9 are almost completely independent of each other.

- **Laplace Transforms**   We provide a detailed chapter on Laplace transforms since this is a recurring topic for engineers. Our treatment emphasizes discontinuous forcing terms and includes a section on the Dirac delta function.

- **Power Series**   Power series solutions is a topic that occasionally causes student anxiety. Possibly, this is due to inadequate preparation in calculus where the more subtle subject of convergent series is (not infrequently) covered at a rapid pace. Our solution has been to provide a thorough treatment of power series solutions that also includes a review of their properties as well as a discussion of real analytic functions. Unlike many texts, we have provided an extensive section on the *method of Frobenius* (Section 8.6) and two sections on the various methods for finding a second linearly independent solution.

   While we have given considerable space to power series solutions, we have also taken great care to allow for the instructor who only wishes to give a basic introduction to the topic. *An introduction to solving differential equations using power series and the method of Frobenius can be accomplished by covering the materials in Section 8.3, Section 8.6, and part of Section 8.7.*

- **Partial Differential Equations**   An introduction to this topic is provided in Chapter 11, which covers the method of separation of variables, Fourier series, the heat equation, the wave equation, and Laplace's equation. Examples in two and three dimensions are included.

- **Syllabi**   As a rough guide in designing a syllabus related to the text, we provide three sample syllabi that can be used for a 15-week course that meets three hours per week: the first emphasizes methods and applications, the second theory and methods, and the third methods and partial differential equations. Chapters 1, 2, and 4 provide the core for any course. The rest of the chapters are, for the most part, independent of each other.

Week	Methods and Applications Sections	Theory and Methods Sections	Methods and Partial Differential Equations Sections
1	1.1, 1.2, 1.3	1.1, 1.2, 2.2	1.1, 1.2, 2.2
2	2.2, 2.3	2.3, 2.4	2.3, 2.4
3	2.4, 2.6, 3.1	2.6, 4.2	4.2, 4.3
4	3.2, 3.4	4.3, 4.4, 4.5	4.4, 4.5
5	4.2, 4.3, 4.5	4.6, 4.7	4.6, 4.7
6	4.6, 4.7	4.8, 4.10	4.8, 4.10
7	4.8, 4.10	6.2	5.1, 5.2, 5.3
8	5.1, 5.2, 5.3	6.3, 6.4	7.2, 7.3
9	7.2, 7.3	6.5, 8.3	7.4, 7.5
10	7.4, 7.5	8.4, 8.5	7.6, 8.2
11	7.6, 8.2	8.6	8.3, 8.6
12	8.3, 8.6	8.7, 10.2	8.7, 9.2
13	8.7, 9.2	10.3, 10.4	9.4, 11.2
14	9.3, 9.4	10.5	11.3, 11.4
15	9.6	10.6	11.5, 11.6, 11.7

FEATURES FOR THE STUDENT:

- **Procedure Boxes**   Step-by-step procedure boxes shaded in color are provided as a convenient summary and a readily accessible reference. However, the student should be cautioned not to memorize or rely solely on these outlines. Indeed, they are no substitute for a solid understanding of the subject.

- **Worked Examples**   A substantial number of worked-out examples illustrate the various difficulties one might encounter in carrying out the procedures. Introductory examples are treated in greater detail. For easy identification, examples are set off by colored disks ●.

- **Numerical Algorithms**   Several numerical methods for approximating solutions to differential equations are presented along with program outlines that are easily implemented on a microcomputer. These methods are introduced early in the text so that teachers and/or students can use them for numerical experimentation and for tackling complicated applications.

- **Computer Graphics**   Most of the figures in the text were generated on a microcomputer with a laser printer. Computer graphics not only ensure greater accuracy in the illustrations, they demonstrate the use of numerical experimentation in studying the behavior of solutions.

- **Historical Footnotes**   Throughout the text historical footnotes are set off by colored daggers (†). These footnotes typically provide the name of the person who developed the technique, the date, and the context of the original research.

- **Motivating Problem**   Most chapters begin with a discussion of a problem from physics or engineering that motivates the topic presented and illustrates the methodology.

- **Modern Applications**   Aside from standard applications such as Newtonian mechanics, electric circuits, and population models, we have included material on mathematical modeling, aquaculture, pollution, heating and cooling of buildings, combat models, frequency response modeling, and difference equations.

- **Chapter Summary and Review Problems**   All of the main chapters contain a set of review problems along with a synopsis of the major concepts presented.

SUPPLEMENTS

Two supplements are available from Addison-Wesley. They are

- *Student's Solutions Manual to Fundamentals of Differential Equations* by John A. Banks, Jessica M. Craig, R. Kent Nagle and Edward B. Saff

- *Instructor's Guide and Answer Book to Fundamentals of Differential Equations* by R. Kent Nagle and Edward B. Saff

An expanded version of this text entitled *Fundamentals of Differential Equations and Boundary Value Problems* contains three additional chapters: Chapter 12, *Boundary Value Problems*; Chapter 13, *Stability of Autonomous Systems*; and Chapter 14, *Existence and Uniqueness Theory*.

ACKNOWLEDGMENTS

The staging of this text involved considerable behind-the-scenes activity. We want to thank Frank Glaser (California State Polytechnic University, Pomona) for many of the historical footnotes. We are indebted to Herbert E. Rauch (Lockheed Research Laboratory) for help with Section 3.3 on heating and cooling of buildings, Project B in Chapter 3 on aquaculture, and other application problems. Our appreciation goes to George Fix and R. Kannan (University of Texas, Arlington) for their useful suggestions concerning Section 3.7. We give special thanks to Richard H. Elderkin (Pomona College), Jerrold Marsden (University of California, Berkeley), T. G. Proctor (Clemson University), and Philip W. Schaefer (University of Tennessee) who read and reread the manuscript for the first edition making numerous suggestions that greatly improved the book. We also extend thanks to the many people who reviewed the manuscripts for the third edition:

Randy Campbell-Wright	University of Tampa
Conduff G. Childress, Jr.	Shaw University
Bruce Edwards	University of Florida
Gurcharan S. Gill	Brigham Young University
Alan Gorfin	Western New England College
Vera Granlund	University of Virginia
Ronald Guenther	Oregon State University
Greg Henderson	University of Tampa
Alan M. Johnson	University of Arkansas at Little Rock
Allan M. Krall	Pennsylvania State University
Melvin D. Lax	California State University at Long Beach
Xin Li	University of Central Florida
Jerrold Marsden	University of California, Berkeley
Leroy F. Meyers	Ohio State University
G. Paul Neitzel, Jr.	Georgia Institute of Technology
Carol S. O'Dell	Elizabeth City State University
Mary Parrott	University of South Florida
Samuel M. Rankin, III	Worcester Polytechnic Institute
Klaus Schmidt	University of Utah
Donald C. Solmon	Oregon State University
Michael Stecher	Texas A&M University
Howell K. Wilson	Southern Illinois University at Edwardsville
Jet Wimp	Drexel University
Yuncheng You	University of South Florida

We are grateful for the suggestions and encouragement we received from our students here at the University of South Florida, especially William Albrecht, Mark Clark, Jessica Craig, Vladamir Veselov, Jun Cao, Yanmu Zhou, Kurt Van Etten, William Hughes, David Kaplan, Laura Kneeberg, Brian Melloy, Rocky Rathgeber, Mehrdad Simkani, and Zachariah Sinkala who helped us obtain correct answers to the problems. The credit for the computer graphics goes to Hao Nguyen, Rafael Muñoz, Charles Godfrey and Ron Yoder. While several typists including Mary Baroli, Selma Canas, Carol Crosson, Jo Ann Dennison, and Loretta Saff were extremely helpful, the majority of the typing of the first edition was done by Sandy Nagle, who spent many late nights working on the manuscript. To Sandy we give our heartfelt thanks.

Finally, we want to thank the staff at Addison-Wesley for their dedicated assistance. Special kudos go to mathematics editor Jerome Grant and production supervisor Peggy McMahon.

R. Kent Nagle
Edward B. Saff
Tampa, Florida

Contents

*Denotes optional sections that can be deleted without compromising the logical flow.

• 3 • MATHEMATICAL MODELS AND NUMERICAL METHODS INVOLVING FIRST ORDER EQUATIONS   70

• 4 • LINEAR SECOND ORDER EQUATIONS   129

•5• APPLICATIONS AND NUMERICAL METHODS FOR SECOND ORDER EQUATIONS   198

•6• HIGHER ORDER LINEAR DIFFERENTIAL EQUATIONS   242

•7• LAPLACE TRANSFORMS   275

•8• SERIES SOLUTION OF DIFFERENTIAL EQUATIONS   344

•9• SYSTEMS OF DIFFERENTIAL EQUATIONS AND THEIR APPLICATIONS 419

●10● MATRIX METHODS FOR LINEAR SYSTEMS   478

●11● PARTIAL DIFFERENTIAL EQUATIONS   533

CHAPTER 1

Introduction

1.1 BACKGROUND

In the sciences and engineering, mathematical models are developed to aid in the understanding of physical phenomena. These models often yield an equation that contains some derivatives of an unknown function. Such an equation is referred to as a **differential equation.** Two examples of models developed in calculus are the free fall of a body and the decay of a radioactive substance.

In the case of free fall, an object is dropped from a certain height above the ground and falls under the force of gravity.[†] Newton's second law, which states that an object's mass times its acceleration equals the total force acting on it, can be applied to the falling object. This leads to the equation

$$m\frac{d^2h}{dt^2} = -mg,$$

where m is the mass of the object, h is its height above the ground, d^2h/dt^2 is its acceleration, g is a constant, and $-mg$ is the force due to gravity.

Fortunately, the above equation is easy to solve for h. All we have to do is divide by m and integrate twice with respect to t. That is,

$$\frac{d^2h}{dt^2} = -g,$$

[†] We are assuming here that gravity is the *only* force acting on the object and that this force is constant. More general models would take into account other forces, such as air resistance.

so

$$\frac{dh}{dt} = -gt + c_1,$$

and

$$h = \frac{-gt^2}{2} + c_1 t + c_2.$$

The constants of integration, c_1 and c_2, can be determined if we know the initial height and the initial velocity of the object. We then have a formula for the height of the object at time t.

In the case of radioactive decay, we begin from the premise that the rate of decay is proportional to the amount of radioactive substance present. This leads to the equation

$$\frac{dA}{dt} = -kA, \qquad k > 0,$$

where A (>0) is the unknown amount of radioactive substance present at time t and k is the proportionality constant. To solve this equation we write it in the form

$$\frac{1}{A} dA = -k \, dt$$

and integrate to obtain

$$\ln A = \int \frac{1}{A} dA = \int -k \, dt = -kt + C_1.$$

Solving for A yields

$$A = e^{\ln A} = e^{-kt} e^{C_1} = Ce^{-kt},$$

where C is the new constant e^{C_1}. The constants C and k can be determined if the initial amount of radioactive substance and the half-life of the substance are given. We then have a formula for the amount of radioactive substance at any future time t.

Even though the above examples were easily solved by methods learned in calculus, they do give us some insight into the study of differential equations. First of all, **integration** is an important tool in solving differential equations. Second, we do not expect to get a unique solution to a differential equation since there will be arbitrary "constants of integration." Third, when a mathematical model involves the **rate of change** of one variable with respect to another, a differential equation is apt to appear. Unfortunately, in contrast to the examples for free fall and radioactive decay, the differential equation may be very complicated and difficult to analyze.

Differential equations arise in a variety of subject areas, which include not only the physical sciences but also such diverse fields as economics, medicine, psychology, and operations research. To list all the occurrences would be a Herculean task, so we shall limit the discussion to a few specific examples.

1. A classical application of differential equations is found in the study of an electric circuit consisting of resistors, inductors, and capacitors, which is driven by an electromotive force. Here an application of Kirchhoff's laws[†] leads to the equation

(1)
$$L\frac{d^2q}{dt^2} + R\frac{dq}{dt} + \frac{1}{C}q = E(t),$$

where L is the inductance, R is the resistance, C is the capacitance, $E(t)$ is the electromotive force, $q(t)$ is the charge, and t is the time.

2. In the study of the gravitational equilibrium of a star, an application of Newton's law of gravity and of the Stefan-Boltzmann law for gases leads to the equilibrium equation

(2)
$$\frac{1}{r^2}\frac{d}{dr}\left(\frac{r^2}{\rho}\frac{dP}{dr}\right) = -4\pi\rho G,$$

where P is the sum of the gas kinetic pressure and the radiation pressure, r is the distance from the center of the star, ρ is the density of matter, and G is the gravitational constant.

3. In psychology, one model of the learning of a task involves the equation

(3)
$$\frac{dy/dt}{y^{3/2}(1 - y)^{3/2}} = \frac{2p}{\sqrt{n}}.$$

Here the variable y represents the state of the learner or the learner's skill level as a function of time t. The constants p and n depend on the individual learner and the nature of the task.

4. In the study of vibrating strings and the propagation of waves, we find the partial differential equation

(4)
$$\frac{\partial^2 u}{\partial t^2} - \frac{\partial^2 u}{\partial x^2} = 0,\text{[††]}$$

where t represents time, x the location along the string, and u the displacement of the string, which is a function of time and location.

To begin our study of differential equations we need some common terminology. If an equation involves the derivative of one variable with respect to another, then the former is called a **dependent variable** and the latter is an **independent variable.** Thus, in the equation

(5)
$$\frac{d^2x}{dt^2} + a\frac{dx}{dt} + kx = 0,$$

t is the independent variable and x is the dependent variable. We refer to a and k as **coefficients** in equation (5).

[†] We will discuss the applications of Kirchhoff's laws in Sections 5.4 and 9.1.

[††] *Historical Footnote:* This partial differential equation was first discovered by Jean le Rond d'Alembert (1717–1783) in 1747.

In the equation

(6) $$\frac{\partial u}{\partial x} - \frac{\partial u}{\partial y} = x - 2y,$$

x and y are independent variables and u is a dependent variable.

A differential equation involving ordinary derivatives with respect to a single independent variable is called an **ordinary differential equation.** A differential equation involving partial derivatives with respect to more than one independent variable is a **partial differential equation.** Notice that equation (5) is an ordinary differential equation, whereas equation (6) is a partial differential equation.

The **order** of a differential equation is the order of the highest-order derivatives present in the equation. Equation (5) is a second order equation because d^2x/dt^2 is the highest order derivative present. Equation (6) is a first order equation because only first order partial derivatives occur.

It will be useful to classify ordinary differential equations as being either linear or nonlinear. A **linear** differential equation is any equation that can be written in the form

(7) $$a_n(x)\frac{d^n y}{dx^n} + a_{n-1}(x)\frac{d^{n-1}y}{dx^{n-1}} + \cdots + a_1(x)\frac{dy}{dx} + a_0(x)y = F(x),$$

where $a_n(x), a_{n-1}(x), \ldots, a_0(x)$, and $F(x)$ depend only on the independent variable x, not on y. If an ordinary differential equation is not linear, then we call it **nonlinear.** For example,

$$\frac{d^2 y}{dx^2} + y = x^2$$

is a linear second order ordinary differential equation, whereas

$$\frac{d^2 y}{dx^2} + \sin y = 0$$

is nonlinear because of the $\sin y$ term. The equation

$$\frac{d^2 y}{dx^2} - y\frac{dy}{dx} = \cos x$$

is nonlinear because of the $y\,dy/dx$ term. As we shall see in Chapter 4, linear equations are more amenable to analytic attack than nonlinear ones, so the distinction is made, even though most phenomena in nature are nonlinear.

EXERCISES 1.1

2, 4, 6, 7, 9, 12

In Problems 1 through 14, a differential equation is given along with the field or problem area in which it arises. Classify each as an ordinary differential equation (ODE) or a partial differential equation (PDE), give the order, and indicate the independent and dependent variables. If the equation is an ordinary differential equation, indicate whether the equation is linear or nonlinear.

1. $5\dfrac{d^2 x}{dt^2} + 2\dfrac{dx}{dt} + 9x = 2\cos 3t$

(mechanical vibrations, electrical circuits, seismology).

2. $\dfrac{d^2 y}{dx^2} - 2x\dfrac{dy}{dx} + 2py = 0$, where p is a constant

(Hermite's equation, quantum mechanics, harmonic oscillator).

3. $\frac{dp}{dt} = kp(P - p)$, where k, P are constants
(logistic curve, epidemiology, economics).

4. $\frac{\partial^2 u}{\partial x^2} + \frac{\partial^2 u}{\partial y^2} = 0$
(Laplace's equation, potential theory, electricity, heat, aerodynamics).

5. $\frac{dy}{dx} = \frac{y(2 - 3x)}{x(1 + 3y)}$
(competition between two species, ecology).

6. $\frac{dx}{dt} = k(4 - x)(1 - x)$, where k is a constant
(chemical reaction rates).

7. $y\left[1 + \left(\frac{dy}{dx}\right)^2\right] = C$, where C is a constant
(brachistochrone problem,† calculus of variations).

8. $x\frac{d^2y}{dx^2} + \frac{dy}{dx} + xy = 0$
(aerodynamics, stress analysis).

9. $\sqrt{1 - \alpha y}\frac{d^2y}{dx^2} + 2x\frac{dy}{dx} = 0$, where α is a constant
(Kidder's equation, flow of gases through a porous medium).

10. $8\frac{d^4y}{dx^4} = x(1 - x)$
(deflection of beams).

11. $\frac{\partial N}{\partial t} = \frac{\partial^2 N}{\partial r^2} + \frac{1}{r}\frac{\partial N}{\partial r} + kN$, where k is a constant
(nuclear fission).

12. $\frac{d^2y}{dx^2} - \varepsilon(1 - y^2)\frac{dy}{dx} + 9y = 0$, where ε is a constant
(Van der Pol's equation, triode vacuum tube).

13. $x\frac{d^2y}{dx^2} + 2\frac{dy}{dx} + x(y^2 - C)^{3/2} = 0$, where C is a constant
(white dwarf equation, gravitational potential in degenerate stars).

14. $-\pi(y \tan\alpha)^2\frac{dy}{dt} = 12(2gy)^{1/2}$, where α, g are constants
(liquid flow from a container).

In Problems 15 and 16 write a differential equation that fits the given description.

15. The rate of change of the mass A of salt at time t is proportional to the square of the mass of salt present at time t.

16. The rate of change in the temperature T of coffee at time t is proportional to the difference between the temperature M of the air at time t and the temperature of the coffee at time t.

17. **Drag Race.** Two drivers, Alison and Kevin, are participating in a drag race. Beginning from a standing start, they each proceed with a constant acceleration. Alison covers the last $\frac{1}{4}$ of the distance in 3 seconds, whereas Kevin covers the last $\frac{1}{3}$ of the distance in 4 seconds. Who wins, and by how much time?

1.2 SOLUTIONS AND INITIAL VALUE PROBLEMS

Any nth order ordinary differential equation can be expressed in the general form

$$\text{(1)} \qquad F\left(x, y, \frac{dy}{dx}, \ldots, \frac{d^ny}{dx^n}\right) = 0,$$

where F is a function of the independent variable x, the dependent variable y, and the derivatives of y up to order n; that is, $x, y, \ldots, d^ny/dx^n$. We assume that x lies in an interval I that can be any of the usual intervals (a, b), $[a, b]$, $[a, b)$, and so on. In many cases we

† *Historical Footnote*: In 1630 Galileo formulated the brachistochrone problem (βράχίστος = shortest, χρόνος = time), that is, to determine a path down which a particle will fall from one given point to another in the shortest time. It was reproposed by John Bernoulli in 1696 and solved by him the following year.

can solve the differential equation for $d^n y/dx^n$. This allows us to write it in the form

(2) $$\frac{d^n y}{dx^n} = f\left(x, y, \frac{dy}{dx}, \ldots, \frac{d^{n-1}y}{dx^{n-1}}\right),$$

which is often preferable to (1) for theoretical and computational reasons.

> ### EXPLICIT SOLUTION
>
> **Definition 1.**   A function $\phi(x)$ that when substituted for y in equation (1) [or (2)] satisfies the equation for all x in the interval I is called an **explicit solution** to the equation on I.

● **EXAMPLE 1**   Show that $\phi(x) = x^2 - x^{-1}$ is an explicit solution to

(3) $$\frac{d^2 y}{dx^2} - \frac{2}{x^2} y = 0.$$

Solution   The functions $\phi(x) = x^2 - x^{-1}$, $\phi'(x) = 2x + x^{-2}$, and $\phi''(x) = 2 - 2x^{-3}$ are defined for all $x \neq 0$. Substitution of $\phi(x)$ for y in equation (3) gives

$$(2 - 2x^{-3}) - \frac{2}{x^2}(x^2 - x^{-1}) = (2 - 2x^{-3}) - (2 - 2x^{-3}) = 0.$$

Since this is valid for any $x \neq 0$, the function $\phi(x) = x^2 - x^{-1}$ is an explicit solution to (3) on $(-\infty, 0)$ and also on $(0, \infty)$.   ●

● **EXAMPLE 2**   Show that for *any* choice of the constants c_1 and c_2, the function

$$\phi(x) = c_1 e^{-x} + c_2 e^{2x}$$

is an explicit solution to

(4) $$y'' - y' - 2y = 0.$$

Solution   We compute $\phi'(x) = -c_1 e^{-x} + 2c_2 e^{2x}$ and $\phi''(x) = c_1 e^{-x} + 4c_2 e^{2x}$. Substitution of ϕ, ϕ', and ϕ'' for y, y', and y'' in equation (4) yields

$$(c_1 e^{-x} + 4c_2 e^{2x}) - (-c_1 e^{-x} + 2c_2 e^{2x}) - 2(c_1 e^{-x} + c_2 e^{2x})$$
$$= (c_1 + c_1 - 2c_1)e^{-x} + (4c_2 - 2c_2 - 2c_2)e^{2x} = 0.$$

Since equality holds for all x in $(-\infty, \infty)$, then $\phi(x) = c_1 e^{-x} + c_2 e^{2x}$ is an explicit solution to (4) on the interval $(-\infty, \infty)$ for any choice of the constants c_1 and c_2.   ●

As we shall see in Chapter 2, the methods for solving differential equations do not always yield an explicit solution for the equation. We may have to settle for a solution that is defined implicitly.

> ### IMPLICIT SOLUTION
>
> **Definition 2.**   A relation $G(x, y) = 0$ is said to be an **implicit solution** to equation (1) on the interval I if it defines one or more explicit solutions on I.

● **EXAMPLE 3**   Show that

(5) $y^2 - x^3 + 8 = 0$

is an implicit solution to

(6) $\dfrac{dy}{dx} = \dfrac{3x^2}{2y}$

on the interval $(2, \infty)$.

Solution   When we solve (5) for y we obtain $y = \pm\sqrt{x^3 - 8}$. Let's try $\phi(x) = \sqrt{x^3 - 8}$ to see if it is an explicit solution. Since $d\phi/dx = 3x^2/2\sqrt{x^3 - 8}$, both ϕ and $d\phi/dx$ are defined on $(2, \infty)$. Substituting them into (6) yields

$$\frac{3x^2}{2\sqrt{x^3 - 8}} = \frac{3x^2}{2(\sqrt{x^3 - 8})},$$

which is valid for all x in $(2, \infty)$. Hence relation (5) is an implicit solution to (6) on $(2, \infty)$. (The reader can check that $\psi(x) = -\sqrt{x^3 - 8}$ is also an explicit solution to (6).)   ●

● **EXAMPLE 4**   Show that

(7) $x + y + e^{xy} = 0$

is an implicit solution to

(8) $(1 + xe^{xy})\dfrac{dy}{dx} + 1 + ye^{xy} = 0.$

Solution   First, we observe that we are unable to solve (7) directly for y in terms of x alone. However, for (7) to hold, we realize that any change in x requires a change in y, so we expect the relation (7) to define implicitly at least one function $y(x)$. This is difficult to show directly but can be rigorously verified using the **implicit function theorem**[†] of advanced calculus, which guarantees that such a function $y(x)$ exists that is also differentiable (see Problem 30).

Once we know that y is a differentiable function of x, we can use the technique of implicit differentiation. Indeed, from (7) we obtain

$$1 + \frac{dy}{dx} + e^{xy}\left(y + x\frac{dy}{dx}\right) = 0$$

or

$$(1 + xe^{xy})\frac{dy}{dx} + 1 + ye^{xy} = 0,$$

[†] See *Vector Calculus*, Third Edition, by J. E. Marsden and A. J. Tromba, W. H. Freeman and Company, San Francisco, 1988, p. 281.

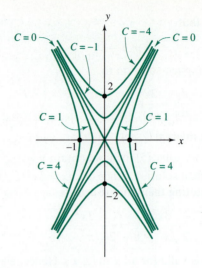

Figure 1.1 Implicit solutions $4x^2 - y^2 = C$

which is identical to the differential equation (8). Thus relation (7) is an implicit solution on some interval guaranteed by the implicit function theorem. ●

● **EXAMPLE 5**   Verify that $4x^2 - y^2 = C$, where C is an arbitrary constant, gives a one-parameter family of implicit solutions to

(9) $$y\frac{dy}{dx} - 4x = 0$$

and graph several of these solution curves.

Solution   When we implicitly differentiate the equation $4x^2 - y^2 = C$ with respect to x, we find

$$8x - 2y\frac{dy}{dx} = 0,$$

which is equivalent to (9). In Figure 1.1 we have sketched the implicit solutions for $C = 0, \pm 1, \pm 4$. The curves are hyperbolas with common asymptotes $y = \pm 2x$. Notice that the implicit solution curves (with C arbitrary) fill the entire plane and are nonintersecting for $C \neq 0$. For $C = 0$, the implicit solution gives rise to the two explicit solutions $y = 2x$, $y = -2x$, both of which pass through the origin. ●

For brevity, we hereafter use the term *solution* to mean either an explicit or an implicit solution.

As we shall see later in the text, the methods for solving nth order differential equations involve n arbitrary constants. In most cases we will be able to determine the n constants if we know the n values $y(x_0), y'(x_0), \dots, y^{(n-1)}(x_0)$.

> ### INITIAL VALUE PROBLEM
>
> **Definition 3.**   By an **initial value problem** for an nth order differential equation
>
> $$F\left(x, y, \frac{dy}{dx}, \ldots, \frac{d^n y}{dx^n}\right) = 0$$
>
> we mean: Find a solution to the differential equation on an interval I that satisfies at x_0 the n initial conditions
>
> $$y(x_0) = y_0,$$
>
> $$\frac{dy}{dx}(x_0) = y_1,$$
>
> $$\vdots$$
>
> $$\frac{d^{n-1}y}{dx^{n-1}}(x_0) = y_{n-1},$$
>
> where $x_0 \in I$ and $y_0, y_1, \ldots, y_{n-1}$ are given constants.

In the case of a first order equation, the initial conditions reduce to the single requirement

$$y(x_0) = y_0,$$

and in the case of a second order equation the initial conditions have the form

$$y(x_0) = y_0, \qquad \frac{dy}{dx}(x_0) = y_1.$$

The terminology *initial conditions* comes from mechanics, in which $y(x_0) = y_0$ usually represents the location of an object at time x_0 and $dy(x_0)/dx = y_1$ gives the velocity of the object at time x_0.

● **EXAMPLE 6**   Show that $\phi(x) = \sin x - \cos x$ is a solution to the initial value problem

(10) $\dfrac{d^2 y}{dx^2} + y = 0; \qquad y(0) = -1, \qquad \dfrac{dy}{dx}(0) = 1.$

Solution   Observe that $\phi(x) = \sin x - \cos x$, $d\phi/dx = \cos x + \sin x$, and $d^2\phi/dx^2 = -\sin x + \cos x$ are all defined on $(-\infty, \infty)$. Substituting into the differential equation gives

$$(-\sin x + \cos x) + (\sin x - \cos x) = 0,$$

which holds for all $x \in (-\infty, \infty)$. Hence $\phi(x)$ is a solution to the differential equation in (10) on $(-\infty, \infty)$. When we check the initial conditions, we find

$$\phi(0) = \sin 0 - \cos 0 = -1,$$

$$\frac{d\phi}{dx}(0) = \cos 0 + \sin 0 = 1,$$

which meets the requirements of (10). Therefore, $\phi(x)$ is a solution to the given initial value problem.   ●

● **EXAMPLE 7**   As shown in Example 2, the function $\phi(x) = c_1 e^{-x} + c_2 e^{2x}$ is a solution to

$$\frac{d^2 y}{dx^2} - \frac{dy}{dx} - 2y = 0,$$

for any choice of the constants c_1 and c_2. Determine c_1 and c_2 so that the initial conditions

$$y(0) = 2 \quad \text{and} \quad \frac{dy}{dx}(0) = -3$$

are satisfied.

Solution   To determine the constants c_1 and c_2, we first compute $d\phi/dx$ to get $d\phi/dx = -c_1 e^{-x} + 2c_2 e^{2x}$. Substituting in our initial conditions gives the following system of equations:

$$\begin{cases} \phi(0) = c_1 e^0 + c_2 e^0 = 2, \\ \dfrac{d\phi}{dx}(0) = -c_1 e^0 + 2c_2 e^0 = -3, \end{cases} \quad \text{or} \quad \begin{cases} c_1 + c_2 = 2, \\ -c_1 + 2c_2 = -3. \end{cases}$$

Adding the last two equations yields $3c_2 = -1$, so $c_2 = -\frac{1}{3}$. Since $c_1 + c_2 = 2$, we find $c_1 = \frac{7}{3}$. Hence the solution to the initial value problem is $\phi(x) = \frac{7}{3} e^{-x} - \frac{1}{3} e^{2x}$.   ●

We now state an existence and uniqueness theorem for first order initial value problems.

EXISTENCE AND UNIQUENESS OF SOLUTION

Theorem 1.   Given the initial value problem

$$\frac{dy}{dx} = f(x, y), \qquad y(x_0) = y_0,$$

assume that f and $\partial f/\partial y$ are continuous functions in a rectangle

$$R = \{(x, y): a < x < b, c < y < d\}$$

that contains the point (x_0, y_0). Then the initial value problem has a unique solution $\phi(x)$ in some interval $x_0 - h < x < x_0 + h$, where h is a positive number.

The preceding theorem tells us two things. First, when an equation satisfies the hypotheses of Theorem 1, we are assured that a solution to the initial value problem exists. Naturally it is desirable to know whether the equation we are trying to solve actually has a solution before we spend too much time trying to solve it. Second, when the hypotheses are satisfied, there is a **unique** solution to the initial value problem. This uniqueness tells us that if we can find a solution, then it is the *only* solution for the initial value problem. Graphically, the theorem says that there is only one solution curve that passes through the point (x_0, y_0); in fact, for this first order equation two solutions can't cross anywhere in the rectangle. Notice that the existence and uniqueness of the solution holds only in *some* neighborhood of x_0. Unfortunately, the theorem does not tell us the size of this neighborhood but only that it has a positive radius denoted by h.

When initial value problems are used to model physical phenomena, the issues of existence and uniqueness are especially important. For the initial value problem to be a reasonable model, we certainly expect it to have a solution, because physically "something does happen." Moreover, the solution should be unique in those cases when repetition of our experiment under identical conditions yields the same results.[†] Certainly, the software user who is trying to obtain a numerical solution to an initial value problem would be well advised to verify its existence and uniqueness since the program may otherwise produce meaningless results.

The proof of Theorem 1 involves converting the initial value problem into an integral equation and then using Picard's method to generate a sequence of successive approximations that converge to the solution. The conversion to an integral equation and Picard's method are discussed in Project B at the end of this chapter. A detailed discussion and proof of the theorem are given in Chapter 14.[††]

● **EXAMPLE 8**   For the initial value problem

(11) $\dfrac{dy}{dx} = x^2 - xy^3, \qquad y(1) = 6,$

does Theorem 1 imply the existence of a unique solution?

Solution   Here $f(x, y) = x^2 - xy^3$ and $\partial f/\partial y = -3xy^2$. Since both of these functions are continuous in any rectangle containing the point $(1, 6)$, the hypotheses of Theorem 1 are satisfied. It then follows from Theorem 1 that the initial value problem (11) has a unique solution in an interval about $x = 1$ of the form $(1 - h, 1 + h)$, where h is some positive number.   ●

● **EXAMPLE 9**   For the initial value problem

(12) $\dfrac{dy}{dx} = 3y^{2/3}, \qquad y(2) = 0,$

does Theorem 1 imply the existence of a unique solution?

Solution   Here $f(x, y) = 3y^{2/3}$ and $\partial f/\partial y = 2y^{-1/3}$. Unfortunately, $\partial f/\partial y$ is not continuous or even defined when $y = 0$. Consequently, there is no rectangle containing $(2, 0)$ in which both f and $\partial f/\partial y$ are continuous. Since the hypotheses of Theorem 1 do not hold, we cannot use Theorem 1 to determine whether the initial value problem does or does not have a unique solution. It turns out that this initial value problem does *not* have a unique solution. We refer the reader to Problem 29 for the details.   ●

In Example 9, suppose the initial condition is changed to $y(2) = 1$. Then, since f and $\partial f/\partial y$ are continuous in any rectangle that contains the point $(2, 1)$ but does not intersect the x-axis, say $R = \{(x, y): 0 < x < 10, 0 < y < 5\}$, it follows from Theorem 1 that this *new* initial value problem has a unique solution in some interval about $x = 2$.

[†] At least this is the case when we are considering a deterministic model, as opposed to a probabilistic model.

[††] All references to Chapters 12–14 refer to the expanded text *Fundamentals of Differential Equations and Boundary Value Problems*.

EXERCISES 1.2

$1, 2, 3, 6, 9, 10, 15, 17, 22, 25, 30$

1. (a) Show that $\phi(x) = x^2$ is an explicit solution to

$$x\frac{dy}{dx} = 2y$$

on the interval $(-\infty, \infty)$.

(b) Show that $\phi(x) = e^x - x$ is an explicit solution to

$$\frac{dy}{dx} + y^2 = e^{2x} + (1 - 2x)e^x + x^2 - 1$$

on the interval $(-\infty, \infty)$.

(c) Show that $\phi(x) = x^2 - x^{-1}$ is an explicit solution to $x^2 d^2y/dx^2 = 2y$ on the interval $(0, \infty)$.

2. (a) Show that $y^2 + x - 3 = 0$ is an implicit solution to $dy/dx = -1/(2y)$ on the interval $(-\infty, 3)$.

(b) Show that $xy^3 - xy^3 \sin x = 1$ is an implicit solution to

$$\frac{dy}{dx} = \frac{(x \cos x + \sin x - 1)y}{3(x - x \sin x)}$$

on the interval $(0, \pi/2)$.

In Problems 3 through 8 determine whether the given function is a solution to the given differential equation.

3. $y = \sin x + x^2,$ $\dfrac{d^2y}{dx^2} + y = x^2 + 2.$

4. $x = \cos t - 2\sin t,$ $x'' + x = 0.$

5. $\theta = 2e^{3t} - e^{2t},$ $\dfrac{d^2\theta}{dt^2} - \theta\dfrac{d\theta}{dt} + 3\theta = -2e^{2t}.$

6. $x = \cos 2t,$ $\dfrac{dx}{dt} + tx = \sin 2t.$

7. $y = e^{2x} - 3e^{-x},$ $\dfrac{d^2y}{dx^2} - \dfrac{dy}{dx} - 2y = 0.$

8. $y = 3\sin 2x + e^{-x},$ $y'' + 4y = 5e^{-x}.$

In Problems 9 through 13 determine whether the given relation is an implicit solution to the differential equation. Assume that the relationship does define y implicitly as a function of x and use implicit differentiation.

9. $x^2 + y^2 = 4,$ $\dfrac{dy}{dx} = \dfrac{x}{y}.$

10. $e^{xy} + y = x - 1,$ $\dfrac{dy}{dx} = \dfrac{e^{-xy} - y}{e^{-xy} + x}.$

11. $y - \ln y = x^2 + 1,$ $\dfrac{dy}{dx} = \dfrac{2xy}{y - 1}.$

12. $x^2 - \sin(x + y) = 1,$ $\dfrac{dy}{dx} = 2x\sec(x + y) - 1.$

13. $\sin y + xy - x^3 = 2,$ $y'' = \dfrac{6xy' + (y')^3 \sin y - 2(y')^2}{3x^2 - y}.$

14. Show that $\phi(x) = c_1 \sin x + c_2 \cos x$ is a solution to $d^2y/dx^2 + y = 0$ for any choice of the constants c_1 and c_2. Thus $c_1 \sin x + c_2 \cos x$ is a two-parameter family of solutions to the differential equation.

15. Show that $\phi(x) = Ce^{3x} + 1$ is a solution to $dy/dx - 3y = -3$ for any choice of the constant C. Thus $Ce^{3x} + 1$ is a one-parameter family of solutions to the differential equation. Graph several of the solution curves using the same coordinate axes.

16. Verify that $x^2 + cy^2 = 1$, where c is an arbitrary nonzero constant, is a one-parameter family of implicit solutions to

$$\frac{dy}{dx} = \frac{xy}{x^2 - 1}$$

and graph several of the solution curves using the same coordinate axes.

17. Verify that $\phi(x) = 2/(1 - ce^x)$, where c is an arbitrary constant, is a one-parameter family of solutions to

$$\frac{dy}{dx} = \frac{y(y - 2)}{2}.$$

Graph the solution curves corresponding to $c = 0$, ± 1, ± 2 using the same coordinate axes.

18. Movable Singular Points. Show that for any choice of the constant C, the function $\phi(x) = (C - x)^{-1}$ is a solution to $dy/dx = y^2$ on the interval (C, ∞) and on $(-\infty, C)$. Moreover, this solution becomes unbounded as x approaches C. (Observe that there is no clue from the differential equation $dy/dx = y^2$ itself that the solution $\phi(x)$ will become unbounded at $x = C$.)

19. Show that the equation $(dy/dx)^2 + y^2 + 3 = 0$ has no (real-valued) solution.

20. Determine for which values of m the function $\phi(x) = e^{mx}$ is a solution to the given equation.

(a) $\dfrac{d^2y}{dx^2} + 6\dfrac{dy}{dx} + 5y = 0.$

(b) $\dfrac{d^3y}{dx^3} + 3\dfrac{d^2y}{dx^2} + 2\dfrac{dy}{dx} = 0.$

21. Determine for which values of m the function $\phi(x) = x^m$ is a solution to the given equation.

(a) $x^2 \dfrac{d^2 y}{dx^2} + x \dfrac{dy}{dx} - y = 0.$

(b) $x^2 \dfrac{d^2 y}{dx^2} - x \dfrac{dy}{dx} - 5y = 0.$

22. The function $\phi(x) = c_1 e^{-x} + c_2 e^{2x}$ is a solution to

$$\frac{d^2 y}{dx^2} - \frac{dy}{dx} - 2y = 0$$

for any choice of the constants c_1 and c_2. Determine c_1 and c_2 so that each of the following initial conditions is satisfied.

(a) $y(0) = 2, \quad y'(0) = 1.$
(b) $y(1) = 1, \quad y'(1) = 0.$

In Problems 23 through 28 determine whether Theorem 1 implies that the given initial value problem has a unique solution.

23. $\dfrac{dy}{dx} = x^3 - y^3, \qquad y(0) = 6.$

24. $\dfrac{dy}{d\theta} - \theta y = \sin^2 \theta, \qquad y(\pi) = 5.$

25. $y \dfrac{dy}{dx} + 4x = 0, \qquad y(2) = -\pi.$

26. $y \dfrac{dy}{dx} = x, \qquad y(1) = 0.$

27. $\dfrac{dy}{dx} = 3x - \sqrt[3]{y - 1}, \qquad y(2) = 1.$

28. $\dfrac{dy}{dx} + \cos y = \sin x, \qquad y(\pi) = 0.$

29. For the initial value problem (12) of Example 9, show that $\phi_1(x) \equiv 0$ and $\phi_2(x) = (x - 2)^3$ are solutions. Hence the initial value problem does not have a *unique* solution.

30. Implicit Function Theorem. *Let $G(x, y)$ have continuous first partial derivatives in the rectangle $R = \{(x, y):$ $a < x < b, c < y < d\}$ containing the point (x_0, y_0). If $G(x_0, y_0) = 0$ and $G_y(x_0, y_0) \neq 0$, then there exists a differentiable function $y = \phi(x)$, defined in some interval $I = (x_0 - h, x_0 + h)$, that satisfies $G(x, \phi(x)) = 0$ for all $x \in I.$*

The implicit function theorem gives conditions under which the relationship $G(x, y) = 0$ defines y implicitly as a function of x. Use the implicit function theorem to show that the relationship $x + y + e^{xy} = 0$, given in Example 4, defines y implicitly as a function of x near the point $(0, -1)$.

31. Consider the equation of Example 5,

$$\textbf{(13)} \quad y \frac{dy}{dx} - 4x = 0.$$

(a) Does Theorem 1 imply the existence of a unique solution to (13) that satisfies $y(x_0) = 0$?
(b) Show that when $x_0 \neq 0$, equation (13) can't possibly have a solution in a neighborhood of $x = x_0$ that satisfies $y(x_0) = 0$.
(c) Show that there are two distinct solutions to (13) satisfying $y(0) = 0$ (see Figure 1.1).

•1.3• DIRECTION FIELDS AND THE APPROXIMATION METHOD OF EULER[†] (Optional)

The existence and uniqueness theorems that we discussed in the preceding sections certainly have great value, but they stop short of telling us anything about the *nature* of the solution to a differential equation. For practical reasons we may need to know the value of the solution at a certain point or the intervals where the solution is increasing or the points where the solution attains a maximum value. Certainly knowing an explicit representation (a formula) for the solution would be a considerable help in answering these questions. However, for many of the differential equations that the reader is likely to encounter in "real-world" applications, it will be impossible to find such a formula. Moreover, even if

---

[†] This optional section will be particularly appropriate in courses emphasizing numerical techniques.

we are lucky enough to obtain an implicit solution, it may be difficult to use this relationship to determine an explicit form. Thus we must rely on other methods that help us to analyze and approximate the solution.

One technique that is useful in visualizing (graphing) the solutions to a differential equation is to sketch the direction field for the equation. To describe this method we need to make a general observation. Namely, a first order equation

$$\frac{dy}{dx} = f(x, y)$$

specifies a slope at each point in the plane where f is defined; in other words, it gives the direction that a solution to the equation must have at each point. A plot of the directions associated with various points in the plane is called the **direction field** for the differential equation. Since the direction field gives the "flow of solutions," it facilitates the drawing of any particular solution (such as the solution to an initial value problem). In Figure 1.2 we have sketched the direction field for the equation

(1) $$\frac{dy}{dx} = x^2 - y$$

and indicated in color several solutions.[†]

A systematic way to construct the direction field for the equation $y' = f(x, y)$ is to first determine all the points in the xy-plane that are associated with the same slope c; that is, all the points where $f(x, y) = c$. Such a locus is called an **isocline.** For equation (1), the

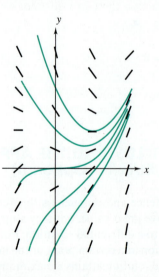

Figure 1.2 Direction field and solutions to $dy/dx = x^2 - y$

† There are several microcomputer programs available for drawing direction fields, such as MACMATH by John H. Hubbard and Beverly H. West available from Springer-Verlag, New York and PHASEPLANE by Brad Ermentrout available from Brooks/Cole Publishing Co., Pacific Groves, California.

isoclines are just the parabolas $x^2 - y = c$. Once the isoclines are determined, we draw short segments along them with slope equal to the corresponding value of c. This procedure is called the **method of isoclines.**

● **EXAMPLE 1**   Use the method of isoclines to sketch the solution to the initial value problem

(2) $\dfrac{dy}{dx} = x - y, \qquad y(0) = 1.$

Solution   We first determine the isoclines. These are the curves

$$x - y = c$$

which, of course, are just straight lines. In Figure 1.3(a) we have sketched (in color) several of these isoclines and indicated the corresponding value for c. We have also drawn short segments (hash marks) with the appropriate slope along each isocline. Enough of the isoclines were drawn so that we can make a rough sketch (see Figure 1.3(b)) of the particular solution $\phi(x)$ to (2). Beginning at (0, 1), the solution is decreasing with slope $c = -1$. When the solution reaches and crosses the line $y = x$, it must have zero slope. After crossing the line $y = x$, the solution begins to increase. As it nears the line $y = x - 1$, its slope tends to 1, and the solution appears to approach this line asymptotically from above. However, the solution $\phi(x)$ can never cross the line $y = x - 1$; this is visually obvious and rigorously follows from the fact that $y = x - 1$ is also a solution to the differential equation in (2). Indeed, the existence and uniqueness theorem (Theorem 1 in Section 1.2) precludes the intersection of these two different solutions.   ●

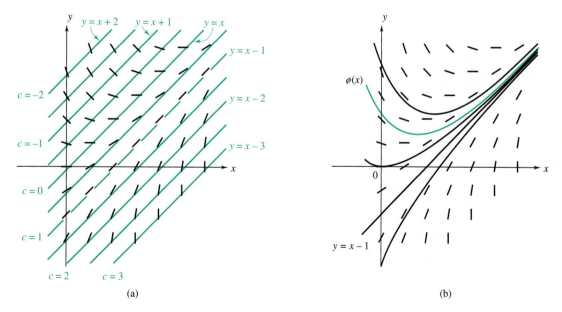

(a)   (b)

Figure 1.3 (a) Isoclines and direction field to $dy/dx = x - y$
(b) Direction field and solutions to $dy/dx = x - y$

When we use the direction field method to sketch a particular solution, we try to visualize the intermediate directions between the isoclines we have drawn. If we follow a finite number of these directions, the sketch becomes a polygonal curve or chain of line segments (see Figure 1.4). This polygonal curve is, visually speaking, an approximation to the solution. Moreover, it is possible to quantify this method, so that we obtain numerical approximations to the solution at specified points. This is the so-called method of Euler (or tangent line method), which we now make precise.

Let's assume that the initial value problem

(3) $y' = f(x, y), \qquad y(x_0) = y_0$

has a unique solution $\phi(x)$ in some interval centered at x_0. Let $h > 0$ be fixed, and consider the equally spaced points[†]

$$x_n := x_0 + nh, \qquad n = 0, 1, 2, \ldots.$$

We can construct values y_n that approximate the solution values $\phi(x_n)$ as follows. At the point (x_0, y_0), the slope of the solution to (3) is given by $dy/dx = f(x_0, y_0)$. Hence the tangent line to the solution curve at the initial point (x_0, y_0) is

$$y = y_0 + (x - x_0)f(x_0, y_0).$$

Using this tangent line to approximate $\phi(x)$, we find that for the point $x_1 = x_0 + h$,

$$\phi(x_1) \approx y_1 := y_0 + hf(x_0, y_0).$$

Next, starting at the point (x_1, y_1), we construct the line with slope given by the direction field at the point (x_1, y_1); that is, with slope equal to $f(x_1, y_1)$. If we follow this line[††] (namely, $y = y_1 + (x - x_1)f(x_1, y_1)$) in stepping from x_1 to $x_2 = x_1 + h$, we arrive at the approximation

$$\phi(x_2) \approx y_2 := y_1 + hf(x_1, y_1).$$

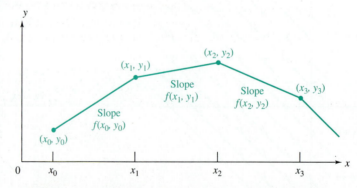

Figure 1.4 Polygonal line approximation given by Euler's method

---

[†] The symbol := means "is defined to be."

[††] Since y_1 is an approximation to $\phi(x_1)$, we cannot assert that this line is tangent to the solution curve $y = \phi(x)$.

Repeating the process (as illustrated in Figure 1.4), we get

$$\phi(x_3) \approx y_3 := y_2 + hf(x_2, y_2),$$
$$\phi(x_4) \approx y_4 := y_3 + hf(x_3, y_3), \qquad \text{etc.}$$

This simple procedure is called **Euler's method** and can be summarized by the recursive formulas

$$(4) \qquad x_{n+1} = x_n + h,$$
$$(5) \qquad y_{n+1} = y_n + hf(x_n, y_n), \qquad n = 0, 1, 2, \ldots .$$

● EXAMPLE 2 Use Euler's method with step size $h = 0.1$ to approximate the solution to the initial value problem

$$(6) \qquad y' = x\sqrt{y}, \qquad y(1) = 4$$

at the points $x = 1.1, 1.2, 1.3, 1.4$, and 1.5.

Solution Here $x_0 = 1$, $y_0 = 4$, $h = 0.1$, and $f(x, y) = x\sqrt{y}$. Thus the recursive formula (5) for y_n is

$$y_{n+1} = y_n + hf(x_n, y_n) = y_n + (0.1)x_n\sqrt{y_n}.$$

Substituting $n = 0$, we get

$$x_1 = x_0 + 0.1 = 1 + 0.1 = 1.1,$$
$$y_1 = y_0 + (0.1)x_0\sqrt{y_0} = 4 + (0.1)(1)\sqrt{4} = 4.2.$$

Putting $n = 1$ yields

$$x_2 = x_1 + 0.1 = 1.1 + 0.1 = 1.2,$$
$$y_2 = y_1 + (0.1)x_1\sqrt{y_1} = 4.2 + (0.1)(1.1)\sqrt{4.2} = 4.42543.$$

Continuing in this manner we obtain the results listed in Table 1.1. For comparison we have included the exact value (to five decimal places) of the solution $\phi(x) = (x^2 + 7)^2/16$ to (6), which can be obtained using separation of variables (see Section 2.2). ●

TABLE 1.1 COMPUTATIONS FOR $y' = x\sqrt{y}$, $y(1) = 4$

n	x_n	Euler's Approximation	Exact Value
0	1	4	4
1	1.1	4.2	4.21276
2	1.2	4.42543	4.45210
3	1.3	4.67787	4.71976
4	1.4	4.95904	5.01760
5	1.5	5.27081	5.34766

Given the initial value problem (3) and a specific point x, how can Euler's method be used to approximate $\phi(x)$? Starting at x_0 we can take one giant step that lands on x, or we can take several smaller steps to arrive at x. If we wish to take n steps, then we set $h = (x - x_0)/n$, so that the step size h and the number of steps n are related in a specific way. It is expected that the more steps we take, the better will be the approximation. But it should be kept in mind that more steps mean more computations and hence greater roundoff error.

● **EXAMPLE 3**   Use Euler's method to approximate the solution to the initial value problem

(7) $y' = y$, $y(0) = 1$

at $x = 1$. Take $n = 1, 2, 4, 8,$ and 16.

Remark.   We know that the solution to (7) is just $\phi(x) = e^x$, so Euler's method will generate algebraic approximations to the transcendental number $e = 2.71828\ldots$.

Solution   Here $f(x, y) = y$, $x_0 = 0$, and $y_0 = 1$. The recursive formula for Euler's method is

$$y_{j+1} = y_j + hy_j = (1 + h)y_j.$$

To obtain approximations at $x = 1$ with n steps, we take the step size $h = 1/n$. For $n = 1$ we have

$$\phi(1) \approx y_1 = (1 + 1)(1) = 2.$$

For $n = 2$, $\phi(x_2 = 1) \approx y_2$. In this case we get

$$y_1 = (1 + 0.5)(1) = 1.5,$$
$$\phi(1) \approx y_2 = (1 + 0.5)(1.5) = 2.25.$$

For $n = 4$, $\phi(x_4 = 1) \approx y_4$, where

$$y_1 = (1 + 0.25)(1) = 1.25,$$
$$y_2 = (1 + 0.25)(1.25) = 1.5625,$$
$$y_3 = (1 + 0.25)(1.5625) = 1.95313,$$
$$\phi(1) \approx y_4 = (1 + 0.25)(1.95313) = 2.44141.$$

(In the above computations we have rounded to five decimal places.) Similarly, taking $n = 8$ and 16, we obtain even better estimates for $\phi(1)$. These approximations are shown in Table 1.2. For comparison, Figure 1.5 displays the polygonal line approximations to e^x using Euler's Method with $h = 1/4$ ($n = 4$) and $h = 1/8$ ($n = 8$). Notice that the smaller step size yields the better approximation.   ●

How good (or bad) is Euler's method? In judging a numerical scheme, we must begin with two fundamental questions. Does the method converge? And, if so, what is the rate of convergence? These important issues are discussed in Section 3.5 where improvements in Euler's method are introduced (see also Problems 20 and 21).

TABLE 1.2 EULER'S METHOD FOR $y' = y$, $y(0) = 1$

n	h	Approximation for $\phi(1) = e$
1	1.0	2.0
2	0.5	2.25
4	0.25	2.44141
8	0.125	2.56578
16	0.0625	2.63793

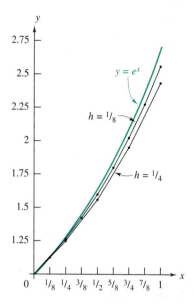

Figure 1.5 Euler's approximations of e^x with $h = 1/4$ and $1/8$

EXERCISES 1.3

In Problems 1 through 6 draw the isoclines with their direction markers and sketch several solution curves, including the curve satisfying the given initial conditions.

1. $dy/dx = 2x, \quad y(0) = -1.$
2. $dy/dx = y, \quad y(0) = 1.$
3. $dy/dx = -x/y, \quad y(0) = 4.$
4. $dy/dx = x/y, \quad y(0) = -1.$
5. $dy/dx = y(2-y), \quad y(0) = 3.$ [Hint: For each $c < 1$ the isocline $y(2-y) = c$ consists of two horizontal lines, which can be obtained using the quadratic formula.]
6. $dy/dx = x + y, \quad y(0) = 1.$

7. From a sketch of the direction field, what can one say about the behavior as x approaches ∞ of a solution to
$$\frac{dy}{dx} = 2 - y + \frac{1}{x}\,?$$

8. From a sketch of the direction field, what can one say about the behavior as x approaches ∞ of a solution to
$$\frac{dy}{dx} = -y\,?$$

In many of the problems below it will be helpful to have a calculator or microcomputer available. The reader may also find it convenient to write a program for solving initial value problems using Euler's method. (Remember, all trigonometric calculations are done in radians.)

In Problems 9 through 12 use Euler's method to approximate the solution to the given initial value problem at the points $x = 0.1, 0.2, 0.3, 0.4,$ and 0.5 using steps of size 0.1 ($h = 0.1$).

9. $dy/dx = -x/y, \quad y(0) = 4 \quad$ (cf. Problem 3).
10. $dy/dx = x/y, \quad y(0) = -1 \quad$ (cf. Problem 4).
11. $dy/dx = y(2-y), \quad y(0) = 3 \quad$ (cf. Problem 5).
12. $dy/dx = x + y, \quad y(0) = 1 \quad$ (cf. Problem 6).

13. Use Euler's method with step size $h = 0.1$ to approximate the solution to the initial value problem
$$y' = x - y^2, \qquad y(1) = 0$$
at the points $x = 1.1, 1.2, 1.3, 1.4,$ and 1.5.

14. Use Euler's method with step size $h = 0.2$ to approximate the solution to the initial value problem
$$y' = \frac{1}{x}(y^2 + y), \qquad y(1) = 1$$

at the points $x = 1.2, 1.4, 1.6,$ and 1.8.

15. Use Euler's method to approximate the solution to the initial value problem

$$\frac{dx}{dt} = 1 + t\sin(tx), \qquad x(0) = 0$$

at $t = 1$. Take $n = 1, 2, 4,$ and 8.

16. Use Euler's method to approximate the solution to the initial value problem

$$y' = 1 - \sin y, \qquad y(0) = 0$$

at $x = \pi$. Take $n = 1, 2, 4,$ and 8.

17. Use Euler's method with $h = 0.1$ to approximate the solution to the initial value problem

$$y' = \frac{1}{x^2} - \frac{y}{x} - y^2, \qquad y(1) = -1$$

on the interval $1 \le x \le 2$. Compare these approximations with the actual solution $y = -1/x$ by graphing the polygonal line approximation and the actual solution on the same coordinate system.

18. Use Euler's method with $h = 0.1$ to approximate the solution to the initial value problem

$$y' = x - y, \qquad y(0) = 0$$

on the interval $0 \le x \le 1$. Compare these approximations with the actual solution $y = e^{-x} + x - 1$ by graphing the polygonal line approximation and the actual solution on the same coordinate system.

19. Use Euler's method with $n = 20$ to approximate the solution to the initial value problem

$$\frac{dx}{dt} = 1 + x^2, \qquad x(0) = 0$$

at $t = 1$. Compare the approximation with the actual solution $x = \tan t$ evaluated at $t = 1$.

20. In Example 3 we approximated the transcendental number e by using Euler's method to solve the initial value problem

$$y' = y, \qquad y(0) = 1.$$

Show that the Euler approximation y_n obtained by

using the step size $1/n$ is given by the formula

$$y_n = \left(1 + \frac{1}{n}\right)^n, \qquad n = 1, 2, \dots.$$

Recall from calculus that

$$\lim_{n \to \infty} \left(1 + \frac{1}{n}\right)^n = e,$$

and hence Euler's method converges (theoretically) to the correct value.

21. Prove that the "rate of convergence" for Euler's method in Problem 20 is like $1/n$ by showing that

$$\lim_{n \to \infty} \frac{e - y_n}{1/n} = \frac{e}{2}.$$

[Hint: Use L'Hôpital's rule and the Maclaurin expansion for $\ln(1 + t)$.]

22. Newton's Law of Cooling. Newton's law of cooling states that the rate of change in the temperature $T(t)$ of a body is proportional to the difference between the temperature of the medium $M(t)$ and the temperature of the body. That is,

$$\frac{dT}{dt} = K[M(t) - T(t)],$$

where K is a constant. Let $K = 1\ (\text{min})^{-1}$ and let the temperature of the medium be constant, $M(t) \equiv 70°$. If the body is initially at $100°$, use Euler's method with $h = 0.1$ to approximate the temperature of the body after **(a)** 1 minute, **(b)** 2 minutes.

23. Stefan's Law of Radiation. Stefan's law of radiation states that the rate of change in temperature of a body at $T(t)$ degrees in a medium at $M(t)$ degrees is proportional to $M^4 - T^4$. That is,

$$\frac{dT}{dt} = K(M^4 - T^4),$$

where K is a constant. Let $K = (40)^{-4}$ and assume the medium temperature is constant, $M(t) \equiv 70°$. If $T(0) = 100°$, use Euler's method with $h = 0.1$ to approximate $T(1)$ and $T(2)$.

● CHAPTER SUMMARY ●

In this chapter some basic terminology for differential equations was introduced. The **order** of a differential equation is the order of the highest derivative present. The subject of this text is **ordinary** differential equations, which involve derivatives with respect to a single independent variable. Such equations are classified as **linear** or **nonlinear.**

An **explicit solution** of a differential equation is a function that satisfies the equation on some interval. An **implicit solution** is a relation that implicitly defines a function that is an explicit solution. A differential equation typically has infinitely many solutions. In contrast, there are theorems that insure that a unique solution exists for certain **initial value problems** in which it is required to find a solution to the differential equation that satisfies given initial conditions. For an nth order equation, these conditions refer to the values of the solution and its first $n - 1$ derivatives at some point.

Whether or not one is successful in finding solutions to a differential equation, several techniques can be used to help analyze the solutions. For first order differential equations, one such technique is the **method of isoclines.** In this method, a differential equation $dy/dx = f(x, y)$ is viewed as specifying directions (slopes) at points on the plane. The conglomerate of such slopes is the **direction field** for the equation. Plotting this direction field is facilitated by first graphing the **isoclines** $f(x, y) = c$ (c a constant). Knowing the "flow of solutions" is helpful in sketching the solution to an initial value problem. Furthermore, carrying out this method algebraically leads to numerical approximations to the desired solution. This numerical process is called **Euler's method.**

TECHNICAL WRITING EXERCISES

1. Select four fields (e.g. Astronomy, Geology, Biology, and Economics) and for each field discuss a situation in which differential equations are used to solve a problem. Select examples that are not covered in Section 1.1.

2. Compare the different types of solutions discussed in this chapter—explicit, implicit, graphical, and numerical. What are advantages and disadvantages of each?

GROUP PROJECTS FOR CHAPTER 1

A. Taylor Series Method

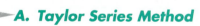

The numerical method of Euler was based on the fact that the tangent line gives a good local approximation for the function. But why restrict ourselves to linear approximants when higher-degree polynomial approximants are available? For example, we can use the **Taylor polynomial** of degree n about $x = x_0$ which is defined by

$$P_n(x) := y(x_0) + y'(x_0)(x - x_0) + \frac{y''(x_0)}{2!}(x - x_0)^2 + \cdots + \frac{y^{(n)}(x_0)}{n!}(x - x_0)^n.$$

This polynomial is the nth partial sum of the Taylor series representation

$$\sum_{k=0}^{\infty} \frac{y^{(k)}(x_0)}{k!}(x - x_0)^k.$$

To determine the Taylor series for the solution $\phi(x)$ to the initial value problem

$$dy/dx = f(x, y), \qquad y(x_0) = y_0,$$

we need only determine the values of the derivatives of ϕ (assuming they exist) at x_0; that is, $\phi(x_0)$, $\phi'(x_0), \dots$. The initial condition gives the first value $\phi(x_0) = y_0$. Using the equation $y' = f(x, y)$, we find $\phi'(x_0) = f(x_0, y_0)$. To determine $\phi''(x_0)$ we differentiate the equation $y' = f(x, y)$ implicitly with respect to x to obtain

$$y'' = \frac{\partial f}{\partial x} + \frac{\partial f}{\partial y}\frac{dy}{dx} = \frac{\partial f}{\partial x} + \frac{\partial f}{\partial y}f$$

and thereby we can compute $\phi''(x_0)$.

(a) Compute the Taylor polynomials of degree 4 for the solutions to the given initial value problems. Use these Taylor polynomials to approximate the solution at $x = 1$.

(i) $\dfrac{dy}{dx} = x - y; \quad y(0) = 1.$ (ii) $\dfrac{dy}{dx} = y(2 - y); \quad y(0) = 3.$

(b) In order to compare the use of Euler's method with that of Taylor series to approximate the solution $\phi(x)$ to the initial value problem

$$\frac{dy}{dx} + y = \sin x - \cos x, \qquad y(0) = 0,$$

fill in Table 1.3. Give the approximations for $\phi(1)$ and $\phi(3)$ to the nearest thousandth. Verify that $\phi(x) = e^{-x} - \cos x$ and use this formula together with a calculator or tables to find the exact values of $\phi(x)$ to the nearest thousandth. Finally, decide which of the first four methods in Table 1.3 will yield the closest approximation to $\phi(10)$ and give the reasons for your choice. (Remember that the computation of trigonometric functions must be done in the radian mode.)

TABLE 1.3

Method	Approximation of $\phi(1)$	Approximation of $\phi(3)$
Euler's method using steps of size 0.1		
Euler's method using steps of size 0.01		
Taylor polynomial of degree 2		
Taylor polynomial of degree 5		
Exact value of $\phi(x)$ to nearest thousandth		

B. Picard's Approximation Method

The initial value problem

(1) $y'(x) = f(x, y), \qquad y(x_0) = y_0$

can be expressed as an **integral** equation. This is obtained by integrating both sides of (1) with respect to x from $x = x_0$ to $x = x_1$:

$$(2) \qquad \int_{x_0}^{x_1} y'(x)\,dx = y(x_1) - y(x_0) = \int_{x_0}^{x_1} f(x, y(x))\,dx.$$

Substituting $y(x_0) = y_0$ and solving for $y(x_1)$ gives

$$(3) \qquad y(x_1) = y_0 + \int_{x_0}^{x_1} f(x, y(x))\,dx.$$

If we use t instead of x as the variable of integration, we can let $x = x_1$ be the upper limit of integration. Equation (3) then becomes

$$(4) \qquad y(x) = y_0 + \int_{x_0}^{x} f(t, y(t))\,dt.$$

Equation (4) can be used to generate successive approximations of a solution to (1). Let the function $\phi_0(x)$ be an initial guess or approximation of a solution to (1); then a new approximation is given by

$$\phi_1(x) := y_0 + \int_{x_0}^{x} f(t, \phi_0(t))\,dt,$$

where we have replaced $y(t)$ by the approximation $\phi_0(t)$ in the argument of f. In a similar fashion we can use $\phi_1(x)$ to generate a new approximation $\phi_2(x)$, and so on. In general, we obtain the $(n + 1)$st approximation from the relation

$$(5) \qquad \phi_{n+1}(x) := y_0 + \int_{x_0}^{x} f(t, \phi_n(t))\,dt.$$

This procedure is called **Picard's method**,[†] and under certain assumptions on f and $\phi_0(x)$, the sequence $\{\phi_n(x)\}$ is known to converge to a solution to (1). These assumptions and the proof of convergence are given in Chapter 14.[††]

Without further information about the solution to (1), it is common practice to take $\phi_0(x) \equiv y_0$.

(a) Use Picard's method with $\phi_0(x) \equiv 1$ to obtain the next four successive approximations of the solution to

$$(6) \qquad y'(x) = y(x), \qquad y(0) = 1.$$

Show in general, that these approximations are just the partial sums of the Maclaurin series for the actual solution e^x.

(b) Use Picard's method with $\phi_0(x) \equiv 0$ to obtain the next three successive approximations of the solution to the nonlinear problem

$$(7) \qquad y'(x) = x - [y(x)]^2, \qquad y(0) = 0.$$

(c) In Problem 29 in Exercises 1.2 we showed that the initial value problem

$$(8) \qquad y'(x) = 3[y(x)]^{2/3}, \qquad y(2) = 0$$

[†] *Historical Footnote*: This approximation method is a by-product of the famous Picard-Lindelöf existence theorem formulated at the end of the nineteenth century.

[††] All references to Chapters 12–14 refer to the expanded text *Fundamentals of Differential Equations and Boundary Value Problems*.

does not have a unique solution. Show that Picard's method beginning with $\phi_0(x) \equiv 0$ converges to the solution $y(x) \equiv 0$, whereas Picard's method beginning with $\phi_0(x) = x - 2$ converges to the second solution $y(x) = (x - 2)^3$. [Hint: For the guess $\phi_0(x) = x - 2$, show that $\phi_n(x)$ has the form $c_n(x - 2)^{r_n}$, where $c_n \to 1$ and $r_n \to 3$ as $n \to \infty$.]

C. Magnetic "Dipole"

A bar magnet is often thought of as a magnetic dipole with one end labeled the north pole N and the opposite end labeled the south pole S. The magnetic field for the magnetic dipole is symmetric with respect to rotation about the axis passing lengthwise through the center of the bar. Hence, we can study the magnetic field by restricting ourselves to a plane containing this axis of symmetry with the origin at the center of the bar (see Figure 1.6).

For a point P that is located a distance r from the origin, where r is much greater than the length of the magnet, the magnetic field lines satisfy the differential equation

$$(9) \qquad \frac{dy}{dx} = \frac{3xy}{2x^2 - y^2},$$

and the equipotential lines satisfy the equation

$$(10) \qquad \frac{dy}{dx} = \frac{y^2 - 2x^2}{3xy}.$$

(a) Show that the two families of curves are perpendicular where they intersect. [Hint: Consider the slopes of the tangent lines of the two curves at a point of intersection.]

(b) Sketch the direction field for equation (9) for $-5 \le x, y \le 5$. You can use a microcomputer to generate the direction field or use the method of isoclines discussed in Section 1.3. The direction field should remind you of the experiment where iron filings are sprinkled on a sheet of paper that is held above a bar magnet. The iron filings are like the hash marks.

(c) Use the direction field found in part (b) to help sketch the magnetic field lines that are solutions to (9).

(d) Use the results of parts (a) and (c) to sketch the equipotential lines that are solutions to (10). The magnetic field lines and the equipotential lines are examples of *orthogonal trajectories*. (See Problem 31 in Exercises 2.3 on page 42.)

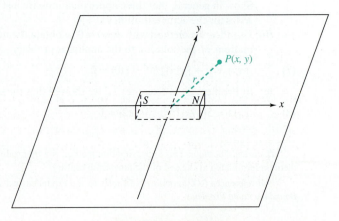

Figure 1.6 Magnetic "dipole"

CHAPTER 2

First Order Differential Equations

2.1 • INTRODUCTION: MOTION OF A FALLING BODY

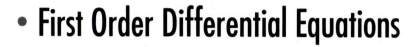

An object falls through the air toward the Earth. Assuming that the only forces acting on the object are gravity and air resistance, determine the velocity of the object as a function of time.

Newton's second law states that force is equal to mass times acceleration, assuming the mass is constant. We can express this by the equation

$$m\frac{dv}{dt} = F,$$

where F represents the total force on the object, m is the mass of the object, and dv/dt is the acceleration. Here we have expressed acceleration as the derivative of velocity with respect to time, dv/dt.

Near the Earth's surface the force due to gravity is just the weight of the object. This force can be expressed by mg, where g is the acceleration due to gravity. There is no general law that exactly determines the air resistance acting on the object, since this force seems to depend upon the velocity of the object, the density of the air, and the shape of the object, among other things. However, in some instances air resistance can be reasonably represented by $-kv$, where k is a positive constant depending on the density of the air and the shape of the object. We use the minus sign because air resistance is a force that opposes the motion. The forces acting on the object are depicted in Figure 2.1 on page 26.

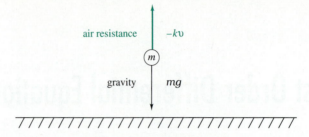

Figure 2.1 Forces on falling object

Applying Newton's law we obtain the first order differential equation

(1) $$m\frac{dv}{dt} = mg - kv.$$

This equation may be solved using **separation of variables** (this method is discussed more fully in Section 2.2). Treating dv and dt as differentials, we separate the variables to obtain

$$\frac{dv}{mg - kv} = \frac{dt}{m}.$$

Integrating, we find

$$\int \frac{1}{mg - kv}\,dv = \int \frac{1}{m}\,dt,$$

$$-\frac{1}{k}\ln(mg - kv) = \frac{t}{m} + C,$$

where C is a "constant of integration." The last equation is solved for v by exponentiating both sides, giving

(2) $$v = \frac{mg}{k} - \frac{1}{k}e^{-kC}e^{-kt/m},$$

which is called a **general solution** to the differential equation because, as we shall see in Section 2.4, every solution to (1) can be expressed in the form given in (2).

In a specific case we would be given the values of m, g, and k. To determine the constant C in the general solution, we can use the initial velocity of the obect v_0. That is, we solve the **initial value problem**

$$m\frac{dv}{dt} = mg - kv, \qquad v(0) = v_0.$$

Substituting $v = v_0$ and $t = 0$ into the general solution to the differential equation, we can solve for C. With this value for C, the solution to the initial value problem is

(3) $$v = \frac{mg}{k} + \left(v_0 - \frac{mg}{k}\right)e^{-kt/m}.$$

The preceding formula gives the velocity of the object falling through the air as a

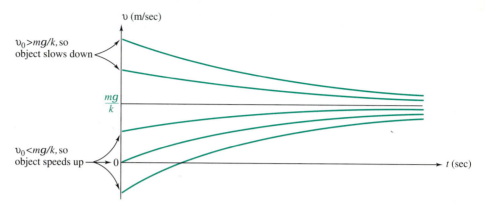

Figure 2.2 Graph of $v(t)$ for six different initial velocities v_0. ($g = 9.8$ m/sec^2, $m/k = 5$ sec)

function of time if the initial velocity of the object is v_0. In Figure 2.2 we have sketched the graph of $v(t)$ for various values of v_0. It appears from Figure 2.2 that the velocity $v(t)$ approaches mg/k *regardless of the initial velocity* v_0. (This is easy to see from formula (3) by letting ($t \to \infty$.) The constant mg/k is referred to as the **limiting** or **terminal velocity** of the object.

From this model for a falling body we can make certain observations. Since $e^{-kt/m}$ rapidly tends to zero, the velocity is approximately the weight, mg, divided by the coefficient of air resistance, k. Thus, in the presence of air resistance, the heavier the object, the faster it will fall, assuming shapes and sizes are the same. Also, when air resistance is lessened (k is made smaller), the object will fall faster. These observations certainly agree with our experience.

Many other physical problems,[†] when given a mathematical formulation, lead to first order differential equations or initial value problems. Several of these are discussed in Chapter 3. In this chapter we shall learn how to recognize and obtain solutions for some special types of first order equations. We begin by studying separable equations, then exact equations, and then linear equations. The methods for solving these are the most basic. In the last two sections we illustrate how devices such as integrating factors, substitutions, and transformations can be used to transform a special type of equation into either a separable, exact, or linear equation that we can solve. Through a discussion of these special types of equations, we hope that the reader will gain insight into the behavior of solutions to more general equations and the possible difficulties in finding these solutions.

•2.2• SEPARABLE EQUATIONS

A simple class of differential equations that can be solved using integration is the class of separable equations.

[†] The physical problem just discussed leads to other mathematical models. For example, more precise models would take into account the variations in the gravitational field of the Earth.

> ### SEPARABLE EQUATION
>
> **Definition 1.**   If the right-hand side of the equation
>
> $$\frac{dy}{dx} = f(x, y)$$
>
> can be expressed as a function that depends only on x times a function that depends only on y, then the differential equation is called **separable.**[†]

Thus a first order equation is separable if it can be written in the form

(1) $$\frac{dy}{dx} = g(x)p(y).$$

For example, the equation

$$\frac{dy}{dx} = \frac{2x + xy}{y^2 + 1}$$

is separable, since

$$\frac{2x + xy}{y^2 + 1} = x\frac{2 + y}{y^2 + 1} = g(x)p(y).$$

However, the equation

$$\frac{dy}{dx} = 1 + xy$$

admits no such factorization of the right-hand side and so is not separable.

To obtain solutions to the separable equation (1), we first write it in the form

(2) $$h(y)\frac{dy}{dx} = g(x),$$

where[††] $h(y) := 1/p(y)$. Letting $H(y)$ and $G(x)$ denote antiderivatives (indefinite integrals) of $h(y)$ and $g(x)$, respectively, that is

$$H'(y) = h(y), \qquad G'(x) = g(x),$$

then equation (2) becomes

$$H'(y)\frac{dy}{dx} = G'(x).$$

[†] *Historical Footnote*: A procedure for solving separable equations was implicitly discovered by Gottfried Leibniz in 1691. The explicit technique called separation of variables was formalized by John Bernoulli in 1694.

[††] The symbol := means "is defined to be."

Now let $y(x)$ be a solution. Recalling the chain rule for differentiation, we then have

$$\frac{d}{dx} H(y(x)) = \frac{d}{dx} G(x).$$

In other words, $H(y(x))$ and $G(x)$ are two functions of x that have the same derivative; thus they must differ by a constant. That is,

(3) $H(y(x)) = G(x) + C,$

which defines the solution $y(x)$ *implicitly*.

This procedure can be considerably streamlined by treating dy and dx as differentials.

▶ **METHOD FOR SOLVING SEPARABLE EQUATIONS**

To solve the equation

$$h(y)\frac{dy}{dx} = g(x)$$

multiply by dx to obtain

$$h(y)\,dy = g(x)\,dx.$$

Then integrate both sides:

$$\int h(y)\,dy = \int g(x)\,dx,$$

$$H(y) = G(x) + C.$$

The last equation is the implicit solution previously obtained in equation (3).

● **EXAMPLE 1** Solve

$$\frac{dy}{dx} = \frac{x-5}{y^2}.$$

Solution Following the streamlined approach, we separate the variables and rewrite the equation in the form

$$y^2\,dy = (x-5)\,dx.$$

Integrating, we have

$$\int y^2\,dy = \int (x-5)\,dx$$

$$\frac{y^3}{3} = \frac{x^2}{2} - 5x + C,$$

and solving this last equation for y gives

$$y = \left(\frac{3x^2}{2} - 15x + 3C\right)^{1/3}.$$

Since C is a constant of integration that can be any real number, $3C$ can also be any real number. Replacing $3C$ by the single symbol K, we then have

$$y = \left(\frac{3x^2}{2} - 15x + K\right)^{1/3}.$$

If we wish to abide by the custom of letting C represent an arbitrary constant, we can go one step further and use C instead of K in the final answer. ●

As Example 1 attests, separable equations are among the easiest to solve. However, the procedure does require a facility for computing integrals. Many of the procedures to be discussed later in the text also require a familiarity with the techniques of integration. For the student who requires it, a brief table of integrals can be found on the inside front cover.

● EXAMPLE 2 Solve the initial value problem

$$\textbf{(4)} \qquad \frac{dy}{dx} = \frac{y-1}{x+3}, \qquad y(-1) = 0.$$

Solution Separating the variables and integrating gives

$$\frac{dy}{y-1} = \frac{dx}{x+3},$$

$$\int \frac{dy}{y-1} = \int \frac{dx}{x+3},$$

$$\textbf{(5)} \qquad \ln|y-1| = \ln|x+3| + C.$$

At this point we can either solve for y explicitly (retaining the constant C) or use the initial condition to first determine C and then solve explicitly for y. Let's try the first approach.

Exponentiating equation (5), we have

$$e^{\ln|y-1|} = e^{\ln|x+3|+C} = e^C e^{\ln|x+3|},$$

$$\textbf{(6)} \qquad |y-1| = e^C|x+3| = C_1|x+3|,$$

where $C_1 := e^C$. Now, depending on the values of y, we have $|y-1| = \pm(y-1)$; and similarly, $|x+3| = \pm(x+3)$. Thus (6) can be written as

$$y - 1 = \pm C_1(x+3) \quad \text{or} \quad y = 1 \pm C_1(x+3),$$

where the choice of sign depends upon the values of x and y. Since C_1 is an arbitrary positive constant (recall that $C_1 = e^C > 0$), we can replace $\pm C_1$ by C, where C now represents an arbitrary nonzero constant. We then obtain

$$\textbf{(7)} \qquad y = 1 + C(x+3).$$

Finally, we determine C such that the initial condition $y(-1) = 0$ is satisfied. Putting $x = -1$ and $y = 0$ in equation (7) gives

$$0 = 1 + C(-1 + 3) = 1 + 2C,$$

and so $C = -\frac{1}{2}$. Thus the solution to the initial value problem is

$$(8) \qquad y = 1 - \tfrac{1}{2}(x + 3) = -\tfrac{1}{2}(x + 1).$$

Alternative Approach.   The second approach is to first set $x = -1$ and $y = 0$ in equation (5) and then solve for C. In this case we obtain

$$\ln|0 - 1| = \ln|-1 + 3| + C,$$
$$0 = \ln 1 = \ln 2 + C,$$

and so $C = -\ln 2$. Thus, from (5), the solution y is given implicitly by

$$\ln(1 - y) = \ln(x + 3) - \ln 2,$$

where we have replaced $|y - 1|$ by $1 - y$ and $|x + 3|$ by $x + 3$ since we are interested in x and y near the initial values $x = -1$, $y = 0$ (for such values, $y - 1 < 0$ and $x + 3 > 0$). Solving for y, we find

$$\ln(1 - y) = \ln(x + 3) - \ln 2 = \ln\left(\frac{x + 3}{2}\right),$$

$$1 - y = \frac{x + 3}{2},$$

$$y = 1 - \tfrac{1}{2}(x + 3) = -\tfrac{1}{2}(x + 1),$$

which agrees with the solution (8) found by the first method. ●

We caution the reader that it is possible to lose solutions to a differential equation when applying the separation of variables technique. Indeed, in solving the equation of Example 2,

$$\frac{dy}{dx} = \frac{y - 1}{x + 3},$$

we obtained $y = 1 + C(x + 3)$ as the set of solutions, where C was a *nonzero* constant. But notice that the constant function $y \equiv 1$ (which in this case corresponds to $C = 0$) is also a solution to the differential equation. The reason we lost this solution can be traced back to a division by $y - 1$ in the separation process. (See Problem 31 for an example where a solution is lost and cannot be retrieved by setting the constant $C = 0$.) In general, if the separation of variables process involves a division, then functions that cause the divisor to be zero are candidates for solutions to the differential equation and should be individually checked.

● **EXAMPLE 3**   Solve

$$(9) \qquad \frac{dy}{dx} = \frac{6x^5 - 2x + 1}{\cos y + e^y}.$$

Solution Separating variables and integrating, we find

$$(\cos y + e^y)\,dy = (6x^5 - 2x + 1)\,dx,$$

$$\int (\cos y + e^y)\,dy = \int (6x^5 - 2x + 1)\,dx,$$

$$\sin y + e^y = x^6 - x^2 + x + C.$$

At this point we reach an impasse. We would like to solve for y explicitly, but we are unable to do so. This is often the case in solving nonlinear first order equations. Consequently, when we say "solve the equation," we must on occasion be content if only an implicit form of the solution has been found. ●

As we shall see in the next section, first order differential equations are sometimes written in the differential form

(10) $\quad M(x, y)\,dx + N(x, y)\,dy = 0.$

Such an equation is separable if both M and N are separable; that is, if each can be expressed as a function that depends only on x times a function that depends only on y (see Problem 28).

EXERCISES 2.2

In Problems 1 through 6 determine whether the given differential equation is separable.

1. $\dfrac{dy}{dx} = y^3 + y.$

2. $\dfrac{dy}{dx} = \sin(x + y).$

3. $\dfrac{dy}{dx} = \dfrac{3e^{x+y}}{x^2 + 2}.$

4. $\dfrac{ds}{dt} = t\ln(s^{2t}) + 8t^2.$

5. $s^2 + \dfrac{ds}{dt} = \dfrac{s + 1}{st}.$

6. $(xy^2 + 3y^2)\,dy - 2x\,dx = 0.$

In Problems 7 through 16 solve the equation.

7. $\dfrac{dy}{dx} = \dfrac{x^2 - 1}{y^2}.$

8. $\dfrac{dy}{dx} = \dfrac{1}{xy^3}.$

9. $\dfrac{dx}{dt} = 3xt^2.$

10. $\dfrac{dy}{dx} = y(2 + \sin x).$

11. $\dfrac{dy}{dx} = \dfrac{\sec^2 y}{1 + x^2}.$

12. $x\dfrac{dv}{dx} = \dfrac{1 - 4v^2}{3v}.$

13. $\dfrac{dy}{dx} + y^2 = y.$

14. $\dfrac{dy}{dx} = 3x^2(1 + y^2).$

15. $y^{-1}\,dy + y e^{\cos x}\sin x\,dx = 0.$

16. $(x + xy^2)\,dx + e^{x^2} y\,dy = 0.$

In Problems 17 through 26 solve the initial value problem.

17. $y' = x^3(1 - y), \quad y(0) = 3.$

18. $\dfrac{dy}{dx} = (1 + y^2)\tan x, \quad y(0) = \sqrt{3}.$

19. $\dfrac{dy}{d\theta} = y\sin\theta, \quad y(\pi) = -3.$

20. $\dfrac{dy}{dx} = \dfrac{3x^2 + 4x + 2}{2y + 1}, \quad y(0) = -1.$

21. $\dfrac{dy}{dx} = 2\sqrt{y + 1}\cos x, \quad y(\pi) = 0.$

22. $x^2\,dx + 2y\,dy = 0, \quad y(0) = 2.$

23. $\dfrac{dy}{dx} = 8x^3 e^{-2y}, \quad y(1) = 0.$

24. $\dfrac{dy}{dx} = 2x\cos^2 y, \quad y(0) = \pi/4.$

25. $\dfrac{dy}{dx} = x^2(1 + y), \quad y(0) = 3.$

26. $\sqrt{y}\,dx + (1 + x)\,dy = 0, \quad y(0) = 1.$

27. Solutions Not Expressible in Terms of Elementary Functions.   As discussed in calculus, certain indefinite integrals (antiderivatives) such as $\int e^{x^2}\,dx$ cannot be expressed in finite terms using elementary functions. When such an integral is encountered while solving a differential equation, it is often helpful to use definite integration (integrals with variable upper limit). For example, consider the initial value problem

$$\frac{dy}{dx} = e^{x^2}y^2, \qquad y(2) = 1.$$

Dividing both sides by y^2 and integrating from $x = 2$ to $x = x_1$ gives

$$\int_2^{x_1} y^{-2}(x)\frac{dy}{dx}\,dx = -y^{-1}(x_1) + y^{-1}(2)$$

$$= \int_2^{x_1} e^{x^2}\,dx.$$

If we let t be the variable of integration and replace x_1 by x and $y(2)$ by 1, then we can express the solution to the initial value problem by

$$y(x) = \left\{1 - \int_2^x e^{t^2}\,dt\right\}^{-1}.$$

Use definite integration to find an explicit solution to the initial value problems in (a)–(c).

(a) $dy/dx = e^{x^2}$, $\quad y(0) = 0.$
(b) $dy/dx = e^{x^2}y^{-2}$, $\quad y(0) = 1.$
(c) $dy/dx = \sqrt{1 + \sin x}(1 + y^2)$, $\quad y(0) = 1.$
(d) Use Simpson's rule (see Appendix B) to help approximate the solution to part (b) at $x = 0.5$ to three decimal places.

28. Differential Form.   Show that equation (10) represents a separable equation if both functions $M(x, y)$ and $N(x, y)$ can be expressed as a function that depends only on x times a function that depends only on y. (For examples, see Problems 6, 15, 16, 22, and 26.)

29. Uniqueness Questions.   In our discussion in Chapter 1 we indicated that in applications most *initial value problems* will have a unique solution. In fact, the existence of unique solutions was so important that we stated an existence and uniqueness theorem, Theorem 1, page 10. The method for separable equations will give us a solution, but it may not give us all the solutions (also see Problem 31). To illustrate this, consider the equation $dy/dx = y^{1/3}$.

(a) Use the method of separation of variables to show

that

$$y = \left(\frac{2x}{3} + C\right)^{3/2}$$

is a solution.
(b) Show that the initial value problem $dy/dx = y^{1/3}$ with $y(0) = 0$ is satisfied for $C = 0$ by $y = (2x/3)^{3/2}$.
(c) Now show that the constant function $y \equiv 0$ also satisfies the initial value problem given in part (b). Hence this initial value problem does not have a unique solution.
(d) Finally, show that the conditions of Theorem 1 on page 10 are not satisfied.
(The solution $y \equiv 0$ was lost because of a division by zero in the separation process; see Problem 31.)

30. Formal Solutions.   The method of separation of variables usually produces an implicit solution to the equation, which in certain cases can be solved to obtain an explicit solution. This problem illustrates that the method actually gives expressions that "formally" satisfy the equation, but some care is required when choosing the arbitrary constant if one is, in fact, going to obtain an implicit solution.

(a) Solve the equation $x + y\,dy/dx = 0$ using separation of variables to get an expression of the form $x^2 + y^2 = C$.
(b) Show that when $C = -1$, there are no real values for x and y that satisfy $x^2 + y^2 = C$.
(c) While C was an arbitrary constant of integration, show that it is only for $C > 0$ that we obtain an implicit relationship between x and y.

31. Division by Zero.   In developing our method of separation of variables, we tacitly assumed, when we divided equation (1) by $p(y)$, that $p(y) \neq 0$. This assumption may cause us to lose solutions.

(a) For the equation

$$\frac{dy}{dx} = (x - 3)(y + 1)^{2/3},$$

use separation of variables to derive the solution

$$y = -1 + (x^2/6 - x + C)^3.$$

(b) Show that $y \equiv -1$ satisfies the original equation $dy/dx = (x - 3)(y + 1)^{2/3}$.
(c) Show that there is no choice of the constant C that will make the solution in part (a) yield the solution $y \equiv -1$. Thus we lost the solution $y \equiv -1$ when we divided by $(y + 1)^{2/3}$.

32. Interval of Definition. By looking at an initial value problem $dy/dx = f(x, y)$ with $y(x_0) = y_0$, it is not always possible to determine the domain of the solution $y(x)$ or the interval over which the function $y(x)$ satisfies the differential equation.

(a) Solve the equation $dy/dx = xy^3$.

(b) Give explicitly the solutions to the initial value problem with $y(0) = 1$; $y(0) = \frac{1}{2}$; $y(0) = 2$.

(c) Determine the domains of the solutions in part (b).

(d) As found in part (c), the domains of the solutions depend upon the initial conditions. For the initial value problem $dy/dx = xy^3$ with $y(0) = a$, $a > 0$, show that *as $a \to 0^+$*, the domain approaches the whole real line $(-\infty, \infty)$, and as $a \to +\infty$, the domain shrinks to a single point.

(e) Sketch the solutions to the initial value problem $dy/dx = xy^3$ with $y(0) = a$ for $a = \pm\frac{1}{2}, \pm 1$, and ± 2.

33. Mixing. Suppose a brine containing 3 kg of salt per liter (L) runs into a tank initially filled with 400 L of water containing 20 kg of salt. If the brine enters at 10 L/min, the mixture is kept uniform by stirring, and the mixture flows out at the same rate, find the mass of salt in the tank after 10 min (see Figure 2.3). [Hint: Let A denote the number of kilograms of salt in the tank at t minutes after the process begins, and use the fact that

rate of increase in A = rate of input−rate of exit.

A further discussion of mixing problems is given in Section 3.2.]

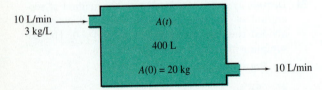

Figure 2.3 Schematic representation of a mixing problem

34. Newton's Law of Cooling. According to Newton's law of cooling, if an object at temperature T is immersed in a medium having the constant temperature M, then the rate of change of T is proportional to the difference of temperature $M - T$. This gives the differential equation

$$dT/dt = k(M - T).$$

(a) Solve the differential equation for T.

(b) A thermometer reading $100°$ is placed in a medium having a constant temperature of $70°$. After 6 min the thermometer reads $80°$. What is the reading after 20 min?

(Further applications of Newton's law of cooling appear in Section 3.3.)

35. Blood plasma is stored at $40°$. Before the plasma can be used, it must be $90°$. When the plasma is placed in an oven at $120°$, it takes 45 min for the plasma to warm to $90°$. Assume that Newton's law of cooling (Problem 34) applies and that the cooling constant k is independent of M, the temperature of the oven. How long will it take the plasma to warm to $90°$ if the oven temperature is set at (a) $100°$, (b) $140°$, (c) $80°$?

36. Free Fall. In Section 2.1 we discussed a model for an object falling toward the Earth. Assuming only air resistance and gravity are acting upon the object, we found that the velocity v must satisfy the equation

$$m\frac{dv}{dt} = mg - kv,$$

where m is the mass, g is the acceleration due to gravity, and $k > 0$ is a constant (See Figure 2.1). If $m = 100$ kg, $g = 9.8$ m/sec^2, $k = 5$ kg/sec, and $v(0) = 10$ m/sec, solve for $v(t)$. What is the limiting (or terminal) velocity of the object?

37. Compound Interest. If $P(t)$ is the amount of dollars in a savings bank account that pays a yearly interest rate of $r\%$ *compounded continuously*, then

$$\frac{dP}{dt} = \frac{r}{100}P, \quad t \text{ in years.}$$

Assume that the interest is 5% annually, $P(0) = \$1000$, and that no monies are withdrawn.

(a) How much will be in the account after 2 years?

(b) When will the account reach $4000?

(c) If $1000 is added to the account every 12 months, how much will be in the account after $3\frac{1}{2}$ years?

•2.3• EXACT EQUATIONS

The first order differential equation

$$\frac{dy}{dx} = f(x, y)$$

may also be expressed in the differential form

(1) $M(x, y)\, dx + N(x, y)\, dy = 0.$

For example, the equation

$$\frac{dy}{dx} = \frac{3x^2 - y}{x - 1}$$

can be expressed as

$$(y - 3x^2)\, dx + (x - 1)\, dy = 0,$$

where $M(x, y) = y - 3x^2$ and $N(x, y) = x - 1$. Notice that there are other ways of expressing the given equation in differential form, such as

$$\left(\frac{y - 3x^2}{x - 1}\right) dx + dy = 0.$$

To solve equation (1), it is helpful to know whether the left-hand side is a **total differential**. Recall that the total differential $dF(x, y)$ of a function $F(x, y)$ of two variables is defined by

$$dF(x, y) := \frac{\partial F}{\partial x}(x, y)\, dx + \frac{\partial F}{\partial y}(x, y)\, dy,$$

where dx and dy are arbitrary increments. For example, if $F(x, y) = \sin(xy)$, then

$$dF(x, y) = y\cos(xy)\, dx + x\cos(xy)\, dy.$$

EXACT DIFFERENTIAL FORM

Definition 2.   The differential form

$$M(x, y)\, dx + N(x, y)\, dy$$

is said to be **exact** in a rectangle R if there is a function $F(x, y)$ such that

(2) $\dfrac{\partial F}{\partial x}(x, y) = M(x, y)$ **and** $\dfrac{\partial F}{\partial y}(x, y) = N(x, y)$

for all (x, y) in R. That is, the total differential of $F(x, y)$ satisfies

$$dF(x, y) = M(x, y)\, dx + N(x, y)\, dy.$$

If $M(x, y)\, dx + N(x, y)\, dy$ is an exact differential form, then the equation

$$M(x, y)\, dx + N(x, y)\, dy = 0$$

is called an **exact equation.**

For example, the equation $y\,dx + x\,dy = 0$ is exact, since

$$d(xy) = y\,dx + x\,dy$$

is the total differential of $F(x, y) = xy$.

Solving exact equations is a simple matter once the function $F(x, y)$ of (2) is found. If we treat y as a function of x on some interval I, the equation $dF(x, y) = 0$ is equivalent to $\dfrac{d}{dx}F(x, y(x)) = 0$ for x in I. Hence $F(x, y(x)) = C$. In other words, the solutions of

$$dF(x, y) = 0$$

are given implicitly by

$$F(x, y) = C.$$

Like indefinite integrals, the function $F(x, y)$ satisfying conditions (2) is not uniquely determined, since we can always add an arbitrary constant K to $F(x, y)$. Because we are ultimately interested in the solution to the equation $dF = 0$, we can absorb the constant K in with the constant C and still express the solution by $F(x, y) = C$. Hence we need only find one function $F(x, y)$ whose total differential dF is $M\,dx + N\,dy$ in order to solve an exact differential equation.

● **EXAMPLE 1**   Show that the equation

(3) $(y - 3x^2)\,dx + (x - 1)\,dy = 0$

is exact and find its solutions.

Solution   If we regroup the terms in (3) as follows:

$$(y\,dx + x\,dy) - 3x^2\,dx - dy = 0,$$

then we see that this equation can be written in the form

$$d(xy) - d(x^3) - dy = 0,$$
$$d(xy - x^3 - y) = 0.$$

Hence $F(x, y) = xy - x^3 - y$ satisfies

$$\frac{\partial F}{\partial x}(x, y) = y - 3x^2 = M(x, y),$$

$$\frac{\partial F}{\partial y}(x, y) = x - 1 = N(x, y),$$

and so (3) is exact. Furthermore, the solutions of (3) are given implicitly by $xy - x^3 - y = C$ or explicitly by $y = (x^3 + C)/(x - 1)$.   ●

In the preceding example we were able to see a regrouping of the terms that made it easy for us to recognize a function $F(x, y)$ whose total differential was $M(x, y)\,dx + N(x, y)\,dy$. Unfortunately, for other equations it may not be simple to determine $F(x, y)$ by inspection.

Consider the equation

$$(3y + x)\,dx + 2x\,dy = 0.$$

For this equation no amount of regrouping helps, since (as we shall soon see) it is *not* an exact equation. What we really need is a test to determine whether an equation is exact, and if it is exact, then a procedure for finding the function $F(x, y)$. These needs are met by the following theorem.

> ### TEST FOR EXACTNESS
>
> **Theorem 1.**   Suppose the first partial derivatives of $M(x, y)$ and $N(x, y)$ are continuous in a rectangle R. Then
>
> $$M(x, y)\,dx + N(x, y)\,dy = 0$$
>
> is an exact equation in R if and only if
>
> **(4)** $$\frac{\partial M}{\partial y}(x, y) = \frac{\partial N}{\partial x}(x, y)$$
>
> for all (x, y) in $R.^{\dagger}$

Proof.   There are two parts to the theorem. First, if the equation is exact, then condition (4) is satisfied. Second, if condition (4) is satisfied, then the equation is exact. We will prove the first part and indicate the proof of the second, leaving the details for the exercises.

($\Rightarrow$) Assume that $M\,dx + N\,dy = 0$ is exact. Then there is a function $F(x, y)$ satisfying

(5) $$\frac{\partial F}{\partial x}(x, y) = M(x, y) \quad \text{and} \quad \frac{\partial F}{\partial y}(x, y) = N(x, y).$$

Using these equations to compute $\partial M/\partial y$ and $\partial N/\partial x$, we obtain

$$\frac{\partial M}{\partial y}(x, y) = \frac{\partial^2 F}{\partial y\,\partial x}(x, y) \quad \text{and} \quad \frac{\partial N}{\partial x}(x, y) = \frac{\partial^2 F}{\partial x\,\partial y}(x, y).$$

Since the first partial derivatives of M and N are continuous in R, the same must be true for the second order mixed partial derivatives of F. From calculus we recall that the continuity of these mixed partial derivatives implies that they are equal. Hence,

$$\frac{\partial M}{\partial y} = \frac{\partial^2 F}{\partial y\,\partial x} = \frac{\partial^2 F}{\partial x\,\partial y} = \frac{\partial N}{\partial x}$$

in R. Consequently, if $M\,dx + N\,dy = 0$ is exact, then condition (4) is satisfied.

Before proceeding with the second part of the proof, let's derive a formula for the function $F(x, y)$ satisfying the equations (5). Holding y constant, we integrate the first

---

† *Historical Footnote:* This theorem was proven by Leonhard Euler in 1734.

equation in (5) with respect to x to obtain

(6) $$F(x, y) = \int M(x, y)\,dx + g(y).$$

Notice that instead of using C to represent the constant of integration, we have written $g(y)$. This is because y is held fixed while integrating with respect to x, and so our "constant" may well depend on y. To determine $g(y)$, we differentiate both sides of (6) with respect to y to obtain

(7) $$\frac{\partial F}{\partial y}(x, y) = \frac{\partial}{\partial y}\int M(x, y)\,dx + \frac{\partial}{\partial y}g(y).$$

As g is a function of y alone, we can write $\partial g/\partial y = g'(y)$, and solving (7) for $g'(y)$ gives

$$g'(y) = \frac{\partial F}{\partial y}(x, y) - \frac{\partial}{\partial y}\int M(x, y)\,dx.$$

Since $\partial F/\partial y = N$, this last equation becomes

(8) $$g'(y) = N(x, y) - \frac{\partial}{\partial y}\int M(x, y)\,dx.$$

Notice that although the right-hand side of (8) indicates a possible dependence on x, the appearances of this variable must cancel because the left-hand side, $g'(y)$, depends only on y. By integrating (8) we can determine $g(y)$ up to a numerical constant and therefore we can determine the function $F(x, y)$ up to a numerical constant from the functions $M(x, y)$ and $N(x, y)$.

($\Leftarrow$) Suppose now that condition (4) holds. We will show that $M\,dx + N\,dy = 0$ is exact by actually exhibiting a function $F(x, y)$ that satisfies $\partial F/\partial x = M$ and $\partial F/\partial y = N$. Fortunately, we needn't look too far for such a function. The discussion in the first part of the proof suggests (6) as a candidate, where $g'(y)$ is given by (8). Namely, we define $F(x, y)$ by

(9) $$F(x, y) := \int_{x_0}^{x} M(t, y)\,dt + g(y),$$

where (x_0, y_0) is a fixed point in the rectangle R and $g(y)$ is determined, up to a numerical constant, by the equation

(10) $$g'(y) := N(x, y) - \frac{\partial}{\partial y}\int_{x_0}^{x} M(t, y)\,dt.$$

Before proceeding, we must address an extremely important question concerning the definition of $F(x, y)$. That is, how can we be sure (in this portion of the proof) that $g'(y)$, as given in equation (10), is really a function of just y? To show that the right-hand side of (10) is independent of x—i.e., that the appearances of the variable x cancel—all we need do is show that its partial derivative with respect to x is zero. This is where condition (4) is utilized. We leave to the reader this computation and the verification that $F(x, y)$ satisfies conditions (2) (see Problems 29 and 30). ◄◄◄

Theorem 1 suggests the following procedure for solving equation (1). If the equation is not separable, then compute $\partial M/\partial y$ and $\partial N/\partial x$. If $\partial M/\partial y = \partial N/\partial x$, then the equation is exact and all we must do is determine an appropriate function $F(x, y)$. Let's review the process.

> ▶ **METHOD FOR SOLVING EXACT EQUATIONS**
>
> **(a)** If $M\,dx + N\,dy = 0$ is exact, then $\partial F/\partial x = M$. Integrate this last equation with respect to x to get
>
> **(11)** $$F(x, y) = \int M(x, y)\,dx + g(y).$$
>
> **(b)** To determine $g(y)$, take the partial derivative with respect to y of both sides of equation (11) and substitute N for $\partial F/\partial y$. We can now solve for $g'(y)$.
> **(c)** Integrate $g'(y)$ to obtain $g(y)$ up to a numerical constant. Substituting $g(y)$ into equation (11) gives $F(x, y)$.
> **(d)** The solution to $M\,dx + N\,dy = 0$ is given implicitly by
>
> $$F(x, y) = C.$$
>
> (Alternatively, starting with $\partial F/\partial y = N$, the implicit solution can be found by first integrating with respect to y; see Example 3.)

● **EXAMPLE 2**   Solve

(12) $$(2xy - \sec^2 x)\,dx + (x^2 + 2y)\,dy = 0.$$

Solution   Here $M(x, y) = 2xy - \sec^2 x$ and $N(x, y) = x^2 + 2y$. Since

$$\frac{\partial M}{\partial y} = 2x = \frac{\partial N}{\partial x},$$

equation (12) is exact. To find $F(x, y)$, we begin by integrating M with respect to x:

(13) $$F(x, y) = \int (2xy - \sec^2 x)\,dx + g(y)$$

$$= x^2 y - \tan x + g(y).$$

Next we take the partial derivative of (13) with respect to y and substitute $x^2 + 2y$ for N:

$$\frac{\partial F}{\partial y}(x, y) = N(x, y),$$

$$x^2 + g'(y) = x^2 + 2y.$$

Thus $g'(y) = 2y$, and since the choice of the constant of integration is not important, we can take $g(y) = y^2$. Hence, from (13), we have $F(x, y) = x^2 y - \tan x + y^2$, and the solution to equation (12) is given implicitly by $x^2 y - \tan x + y^2 = C$.   ●

Remark.   The procedure for solving exact equations requires several steps. As a check on our work, we observe that when we solve for $g'(y)$, we must obtain a function that is independent of x. If this is not the case, then we have erred either in our computation of $F(x, y)$ or in computing $\partial M/\partial y$ or $\partial N/\partial x$.

In the construction of $F(x, y)$ we can first integrate $N(x, y)$ with respect to y to get

$$F(x, y) = \int N(x, y)\,dy + h(x) \tag{14}$$

and then proceed to find $h(x)$. We illustrate this alternative method in the next example.

EXAMPLE 3 Solve

$$(1 + e^x y + xe^x y)\,dx + (xe^x + 2)\,dy = 0. \tag{15}$$

Solution Here $M = 1 + e^x y + xe^x y$ and $N = xe^x + 2$. Since

$$\frac{\partial M}{\partial y} = e^x + xe^x = \frac{\partial N}{\partial x},$$

equation (15) is exact. If we now integrate $N(x, y)$ with respect to y, we obtain

$$F(x, y) = \int (xe^x + 2)\,dy + h(x) = xe^x y + 2y + h(x).$$

When we take the partial derivative with respect to x and substitute for M, we get

$$\frac{\partial F}{\partial x}(x, y) = M(x, y)$$

$$e^x y + xe^x y + h'(x) = 1 + e^x y + xe^x y.$$

Thus $h'(x) = 1$, and so we take $h(x) = x$. Hence $F(x, y) = xe^x y + 2y + x$, and the solution to equation (15) is given implicitly by $xe^x y + 2y + x = C$. In this case, we can solve explicitly for y to obtain $y = (C - x)/(2 + xe^x)$. ●

Remark. Since we can use either procedure for finding $F(x, y)$, it may be worthwhile to consider each of the integrals $\int M(x, y)\,dx$ and $\int N(x, y)\,dy$. If one is easier to evaluate than the other, this would be sufficient reason for us to use one method over the other. (The skeptical reader should try solving equation (15) by first integrating $M(x, y)$.)

EXAMPLE 4 Show that

$$(x + 3x^3 \sin y)\,dx + (x^4 \cos y)\,dy = 0 \tag{16}$$

is *not* exact, but that multiplying this equation by the factor x^{-1} yields an exact equation. Use this fact to solve (16).

Solution In equation (16), $M = x + 3x^3 \sin y$ and $N = x^4 \cos y$. Since

$$\frac{\partial M}{\partial y} = 3x^3 \cos y \not\equiv 4x^3 \cos y = \frac{\partial N}{\partial x},$$

then (16) is not exact. When we multiply equation (16) by the factor x^{-1}, we obtain

$$(1 + 3x^2 \sin y)\,dx + (x^3 \cos y)\,dy = 0. \tag{17}$$

For this new equation, $M = 1 + 3x^2 \sin y$ and $N = x^3 \cos y$. If we test for exactness we now find that

$$\frac{\partial M}{\partial y} = 3x^2 \cos y = \frac{\partial N}{\partial x},$$

and hence (17) is exact. Upon solving (17) we find that the solution is given implicitly by $x + x^3 \sin y = C$. Since equations (16) and (17) differ only by a factor of x, then any solution to one will be a solution for the other whenever $x \neq 0$. Hence the solution to equation (16) is given implicitly by $x + x^3 \sin y = C$. ●

In the preceding example the function $\mu = x^{-1}$ is called an **integrating factor** for equation (16). In general, any factor $\mu(x, y)$ that changes a nonexact equation into an exact equation is called an integrating factor. We discuss how to find such integrating factors in Sections 2.4 and 2.5 (see also Problems 27 and 28).

EXERCISES 2.3

In Problems 1 through 6 determine whether the given equation is separable, exact, neither, or both.

1. $(ye^{xy} + 2x)\,dx + (xe^{xy} - 2y)\,dy = 0.$

2. $(6xy - \cos x)\,dx + (3x^2)\,dy = 0.$

3. $[\cos(x - y) + 1]\,dx + [\cos(x - y) + 2y]\,dy = 0.$

4. $x\,dx + (\sin y + x^2 \sin y)\,dy = 0.$

5. $[\arctan y + \cos(x + 2y)]\,dx$
$$+ \left[\frac{x}{1 + y^2} + 2\cos(x + 2y) + y\right]dy = 0.$$

6. $\sec^2 x\,dx + \sqrt{1 - y}\,dy = 0.$

In Problems 7 through 18 determine whether the equation is exact. If it is exact, then solve it.

7. $(2xy + 3)\,dx + (x^2 - 1)\,dy = 0.$

8. $(2x + y)\,dx + (x - 2y)\,dy = 0.$

9. $(e^x \sin y - 3x^2)\,dx + (e^x \cos y + y^{-2/3}/3)\,dy = 0.$

10. $(\cos x \cos y + 2x)\,dx - (\sin x \sin y + 2y)\,dy = 0.$

11. $(1 + \ln y)\,dt + (t/y)\,dy = 0.$

12. $e^t(y - t)\,dt + (1 + e^t)\,dy = 0.$

13. $\cos\theta\,dr - (r \sin\theta - e^\theta)\,d\theta = 0.$

14. $(ye^{xy} - 1/y)\,dx + (xe^{xy} + x/y^2)\,dy = 0.$

15. $(1/y)\,dx - (2y - x/y^2)\,dy = 0.$

16. $[2x + y^2 - \cos(x + y)]\,dx$
$$+ [2xy - \cos(x + y) - e^y]\,dy = 0.$$

17. $\left(2x + \dfrac{y}{1 + x^2y^2}\right)dx + \left(\dfrac{x}{1 + x^2y^2} - 2y\right)dy = 0.$

18. $\left[\dfrac{2}{\sqrt{1 - x^2}} + y\cos(xy)\right]dx + [x\cos(xy) - y^{-1/3}]\,dy = 0.$

In Problems 19 through 24 solve the initial value problem.

19. $(1/x + 2y^2x)\,dx + (2yx^2 - \cos y)\,dy = 0, \quad y(1) = \pi.$

20. $(ye^{xy} - 1/y)\,dx + (xe^{xy} + x/y^2)\,dy = 0, \quad y(0) = 1.$

21. $(e^t y + te^t y)\,dt + (te^t + 2)\,dy = 0, \quad y(0) = -1.$

22. $(e^t x + 1)\,dt + (e^t - 1)\,dx = 0, \quad x(1) = 1.$

23. $(y^2 \sin x)\,dx + (1/x - y/x)\,dy = 0, \quad y(\pi) = 1.$

24. $(\tan y - 2)\,dx + (x\sec^2 y + 1/y)\,dy = 0, \quad y(0) = 1.$

25. For each of the following equations find the most general function $M(x, y)$ so that the equation is exact.
- **(a)** $M(x, y)\,dx + (\sec^2 y - x/y)\,dy = 0.$
- **(b)** $M(x, y)\,dx + (\sin x \cos y - xy - e^{-y})\,dy = 0.$

26. For each of the following equations find the most general function $N(x, y)$ so that the equation is exact.
- **(a)** $[y\cos(xy) + e^x]\,dx + N(x, y)\,dy = 0.$
- **(b)** $[ye^{xy} - 4x^3y + 2]\,dx + N(x, y)\,dy = 0.$

27. Integrating Factors. Consider the equation
$$(y^2 + 2xy)\,dx - x^2\,dy = 0.$$
- **(a)** Show that this equation is not exact.
- **(b)** Show that multiplying both sides of the equation by y^{-2} yields a new equation that is exact; that is, show that y^{-2} is an integrating factor.
- **(c)** Use the solution of the resulting exact equation to solve the original equation.
- **(d)** Were any solutions "lost" in the process?

28. Consider the equation

$$(5x^2y + 6x^3y^2 + 4xy^2)\,dx$$
$$+ (2x^3 + 3x^4y + 3x^2y)\,dy = 0.$$

(a) Show that the equation is not exact.

(b) Find an integrating factor of the form x^ny^m by multiplying the equation by x^ny^m and determining values for n and m that make the resulting equation exact.

(c) Use the solution of the resulting exact equation to solve the original equation.

29. Using condition (4), show that the right-hand side of (10) is independent of x by showing that its partial derivative with respect to x is zero. [Hint: Since the partial derivatives of M are continuous, Leibniz's theorem allows you to interchange the operations of integration and differentiation.]

30. Verify that $F(x, y)$ as defined by (9) and (10) satisfies conditions (2).

31. Orthogonal Trajectories. The geometric problem of finding a family of curves (orthogonal trajectories) that intersects a given family of curves orthogonally at each point occurs frequently in engineering. For example, we may be given the lines of force and want to find the equation for the equipotential lines. Consider the family of curves described by $F(x, y) = k$, where k is a parameter.

(a) Using implicit differentiation, show that for each curve in the family its slope is given by

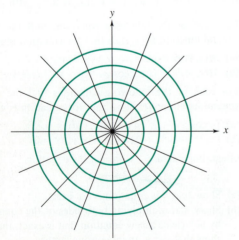

Figure 2.4 Orthogonal trajectories for concentric circles are lines through the center

$$\frac{dy}{dx} = -\frac{\partial F}{\partial x}\bigg/\frac{\partial F}{\partial y}.$$

(b) Using the fact that the slope of a curve that is orthogonal (perpendicular) to a given curve is just the negative reciprocal of the slope of the given curve, show that the curves orthogonal to the family $F(x, y) = k$ satisfy the differential equation

$$\frac{\partial F}{\partial y}(x, y)\,dx - \frac{\partial F}{\partial x}(x, y)\,dy = 0.$$

(c) Using the preceding differential equation, show that the orthogonal trajectories to the family of circles $x^2 + y^2 = k$ are just straight lines through the origin (see Figure 2.4).

(d) Show that the orthogonal trajectories to the family of hyperbolas $xy = k$ are the hyperbolas $x^2 - y^2 = k$ (see Figure 2.5).

32. Use the method in Problem 31 to find the orthogonal trajectories for each of the given families of curves.

(a) $2x^2 + y^2 = k.$ **(b)** $y = kx^4.$

(c) $y = e^{kx}.$ **(d)** $y^2 = kx.$

[Hint: First express the family in the form $F(x, y) = k$.]

33. Use the method described in Problem 31 to show that the orthogonal trajectories to the family of curves $x^2 + y^2 = kx$, k a parameter, satisfy

$$(2yx^{-1})\,dx + (y^2x^{-2} - 1)\,dy = 0.$$

Find the orthogonal trajectories by solving the above equation. Sketch the family of curves, along with their orthogonal trajectories. [Hint: Look for an integrating factor of the form $x^m y^n$.]

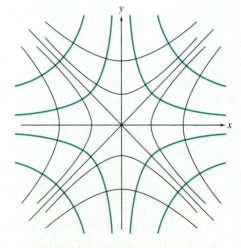

Figure 2.5 Families of orthogonal hyperbolas

2.4 LINEAR EQUATIONS

A type of first order differential equation that occurs frequently in applications is the linear equation. Recall from Section 1.1 that a **linear first order equation** is an equation that can be expressed in the form

$$(1) \qquad a_1(x)\frac{dy}{dx} + a_0(x)y = b(x),$$

where $a_1(x)$, $a_0(x)$, and $b(x)$ depend only on the independent variable x, not on y.

For example, the equation

$$x^2 \sin x - (\cos x)y = (\sin x)\frac{dy}{dx}$$

is linear since it can be rewritten in the form

$$(\sin x)\frac{dy}{dx} + (\cos x)y = x^2 \sin x.$$

However, thc equation

$$y\frac{dy}{dx} + (\sin x)y^3 = e^x + 1$$

is not linear, because it cannot be put in the form of equation (1). This is due to the presence of the y^3 and $y\,dy/dx$ terms.

In this section we assume that the functions $a_1(x)$, $a_0(x)$, and $b(x)$ are continuous on an interval and that $a_1(x) \neq 0$ on that interval. Then, on dividing by $a_1(x)$, we can rewrite equation (1) in the **standard form**

$$(2) \qquad \frac{dy}{dx} + P(x)y = Q(x),$$

where $P(x)$ and $Q(x)$ are continuous functions on the interval.

Let's express equation (2) in the differential form

$$(3) \qquad [P(x)y - Q(x)]\,dx + dy = 0.$$

If we test this equation for exactness, we find $\partial M/\partial y = P(x)$ and $\partial N/\partial x = 0$. Consequently, equation (3) is exact only when $P(x) \equiv 0$. While this is disappointing news, it is no cause for despair, for it turns out that an **integrating factor** μ, which depends only on x, can be easily obtained for the general equation (3).[†]

So let's multiply (3) by a function $\mu(x)$ and try to determine $\mu(x)$ so that the resulting equation

$$(4) \qquad [\mu(x)P(x)y - \mu(x)Q(x)]\,dx + \mu(x)\,dy = 0$$

is exact.

[†] *Historical Footnote*: This method was discovered by Gottfried Leibniz in 1694.

From the equations

$$\frac{\partial M}{\partial y}(x, y) = \mu(x)P(x) \quad \text{and} \quad \frac{\partial N}{\partial x}(x, y) = \frac{d\mu}{dx}(x),$$

we see that (4) is exact if μ satisfies the differential equation

(5) $$\frac{d\mu}{dx}(x) = P(x)\mu(x),$$

Fortunately this equation is separable and, for *any* antiderivative $\int P(x)\,dx$, has the solution

(6) $$\mu(x) = \exp\left(\int P(x)\,dx\right),$$

which is our desired integrating factor. With this choice, we can, of course, proceed to solve equation (4) by the method discussed in Section 2.3. However, there is a shorter path to finding the solution, which we now describe.

Returning to equation (2), we multiply by $\mu(x)$ defined in (6) to obtain

(7) $$\mu(x)\frac{dy}{dx} + P(x)\mu(x)y = \mu(x)Q(x).$$

We know from (5) that $P(x)\mu(x) = d\mu/dx$, and so (7) can be written in the form

$$\underbrace{\mu(x)\frac{dy}{dx} + \frac{d\mu}{dx}(x)\,y}_{} = \mu(x)Q(x),$$

$$\frac{d}{dx}(\mu(x)y) = \mu(x)Q(x),$$

where we have made use of the product rule for differentiation. Integrating with respect to x gives

$$\mu(x)y = \int \mu(x)Q(x)\,dx + C,$$

and solving for y yields

(8) $$y = \mu(x)^{-1}\left(\int \mu(x)Q(x)\,dx + C\right),$$

where $\mu(x)^{-1} = \exp(-\int P(x)\,dx)$. The function $y(x)$ given by equation (8) is referred to as the **general solution** to equation (2).

The above discussion suggests the following theorem.

> ### ◢ EXISTENCE AND UNIQUENESS OF SOLUTION
>
> **Theorem 2.**   Suppose $P(x)$ and $Q(x)$ are continuous on an interval (a, b) that contains the point x_0. Then for any choice of initial value y_0, there exists a unique solution $y(x)$ on (a, b) to the initial value problem
>
> **(9)** $$\frac{dy}{dx} + P(x)y = Q(x), \qquad y(x_0) = y_0.$$
>
> In fact, the solution is given by (8) for a suitable value of C.

The proof of Theorem 2 is based on the previous discussion, that led to equation (8). (See Problems 32 and 33 for the details.) Like Theorem 1 of Chapter 1, Theorem 2 is an existence and uniqueness theorem. It asserts that the initial value problem (9) *always has a solution* on the interval (a, b) *for any choice of initial value* y_0. Furthermore, the initial value problem (9) has *only one solution* on (a, b) for a given value y_0. Theorem 2 does differ from Theorem 1 on page 10 in that for the linear initial value problem (9) we have the existence and uniqueness of the solution on the *whole* interval (a, b), rather than on some smaller interval about x_0.

We can summarize the method for solving linear equations as follows:

> ### ◢ METHOD FOR SOLVING LINEAR EQUATIONS
>
> **(a)** Write the equation in the standard form
>
> $$\frac{dy}{dx} + P(x)y = Q(x).$$
>
> **(b)** Calculate the integrating factor $\mu(x)$ by the formula
>
> $$\mu(x) = \exp\left(\int P(x)\,dx\right).$$
>
> **(c)** Multiply the equation in standard form by $\mu(x)$ and, recalling that the left-hand side is just $\frac{d}{dx}(\mu(x)y)$, obtain
>
> $$\underbrace{\mu(x)\frac{dy}{dx} + P(x)\mu(x)y}_{} = \mu(x)Q(x),$$
>
> $$\frac{d}{dx}(\mu(x)y) \qquad = \mu(x)Q(x).$$
>
> **(d)** Integrate the last equation and solve for y by dividing by $\mu(x)$ to obtain (8).

● EXAMPLE 1 Find the general solution to

$$\textbf{(10)} \qquad \frac{1}{x}\frac{dy}{dx} - \frac{2y}{x^2} = x\cos x, \qquad x > 0.$$

Solution To put this linear equation in standard form, we multiply by x to obtain

$$\textbf{(11)} \qquad \frac{dy}{dx} - \frac{2}{x}y = x^2\cos x.$$

Here $P(x) = -2/x$, and so

$$\int P(x)\,dx = \int \frac{-2}{x}\,dx = -2\ln|x|.$$

Thus an integrating factor is

$$\mu(x) = e^{-2\ln|x|} = e^{\ln(x^{-2})} = x^{-2}.$$

Multiplying equation (11) by $\mu(x)$ yields

$$\underbrace{x^{-2}\frac{dy}{dx} - 2x^{-3}y}_{} = \cos x,$$

$$\frac{d}{dx}(x^{-2}y) \quad = \cos x.$$

We now integrate both sides and solve for y to find

$$x^{-2}y = \int \cos x\,dx = \sin x + C.$$

$$\textbf{(12)} \qquad y = x^2\sin x + Cx^2.$$

This solution is valid for $x > 0$ since both $P(x) = -2/x$ and $Q(x) = x^2\cos x$ are continuous for $x > 0$. In Figure 2.6 we have sketched solutions for various values of the constant C in (12). ●

In the next example we encounter a linear equation that arises in the study of the radioactive decay of an isotope.

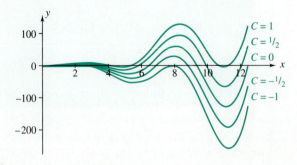

Figure 2.6 Graph of $y = x^2\sin x + Cx^2$ for five values of the constant C

● **EXAMPLE 2**   A rock contains two radioactive isotopes RA_1 and RA_2 that belong to the same radioactive series; that is, RA_1 decays into RA_2 which then decays into stable atoms. Assume that the rate that RA_1 decays into RA_2 is $50e^{-10t}$ kg/sec. Since the rate of decay of RA_2 is proportional to the mass $y(t)$ of RA_2 present, then the rate of change in RA_2 is

$$\frac{dy}{dt} = \text{rate of creation} - \text{rate of decay},$$

(13) $$\frac{dy}{dt} = 50e^{-10t} - ky,$$

where $k > 0$ is the decay constant. If $k = 2/\text{sec}$ and initially $y(0) = 40$ kg, find the mass $y(t)$ of RA_2 for $t \geq 0$.

Solution   Equation (13) is linear so we begin by writing it in standard form

(14) $$\frac{dy}{dt} + 2y = 50e^{-10t}, \qquad y(0) = 40,$$

where we have substituted $k = 2$ and displayed the initial condition. We now see that $P(t) = 2$ and so $\int P(t)\,dt = \int 2\,dt = 2t$. Thus an integrating factor is $\mu(t) = e^{2t}$. Multiplying equation (14) by $\mu(t)$ yields

$$\underbrace{e^{2t}\frac{dy}{dt} + 2e^{2t}y}_{} = 50e^{-8t},$$

$$\frac{d}{dt}(e^{2t}y) = 50e^{-8t}.$$

Integrating both sides and solving for y, we find

$$e^{2t}y = -\frac{25}{4}e^{-8t} + C,$$

$$y = -\frac{25}{4}e^{-10t} + Ce^{-2t}.$$

Substituting $t = 0$ and $y(0) = 40$ gives

$$40 = -\frac{25}{4}e^0 + Ce^0 = -\frac{25}{4} + C,$$

and so $C = 40 + 25/4 = 185/4$. Thus the mass $y(t)$ of RA_2 is given by

(15) $$y(t) = \left(\frac{185}{4}\right)e^{-2t} - \left(\frac{25}{4}\right)e^{-10t}, \qquad t \geq 0. \quad ●$$

The theory of linear differential equations is an important branch of mathematics not only because these equations occur in applications, but because of the elegant structure associated with them. For example, first order linear equations always have a general solution given by equation (8). Some further properties of first order linear equations are described in Problems 26 and 34. Higher order linear equations will be treated in Chapters 4, 6, and 8.

EXERCISES 2.4

1, 3, 5, 9, 19, 21

In Problems 1 through 8 classify the equation as separable, exact, or linear. Notice that some equations may have more than one classification.

1. $(x^{10/3} - 2y)\,dx + x\,dy = 0.$

2. $(x^2 y + x^4 \cos x)\,dx - x^3\,dy = 0.$

3. $(ye^{xy} + 2x)\,dx + (xe^{xy} - 2y)\,dy = 0.$

4. $\sqrt{-2y - y^2}\,dx + (3 + 2x - x^2)\,dy = 0.$

5. $y^2\,dx + (2xy + \cos y)\,dy = 0.$

6. $xy\,dx + dy = 0.$

7. $(3r - \theta - 1)\,d\theta + \theta\,dr = 0.$

8. $[2x + y\cos(xy)]\,dx + [x\cos(xy) - 2y]\,dy = 0.$

In Problems 9 through 18 obtain the general solution to the equation.

9. $\dfrac{dy}{dx} = \dfrac{y}{x} + 2x + 1.$

10. $\dfrac{dy}{dx} - y = e^{3x}.$

11. $(t + y + 1)\,dt - dy = 0.$

12. $x\dfrac{dy}{dx} + 2y = x^{-3}.$

13. $\dfrac{dr}{d\theta} + r\tan\theta = \sec\theta.$

14. $\dfrac{dy}{dx} = x^2 e^{-4x} - 4y.$

15. $y\dfrac{dx}{dy} + 2x = 5y^3.$

16. $x\dfrac{dy}{dx} + 3y + 2x^2 = x^3 + 4x.$

17. $(x^2 + 1)\dfrac{dy}{dx} + xy = x.$

18. $(x^2 + 1)\dfrac{dy}{dx} = x^2 + 2x - 1 - 4xy.$

In Problems 19 through 24 solve the initial value problem.

19. $\dfrac{dy}{dx} - \dfrac{y}{x} = xe^x, \qquad y(1) = e - 1.$

20. $\dfrac{dy}{dx} + 4y - e^{-x} = 0, \qquad y(0) = \dfrac{4}{3}.$

21. $x^3\dfrac{dy}{dx} + 3x^2 y = x, \qquad y(2) = 0.$

22. $\dfrac{dy}{dx} + \dfrac{3y}{x} + 2 = 3x, \qquad y(1) = 1.$

23. $\cos x\dfrac{dy}{dx} + y\sin x = 2x\cos^2 x, \qquad y\!\left(\dfrac{\pi}{4}\right) = \dfrac{-15\sqrt{2}\pi^2}{32}.$

24. $\sin x\dfrac{dy}{dx} + y\cos x = x\sin x, \qquad y\!\left(\dfrac{\pi}{2}\right) = 2.$

25. Solve the equation
$$\frac{dy}{dx} = \frac{1}{e^{4y} + 2x}.$$

26. Constant Multiples of Solutions.

(a) Show that $y = e^{-x}$ is a solution of the linear equation

(16) $\qquad \dfrac{dy}{dx} + y = 0,$

and $y = x^{-1}$ is a solution of the nonlinear equation

(17) $\qquad \dfrac{dy}{dx} + y^2 = 0.$

(b) Show that for any constant C, Ce^{-x} is a solution of equation (16), while Cx^{-1} is a solution of equation (17) only when $C = 0$ or 1.

(c) Show that for any linear equation of the form
$$\frac{dy}{dx} + P(x)y = 0,$$
if $\hat{y}(x)$ is a solution, then for any constant C, the function $C\hat{y}(x)$ is also a solution.

27. Solutions Not Expressible in Terms of Elementary Functions. Solve the following initial value problems using definite integration (see Problem 27 in Exercises 2.2).

(a) $\dfrac{dy}{dx} + 2xy = 1, \qquad y(2) = 1.$

(b) $\dfrac{dy}{dx} + \dfrac{\sin 2x}{2(1 + \sin^2 x)}y = 1, \qquad y(0) = 0.$

(c) Use Simpson's rule (see Appendix B) to help approximate the solution to part (a) at $x = 2.5$ to three decimal places.

28. Bernoulli Equations. The equation

(18) $\qquad \dfrac{dy}{dx} + 2y = xy^{-2},$

is an example of a Bernoulli equation. (Further discussion of Bernoulli equations appears in Section 2.6.)

(a) Show that the substitution $v = y^3$ reduces equation (18) to the equation

(19) $\qquad \dfrac{dv}{dx} + 6v = 3x.$

(b) Solve equation (19) for v; then make the substitution $v = y^3$ to obtain the solution to equation (18).

29. Discontinuous Coefficients. As we will see in Chapter 3, there are times when the coefficient $P(x)$ in a linear equation may not be continuous, but may have a jump discontinuity. Fortunately, we may still obtain a reasonable solution. For example, consider the initial value problem

$$\frac{dy}{dx} + P(x)y = x, \qquad y(0) = 1,$$

where

$$P(x) := \begin{cases} 1, & 0 \le x \le 2, \\ 3, & x > 2. \end{cases}$$

(a) Find the general solution for $0 \le x \le 2$.
(b) Choose the constant in the solution of part (a) so that the initial condition is satisfied.
(c) Find the general solution for $x > 2$.
(d) Now choose the constant in the general solution from part (c) so that the solution from part (b) and the solution from part (c) agree at $x = 2$. By patching the two solutions together we are able to obtain a continuous function that satisfies the differential equation except at $x = 2$, where its derivative is undefined.
(e) Sketch the graph of the solution from $x = 0$ to $x = 5$.

30. Discontinuous Forcing Terms. There are times when the forcing term $Q(x)$ in a linear equation may not be continuous, but may have a jump discontinuity. Fortunately, we may still obtain a reasonable solution using the procedure discussed in Problem 29. Use this procedure to find the continuous solution to the initial value problem

$$\frac{dy}{dx} + 2y = Q(x), \qquad y(0) = 0,$$

where

$$Q(x) := \begin{cases} 2, & 0 \le x \le 3, \\ -2, & x > 3. \end{cases}$$

Sketch the graph of the solution from $x = 0$ to $x = 7$.

31. Singular Points. Those values of x for which $P(x)$ in equation (2) is not defined are called **singular points** of the equation. For example, $x = 0$ is a singular point of the equation $xy' + 2y = 3x$, since when the equation is written in the standard form $y' + (2/x)y = 3$, we see that $P(x) = 2/x$ is not defined at $x = 0$. On an interval containing a singular point the questions of the existence and uniqueness of a solution are left unanswered since Theorem 2 does not apply. To show the possible behavior of solutions near a singular point consider the following equations.

(a) Show that $xy' + 2y = 3x$ has only one solution defined at $x = 0$. Then show that the initial value problem for this equation with initial condition $y(0) = y_0$ has a unique solution when $y_0 = 0$ and no solution when $y_0 \ne 0$.

(b) Show that $xy' - 2y = 3x$ has an infinite number of solutions defined at $x = 0$. Then show that the initial value problem for this equation with initial condition $y(0) = 0$ has an infinite number of solutions.

32. Existence. Under the assumptions of Theorem 2, we will show that equation (8) gives a solution to equation (2) on (a, b). We can then choose the constant C in equation (8) so that the initial value problem (9) has a solution.

(a) Show that since $P(x)$ is continuous on (a, b), then $\mu(x)$ defined in (6) is a positive, continuous function satisfying $d\mu/dx = P(x)\mu(x)$ on (a, b).
(b) Since

$$\frac{d}{dx} \int \mu(x)Q(x)\, dx = \mu(x)Q(x),$$

verify that y given in equation (8) satisfies equation (2) by differentiating both sides of equation (8).
(c) Show that when we let $\int \mu(x)Q(x)\, dx$ be the antiderivative whose value at x_0 is 0, (i.e., $\int_{x_0}^{x} \mu(t)Q(t)\, dt$), then by choosing C to be $y_0\mu(x_0)$, the initial condition $y(x_0) = y_0$ is satisfied.

33. Uniqueness. Under the assumptions of Theorem 2, we will show that the initial value problem (9) has only one solution. Let $y_1(x)$ and $y_2(x)$ be solutions to (9).

(a) Show that $\hat{y}(x) = y_1(x) - y_2(x)$ satisfies the initial value problem

$$\frac{dy}{dx} + P(x)y = 0, \qquad y(x_0) = 0.$$

(b) Let $\mu(x)$ be the integrating factor defined in equation (6). Show that

$$\frac{d}{dx}(\mu(x)\hat{y}(x)) = 0,$$

and thus $\mu(x)\hat{y}(x) = C$, for some constant C.
(c) Since $\hat{y}(x_0) = 0$ and $\mu(x) > 0$ for x in (a, b), show

that $\hat{y}(x) = 0$ for all x in (a, b). Hence $y_1(x) = y_2(x)$ for all x in (a, b).

34. Variation of Parameters. Here is another procedure for solving linear equations which is particularly useful for higher order linear equations. This method is called **variation of parameters.** It is based on the idea that just knowing the *form* of the solution, we can substitute into the given equation and solve for any unknowns. Here we illustrate the method for first order equations (see Sections 4.10 and 6.5 for the generalization to higher order equations).

(a) Show that the general solution to

(20) $\qquad \dfrac{dy}{dx} + P(x)y = Q(x)$

has the form

$$y(x) = Cy_h(x) + y_p(x),$$

where y_h is a solution to equation (20) when $Q(x) \equiv 0$, C is a constant, and $y_p(x) = v(x)y_h(x)$ for a suitable function $v(x)$. [Hint: Show that we can take $y_h = \mu^{-1}(x)$ and then use equation (8).]

Knowing this form, we can determine the unknown function y_h by solving a separable equation; then direct substitution into the original equation will give a simple equation that can be solved for v.

Use this procedure to find the general solution to

(21) $\qquad \dfrac{dy}{dx} + \dfrac{3}{x}y = x^2, \qquad x > 0,$

by completing the following steps:

(b) Find a nontrivial solution y_h to the separable equation

(22) $\qquad \dfrac{dy}{dx} + \dfrac{3}{x}y = 0, \qquad x > 0.$

(c) Assuming that (21) has a solution of the form $y_p(x) = v(x)y_h(x)$, substitute this into equation (21) and simplify to obtain $v'(x) = x^2/y_h(x)$.

(d) Now integrate to get $v(x)$.

(e) Verify that $y(x) = Cy_h(x) + v(x)y_p(x)$ is a general solution to (21).

35. Secretion of Hormones. The secretion of hormones into the blood is often a periodic activity. If a hormone is secreted on a 24-hr cycle, then the rate of change of the level of the hormone in the blood may be represented by the initial value problem

$$\dfrac{dx}{dt} = \alpha - \beta \cos\dfrac{\pi t}{12} - kx, \qquad x(0) = x_0,$$

where $x(t)$ is the amount of the hormone in the blood at time t, α is the average secretion rate, β is the amount of variation in the secretion, and k is a positive constant reflecting the rate at which the body removes the hormone from the blood. If $\alpha = \beta = 1$, $k = 2$, and $x_0 = 10$, solve for $x(t)$.

36. Mixing. Suppose a brine containing 2 kg of salt per liter runs into a tank initially filled with 500 L of water containing 50 kg of salt. The brine enters the tank at a rate of 5 L/min. The mixture, kept uniform by stirring, is flowing out at the rate of 5 L/min (see Figure 2.7).

(a) Find the concentration, in kilograms per liter, of salt in the tank after 10 min. [Hint: Let A denote the number of kilograms of salt in the tank at t minutes after the process begins, and use the fact that

rate of increase in A = rate of input−rate of exit.

A further discussion of mixing problems is given in Section 3.2.]

(b) After 10 min a leak develops in the tank and an additional liter per minute of mixture flows out of the tank (see Figure 2.8). What will be the concentration, in kilograms per liter, of salt in the tank 20 min after the leak develops? [Hint: Use the method discussed in Problems 29 and 30.]

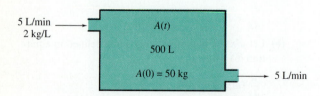

Figure 2.7 Mixing problem with equal flow rates

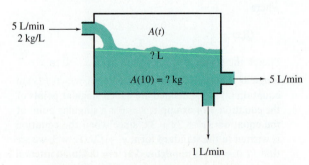

Figure 2.8 Mixing problem with unequal flow rates

37. Radioactive Decay. In Example 2 assume that the rate at which RA_1 decays into RA_2 is $40e^{-20t}$ kg/sec and the decay constant in equation (13) is $k = 5/$sec. Find the mass $y(t)$ of RA_2 for $t \geq 0$ if initially $y(0) = 10$ kg.

38. In Example 2 the decay constant for isotope RA_1 was $10/$sec which expresses itself in the exponent of the rate term $50e^{-10t}$ kg/sec. When the decay constant for RA_2 is $k = 2/$sec, we see that in formula (15) for y the term $(185/4)e^{-2t}$ eventually dominates (has greater magnitude for t large).

(a) Redo Example 2 taking $k = 20/$sec. Which term in the solution now eventually dominates?

(b) Redo Example 2 taking $k = 10/$sec.

• 2.5 • SPECIAL INTEGRATING FACTORS (Optional)

In solving linear equations written in differential form, we found a multiplier μ that made the resulting equation exact. Do such multiplicative factors μ exist for other classes of differential equations? If so, how are they computed? The present section is devoted to these questions.[†]

INTEGRATING FACTOR

Definition 3.   If the equation

(1) $M(x, y)\,dx + N(x, y)\,dy = 0$

is not exact, but the equation

(2) $\mu(x, y)M(x, y)\,dx + \mu(x, y)N(x, y)\,dy = 0,$

which results from multiplying equation (1) by the function $\mu(x, y)$, is exact, then $\mu(x, y)$ is called an **integrating factor** of the equation (1).

● **EXAMPLE 1**   Show that $\mu(x, y) = xy^2$ is an integrating factor for

(3) $(2y - 6x)\,dx + (3x - 4x^2y^{-1})\,dy = 0.$

Use this integrating factor to solve the equation.

Solution   We leave it for the reader to show that (3) is not exact. Multiplying (3) by $\mu(x, y) = xy^2$, we obtain

(4) $(2xy^3 - 6x^2y^2)\,dx + (3x^2y^2 - 4x^3y)\,dy = 0.$

For this equation we have $M = 2xy^3 - 6x^2y^2$ and $N = 3x^2y^2 - 4x^3y$. Since

$$\frac{\partial M}{\partial y}(x, y) = 6xy^2 - 12x^2y = \frac{\partial N}{\partial x}(x, y),$$

equation (4) is exact, and so $\mu(x, y) = xy^2$ is indeed an integrating factor of equation (3).

[†] *Historical Footnote*: A general theory of integrating factors was developed by Alexis Clairaut in 1739. Leonhard Euler also studied classes of equations that could be solved using a specific integrating factor.

Let us now solve equation (4) using the procedure of Section 2.3. To find $F(x, y)$ we begin by integrating M with respect to x:

$$F(x, y) = \int (2xy^3 - 6x^2y^2)\, dx + g(y) = x^2y^3 - 2x^3y^2 + g(y).$$

When we take the partial derivative with respect to y and substitute for N, we find

$$\frac{\partial F}{\partial y}(x, y) = N(x, y)$$

$$3x^2y^2 - 4x^3y + g'(y) = 3x^2y^2 - 4x^3y.$$

Thus $g'(y) = 0$, and so we can take $g(y) \equiv 0$. Hence $F(x, y) = x^2y^3 - 2x^3y^2$, and the solution to equation (4) is given implicitly by

$$x^2y^3 - 2x^3y^2 = C.$$

While equations (3) and (4) have essentially the same solutions, *it is possible to lose or gain solutions when multiplying by $\mu(x, y)$.* In this case $y \equiv 0$ is a solution of equation (4) but not of equation (3). The extraneous solution arises because when we multiply (3) by $\mu = xy^2$ to obtain (4), we are actually multiplying both sides of (3) by zero if $y \equiv 0$. This gives us $y \equiv 0$ as a solution to (4), while it is not a solution to (3). ●

How do we find an integrating factor? If $\mu(x, y)$ is an integrating factor of (1) with continuous first partial derivatives, then testing (2) for exactness we must have

$$\frac{\partial}{\partial y}[\mu(x, y)M(x, y)] = \frac{\partial}{\partial x}[\mu(x, y)N(x, y)].$$

By use of the product rule, this reduces to the equation

(5) $$M\frac{\partial \mu}{\partial y} - N\frac{\partial \mu}{\partial x} = \left(\frac{\partial N}{\partial x} - \frac{\partial M}{\partial y}\right)\mu.$$

But solving the partial differential equation (5) for μ is usually more difficult than solving the original equation (1). There are, however, two important exceptions.

Let's assume equation (1) has an integrating factor that depends only on x; that is, $\mu = \mu(x)$. In this case equation (5) reduces to the separable equation

(6) $$\frac{d\mu}{dx} = \left\{\frac{\partial M/\partial y - \partial N/\partial x}{N}\right\}\mu,$$

where $(\partial M/\partial y - \partial N/\partial x)/N$ is just a function of x. In a similar fashion, if equation (1) has an integrating factor that depends only on y, then equation (5) reduces to the separable equation

(7) $\qquad \dfrac{d\mu}{dy} = \left\{ \dfrac{\partial N/\partial x - \partial M/\partial y}{M} \right\} \mu,$

where $(\partial N/\partial x - \partial M/\partial y)/M$ is a function of just y.

    We can reverse the above argument. In particular, if $(\partial M/\partial y - \partial N/\partial x)/N$ is a function that depends only on x, then we can solve the separable equation (6) to obtain the integrating factor

$$\mu(x) = \exp\left(\int \left\{ \dfrac{\partial M/\partial y - \partial N/\partial x}{N} \right\} dx \right)$$

for equation (1). We summarize these observations in the following theorem.

SPECIAL INTEGRATING FACTORS

Theorem 3.   If $(\partial M/\partial y - \partial N/\partial x)/N$ is continuous and depends only on x, then

(8) $\qquad \mu(x) = \exp\left(\int \left\{ \dfrac{\partial M/\partial y - \partial N/\partial x}{N} \right\} dx \right)$

is an integrating factor for equation (1).

    If $(\partial N/\partial x - \partial M/\partial y)/M$ is continuous and depends only on y, then

(9) $\qquad \mu(y) = \exp\left(\int \left\{ \dfrac{\partial N/\partial x - \partial M/\partial y}{M} \right\} dy \right)$

is an integrating factor for equation (1).

Theorem 3 suggests the following procedure:

METHOD FOR FINDING SPECIAL INTEGRATING FACTORS

If $M\,dx + N\,dy = 0$ is neither separable nor linear, compute $\partial M/\partial y$ and $\partial N/\partial x$. If $\partial M/\partial y = \partial N/\partial x$, then the equation is exact. If it is not exact, consider

(10) $\qquad \dfrac{\partial M/\partial y - \partial N/\partial x}{N}.$

If (10) is a function of just x, then an integrating factor is given by formula (8). If not, consider

(11) $\qquad \dfrac{\partial N/\partial x - \partial M/\partial y}{M}.$

If (11) is a function of just y, then an integrating factor is given by formula (9).

● EXAMPLE 2 Solve

$$(2x^2 + y)\,dx + (x^2y - x)\,dy = 0. \tag{12}$$

Solution A quick inspection shows that equation (12) is neither separable nor linear. We also note that

$$\frac{\partial M}{\partial y} = 1 \not\equiv (2xy - 1) = \frac{\partial N}{\partial x}.$$

Since (12) is not exact, we compute

$$\frac{\partial M/\partial y - \partial N/\partial x}{N} = \frac{1 - (2xy - 1)}{x^2y - x} = \frac{2(1 - xy)}{-x(1 - xy)} = \frac{-2}{x}.$$

Since we obtain a function of only x, an integrating factor for (12) is given by formula (8). That is,

$$\mu(x) = \exp\left(\int \frac{-2}{x}\,dx\right) = x^{-2}.$$

When we multiply (12) by $\mu = x^{-2}$ we get the exact equation

$$(2 + yx^{-2})\,dx + (y - x^{-1})\,dy = 0.$$

Solving this equation, we ultimately derive the implicit solution

$$2x - yx^{-1} + \frac{y^2}{2} = C. \tag{13}$$

Notice that the solution $x \equiv 0$ was lost in multiplying by $\mu = x^{-2}$. Hence (13) and $x \equiv 0$ are solutions to equation (12). ●

There are many differential equations that are not covered by Theorem 3 but for which an integrating factor nevertheless exists. The major difficulty, however, is in finding an explicit formula for these integrating factors, which in general will depend on both x and y.

EXERCISES 2.5

In Problems 1 through 6 identify the equation as separable, linear, exact, or having an integrating factor that is a function of either x alone or y alone.

1. $(2x + yx^{-1})\,dx + (xy - 1)\,dy = 0.$
2. $(2y^3 + 2y^2)\,dx + (3y^2x + 2xy)\,dy = 0.$
3. $(y^2 + 2xy)\,dx - x^2\,dy = 0.$
4. $(2x + y)\,dx + (x - 2y)\,dy = 0.$
5. $(2y^2x - y)\,dx + x\,dy = 0.$
6. $(x^2 \sin x + 4y)\,dx + x\,dy = 0.$

In Problems 7 through 12 solve the equation.

7. $(3x^2 + y)\,dx + (x^2y - x)\,dy = 0.$
8. $(2xy)\,dx + (y^2 - 3x^2)\,dy = 0.$
9. $(2y^2 + 2y + 4x^2)\,dx + (2xy + x)\,dy = 0.$
10. $(x^4 - x + y)\,dx - x\,dy = 0.$
11. $(y^2 + 2xy)\,dx - x^2\,dy = 0.$
12. $(2xy^3 + 1)\,dx + (3x^2y^2 - y^{-1})\,dy = 0.$

In Problems 13 and 14 find an integrating factor of the form x^ny^m and solve the equation.

13. $(2y^2 - 6xy) dx + (3xy - 4x^2) dy = 0.$

14. $(12 + 5xy) dx + (6xy^{-1} + 3x^2) dy = 0.$

15. Show that if $(\partial N/\partial x - \partial M/\partial y)/(xM - yN)$ depends only on the product xy, that is

$$\frac{\partial N/\partial x - \partial M/\partial y}{xM - yN} = H(xy),$$

then the equation $M(x, y) dx + N(x, y) dy = 0$ has an integrating factor of the form $\mu(xy)$. Give the general formula for $\mu(xy)$.

16. If $xM(x, y) + yN(x, y) \equiv 0$, find the solution to the equation

$$M(x, y) dx + N(x, y) dy = 0.$$

17. Fluid Flow.   The streamlines associated with a certain fluid flow are represented by the family of curves $y = x - 1 + ke^{-x}$. The velocity potentials of the flow are just the orthogonal trajectories of this family.

(a) Use the method described in Problem 31 of Exercises 2.3 to show that the velocity potentials satisfy

$$dx + (x - y) dy = 0.$$

[Hint: First express the family $y = x - 1 + ke^{-x}$ in the form $F(x, y) = k$.]

(b) Find the velocity potentials by solving the equation obtained in part (a).

•2.6 • SUBSTITUTIONS AND TRANSFORMATIONS (Optional)

When the equation

$$M(x, y) dx + N(x, y) dy = 0$$

is not a separable, exact, or linear equation, it may still be possible to transform it into one that we know how to solve. This, in fact, was our approach in Section 2.5, where we used an integrating factor to transform our original equation into an exact equation.

In this section we study four types of equations that can be transformed into either a separable or linear equation by means of a suitable substitution or transformation.

► SUBSTITUTION PROCEDURE

(a) Identify the type of equation and determine the appropriate substitution or transformation.
(b) Rewrite the original equation in terms of new variables.
(c) Solve the transformed equation.
(d) Express the solution in terms of the original variables.

Homogeneous Equations

► HOMOGENEOUS EQUATION

Definition 4.   If the right-hand side of the equation

(1) $$\frac{dy}{dx} = f(x, y)$$

can be expressed as a function of the ratio y/x alone, then we say the equation is **homogeneous.**

For example, the equation

(2) $(x - y)\,dx + x\,dy = 0$

can be written in the form

$$\frac{dy}{dx} = \frac{y - x}{x} = \frac{y}{x} - 1.$$

Since we have expressed $(y - x)/x$ as a function of the ratio y/x; that is, $(y - x)/x = G(y/x)$, where $G(v) := v - 1$, then equation (2) is homogeneous.

The equation

(3) $(x - 2y + 1)\,dx + (x - y)\,dy = 0.$

can be written in the form

$$\frac{dy}{dx} = \frac{x - 2y + 1}{y - x} = \frac{1 - 2(y/x) + (1/x)}{(y/x) - 1}.$$

Here the right-hand side cannot be expressed as a function of y/x alone because of the term $1/x$ in the numerator. Hence equation (3) is not homogeneous.

One test for the homogeneity of equation (1) is to replace x by tx and y by ty; then (1) is homogeneous if and only if

$$f(tx, ty) = f(x, y),$$

for all $t \neq 0$ (see Problem 43(a)).

To solve a homogeneous equation we make a rather obvious substitution. Let

$$v = \frac{y}{x}.$$

Our homogeneous equation now has the form

(4) $\dfrac{dy}{dx} = G(v),$

and all we need is to express dy/dx in terms of x and v. Since $v = y/x$, then $y = vx$. Keeping in mind that both v and y are functions of x, we use the product rule for differentiation to obtain from $y = vx$ that

$$\frac{dy}{dx} = v + x\frac{dv}{dx}.$$

Substituting the above expression for dy/dx into equation (4) yields

(5) $v + x\dfrac{dv}{dx} = G(v).$

Fortunately, the new equation (5) is separable, and we can obtain its implicit solution from

$$\int \frac{1}{G(v) - v}\,dv = \int \frac{1}{x}\,dx.$$

All that remains is to express the solution in terms of the original variables x and y.

● **EXAMPLE 1**   Solve

(6) $(xy + y^2 + x^2)\,dx - x^2\,dy = 0.$

Solution   A check will show that equation (6) is not separable, exact, or linear. If we express (6) in the derivative form

(7) $\dfrac{dy}{dx} = \dfrac{xy + y^2 + x^2}{x^2} = \dfrac{y}{x} + \left(\dfrac{y}{x}\right)^2 + 1,$

then we see that the right-hand side of (7) is a function of just y/x. Thus equation (6) is homogeneous.

Now let $v = y/x$ and recall that $dy/dx = v + x(dv/dx)$. With these substitutions equation (7) becomes

$$v + x\frac{dv}{dx} = v + v^2 + 1.$$

The above equation is separable, and, on separating the variables and integrating, we obtain

$$\int \frac{1}{v^2 + 1}\,dv = \int \frac{1}{x}\,dx,$$

$$\arctan v = \ln|x| + C.$$

Hence

$$v = \tan(\ln|x| + C).$$

Finally, we substitute y/x for v and solve for y to get

$$y = x\tan(\ln|x| + C)$$

as an explicit solution to equation (6).   ●

Equations of the Form $dy/dx = G(ax + by)$

When the right-hand side of the equation $dy/dx = f(x, y)$ can be expressed as a function of $ax + by$, where a and b are constants, that is

$$\frac{dy}{dx} = G(ax + by),$$

then the substitution

$$z = ax + by$$

transforms the equation into a separable one. The method is illustrated in the next example.

● **EXAMPLE 2**   Solve

(8) $\dfrac{dy}{dx} = y - x - 1 + (x - y + 2)^{-1}.$

Solution   Since the right-hand side can be expressed as a function of $x - y$, that is,

$$y - x - 1 + (x - y + 2)^{-1} = -(x - y) - 1 + [(x - y) + 2]^{-1},$$

let $z = x - y$. To solve for dy/dx we differentiate $z = x - y$ with respect to x to obtain $dz/dx = 1 - dy/dx$, and so $dy/dx = 1 - dz/dx$. Substituting into (8) yields

$$1 - \dfrac{dz}{dx} = -z - 1 + (z + 2)^{-1},$$

or

$$\dfrac{dz}{dx} = (z + 2) - (z + 2)^{-1}.$$

Solving this separable equation, we obtain

$$\int \dfrac{z + 2}{(z + 2)^2 - 1}\, dz = \int dx,$$

$$\dfrac{1}{2}\ln|(z + 2)^2 - 1| = x + C_1,$$

from which it follows that

$$(z + 2)^2 = Ce^{2x} + 1.$$

Finally, replacing z by $x - y$ yields

$$(x - y + 2)^2 = Ce^{2x} + 1$$

as an implicit solution to equation (8).   ●

Bernoulli Equations

> **BERNOULLI EQUATION**
>
> **Definition 5.**   A first order equation that can be written in the form
>
> (9) $\dfrac{dy}{dx} + P(x)y = Q(x)y^n,$
>
> where $P(x)$ and $Q(x)$ are continuous on an interval (a, b) and n is a real number, is called a **Bernoulli equation.**[†]

[†] *Historical Footnote*: This equation was proposed for solution by James Bernoulli in 1695. It was solved by his brother John Bernoulli. (James and John were two of eight mathematicians in the Bernoulli family.) In 1696 Gottfried Leibniz showed that the Bernoulli equation can be reduced to a linear equation by making the substitution $v = y^{1-n}$.

Notice that when $n = 0$ or 1, equation (9) is also a linear equation and can be solved by the method discussed in Section 2.4. For other values of n, the substitution

$$v = y^{1-n}$$

transforms the Bernoulli equation into a linear equation as we now show.

Dividing equation (9) by y^n yields

(10) $y^{-n}\dfrac{dy}{dx} + P(x)y^{1-n} = Q(x).$

Taking $v = y^{1-n}$, we find via the chain rule that

$$\frac{dv}{dx} = (1 - n)y^{-n}\frac{dy}{dx},$$

and so equation (10) becomes

$$\frac{1}{1 - n}\frac{dv}{dx} + P(x)v = Q(x).$$

Since $1/(1 - n)$ is just a constant, the last equation is indeed linear.

● **EXAMPLE 3**   Solve

(11) $\dfrac{dy}{dx} - 5y = -\dfrac{5}{2}xy^3.$

Solution   This is a Bernoulli equation with $n = 3$, $P(x) = -5$, and $Q(x) = -5x/2$. To transform (11) into a linear equation, we first divide by y^3 to obtain

$$y^{-3}\frac{dy}{dx} - 5y^{-2} = -\frac{5}{2}x.$$

Next we make the substitution $v = y^{-2}$. Since $dv/dx = -2y^{-3}dy/dx$, the transformed equation is

$$-\frac{1}{2}\frac{dv}{dx} - 5v = -\frac{5}{2}x,$$

(12) $\dfrac{dv}{dx} + 10v = 5x.$

Equation (12) is linear, so we can solve it for v using the method discussed in Section 2.4. When we do this, it turns out that

$$v = \frac{x}{2} - \frac{1}{20} + Ce^{-10x}.$$

Substituting $v = y^{-2}$ gives the solution

$$y^{-2} = \frac{x}{2} - \frac{1}{20} + Ce^{-10x}.$$

Not included in the last equation is the solution $y \equiv 0$ that was lost in the process of dividing (11) by y^3.   ●

Equations with Linear Coefficients

We have used various substitutions for y to transform the original equation into a new equation that we could solve. In some cases we must transform *both* x and y into new variables, say u and v. This is the situation for **equations with linear coefficients,** that is, equations of the form

(13) $(a_1 x + b_1 y + c_1)\,dx + (a_2 x + b_2 y + c_2)\,dy = 0,$

where the a_i's, b_i's, and c_i's are constants. We leave it as an exercise to show that when $a_1 b_2 = a_2 b_1$, equation (13) can be put in the form $dy/dx = G(ax + by)$, which we solved via the substitution $z = ax + by$.

Before considering the general case when $a_1 b_2 \neq a_2 b_1$, let's first look at the special situation when $c_1 = c_2 = 0$. Equation (13) then becomes

$(a_1 x + b_1 y)\,dx + (a_2 x + b_2 y)\,dy = 0,$

which can be rewritten in the form

$$\frac{dy}{dx} = -\frac{a_1 x + b_1 y}{a_2 x + b_2 y} = -\frac{a_1 + b_1(y/x)}{a_2 + b_2(y/x)}.$$

Since this equation is homogeneous, we can solve it using the method discussed earlier in this section.

The above discussion suggests the following procedure for solving (13). If $a_1 b_2 \neq a_2 b_1$, then we seek a translation of axes of the form

$$x = u + h \quad \text{and} \quad y = v + k,$$

where h, k are constants, that will change $a_1 x + b_1 y + c_1$ into $a_1 u + b_1 v$ and change $a_2 x + b_2 y + c_2$ into $a_2 u + b_2 v$. Some elementary algebra shows that such a transformation exists if the system of equations

(14) $\begin{aligned} a_1 h + b_1 k + c_1 &= 0, \\ a_2 h + b_2 k + c_2 &= 0 \end{aligned}$

has a solution. This is assured by the assumption $a_1 b_2 \neq a_2 b$, which is geometrically equivalent to assuming that the two lines described by the system (14) intersect. Now if (h, k) satisfies (14), then the substitutions $x = u + h$ and $y = v + k$ transform equation (13) into the homogeneous equation

(15) $$\frac{dv}{du} = -\frac{a_1 u + b_1 v}{a_2 u + b_2 v} = -\frac{a_1 + b_1(v/u)}{a_2 + b_2(v/u)},$$

which we know how to solve.

● **EXAMPLE 4**   Solve

(16) $(-3x + y + 6)\,dx + (x + y + 2)\,dy = 0.$

Solution   Since $a_1 b_2 = (-3)(1) \neq (1)(1) = a_2 b_1$, we will use the translation of axes $x = u + h$, $y = v + k$, where h and k satisfy the system

$$-3h + k + 6 = 0,$$
$$h + k + 2 = 0.$$

Solving the above system for h and k gives $h = 1$, $k = -3$. Hence, we let $x = u + 1$ and $y = v - 3$. Since $dy = dv$ and $dx = du$, substituting in equation (16) for x and y yields

$$(-3u + v)\, du + (u + v)\, dv = 0,$$

or

$$\frac{dv}{du} = \frac{3 - (v/u)}{1 + (v/u)}.$$

The above equation is homogeneous, so we let $z = v/u$. Then $dv/du = z + u(dz/du)$, and, substituting for v/u, we obtain

$$z + u\frac{dz}{du} = \frac{3 - z}{1 + z}.$$

Separating variables gives

$$\int \frac{z + 1}{z^2 + 2z - 3}\, dz = -\int \frac{1}{u}\, du,$$

$$\frac{1}{2}\ln|z^2 + 2z - 3| = -\ln|u| + C_1,$$

from which it follows that

$$z^2 + 2z - 3 = Cu^{-2}.$$

When we substitute back in for z, u, and v, we find

$$(v/u)^2 + 2(v/u) - 3 = Cu^{-2},$$
$$v^2 + 2uv - 3u^2 = C,$$
$$(y + 3)^2 + 2(x - 1)(y + 3) - 3(x - 1)^2 = C.$$

This last equation gives an implicit solution to (16).   ●

EXERCISES 2.6

In Problems 1 through 8 identify the equation as homogeneous, Bernoulli, linear coefficients, or of the form $y' = G(ax + by)$.

1. $(y - 4x - 1)^2\, dx - dy = 0$.

2. $(t^2 - x^2)\, dt + 2tx\, dx = 0$.

3. $(3t - x - 6)\, dt + (t + x + 2)\, dx = 0$.

4. $dy/dx + y/x = x^3 y^2$.

5. $(ye^{-2x} + y^3)\, dx - e^{-2x}\, dy = 0$.

6. $\theta\, dy - y\, d\theta = \sqrt{\theta y}\, d\theta$.

7. $(y^3 - \theta y^2)\,d\theta + 2\theta^2 y\,dy = 0.$

8. $\cos(x + y)\,dy = \sin(x + y)\,dx.$

Use the method discussed under "Homogeneous Equations" to solve Problems 9 through 16.

9. $(3x^2 - y^2)\,dx + (xy - x^3 y^{-1})\,dy = 0.$

10. $(xy + y^2)\,dx - x^2\,dy = 0.$

11. $(y^2 - xy)\,dx + x^2\,dy = 0.$

12. $(x^2 + y^2)\,dx + 2xy\,dy = 0.$

13. $\dfrac{dx}{dt} = \dfrac{x^2 + t\sqrt{t^2 + x^2}}{tx}.$ **14.** $\dfrac{dy}{d\theta} = \dfrac{\theta\sec(y/\theta) + y}{\theta}.$

15. $\dfrac{dy}{dx} = \dfrac{x^2 - y^2}{3xy}.$ **16.** $\dfrac{dy}{dx} = \dfrac{y(\ln y - \ln x + 1)}{x}.$

Use the method discussed under "Equations of the Form $dy/dx = G(ax + by)$" to solve Problems 17 through 20.

17. $dy/dx = \sqrt{x + y} - 1.$ **18.** $dy/dx = (x + y + 2)^2.$

19. $dy/dx = (x - y + 5)^2.$ **20.** $dy/dx = \sin(x - y).$

Use the method discussed under "Bernoulli Equations" to solve Problems 21 through 28.

21. $\dfrac{dy}{dx} + \dfrac{y}{x} = x^2 y^2.$

22. $\dfrac{dy}{dx} - y = e^{2x} y^3.$

23. $\dfrac{dy}{dx} = \dfrac{2y}{x} - x^2 y^2.$

24. $\dfrac{dy}{dx} + \dfrac{y}{(x - 2)} = 5(x - 2)y^{1/2}.$

25. $\dfrac{dx}{dt} + tx^3 + \dfrac{x}{t} = 0.$ **26.** $\dfrac{dy}{dx} + y = e^x y^{-2}.$

27. $\dfrac{dr}{d\theta} = \dfrac{r^2 + 2r\theta}{\theta^2}.$ **28.** $\dfrac{dy}{dx} + y^3 x + y = 0.$

Use the method discussed under "Equations with Linear Coefficients" to solve Problems 29 through 32.

29. $(-3x + y - 1)\,dx + (x + y + 3)\,dy = 0.$

30. $(x + y - 1)\,dx + (y - x - 5)\,dy = 0.$

31. $(2x - y)\,dx + (4x + y - 3)\,dy = 0.$

32. $(2x + y + 4)\,dx + (x - 2y - 2)\,dy = 0.$

In Problems 33 through 40 solve the equation given in:

33. Problem 1.   **34.** Problem 2.

35. Problem 3.   **36.** Problem 4.

37. Problem 5.   **38.** Problem 6.

39. Problem 7.   **40.** Problem 8.

41. Use the substitution $v = x - y + 2$ to solve equation (8).

42. Use the substitution $y = vx^2$ to solve
$$\frac{dy}{dx} = \frac{2y}{x} + x\cos(y/x^2).$$

43. (a) Show that the equation $dy/dx = f(x, y)$ is homogeneous if and only if $f(tx, ty) = f(x, y)$. [Hint: Let $t = 1/x$.]

(b) A function $H(x, y)$ is called **homogeneous of order n** if $H(tx, ty) = t^n H(x, y)$. Show that the equation
$$M(x, y)\,dx + N(x, y)\,dy = 0$$
is homogeneous if $M(x, y)$ and $N(x, y)$ are both homogeneous of the same order.

44. Show that equation (13) reduces to an equation of the form
$$\frac{dy}{dx} = G(ax + by),$$
when $a_1 b_2 = a_2 b_1$. [Hint: If $a_1 b_2 = a_2 b_1$, then $a_2/a_1 = b_2/b_1 = k$, so that $a_2 = ka_1$ and $b_2 = kb_1$.]

45. Ecological Systems. In modeling ecological systems, there often arise coupled equations of the form
$$\frac{dy}{dt} = ax + by,$$
$$\frac{dx}{dt} = \alpha x + \beta y,$$
where a, b, α, and β are constants. The quantities x and y represent the populations at time t of two competing species. In this problem we wish to determine the relationship between x and y rather than the individual populations as functions of t. For this purpose, divide the first equation by the second to obtain

(17) $\qquad \dfrac{dy}{dx} = \dfrac{ax + by}{\alpha x + \beta y}.$

This new equation is homogeneous, so we can solve it via the substitution $v = y/x$. We refer to the solutions of (17) as **integral curves** (see Section 9.6). Determine the integral curves for the system
$$\frac{dy}{dt} = -4x - y,$$
$$\frac{dx}{dt} = 2x - y.$$

46. Riccati Equation. An equation of the form

(18) $$\frac{dy}{dx} = P(x)y^2 + Q(x)y + R(x)$$

is called a generalized Riccati equation.[†]

(a) If one solution, say $u(x)$, of (18) is known, show that the substitution $y = u + 1/v$ reduces (18) to a linear equation in v.

(b) Given that $u(x) = x$ is a solution to

$$\frac{dy}{dx} = x^3(y-x)^2 + \frac{y}{x},$$

use the result of part (a) to find all the other solutions to this equation. (The particular solution $u(x) = x$ can be found by inspection or by using a Taylor series method; see Section 8.2.)

47. Guerrilla Combat Model[††]. In modeling a pair of guerrilla forces in combat, the following system arises

$$\frac{dy}{dt} = -axy,$$

$$\frac{dx}{dt} = -bxy,$$

where $x(t)$ and $y(t)$ are the strengths of opposing forces at time t and a and b are positive constants. The terms $-axy$ and $-bxy$ represent the *combat loss rate* for the troops y and x, respectively. This model assumes no reinforcements.

(a) Show that x and y satisfy the equation

$$\frac{dy}{dx} = \frac{a}{b}.$$

(b) Let $y(0) = y_0$ and $x(0) = x_0$. Solving the equation in part (a), derive the **linear combat law**

$$by - ax = c,$$

where $c = by_0 - ax_0$.

(c) Use the linear combat law to show that the y troops win if $c > 0$ and lose if $c < 0$.

48. Mixed Combat Model. In modeling a conflict between a guerrilla force and a conventional force, the following system arises

$$\frac{dy}{dt} = -ax,$$

$$\frac{dx}{dt} = -bxy,$$

where $x(t)$ is the strength of the guerrilla force, $y(t)$ is the strength of the conventional force at time t, and a and b are positive constants. The terms $-ax$ and $-bxy$ represent the *combat loss rate* for troops y and x, respectively. This model assumes no reinforcements.

(a) Show that x and y satisfy the equation

$$\frac{dy}{dx} = \frac{a}{by}.$$

(b) Let $y(0) = y_0$ and $x(0) = x_0$. Solving the equation in part (a), derive the **parabolic combat law**

$$by^2 - 2ax = c,$$

where $c = by_0^2 - 2ax_0$.

(c) Use the parabolic combat law to show that the y troops win if $c > 0$ and lose if $c < 0$.

• CHAPTER SUMMARY •

In this chapter we have discussed various types of first order differential equations. The most important were the separable, exact, and linear equations. Their principal features and method of solution are outlined below.

Separable Equations: $dy/dx = g(x)p(y)$. Separate the variables and integrate.

[†] *Historical Footnote*: Count Jacopo Riccati studied a particular case of this equation in 1724 during his investigation of curves whose radii of curvature depend only on the variable y and not the variable x.

[††] For a discussion of combat models see *Differential Equation Models*, M. Braun, C. S. Coleman, and D. A. Drew, eds., Springer-Verlag, New York, 1983, Chapter 8.

Exact Equations: $dF(x, y) = 0$. Solutions are given implicitly by $F(x, y) = C$. If $\partial M/\partial y = \partial N/\partial x$, then $M\,dx + N\,dy = 0$ is exact and F is given by

$$F = \int M\,dx + g(y), \quad \text{where} \quad g'(y) = N - \frac{\partial}{\partial y}\int M\,dx,$$

or

$$F = \int N\,dy + h(x), \quad \text{where} \quad h'(x) = M - \frac{\partial}{\partial x}\int N\,dy.$$

Linear Equations: $dy/dx + P(x)y = Q(x)$. The integrating factor $\mu = \exp(\int P(x)\,dx)$ reduces the equation to $d(\mu y)/dx = \mu Q$, so that $\mu y = \int \mu Q\,dx + C$.

When an equation is not separable, exact, or linear, it may be possible to find an integrating factor or perform a substitution that will enable us to solve the equation.

Special Integrating Factors: $\mu M\,dx + \mu N\,dy = 0$ is exact. If $(\partial M/\partial y - \partial N/\partial x)/N$ depends only on x, then

$$\mu(x) = \exp\left(\int \left\{\frac{\partial M/\partial y - \partial N/\partial x}{N}\right\} dx\right)$$

is an integrating factor. If $(\partial N/\partial x - \partial M/\partial y)/M$ depends only on y, then

$$\mu(y) = \exp\left(\int \left\{\frac{\partial N/\partial x - \partial M/\partial y}{M}\right\} dy\right)$$

is an integrating factor.

Homogeneous Equations: $dy/dx = G(y/x)$. Let $v = y/x$. Then $dy/dx = v + x(dv/dx)$, and the transformed equation in the variables v and x is separable.

Equations of the Form: $dy/dx = G(ax + by)$. Let $z = ax + by$. Then $dz/dx = a + b(dy/dx)$, and the transformed equation in the variables z and x is separable.

Bernoulli Equations: $dy/dx + P(x)y = Q(x)y^n$. For $n \neq 0$ or 1, let $v = y^{1-n}$. Then $dv/dx = (1 - n)y^{-n}(dy/dx)$, and the transformed equation in the variables v and x is linear.

Linear Coefficients: $(a_1x + b_1y + c_1)\,dx + (a_2x + b_2y + c_2)\,dy = 0$. For $a_1b_2 \neq a_2b_1$, let $x = u + h$ and $y = v + k$, where h and k satisfy

$$a_1h + b_1k + c_1 = 0,$$
$$a_2h + b_2k + c_2 = 0.$$

Then the transformed equation in the variables u and v is homogeneous.

REVIEW PROBLEMS

In Problems 1 through 30 solve the equation.

1. $\dfrac{dy}{dx} = \dfrac{e^{x+y}}{y-1}$.

2. $\dfrac{dy}{dx} - 4y = 32x^2$.

3. $(2xy - 3x^2)\,dx + (x^2 - 2y^{-3})\,dy = 0$.

4. $\dfrac{dy}{dx} + \dfrac{3y}{x} = x^2 - 4x + 3$.

5. $[\sin(xy) + xy\cos(xy)]\,dx + [1 + x^2\cos(xy)]\,dy = 0$.

6. $2xy^3\,dx - (1 - x^2)\,dy = 0$.

7. $x^3y^2\,dx + x^4y^{-6}\,dy = 0$.

8. $\dfrac{dy}{dx} + \dfrac{2y}{x} = 2x^2y^2$.

9. $(x^2 + y^2)\,dx + 3xy\,dy = 0$.

10. $[1 + (1 + x^2 + 2xy + y^2)^{-1}]\,dx$
$+ [y^{-1/2} + (1 + x^2 + 2xy + y^2)^{-1}]\,dy = 0$.

11. $\dfrac{dx}{dt} = 1 + \cos^2(t - x)$.

12. $(y^3 + 4e^x y)\,dx + (2e^x + 3y^2)\,dy = 0$.

13. $\dfrac{dy}{dx} - \dfrac{y}{x} = x^2\sin 2x$.

14. $\dfrac{dx}{dt} - \dfrac{x}{t-1} = t^2 + 2$.

15. $\dfrac{dy}{dx} = 2 - \sqrt{2x - y + 3}$.

16. $\dfrac{dy}{dx} + y\tan x + \sin x = 0$.

17. $\dfrac{dy}{d\theta} + 2y = y^2$.

18. $\dfrac{dy}{dx} = (2x + y - 1)^2$.

19. $(x^2 - 3y^2)\,dx + 2xy\,dy = 0$.

20. $\dfrac{dy}{d\theta} + \dfrac{y}{\theta} = -4\theta y^{-2}$.

21. $(y - 2x - 1)\,dx + (x + y - 4)\,dy = 0$.

22. $(2x - 2y - 8)\,dx + (x - 3y - 6)\,dy = 0$.

23. $(y - x)\,dx + (x + y)\,dy = 0$.

24. $(\sqrt{y/x} + \cos x)\,dx + (\sqrt{x/y} + \sin y)\,dy = 0$.

25. $y(x - y - 2)\,dx + x(y - x + 4)\,dy = 0$.

26. $\dfrac{dy}{dx} + xy = 0$.

27. $(3x - y - 5)\,dx + (x - y + 1)\,dy = 0$.

28. $\dfrac{dy}{dx} = \dfrac{x - y - 1}{x + y + 5}$.

29. $(4xy^3 - 9y^2 + 4xy^2)\,dx + (3x^2y^2 - 6xy + 2x^2y)\,dy = 0$.

30. $\dfrac{dy}{dx} = (x + y + 1)^2 - (x + y - 1)^2$.

In Problems 31 through 40 solve the initial value problem.

31. $(x^3 - y)\,dx + x\,dy = 0, \quad y(1) = 3$.

32. $\dfrac{dy}{dx} = \left(\dfrac{x}{y} + \dfrac{y}{x}\right), \quad y(1) = -4$.

33. $(t + x + 3)\,dt + dx = 0, \quad x(0) = 1$.

34. $\dfrac{dy}{dx} - \dfrac{2y}{x} = x^2\cos x, \quad y(\pi) = 2$.

35. $(2y^2 + 4x^2)\,dx - xy\,dy = 0, \quad y(1) = -2$.

36. $[2\cos(2x + y) - x^2]\,dx + [\cos(2x + y) + e^y]\,dy = 0,$
$y(1) = 0$.

37. $(2x - y)\,dx + (x + y - 3)\,dy = 0, \quad y(0) = 2$.

38. $\sqrt{y}\,dx + (x^2 + 4)\,dy = 0, \quad y(0) = 4$.

39. $\dfrac{dy}{dx} - \dfrac{2y}{x} = x^{-1}y^{-1}, \quad y(1) = 3$.

40. $\dfrac{dy}{dx} - 4y = 2xy^2, \quad y(0) = -4$.

TECHNICAL WRITING EXERCISES

1. An instructor at Ivey U. asserted: "All you need to know about first order differential equations is how to solve those that are exact." Give arguments that support and arguments that refute the instructor's claim.

2. What properties do solutions to linear equations have that are not shared by solutions to either separable or exact equations? Give some specific examples to support your conclusions.

3. Consider the differential equation

$$\frac{dy}{dx} = ay + be^{-x}, \qquad y(0) = c,$$

where a, b, and c are constants. Describe what happens to the asymptotic behavior as $x \to +\infty$ of the solution as the constants a, b, and c are varied. Illustrate with figures and/or graphs.

GROUP PROJECTS FOR CHAPTER 2

A. The Snowplow Problem

To apply the techniques discussed in this chapter to real-world problems, it is necessary to "translate" these problems into questions that can be answered mathematically. The process of reformulating a real-world problem as a mathematical one often requires making certain simplifying assumptions. To illustrate this, consider the following snowplow problem.

One morning it began to snow very hard and continued snowing steadily throughout the day. A snowplow set out at 9:00 A.M. to clear a road, clearing 2 mi by 11:00 A.M. and an additional mile by 1:00 P.M. *At what time did it start snowing?*

To solve this problem, you can make two physical assumptions concerning the rate at which it is snowing and the rate at which the snowplow can clear the road. Since it is snowing steadily, it is reasonable to assume that it is snowing at a constant rate. From the data given (and from our experience) the snowplow moves slower, the deeper the snow. With this in mind, assume that the rate (in mi/hr) at which a snowplow can clear a road is inversely proportional to the depth of the snow.

B. Asymptotic Behavior of Solutions to Linear Equations

To illustrate how the asymptotic behavior of the forcing term $Q(x)$ affects the solution to a linear equation, consider the equation

(1) $$\frac{dy}{dx} + ay = Q(x),$$

where the constant a is positive and $Q(x)$ is continuous on $(0, \infty)$.

(a) Show that the general solution to equation (1) can be written in the form

(2) $$y(x) = y(x_0)e^{-a(x-x_0)} + e^{-ax} \int_{x_0}^{x} e^{at}Q(t)\, dt,$$

where x_0 is a nonnegative constant.

(b) If $|Q(x)| \leq k$ for $x \geq x_0$, where k and x_0 are nonnegative constants, show that

$$|y(x)| \leq |y(x_0)|e^{-a(x-x_0)} + \frac{k}{a}[1 - e^{-a(x-x_0)}]$$

for $x \geq x_0$.

(c) Let $z(x)$ satisfy the same equation as (1) but with forcing function $\tilde{Q}(x)$, that is,

$$\frac{dz}{dx} + az = \tilde{Q}(x),$$

where $\tilde{Q}(x)$ is continuous on $(0, \infty)$. Show that if

$$|\tilde{Q}(x) - Q(x)| \leq K \quad \text{for} \quad x \geq x_0,$$

then

$$|z(x) - y(x)| \leq |z(x_0) - y(x_0)|e^{-a(x-x_0)} + \frac{K}{a}[1 - e^{-a(x-x_0)}]$$

for $x \geq x_0$.

(d) Now show that if $Q(x) \to \beta$ as $x \to \infty$, then any solution $y(x)$ of equation (1) satisfies $y(x) \to \beta/a$ as $x \to \infty$. [Hint: Take $\tilde{Q}(x) \equiv \beta$ and $z(x) \equiv \beta/a$ in part (c).]

(e) As an application of part (d), suppose that a brine solution containing $q(t)$ kilograms of salt per liter at time t runs into a tank of water at a fixed rate and that the mixture, kept uniform by stirring, flows out at the same rate. Given that $q(t) \to \beta$ as $t \to \infty$, use the result of part (d) to determine the limiting concentration of the salt in the tank as $t \to \infty$ (see Problem 36, Exercises 2.4).

C. Designing a Solar Collector

You want to design a solar collector that will concentrate the sun's rays at a point. By symmetry this surface will have a shape that is a surface of revolution obtained by revolving a curve about an axis. Without loss of generality, you can assume that this axis is the x-axis and the rays parallel to this axis are focused at the origin (see Figure 2.9). To derive the equation for the curve, proceed as follows:

(a) The law of reflection says that the angles γ and δ are equal. Use this and results from geometry to show that $\beta = 2\alpha$.

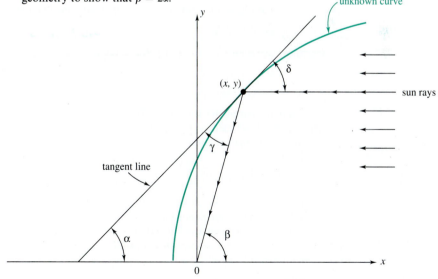

Figure 2.9 Curve that generates a solar collector

(b) From calculus recall that $dy/dx = \tan\alpha$. Use this, the fact that $y/x = \tan\beta$, and the double angle formula to show that

$$\frac{y}{x} = \frac{2\,dy/dx}{1 - (dy/dx)^2}.$$

(c) Now show that the curve satisfies the differential equation

(3) $$\frac{dy}{dx} = \frac{-x + \sqrt{x^2 + y^2}}{y}.$$

(d) Solve equation (3).

(e) Describe the solutions and identify the type of collector obtained.

D. Clairaut Equations and Singular Solutions

An equation of the form

(4) $$y = x\frac{dy}{dx} + f(dy/dx),$$

where the continuously differentiable function $f(t)$ is evaluated at $t = dy/dx$, is called a **Clairaut equation.**[†] Interest in these equations is due to the fact that (4) has a one-parameter family of solutions that consist of *straight lines*. Furthermore, the **envelope** of this family—that is, the curve whose tangent lines are given by the family—is also a solution to (4) and is called the **singular solution.**
To solve a Clairaut equation:

(a) Differentiate equation (4) with respect to x and simplify to show that

(5) $$[x + f'(dy/dx)]\frac{d^2y}{dx^2} = 0, \quad \text{where} \quad f'(t) = \frac{d}{dt}f(t).$$

(b) From (5), conclude that $dy/dx = c$ or $f'(dy/dx) = -x$. Assume that $dy/dx = c$ and substitute back into (4) to obtain the family of *straight-line solutions*

$$y = cx + f(c).$$

(c) Show that another solution to (4) is given parametrically by

$$x = -f'(p),$$
$$y = f(p) - pf'(p),$$

where the parameter $p = dy/dx$. This solution is the *singular solution*.

(d) Use the above method to find the family of straight-line solutions and the singular solution to the equation

$$y = x\left(\frac{dy}{dx}\right) + 2\left(\frac{dy}{dx}\right)^2.$$

Here $f(t) = 2t^2$. Sketch several of the straight-line solutions along with the singular solution on the same coordinate system. Observe that the straight-line solutions are all tangent to the singular solution.

[†] *Historical Footnote*: These equations were studied by Alexis Clairaut in 1734.

(e) Repeat part (d) for the equation

$$x\left(\frac{dy}{dx}\right)^3 - y\left(\frac{dy}{dx}\right)^2 + 1 = 0.$$

E. Analytic Functions and Orthogonal Trajectories

A complex-valued function $f(z)$ of the complex variable $z = x + iy$, where $i = \sqrt{-1}$, can be written in the form $f(z) = u(x, y) + iv(x, y)$, where u and v are real-valued functions. If $f(z)$ has a derivative at each point of a region $\mathcal{D}$ in the plane, then $f(z)$ is said to be **analytic**[†] in $\mathcal{D}$. For example, any polynomial function of z is analytic in the whole plane. It is known that the real and imaginary parts of an analytic function satisfy the **Cauchy-Riemann equations**

$$\frac{\partial u}{\partial x} = \frac{\partial v}{\partial y}, \qquad \frac{\partial u}{\partial y} = -\frac{\partial v}{\partial x}.$$

(a) For an analytic function $f(z) = u(x, y) + iv(x, y)$, show that the level curves

$$u(x, y) = \text{constant}$$

are orthogonal trajectories (see Problem 31, Exercises 2.3) for the level curves

$$v(x, y) = \text{constant}. \qquad [\text{Hint: Use the Cauchy-Riemann equations.}]$$

(b) Let $v(x, y) = y - 2xy$. Find $u(x, y)$ so that $f(z) = u(x, y) + iv(x, y)$ satisfies the Cauchy-Riemann equations. [Hint: Start with $\partial u/\partial x = \partial v/\partial y = 1 - 2x$.] Now use the result of part (a) to determine the orthogonal trajectories of the family of curves $y - 2xy = C$.

(c) The function $f(z) = z + 1/z$ is analytic for $z \neq 0$. Using complex arithmetic, write $f(z)$ in the form $u(x, y) + iv(x, y)$ and show that the level curve $v(x, y) = 0$ consists of the x-axis and the unit circle $x^2 + y^2 = 1$.

(d) The level curves $v(x, y) = C$ from part (c) that lie outside the unit circle can be interpreted as the stream lines of a fluid flow around a cylindrical obstacle. Sketch these level curves for various values of C. [Hint: Solve for x in terms of y and C and plot points.]

[†] For a discussion of analytic functions of a complex variable see, for example, *Fundamentals of Complex Analysis*, Second Edition, by E. B. Saff and A. D. Snider, Prentice-Hall, Inc., Englewood Cliffs, New Jersey, 1993.

CHAPTER 3

Mathematical Models and Numerical Methods Involving First Order Equations

• 3.1 • MATHEMATICAL MODELING

Adopting the Babylonian practices of careful measurement and detailed observations, the ancient Greeks sought to comprehend nature by logical analysis. Aristotle's convincing arguments that the world was not flat, but spherical, led the intellectuals of that day to ponder the question: What is the circumference of the Earth? And it was astonishing that Eratosthenes managed to obtain a fairly accurate answer to this problem without having to set foot beyond the ancient city of Alexandria. His method involved certain assumptions and simplifications: The Earth is a perfect sphere, the Sun's rays travel parallel paths, the city of Syene was 5000 stadia due south of Alexandria, etc. With these idealizations, Eratosthenes created a mathematical context in which the principles of geometry could be applied.[†]

Today, as scientists seek to further our understanding of nature and as engineers seek, on a more pragmatic level, to find answers to technical problems, the technique of representing our "real world" in mathematical terms has become an invaluable tool. This process of mimicking reality by using the language of mathematics is known as **mathematical modeling**.

Formulating problems in mathematical terms has several benefits. First, it requires that we clearly state our premises. Real-world problems are often complex, involving several different and possibly interrelated processes. Before mathematical treatment can proceed, one must determine which variables are significant and which can be ignored. Often, for the relevant variables, relationships are postulated in the form of laws, formulas, theories, etc. These assumptions constitute the **idealizations** of the model.

Mathematics contains a wealth of theorems and techniques for making logical deductions and manipulating equations. Hence it provides a context in which analysis can

---

[†] For further reading see, for example, *The Mapmakers*, by John Noble Wilford, Vintage Books, New York, 1982, Chapter 2.

take place free of any preconceived notions of the outcome. It is also of great practical importance that mathematics provides a format for obtaining numerical answers via a computer.

The process of building an effective mathematical model takes skill, imagination, and objective evaluation. Certainly an exposure to several existing models that illustrate various aspects of modeling can lead to a better feel for the process. Several excellent books and articles are devoted exclusively to the subject.[†] In this chapter we concentrate on examples of models that involve first order differential equations. In studying these and in building your own models, the following broad outline of the process may be helpful.

Formulate the Problem

Here you must pose the problem in such a way that it can be "answered" mathematically. This requires an understanding of the problem area as well as the mathematics. At this stage you may need to spend time talking with nonmathematicians and reading the relevant literature.

Develop the Model

There are two things to be done here. First you must decide which variables are important and which are not. The former are then classified as independent variables or dependent variables. The unimportant variables are those that have very little or no effect on the process. (For example, in studying the motion of a falling body, its color is usually of little interest.) The independent variables are those whose effect is significant and that will serve as input for the model.[††] For the falling body, its shape, mass, initial position, initial velocity, and time from release are possible independent variables. The dependent variables are those that are affected by the independent variables and that are important to solving the problem. Again for a falling body, its velocity, location, and time of impact are all possible dependent variables.

Second, you must determine or specify the relationships (e.g., a differential equation) that exist among the relevant variables. This requires a good background in the area and insight into the problem. You may begin with a crude model; then, based upon testing, refine the model as needed. For example, you might begin by ignoring any friction acting on the falling body. Then, if necessary to obtain a more acceptable answer, try to take into account any frictional forces that may affect the motion.

Test the Model

Before attempting to "verify" a model by comparing its output with experimental data, the following questions should be considered:

[†] See, for example, *A First Course in Mathematical Modeling*, by F. R. Giordano and M. D. Weir, Brooks/Cole Publishing Company, Monterey, California, 1985, or *Concepts of Mathematical Modeling*, by W. J. Meyer, McGraw-Hill Book Company, New York, 1984.

[††] In the mathematical formulation of the model, certain of the independent variables may be called **parameters**.

Are the assumptions reasonable?

Are the equations dimensionally consistent? (For example, we don't want to add units of force to units of velocity.)

Is the model internally consistent in the sense that equations do not contradict one another?

Do the relevant equations have solutions?

How difficult is it to obtain the solutions?

Do the solutions provide an answer for the problem being studied?

When possible, try to validate the model by comparing its predictions with any experimental data. Begin with rather simple predictions that involve little computation or analysis. Then, as the model is refined, check to see that the accuracy of the model's predictions are acceptable to you. In some cases validation is impossible or socially, politically, economically, or morally unreasonable. For example, how does one validate a model that predicts when our sun will die out?

Each time the model is used to predict the outcome of a process and hence solve a problem, it represents a further test of the model that may lead to further refinements or simplifications. In many cases a model is simplified to give a quicker or less expensive answer—provided, of course, that sufficient accuracy is maintained.

One should always keep in mind that a model is *not* reality, but only a representation of reality. The more refined models *may* provide an understanding of the underlying processes of nature. For this reason applied mathematicians strive for better, more refined models. Still, the real test of a model is its ability to find an acceptable answer for the posed problem.

In this chapter we discuss various models that involve differential equations. In Section 3.2, Compartmental Analysis, and Section 3.4, Newtonian Mechanics, we consider two general models that result from interpreting the derivative as a rate of change of one variable with respect to another. Section 3.3 describes a model for the heating and cooling of a building. Sections 3.5 and 3.6 contain a brief introduction to some numerical techniques for solving first order initial value problems. This will enable us to study more realistic models that cannot be solved using the techniques of Chapter 2.

•3.2• COMPARTMENTAL ANALYSIS

Many complicated processes can be broken down into distinct stages and the entire system modeled by describing the interactions between the various stages. Such systems are called **compartmental** and are graphically depicted by **block diagrams.** In this section we study the basic unit of these systems, a single compartment, and analyze some simple processes that can be handled by such a model.

The basic one-compartment system consists of a function $x(t)$ that represents the amount of a substance in the compartment at time t, an input rate at which the substance enters the compartment, and an output rate at which the substance leaves the compartment (see Figure 3.1).

Since the derivative of x with respect to t can be interpreted as the rate of change in the amount of the substance in the compartment with respect to time, the one-compartment system suggests

(1) $$\frac{dx}{dt} = \text{input rate} - \text{output rate},$$

as a mathematical model for the process.

Mixing Problems

A problem for which the one-compartment system provides a useful representation is the mixing of fluids in a tank. Let $x(t)$ represent the amount of a substance in a tank (compartment) at time t. In order to use the compartmental analysis model we must be able to determine the rates at which this substance enters and leaves the tank. In mixing problems, one is often given the rate at which a fluid containing the substance flows into the tank along with the concentration of the substance in that fluid. Hence multiplying the flow rate (volume/time) by the concentration (amount/volume) yields the input rate (amount/time).

The output rate of the substance is usually more difficult to determine. If we are given the exit rate of the mixture of fluids in the tank, then how do we determine the concentration of the substance in this mixture? One simplifying assumption that we might make is that the concentration is kept uniform in the mixture. Then, we can compute the concentration of the substance in the mixture by dividing the amount $x(t)$ by the volume of the mixture in the tank at time t. Multiplying this concentration by the exit rate of the mixture then gives the desired output rate of the substance. This model is used in the next two examples.

● **EXAMPLE 1**   Consider a large tank holding 1000 L of water into which a brine solution of salt begins to flow at a constant rate of 6 L/min. The solution inside the tank is kept well stirred and is flowing out of the tank at a rate of 6 L/min. If the concentration of salt in the brine entering the tank is 1 kg/L, determine when the concentration of salt in the tank will reach $\frac{1}{2}$ kg/L (see Figure 3.2).

Solution   We can view the tank as a compartment containing salt. If we let $x(t)$ denote the mass of salt in the tank at time t, we can determine the concentration of salt in the tank by dividing $x(t)$ by the volume of fluid in the tank at time t. We use the mathematical model described by equation (1) to solve for $x(t)$.

First we must determine the rate at which salt enters the tank. We are given that brine flows into the tank at a rate of 6 L/min. Since the concentration of this brine is 1 kg/L, we

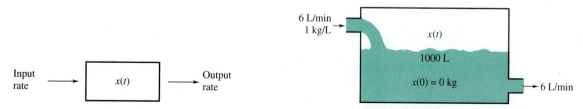

Figure 3.1 Schematic representation of a one-compartment system

Figure 3.2 Mixing problem with equal flow rates

conclude that the input rate of salt into the tank is

(2)   (6 L/min)(1 kg/L) = 6 kg/min.

We must now determine the output rate of salt from the tank. Since the brine solution in the tank is kept well stirred, let's assume that the concentration of salt in the tank is uniform. That is, the concentration of salt in any part of the tank at time t is just $x(t)$ divided by the volume of fluid in the tank. Since the tank initially contains 1000 L and the rate of flow into the tank is the same as the rate of flow out, the volume is a constant 1000 L. Hence the output rate of salt is

(3) $(6 \text{ L/min})\left(\dfrac{x(t)}{1000} \text{ kg/L}\right) = \dfrac{3x(t)}{500} \text{ kg/min.}$

Since the tank initially contained just water, we set $x(0) = 0$. Substituting the rates in (2) and (3) into equation (1) then gives the initial value problem

(4) $\dfrac{dx}{dt} = 6 - \dfrac{3x}{500}, \qquad x(0) = 0,$

as a mathematical model for the mixing problem.

The equation in (4) is separable (and linear) and easy to solve. Using the initial condition $x(0) = 0$ to evaluate the arbitrary constant, we obtain

(5) $x(t) = 1000(1 - e^{-3t/500}).$

Thus the concentration of salt in the tank at time t is

$$\frac{x(t)}{1000} = 1 - e^{-3t/500} \text{ kg/L.}$$

To determine when the concentration of salt is $\frac{1}{2}$ kg/L, we set the right hand side equal to $\frac{1}{2}$ and solve for t. This gives

$$1 - e^{-3t/500} = \frac{1}{2} \quad \text{or} \quad e^{-3t/500} = \frac{1}{2},$$

and hence

$$t = \frac{500 \ln 2}{3} \approx 115.52 \text{ min.}$$

Consequently the concentration of salt in the tank will be $\frac{1}{2}$ kg/L after 115.52 min.   ●

From equation (5) we observe that the mass of salt in the tank steadily increases and has the limiting value

$$\lim_{t \to \infty} x(t) = \lim_{t \to \infty} 1000(1 - e^{-3t/500}) = 1000 \text{ kg.}$$

Thus, the limiting concentration of salt in the tank is 1 kg/L, which is the same as the concentration of salt in the brine flowing into the tank. This certainly agrees with our expectations!

It might be interesting to see what would happen to the concentration if the flow rate into the tank is greater than the flow rate out.

● **EXAMPLE 2**   For the mixing problem described in Example 1, assume now that the brine leaves the tank at a rate of 5 L/min instead of 6 L/min, with all else being the same (see Figure 3.3). Determine the concentration of salt in the tank as a function of time.

Solution   Since the difference between the rate of flow into the tank and the rate of flow out is $6 - 5 = 1$ L/min, the volume of fluid in the tank after t minutes is $(1000 + t)$ L. Hence the rate at which salt leaves the tank is

$$(5 \text{ L/min})\left[\frac{x(t)}{1000 + t} \text{ kg/L}\right] = \frac{5x(t)}{1000 + t} \text{ kg/min}.$$

Using this in place of (3) for the output rate gives the initial value problem

(6) $$\frac{dx}{dt} = 6 - \frac{5x}{1000 + t}, \qquad x(0) = 0,$$

as a mathematical model for the mixing problem.

Since the differential equation in (6) is linear, we can use the procedure outlined on page 45 to solve for $x(t)$. The integrating factor is $\mu(t) = (1000 + t)^5$. Thus

$$\frac{d}{dt}[(1000 + t)^5 x] = 6(1000 + t)^5,$$

$$(1000 + t)^5 x = (1000 + t)^6 + c,$$

$$x(t) = (1000 + t) + c(1000 + t)^{-5}.$$

Using the initial condition $x(0) = 0$, we find $c = -(1000)^6$, and thus the solution to (6) is

$$x(t) = (1000 + t) - (1000)^6 (1000 + t)^{-5}.$$

Hence, the concentration of salt in the tank at time t is

(7) $$\frac{x(t)}{1000 + t} = 1 - (1000)^6 (1000 + t)^{-6} \text{ kg/L}. \qquad ●$$

As in Example 1, the concentration given by (7) approaches 1 kg/L as $t \to \infty$. However, in Example 2 the volume of fluid in the tank becomes unbounded, and when the tank begins to overflow, the model in (6) is no longer appropriate.

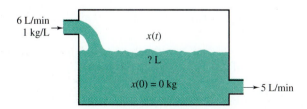

Figure 3.3 Mixing problem with unequal flow rates

For mixing problems, the one-compartment model is extremely useful. The only assumption made that is the least bit questionable is that the well-stirred mixture in the tank has a uniform concentration. Even with stirring, the salt entering the tank at the top has a delayed effect on the concentration of the brine leaving at the bottom. It is more realistic to assume that the concentration of brine leaving the tank at time t is equal to the average concentration at an earlier instant, say at time $t - t_0$, where t_0 is a positive constant. The original problem of Example 1 would now be described by

$$x'(t) = 6 - \frac{3}{500}x(t - t_0), \qquad x(t) = 0 \quad \text{for} \quad t \in [-t_0, 0].$$

Such equations are called **delay differential equations** due to the "delay" in the argument of $x(t - t_0)$. These equations are discussed further in the projects at the end of this chapter (see Project A).

Population Models

How does one predict the growth of a population? If we are interested in a single population, we can think of the species as being contained in a compartment (a Petrie dish, an island, a country, etc.) and study the growth process as a one-compartment system.

Let $p(t)$ be the population at time t. While the population is always an integer, it is usually large enough so that very little error is introduced in assuming that $p(t)$ is a continuous function. We now need to determine the growth (input) rate and the death (output) rate for the population.

Let's consider a population of bacteria that reproduce by simple cell division. In our model we assume that the growth rate is proportional to the population present. This assumption is consistent with observations of bacteria growth. As long as there are sufficient space and ample food supply for the bacteria, we can also assume that the death rate is zero. (Remember that in cell division, the parent cell does not die, but becomes two new cells.) Hence a mathematical model for a population of bacteria is

$$(8) \qquad \frac{dp}{dt} = k_1 p, \qquad p(0) = p_0,$$

where $k_1 > 0$ is the proportionality constant for the growth rate and p_0 is the population at time $t = 0$. For human populations, the assumption that the death rate is zero is certainly wrong! However, if we assume that people die only of natural causes, we might expect the death rate also to be proportional to the size of the population. That is, we revise (8) to read

$$(9) \qquad \frac{dp}{dt} = k_1 p - k_2 p = (k_1 - k_2)p = kp,$$

where $k := k_1 - k_2$ and k_2 is the proportionality constant for the death rate. Let's assume $k_1 > k_2$, so that $k > 0$. This gives the mathematical model

$$(10) \qquad \frac{dp}{dt} = kp, \qquad p(0) = p_0,$$

which is called the **Malthusian** or **exponential law** of population growth. This equation is separable, and solving the initial value problem for $p(t)$, gives

(11) $p(t) = p_0 e^{kt}$.

To test the Malthusian model, let's apply it to the demographic history of the United States.

● **EXAMPLE 3**   In 1790 the population of the United States was 3.93 million, and in 1890 the population was 62.95 million. Using the Malthusian model, estimate the population of the United States as a function of time.

Solution   If we set $t = 0$ to be the year 1790, then by formula (11), we have

(12) $p(t) = (3.93)e^{kt}$,

where $p(t)$ is the population in millions. To determine k, we observe that in 1890, when $t = 100$ years, we have

$$p(100) = 62.95 = (3.93)e^{100k}.$$

Solving for k yields

$$k = \frac{\ln(62.95) - \ln(3.93)}{100} \approx 0.027737,$$

and substituting this value in equation (12), we find

(13) $p(t) = (3.93)e^{(0.027737)t}$.   ●

In Table 3.1 on page 78 we have listed the United States population as given by the U.S. Bureau of the Census, and the population predicted by the Malthusian model using equation (13). Looking at Table 3.1 we see that the predictions based on the Malthusian model are in reasonable agreement with the census data until about 1900. After 1900 the predicted population is too large, and the Malthusian model is unacceptable.

We remark that a Malthusian model can be generated using the census data for any two different years. We selected 1790 and 1890 for purposes of comparison with the logistic model which we now describe.

The Malthusian model considered only death by natural causes. What about premature deaths due to malnutrition, inadequate medical supplies, communicable diseases, violent crimes, etc.? Since these factors involve a competition within the population, we might assume that there is another component of the death rate that is proportional to the number of two-party interactions. For a population of size p, there are $p(p-1)/2$ such interactions. Thus we will assume that the death rate due to intraspecies competition is

$$k_3 \frac{p(p-1)}{2},$$

where k_3 is the proportionality constant. Together with equation (9), this leads to the **logistic model**

(14) $\dfrac{dp}{dt} = ap - bp^2$, $p(0) = p_0$, where $a = k_1 - k_2 + k_3/2$ and $b = k_3/2$.

The equation (14) is separable and can be solved using a partial fractions expansion:

$$\int \frac{1}{p(a - bp)} \, dp = \int dt,$$

$$\frac{1}{a} \int \frac{1}{p} \, dp - \frac{1}{a} \int \frac{-b}{a - bp} \, dp = \int dt,$$

$$\frac{1}{a} \ln|p| - \frac{1}{a} \ln|a - bp| = t + c_1,$$

$$\ln \left| \frac{p}{a - bp} \right| = at + c_2,$$

$$\frac{p}{a - bp} = c_3 e^{at}.$$

TABLE 3.1	A COMPARISON OF THE MALTHUSIAN AND LOGISTIC MODELS WITH U.S. CENSUS DATA (POPULATION IS GIVEN IN MILLIONS)		
Year	**U.S. Census**	**Malthusian (Example 3)**	**Logistic (Example 4)**
1790	3.93	3.93	3.93
1800	5.31	5.19	5.30
1810	7.24	6.84	7.13
1820	9.64	9.03	9.58
1830	12.87	11.92	12.82
1840	17.07	15.73	17.07
1850	23.19	20.76	22.60
1860	31.44	27.39	29.70
1870	39.82	36.15	38.65
1880	50.16	47.70	49.69
1890	62.95	62.95	62.95
1900	75.99	83.07	78.37
1910	91.97	109.63	95.64
1920	105.71	144.67	114.21
1930	122.78	190.91	133.28
1940	131.67	251.94	152.00
1950	151.33	332.47	169.56
1960	179.32	438.75	185.35
1970	203.21	579.00	199.01
1980	226.50	764.08	210.46
1990	249.63	1008.32	219.77
2000	?	1330.63	227.19

Solving for p we have

$$p(t) = \frac{ac_3 e^{at}}{bc_3 e^{at} + 1} = \frac{ac_3}{bc_3 + e^{-at}}.$$

Using the initial condition in (14), we find

$$c_3 = \frac{p_0}{a - bp_0}, \quad \text{if} \quad p_0 \neq 0 \quad \text{or} \quad a/b.$$

Thus, on substituting for c_3 and simplifying, we get

(15) $$p(t) = \frac{ap_0}{bp_0 + (a - bp_0)e^{-at}}.$$

The function $p(t)$ given in (15) is called the **logistic function.** In Figure 3.4, we have sketched the graph of this function, which is called the **logistic curve.** The logistic model for population growth was first developed by P. F. Verhulst around 1840.

An important property of the logistic function is that

$$\lim_{t \to \infty} p(t) = \frac{a}{b} \quad \text{if} \quad p_0 > 0.$$

Consequently, we see that any population following the logistic model will remain **bounded.** Moreover, when the initial population p_0 is greater than the ratio a/b, the population will decrease toward a/b, and when the initial population is less than a/b, it will increase toward a/b as illustrated in Figure 3.4.

In deriving the solution given in (15), we excluded the two cases when $p_0 = a/b$ or 0. These two populations are referred to as **equilibrium populations.** This terminology follows from the observation that when $p(t) \equiv a/b$ (or 0), then

$$0 \equiv \frac{dp}{dt} \quad \text{and} \quad ap - bp^2 = p(a - bp) \equiv 0.$$

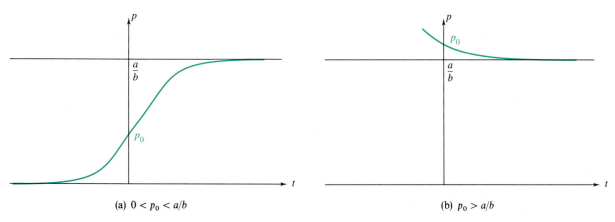

(a) $0 < p_0 < a/b$ (b) $p_0 > a/b$

Figure 3.4 The logistic curves

Hence, $p(t) \equiv a/b$ (or 0) are constant solutions to (14). Notice that the graph of these equilibrium populations are the horizontal asymptotes of the logistic curves in Figure 3.4.

Let's test the logistic model on the population growth of the United States.

● **EXAMPLE 4**   Taking the 1790 population of 3.93 million as the initial population and given the 1840 and 1890 populations of 17.07 and 62.95 million, respectively, use the logistic model to estimate the population at time t.

Solution   With $t = 0$ corresponding to the year 1790, we know that $p_0 = 3.93$. We must now determine the parameters a, b in equation (15). For this purpose we use the given facts that $p(50) = 17.07$ and $p(100) = 62.95$; that is,

(16) $$17.07 = \frac{3.93a}{3.93b + (a - 3.93b)e^{-50a}},$$

(17) $$62.95 = \frac{3.93a}{3.93b + (a - 3.93b)e^{-100a}}.$$

Equations (16) and (17) are two nonlinear equations in the two unknowns a, b. To solve such a system we would usually resort to a numerical approximation scheme such as Newton's method. However, for the case at hand, it is possible to find the solutions directly because the data is given at times t_1 and t_2 with $t_2 = 2t_1$. Indeed, from equation (16), we find that

$$3.93b = \frac{(3.93 - 17.07e^{-50a})a}{17.07(1 - e^{-50a})},$$

and substituting this expression into (17) leads to a linear equation in $x := e^{-50a}$ (see Problem 12). Carrying out the algebra, we ultimately find that

(18) $a = 0.0304667$ and $b = 0.0001214$.

Thus the logistic model for the given data is

(19) $$p(t) = \frac{0.1197340}{0.0004771 + (0.0299896)e^{-(0.0304667)t}}. \qquad ●$$

The model (19) predicts a limit on the future population of the United States, namely,

$$\lim_{t \to \infty} p(t) = \frac{a}{b} = \frac{0.0304667}{0.0001214} \approx 251 \text{ million}.$$

In Table 3.1 we have also listed the U.S. population as predicted by the logistic function in (19). As you can see, these predictions are in better agreement with the census data than the Malthusian model. We have still not considered the effect of major wars, immigration, social and technological changes, etc. A more refined population model would take these into account as well as the number of females in various age groups, especially the number of females of child-bearing age. Some sophisticated models are probabilistic in nature and involve what are called **Markov chains.**

EXERCISES 3.2

1. A brine solution of salt flows at a constant rate of 8 L/min into a large tank that initially held 100 L of brine solution in which was dissolved 5 kg of salt. The solution inside the tank is kept well stirred and flows out of the tank at the same rate. If the concentration of salt in the brine entering the tank is 0.5 kg/L, determine the mass of salt in the tank after t minutes. When will the concentration of salt in the tank reach 0.2 kg/L?

2. A brine solution of salt flows at a constant rate of 6 L/min into a large tank that initially held 50 L of brine solution in which was dissolved 5 kg of salt. The solution inside the tank is kept well stirred and flows out of the tank at the same rate. If the concentration of salt in the brine entering the tank is 0.5 kg/L, determine the mass of salt in the tank after t minutes. When will the concentration of salt in the tank reach 0.3 kg/L?

3. A nitric acid solution flows at a constant rate of 6 L/min into a large tank that initially held 200 L of a 0.5% nitric acid solution. The solution inside the tank is kept well stirred and flows out of the tank at a rate of 8 L/min. If the solution entering the tank is 20% nitric acid, determine the volume of nitric acid in the tank after t minutes. When will the percentage of nitric acid in the tank reach 10%?

4. A brine solution of salt flows at a constant rate of 4 L/min into a large tank that initially held 100 L of water. The solution inside the tank is kept well stirred and flows out of the tank at a rate of 3 L/min. If the concentration of salt in the brine entering the tank is 0.2 kg/L, determine the mass of salt in the tank after t minutes. When will the concentration of salt in the tank reach 0.1 kg/L?

5. A swimming pool whose volume is 10,000 gallons (gal) contains water that is 0.01% chlorine. Starting at $t = 0$, city water containing 0.001% chlorine is pumped into the pool at a rate of 5 gal/min, and the pool water flows out at the same rate. What is the percentage of chlorine in the pool after 1 hr? When will the pool water be 0.002% chlorine?

6. The air in a small room 12 ft by 8 ft by 8 ft is 3% carbon monoxide. Starting at $t = 0$, fresh air containing no carbon monoxide is blown into the room at a rate of 100 ft^3/min. If air in the room flows out through a vent at the same rate, when will the air in the room be 0.01% carbon monoxide?

7. Blood carries a drug into an organ at a rate of 3 cm^3/sec and leaves at the same rate. The organ has a liquid volume of 125 cm^3. If the concentration of the drug in the blood entering the organ is 0.2 g/cm^3, what is the concentration of the drug in the organ at time t if there was no trace of the drug initially? When will the concentration of the drug in the organ reach 0.1 g/cm^3?

8. Water flows into Lake Magdalene from Sweetwater Creek at a rate of 300 gal/min. Lake Magdalene contains about 100 million gal of water. The spraying of nearby orange groves has caused the concentration of pesticides in the lake to reach 0.000035, or 35 parts per million. If pesticides are banned, how long will it be before the concentration of pesticides in Lake Magdalene is below 10 parts per million? (Assume that Sweetwater Creek contains no pesticides and the volume of Lake Magdalene stays constant.)

9. In 1970 the Department of Natural Resources released 1000 splake (a crossbreed of fish) into a lake. In 1977 the population of splake in the lake was estimated to be 3000. Using a Malthusian law for population growth, estimate the population of splake in the lake in 1980. What would the estimate be for the year 1991 using the Malthusian law?

10. From a sketch of the direction field, (see Section 1.3), argue that any solution to the mixing problem model

$$\frac{dx}{dt} = a - bx; \qquad a, b > 0,$$

approaches the equilibrium solution $x(t) \equiv b/a$ as t approaches ∞.

11. From a sketch of the direction field (see Section 1.3), argue that any solution to the logistic model

$$\frac{dp}{dt} = (a - bp)p; \qquad p(t_0) = p_0,$$

where a, b, and p_0 are positive constants, approaches the equilibrium solution $p(t) \equiv b/a$ as t approaches ∞.

12. For the logistic curve (15), assume that $p_1 := p(t_1)$ and $p_2 := p(t_2)$ are given with $t_2 = 2t_1(t_1 > 0)$. Show that

$$a = \frac{1}{t_1} \ln \left[\frac{p_2(p_1 - p_0)}{p_0(p_2 - p_1)} \right],$$

$$b = \frac{a}{p_1} \left[\frac{p_1^2 - p_0 p_2}{p_1 p_2 - 2p_0 p_2 + p_0 p_1} \right].$$

[Hint: Solve, separately, the equations $p_1 = p(t_1)$ and

$p_2 = p(t_2)$ for bp_0 and set the answers equal.] Use these formulas to verify the numerical values given in (18).

13. In Problem 9, suppose we have the additional information that the population of splake in 1984 was estimated to be 5000. Use a logistic model to estimate the population of splake in the year 1991. What is the predicted limiting population? [Hint: Use the formulas in Problem 12.]

14. In 1970 the population of alligators on the Kennedy Space Center grounds was estimated to be just 300. In 1980 the population had grown to an estimated 1500. Using a Malthusian law for population growth, estimate the alligator population on the Kennedy Space Center grounds in the year 2000.

15. In Problem 14 suppose we have the additional information that the population of alligators on the grounds of the Kennedy Space Center in 1975 was estimated to be 1200. Use a logistic model to estimate the population of alligators in the year 2000. What is the predicted limiting population? [Hint: Use the formulas in Problem 12.]

16. A population model used in actuarial predictions is based on the **Gompertz equation**

$$(20) \qquad \frac{dP}{dt} = P(a - b \ln P),$$

where a and b are constants.

(a) Solve the Gompertz equation for $P(t)$.
(b) If $P(0) = P_0 > 0$, give a formula for $P(t)$ in terms of a, b, P_0, and t.
(c) Describe the behavior of $P(t)$ as $t \to \infty$. [Hint: Consider cases for $b > 0$ and $b < 0$.]

17. The initial mass of a certain species of fish is 7 million tons. The mass of fish, if left alone, would increase at a rate proportional to the mass, with a proportionality constant of 2/year. However, commercial fishing removes fish mass at a constant rate of 15 million tons per year. When will all the fish be gone? If the fishing rate is changed so that the initial mass of fish remains constant, what should that rate be?

18. From theoretical considerations, it is known that light from a certain star should reach the Earth with intensity I_0. However, the path taken by the light from the star to the Earth passes through a dust cloud, with absorption coefficient 0.1 per light year, and the light reaching the Earth has intensity $\frac{1}{2}I_0$. How thick is the dust cloud? (The rate of change of light intensity with respect to thickness is proportional to the intensity. One light year is the distance traveled by light during one year.)

19. A snowball melts in such a way that the rate of change in its volume is proportional to its surface area. If the snowball was initially 4 in. in diameter and after 30 min its diameter is 3 in., when will its diameter be 2 in.? Mathematically speaking, when will the snowball disappear?

20. Suppose that the snowball in Problem 19 melts so that the rate of change in its *diameter* is proportional to its surface area. With the same given data, when will its diameter be 2 in.? Mathematically speaking, when will the snowball disappear?

In Problems 21 through 25 assume that the rate of decay of a radioactive substance is proportional to the amount of the substance present. The half-life of a radioactive substance is the time it takes for one-half of the substance to disintegrate.

21. If initially there are 50 grams (g) of a radioactive substance and after 3 days there are only 10 g remaining, what percentage of the original amount remains after 4 days?

22. If initially there are 300 g of a radioactive substance and after 5 years there are 200 g remaining, how much time must elapse before only 10 g remain?

23. Carbon dating is often used to determine the age of a fossil. For example, a humanoid skull was found in a cave in South Africa along with the remains of a campfire. Archaeologists believe the age of the skull to be the same age as the campfire. It is determined that only 2% of the original amount of carbon-14 remains in the burnt wood of the campfire. Estimate the age of the skull if the half-life of carbon-14 is about 5600 years.

24. To see how sensitive the technique of carbon dating of Problem 23 is,

(a) Redo Problem 23 assuming the half-life of carbon-14 is 5550 years.
(b) Redo Problem 23 assuming 3% of the original mass remains.
(c) If each of the figures in parts (a) and (b) represents a "1%" error in measuring the two parameters of half-life and percent of mass remaining, to which parameter is the model more sensitive?

25. The only undiscovered isotopes of the two unknown

elements hohum and inertium (symbols Hh and It) are radioactive. Hohum decays into inertium with a decay constant of 2/year, and inertium decays into the non-radioactive isotope of bunkum (symbol Bu) with a decay constant of 1/year. An initial mass of 1 kilogram of hohum is put into a non-radioactive container, with no other source of hohum, inertium, or bunkum. How much of each of the three elements is in the container after t years? (The **decay constant** is the constant of proportionality in the statement that the rate of loss of mass of the element at any time is proportional to the mass of the element at that time.)

•3.3• HEATING AND COOLING OF BUILDINGS

Our goal is to formulate a mathematical model that describes the 24-hr temperature profile inside a building as a function of the outside temperature, the heat generated inside the building, and the furnace heating or air conditioning cooling. From this model we would like to answer the following three questions:

(a) How long does it take to change the building temperature substantially?

(b) How does the building temperature vary during spring and fall when there is no furnace heating or air conditioning?

(c) How does the building temperature vary in summer when there is air conditioning or in the winter when there is furnace heating?

A natural approach to modeling the temperature inside a building is to use compartmental analysis. If we let $T(t)$ represent the temperature inside the building at time t and view the building as a single compartment, then the rate of change in the temperature is just the difference between the rate at which the temperature increases and the rate at which the temperature decreases.

We will consider three main factors affecting the temperature inside the building. The first factor is the heat produced by people, lights, and machines inside the building. This causes a rate of increase in temperature that we will denote by $H(t)$. The second factor is the heating (or cooling) supplied by the furnace (or air conditioning). This rate of increase (or decrease) in temperature will be represented by $U(t)$. In general, the additional heating rate $H(t)$ and the furnace (or air conditioning) rate $U(t)$ are described in terms of energy per unit time (such as British thermal units per hour). However, by multiplying by the heat capacity of the building (in units of degrees temperature change per heat energy) we can express the two quantities $H(t)$ and $U(t)$ in terms of temperature per unit time.

The third factor is the effect of the outside temperature $M(t)$ on the temperature inside the building. Experimental evidence has shown that this factor can be modeled using **Newton's law of cooling,** which states that there is a rate of change in the temperature $T(t)$ that is proportional to the difference between the outside temperature $M(t)$ and the inside temperature $T(t)$. That is, the rate of change in the building temperature due to $M(t)$ is

$$K[M(t) - T(t)].$$

The positive constant K depends on the physical properties of the building, such as the number of doors and windows and the type of insulation, but K does not depend on M, T, or t. Hence, when the outside temperature is greater than the inside temperature, then $M(t) - T(t) > 0$, and there is an increase in the rate of change of the building temperature due to $M(t)$. On the other hand, when the outside temperature is less than the inside temperature, then $M(t) - T(t) < 0$, and there is a decrease in this rate of change.

Summarizing, we find

(1) $\qquad \dfrac{dT}{dt} = K[M(t) - T(t)] + H(t) + U(t),$

where the additional heating rate $H(t)$ is always nonnegative and $U(t)$ is positive for furnace heating and negative for air conditioner cooling. A more detailed model of the temperature dynamics of the building could involve more variables to represent different temperatures in different rooms or zones. Such an approach would use compartmental analysis, with the rooms as different compartments. (See Section 9.4.)

Since equation (1) is linear, it can be solved using the method discussed in Section 2.4. Rewriting (1) in the standard form

(2) $\qquad \dfrac{dT}{dt}(t) + P(t)T(t) = Q(t),$

where

$\qquad P(t) := K,$

(3) $\qquad Q(t) := KM(t) + H(t) + U(t),$

we find that the integrating factor is

$$\mu(t) = \exp\left(\int K \, dt \right) = e^{Kt}.$$

To solve (2), multiply each side by e^{Kt} and integrate:

$$e^{Kt}\dfrac{dT}{dt}(t) + Ke^{Kt}T(t) = e^{Kt}Q(t),$$

$$e^{Kt}T(t) = \int e^{Kt}Q(t) \, dt + C.$$

Solving for $T(t)$ gives

(4) $\qquad T(t) = e^{-Kt} \int e^{Kt}Q(t) \, dt + Ce^{-Kt}$

$$= e^{-Kt}\left\{ \int e^{Kt}[KM(t) + H(t) + U(t)] \, dt + C \right\}.$$

● **EXAMPLE 1**   Suppose that, at the end of the day (at time t_0), when people leave the building, the outside temperature stays constant at M_0, the additional heating rate H inside the building is zero, and the furnace/air conditioner rate U is zero. Determine $T(t)$, given the initial condition $T(t_0) = T_0$.

Solution   With $M = M_0$, $H = 0$, and $U = 0$, equation (4) becomes

$$T(t) = e^{-Kt}\left\{ \int e^{Kt}KM_0 \, dt + C \right\} = e^{-Kt}[M_0 e^{Kt} + C]$$

$$= M_0 + Ce^{-Kt}.$$

Setting $t = t_0$ and using the initial value T_0 of the temperature, we find that the constant C is $(T_0 - M_0)\exp(Kt_0)$. Hence,

(5) $T(t) = M_0 + (T_0 - M_0)e^{-K(t - t_0)}.$ ●

When $M_0 < T_0$, the solution in (5) decreases exponentially from the initial temperature T_0 to the final temperature M_0. To determine a measure of the time it takes for the temperature to change "substantially," let's consider the simple linear equation $dA/dt = -\alpha A$, whose solutions have the form $A(t) = A(0)e^{-\alpha t}$. Now as $t \to \infty$, the function $A(t)$ either decays exponentially ($\alpha > 0$) or grows exponentially ($\alpha < 0$). In either case, the time it takes for $A(t)$ to change from $A(0)$ to $A(0)/e$ ($\approx 0.368\, A(0)$) is just $1/\alpha$, because

$$A\left(\frac{1}{\alpha}\right) = A(0)e^{-\alpha(1/\alpha)} = \frac{A(0)}{e}.$$

The quantity $1/|\alpha|$, which is independent of $A(0)$, is called the **time constant** for the equation. For linear equations of the more general form $dA/dt = -\alpha A + g(t)$, we again refer to $1/|\alpha|$ as the time constant.

Returning to Example 1, we see that the temperature $T(t)$ satisfies the equations

$$\frac{dT}{dt}(t) = -KT(t) + KM_0, \qquad \frac{d(T - M_0)}{dt}(t) = -K[T(t) - M_0]$$

for M_0 a constant. In either case, the time constant is just $1/K$, which represents the time it takes for the temperature difference $T - M_0$ to change from $T_0 - M_0$ to $(T_0 - M_0)/e$. We also call $1/K$ the **time constant for the building** (without heating or air conditioning). A typical value for the time constant of a building is 2 to 4 hr, but the time constant can be much shorter if windows are open or if there is a fan circulating air, or it can be much longer if the building is well insulated.

In the context of Example 1, we can use the notion of time constant to answer our initial question (a): The building temperature changes exponentially with a time constant of $1/K$. An answer to question (b) about the temperature inside the building during spring and fall is given in the next example.

● **EXAMPLE 2**   Find the building temperature $T(t)$ if the additional heating rate $H(t)$ is equal to the constant H_0, there is no heating or cooling ($U(t) \equiv 0$), and the outside temperature M varies as a sine wave over a 24-hr period, with its minimum at $t = 0$ (midnight) and its maximum at $t = 12$ (noon); that is,

$$M(t) = M_0 - B\cos\omega t,$$

where B is a positive constant, M_0 is the average outside temperature, and $\omega = 2\pi/24 = \pi/12$ radians/hr. (This could be the situation during the spring or fall when there is neither furnace heating nor air conditioning.)

Solution   The function $Q(t)$ in (3) is now

$$Q(t) = K(M_0 - B\cos\omega t) + H_0.$$

Setting $B_0 := M_0 + H_0/K$, we can rewrite Q as

(6) $Q(t) = K(B_0 - B \cos \omega t)$,

where KB_0 represents the daily average value of $Q(t)$; that is,

$$KB_0 = \frac{1}{24} \int_0^{24} Q(t) \, dt.$$

When the forcing function $Q(t)$ in (6) is substituted into the expression for the temperature in equation (4), the result (after using integration by parts) is

$$T(t) = e^{-Kt} \left\{ \int e^{Kt}(KB_0 - KB \cos \omega t) \, dt + C \right\}$$

(7) $T(t) = B_0 - BF(t) + Ce^{-Kt}$,

where

$$F(t) := \frac{\cos \omega t + (\omega/K) \sin \omega t}{1 + (\omega/K)^2}.$$

The constant C is chosen so that at midnight $(t_0 = 0)$ the value of the temperature T is equal to some initial temperature T_0. Thus,

$$C = T_0 - B_0 + BF(0) = T_0 - B_0 + \frac{B}{1 + (\omega/K)^2}. \quad \bullet$$

Notice that the third term in solution (7) involving the constant C tends to zero exponentially. The constant term B_0 in (7) is equal to $M_0 + H_0/K$ and represents the daily average temperature inside the building (neglecting the exponential term). When there is no additional heating rate inside the building $(H_0 = 0)$, this average temperature is equal to the average outside temperature M_0. The term $BF(t)$ in (7) represents the sinusoidal variation of temperature inside the building responding to the outside temperature variation. Since $F(t)$ can be written in the form

(8) $F(t) = [1 + (\omega/K)^2]^{-1/2} \cos(\omega t - \alpha)$,

where $\tan \alpha = \omega/K$ (see Problem 16), the sinusoidal variation inside the building lags behind the outside variation by α hours. Furthermore, the magnitude of the variation inside the building is slightly less, by a factor of $[1 + (\omega/K)^2]^{-1/2}$, than the outside variation. The angular frequency of variation ω is $\pi/12$ radians/hr (which is about $\frac{1}{4}$). Typical values for the dimensionless ratio ω/K lie between $\frac{1}{2}$ and 1. For this range, the lag between inside and outside temperature is approximately 1.8 to 3 hr and the magnitude of the inside variation is between 89% and 71% of the variation outside. Figure 3.5 shows the 24-hr sinusoidal variation of the outside temperature for a typical moderate day as well as the temperature variations inside the building for a dimensionless ratio ω/K of unity, which corresponds to a time constant $1/K$ of approximately 4 hr. In sketching the latter curve, we have assumed that the exponential term has died out.

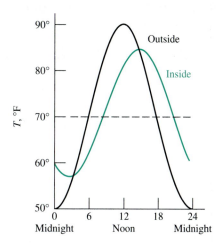

Figure 3.5 Temperature variation inside and outside an unheated building

● **EXAMPLE 3**   Suppose that, in the building of Example 2, a simple thermostat is installed that is used to compare the actual temperature inside the building with a desired temperature T_D. If the actual temperature is below the desired temperature, the furnace supplies heating; otherwise it is turned off. If the actual temperature is above the desired temperature, the air conditioner supplies cooling; otherwise it is off. (In practice there is some dead zone around the desired temperature in which the temperature difference is not sufficient to activate the thermostat, but that is to be ignored here.) Assuming that the amount of heating or cooling supplied is proportional to the difference in temperature—that is,

$$U(t) = K_U[T_D - T(t)],$$

where K_U is the (positive) proportionality constant—find $T(t)$.

Solution   If the proportional control $U(t)$ is substituted directly into the differential equation (1) for the building temperature, we get

(9) $$\frac{dT(t)}{dt} = K[M(t) - T(t)] + H(t) + K_U[T_D - T(t)].$$

A comparison of equation (9) with the first order linear differential equation (2) shows that for this example the quantity P is equal to $K + K_U$, while the quantity $Q(t)$ representing the forcing function includes the desired temperature T_D. That is,

$$P = K + K_U,$$
$$Q(t) = KM(t) + H(t) + K_U T_D.$$

When the additional heating rate is a constant H_0 and the outside temperature M varies as a sine wave over a 24-hr period in the same way as it did in Example 2, the forcing function is

$$Q(t) = K(M_0 - B\cos\omega t) + H_0 + K_U T_D.$$

The function $Q(t)$ has a constant term and a cosine term just as in equation (6), and this equivalence becomes more apparent after the following substitution:

(10) $Q(t) = K_1(B_2 - B_1 \cos \omega t)$,

where

$$\omega := \frac{2\pi}{24} = \frac{\pi}{12}, \qquad\qquad K_1 := K + K_U,$$

(11) $B_2 := \dfrac{K_U T_D + K M_0 + H_0}{K_1}, \qquad B_1 := \dfrac{BK}{K_1}.$

The expressions for the constant P and the forcing function $Q(t)$ of equation (10) are the same as the expressions in Example 2, except that the constants K, B_0, and B are replaced, respectively, by the constants K_1, B_2, and B_1. Hence, the solution to the differential equation (9) will be the same as the temperature solution in Example 2, except that the three constant terms are changed. Thus,

(12) $T(t) = B_2 - B_1 F_1(t) + C \exp(-K_1 t)$,

where

$$F_1(t) := \frac{\cos \omega t + (\omega / K_1) \sin \omega t}{1 + (\omega / K_1)^2}.$$

The constant C is chosen so that at time $t_0 = 0$ the value of the temperature T equals T_0. Thus,

$$C = T_0 - B_2 + B_1 F_1(0). \quad \bullet$$

In the above example the time constant for equation (9) is $1/P = 1/K_1$, where $K_1 = K + K_U$. Here $1/K_1$ is referred to as the **time constant for the building with heating and air conditioning.** For a typical heating and cooling system K_U is somewhat less than 2, and for a typical building the constant K is between $\frac{1}{2}$ and $\frac{1}{4}$, so the sum gives a value for K_1 of about 2. Hence, the time constant for the building with heating and air conditioning is about $\frac{1}{2}$ hr.

When the heating or cooling is turned on, it takes about 30 min for the exponential term in (12) to die off. If we neglect this exponential term, the average temperature inside the building is B_2. Since K_1 is much larger than K and H_0 is small, it follows from (11) that B_2 is roughly T_D, the desired temperature. In other words, after a certain period of time, the temperature inside the building is roughly T_D, with a small sinusoidal variation. (The outside average M_0 and inside heating rate H_0 have only a small effect.) Thus, to save energy, the heating or cooling system may be left off during the night. When it is turned on in the morning, it will take roughly 30 min for the inside of the building to attain the desired temperature. These observations provide an answer to the question regarding the temperature inside the building during summer and winter that was posed at the beginning of this section.

The assumption made in Example 3 that the amount of heating or cooling is $U(t) = K_u[T_D - T(t)]$ may not always be suitable. We have used it here and in the exercises to illustrate the use of the time constant. More adventuresome students may want to experiment with other models for $U(t)$, especially if they have available the numerical techniques discussed in Sections 3.5 and 3.6.

EXERCISES 3.3

1. On a hot Saturday morning while people are working, the air conditioner keeps the temperature inside the building at 24°C. At noon the air conditioner is turned off and the people go home. The temperature outside is a constant 35°C the rest of the afternoon. If the time constant for the building is 4 hr, what will be the temperature inside the building at 2:00 P.M.? At 6:00 P.M.? When will the temperature inside the building reach 27°C?

2. On a mild Saturday morning while people are working, the furnace keeps the temperature inside the building at 21°C. At noon the furnace is turned off and the people go home. The temperature outside is a constant 12°C the rest of the afternoon. If the time constant for the building is 3 hr, when will the temperature inside the building reach 16°C? If some windows are left open and the time constant drops to 2 hr, when will the temperature inside reach 16°C?

3. A warehouse is being built that will have neither heating nor cooling. Depending upon the amount of insulation, the time constant for the building may range from 1 to 5 hr. To illustrate the effect insulation will have on the temperature inside the warehouse, assume that the outside temperature varies as a sine wave, with a minimum of 16°C at 2:00 A.M. and a maximum of 32°C at 2:00 P.M. Assuming that the exponential term (which involves the initial temperature T_0) has died off, what is the lowest temperature inside the building if the time constant is 1 hr? If the time constant is 5 hr? What is the highest temperature inside the building if the time constant is 1 hr? If it is 5 hr?

4. A garage with no heating or cooling has a time constant of 2 hr. If the outside temperature varies as a sine wave with a minimum of 50°F at 2:00 A.M. and a maximum of 80°F at 2:00 P.M., determine the times at which the building reaches its lowest temperature and its highest temperature, assuming that the exponential term has died off.

5. During the summer the temperature inside a van reaches 55°C while the outside temperature is a constant 35°C. When the driver gets into the van, she turns on the air conditioner with the thermostat set at 16°C. If the time constant for the van is $1/K = 2$ hr and for the van with its air conditioning system is $1/K_1 = \frac{1}{3}$ hr, when will the temperature inside the van reach 27°C?

6. Early Monday morning the temperature in the lecture hall has fallen to 40°F, the same as the temperature outside. At 7:00 A.M. the janitor turns on the furnace with the thermostat set at 70°F. The time constant for the building is $1/K = 2$ hr and for the building along with its heating system is $1/K_1 = \frac{1}{2}$ hr. Assuming that the outside temperature remains constant, what will be the temperature inside the lecture hall at 8:00 A.M.? When will the temperature inside the lecture hall reach 65°F?

7. A solar hot water heating system consists of a hot water tank and a solar panel. The tank is well insulated and has a time constant of 64 hr. The solar panel generates 2000 Btu/hr during the day, and the tank has a heat capacity of 2°F per thousand Btu. If the water in the tank is initially 110°F and the room temperature outside the tank is 80°F, what will be the temperature in the tank after 12 hr of sunlight?

8. In Problem 7, if a larger tank with a heat capacity of 1°F per thousand Btu and a time constant of 72 hr is used instead (with all other factors being the same), what would be the temperature in the tank after 12 hr?

9. A cup of hot coffee initially at 95°C cools to 80°C in 5 min while sitting in a room of temperature 21°C. Using just Newton's law of cooling, determine when the temperature of the coffee will be a nice 50°C.

10. A cold beer initially at 35°F warms up to 40°F in 3 min while sitting in a room of temperature 70°F. How warm will the beer be if left out for 20 min?

11. A white wine at room temperature 70°F is chilled in ice (32°F). If it takes 15 min for the wine to chill to 60°F, how long will it take for the wine to reach 56°F?

12. A red wine is brought up from the wine cellar, which is a cool 10°C, and left to breathe in a room of temperature 23°C. When will the temperature of the wine reach 18°C if it takes 10 min for the wine to reach 15°C?

13. Stefan's law of radiation states that the rate of change of temperature of a body at T degrees Kelvin in a medium at M degrees Kelvin is proportional to $M^4 - T^4$. That is,

$$\frac{dT}{dt} = k(M^4 - T^4),$$

where k is a positive constant. Solve this equation using separation of variables. Explain why Newton's law and Stefan's law are nearly the same when T is close to M and M is constant. [Hint: Factor $M^4 - T^4$.]

14. Two friends sit down to talk and enjoy a cup of coffee. When the coffee is served, the impatient friend immediately adds a teaspoon of cream to his coffee. The relaxed friend waits 5 min before adding a teaspoon of cream (which has been kept at a constant tempera-

ture). The two now begin to drink their coffee. *Who has the hotter coffee?* Assume that the cream is cooler than the air and use Newton's law of cooling.

15. It was noon on a cold December day in Tampa; 16°C. Detective Taylor arrived at the crime scene to find the sergeant leaning over the body. The sergeant said that there were several suspects. If only they knew the exact time of death, then they could narrow the list. Detective Taylor took out a thermometer and measured the temperature of the body; 34.5°C. He then left for lunch. Upon returning at 1:00 P.M., he found the body temperature to be 33.7°C. When did the murder occur? [Hint: Normal body temperature is 37°C.]

16. Show that $C_1 \cos \omega t + C_2 \sin \omega t$ can be written in the form $A \cos(\omega t - \alpha)$, where $A = \sqrt{C_1^2 + C_2^2}$ and $\tan \alpha = C_2/C_1$. [Hint: Use a standard trigonometric identity with $C_1 = A \cos \alpha$, $C_2 = A \sin \alpha$.] Use this fact to verify the alternate representation (8) of $F(t)$ discussed in Example 2.

•3.4• NEWTONIAN MECHANICS

Mechanics is the study of the motion of objects and the effect of forces acting on these objects. It is the foundation of several branches of physics and engineering. **Newtonian** or **classical mechanics** deals with the motion of **ordinary** objects; that is, objects that are large compared to an atom and slow moving compared with the speed of light. A model for Newtonian mechanics can be based on **Newton's laws of motion:**[†]

1. When a body is subject to no resultant external force, it moves with a constant velocity.
2. When a body is subject to one or more external forces, the time rate of change of the body's momentum is equal to the vector sum of the external forces acting on it.
3. When one body interacts with a second body, the force of the first body on the second is equal in magnitude, but opposite in direction, to the force of the second body on the first.

Experimental results for more than two centuries verify that these laws are extremely useful for studying the motion of ordinary objects in an **inertial reference frame;** that is, a reference frame in which an undisturbed body moves with a constant velocity. It is Newton's second law, which applies only to inertial reference frames, that enables us to formulate the equations of motion for a moving body. We can express Newton's second

---

[†] For a discussion of Newton's laws of motion see *University Physics*, by F. W. Sears, M. W. Zemansky, and H. D. Young, Addison-Wesley Publishing Co., Reading, Massachusetts, Sixth Edition, 1982.

law by

(1) $$\frac{dp}{dt} = F\left(t, x, \frac{dx}{dt}\right),$$

where $F(t, x, dx/dt)$ is the resultant force on the body at time t, location x, and velocity dx/dt, and $p(t)$ is the momentum of the body at time t. The momentum is the product of the mass of the body and its velocity, that is,

$$p(t) = m(t)v(t).$$

In most applications the mass remains constant, so we can express Newton's second law as

(2) $$m\frac{dv}{dt} = ma = F(t, x, dx/dt),$$

where $v = dx/dt$ is the velocity and $a = dv/dt$ is the acceleration of the body at time t. Observe that equation (2) will be a first order equation in $v(t)$ provided that F does not depend upon x.

　　To apply Newton's laws of motion to a problem in mechanics, the following general procedure may be useful.

 PROCEDURE FOR NEWTONIAN MODELS

(a) Determine *all* relevant forces acting on the object being studied. It is helpful to draw a simple diagram of the object that depicts these forces.
(b) Choose an appropriate axis or coordinate system in which to represent the motion of the object and the forces acting on it. Keep in mind that this coordinate system must be an inertial reference frame.
(c) Apply Newton's second law as expressed in equation (1) or equation (2), whichever is appropriate, to determine the equations of motion for the object.

　　In this section we consider examples in which the motion of the object lies in one dimension and Newton's second law gives rise to a first order differential equation. We use either of two systems of units: the U.S. Customary system or the meter-kilogram-second (MKS) system. The various units in these systems are summarized in Table 3.2, along with approximate values for the Earth's gravitational constant.

TABLE 3.2	MECHANICAL UNITS IN THE U.S. CUSTOMARY AND METRIC SYSTEMS	
Unit	**U.S. Customary System**	**MKS System**
Distance	feet ft	meters m
Mass	slugs	kilograms kg
Time	seconds sec	seconds sec
Force	pounds lb	newtons N
g (Earth)	32 ft/sec^2	9.81 m/sec^2

● **EXAMPLE 1** An object of mass m is given an initial downward velocity v_0 and allowed to fall under the influence of gravity. Assuming that the gravitational force is constant and the force due to air resistance is proportional to the velocity of the object, determine the equation of motion for this object.

Solution We are told that there are two forces acting on the object: a constant force due to the downward pull of gravity and a force due to air resistance that is proportional to the velocity of the object and acts in opposition to the motion of the object. Hence the motion of the object will take place along a vertical axis. On this axis we choose the origin to be the point where the object was initially dropped and let $x(t)$ denote the distance the object has fallen in time t (see Figure 3.6).

The forces acting on the object can be expressed in terms of this axis. The force due to gravity is

$$F_1 = mg,$$

where g is the acceleration due to gravity near the Earth (see Table 3.2). The force due to air resistance is

$$F_2 = -kv(t) = -kx'(t),$$

where k (>0) is the proportionality constant and the negative sign is present because air resistance acts in opposition to the motion of the object. Hence the net force acting on the object (see Figure 3.6) is

(3) $\quad F = F_1 + F_2 = mg - kv(t).$

We now apply Newton's second law by substituting (3) into (2) to obtain

$$m\frac{dv}{dt} = mg - kv.$$

Since the initial velocity of the object is v_0, a model for the velocity of the falling body is expressed by the initial value problem

(4) $\quad m\dfrac{dv}{dt} = mg - kv, \qquad v(0) = v_0,$

where g and k are positive constants.

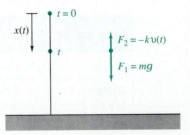

Figure 3.6 Forces on a falling object

The model (4) is the same as the one we obtained in Section 2.1. Using separation of variables (as we did previously), we get

(5) $$v(t) = \frac{mg}{k} + \left(v_0 - \frac{mg}{k}\right)e^{-kt/m}.$$

Since we have taken $x = 0$ when $t = 0$, we can determine the equation of motion of the object by integrating $v = dx/dt$ with respect to t. Thus, from (5) we obtain

$$x(t) = \int v(t)\,dt = \frac{mg}{k}t - \frac{m}{k}\left(v_0 - \frac{mg}{k}\right)e^{-kt/m} + c,$$

and setting $x = 0$ when $t = 0$, we find

$$0 = -\frac{m}{k}\left(v_0 - \frac{mg}{k}\right) + c,$$

$$c = \frac{m}{k}\left(v_0 - \frac{mg}{k}\right).$$

Hence the equation of motion is

(6) $$x(t) = \frac{mg}{k}t + \frac{m}{k}\left(v_0 - \frac{mg}{k}\right)(1 - e^{-kt/m}). \quad \bullet$$

In Figure 3.7 we have sketched the graphs of the velocity and the position as functions of t. Observe that the velocity $v(t)$ approaches the horizontal asymptote $v = mg/k$ as $t \to \infty$, and that the position $x(t)$ asymptotically approaches the line

$$x = \frac{mg}{k}t - \frac{m^2 g}{k^2} + \frac{m}{k}v_0$$

as $t \to +\infty$. The value mg/k of the horizontal asymptote for $v(t)$ is called the **limiting** or **terminal velocity** of the object.

Now that we have obtained the equation of motion for a falling object with air resistance proportional to v, we can answer a variety of questions.

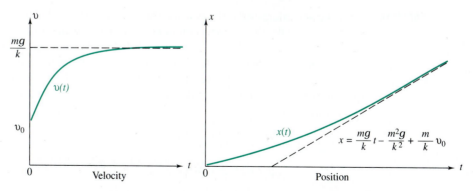

Figure 3.7 Graphs of the position and velocity of a falling object

● **EXAMPLE 2**   An object of mass 3 kg is released from rest 500 m above the ground and allowed to fall under the influence of gravity. Assuming that the gravitational force is constant, with $g = 9.81$ m/sec^2, and the force due to air resistance is proportional to the velocity of the object with proportionality constant $k = 3$ kg/sec, determine when the object will strike the ground.

Solution   We can use the model discussed in Example 1 with $v_0 = 0$, $m = 3$, $k = 3$, and $g = 9.81$. From (6), the equation of motion in this case is

$$x(t) = \frac{(3)(9.81)}{(3)}t - \frac{(3)^2(9.81)}{(3)^2}(1 - e^{-3t/3}) = (9.81)t - (9.81)(1 - e^{-t}).$$

Since the object is released 500 m above the ground, we can determine when the object strikes the ground by setting $x(t) = 500$ and solving for t. Thus we put

$$500 = (9.81)t - 9.81 + (9.81)e^{-t}$$

or

$$t + e^{-t} = \frac{509.81}{9.81} = 51.97,$$

where we have rounded the computations to two decimal places. Unfortunately this equation cannot be solved explicitly for t. We might try to approximate t using Newton's approximation method (see Appendix A), but in this case it is not necessary. Since e^{-t} will be very small for t near 51.97 ($e^{-51.97} \approx 10^{-22}$), we simply ignore the term e^{-t} and obtain as our approximation $t = 51.97$ sec.   ●

● **EXAMPLE 3**   A parachutist whose mass is 75 kg drops from a helicopter hovering 4000 m above the ground and falls toward the earth under the influence of gravity. Assume that the gravitational force is constant and that the force due to air resistance is proportional to the velocity of the parachutist, with the proportionality constant $k_1 = 15$ kg/sec when the chute is closed and with constant $k_2 = 105$ kg/sec when the chute is open. If the chute does not open until 1 min after the parachutist leaves the helicopter, after how many seconds will she hit the ground?

Solution   Since we are interested only in when the parachutist will hit the ground and not where, we consider only the vertical component of her descent. For this we need to use two equations—one to describe the motion before the chute opens and the other to apply after it opens. Before the chute opens, the model is the same as in Example 1 with $v_0 = 0$, $m = 75$ kg, $k = k_1 = 15$ kg/sec, and $g = 9.81$ m/sec^2. If we let $x_1(t)$ be the distance the parachutist has fallen in t seconds and let $v_1 = dx_1/dt$, then substituting into equations (5) and (6) we have

$$v_1(t) = \frac{(75)(9.81)}{15}(1 - e^{-(15/75)t})$$

$$= (49.05)(1 - e^{-0.2t}),$$

and

$$x_1(t) = \frac{(75)(9.81)}{15}t - \frac{(75)^2(9.81)}{(15)^2}(1 - e^{-(15/75)t})$$
$$= 49.05t - 245.25(1 - e^{-0.2t}).$$

Hence, after 1 min, when $t = 60$, the parachutist is falling at a rate

$$v_1(60) = (49.05)(1 - e^{-0.2(60)}) = 49.05 \text{ m/sec},$$

and has fallen

$$x_1(60) = (49.05)(60) - (245.25)(1 - e^{-0.2(60)}) = 2697.75 \text{ m}.$$

(In these and other computations for this problem we round our answers to two decimal places.)

Now when the chute opens, the parachutist is $4000 - 2697.75$ or 1302.25 m above the ground and traveling at a velocity of 49.05 m/sec. To determine the equation of motion after the chute opens, let $x_2(T)$ denote the position of the parachutist T seconds after the chute opens (so that $T = t - 60$), taking $x_2(0) = 0$ at $x_1(60)$ (see Figure 3.8). Furthermore, assume that the initial velocity of the parachutist after the chute opens is the same as the final velocity before it opens, that is, $x_2'(0) = x_1'(60) = 49.05$ m/sec. Since the forces acting on the parachutist are the same as those acting on the object in Example 1, we can again use equations (5) and (6). With $v_0 = 49.05$, $m = 75$, $k = k_2 = 105$, and $g = 9.81$, we find from (6)

$$x_2(T) = \frac{(75)(9.81)}{105}T + \frac{75}{105}\left[49.05 - \frac{(75)(9.81)}{105}\right](1 - e^{-(105/75)T})$$
$$= 7.01T + 30.03(1 - e^{-1.4T}).$$

Now to determine when the parachutist will hit the ground, we just set $x_2(T) = 1302.25$, the height the parachutist was above the ground when her parachute opened. This gives

$$7.01T + 30.03 - 30.03e^{-1.4T} = 1302.25$$

(7) $$T - 4.28e^{-1.4T} - 181.49 = 0.$$

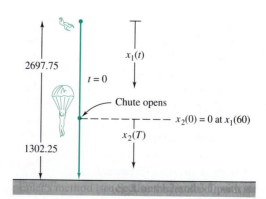

Figure 3.8 The fall of the parachutist

Again we cannot solve (7) explicitly for T. However, since $e^{-1.4T}$ is very small for T near 181.49, we ignore the exponential term and obtain $T = 181.49$. Hence, the parachutist will strike the earth 181.49 sec after the parachute opens, or 241.49 sec after dropping from the helicopter. ●

In the computation for T in equation (7), we found that the exponential $e^{-1.4T}$ was negligible. Consequently, ignoring the corresponding exponential term in equation (5), we see that the parachutist's velocity at impact is

$$\frac{mg}{k_2} = \frac{(75)(9.81)}{105} = 7.01 \text{ m/sec},$$

which is the limiting velocity for her fall with the chute open.

EXERCISES 3.4

Unless otherwise stated, in the following problems we assume that the gravitational force is constant with $g = 9.81 \text{ m/sec}^2$ in the MKS system and $g = 32 \text{ ft/sec}^2$ in the U.S. Customary system.

1. An object of mass 5 kg is released from rest 1000 m above the ground and allowed to fall under the influence of gravity. Assuming that the force due to air resistance is proportional to the velocity of the object with proportionality constant $k = 50$ kg/sec, determine the equation of motion of the object. When will the object strike the ground?

2. A 400-lb object is released from rest 500 ft above the ground and allowed to fall under the influence of gravity. Assuming that the force in pounds due to air resistance is $-10v$, where v is the velocity of the object in ft/sec, determine the equation of motion of the object. When will the object hit the ground?

3. If the object in Problem 1 has a mass of 500 kg instead of 5 kg, when will it strike the ground? [Hint: Here the exponential term is too large to ignore. Use Newton's method to approximate the time t when the object strikes the ground (see Appendix A).]

4. If the object in Problem 2 is released from rest 30 ft above the ground instead of 500 ft, when will it strike the ground? [Hint: Use Newton's method to solve for t.]

5. An object of mass 5 kg is given an initial downward velocity of 50 m/sec and then allowed to fall under the influence of gravity. Assume that the force in newtons (N) due to air resistance is $-10v$, where v is the velocity of the object in m/sec. Determine the equation of motion of the object. If the object is initially 500 m above the ground, determine when the object will strike the ground.

6. An object of mass 8 kg is given an upward initial velocity of 20 m/sec and then allowed to fall under the influence of gravity. Assume that the force in newtons due to air resistance is $-16v$, where v is the velocity of the object in m/sec. Determine the equation of motion of the object. If the object is initially 100 m above the ground, determine when the object will strike the ground.

7. A parachutist whose mass is 75 kg drops from a helicopter hovering 2000 m above the ground and falls toward the ground under the influence of gravity. Assume that the force due to air resistance is proportional to the velocity of the parachutist, with the proportionality constant $k_1 = 30$ kg/sec when the chute is closed and $k_2 = 90$ kg/sec when the chute is open. If the chute does not open until the velocity of the parachutist reaches 20 m/sec, after how many seconds will she reach the ground?

8. A parachutist whose mass is 100 kg drops from a helicopter hovering 3000 m above the ground and falls under the influence of gravity. Assume that the force due to air resistance is proportional to the velocity of the parachutist, with the proportionality constant $k_3 = 20$ kg/sec when the chute is closed and $k_4 = 100$ kg/sec when the chute is open. If the chute does

not open until 30 sec after the parachutist leaves the helicopter, after how many seconds will he hit the ground? If the chute does not open until 1 min after he leaves the helicopter, after how many seconds will he hit the ground?

9. An object of mass 100 kg is released from rest from a boat into the water and allowed to sink. While gravity is pulling the object down, a buoyancy force of $\frac{1}{40}$ times the weight of the object is pushing the object up (weight = mg). If we assume that water resistance exerts a force on the object that is proportional to the velocity of the object, with proportionality constant 10 kg/sec, find the equation of motion of the object. After how many seconds will the velocity of the object be 70 m/sec?

10. An object of mass 2 kg is released from rest from a platform 30 m above the water and allowed to fall under the influence of gravity. After the object strikes the water, it begins to sink with gravity pulling down and a buoyancy force pushing up. Assuming that the force of gravity is constant, that no change in momuntum occurs on impact with the water, that the buoyancy force is $\frac{1}{2}$ the weight (weight = mg), and that the force due to air resistance or water resistance is proportional to the velocity, with proportionality constant $k_1 = 10$ kg/sec in the air and $k_2 = 100$ kg/sec in the water, find the equation of motion of the object. What is the velocity of the object 1 min after it is released?

11. In Example 1 we solved for the velocity of the object as a function of time (equation (5)). In some cases it is useful to have an expression, independent of t, that relates v and x. Find this relation for the motion in Example 1. [Hint: Let $v(t) = V(x(t))$, then $dv/dt = (dV/dx)V$.]

12. A shell of mass 2 kg is shot upward with an initial velocity of 200 m/sec. The magnitude of the force on the shell due to air resistance is $|v|/20$. When will the shell reach its maximum height above the ground? What is the maximum height?

13. When the velocity v of an object is very large, the magnitude of the force due to air resistance is proportional to v^2 with the force acting in opposition to the motion of the object. A shell of mass 3 kg is shot upward from the ground with an initial velocity of 500 m/sec. If the magnitude of the force due to air resistance is $(0.1)v^2$, when will the shell reach its maximum height above the ground? What is the maximum height?

14. An object of mass m is released from rest and falls

under the influence of gravity. If the magnitude of the force due to air resistance is kv^n, where k and n are positive constants, find the limiting velocity of the object (assuming this limit exists). [Hint: Argue that the existence of a (finite) limiting velocity implies that $dv/dt \to 0$ as $t \to \infty$.]

15. A rotating flywheel is being turned by a motor that exerts a constant torque T (see Figure 3.9). A retarding torque due to friction is proportional to the angular velocity ω. If the moment of inertia of the flywheel is I and its initial angular velocity is ω_0, find the equation for the angular velocity ω as a function of time. [Hint: Use Newton's second law for rotational motion, that is, moment of inertia × angular acceleration = sum of the torques.]

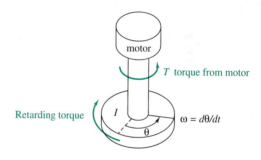

Figure 3.9 Motor-driven flywheel

16. Find the equation for the angular velocity ω in Problem 15, assuming that the retarding torque is proportional to $\sqrt{\omega}$.

17. In Problem 16 let $I = 50$ kg-m^2 and let the retarding torque be $5\sqrt{\omega}$ N-m. If the motor is turned off with the angular velocity at 225 rad/sec, determine how long it will take for the flywheel to come to rest.

18. When an object moves on a surface, it encounters a resistance force called **friction**. This force has a magnitude of μN, where μ is the **coefficient of friction** and N is the magnitude of the normal force that the surface applies to the object. Friction acts in opposition to motion, but cannot initiate motion. Suppose that an object of mass 30 kg is released from the top of an inclined plane that is inclined 30° to the horizontal (see Figure 3.10 on page 98). Assume that the gravitational force is constant, air resistance is negligible, and the coefficient of friction $\mu = 0.2$. Determine the equation of motion for the object as it slides down the

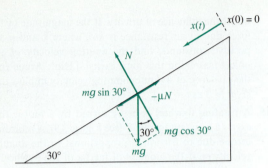

Figure 3.10 Forces on an object on an inclined plane

plane. If the top surface of the plane is 5 m long, what is the velocity of the object when it reaches the bottom?

19. An object of mass 60 kg starts from rest at the top of a 45° inclined plane. Assume that the coefficient of friction is 0.05 (see Problem 18). If the force due to air resistance is proportional to the velocity of the object, say $-3v$, find the equation of motion of the object. How long will it take the object to reach the bottom of the inclined plane if the incline is 10 m long?

20. Recall that an object will not move down an inclined plane until the component of the gravitational force down the incline is greater than the force due to friction (see Problem 18). If μ is the coefficient of friction and α is the angle at which the plane is inclined, determine the critical angle α_0 for which an object will begin to move down the plane if $\alpha > \alpha_0$ but will not move if $\alpha < \alpha_0$.

21. A sailboat has been running (on a straight course) under a light wind at 1 m/sec. Suddenly the wind picks up, blowing hard enough to apply a constant force of 600 N to the sailboat. The only other force acting on the boat is water resistance that is proportional to the velocity of the boat. If the proportionality constant for water resistance is $k = 100$ kg/sec and the mass of the sailboat is 50 kg, find the equation of motion of the sailboat. What is the limiting velocity of the sailboat under this wind?

22. In Problem 21 it is observed that when the velocity of the sailboat reaches 5 m/sec, the boat begins to rise out of the water and "plane." When this happens, the proportionality constant for the water resistance drops to $k_0 = 60$ kg/sec. Now find the equation of motion of the sailboat. What is the limiting velocity of the sailboat under this wind as it is planing?

23. Sailboats A and B each have a mass of 60 kg and cross the starting line at the same time on the first leg of a race. Each has an initial velocity of 2 m/sec. The wind applies a constant force of 650 N to each boat, and the force due to water resistance is proportional to the velocity of the boat. For sailboat A, the proportionality constants are $k_1 = 80$ kg/sec before planing when the velocity is less than 5 m/sec, and $k_2 = 60$ kg/sec when the velocity is above 5 m/sec. For sailboat B, the proportionality constants are $k_3 = 100$ kg/sec before planing when the velocity is less than 6 m/sec and $k_4 = 50$ kg/sec when the velocity is above 6 m/sec. If the first leg of the race is 500 m long, which sailboat will be leading at the end of the first leg?

24. Rocket Flight. A model rocket having initial mass m_0 kg is launched vertically from the ground. The rocket expels gas at a constant rate of α kg/sec and at a constant velocity of β m/sec relative to the rocket. Assume that the magnitude of the gravitational force is proportional to the mass with proportionality constant g. Since the mass is not constant, Newton's second law, which states that force is equal to the time rate of change of the momentum, leads to the equation

$$(m_0 - \alpha t)\frac{dv}{dt} - \alpha\beta = -g(m_0 - \alpha t),$$

where $v = dx/dt$ is the velocity of the rocket, x is its height above the ground, and $m_0 - \alpha t$ is the mass of the rocket at t seconds after launch. If the initial velocity is zero, solve the above equation to determine the velocity of the rocket and its height above ground for $0 \le t < m_0/\alpha$.

25. Escape Velocity. According to Newton's law of gravitation, the attractive force between two objects varies directly as the square of the distance between them. That is, $F_g = GM_1M_2/r^2$ where M_1 and M_2 are the masses of the objects, r is the distance between them (center to center), F_g is the attractive force, and G is the constant of proportionality. Consider a projectile of constant mass m being fired vertically from the Earth (see Figure 3.11). Let t represent time and v the velocity of the projectile.

(a) Show that the motion of the projectile, under the gravitational force of the Earth, is governed by the equation

$$\frac{dv}{dt} = -\frac{gR^2}{r^2},$$

where r is the distance between the projectile and

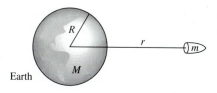

Figure 3.11 Projectile escaping from Earth

the center of the Earth, R is the radius of the Earth, and $g = GM/R^2$, where M is the mass of the Earth.

(b) Use the fact that $dr/dt = v$ to obtain

$$v\frac{dv}{dr} = -\frac{gR^2}{r^2}.$$

(c) If the projectile leaves the Earth's surface with velocity v_0, show that

$$v^2 = \frac{2gR^2}{r} + v_0^2 - 2gR.$$

(d) Use the result of part (c) to show that the velocity of the projectile remains positive if and only if $v_0^2 - 2gR > 0$. The velocity $v_e = \sqrt{2gR}$ is called the **escape velocity** of the Earth.

(e) If $g = 9.81$ m/sec^2 and $R = 6370$ km for the Earth, what is the escape velocity of the Earth?

(f) If the acceleration due to gravity for the moon $g_m = g/6$ and the radius of the moon $R_m = 1738$ km, what is the escape velocity of the moon?

•3.5• IMPROVED EULER'S METHOD

While the analytical techniques presented in Chapter 2 were useful for the variety of mathematical models presented earlier in this chapter, **the majority of the differential equations encountered in applications cannot be solved either implicitly or explicitly.** This is especially true of higher order equations and systems of equations, which we study in later chapters. In this section and the next, we discuss methods for obtaining a numerical *approximation* of the solution to an initial value problem for a first order differential equation. Our goal is to develop algorithms that you can use with a calculator or microcomputer. These algorithms also extend naturally to higher order equations (see Sections 5.6 and 9.5). We describe the rationale behind each method, but leave the more detailed discussion to texts on numerical analysis.[†]

Consider the initial value problem

(1) $\qquad y' = f(x, y) \qquad y(x_0) = y_0.$

To guarantee that (1) has a unique solution, we assume that f and $\partial f/\partial y$ are continuous in a rectangle $R := \{(x, y): a < x < b, c < y < d\}$ containing (x_0, y_0). It follows from Theorem 1 in Chapter 1 that the initial value problem (1) has a unique solution $\phi(x)$ in some interval $x_0 - \delta < x < x_0 + \delta$, where δ is a positive number. Since δ is not known a priori, there is no assurance that the solution will exist at a particular point $x(\neq x_0)$, even if x is in the interval (a, b). However, if $\partial f/\partial y$ is continuous and *bounded*[††] on the vertical strip

$$S := \{(x, y): a < x < b, -\infty < y < \infty\},$$

then it turns out that (1) has a unique solution on the whole interval (a, b). In describing numerical methods we will assume that this last condition is satisfied and that f possesses as many continuous partial derivatives as needed.

In the optional Section 1.3 we used the method of isoclines to motivate a scheme for

---

[†] See, for example, *Elements of Numerical Analysis*, by P. Henrici, John Wiley and Sons, Inc., New York, 1964, or *Numerical Analysis*, Third Edition, by R. L. Burden and J. D. Faires, Prindle, Weber & Schmidt, Boston, 1985.

[††] A function $g(x, y)$ is bounded on S if there exists a number M such that $|g(x, y)| \leq M$ for all (x, y) in S.

approximating the solution to the initial value problem (1). This scheme, called **Euler's method,** is one of the most basic and so it is worthwhile to discuss its advantages, disadvantages, and possible improvements. We begin with a derivation of Euler's method that is somewhat different from that presented in Section 1.3.

Let $h > 0$ be fixed (h is called the **step size**), and consider the equally spaced points

(2) $x_n := x_0 + nh, \qquad n = 0, 1, 2, \dots .$

Our goal is to obtain an approximation to the solution $\phi(x)$ of the initial value problem (1) at those points x_n that lie in the interval (a, b). Namely, we will describe a method that generates values $y_0, y_1, y_2, \dots$ that approximate $\phi(x)$ at the respective points $x_0, x_1, x_2, \dots$; that is,

$y_n \approx \phi(x_n), \qquad n = 0, 1, 2, \dots .$

Of course, the first "approximant" y_0 is exact, since $y_0 = \phi(x_0)$ is given. Thus we must describe how to compute $y_1, y_2, \dots .$

For Euler's method we begin by integrating both sides of equation (1) from x_n to x_{n+1} to obtain

$$\phi(x_{n+1}) - \phi(x_n) = \int_{x_n}^{x_{n+1}} \phi'(t)\, dt = \int_{x_n}^{x_{n+1}} f(t, \phi(t))\, dt,$$

where we have substituted $\phi(x)$ for y. Solving for $\phi(x_{n+1})$, we have

(3) $$\phi(x_{n+1}) = \phi(x_n) + \int_{x_n}^{x_{n+1}} f(t, \phi(t))\, dt.$$

Without knowing $\phi(t)$, we cannot integrate $f(t, \phi(t))$. Hence, we must approximate the integral in (3). Assuming we have already found $y_n \approx \phi(x_n)$, the simplest approach is to appproximate the area under the function $f(t, \phi(t))$ by the rectangle with base $[x_n, x_{n+1}]$ and height $f(x_n, \phi(x_n))$ (see Figure 3.12). This gives

$$\phi(x_{n+1}) \approx \phi(x_n) + (x_{n+1} - x_n) f(x_n, \phi(x_n)).$$

Substituting h for $x_{n+1} - x_n$ and the approximation y_n for $\phi(x_n)$, we arrive at the numerical scheme

(4) $y_{n+1} = y_n + hf(x_n, y_n), \qquad n = 0, 1, 2, \dots,$

which is Euler's method.

Starting with the given value y_0, we use (4) to compute $y_1 = y_0 + hf(x_0, y_0)$, then use y_1 to compute $y_2 = y_1 + hf(x_1, y_1)$, etc. Several examples of Euler's method can be found in Section 1.3.

As discussed in Section 1.3, if we wish to use Euler's method to approximate the solution to the initial value problem (1) at a particular value of x, say $x = c$, then we must first determine a suitable step size h so that $x_0 + nh = c$ for some integer n. For example, we can take $n = 1$ and $h = c - x_0$ in order to arrive at the approximation after just one step:

$$\phi(c) = \phi(x_0 + h) \approx y_1.$$

If, instead, we wish to take 10 steps ($n = 10$) in Euler's method, we choose $h = (c - x_0)/10$

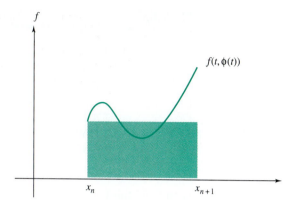

Figure 3.12 Approximation by a rectangle

and ultimately obtain

$$\phi(c) = \phi(x_0 + 10h) = \phi(x_{10}) \approx y_{10}.$$

In general, depending on the size of h, we will get different approximations to $\phi(c)$. It is reasonable to expect that as h gets smaller (or, equivalently, as n gets larger), then the Euler approximations approach the exact value $\phi(c)$. On the other hand, as h gets smaller, the number (and cost) of computations increases and hence so do machine errors that arise from round-off. Thus, it is important to analyze how the error in the approximation scheme varies with h.

If Euler's method is used to approximate the solution $\phi(x) = e^x$ to the problem

(5) $y' = y, \qquad y(0) = 1,$

at $x = 1$, then we obtain approximations to the constant $e = \phi(1)$. It turns out that these approximations take a particularly simple form that enables us to compare the error in the approximation with the step size h. Indeed, setting $f(x, y) = y$ in (4) yields

$$y_{n+1} = y_n + hy_n = (1 + h)y_n, \qquad n = 0, 1, 2, \dots.$$

Since $y_0 = 1$, we get

$$y_1 = (1 + h)y_0 = 1 + h$$
$$y_2 = (1 + h)y_1 = (1 + h)(1 + h) = (1 + h)^2$$
$$y_3 = (1 + h)y_2 = (1 + h)(1 + h)^2 = (1 + h)^3,$$

and, in general,

(6) $y_n = (1 + h)^n, \qquad n = 0, 1, 2, \dots.$

For the problem in (5) we have $x_0 = 0$; so to obtain approximations at $x = 1$, we must set $nh = 1$. That is, h must be the reciprocal of an integer ($h = 1/n$). Replacing n by $1/h$ in (6), we see that Euler's method gives the (familiar) approximation $(1 + h)^{1/h}$ to the constant e. In Table 3.3 on page 102 we list this approximation for $h = 1, 10^{-1}, 10^{-2}, 10^{-3}, 10^{-4}$, along with the corresponding errors

$$e - (1 + h)^{1/h}.$$

TABLE 3.3 EULER'S APPROXIMATIONS TO $e = 2.71828\ldots$

h	Euler's Approximation $(1+h)^{1/h}$	Error $e-(1+h)^{1/h}$	$\dfrac{\text{Error}}{h}$
1	2.0000	0.71828	0.71828
10^{-1}	2.59374	0.12454	1.24539
10^{-2}	2.70481	0.01347	1.34680
10^{-3}	2.71692	0.00136	1.35790
10^{-4}	2.71815	0.00014	1.35902

From the second and third columns in Table 3.3 we see that the approximation gains roughly one decimal place in accuracy as h decreases by a factor of 10. That is, the error is roughly proportional to h. This observation is further confirmed by the entries in the last column of Table 3.3. In fact, using methods of calculus (see Problem 21, Section 1.3, page 20) it can be shown that

(7) $$\lim_{h\to 0}\frac{\text{error}}{h} = \lim_{h\to 0}\frac{e-(1+h)^{1/h}}{h} = \frac{e}{2} \approx 1.35914.$$

The general situation is similar: When Euler's method is used to approximate the solution to the initial value problem (1), the error in the approximation is at worst a constant times the step size h. Moreover, in view of (7), this is the best one can say.

Numerical analysts have a convenient notation for describing the convergence behavior of a numerical scheme. For fixed x, we denote by $y(x;h)$ the approximation to the solution $\phi(x)$ of (1) obtained via the scheme when using a step size of h. We say that the numerical scheme **converges** at x if

$$\lim_{h\to 0} y(x;h) = \phi(x).$$

In other words, as the step size h decreases to zero, the approximations approach the exact value $\phi(x)$. The rate at which $y(x;h)$ tends to $\phi(x)$ is often expressed in terms of a suitable power of h. If the error $\phi(x) - y(x;h)$ tends to zero like a constant times h^p, we write

$$\phi(x) - y(x;h) = O(h^p),$$

and say that the **method is of order p.** Of course, the higher the power p, the faster is the rate of convergence.

As seen from our earlier discussion, the rate of convergence of Euler's method is $O(h)$; that is, *Euler's method is of order* $p = 1$. This means that to have an error less than 10^{-2} generally requires more than 100 steps ($n > 10^2$). Thus Euler's method converges too slowly to be of practical use.

How can we improve Euler's method? To answer this, let's return to the derivation expressed in formulas (3) and (4), and analyze the "errors" that were introduced to get the approximation. A crucial step in the process was to approximate the integral

$$\int_{x_n}^{x_{n+1}} f(t,\phi(t))\,dt$$

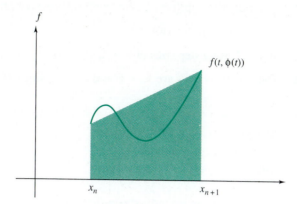

Figure 3.13 Approximation by a trapezoid

by using a rectangle (recall Figure 3.12). This step gives rise to what is called the local truncation error in the method. From calculus we know that a better (more accurate) approach to approximating the integral is to use a trapezoid; that is, to apply the trapezoidal rule (see Figurc 3.13). This gives

$$\int_{x_n}^{x_{n+1}} f(t, \phi(t))\,dt \approx \frac{h}{2}[f(x_n, \phi(x_n)) + f(x_{n+1}, \phi(x_{n+1}))]$$

which leads to the numerical scheme

$$\textbf{(8)} \qquad y_{n+1} = y_n + \frac{h}{2}[f(x_n, y_n) + f(x_{n+1}, y_{n+1})], \qquad n = 0, 1, 2, \ldots.$$

We call equation (8) the **trapezoid scheme.** It is an example of an **implicit method;** that is, unlike Euler's method, equation (8) gives only an implicit formula for y_{n+1}. Assuming that we have already computed y_n, some root-finding technique such as Newton's method (see Appendix A) might be needed to compute y_{n+1}. Despite the inconvenience of working with an implicit method, the trapezoid scheme has two advantages over Euler's method. First, it is a method of order $p = 2$; that is, it converges at a rate that is proportional to h^2 and hence is faster than Euler's method. Second, as illustrated in Project F, the trapezoid scheme has the desirable feature of being **stable.**

Can we somehow modify the trapezoid scheme in order to obtain an explicit method? One idea is to first get an estimate (say, y_{n+1}^*) of the value y_{n+1} using Euler's method and then use formula (8) with y_{n+1} replaced by y_{n+1}^* on the right-hand side. This two-step process is an example of a **predictor-corrector method.** That is, we predict y_{n+1} using Euler's method and then use that value in (8) to obtain a "more correct" approximation. Setting $y_{n+1} = y_n + hf(x_n, y_n)$ in the right-hand side of (8), we obtain

$$\textbf{(9)} \qquad y_{n+1} = y_n + \frac{h}{2}[f(x_n, y_n) + f(x_n + h, y_n + hf(x_n, y_n))], \qquad n = 0, 1, \ldots,$$

where $x_{n+1} = x_n + h$. This explicit scheme is referred to as the **improved Euler's method.**

● EXAMPLE 1 Compute the improved Euler's method approximation to the solution $\phi(x) = e^x$ of

$$y' = y, \qquad y(0) = 1$$

at $x = 1$ using step sizes of $h = 1, 10^{-1}, 10^{-2}, 10^{-3}, 10^{-4}$.

Solution The starting values are $x_0 = 0$ and $y_0 = 1$. Since $f(x, y) = y$, formula (9) becomes

$$y_{n+1} = y_n + \frac{h}{2}[y_n + (y_n + hy_n)] = y_n + hy_n + \frac{h^2}{2}y_n,$$

that is,

$$\textbf{(10)} \qquad y_{n+1} = \left(1 + h + \frac{h^2}{2}\right)y_n.$$

Since $y_0 = 1$, we see inductively that

$$y_n = \left(1 + h + \frac{h^2}{2}\right)^n, \qquad n = 0, 1, 2, \ldots.$$

To obtain approximations at $x = 1$, we must have $1 = x_0 + nh = nh$, and so $n = 1/h$. Hence the improved Euler's approximations to $e = \phi(1)$ are just

$$\left(1 + h + \frac{h^2}{2}\right)^{1/h}.$$

In Table 3.4 we have computed this approximation for the specified values of h, along with the corresponding errors

$$e - \left(1 + h + \frac{h^2}{2}\right)^{1/h}.$$

Comparing the entries of this table with those of Table 3.3, we observe that the improved Euler's method converges much more rapidly than the original Euler's method. In fact, from the first few entries in the second and third columns of Table 3.4, it appears that the approximation gains two decimal places in accuracy each time h is decreased by a factor of 10. In other words, the error is roughly proportional to h^2 (see the last column of the

TABLE 3.4 IMPROVED EULER'S APPROXIMATION TO $e = 2.71828\ldots$

h	Approximation $\left(1 + h + \frac{h^2}{2}\right)^{1/h}$	Error	$\frac{\text{Error}}{h^2}$
1	2.50000	0.21828	21.828
10^{-1}	2.71408	0.00420	0.4201
10^{-2}	2.71824	0.00004	0.4496
10^{-3}	2.71828	0.00000	0.5430
10^{-4}	2.71828	0.00000	0.0000

table and also Problem 4). The entries in the last two rows of the table must be regarded with caution. Indeed, when $h = 10^{-3}$ or $h = 10^{-4}$, the true error is so small that the calculator used rounded this error to zero in its display (but internally maintained additional digits). The entries in color in the last column are inaccurate due to the loss of significant figures in the calculator arithmetic.   ●

As the preceding example suggests, the improved Euler's method converges at the rate $O(h^2)$ and, indeed, it can be proved that, in general, *this method is of order p = 2.*

A step-by-step outline for an algorithm that implements the improved Euler's method is given below. Notice that it halves the size of h in going from one approximation to the next. For binary arithmetic and certain extrapolation methods (see Project E), it is more convenient to use powers of 2 rather than powers of 10.

IMPROVED EULER'S METHOD ALGORITHM

Purpose   To approximate the solution to the initial value problem

$$y' = f(x, y), \qquad y(x_0) = y_0,$$

at $x = c$.

INPUT x_0, y_0, c
ε (tolerance)
M (maximum number of iterations)

Step 1   Set $z = y_0$

Step 2   For $m = 0$ to M do Steps 3–9[†]

Step 3   Set $h = (c - x_0)2^{-m}$, $x = x_0$, $y = y_0$

Step 4   For $i = 1$ to 2^m do Steps 5 and 6

Step 5   Set

$$F = f(x, y)$$
$$G = f(x + h, y + hF)$$

Step 6   Set

$$x = x + h$$
$$y = y + \frac{h}{2}(F + G)$$

Step 7   Print h, y

Step 8   If $|y - z| < \varepsilon$, go to Step 12

Step 9   Set $z = y$

Step 10   Print "$\phi(c)$ is approximately"; y; "but may not be within the tolerance"; ε

Step 11   Go to Step 13

Step 12   Print "$\phi(c)$ is approximately"; y; "with tolerance"; ε

Step 13   Stop

OUTPUT   Approximations of the solution to the initial value problem at $x = c$ using 2^m steps.

[†] To save time, one can start with $m = K < M$ rather than with $m = 0$.

Notice that one of the inputs to this algorithm is the "tolerance." In practice we want to estimate $\phi(x)$ to some desired accuracy, say with a permissible error of ε. To save time and money, we will want to stop the algorithm when we suspect that our estimate is within ε of $\phi(x)$. One such procedure is to terminate the computations when

(11) $|y(c;(c-x_0)2^{-m}) - y(c;(c-x_0)2^{-(m-1)})| < \varepsilon,$

where $y(x;h)$ denotes the approximation of $\phi(x)$ using step size h. The preceding algorithm repeatedly uses the improved Euler's method to approximate $\phi(c)$, halving the step size with each iteration. When two successive iterations differ by less than ε, the process terminates. Using $y(c;(c-x_0)2^{-m})$ to approximate $\phi(c)$, inequality (11) simulates the **absolute error**

$$|y(c;(c-x_0)2^{-m}) - \phi(c)|.$$

If one desires a stopping procedure that simulates the **relative error** $|[y(c;(c-x_0)2^{-m}) - \phi(c)]/\phi(c)|$, then replace Step 8 by

Step 8′   If $\left|\dfrac{y-z}{z}\right| < \varepsilon,$ go to Step 12.

● **EXAMPLE 2**   Use the algorithm for improved Euler's method to approximate the solution to the initial value problem

(12) $y' = x + 2y,$ $y(0) = 0.25,$

at $x = 2$. For a tolerance of $\varepsilon = 0.001$, use a stopping procedure based on the absolute error.

Solution   The starting values are $x_0 = 0$, $y_0 = 0.25$. Since we are computing the approximations for $c = 2$, the initial value for h in Step 3 of the algorithm is

$$h = (2-0)2^{-0} = 2.$$

For equation (12) we have $f(x, y) = x + 2y$, and so the numbers F and G in Step 5 are

$$F = x + 2y$$
$$G = (x+h) + 2(y+hF) = x + 2y + h(1 + 2x + 4y),$$

and Step 6 becomes

$$x = x + h$$
$$y = y + \frac{h}{2}(F + G) = y + \frac{h}{2}(2x + 4y) + \frac{h^2}{2}(1 + 2x + 4y).$$

Thus, with $x_0 = 0$, $y = 0.25$, and $h = 2$, we get for the first approximation

$$y = 0.25 + (0 + 1) + 2(1 + 1) = 5.25.$$

To describe the further outputs of the algorithm, we use the notation $y(2;h)$ for the approximation obtained with step size h. Thus $y(2;2) = 5.25$, and we find from the algorithm

$$y(2; 1) = 11.25000 \qquad y(2; 2^{-5}) = 25.98132$$
$$y(2; 2^{-1}) = 18.28125 \qquad y(2; 2^{-6}) = 26.03172$$
$$y(2; 2^{-2}) = 23.06067 \qquad y(2; 2^{-7}) = 26.04468$$
$$y(2; 2^{-3}) = 25.12012 \qquad y(2; 2^{-8}) = 26.04797$$
$$y(2; 2^{-4}) = 25.79127 \qquad y(2; 2^{-9}) = 26.04880.$$

Since $|y(2; 2^{-9}) - y(2; 2^{-8})| = 0.00083$, which is less than $\varepsilon = 0.001$, we stop.

The exact solution to (12) is $\phi(x) = \frac{1}{2}(e^{2x} - x - \frac{1}{2})$, and so we have determined that

$$\phi(2) = \frac{1}{2}(e^4 - \frac{5}{2}) \approx 26.04880. \quad \bullet$$

In the next section we discuss methods with higher rates of convergence than either Euler's or the improved Euler's methods.

EXERCISES 3.5

In many of the following problems it will be helpful to have a calculator or microcomputer available. The reader may also find it convenient to write a program for solving initial value problems using the improved Euler's method algorithm on page 105. (Remember, all trigonometric calculations are done in radians.)

1. Show that when Euler's method is used to approximate the solution of the initial value problem

   $$y' = 5y, \qquad y(0) = 1,$$

   at $x = 1$, then the approximation with step size h is $(1 + 5h)^{1/h}$.

2. Show that when Euler's method is used to approximate the solution of the initial value problem

   $$y' = -\frac{1}{2}y, \qquad y(0) = 3,$$

   at $x = 2$, then the approximation with step size h is

   $$3\left(1 - \frac{h}{2}\right)^{2/h}.$$

3. Show that when the trapezoid scheme given in formula (8) is used to approximate the solution $\phi(x) = e^x$ of

   $$y' = y, \qquad y(0) = 1,$$

   at $x = 1$, then we get

   $$y_{n+1} = \left(\frac{1 + h/2}{1 - h/2}\right)y_n, \qquad n = 0, 1, 2, \ldots,$$

   which leads to the approximation

   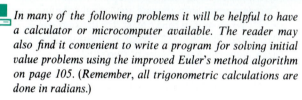

   $$\left(\frac{1 + h/2}{1 - h/2}\right)^{1/h}$$

   for the constant e. Compute this approximation for

$h = 1$, 10^{-1}, 10^{-2}, 10^{-3}, and 10^{-4} and compare your results with those in Tables 3.3 and 3.4.

4. In Example 1, the improved Euler's method approximation to e with step size h was shown to be

   $$\left(1 + h + \frac{h^2}{2}\right)^{1/h}$$

   First prove that error $:= e - (1 + h + h^2/2)^{1/h}$ approaches zero as $h \to 0$. Then use L'Hôpital's rule to show that

   $$\lim_{h \to 0} \frac{\text{error}}{h^2} = \frac{e}{6} \approx 0.45304.$$

   Compare this constant with the entries in the last column of Table 3.4.

5. Show that when the improved Euler's method is used to approximate the solution of the initial value problem

   $$y' = 4y, \qquad y(0) = \frac{1}{3},$$

   at $x = \frac{1}{2}$, then the approximation with step size h is

   $$\frac{1}{3}(1 + 4h + 8h^2)^{1/(2h)}.$$

6. Since the integral $y(x) := \int_0^x f(t)\,dt$ with variable upper limit satisfies (for continuous f) the initial value problem

   $$y' = f(x), \qquad y(0) = 0,$$

   any numerical scheme that is used to approximate the solution at $x = 1$ will give an approximation to the definite integral

   $$\int_0^1 f(t)\,dt.$$

Derive a formula for this approximation of the integral using

(a) Euler's method.
(b) Trapezoid scheme.
(c) Improved Euler's method.

7. Use formula (9) for the improved Euler's method with step size $h = 0.1$ to approximate the solution to the initial value problem

$$y' = x - y^2, \qquad y(1) = 0,$$

at the points $x = 1.1, 1.2, 1.3, 1.4,$ and 1.5. Compare these approximations with those obtained using Euler's method (see Problem 13 in Exercises 1.3).

8. Use formula (9) for the improved Euler's method with step size $h = 0.2$ to approximate the solution to the initial value problem

$$y' = \frac{1}{x}(y^2 + y), \qquad y(1) = 1,$$

at the points $x = 1.2, 1.4, 1.6,$ and 1.8. Compare these approximations with those obtained using Euler's method (see Problem 14 in Exercises 1.3).

9. Use the improved Euler's method with step size $h = 0.2$ to approximate the solution to

$$y' = x + 3\cos(xy), \qquad y(0) = 0,$$

at the points $x = 0, 0.2, 0.4, \dots, 2.0$. Use your answers to make a rough sketch of the solution on $[0, 2]$.

10. Use the improved Euler's method with step size $h = 0.1$ to approximate the solution to

$$y' = 4\cos(x + y), \qquad y(0) = 1,$$

at the points $x = 0, 0.1, 0.2, \dots, 1.0$. Use your answers to make a rough sketch of the solution on $[0, 1]$.

11. Use the algorithm for the improved Euler's method to approximate the solution to

$$\frac{dx}{dt} = 1 + t\sin(tx), \qquad x(0) = 0,$$

at $t = 1$. For a tolerance of $\varepsilon = 0.01$, use a stopping procedure based on the absolute error.

12. Use the algorithm for the improved Euler's method to approximate the solution to

$$y' = 1 - \sin y, \qquad y(0) = 0,$$

at $x = \pi$. For a tolerance of $\varepsilon = 0.01$, use a stopping procedure based on the absolute error.

13. Use the algorithm for the improved Euler's method to approximate the solution to

$$y' = 1 - y + y^3, \qquad y(0) = 0,$$

at $x = 1$. For a tolerance of $\varepsilon = 0.003$, use a stopping procedure based on the absolute error.

14. Use the algorithm for the improved Euler's method to approximate the solution to

$$\frac{dx}{dt} = 1 + x^2, \qquad x(0) = 0,$$

at $t = 1.2$. For a tolerance of $\varepsilon = 0.1$, use a stopping procedure based on the absolute error.

15. Redo Problem 13 for a tolerance of $\varepsilon = 0.003$ using a stopping procedure based on the relative error.

16. Redo Problem 14 for a tolerance of $\varepsilon = 0.1$ using a stopping procedure based on the relative error.

17. Use Euler's method (4) with $h = 0.1$ to approximate the solution to the initial value problem

$$y' = -20y, \qquad y(0) = 1,$$

on the interval $0 \le x \le 1$ (that is, at $x = 0, 0.1, \dots, 1.0$). Compare your answers with the actual solution $y = e^{-20x}$. What went wrong? Next, try the step size $h = 0.025$ and also $h = 0.2$. What conclusions can you draw concerning the choice of step size?

18. **Local versus Global Error.** In deriving formula (4) for Euler's method, a rectangle was used to approximate the area under a curve (see Figure 3.12). With $g(t) := f(t, \phi(t))$, this approximation can be written as

$$\int_{x_n}^{x_{n+1}} g(t)\,dt \approx hg(x_n), \qquad \text{where} \quad h = x_{n+1} - x_n.$$

(a) Show that if g has a continuous derivative that is bounded in absolute value by B, then the rectangle approximation has error $O(h^2)$, that is, for some constant M,

$$\left| \int_{x_n}^{x_{n+1}} g(t)\,dt - hg(x_n) \right| \le Mh^2.$$

This is called the *local truncation error* of the scheme. [Hint: Write

$$\int_{x_n}^{x_{n+1}} g(t)\,dt - hg(x_n) = \int_{x_n}^{x_{n+1}} [g(t) - g(x_n)]\,dt.$$

Next, using the mean value theorem, show that $|g(t) - g(x_n)| \le B|t - x_n|$. Now integrate to obtain the error bound $(B/2)h^2$.]

(b) In applying Euler's method, local truncation errors occur in each step of the process and are propagated throughout the further computations. Show that the *sum* of the local truncation errors in part (a) that arise after n steps is $O(h)$. This is the *global error*, which is the same as the convergence rate of Euler's method.

19. Logistic Model.   In Section 3.2 we discussed the logistic equation

$$\frac{dp}{dt} = ap - bp^2, \qquad p(0) = p_0$$

and its use in modeling population growth. A more general model might involve the equation

(13) $\dfrac{dp}{dt} = ap - bp^r, \qquad p(0) = p_0,$

where $r > 1$. To see the effect of changing the parameter r in (13), take $a = 3$, $b = 1$, and $p_0 = 1$. Now use the improved Euler's method with $h = 0.25$ to approximate the solution to (13) on the interval $0 \le t \le 5$ for $r = 1.5$, 2, and 3.

20. Falling Body.   In Example 1 of Section 3.4 we modeled the velocity of a falling body by the initial value problem

$$m\frac{dv}{dt} = mg - kv, \qquad v(0) = v_0,$$

under the assumption that the force due to air resistance is $-kv$. However, in certain cases the force due to air resistance behaves more like $-kv^r$, where $r\,(>1)$ is some constant. This leads to the model

(14) $m\dfrac{dv}{dt} = mg - kv^r, \qquad v(0) = v_0.$

To study the effect of changing the parameter r in (14), take $m = 1$, $g = 9.81$, $k = 2$, and $v_0 = 0$. Now use the improved Euler's method with $h = 0.2$ to approximate the solution to (14) on the interval $0 \le t \le 5$ for $r = 1.0$, 1.5, and 2.0.

21. Building Temperature.   In Section 3.3 we modeled the temperature inside a building by the initial value problem

(15)
$$\frac{dT}{dt} = K[M(t) - T(t)] + H(t) + U(t),$$

$$T(t_0) = T_0,$$

where M is the temperature outside the building, T is the temperature inside the building, H is the additional heating rate, U is the furnace heating or air conditioner cooling rate, K is a positive constant, and T_0 is the initial temperature at time t_0. In a typical model, $t_0 = 0$ (midnight), $T_0 = 65°F$, $H(t) = 0.1$, $U(t) = 1.5[70 - T(t)]$, and $M(t) = 75 - 20\cos(\pi t/12)$. The constant K is usually between $\frac{1}{4}$ and $\frac{1}{2}$, depending on such things as insulation. To study the effect of insulating this building, consider the typical building described above and use the improved Euler's method with $h = \frac{2}{3}$ to approximate the solution to (15) on the interval $0 \le t \le 24$ (1 day) for $K = 0.2$, 0.4, and 0.6.

● 3.6 ● HIGHER ORDER NUMERICAL METHODS: TAYLOR AND RUNGE-KUTTA

In Sections 1.3 and 3.5 we discussed a simple numerical procedure, Euler's method, for obtaining a numerical approximation of the solution $\phi(x)$ to the initial value problem

(1) $y' = f(x, y), \qquad y(x_0) = y_0.$

Euler's method is easy to implement because it just involves linear approximations to the solution $\phi(x)$. But it suffers from slow convergence, being a method of order 1; that is, the error is $O(h)$. Even the improved Euler's method discussed in Section 3.5 has order of only 2. In this section we present numerical methods that have faster rates of convergence. These include **Taylor methods,** which are natural extensions of the Euler procedure, and **Runge-Kutta methods,** which are the more popular schemes for solving initial value problems because they have fast rates of convergence and are easy to program.

As in the previous section we assume that f and $\partial f/\partial y$ are continuous and bounded on the vertical strip $\{(x, y): a < x < b, -\infty < y < \infty\}$ and that f possesses as many continuous partial derivatives as needed.

Let $\phi_n(x)$ be the exact solution of the related initial value problem

(2) $\phi'_n = f(x, \phi_n)$ $\phi_n(x_n) = y_n$.

The Taylor series for $\phi_n(x)$ about the point x_n is

$$\phi_n(x) = \phi_n(x_n) + h\phi'_n(x_n) + \frac{h^2}{2!}\phi''_n(x_n) + \cdots,$$

where $h = x - x_n$. Since ϕ_n satisfies (2), we can write this series in the form

(3) $$\phi_n(x) = y_n + hf(x_n, y_n) + \frac{h^2}{2!}\phi''_n(x_n) + \cdots.$$

Observe that the recursive formula for y_{n+1} in Euler's method is obtained by truncating the Taylor series after the linear term. For a better approximation, we will use more terms in the Taylor series. This requires that we express the higher order derivatives of the solution in terms of the function $f(x, y)$.

If y satisfies $y' = f(x, y)$, we can compute y'' by using the chain rule:

(4) $$y'' = \frac{\partial f}{\partial x}(x, y) + \frac{\partial f}{\partial y}(x, y)y'$$

$$= \frac{\partial f}{\partial x}(x, y) + \frac{\partial f}{\partial y}(x, y)f(x, y)$$

$$=: f_2(x, y).$$

In a similar fashion, define f_3, f_4, etc., that correspond to the expressions for $y'''(x)$, $y^{(4)}(x)$, etc. If we truncate the expansion in (3) after the h^p term, then, with the above notation, the recursive formulas for the **Taylor method of order p** are

(5) $x_{n+1} = x_n + h,$

(6) $y_{n+1} = y_n + hf(x_n, y_n) + \frac{h^2}{2!}f_2(x_n, y_n) + \cdots + \frac{h^p}{p!}f_p(x_n, y_n).$

As before, $y_n \approx \phi(x_n)$, where $\phi(x)$ is the solution to the initial value problem (1). It can be shown[†] that **the Taylor method of order p has the rate of convergence $O(h^p)$.**

● **EXAMPLE 1**   Determine the recursive formulas for the Taylor method of order 2 for the initial value problem

(7) $y' = \sin(xy),$ $y(0) = \pi.$

Solution   We must compute $f_2(x, y)$ as defined in (4). Since $f(x, y) = \sin(xy)$,

$$\frac{\partial f}{\partial x}(x, y) = y\cos(xy), \qquad \frac{\partial f}{\partial y}(x, y) = x\cos(xy).$$

Substituting into (4) we have

[†] See P. Henrici, *Elements of Numerical Analysis*, John Wiley and Sons, Inc., New York, 1964, Chapter 14.

$$f_2(x, y) = \frac{\partial f}{\partial x}(x, y) + \frac{\partial f}{\partial y}(x, y)f(x, y)$$

$$= y\cos(xy) + x\cos(xy)\sin(xy)$$

$$= y\cos(xy) + \frac{x}{2}\sin(2xy),$$

and the recursive formulas (5) and (6) become

$$x_{n+1} = x_n + h,$$

$$y_{n+1} = y_n + h\sin(x_n y_n) + \frac{h^2}{2}\left[y_n\cos(x_n y_n) + \frac{x_n}{2}\sin(2x_n y_n) \right],$$

where $x_0 = 0$, $y_0 = \pi$ are the starting values.   ●

The difficulty in employing higher order Taylor methods is the tedious computation of the partial derivatives needed to determine f_p; these computations grow exponentially with p. Fortunately, this difficulty can be circumvented by using one of the **Runge-Kutta methods.**[†]

Observe that the general Taylor method has the form

(8) $\qquad y_{n+1} = y_n + hF(x_n, y_n; h),$

where the choice of F depends on p. In particular (cf. (6)), for

$$p = 1, \qquad F = T_1(x, y; h) := f(x, y),$$

(9) $\qquad p = 2, \qquad F = T_2(x, y; h) := f(x, y) + \frac{h}{2}\left[\frac{\partial f}{\partial x}(x, y) + \frac{\partial f}{\partial y}(x, y)f(x, y) \right].$

The idea behind the Runge-Kutta method of order 2 is to choose F in (8) of the form

(10) $\qquad F = K_2(x, y; h) := f(x + \alpha h, y + \beta hf(x, y)),$

where the constants α, β are to be selected so that (8) has the rate of convergence $O(h^2)$. The advantage here is that K_2 is computed by two evaluations of the original function $f(x, y)$ and does not involve the derivatives of $f(x, y)$.

To ensure $O(h^2)$ convergence, we compare this new scheme with the Taylor method of order 2 and require

$$T_2(x, y; h) - K_2(x, y; h) = O(h^2), \quad \text{as} \quad h \to 0.$$

That is, we choose α, β so that the Taylor expansions for T_2 and K_2 agree through terms of order h. For (x, y) fixed, when we expand $K_2 = K_2(h)$ as given in (10) about $h = 0$, we find

(11) $\qquad K_2(h) = K_2(0) + \frac{dK_2}{dh}(0)h + O(h^2)$

$$= f(x, y) + \left[\alpha\frac{\partial f}{\partial x}(x, y) + \beta\frac{\partial f}{\partial y}(x, y)f(x, y) \right]h + O(h^2),$$

[†] *Historical Footnote*: These methods were developed by C. Runge in 1895 and W. Kutta in 1901.

where the expression in brackets for dK_2/dh follows from the chain rule. Comparing (11) with (9), we see that for T_2 and K_2 to agree through terms of order h, we must have $\alpha = \beta = \frac{1}{2}$. Thus

$$K_2(x, y; h) = f\left(x + \frac{h}{2}, y + \frac{h}{2}f(x, y)\right).$$

The Runge-Kutta method we have derived is called the **midpoint method** and it has the recursive formulas

(12) $x_{n+1} = x_n + h,$

(13) $y_{n+1} = y_n + hf\left(x_n + \frac{h}{2}, y_n + \frac{h}{2}f(x_n, y_n)\right).$

By construction, the midpoint method has the same rate of convergence as the Taylor method of order 2; that is, $O(h^2)$. This is the same rate as the improved Euler's method.

In a similar fashion one can work with the Taylor method of order 4, and after some elaborate calculations, obtain the **classical** or **fourth order Runge-Kutta method.** The recursive formulas for this method are

(14)
$$x_{n+1} = x_n + h,$$
$$y_{n+1} = y_n + \tfrac{1}{6}(k_1 + 2k_2 + 2k_3 + k_4),$$

where

$$k_1 = hf(x_n, y_n),$$

$$k_2 = hf\left(x_n + \frac{h}{2}, y_n + \frac{k_1}{2}\right),$$

$$k_3 = hf\left(x_n + \frac{h}{2}, y_n + \frac{k_2}{2}\right),$$

$$k_4 = hf(x_n + h, y_n + k_3).$$

The fourth order Runge-Kutta method is one of the more popular methods because its rate of convergence is $O(h^4)$ and it is easy to program. When combined with extrapolation (see Project E), it produces very accurate approximations when the number of iterations is reasonably small. However, as the number of iterations becomes large, other types of errors may creep in.

A program outline for the fourth order Runge-Kutta method is given on the next page. In this algorithm (as in the improved Euler's algorithm) the step sizes are successively halved, and a stopping procedure is included that compares the two approximations $y(x; h)$ and $y(x; h/2)$ for $\phi(x)$. In particular, if ε is the prescribed tolerance, then the computations end when these consecutive approximations differ by less than ε.

● **EXAMPLE 2**   Use the fourth order Runge-Kutta algorithm outlined below to approximate the solution $\phi(x)$ of the initial value problem

$$y' = y, \qquad y(0) = 1,$$

at $x = 1$ with a tolerance of 0.001.

◤ FOURTH ORDER RUNGE-KUTTA ALGORITHM

Purpose   To approximate the solution to the initial value problem

$$y' = f(x, y), \qquad y(x_0) = y_0$$

at $x = c$.

INPUT x_0, y_0, c
ε (tolerance)
M (maximum number of iterations)

Step 1   Set $z = y_0$
Step 2   For $m = 0$ to M do Steps 3–9[†]
Step 3   Set $h = (c - x_0)2^{-m}, \quad x = x_0, \quad y = y_0$
Step 4   For $i = 1$ to 2^m do Steps 5 and 6
Step 5   Set

$$k_1 = hf(x, y)$$

$$k_2 = hf\left(x + \frac{h}{2}, y + \frac{k_1}{2}\right)$$

$$k_3 = hf\left(x + \frac{h}{2}, y + \frac{k_2}{2}\right)$$

$$k_4 = hf(x + h, y + k_3)$$

Step 6   Set

$$x = x + h$$
$$y = y + \tfrac{1}{6}(k_1 + 2k_2 + 2k_3 + k_4)$$

Step 7   Print h, y
Step 8   If $|z - y| < \varepsilon$ go to Step 12
Step 9   Set $z = y$
Step 10   Print "$\phi(c)$ is approximately,"; y; "but may not be within the tolerance"; ε
Step 11   Go to Step 13
Step 12   Print "$\phi(c)$ is approximately,"; y; "with tolerance"; ε
Step 13   STOP
OUTPUT   Approximations of the solution to the initial value problem at $x = c$, using 2^m steps.

[†] To save time, one can start with $m = K < M$ rather than with $m = 0$.

Solution   The inputs are $x_0 = 0$, $y_0 = 1$, $c = 1$, $\varepsilon = 0.001$, and $M = 100$ (say). Since $f(x, y) = y$, the formulas in Step 5 of the algorithm become

$$k_1 = hy, \qquad k_2 = h\left(y + \frac{k_1}{2}\right), \qquad k_3 = h\left(y + \frac{k_2}{2}\right), \qquad k_4 = h(y + k_3).$$

The initial value for h in Step 3 is

$$h = (1 - 0)2^{-0} = 1.$$

Thus, in Step 5, we compute

$$k_1 = (1)(1) = 1, \qquad k_2 = (1)(1 + 0.5) = 1.5,$$
$$k_3 = (1)(1 + 0.75) = 1.75, \qquad k_4 = (1)(1 + 1.75) = 2.75,$$

and, in Step 6, we get for the first approximation

$$\begin{aligned}
y &= y_0 + \tfrac{1}{6}(k_1 + 2k_2 + 2k_3 + k_4) \\
&= 1 + \tfrac{1}{6}[1 + 2(1.5) + 2(1.75) + 2.75] \\
&= 2.70833,
\end{aligned}$$

where we have rounded to five decimal places. Since

$$|z - y| = |y_0 - y| = |1 - 2.70833| = 1.70833 > \varepsilon,$$

we return to Step 3 and set $h = 0.5$.
    Doing Steps 5 and 6 for $i = 1$ and 2, we ultimately obtain (for $i = 2$) the approximation

$$y = 2.71735.$$

Since $|z - y| = |2.70833 - 2.71735| = 0.00902 > \varepsilon$, we return to Step 3 and set $h = 0.25$. This leads to the approximation

$$y = 2.71821,$$

so that

$$|z - y| = |2.71735 - 2.71821| = 0.00086,$$

which is less than $\varepsilon = 0.001$. Hence $\phi(1) = e \approx 2.71821$.   ●

    In the previous example we were able to obtain a better approximation for $\phi(1) = e$ with $h = 0.25$ than we obtained in Section 3.5 using Euler's method with $h = 0.001$ (see Table 3.3) and roughly the same accuracy as we obtained in Section 3.5 using the improved Euler's method with $h = 0.01$ (see Table 3.4).

● **EXAMPLE 3**   Use the fourth order Runge-Kutta method (14) to approximate the solution $\phi(x)$ of the initial value problem

(15) $y' = y^2$, $y(0) = 1$,

at $x = 2$ using $h = 0.25$.

Solution   Here the starting values are $x_0 = 0$ and $y_0 = 1$. Since $f(x, y) = y^2$, we use formulas (14) with

$$k_1 = h(y_n)^2, \qquad k_2 = h\left(y_n + \frac{k_1}{2}\right)^2,$$

$$k_3 = h\left(y_n + \frac{k_2}{2}\right)^2, \qquad k_4 = h(y_n + k_3)^2.$$

For $h = 0.25$, we have $n = 8$. From the preceding formulas, we find

$$y_1 = 1.33322,$$
$$y_2 = 1.99884,$$
$$y_3 = 3.97238,$$
$$y_4 = 32.82820,$$
$$y_5 = 4.09664 \times 10^{11},$$
$$y_6 = \text{overflow}.$$

What happened? Fortunately the equation in (15) is separable, and, solving for $\phi(x)$, we obtain $\phi(x) = (1 - x)^{-1}$. It is now obvious where the problem lies; the true solution $\phi(x)$ is not defined at $x = 1$. If we had been more cautious, we would have realized that $\partial f/\partial y = 2y$ is *not* bounded for all y. Hence the existence of a unique solution is not guaranteed for all x between 0 and 2, and in this case the method does *not* converge.   ●

● **EXAMPLE 4**   Use the fourth order Runge-Kutta algorithm to approximate the solution $\phi(x)$ of the initial value problem

$$y' = x - y^2, \qquad y(0) = 1,$$

at $x = 2$ with a tolerance of 0.0001.

Solution   This time we check to see whether $\partial f/\partial y$ is bounded. Here $\partial f/\partial y = -2y$, which is certainly unbounded in any vertical strip. However, let's consider the qualitative behavior of the solution $\phi(x)$. The solution curve starts at $(0, 1)$, where $\phi'(0) = 0 - 1 < 0$, so $\phi(x)$ begins decreasing and continues to decrease until it crosses the curve $y = \sqrt{x}$. After crossing this curve, $\phi(x)$ begins to increase, since $\phi'(x) = x - \phi^2(x) > 0$. As $\phi(x)$ increases, it remains below the curve $y = \sqrt{x}$, because if the solution were to get "close" to the curve $y = \sqrt{x}$, then the slope of $\phi(x)$ would approach zero, so that overtaking the function $\sqrt{x}$ is impossible.

The above argument shows that $\phi(x)$ exists for $x > 0$, and since it exists, we feel reasonably sure that the fourth order Runge-Kutta method will give a good approximation of the true solution $\phi(x)$. Proceeding with the algorithm, we use the starting values $x_0 = 0$ and $y_0 = 1$. Since $f(x, y) = x - y^2$, the formulas in Step 5 become

$$k_1 = h(x - y^2), \qquad k_2 = h\left[\left(x + \frac{h}{2}\right) - \left(y + \frac{k_1}{2}\right)^2\right],$$

$$k_3 = h\left[\left(x + \frac{h}{2}\right) - \left(y + \frac{k_2}{2}\right)^2\right], \qquad k_4 = h[(x + h) - (y + k_3)^2].$$

TABLE 3.5 FOURTH ORDER RUNGE-KUTTA APPROXIMATION FOR $\phi(2)$

m	h	Approximation for $\phi(2)$	$\lvert y(2; 2^{-m+1}) - y(2; 2^{-m+2}) \rvert$
0	2.0	-8.33334	
1	1.0	1.27504	9.60838
2	0.5	1.25170	0.02334
3	0.25	1.25132	0.00038
4	0.125	1.25132	0.00000

In Table 3.5 we give the approximations $y(2; 2^{-m+1})$ for $\phi(2)$ for $m = 0, 1, 2, 3$, and 4. The algorithm stops at $m = 4$ since

$$|y(2; 0.125) - y(2; 0.25)| = 0.00000.$$

Hence $\phi(2) \approx 1.25132$, with a tolerance of 0.0001. ●

EXERCISES 3.6

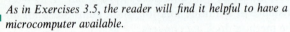

As in Exercises 3.5, the reader will find it helpful to have a microcomputer available.

1. Determine the recursive formulas for the Taylor method of order 2 for the initial value problem

$$y' = \cos(x + y), \qquad y(0) = \pi.$$

2. Determine the recursive formulas for the Taylor method of order 2 for the initial value problem

$$y' = xy - y^2, \qquad y(0) = -1.$$

3. Determine the recursive formulas for the Taylor method of order 4 for the initial value problem

$$y' = x - y, \qquad y(0) = 0.$$

4. Determine the recursive formulas for the Taylor method of order 4 for the initial value problem

$$y' = x^2 + y, \qquad y(0) = 0.$$

5. Use the Taylor methods of orders 2 and 4 with $h = 0.25$ to approximate the solution to the initial value problem

$$y' = x + 1 - y, \qquad y(0) = 1,$$

at $x = 1$. Compare these approximations to the actual solution $y = x + e^{-x}$ evaluated at $x = 1$.

6. Use the Taylor methods of orders 2 and 4 with $h = 0.25$ to approximate the solution to the initial value problem

$$y' = 1 - y, \qquad y(0) = 0,$$

at $x = 1$. Compare these approximations to the actual solution $y = 1 - e^{-x}$ evaluated at $x = 1$.

7. Use the fourth order Runge-Kutta method (14) with $h = 0.25$ to approximate the solution to the initial value problem

$$y' = 2y - 6, \qquad y(0) = 1,$$

at $x = 1$. Compare this approximation to the actual solution $y = 3 - 2e^{2x}$ evaluated at $x = 1$.

8. Use the fourth order Runge-Kutta method (14) with $h = 0.25$ to approximate the solution to the initial value problem

$$\frac{dx}{dt} = 1 + x^2, \qquad x(0) = 0,$$

at $t = 1$.

9. Use the fourth order Runge-Kutta method (14) with $h = 0.5$ and $h = 0.25$ to approximate the solution to the initial value problem

$$y' = x^2 - y^2, \qquad y(0) = 1,$$

at $x = 1$.

10. Use the fourth order Runge-Kutta method (14) with $h = 0.5$ and $h = 0.25$ to approximate the solution to the initial value problem

$$y' = \frac{2y}{1 + x}, \qquad y(0) = 1,$$

at $x = 1$.

11. Use the fourth order Runge-Kutta method (14) with $h = 0.25$ to approximate the solution to the initial

value problem

$$y' = x + 1 - y, \qquad y(0) = 1,$$

at $x = 1$. Compare this approximation with the one obtained in Problem 5 using the Taylor method of order 4.

12. Use the fourth order Runge-Kutta method (14) with $h = 0.25$ to approximate the solution to the initial value problem

$$y' = 1 - y, \qquad y(0) = 0,$$

at $x = 1$. Compare this approximation with the one obtained in Problem 6 using the Taylor method of order 4.

13. Use the fourth order Runge-Kutta algorithm to approximate the solution to the initial value problem

$$y' = 1 - xy, \qquad y(1) = 1,$$

at $x = 2$. For a tolerance of $\varepsilon = 0.001$, use a stopping procedure based on the absolute error.

14. Use the fourth order Runge-Kutta algorithm to approximate the solution to the initial value problem

$$y' = y \cos x, \qquad y(0) = 1,$$

at $x = \pi$. For a tolerance of $\varepsilon = 0.01$, use a stopping procedure based on the absolute error.

15. Use the fourth order Runge-Kutta method (14) with $h = 0.1$ to approximate the solution to

$$y' = \cos(5y) - x, \qquad y(0) = 0,$$

at the points $x = 0, 0.1, 0.2, \ldots, 3.0$. Use your answers to make a rough sketch of the solution on $[0, 3]$.

16. Use the fourth order Runge-Kutta method (14) with $h = 0.1$ to approximate the solution to

$$y' = 3 \cos(y - 5x), \qquad y(0) = 0,$$

at the points $x = 0, 0.1, 0.2, \ldots, 4.0$. Use your answers to make a rough sketch of the solution on $[0, 4]$.

17. The Taylor method of order 2 can be used to approximate the solution to the initial value problem

$$y' = y, \qquad y(0) = 1,$$

at $x = 1$. Show that the approximation y_n obtained by using the Taylor method of order 2 with the step size $1/n$ is given by the formula

$$y_n = \left(1 + \frac{1}{n} + \frac{1}{2n^2}\right)^n, \qquad n = 1, 2, \ldots.$$

Since the solution to the initial value problem is $y = e^x$, then y_n is an approximation to the constant e.

18. If the Taylor method of order p is used in Problem 17,

show that

$$y_n = \left(1 + \frac{1}{n} + \frac{1}{2n^2} + \frac{1}{6n^3} + \cdots + \frac{1}{p!n^p}\right)^n,$$

$$n = 1, 2, \ldots.$$

19. **Fluid Flow.**   In the study of the nonisothermal flow of a Newtonian fluid between parallel plates, the equation

$$\frac{d^2 y}{dx^2} + x^2 e^y = 0, \qquad x > 0,$$

was encountered. By a series of substitutions this equation can be transformed into the first order equation

$$\frac{dv}{du} = u \left(\frac{u}{2} + 1\right) v^3 + \left(u + \frac{5}{2}\right) v^2.$$

Use the fourth order Runge-Kutta algorithm to approximate $v(3)$ if $v(t)$ satisfies $v(2) = 0.1$. For a tolerance of $\varepsilon = 0.0001$, use a stopping procedure based on the relative error.

20. **Chemical Reactions.**   The reaction between nitrous oxide and oxygen to form nitrogen dioxide is given by the balanced chemical equation $2NO + O_2 = 2NO_2$. At high temperatures the dependence of the rate of this reaction on the concentrations of NO, O_2, and NO_2 is complicated. However, at 25°C the rate at which NO_2 is formed obeys the law of mass action and is given by the rate equation

$$\frac{dx}{dt} = k(\alpha - x)^2 \left(\beta - \frac{x}{2}\right),$$

where $x(t)$ denotes the concentration of NO_2 at time t, k is the rate constant, α is the initial concentration of NO, and β is the initial concentration of O_2. At 25°C, the constant k is 7.13×10^3 (liter)^2/(mole)^2(second). Let $\alpha = 0.0010$ mole/L, $\beta = 0.0041$ mole/L, and $x(0) = 0$ mole/L. Use the fourth order Runge-Kutta algorithm to approximate $x(10)$. For a tolerance of $\varepsilon = 0.000001$, use a stopping procedure based on the relative error.

21. **Transmission Lines.**   In studying the electric field that is induced by two nearby transmission lines, an equation of the form

$$\frac{dz}{dx} + g(x)z^2 = f(x)$$

arises. Let $f(x) = 5x + 2$ and $g(x) = x^2$. If $z(0) = 1$, use the fourth order Runge-Kutta algorithm to approximate $z(1)$. For a tolerance of $\varepsilon = 0.0001$, use a stopping procedure based on the absolute error.

•3.7• SOME AVAILABLE CODES FOR INITIAL VALUE PROBLEMS

Before describing some commercially available codes for solving initial value problems, we must first introduce multistep methods and the concept of step size control.

In the previous two sections we discussed numerical methods for solving initial value problems for a first order differential equation. These techniques are examples of **one-step methods** because the approximation y_{n+1} of $y(x_{n+1})$ is obtained using the differential equation and the previous approximation y_n of $y(x_n)$. More generally, we can use **multistep methods.** For such schemes the differential equation and the previous k approximations

$$y_n, y_{n-1}, \dots, y_{n-(k-2)}, y_{n-(k-1)}$$

are used to obtain the approximation y_{n+1}. (In this case the method is called a **k-step method.**)

A two-step method that is easy to derive is based on Simpson's rule for approximating integrals (see Appendix B). Let y_n be the approximation of the solution $y(x)$ to the initial value problem

(1) $\qquad y' = f(x, y), \qquad y(x_0) = y_0,$

at $x_n = x_0 + nh$, where $h > 0$ is the step size. If we integrate both sides of the equation in (1) from x_{n-1} to x_{n+1}, then we have

(2) $\qquad y(x_{n+1}) - y(x_{n-1}) = \int_{x_{n-1}}^{x_{n+1}} f(x, y(x)) \, dx.$

We now approximate the integral in (2) using Simpson's rule. This gives

$$y(x_{n+1}) - y(x_{n-1}) \approx \frac{h}{3} [f(x_{n-1}, y(x_{n-1})) + 4f(x_n, y(x_n)) + f(x_{n+1}, y(x_{n+1}))].$$

Substituting the approximations y_{n-1}, y_n, and y_{n+1} for $y(x_{n-1})$, $y(x_n)$, and $y(x_{n+1})$, respectively, we obtain the numerical scheme

(3) $\qquad y_{n+1} - y_{n-1} = \frac{h}{3} [f(x_{n-1}, y_{n-1}) + 4f(x_n, y_n) + f(x_{n+1}, y_{n+1})].$

This method is called **Simpson's rule.** As you can see, it is an implicit, two-step method. Starting with y_0 and y_1, formula (3) determines y_2; knowing y_1 and y_2 we again use (3) to determine y_3; etc. Formula (3) can also be used in a predictor-corrector procedure, where y_{n+1}^*, a predicted value for y_{n+1}, is used in the right-hand side of (3) to obtain a "corrected" estimate y_{n+1}.

Multistep methods were introduced by J. C. Adams prior to the Runge-Kutta methods. In fact, two important classes of multistep methods are the explicit Adams methods called the **Adams-Bashford methods** and the implicit Adams methods called the **Adams-Moulton methods.** Like the Runge-Kutta methods, there exist Adams methods with different orders of convergence.

A disadvantage of the multistep methods is that they are more susceptible to instability. (Stability is discussed in Project F and Example 4, Section 5.5) From a programming viewpoint, the multistep methods also have the disadvantage of not being self-starting;

that is, the first k approximations $y_0, y_1, \ldots, y_{k-1}$ are needed in a k-step method in order to determine y_k. Therefore, one must use some other method to generate the starting values $y_1, y_2, \ldots, y_{k-1}$ (recall that y_0 is given). This means the Runge-Kutta formulas are easier to code. However, the Runge-Kutta formulas require several evaluations of the function $f(x, y)$ at each step, whereas the multistep formulas require only one evaluation of $f(x, y)$ at each stage. This makes the multistep methods less expensive (and faster) when the function $f(x, y)$ is complicated. For a discussion of multistep methods, we refer the reader to the texts *Numerical Initial Value Problems in Ordinary Differential Equations*, by C. W. Gear, Prentice-Hall, 1971, and *Discrete Variable Methods in Ordinary Differential Equations*, by P. Henrici, John Wiley and Sons, 1962.

In practice, one is confronted with the problem of approximating the solution to within a desired tolerance and obtaining this approximation as inexpensively or as quickly as possible. This is seldom achieved using a *constant* step size h. To illustrate the problem, consider the periodic function $y(x)$ whose graph is given in Figure 3.14. (The solutions to the "Brusselator" equation, which models a chemical reaction that possesses periodic solutions, have graphs similar to the one in Figure 3.14.) If we are trying to approximate $y(x)$, then, over most of its period, we can use a step size h that is relatively large. However, at the "spike," we must take h to be very small or else we will miss the spike entirely or poorly approximate the values of y on the spike. Therefore, it is advantageous to be able to control the step size h.

Much work has been done on writing codes that vary the step size h. These codes can choose h at each step so that the local error is within the given tolerance; shorten the step size, if necessary, to keep the error within tolerance; and decide what the next step size should be. The codes using multistep methods can vary the order of approximation by shifting from one type of multistep method to another. This allows them to begin with a one-step method, which resolves the difficulty of determining the starting values y_1, $y_2, \ldots, y_{k-1}$ in a k-step method. Even when using a code with variable step size and order strategies, it is important to use whatever information one might have to assist in approximating the solution on those intervals where it is changing rapidly.

Because of the considerable effort involved in developing and testing codes for numerically solving initial value problems, a scientist or engineer will often use one of the commercially available packages for "real life" applications. Some packages have been specifically written to handle what are called **stiff problems.** Roughly speaking, this refers

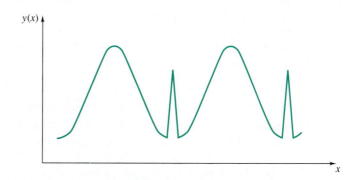

Figure 3.14 Periodic function with spike

to problems that involve one or more fast decaying processes in time, with a time constant (cf. Section 3.3) that is short compared to the time span of interest. For example, the problem

$$\frac{dy}{dt} = -1000y + 999e^{-t}, \qquad y(0) = 0,$$

which has the solution

$$y(t) = e^{-t} - e^{-1000t},$$

would be considered stiff on the interval $[0, 2]$ since the time constant for the term e^{-1000t} is $1/1000$, which is much less than the length of the interval $[0, 2]$.

We now briefly describe several different codes. The double precision versions of any of these will work fine on a PC.

DEABM was developed by L. F. Shampine and H. A. Watts and is a modification of the code **DE** that is described in the book *Computer Solution of Ordinary Differential Equations, the Initial Value Problem,* by L. F. Shampine and M. K. Gordon, Freeman and Company, San Francisco, 1975. This code uses a predictor-corrector approach involving the Adams methods. In this program the step size is adjusted to maintain accuracy in the intervals where the solution varies rapidly and the order is changed to maintain accuracy over the intervals where the solution is slowly varying.

DOPRI8 is a one-step method that uses an eighth order Runge-Kutta formula with a seventh order error estimate and step size control. The formulas were obtained by P. J. Prince and J. R. Dormand. The code is given in the appendix to the book *Solving Ordinary Differential Equations I—Nonstiff Problems,* by E. Hairer, S. P. Nørsett, and G. Wanner, which is published by Springer-Verlag, Berlin, 1987.

DVERK is an initial value problem solver written by T. E. Hull, W. H. Enright, and K. R. Jackson. It uses an explicit Runge-Kutta method that is very effective for nonstiff problems. It is currently being updated. DVERK is available on NETLIB.

ODEPACK is a collection of six initial value problem solvers for stiff and nonstiff problems. The basic member is LSODE which also uses Adams methods. Other component routines are designed for special types of problems such as those involving implicit equation forms and sparse Jacobians. ODEPACK is available from the National Energy Software Center, Building 221, Argonne National Laboratory, 9700 South Cass Avenue, Argonne, IL 60439.

RKF45 is an initial value problem solver written by L. F. Shampine and H. A. Watts. It uses an explicit Runge-Kutta method. It is also available on NETLIB.

SDRIV is an initial value problem solver that works on a great variety of machines (including PCs) and can solve stiff problems. The algorithm has the "G-stop" capability; that is, it can stop the integration based on a user-supplied subroutine. It is included with the book *Numerical Methods and Software,* by David Kahaner, Cleve Moler, and Stephen Nash, Prentice-Hall, Englewood Cliffs, New Jersey, 1989.

VODE is an initial value problem solver for stiff as well as nonstiff problems. It was developed by P. N. Brown, G. D. Byrne, and A. C. Hindmarsh. It uses variable-coefficient Adams-Moulton and Backward Differentiation Formula methods. VODE supersedes EPISODE and is available from the National Energy Software Center (see ODEPACK).

Three software packages that are written for IBM-compatible PCs to be used for both teaching and research are:

PHASEPLANE is an interactive software package for studying dynamical systems. It allows the user to choose between Euler's, Modified Euler's, Runge-Kutta, Adams, and Gear's methods. It was developed by B. Ermentrout and is available from Brooks/Cole Publishing Company, Pacific Grove, California. The software comes with a user's guide.

PHASER is a simulation program written by H. Koçak. It allows the user to choose between Euler's, Improved Euler's, and fourth order Runge-Kutta methods. PHASER comes with the book *Differential and Difference Equations through Computer Experiments*, by H. Koçak, Springer-Verlag, New York, 1986, which describes how to use it.

SIL is a simulation language written by N. Houbak. It uses an implicit Euler method with variable order and variable step size. It can handle discontinuities (switching between several states) and combined discrete/continuous systems. SIL is described in *SIL — a Simulation Language*, by N. Houbak, Lecture Notes in Computer Science No. 426, Springer-Verlag, Berlin, 1990.

GROUP PROJECTS FOR CHAPTER 3

A. Delay Differential Equations

In our discussion of mixing problems in Section 3.2, we encountered the initial value problem

(1) $x'(t) = 6 - \dfrac{3}{500}x(t - t_0),$ $x(t) = 0$ for $t \in [-t_0, 0],$

where t_0 is a positive constant. The equation in (1) is an example of a **delay differential equation.** These equations differ from the usual differential equations by the presence of the shift $(t - t_0)$ in the argument of the unknown function $x(t)$. In general these equations are more difficult to work with than regular differential equations, but quite a bit is known about them.[†]

   (a) Show that the simple linear delay differential equation

(2) $u'(t) = au(t - b),$

[†] See, for example, *Differential-Difference Equations*, by R. Bellman and K. L. Cooke, Academic Press, New York, 1963, or *Ordinary and Delay Differential Equations*, by R. D. Driver, Springer-Verlag, New York, 1977.

where a and b are constants, has a solution of the form $u = Ce^{st}$, for any constant C, provided that s satisfies the transcendental equation $s = ae^{-bs}$.

(b) A solution to (2) for $t > 0$ can also be found using the **method of steps.** Assume that $u(t) = f(t)$ for $-b \le t \le 0$. For $0 \le t \le b$, equation (2) becomes

$$u'(t) = au(t - b) = af(t - b),$$

and so

$$u(t) = \int_0^t af(v - b)\,dv + u(0).$$

Now that we know $u(t)$ on $[0, b]$, we can repeat this procedure to obtain

$$u(t) = \int_b^t au(v - b)\,dv + u(b)$$

for $b \le t \le 2b$. This process can be continued indefinitely.

Use the method of steps to show that the solution to the initial value problem

$$u'(t) = u(t - 1), \qquad u(t) = 1 \quad \text{on} \quad [-1, 0]$$

is given by

$$u(t) = \sum_{k=0}^n \frac{[t - (k - 1)]^k}{k!}, \qquad n - 1 \le t \le n,$$

where n is a nonnegative integer. (This problem can also be solved using Laplace transforms.)

(c) Use the method of steps to compute the solution to the initial value problem given in (1) on the interval $0 \le t \le 10$ for $t_0 = 2$.

B. Aquaculture

Aquaculture is the art of cultivating the plants and animals indigenous to water. In the example considered here, it is assumed that a batch of catfish are raised in a pond. We are interested in determining the best time for harvesting the fish so that the cost per pound for raising the fish is minimized. A differential equation describing the growth of fish may be expressed as

(3) $$\frac{dW}{dt} = KW^\alpha,$$

where $W(t)$ is the weight of the fish at time t, and K and α are empirically determined growth constants. The functional form of this relationship is similar to that of the growth models for other species. Modeling the growth rate or metabolic rate by a term like W^α is a common assumption. Biologists often refer to equation (3) as the **allometric equation.** It can be supported by plausibility arguments such as growth rate depending on the surface area of the gut (which varies like $W^{2/3}$) or depending on the volume of the animal (which varies like W).

(a) Solve equation (3) when $\alpha \ne 1$.

(b) The solution obtained in part (a) grows large without bound, but in practice there is some limiting maximum weight W_{MAX} for the fish. This limiting weight may be included in the differential equation describing growth by inserting a dimensionless variable S that can range between 0 and 1 and involves an empirically determined dimensionless parameter μ.

Namely, we now assume

(4) $$\frac{dW}{dt} = KW^{\alpha}S,$$

where $S := 1 - (W/W_{\text{MAX}})^{\mu}$. When $\mu = 1 - \alpha$, equation (4) has a closed form solution.

Solve equation (4) when $K = 12$, $\alpha = \frac{2}{3}$, $\mu = \frac{1}{3}$, $W_{\text{MAX}} = 64$ (ounces), and $W(0) = 1$ (ounce). The constants are given for t measured in months.

(c) The differential equation describing the total cost in dollars $C(t)$ of raising a fish for t months has one constant term K_1 that specifies the cost per month (due to costs such as interest, depreciation, and labor) and a second constant K_2 that multiplies the growth rate (because the amount of food consumed by the fish is approximately proportional to the growth rate). That is,

(5) $$\frac{dC}{dt} = K_1 + K_2 \frac{dW}{dt}.$$

Solve equation (5) when $K_1 = 0.5$, $K_2 = 0.1$, $C(0) = 1.1$ (dollars), and $W(t)$ is as determined in part (b).

(d) Sketch the curve obtained in part (b) that represents the weight of the fish as a function of time. Next, sketch the curve obtained in part (c) that represents the total cost of raising the fish as a function of time.

(e) To determine the optimal time for harvesting the fish, sketch the ratio $C(t)/W(t)$. This ratio represents the total cost per ounce as a function of time. When this ratio reaches its minimum—that is, when the total cost per ounce is at its lowest—this is the optimal time to harvest the fish. Determine this optimal time to the nearest month.

C. Curve of Pursuit

An interesting geometric model arises when one tries to determine the path of a pursuer chasing its prey. This path is called a *curve of pursuit*. These problems were analyzed using methods of calculus circa 1730 (more than two centuries after Leonardo da Vinci had considered them). The simplest problem is to find the curve along which a vessel moves in pursuing another vessel that flees along a straight line, assuming that the speeds of the two vessels are constant.

Let's assume vessel A, traveling at a speed α, is pursuing vessel B, which is traveling at a speed β. In addition, assume that vessel A begins (at time $t = 0$) at the origin and pursues vessel B, which begins at the point $(b, 0)$, $b > 0$, and travels up the line $x = b$. After t hours, vessel A is located at the

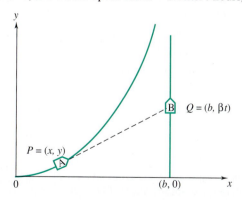

Figure 3.15 The path of vessel A as it pursues vessel B

point $P = (x, y)$ and vessel B is located at the point $Q = (b, \beta t)$ (see Figure 3.15). The goal is to describe the locus of points P; that is, to find y as a function of x.

(a) Since vessel A is pursuing vessel B, then at the time t, vessel A must be heading right at vessel B. That is, the tangent line to the curve of pursuit at P must pass through the point Q (see Figure 3.15). For this to be true, show that

(6)
$$\frac{dy}{dx} = \frac{y - \beta t}{x - b}.$$

(b) Since we know the speed at which vessel A is traveling, we know that the distance it travels in time t is αt. This distance is also the length of the pursuit curve from $(0, 0)$ to (x, y). Using the arclength formula from calculus, show that

(7)
$$\alpha t = \int_0^x \sqrt{1 + [y'(u)]^2}\, du.$$

Solving for t in equations (6) and (7), conclude that

(8)
$$\frac{y - (x - b)(dy/dx)}{\beta} = \frac{1}{\alpha} \int_0^x \sqrt{1 + [y'(u)]^2}\, du.$$

(c) Differentiating both sides of (8) with respect to x, derive the first order equation

$$(x - b)\frac{dw}{dx} = -\frac{\beta}{\alpha}\sqrt{1 + w^2},$$

where $w := dy/dx$.

(d) Using separation of variables and the initial conditions $x = 0$ and $w = dy/dx = 0$ when $t = 0$, show that

(9)
$$\frac{dy}{dx} = w = \frac{1}{2}\left[\left(1 - \frac{x}{b}\right)^{-\beta/\alpha} - \left(1 - \frac{x}{b}\right)^{\beta/\alpha}\right].$$

(e) For $\alpha > \beta$, that is, the pursuing vessel A travels faster than the vessel B that is being pursued, use equation (9) and the initial conditions $x = 0$ and $y = 0$ when $t = 0$ to derive the curve of pursuit

$$y = \frac{b}{2}\left[\frac{(1 - x/b)^{1 + \beta/\alpha}}{1 + \beta/\alpha} - \frac{(1 - x/b)^{1 - \beta/\alpha}}{1 - \beta/\alpha}\right] + \frac{b\alpha\beta}{\alpha^2 - \beta^2}.$$

(f) Find the location where vessel A intercepts vessel B if $\alpha > \beta$.

(g) Show that if $\alpha = \beta$, then the curve of pursuit is given by

$$y = \frac{b}{2}\left\{\frac{1}{2}\left[\left(1 - \frac{x}{b}\right)^2 - 1\right] - \ln\left(1 - \frac{x}{b}\right)\right\}.$$

Will vessel A ever reach vessel B?

D. A Linear Difference Equation

The linear first order difference equation

(10) $y_{n+1} = ay_n + b, \qquad n = 0, 1, 2, \ldots,$

where $a \neq 0$ and b are constants, arises in such applications as computing simple and compound interest and amortizations. One seeks a sequence $\{y_n\}_{n=0}^\infty$ that satisfies relation (10) with y_0 a prescribed "initial condition."

(a) After computing y_1, y_2, y_3, and y_4, deduce the general formula

$$y_n = a^n y_0 + b(1 + a + \cdots + a^{n-1}), \qquad n = 1, 2, 3, \ldots.$$

(b) Show that when $a \neq 1$, we can express y_n in the form

$$y_n = a^n(y_0 - y^*) + y^*, \qquad n = 1, 2, 3, \ldots,$$

where $y^* := b/(1 - a)$.

(c) By considering separately the various cases, describe the behavior (as $n \to \infty$) of the solution sequence $\{y_n\}_{n=0}^{\infty}$. For example, show that when $0 < a < 1$ and $y_0 < y^*$, then the sequence increases monotonically to the limit y^*. [Hint: There are up to ten cases corresponding to the different assumptions on a, b, and y_0.]

◢ E. Extrapolation

▮ When precise information about the *form* of the error in an approximation method is known, a technique called **extrapolation** can be used to improve the rate of convergence.

Suppose the approximation method converges with rate $O(h^p)$ as $h \to 0$ (cf. Section 3.5). From theoretical considerations assume we know, more precisely, that

$$(11) \qquad y(x; h) = \phi(x) + h^p a_p(x) + O(h^{p+1}),$$

where $y(x; h)$ is the approximation to $\phi(x)$ using step size h and $a_p(x)$ is some function that is independent of h (typically we do not know a formula for $a_p(x)$, only that it exists). Our goal is to obtain approximations that converge at the *faster* rate $O(h^{p+1})$.

We start by replacing h by $h/2$ in (11) to get

$$y(x; h/2) = \phi(x) + \frac{h^p}{2^p} a_p(x) + O(h^{p+1}).$$

If we multiply both sides by 2^p and subtract equation (11), we find

$$2^p y(x; h/2) - y(x; h) = (2^p - 1)\phi(x) + O(h^{p+1}).$$

Solving for $\phi(x)$ yields

$$\phi(x) = \frac{2^p y(x; h/2) - y(x; h)}{2^p - 1} + O(h^{p+1}).$$

Hence

$$y^*(x; h/2) := \frac{2^p y(x; h/2) - y(x; h)}{2^p - 1}$$

has a rate of convergence of $O(h^{p+1})$.

(a) Assuming $y^*(x; h/2) = \phi(x) + h^{p+1} a_{p+1}(x) + O(h^{p+2})$, show that

$$y^{**}(x; h/4) := \frac{2^{p+1} y^*(x; h/4) - y^*(x; h/2)}{2^{p+1} - 1}$$

has a rate of convergence of $O(h^{p+2})$.

(b) Assuming $y^{**}(x; h/4) = \phi(x) + h^{p+2} a_{p+2}(x) + O(h^{p+3})$, show that

$$y^{***}(x; h/8) := \frac{2^{p+2} y^{**}(x; h/8) - y^{**}(x; h/4)}{2^{p+2} - 1}$$

has a rate of convergence of $O(h^{p+3})$.

(c) The results of using Euler's method (with $h = 1, \frac{1}{2}, \frac{1}{4}, \frac{1}{8}$) to approximate the solution to the initial value problem

$$y' = y, \qquad y(0) = 1,$$

at $x = 1$ are given in Table 1.2, page 19. For Euler's method, the extrapolation procedure applies with $p = 1$. Use the results in Table 1.2 to find an approximation to $e = y(1)$ by computing $y^{***}(1, \frac{1}{8})$. [Hint: Compute $y^{*}(1; \frac{1}{2})$, $y^{*}(1; \frac{1}{4})$, and $y^{*}(1; \frac{1}{8})$; then compute $y^{**}(1; \frac{1}{4})$ and $y^{**}(1; \frac{1}{8})$.]

(d) Table 1.2 also contains Euler's approximation for $y(1)$ when $h = \frac{1}{16}$. Use this additional information to compute the next step in the extrapolation procedure; that is, compute $y^{****}(1, \frac{1}{16})$.

F. Stability of Numerical Methods

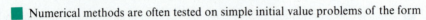

Numerical methods are often tested on simple initial value problems of the form

(12) $\qquad y' + \lambda y = 0, \qquad y(0) = 1, \qquad (\lambda = \text{constant}),$

which has the solution $\phi(x) = e^{-\lambda x}$. Notice that, for each $\lambda > 0$, the solution $\phi(x)$ tends to zero as $x \to \infty$. Thus a desirable property for any numerical scheme that generates approximations $y_0, y_1, y_2, y_3, \ldots$ to $\phi(x)$ at the points $0, h, 2h, 3h, \ldots$ is that, for $\lambda > 0$,

(13) $\qquad y_n \to 0 \quad \text{as} \quad n \to \infty.$

For single-step linear methods, property (13) is called **absolute stability**.

(a) Show that for $x_n = nh$, Euler's method applied to the initial value problem (12) yields the approximations

$$y_n = (1 - \lambda h)^n, \qquad n = 0, 1, 2, \ldots,$$

and deduce that this method is absolutely stable only when $0 < \lambda h < 2$. (This means that for a given $\lambda > 0$, we must choose the step size h sufficiently small in order for property (13) to hold.) Furthermore, show that for $h > 2/\lambda$, the error $y_n - \phi(x_n)$ grows large exponentially!

(b) Show that for $x_n = nh$, the trapezoid scheme of Section 3.5 applied to problem (12) yields the approximations

$$y_n = \left(\frac{1 - \dfrac{\lambda h}{2}}{1 + \dfrac{\lambda h}{2}} \right)^n, \qquad n = 0, 1, 2, \ldots,$$

and deduce that this scheme is absolutely stable for all $\lambda > 0, h > 0$.

(c) Show that the improved Euler's method applied to problem (12) is absolutely stable for $0 < \lambda h < 2$.

Multistep Methods. When multistep numerical methods are used, instability problems may arise that cannot be circumvented by simply choosing the step size h sufficiently small. This is due to the fact that multistep methods yield "extraneous solutions," which may dominate the calculations. To illustrate what can happen, consider the two-step method

(14) $\qquad y_{n+1} = y_{n-1} + 2hf(x_n, y_n), \qquad n = 1, 2, \ldots,$

for the equation $y' = f(x, y)$.

(d) Show that for the initial value problem

(15) $y' + 2y = 0,$ $y(0) = 1,$

the recurrence formula (14), with $x_n = nh$, becomes

(16) $y_{n+1} + 4hy_n - y_{n-1} = 0.$

Equation (16), which is called a **difference equation,** can be solved by using an approach that is similar to solving linear differential equations with constant coefficients. Namely, we try a solution of the form $y_n = r^n$, where r is a constant to be determined.

(e) Show that substituting $y_n = r^n$ in (16) leads to the "characteristic equation"

$$r^2 + 4hr - 1 = 0,$$

which has roots

$$r_1 = -2h + \sqrt{1 + 4h^2} \quad \text{and} \quad r_2 = -2h - \sqrt{1 + 4h^2}.$$

By analogy with the theory for second order differential equations, it can be shown (see Section 5.5) that a general solution of (16) is

$$y_n = c_1 r_1^n + c_2 r_2^n,$$

where c_1 and c_2 are arbitrary constants. Thus the difference equation (16) has two linearly independent solutions, whereas the differential equation (15) has only one—namely, $\phi(x) = e^{-2x}$.

(f) Show that for each $h > 0$,

$$\lim_{n \to \infty} r_1^n = 0, \quad \text{but} \quad \lim_{n \to \infty} |r_2^n| = \infty.$$

Hence the term r_1^n behaves like the solution $\phi(x_n) = e^{-2x_n}$ as $n \to \infty$. However, the extraneous solution r_2^n grows large without bound.

(g) Applying the scheme of (14) to the initial value problem (15) requires two starting values y_0, y_1. The exact values are $y_0 = 1$, $y_1 = e^{-2h}$. However, regardless of the choice of starting values and the size of h, the term $c_2 r_2^n$ will eventually dominate the full solution to the recurrence equation as x_n increases. Illustrate this instability by taking $y_0 = 1$, $y_1 = e^{-2h}$, and using a calculator or microcomputer to compute $y_2, y_3, \ldots, y_{100}$ from the recurrence formula (16) for (i) $h = 0.5$, (ii) $h = 0.05$.

G. Period Doubling and Chaos

In the study of dynamical systems the phenomena of *period doubling* and *chaos* are observed. These phenomena can be seen when one uses a numerical scheme to approximate the solution to an initial value problem for a nonlinear differential equation such as the following logistic model for population growth:

(17) $\dfrac{dp}{dt} = 10p(1 - p),$ $p(0) = 0.1$

(see Section 3.2).

(a) Solve the initial value problem (17) and show that $p(t)$ approaches 1 as t approaches infinity.

(b) Show that using Euler's method (see Sections 1.3 and 3.5) with step size h to approximate the solution to (17) gives

(18) $p_{n+1} = (1 + 10h)p_n - (10h)p_n^2,$ $p_0 = 0.1.$

(c) For $h = 0.18, 0.23, 0.25$, and 0.3, show that the first 40 iterations of (18) appear to: converge to 1 when $h = 0.18$; jump between 1.18 and 0.69 when $h = 0.23$; jump between 1.23, 0.54, 1.16, and 0.70 when $h = 0.25$; and display no discernible pattern when $h = 0.3$.

The transitions from convergence to jumping between two numbers, then four numbers, etc., are called **period doubling.** The phenomenon displayed when $h = 0.3$ is referred to as **chaos.** This transition from period doubling to chaos as h increases is frequently observed in dynamical systems.

The transition to chaos is nicely illustrated in the bifurcation diagram (see Figure 3.16). This diagram is generated for equation (18) as follows. Beginning at $h = 0.18$, compute the sequence $\{p_n\}$ using (18) and, starting at $n = 201$, plot the next 30 values, that is, $p_{201}, p_{202}, \ldots, p_{230}$. Now increment h by 0.001 to 0.181 and repeat. Continue this process until $h = 0.30$. Notice how the figure splits from one branch to two, then four, and then finally gives way to chaos.

Our concern is with the instabilities of the numerical procedure when h is not chosen small enough. Fortunately, the instability observed for Euler's method—the period doubling and chaos—was immediately recognized because we know that this type of behavior is not expected of a solution to the logistic equation. Consequently, if we had tried Euler's method with $h = 0.23, 0.25$, or 0.3 to numerically solve (17), we would have realized that h was not chosen small enough.

The situation for the fourth order Runge-Kutta method (see Section 3.6) is more troublesome. It may happen that for a certain choice of h period doubling occurs, but it is also possible that for other choices of h the numerical solution actually converges to a limiting value that is *not* the limiting value for any solution to the logistic equation in (17).

(d) Approximate the solution to (17) by computing the first 60 iterations of the fourth order Runge-Kutta method using the step size $h = 0.3$. Repeat with $h = 0.325$ and $h = 0.35$. Which values of h (if any) do you feel are giving the "correct" approximation to the solution? Why?

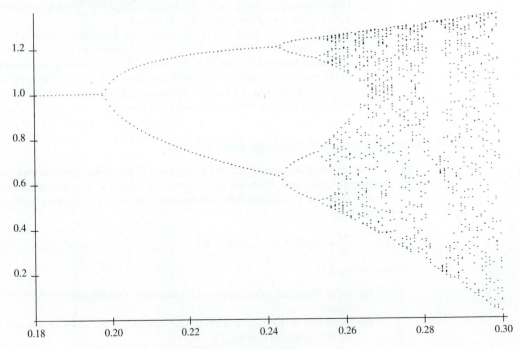

Figure 3.16 Period doubling to chaos

CHAPTER 4

Linear Second Order Equations

4.1 INTRODUCTION: THE SIMPLE PENDULUM

A simple pendulum consists of a mass m suspended by a cable of length l and negligible mass (see Figure 4.1). If the cable is always straight and the mass is free to swing in a vertical plane, find the period of oscillation.

We can use Newton's law to determine the equation that governs the motion of the pendulum. If we can solve this equation, then we can find the period of oscillation.

Newton's second law states that

$$\mathbf{F} = m\mathbf{a},$$

where $\mathbf{F}$ is the net (total) force vector applied to the mass m and $\mathbf{a}$ is the acceleration vector of the mass. Since the mass travels along a circle with radius l, we shall see that the net force $\mathbf{F}$ is a vector acting along (tangent to) this circle. Furthermore the acceleration can be expressed as d^2s/dt^2, where s is the arclength that gives the position of the mass on its circular path.

There are two forces acting on the pendulum. One force is the weight of the mass

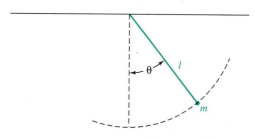

Figure 4.1 The simple pendulum

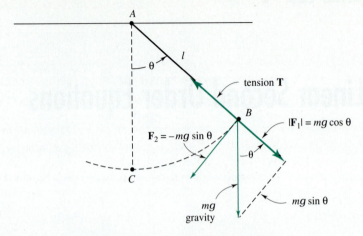

Figure 4.2 Forces on the simple pendulum

pulling down and the second is the tension pulling back up the cable. In Figure 4.2 let AB denote the cable with A the point where the cable is attached and B the end where the mass m is attached. Let AC denote the vertical line through A and let θ be the angle between AB and AC.

The force exerted by the weight mg of the mass at B has two components: a component $\mathbf{F_1}$ along the straight line through AB and a component $\mathbf{F_2}$ perpendicular to AB and in a downward direction tangent to the path of the pendulum. The magnitude of the force $\mathbf{F_1}$ is just

$$|\mathbf{F}_1| = mg \cos \theta,$$

and the magnitude of the force $\mathbf{F_2}$ is

$$|\mathbf{F}_2| = |mg \sin \theta|.$$

Now the tension $\mathbf{T}$ on the cable is equal in magnitude but opposite in direction to the component $\mathbf{F_1}$. Hence $\mathbf{F_1}$ and $\mathbf{T}$ cancel each other. Therefore, the resulting force acting on the pendulum is the force $\mathbf{F_2}$, which is tangent to the path of the pendulum and toward the vertical AC. Let's agree to take $\theta > 0$ for B to the right of C. Then, relative to θ, we can set

$$\mathbf{F}_2 = -mg \sin \theta.$$

Since arclength $s = l\theta$, the acceleration of the mass along the tangent to the path at B is just

$$\frac{d^2 s}{dt^2} = l \frac{d^2 \theta}{dt^2}.$$

Finally, applying Newton's second law, we have

$$m \frac{d^2 s}{dt^2} = \mathbf{F}_2,$$

$$ml \frac{d^2 \theta}{dt^2} = -mg \sin \theta,$$

which reduces to

(1) $$\frac{d^2\theta}{dt^2} + \frac{g}{l}\sin\theta = 0.$$

Equation (1), which governs the motion of the simple pendulum, is an example of a nonlinear equation. In most cases one cannot find a solution to a nonlinear equation, and even when a solution is found it often cannot be expressed using elementary functions. (Two special cases are described in Project D.) In practice, when one encounters a nonlinear equation, it is helpful to study the **linearization** of the equation. This is an approximation of the nonlinear equation by a linear one. In our case, we recall from calculus that for θ small, $\sin\theta \approx \theta$. Using this approximation in equation (1), we obtain the linear equation

(2) $$\frac{d^2\theta}{dt^2} + \frac{g}{l}\theta = 0.$$

Because of the difficulty in solving the pendulum equation (1), we focus instead on its linearization (2). For small oscillations of the pendulum we expect that the period as determined from (2) will be a good approximation to the actual period.

To solve equation (2) we observe that we are looking for a function $\theta(t)$ with the property that its second derivative is just a negative constant times itself. Recalling that this property holds for sine and cosine functions, we are led to guess that our desired function has the form

$$\theta(t) = \cos\omega t \quad \text{or} \quad \theta(t) = \sin\omega t,$$

for some suitable choice of the constant ω. Substituting $\theta(t) = \cos\omega t$ into equation (2) gives

$$-\omega^2\cos\omega t + \frac{g}{l}\cos\omega t = 0,$$

or

$$\left(\frac{g}{l} - \omega^2\right)\cos\omega t = 0.$$

Hence, if we choose $\omega = \sqrt{g/l}$, then $\theta(t) = \cos(\sqrt{g/l}\,t)$ is a solution. Similarly, substituting $\theta(t) = \sin\omega t$, we again find $\omega = \sqrt{g/l}$. Therefore, both

$$\theta_1(t) := \cos\left(\sqrt{\frac{g}{l}}\,t\right) \quad \text{and} \quad \theta_2(t) := \sin\left(\sqrt{\frac{g}{l}}\,t\right)$$

are solutions to (2). In fact, because (2) is a *linear* differential equation, a short computation shows that any function of the form

(3) $$\theta(t) = c_1\cos\left(\sqrt{\frac{g}{l}}\,t\right) + c_2\sin\left(\sqrt{\frac{g}{l}}\,t\right),$$

where c_1 and c_2 are arbitrary constants, is also a solution to (2) (see Section 4.2). Moreover (as we shall see in Section 4.3), every solution to (2) must be of the form (3).

If we are given the initial angular displacement $\theta(0)$ and initial angular velocity $\theta'(0)$, then we can determine c_1 and c_2 in equation (3). However, to calculate the period of oscillation, it is not necessary to know c_1 and c_2. It is clear from (3) that since the period of $\cos \omega t$ and $\sin \omega t$ is $2\pi/\omega$, then the period of oscillation is

$$P = \frac{2\pi}{\sqrt{g/l}} = 2\pi\sqrt{\frac{l}{g}}.$$

This formula provides an approximation to the period of oscillation for the pendulum.

The oscillation described by equation (3) is called **simple harmonic motion.** We will study this motion in Section 5.1 in connection with mechanical vibrations. The nonlinear equation (1) that governs the motion of the simple pendulum is studied in more detail in Projects E and F. Linearization of nonlinear problems is discussed in Project C.

In this chapter we concentrate on the theory of linear second order equations and methods for solving them.

• 4.2 • LINEAR DIFFERENTIAL OPERATORS

Recall that a **second order** differential equation is an equation of the form

$$F\left(x, y, \frac{dy}{dx}, \frac{d^2y}{dx^2}\right) = 0.$$

For example,

$$(x^3 - 1)\frac{d^2y}{dx^2} - x\frac{dy}{dx} + \sin y + e^x = 0$$

is a second order equation. In general, such equations are very difficult to solve. However, for special types of these equations, substitutions are known that transform the original equation into one that can be solved by elementary functions, elliptic functions, or some other type of special function. A class of second order equations for which an extensive theory exists, and that arises quite frequently in applications, is the class of linear equations.

A **linear** second order equation is an equation that can be written in the form

(1) $\qquad a_2(x)\dfrac{d^2y}{dx^2} + a_1(x)\dfrac{dy}{dx} + a_0(x)y = b(x).$

We will assume that $a_0(x)$, $a_1(x)$, $a_2(x)$, and $b(x)$ are continuous functions of x on an interval I. When a_0, a_1, and a_2 are constants, we say that the equation has **constant coefficients;** otherwise it has **variable coefficients.**

For now we are interested only in those linear equations for which $a_2(x) \neq 0$ on I. In this case we can rewrite (1) in the **standard form**

(2) $\qquad \dfrac{d^2y}{dx^2} + p(x)\dfrac{dy}{dx} + q(x)y = g(x),$

where $p(x) := a_1(x)/a_2(x)$, $q(x) := a_0(x)/a_2(x)$, and $g(x) := b(x)/a_2(x)$ are continuous on I.
Associated with equation (2) is the equation

$$(3) \qquad \frac{d^2y}{dx^2} + p(x)\frac{dy}{dx} + q(x)y = 0,$$

which is obtained from (2) by replacing $g(x)$ with the zero function. We say that (2) is a
nonhomogeneous equation and that (3) is the corresponding **homogeneous** equation. Here
the meaning of *homogeneous* is not related to the use of the term for first order equations
but is used more in the sense of a homogeneous system of linear equations as studied in
linear algebra.

Let's consider the expression on the left-hand side of equation (3), which, for convenience, we now write with prime notation:

(4) $\qquad y''(x) + p(x)y'(x) + q(x)y(x).$

Given any function y with a continuous second derivative on the interval I, then (4) generates a new function that we will denote by $L[y]$. That is,

(5) $\qquad L[y] := y'' + py' + qy.$

What we have done is to associate with each function y the function $L[y]$.

The term *function*, in its general sense, means a mapping that associates with each
element in its domain a unique element in its range. Thus L can be interpreted as a function that is defined *on a set of functions*. Its domain is the collection of functions with a
continuous second derivative; its range consists of continuous functions; and the rule of
correspondence is given by (5). To avoid confusion, such mappings are called **operators**.
Since L involves differentiation, we refer to L as a **differential operator.**

The image of a function y under the operator L is the function $L[y]$. If we wish to
evaluate this image function at some point x, we write $L[y](x)$.

For example, let $p(x) = x$ and $q(x) = x - 1$; then

$$L[y](x) = y''(x) + xy'(x) + (x - 1)y(x).$$

If $y_1(x) = x^3$, we get

$$L[y_1](x) = 6x + x(3x^2) + (x - 1)x^3 = x^4 + 2x^3 + 6x.$$

Thus L maps the function x^3 to the function $x^4 + 2x^3 + 6x$. If $y_2(x) = \sin 2x$, then we
similarly find

$$L[y_2](x) = -4\sin 2x + x(2\cos 2x) + (x - 1)\sin 2x$$
$$= x\sin 2x - 5\sin 2x + 2x\cos 2x.$$

The differential operator L defined by (5) has two very important properties that are
extensions of the familiar facts that "the derivative of the sum is the sum of the derivatives" and "the derivative of a constant times a function is the constant times the derivative
of the function."

> ### LINEARITY OF THE DIFFERENTIAL OPERATOR L
>
> **Lemma 1.**   Let $L[y] := y'' + py' + qy$. If y_1 and y_2 are any two functions with continuous second derivatives on the interval I, and if c is any constant, then
>
> (6) $L[y_1 + y_2] = L[y_1] + L[y_2]$,
>
> (7) $L[cy_1] = cL[y_1]$.
>
> In (6) and (7) equality is meant in the sense of equal functions on I.

Proof.   On I we have

$$L[y_1 + y_2] = (y_1 + y_2)'' + p(y_1 + y_2)' + q(y_1 + y_2)$$
$$= (y_1'' + y_2'') + p(y_1' + y_2') + q(y_1 + y_2).$$

On regrouping, this gives

$$L[y_1 + y_2] = (y_1'' + py_1' + qy_1) + (y_2'' + py_2' + qy_2)$$
$$= L[y_1] + L[y_2],$$

which verifies property (6). We leave the proof of property (7) for the exercises (see Problem 28).   ◄◄◄

An operator that satisfies properties (6) and (7) for any constant c and any functions y_1 and y_2 in its domain is called a **linear operator.** If (6) or (7) fails to hold, the operator is **nonlinear.**

As we have shown in Lemma 1, the operator L defined by (5) is linear. We now give an example of a nonlinear operator.

● **EXAMPLE 1**   Show that the operator T defined by

$$T[y](x) := y''(x) + \sin(y(x)),$$

where y is any function whose second derivative is continuous for all real x, is nonlinear.

Solution   To see that T is a nonlinear operator, it suffices to show that property (7) is *not always* satisfied. Let's choose $y_1(x) = x$. Since $y_1''(x) \equiv 0$, we have

$$T[cy_1](x) = \sin(cx), \qquad cT[y_1](x) = c\sin x.$$

But, in general, $\sin(cx) \neq c\sin x$ (take, for example, $c = 2$ and $x = \pi/2$), so property (7) is violated. Hence T is a nonlinear operator.   ●

The linearity of the differential operator L in (5) can be used to prove the following theorem concerning homogeneous equations.

> ### LINEAR COMBINATIONS OF SOLUTIONS
>
> **Theorem 1.**   Let y_1 and y_2 be solutions to the homogeneous equation
>
> **(8)** $y'' + py' + qy = 0.$
>
> Then any linear combination $c_1 y_1 + c_2 y_2$ of y_1 and y_2, where c_1 and c_2 are constants, is also a solution to (8).

Proof.   If we let $L[y] := y'' + py' + qy$, then $L[y_1] = 0$ and $L[y_2] = 0$, since y_1 and y_2 are solutions to (8). Using the linearity of L as expressed in properties (6) and (7), we have

$$L[c_1 y_1 + c_2 y_2] = L[c_1 y_1] + L[c_2 y_2]$$
$$= c_1 L[y_1] + c_2 L[y_2]$$
$$= 0 + 0 = 0.$$

Thus $c_1 y_1 + c_2 y_2$ is a solution to equation (8).   ◄◄◄

● **EXAMPLE 2**   Given that $y_1(x) = e^{2x} \cos 3x$ and $y_2(x) = e^{2x} \sin 3x$ are solutions to the homogeneous equation

(9) $y'' - 4y' + 13y = 0,$

find a solution to (9) that satisfies the initial conditions

(10) $y(0) = 2$ and $y'(0) = -5.$

Solution   As a consequence of Theorem 1, any linear combination

(11) $y(x) = c_1 e^{2x} \cos 3x + c_2 e^{2x} \sin 3x,$

with c_1 and c_2 arbitrary constants, will be a solution to (9). Thus we try to select c_1 and c_2 so as to satisfy the initial conditions. On differentiating (11) we find

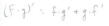

(12) $y'(x) = c_1(2e^{2x} \cos 3x - 3e^{2x} \sin 3x) + c_2(2e^{2x} \sin 3x + 3e^{2x} \cos 3x).$

Substituting (11) and (12) into the initial conditions (10) yields the system of equations

$$c_1 = 2, \qquad 2c_1 + 3c_2 = -5,$$

whose solution is given by $c_1 = 2, c_2 = -3$. Hence, the solution to (9) that satisfies conditions (10) is

$$y(x) = 2e^{2x} \cos 3x - 3e^{2x} \sin 3x.$$ ●

As was the case for linear first order equations, the initial value problem for linear second order equations also has a unique solution.

> ### EXISTENCE AND UNIQUENESS OF SOLUTION
>
> **Theorem 2.** Suppose $p(x)$, $q(x)$, and $g(x)$ are continuous on an interval (a, b) that contains the point x_0. Then, for any choice of the initial values y_0 and y_1, there exists a unique solution $y(x)$ on the whole interval (a, b) to the initial value problem
>
> $$\frac{d^2 y}{dx^2} + p(x)\frac{dy}{dx} + q(x)y = g(x);$$
>
> $$y(x_0) = y_0, \qquad y'(x_0) = y_1.$$

Remark. Geometrically, the uniqueness part of Theorem 2 asserts that two solution curves passing through the same point and *with the same slope* must be identical throughout the interval (a, b).

The proof of Theorem 2 is more complicated than the proof of Theorem 2 in Section 2.4 for linear first order equations; it is given in Chapter 14.[†]

● **EXAMPLE 3**   Determine the largest interval for which Theorem 2 ensures the existence and uniqueness of a solution to the initial value problem

(13) $$\frac{d^2 y}{dx^2} + \frac{1}{x - 3}\frac{dy}{dx} + \sqrt{x}\,y = \ln x;$$

$$y(1) = 3, \qquad y'(1) = -5.$$

Solution   Here $p(x) = 1/(x - 3)$, $q(x) = \sqrt{x}$, $g(x) = \ln x$, and $x_0 = 1$. Now $p(x)$ is continuous for $x \neq 3$, and both $q(x)$ and $g(x)$ are continuous for $x > 0$. Hence the largest open interval containing $x_0 = 1$ for which $p(x)$, $q(x)$, and $g(x)$ are simultaneously continuous is the interval $(0, 3)$. From Theorem 2 we can conclude that the initial value problem (13) has a unique solution on the interval $(0, 3)$. ●

Let D denote differentiation with respect to x and D^2 differentiation with respect to x twice. That is, $Dy := dy/dx$, $D^2 y := d^2 y/dx^2$. In general, let $D^n y := d^n y/dx^n$. Using this notation,[††] we can express L as defined in (5) by

$$L[y] = D^2 y + p\,Dy + qy = (D^2 + p D + q)y.$$

For example, if

$$L[y] = y'' + 4y' + 3y,$$

then we can write L in the form

$$L[y] = D^2 y + 4\,Dy + 3y = (D^2 + 4D + 3)y.$$

[†]All references to Chapters 12–14 refer to the expanded text *Fundamentals of Differential Equations and Boundary Value Problems*.

[††] *Historical Footnote*: The symbol D was introduced by the French mathematician F. J. Servois. The notion of a symbolic operation dates from B. Brisson in 1808, and was extended by Augustin Cauchy in 1827.

When p and q are constants, we can even treat $D^2 + pD + q$ as a polynomial in D and factor it. This is illustrated in the next example.

EXAMPLE 4 Show that the operator $D^2 + 4D + 3$ is the same as the product $(D + 1)(D + 3)$, where by *product* we mean the composition of operators.

Solution For an arbitrary twice-differentiable function y, we have

$$\begin{aligned}(D + 1)(D + 3)y &= (D + 1)[(D + 3)y] = (D + 1)[y' + 3y] \\ &= D[y' + 3y] + 1[y' + 3y] = (y'' + 3y') + (y' + 3y) \\ &= y'' + 4y' + 3y = (D^2 + 4D + 3)y.\end{aligned}$$

Hence $(D + 1)(D + 3) = D^2 + 4D + 3$. ●

Further discussion of the algebra of differential operators is given in Problems 32–36.

EXERCISES 4.2

In Problems 1 through 8 determine whether the equation is linear or nonlinear. If it is linear, classify it as being homogeneous or nonhomogeneous, with constant coefficients or with variable coefficients.

1. $3\dfrac{d^2y}{dx^2} + \dfrac{dy}{dx} = 2y$.

2. $y'' + (1 - x)y' + xy = \sin x$.

3. $xy'' - yy' = \sin x$.

4. $\dfrac{d^2y}{dx^2} - x\dfrac{dy}{dx} + y = 0$.

5. $t^2\dfrac{d^2x}{dt^2} = t\dfrac{dx}{dt} + 4x - \ln t$.

6. $\dfrac{d^2\theta}{dx^2} = \tan\theta$.

7. $\dfrac{d^2y}{d\theta^2} + y = \tan\theta$.

8. $\dfrac{d^2s}{dx^2} + 3\dfrac{ds}{dx} - s^{1/2} = x^2$.

9. Let $L[y](x) = x^2y''(x) - 3xy'(x) - 5y(x)$. Compute:

(a) $L[\cos x]$. **(b)** $L[x^{-1}]$.
(c) $L[x^r]$, r a constant.

10. Let $L[y](x) = y''(x) - 4y'(x) + 3y(x)$. Compute:

(a) $L[x^2]$. **(b)** $L[e^{3x}]$.
(c) $L[e^{rx}]$, r a constant.

11. Show that T defined by

$$T[y](x) := y''(x) - y'(x) + y^2(x),$$

where y is any function whose second derivative is continuous for all real x, is a nonlinear operator.

12. Show that T defined by

$$T[y](x) := y''(x) + \{y'(x)y^2(x)\}^{1/3}$$

is a nonlinear operator.

13. Given that $y_1(x) = e^{2x}\cos x$ and $y_2(x) = e^{2x}\sin x$ are solutions to the homogeneous equation

$$y'' - 4y' + 5y = 0,$$

find solutions to this equation that satisfy the following initial conditions:

(a) $y(0) = 1$, $y'(0) = -1$.
(b) $y(\pi) = 4e^{2\pi}$, $y'(\pi) = 5e^{2\pi}$.

14. Given that $y_1(x) = e^{2x}$ and $y_2(x) = e^{-x}$ are solutions to the homogeneous equation

$$y'' - y' - 2y = 0,$$

find solutions to this equation that satisfy the following initial conditions:

(a) $y(0) = -1$, $y'(0) = 4$.
(b) $y(0) = 3/2$, $y'(0) = 0$.

In Problems 15 through 18 use Theorem 2 to discuss the existence and uniqueness of a solution to the differential *equation satisfying the initial conditions $y(1) = y_0$, $y'(1) = y_1$, where y_0 and y_1 are real constants.*

15. $(1 + x^2)y'' + xy' - y = \tan x$.

16. $x(x - 3)y'' + 2xy' - y = x^2$.

17. $e^x y'' - \dfrac{y'}{x - 3} + y = \ln x$.

18. $x^2 y'' + y = \cos x$.

In Problems 19 through 22 determine whether Theorem 2 applies. If it applies, then discuss what conclusions can be drawn. If it does not apply, explain why.

19. $y'' + yy' = x^2 - 1$; $y(0) = 1$, $y'(0) = -1$.

20. $x^2 z'' + xz' + z = \cos x$; $z(0) = 1$, $z'(0) = 0$.

21. $(1 - t)x'' + tx' - 2x = \sin t$; $x(0) = 1$, $x'(0) = 1$.

22. $y'' + xy' - x^2 y = 0$; $y(0) = 0$, $y(1) = 0$.

In Problems 23 through 26 express the given operator L using the differential operator D.

23. $L[y] := y'' + 7y' + 12y$.

24. $L[z] := 2z'' + z' - z$.

25. $L[y] := y'' + xy$.

26. $L[y] := x^2 y'' - xy' + y$.

27. Let L be defined by $L[y] := (D^2 - xD + 2)y$. Compute:

(a) $L[x^2]$.   (b) $L[\cos x]$.

28. Complete the proof of Lemma 1 by verifying that the differential operator L defined by (5) satisfies property (7).

29. Show that the integro-differential operator L defined by

$$L[y](x) := \int_0^1 (x - t)^2 y(t)\, dt + x^2 y'(x),$$

for any function y having a continuous derivative on $[0, 1]$ is linear; that is, that L satisfies properties (6) and (7).

30. Boundary Value Problems. When the values of a solution to a differential equation are specified at two different points, these conditions are called **boundary conditions.** (In contrast, initial conditions specify the values of a function and its derivative at the same point.) Given that every solution to

(14) $y'' + y = 0$

is of the form

$$y(x) = c_1 \cos x + c_2 \sin x,$$

where c_1 and c_2 are arbitrary constants, show that:

(a) there is a unique solution to (14) that satisfies the boundary conditions $y(0) = 2$ and $y(\pi/2) = 0$;

(b) there is no solution to (14) that satisfies $y(0) = 2$ and $y(\pi) = 0$;

(c) there are infinitely many solutions to (14) that satisfy $y(0) = 2$ and $y(\pi) = -2$.

31. Let $y = \phi(x)$ be the solution to the initial value problem

$$y'' + x^2 y + e^x y = x - 1;$$

$$y(0) = 1, \qquad y'(0) = -1.$$

Assuming that the differential equation can be repeatedly differentiated, determine $\phi''(0)$, $\phi'''(0)$, and $\phi^{(4)}(0)$.

32. Differential Operators with Constant Coefficients. Let A and B represent two linear differential operators with constant coefficients; that is,

$$A := a_2 D^2 + a_1 D + a_0,$$

$$B := b_2 D^2 + b_1 D + b_0,$$

where a_0, a_1, a_2, b_0, b_1, and b_2 are constants. We define the **sum** $A + B$ by

$$(A + B)[y] := A[y] + B[y]$$

and the **product** AB by

$$(AB)[y] := A[B[y]].$$

Notice that the product AB is the composite operator obtained by applying first B, then A. We say two operators A and B are **equal** if $A[y] = B[y]$ for all functions y with the necessary derivatives. With these definitions, verify the following properties.

(a) Commutative laws: $A + B = B + A$, $AB = BA$.

(b) Associative laws: $(A + B) + C = A + (B + C)$, $(AB)C = A(BC)$.

(c) Distributive law: $A(B + C) = AB + AC$.

33. Let $A = (D - 1)$, $B = (D + 2)$, and $y = x^3$. Compute:

(a) $A[y]$.   (b) $B[A[y]]$.

(c) $B[y]$.   (d) $A[B[y]]$.

(e) AB.   (f) $(AB)[y]$.

34. Verify, as in Example 4, each of the following operator equations:

(a) $(D - 1)(D + 6) = D^2 + 5D - 6$.

(b) $(D + 1)(2D^2 - 1) = 2D^3 + 2D^2 - D - 1$.

35. Factor the following differential operators:

(a) $D^2 + 3D - 4$.   (b) $D^2 + D - 6$.

(c) $2D^2 + 9D - 5$.   (d) $D^2 - 2$.

36. Let $A = (xD - 1)$ and $B = (D + 1)$. Show for the function $y = x$ that $(AB)[y] \neq (BA)[y]$. Hence, for operators with **variable** coefficients, the commutative law need not hold.

•4.3• FUNDAMENTAL SOLUTIONS OF HOMOGENEOUS EQUATIONS

In this section we discuss those properties of homogeneous equations that help us obtain *all* solutions of these equations. First, let's examine the particular equation

(1) $L[y] = \dfrac{d^2 y}{dx^2} - y = 0.$

It is easy to verify that the two functions $y_1 = e^x$ and $y_2 = e^{-x}$ are solutions to this homogeneous equation. Furthermore, since L is linear, we know from Theorem 1 in Section 4.2 that $c_1 e^x + c_2 e^{-x}$ is also a solution, for any choice of constants c_1 and c_2. The question is, can all solutions of (1) be represented by

$$c_1 e^x + c_2 e^{-x},$$

with appropriate choices for c_1 and c_2? As we now show, the answer lies in applying the existence and uniqueness theorem (Theorem 2, page 136).

Let $\phi(x)$ be a solution to (1) and let x_0 be a fixed real number. If we can choose c_1 and c_2 so that

(2) $c_1 e^{x_0} + c_2 e^{-x_0} = \phi(x_0),$

(3) $c_1 e^{x_0} - c_2 e^{-x_0} = \phi'(x_0),$

then, since $\phi(x)$ and $c_1 e^x + c_2 e^{-x}$ satisfy the same initial conditions at x_0, the uniqueness conclusion of Theorem 2 guarantees that $\phi(x) = c_1 e^x + c_2 e^{-x}$ for all x.

To solve the system (2)–(3) for c_1, we add the two equations and then divide by $2e^{x_0}$ to obtain

(4) $c_1 = \dfrac{\phi(x_0) + \phi'(x_0)}{2e^{x_0}}.$

Subtracting (3) from (2) and dividing by $2e^{-x_0}$ gives

(5) $c_2 = \dfrac{\phi(x_0) - \phi'(x_0)}{2e^{-x_0}}.$

Notice that since e^{x_0} and e^{-x_0} are never zero, we can always carry out the above procedure for finding c_1 and c_2. Thus, by uniqueness, we have

$$\phi(x) = c_1 e^x + c_2 e^{-x}$$

for all x, where c_1 and c_2 are given in (4) and (5).

We have shown that every solution of (1) can be expressed as a linear combination of two particular solutions of (1). A similar result holds for general linear homogeneous equations, provided that the two solutions y_1 and y_2 satisfy a certain property.

> ### REPRESENTATION OF SOLUTIONS (HOMOGENEOUS CASE)
>
> **Theorem 3.** Let y_1 and y_2 denote two solutions on (a, b) of
>
> **(6)** $\qquad y'' + p(x)y' + q(x)y = 0,$
>
> where p and q are continuous on (a, b). If at some point x_0 in (a, b) these solutions satisfy
>
> **(7)** $\qquad y_1(x_0)y_2'(x_0) - y_1'(x_0)y_2(x_0) \neq 0,$
>
> then every solution of (6) on (a, b) can be expressed in the form
>
> **(8)** $\qquad y(x) = c_1 y_1(x) + c_2 y_2(x),$
>
> where c_1 and c_2 are constants.

The linear combination of y_1 and y_2 in (8), written with arbitrary constants c_1 and c_2, is referred to as a **general solution** to (6).

Proof of Theorem 3. Let $\phi(x)$ be any solution to (6). As before, we consider the system

(9) $\qquad c_1 y_1(x_0) + c_2 y_2(x_0) = \phi(x_0),$

(10) $\qquad c_1 y_1'(x_0) + c_2 y_2'(x_0) = \phi'(x_0),$

where x_0 is a point at which (7) holds. To solve for c_1, multiply (9) by $y_2'(x_0)$, multiply (10) by $y_2(x_0)$, and subtract to obtain

$$c_1[y_1(x_0)y_2'(x_0) - y_1'(x_0)y_2(x_0)] = \phi(x_0)y_2'(x_0) - \phi'(x_0)y_2(x_0).$$

Because of condition (7), the expression in brackets is not zero. Hence,

$$c_1 = \frac{\phi(x_0)y_2'(x_0) - \phi'(x_0)y_2(x_0)}{y_1(x_0)y_2'(x_0) - y_1'(x_0)y_2(x_0)}.$$

Similarly, we find

$$c_2 = \frac{\phi'(x_0)y_1(x_0) - \phi(x_0)y_1'(x_0)}{y_1(x_0)y_2'(x_0) - y_1'(x_0)y_2(x_0)}.$$

With these choices for c_1 and c_2, the functions $c_1 y_1 + c_2 y_2$ and ϕ are solutions to (6) that satisfy the same initial conditions. Hence they must be the same function on (a, b), according to the uniqueness result in Theorem 2. ◄◄◄

Since the expression on the left-hand side of (7) plays an important role in the theory, we adopt the following terminology.

> ### WRONSKIAN
>
> **Definition 1.**   For any two differentiable functions y_1 and y_2, the function
>
> (11) $W[y_1, y_2](x) := y_1(x)y_2'(x) - y_1'(x)y_2(x)$
>
> is called the **Wronskian** of y_1 and y_2.[†]

A convenient way of writing the Wronskian $W[y_1, y_2]$, which will generalize to higher order linear equations, is in terms of the determinant

$$W[y_1, y_2](x) = \begin{vmatrix} y_1(x) & y_2(x) \\ y_1'(x) & y_2'(x) \end{vmatrix}.$$

> ### FUNDAMENTAL SOLUTION SET
>
> **Definition 2.**   A pair of solutions $\{y_1, y_2\}$ of $y'' + py' + qy = 0$ on (a, b) is called a **fundamental solution set** if
>
> $$W[y_1, y_2](x_0) \neq 0$$
>
> at some x_0 in (a, b).[††]

For example, the solutions $y_1 = e^x$, $y_2 = e^{-x}$ of the differential equation (1) form a fundamental solution set for this equation on $(-\infty, \infty)$, since

$$W[y_1, y_2](x) = e^x(-e^{-x}) - e^x e^{-x} = -2 \neq 0$$

for any x.

Using the terminology of Definition 2, Theorem 3 can be summarized as follows.

> ### PROCEDURE FOR SOLVING HOMOGENEOUS EQUATIONS
>
> To determine all solutions to $y'' + py' + qy = 0$:
>
> **(a)** Find two solutions y_1, y_2 that form a fundamental solution set.
> **(b)** A general solution is
>
> $$y(x) = c_1 y_1(x) + c_2 y_2(x),$$
>
> where c_1 and c_2 are arbitrary constants.

● **EXAMPLE 1**   Given that $y_1(x) = \cos 3x$ and $y_2(x) = \sin 3x$ are solutions to

(12) $y'' + 9y = 0$

on $(-\infty, \infty)$, find a general solution to (12).

[†] *Historical Footnote*: The Wronskian was named after the Polish mathematician H. Wronski (1778–1853).

[††] *Historical Footnote*: Lazarus Fuchs introduced the terminology *fundamental system* in 1866.

Solution   First we verify that $\{\cos 3x, \sin 3x\}$ is a fundamental solution set. Since we are given that both $y_1(x) = \cos 3x$ and $y_2(x) = \sin 3x$ are solutions to (12) (the reader can quickly verify this), we need show only that $W[y_1, y_2](x) \neq 0$ for some x in $(-\infty, \infty)$. Substituting y_1 and y_2 into formula (11) yields

$$W[y_1, y_2](x) = (\cos 3x)(3 \cos 3x) - (-3 \sin 3x)(\sin 3x)$$
$$= 3(\cos^2 3x + \sin^2 3x)$$
$$= 3 \neq 0.$$

Thus $\{\cos 3x, \sin 3x\}$ forms a fundamental solution set, and a general solution to (12) is

$$y(x) = c_1 \cos 3x + c_2 \sin 3x. \quad \bullet$$

It is easy to see that a fundamental solution set for $y'' + py' + qy = 0$ always exists. Indeed, let $x_0 \in (a, b)$ and take $y_1(x)$ and $y_2(x)$ to be the solutions that satisfy the initial conditions

$$y_1(x_0) = 1, \quad y_2(x_0) = 0,$$
$$y_1'(x_0) = 0, \quad y_2'(x_0) = 1.$$

Then $W[y_1, y_2](x_0) = 1 \neq 0$, and so $\{y_1, y_2\}$ is a fundamental solution set.

There is another method that can be used to determine quickly whether a pair of solutions form a fundamental solution set. To describe this test, we need to recall some basic facts about **vectors** (directed line segments) in the plane.

A planar vector is prescribed by an ordered pair of real numbers, say (u_1, u_2). Recall that vector addition is done componentwise as is scalar multiplication; that is,

$$(u_1, u_2) + (v_1, v_2) = (u_1 + v_1, u_2 + v_2), \qquad c(u_1, u_2) = (cu_1, cu_2).$$

Two vectors are parallel (point in the same or opposite directions) if and only if one vector is a scalar multiple of the other. This is equivalent to saying that there exist two constants c_1, c_2 *not both zero* such that

$$c_1(u_1, u_2) + c_2(v_1, v_2) = (0, 0).$$

Parallel vectors are said to be **linearly dependent,** whereas two nonparallel vectors are **linearly independent.**

These notions also apply to functions.

▶ LINEAR DEPENDENCE OF FUNCTIONS

Definition 3.   Two functions y_1 and y_2 are said to be **linearly dependent on an interval** *I* if there exist constants c_1 and c_2, not both zero, such that

$$c_1 y_1(x) + c_2 y_2(x) = 0$$

for all x in *I*. If two functions are not linearly dependent, they are said to be **linearly independent.**

The preceding definition extends in a natural way to sets of more than two functions: this is studied in Chapter 6 (see also Problems 24, 25, and 26 in Exercises 4.3). **In the case of just two functions, linear dependence on *I* is equivalent to one function being a constant multiple of the other function on *I*** (see Problem 14).

● **EXAMPLE 2**   Determine whether the following pairs of functions y_1 and y_2 are linearly dependent on $(-5, 5)$:

(a) $y_1(x) = e^{3x}$, $y_2(x) = x + 1$.
(b) $y_1(x) = \sin 2x$, $y_2(x) = \cos x \sin x$.
(c) $y_1(x) = x$, $y_2(x) = |x|$.

Solution   (a) A glance at the functions $y_1(x) = e^{3x}$ and $y_2(x) = x + 1$ indicates that neither is a constant multiple of the other. Indeed, if a constant c exists such that

$$e^{3x} = c(x + 1) \quad \text{for all } x \text{ in } (-5, 5),$$

then we arrive at a contradiction by setting $x = 0$ and $x = 1$:

$$e^0 = c(0 + 1) \Rightarrow c = 1,$$
$$e^3 = c(1 + 1) \Rightarrow c = e^3/2 \neq 1.$$

Hence e^{3x} and $x + 1$ are linearly independent.

(b) Since $y_1(x) = \sin 2x = 2 \sin x \cos x$, we see that $y_1(x) = 2y_2(x)$. Hence y_1 and y_2 are linearly dependent on $(-5, 5)$.

(c) Here the two functions $y_1(x) = x$ and $y_2(x) = |x|$ are identical on the subinterval $(0, 5)$. (In particular, they are linearly dependent on $(0, 5)$.) However, neither function is a fixed constant multiple of the other on the *whole* interval $(-5, 5)$. (On $(0, 5)$, $y_1(x) = 1 \cdot y_2(x)$, but on $(-5, 0)$, $y_1(x) = (-1) \cdot y_2(x)$.) Thus y_1 and y_2 are linearly independent on $(-5, 5)$.   ●

> **CONDITION FOR LINEAR DEPENDENCE OF SOLUTIONS**
>
> **Theorem 4.**   Let y_1, y_2 be solutions to $y'' + py' + qy = 0$ on (a, b), and let $x_0 \in (a, b)$. Then y_1 and y_2 are linearly dependent on (a, b) if and only if the initial vectors[†]
>
> $$\begin{bmatrix} y_1(x_0) \\ y_1'(x_0) \end{bmatrix} \quad \text{and} \quad \begin{bmatrix} y_2(x_0) \\ y_2'(x_0) \end{bmatrix}$$
>
> are linearly dependent.

Remark.   Although not explicitly mentioned, we assume in Theorem 4 and in subsequent results that the coefficient functions p and q are continuous.

Proof of Theorem 4.   ($\Rightarrow$) Assume that y_1 and y_2 are linearly dependent. Then one of these functions is a constant multiple of the other, say $y_1 = cy_2$ on (a, b). Thus $y_1' = cy_2'$,

---

[†] Here and in later discussions we represent planar vectors in column rather than row form.

and we have

$$\begin{bmatrix} y_1(x_0) \\ y_1'(x_0) \end{bmatrix} = \begin{bmatrix} cy_2(x_0) \\ cy_2'(x_0) \end{bmatrix} = c\begin{bmatrix} y_2(x_0) \\ y_2'(x_0) \end{bmatrix}.$$

Since one initial vector is a constant multiple of the other, they are linearly dependent.
($\Leftarrow$) Assume now that the initial vectors are linearly dependent, say

$$\begin{bmatrix} y_1(x_0) \\ y_1'(x_0) \end{bmatrix} = c\begin{bmatrix} y_2(x_0) \\ y_2'(x_0) \end{bmatrix}.$$

This implies that the functions y_1 and cy_2 satisfy the same initial conditions; that is,

$$y_1(x_0) = cy_2(x_0) \quad \text{and} \quad y_1'(x_0) = cy_2'(x_0).$$

Since y_1 and cy_2 are solutions to the same initial value problem, by the uniqueness property they must be the same function on (a, b). Thus y_1 is a constant multiple of y_2 on (a, b); that is, y_1 and y_2 are linearly dependent on (a, b).   ◀◀◀

A simple algebraic test for the linear independence of two column vectors

(13) $\begin{bmatrix} u_1 \\ u_2 \end{bmatrix}$ and $\begin{bmatrix} v_1 \\ v_2 \end{bmatrix}$

is to compute the determinant

$$\begin{vmatrix} u_1 & v_1 \\ u_2 & v_2 \end{vmatrix} = u_1 v_2 - v_1 u_2.$$

The vectors (13) *are linearly independent if and only if this determinant is not zero* (see Problem 16). Using this fact, we obtain the following consequence of Theorem 4.

FUNDAMENTAL SETS AND LINEAR INDEPENDENCE

Corollary 1.   Let y_1, y_2 be solutions to $y'' + py' + qy = 0$ on (a, b). Then $\{y_1, y_2\}$ is a fundamental solution set on (a, b) if and only if the functions y_1 and y_2 are linearly independent on (a, b).

Proof.   Let $x_0 \in (a, b)$. Theorem 4, restated in terms of linear *in*dependence, asserts that the two solutions y_1 and y_2 are linearly independent on (a, b) if and only if the initial vectors

$$\begin{bmatrix} y_1(x_0) \\ y_1'(x_0) \end{bmatrix} \quad \text{and} \quad \begin{bmatrix} y_2(x_0) \\ y_2'(x_0) \end{bmatrix}$$

are linearly independent. By the determinant test, this last condition is equivalent to

$$W[y_1, y_2](x_0) = \begin{vmatrix} y_1(x_0) & y_2(x_0) \\ y_1'(x_0) & y_2'(x_0) \end{vmatrix} \neq 0,$$

which, by Definition 2, is the same as saying that $\{y_1, y_2\}$ is a fundamental solution set.
◀◀◀

Notice that in the proof of Corollary 1, the point x_0 was taken *arbitrarily* in (a, b). In other words, if x_1 is another point in (a, b), then the linear independence of two solutions y_1 and y_2 on (a, b) is also equivalent to the condition $W[y_1, y_2](x_1) \neq 0$. Thus, whenever linear independence on (a, b) holds for these solutions, we must have $W[y_1, y_2](x) \neq 0$ for *every* x in (a, b). On the other hand, if y_1 and y_2 are linearly dependent on (a, b), then $W[y_1, y_2](x) \equiv 0$ on (a, b). We summarize these observations in the following result.

> ### ▶ A PROPERTY OF THE WRONSKIAN OF SOLUTIONS
>
> **Corollary 2.**   Let y_1, y_2 be solutions to $y'' + py' + qy = 0$ on (a, b). Then the Wronskian $W[y_1, y_2](x)$ of the two solutions is either identically zero or never zero on (a, b). Furthermore, the Wronskian of two **solutions** is identically zero if and only if the solutions are linearly dependent.

● **EXAMPLE 3**   Show that $y_1(x) = x^{-1}$ and $y_2(x) = x^3$ are solutions to

$$(14) \qquad x^2 y'' - xy' - 3y = 0$$

on the interval $(0, \infty)$ and give a general solution.

Solution   The verification that y_1 and y_2 are solutions to (14) is straightforward. Substituting $y = x^{-1}$ and $y = x^3$ in (14) gives, respectively, the identities

$$x^2(2x^{-3}) - x(-x^{-2}) - 3(x^{-1}) = 0 \quad \text{and} \quad x^2(6x) - x(3x^2) - 3(x^3) = 0.$$

Furthermore, the solution functions x^{-1} and x^3 are linearly independent on $(0, \infty)$ (neither is a constant multiple of the other on $(0, \infty)$). Hence, by Corollary 1, $\{x^{-1}, x^3\}$ is a fundamental solution set on $(0, \infty)$, and so a general solution is

$$y(x) = c_1 x^{-1} + c_2 x^3. \quad ●$$

● **EXAMPLE 4**   Can the function $w(x) = 3(x - 1)^2$ be the Wronskian on $(0, 2)$ for some homogeneous linear second order equation (with p and q continuous)?

Solution   Since $w(x) = 0$ only for $x = 1$, then by Corollary 2, $w(x)$ cannot be such a Wronskian on any open interval containing $x = 1$.   ●

If two differentiable functions are linearly independent on I, can their Wronskian be identically zero? Surprisingly, the answer is yes. The reader can verify (see Problem 18) that the Wronskian of the functions $y_1(x) = x^3$ and $y_2(x) = |x^3|$ is identically zero on $(-\infty, \infty)$; yet these functions are certainly linearly independent on $(-\infty, \infty)$ because one is not a constant multiple of the other for all x in $(-\infty, \infty)$. Does this example contradict Corollary 2? No, because Corollary 2 refers only to functions $y_1(x)$ and $y_2(x)$ that are solutions to the **same** homogeneous linear second order differential equation. What we can conclude from Corollary 2 is that $y_1(x) = x^3$ and $y_2(x) = |x^3|$ cannot be solutions to the same equation on $(-\infty, \infty)$.

Another representation of the Wronskian for two solutions y_1, y_2 to the equation $y'' + py'' + qy = 0$ on (a, b) is **Abel's identity** [†]:

$$(15) \qquad W[y_1, y_2](x) = C \exp\left(-\int_{x_0}^{x} p(t)\, dt \right),$$

where $x_0 \in (a, b)$ and C is a constant that depends upon y_1 and y_2. This formula reaffirms our earlier observation that the Wronskian of two solutions is always zero ($C = 0$) or never zero ($C \neq 0$) since the exponential factor never vanishes. We leave the proof of (15) as an exercise (see Problem 20).

The properties of the Wronskian and linear independence can be extended to sets of n functions $y_1(x), \ldots, y_n(x)$. This is done in Chapter 6, where we consider higher order linear equations. Although the proofs in Chapter 6 are a bit more complicated, the ideas are essentially the same as in the case of linear second order equations.

To summarize, we have shown in this section that to find *all* the solutions to a homogeneous second order linear differential equation, it suffices to obtain just two linearly independent solutions and then take linear combinations of them. In Sections 4.5 and 4.6 we concentrate on methods for actually finding such a pair of functions.

EXERCISES 4.3

1, 3, 5, 7, 9, 11, 13, 14, 15, 19

In Problems 1 through 6 determine whether the functions y_1 and y_2 are linearly dependent on $(0, 1)$. Also compute the Wronskian $W[y_1, y_2](x)$.

1. $y_1(x) = e^{-x} \cos 2x, \qquad y_2(x) = e^{-x} \sin 2x.$

2. $y_1(x) = e^{3x}, \qquad y_2(x) = e^{-4x}.$

3. $y_1(x) = x e^{2x}, \qquad y_2(x) = e^{2x}.$

4. $y_1(x) = x^2 \cos(\ln x), \qquad y_2(x) = x^2 \sin(\ln x).$

5. $y_1(x) = \tan^2 x - \sec^2 x, \qquad y_2(x) = 3.$

6. $y_1(x) = 0, \qquad y_2(x) = e^x.$

In Problems 7 through 12: (a) verify that the functions y_1 and y_2 are linearly independent solutions of the given differential equation; (b) find a general solution to the given differential equation; (c) find the solution that satisfies the given initial conditions.

7. $y'' - 2y' + 5y = 0; \qquad y_1(x) = e^x \cos 2x,$
$y_2(x) = e^x \sin 2x; \qquad y(0) = 2, \qquad y'(0) = 0.$

8. $y'' - 5y' + 6y = 0; \qquad y_1(x) = e^{2x}, \qquad y_2(x) = e^{3x};$
$y(0) = -1, \qquad y'(0) = -4.$

9. $x^2 y'' - 2y = 0; \qquad y_1(x) = x^2, \qquad y_2(x) = x^{-1};$
$y(1) = -2, \qquad y'(1) = -7.$

10. $y'' - 5y' = 0; \qquad y_1(x) = 2, \qquad y_2(x) = e^{5x};$
$y(0) = 2, \qquad y'(0) = 5.$

11. $xy'' - (x + 2)y' + 2y = 0; \qquad y_1(x) = e^x,$
$y_2(x) = x^2 + 2x + 2; \qquad y(1) = 0, \qquad y'(1) = 1.$

12. $y'' - y = 0; \qquad y_1(x) = \cosh x, \qquad y_2(x) = \sinh x;$
$y(0) = 1, \qquad y'(0) = -1.$

13. Consider the differential equation

$$(16) \qquad y'' + 5y' - 6y = 0.$$

(a) Show that $S_1 := \{e^x, e^x - e^{-6x}\}$ is a fundamental solution set for (16).

(b) Show that $S_2 := \{e^x, 3e^x + e^{-6x}\}$ is another fundamental solution set for (16).

(c) Verify that $\phi(x) = e^{-6x}$ is a solution to (16); then express ϕ as a linear combination of functions in S_1. Likewise, express ϕ as a linear combination of functions in S_2.

14. Prove that two functions are linearly dependent on an interval I if and only if one is a constant times the other on I.

15. Determine whether the following functions can be Wronskians on $(-1, 1)$ for some homogeneous linear

---

[†]*Historical Footnote:* Abel's identity was derived by Niels Abel in 1827.

second order equation (with p and q continuous).

(a) $w(x) = 6e^{4x}$.   **(b)** $w(x) = x^3$.
(c) $w(x) = (x + 1)^{-1}$.   **(d)** $w(x) \equiv 0$.

16. Show that the determinant

$$\begin{vmatrix} u_1 & v_1 \\ u_2 & v_2 \end{vmatrix} \quad (u_1, u_2, v_1, v_2 \text{ constants})$$

is not zero:

(a) if and only if the two column vectors

$$\begin{bmatrix} u_1 \\ u_2 \end{bmatrix} \quad \text{and} \quad \begin{bmatrix} v_1 \\ v_2 \end{bmatrix} \text{ are linearly independent.}$$

(b) if and only if the system with unknowns c_1 and c_2

$$u_1 c_1 + v_1 c_2 = a,$$
$$u_2 c_1 + v_2 c_2 = b,$$

has a unique solution for all a, b.

17. Let y_1 and y_2 be two functions defined on $(-\infty, \infty)$:

(a) (True or False) If y_1 and y_2 are linearly dependent on the interval $[a, b]$, then y_1 and y_2 are linearly dependent on the smaller interval $[c, d] \subset [a, b]$.

(b) (True or False) If y_1 and y_2 are linearly dependent on the interval $[a, b]$, then y_1 and y_2 are linearly dependent on the larger interval $[C, D] \supset [a, b]$.

18. Let $y_1(x) = x^3$ and $y_2(x) = |x^3|$. Are y_1 and y_2 linearly independent on the interval:

(a) $[0, \infty)$?   **(b)** $(-\infty, 0]$?
(c) $(-\infty, \infty)$?
(d) Compute the Wronskian $W[y_1, y_2](x)$ on the interval $(-\infty, \infty)$.

19. Use Abel's identity (15) to determine (up to a constant multiple) the Wronskian of two solutions on $(0, \infty)$ to

$$xy'' + (x - 1)y' + 3y = 0.$$

20. Prove Abel's identity (15) by completing the following steps:

(a) Show that the Wronskian W satisfies the equation $W' + p(x)W = 0$.

(b) Solve the linear equation in part (a) using an appropriate integrating factor.

21. Prove that if y_1 and y_2 are linearly independent solutions of $y'' + py' + qy = 0$ on (a, b), then they cannot both be zero at the same point x_0 in (a, b).

22. Show that if y_1 and y_2 are linearly independent solutions of $y'' + py' + qy = 0$ on (a, b), then they cannot both have an extremum at the same point x_0 in (a, b).

23. Normal Form. Show that the substitution $y(x) = u(x)v(x)$, where

$$v(x) := \exp\left(-\frac{1}{2}\int p(x)\, dx\right),$$

transforms the differential equation

$$y'' + p(x)y' + q(x)y = 0,$$

into an equation of the form

$$u'' + f(x)u = 0.$$

The last equation is called the **normal form** of a homogeneous linear second order equation.

24. Linear Dependence of Three Functions. Three functions $y_1(x)$, $y_2(x)$, and $y_3(x)$ are said to be **linearly dependent on an interval** I if there exist constants C_1, C_2, C_3, not all zero, such that

$$C_1 y_1(x) + C_2 y_2(x) + C_3 y_3(x) = 0, \quad \text{all } x \text{ in } I.$$

Otherwise, we say they are **linearly independent.**

(a) Show that if y_1 and y_2 are two linearly dependent functions on I, then y_1, y_2, and y_3 are linearly dependent on I for any function y_3.

(b) Show that $y_1(x) = e^x$, $y_2(x) = e^{2x}$, and $y_3(x) = e^{-3x}$ are linearly independent on $(-\infty, \infty)$.

(c) Show that $y_1(x) = \cos 2x$, $y_2(x) = \sin^2 x$, and $y_3(x) = \cos^2 x$ are linearly dependent on $(-\infty, \infty)$.

25. Using the definitions in Problem 24, determine whether the given three functions are linearly dependent or linearly independent on $(-\infty, \infty)$:

(a) $y_1(x) = 1$, $y_2(x) = x$, $y_3(x) = x^2$.
(b) $y_1(x) = -3$, $y_2(x) = 5\sin^2 x$, $y_3(x) = \cos^2 x$.
(c) $y_1(x) = e^x$, $y_2(x) = xe^x$, $y_3(x) = x^2 e^x$.

26. Using the definition in Problem 24, prove that $y_1(x) = e^{r_1 x}$, $y_2(x) = e^{r_2 x}$, and $y_3(x) = e^{r_3 x}$ are linearly independent on $(-\infty, \infty)$ if and only if the real numbers r_1, r_2, and r_3 are distinct.

27. Wronskian of Three Functions. The Wronskian of three functions is defined in terms of a determinant as follows.

$$W[y_1, y_2, y_3](x) := \begin{vmatrix} y_1(x) & y_2(x) & y_3(x) \\ y_1'(x) & y_2'(x) & y_3'(x) \\ y_1''(x) & y_2''(x) & y_3''(x) \end{vmatrix}.$$

Find the Wronskian of the three functions given in:

(a) Part (a) of Problem 25.
(b) Part (b) of Problem 25.
(c) Part (c) of Problem 25.
(d) What do you think is the connection between the Wronskian of three functions and their linear dependence?

• **4.4** • REDUCTION OF ORDER (Optional)

As we found in the previous section, a general solution to a linear second order homogeneous equation is given by a linear combination of two linearly independent solutions. But what can be done if a method for solving such an equation produces (apart from a multiplicative constant) only *one* nontrivial solution? It seems reasonable that knowing one solution should help us to find a second linearly independent solution. This is, in fact, the case. We can use the known solution to reduce a homogeneous linear second order equation to a separable first order equation—a type that we discussed in Section 2.2. This method is referred to as **reduction of order.**[†]

Let f be a nontrivial (i.e., not identically zero) solution to the homogeneous equation

(1) $y'' + p(x)y' + q(x)y = 0.$

To find a second solution that is linearly independent of f, the simplest way would be to use function addition or multiplication to modify f. However, trying a solution of the form $y(x) = v(x) + f(x)$ leads us right back where we started; namely, to the equation $v'' + pv' + qv = 0.$ (Check this!) So, let's try to find a solution of the form

(2) $y(x) = v(x)f(x),$

where the nonconstant function v is to be determined. Differentiating the product vf, we have

$$y' = vf' + v'f,$$
$$y'' = vf'' + 2v'f' + v''f.$$

Substituting these expressions into (1) yields

$$(vf'' + 2v'f' + v''f) + p(vf' + v'f) + qvf = 0,$$

or, on regrouping,

(3) $(f'' + pf' + qf)v + fv'' + (2f' + pf)v' = 0.$

Since f is a solution to (1), the factor in front of v is zero. Thus (3) reduces to

(4) $fv'' + (2f' + pf)v' = 0.$

This is a separable first order equation in the variable $w = v'$. Indeed, on substituting w for v' in (4), we have

$$fw' + (2f' + pf)w = 0.$$

Then, separating the variables and integrating gives

$$\int \frac{dw}{w} = -2\int \frac{f'}{f}\,dx - \int p\,dx,$$

$$\ln w = \ln(f^{-2}) - \int p\,dx,$$

$$w = [e^{-\int p\,dx}]/f^2.$$

[†] *Historical Footnote*: The method for the reduction of order is credited to Jean le Rond d'Alembert circa 1760.

If we replace w by v' in the last equation and integrate, we find

(5) $$v = \int \frac{e^{-\int p(x)\,dx}}{[f(x)]^2}\,dx.$$

Hence, a second linearly independent solution is $y(x) = v(x)f(x)$, where $v(x)$ is given in equation (5).

The above discussion can be summarized as follows.

▶ **REDUCTION OF ORDER PROCEDURES**

Given a nontrivial solution $f(x)$ to $y'' + py' + qy = 0$, a second linearly independent solution $y(x)$ can be determined in either of the following ways.

1. Set $y(x) = v(x)f(x)$ and substitute for y, y', and y'' in the given equation. This gives a separable equation for v'. Solve for v' and integrate v' to obtain v. The desired second solution is given by $v(x)f(x)$.
2. The solution $y(x)$ can also be obtained by plugging $p(x)$ and $f(x)$ directly into the **reduction of order formula:**

(6) $$y(x) = f(x) \int \frac{e^{-\int p(x)dx}}{[f(x)]^2}\,dx.$$

Caution: If the equation is $a_2(x)y'' + a_1(x)y' + a_0(x)y = 0$, then it must first be put in the form $y'' + p(x)y' + q(x)y = 0$, so $p(x) = a_1(x)/a_2(x)$.

The first procedure is a technique for which we only have to remember the substitution $y = v(x)f(x)$. The second procedure is quicker, but we must remember precisely the reduction of order formula. In the next example we illustrate both of these methods.

● **EXAMPLE 1**   Given that $f(x) = e^x$ is a solution to

(7) $$y'' - 2y' + y = 0,$$

determine a second linearly independent solution.

Solution   Let $y = v(x)f(x) = ve^x$. Then

$$y' = ve^x + v'e^x,$$
$$y'' = ve^x + 2v'e^x + v''e^x.$$

Substituting these representations into (7) gives

$$(ve^x + 2v'e^x + v''e^x) - 2(ve^x + v'e^x) + ve^x = 0,$$

which simplifies to $v'' = 0$. Hence $v = c_1 x + c_2$. Since we want y to be linearly independent of e^x, we must take $c_1 \neq 0$; so for simplicity let $c_1 = 1$ and $c_2 = 0$. Then a second linearly independent solution is

$$y = vf = xe^x.$$

The alternative approach is to use the reduction of order formula given by equation (6). In this example $p(x) = -2$ and $f(x) = e^x$. Hence

$$-\int p(x)\,dx = \int 2\,dx = 2x + c.$$

We will take $c = 0$. Now

$$y(x) = f(x)\int \frac{e^{-\int p(x)\,dx}}{[f(x)]^2}\,dx = e^x\int \frac{e^{2x}}{(e^x)^2}\,dx$$

$$= e^x\int dx = e^x(x + C).$$

Again, we set $C = 0$. This gives us the same linearly independent solution, $y = xe^x$. ●

● **EXAMPLE 2**   Given that $f(x) = x$ is a solution to

(8) $y'' - 2xy' + 2y = 0,$

determine a second linearly independent solution.

Solution   Here $p(x) = -2x$, and so

$$-\int p(x)\,dx = \int 2x\,dx = x^2,$$

where we have taken the constant of integration to be zero. The reduction of order formula in (6) gives

(9) $y(x) = x\int \frac{e^{x^2}}{x^2}\,dx$

as a second linearly independent solution. Unfortunately, we are not able to evaluate the integral in (9) in terms of elementary functions. One alternative is to write $y(x)$ in terms of a definite integral, say,

$$y(x) = x\int_1^x \frac{e^{t^2}}{t^2}\,dt.$$

Then $y(x)$ can be approximated for various values of x by using a numerical integration method such as Simpson's rule (see Problem 17). Another approach is to obtain a **power series expansion** for the solution in (9). This can be derived as follows.

Recall that the Maclaurin expansion for e^x is

$$e^x = 1 + x + \frac{x^2}{2!} + \frac{x^3}{3!} + \cdots + \frac{x^n}{n!} + \cdots.$$

Hence, the expansion for e^{x^2} is just

$$e^{x^2} = 1 + x^2 + \frac{x^4}{2!} + \frac{x^6}{3!} + \cdots + \frac{x^{2n}}{n!} + \cdots.$$

Substituting this series for e^{x^2} into (9), we integrate term by term to obtain

$$y(x) = x \int x^{-2} \left\{ 1 + x^2 + \frac{x^4}{2} + \cdots + \frac{x^{2n}}{n!} + \cdots \right\} dx$$

$$= x \int \left\{ x^{-2} + 1 + \frac{x^2}{2} + \cdots + \frac{x^{2n-2}}{n!} + \cdots \right\} dx$$

$$= x \left\{ -x^{-1} + x + \frac{x^3}{6} + \cdots + \frac{x^{2n-1}}{n!(2n-1)} + \cdots \right\},$$

where we have taken the constant of integration to be zero. Hence

(10) $$y(x) = -1 + x^2 + \frac{x^4}{6} + \cdots + \frac{x^{2n}}{n!(2n-1)} + \cdots. \quad \bullet$$

In the last example we did not bother to justify the use of term-by-term integration to obtain equation (10). Such questions are addressed in Chapter 8.

The reduction of order procedure can also be used to reduce a linear *nonhomogeneous* second order equation to a linear first order equation (see Problem 18 in Exercises 4.7).

EXERCISES 4.4

In Problems 1 through 6 a differential equation and a non-trivial solution f are given. Find a second linearly independent solution.

1. $y'' + 2y' - 15y = 0$; $f(x) = e^{3x}$.

2. $z'' - 3z' + 2z = 0$; $f(x) = e^x$.

3. $x^2 y'' + 6xy' + 6y = 0$, $x > 0$; $f(x) = x^{-2}$.

4. $x^2 y'' - 2xy' - 4y = 0$, $x > 0$; $f(x) = x^{-1}$.

5. $tx'' - (t+1)x' + x = 0$, $t > 0$; $f(t) = e^t$.

6. $xy'' + (1-2x)y' + (x-1)y = 0$, $x > 0$; $f(x) = e^x$.

7. The reduction of order procedure can be used, more generally, to reduce a homogeneous linear *n*th order equation to a homogeneous linear $(n-1)$th order equation. For the equation

$$xy''' - xy'' + y' - y = 0,$$

which has $f(x) = e^x$ as a solution, use the substitution $y(x) = v(x)f(x)$ to reduce this third order equation to a homogeneous linear second order equation in the variable $w = v'$.

8. The equation

$$xy''' + (1-x)y'' + xy' - y = 0$$

has $f(x) = x$ as a solution. Use the substitution $y(x) = v(x)f(x)$ to reduce this third order equation to a homogeneous linear second order equation in the variable $w = v'$.

9. Given that $f(x) = x$ is a solution to

$$y'' - xy' + y = 0:$$

(a) Obtain an integral representation for a second linearly independent solution.

(b) Using the integral representation from part (a), obtain a Maclaurin series expansion for this solution.

10. Given that $f(x) = x$ is a solution to

$$(1 - x^2)y'' - 2xy' + 2y = 0, \qquad -1 < x < 1:$$

(a) Obtain an integral representation for a second linearly independent solution.

(b) Using the integral representation from part (a), obtain a Maclaurin series expansion for this solution. [Hint: $(1-x)^{-1} = 1 + x + x^2 + \cdots$.]

(c) Use partial fractions to evaluate the integral in part (a), and give a general solution to the differential equation.

11. The reduction of order formula (6) can also be derived from Abel's identity [formula (15) of Section 4.3]. Let

$f(x)$ be a nontrivial solution to (1) and let $y(x)$ be a second linearly independent solution. Show that

$$\left(\frac{y}{f}\right)' = \frac{W[f, y]}{f^2},$$

and then use Abel's identity for the Wronskian $W[f, y]$ to obtain the reduction of order formula.

12. Verify, by direct substitution, that the power series (10) is a solution to the differential equation (8).

13. In quantum mechanics the study of the Schrödinger equation for the case of a harmonic oscillator leads to a consideration of **Hermite's equation,**

$$y'' - 2xy' + \lambda y = 0,$$

where λ is a parameter. Use the reduction of order formula to obtain an integral representation of a second linearly independent solution to Hermite's equation for the given value of λ and corresponding solution $f(x)$.

   (a) $\lambda = 4$, $f(x) = 1 - 2x^2$.
   (b) $\lambda = 6$, $f(x) = 3x - 2x^3$.

14. In mathematical physics, many problems with spherical symmetry involve the study of **Legendre's equation,**

$$(1 - x^2)y'' - 2xy' + \lambda(\lambda + 1)y = 0, \qquad -1 < x < 1,$$

where λ is a parameter. Use the reduction of order formula to obtain an integral representation of a second linearly independent solution to Legendre's equation for the given value of λ and corresponding solution $f(x)$.

   (a) $\lambda = 1$, $f(x) = x$.
   (b) $\lambda = 2$, $f(x) = 3x^2 - 1$.
   (c) $\lambda = 3$, $f(x) = 5x^3 - 3x$.

15. In quantum mechanics the study of the Schrödinger equation for the hydrogen atom leads to a consideration of **Laguerre's equation,**

$$xy'' + (1 - x)y' + \lambda y = 0, \qquad x > 0,$$

where λ is a parameter. Use the reduction of order formula to obtain an integral representation of a second linearly independent solution to Laguerre's equation for the given value of λ and corresponding solution $f(x)$.

   (a) $\lambda = 1$, $f(x) = x - 1$.
   (b) $\lambda = 2$, $f(x) = x^2 - 4x + 2$.

16. For equation (8) we showed in Example 2 that

$$f(x) = x \quad \text{and}$$

$$h(x) = -1 + x^2 + \frac{x^4}{6} + \cdots + \frac{x^{2n}}{n!(2n - 1)} + \cdots$$

are two linearly independent solutions. Show that the solution to the initial value problem

(11) $y'' - 2xy' + 2y = 0;$

$$y(0) = -2,$$

$$y'(0) = -1$$

is given by $y(x) = 2h(x) - f(x)$. Obtain an approximation for the solution to the initial value problem (11) at $x = 1$ by using the first five nonzero terms of the power series for $h(x)$.

17. In Example 2 we showed that a solution to equation (8) is given by

$$g(x) = x \int_1^x \frac{e^{t^2}}{t^2}\, dt.$$

   (a) Show that this solution satisfies $g(1) = 0$, $g'(1) = e$.
   (b) Using Simpson's rule (see Appendix B), approximate the solution to the initial value problem

$$y'' - 2xy' + y = 0; \qquad y(1) = 0, \qquad y'(1) = e,$$

   at $x = 2$.

18. In Example 2 we showed that

$$f(x) = x, \qquad g(x) = x \int_1^x \frac{e^{t^2}}{t^2}\, dt,$$

and

$$h(x) = -1 + x^2 + \frac{x^4}{6} + \cdots + \frac{x^{2n}}{n!(2n - 1)} + \cdots$$

are all solutions to equation (8). Hence, by the representation theorem for homogeneous equations, we know that $h(x)$ can be expressed as a linear combination of $f(x)$ and $g(x)$. This representation can be determined as follows:

   (a) Using integration by parts, show that

$$g(x) = ex - e^{x^2} + 2x \int_1^x e^{t^2}\, dt.$$

   (b) Prove that

$$h(x) = g(x) + \left(2 \int_0^1 e^{t^2}\, dt - e\right) f(x).$$

•4.5• HOMOGENEOUS LINEAR EQUATIONS WITH CONSTANT COEFFICIENTS

Let's consider the homogeneous linear second order differential equation with constant coefficients

(1) $ay'' + by' + cy = 0,$

where $a\,(\neq 0)$, b, and c are real constants. Since constant functions are everywhere continuous, Theorem 2 guarantees that equation (1) has solutions defined for all x in $(-\infty, \infty)$. If we can find two linearly independent solutions to (1), say y_1 and y_2, then we can express a general solution in the form

$$y = c_1 y_1 + c_2 y_2,$$

where c_1 and c_2 are arbitrary constants.

A look at equation (1) tells us that a solution of (1) must have the property that a constant times its second derivative plus a constant times its first derivative plus a constant times itself must sum to zero. This suggests that we try to find a solution of the form $y = e^{rx}$, since derivatives of e^{rx} are just constants times e^{rx}. If we substitute $y = e^{rx}$ into (1), we obtain

$$ar^2 e^{rx} + bre^{rx} + ce^{rx} = 0,$$
$$e^{rx}(ar^2 + br + c) = 0.$$

Because e^{rx} is never zero, we can divide by it to obtain

(2) $ar^2 + br + c = 0.$

Consequently, $y = e^{rx}$ is a solution to (1) if and only if r satisfies equation (2). Equation (2) is called the **auxiliary equation** associated with the homogeneous equation (1).

Now the auxiliary equation is just a quadratic, and its roots are

$$r_1 = \frac{-b + \sqrt{b^2 - 4ac}}{2a} \quad \text{and} \quad r_2 = \frac{-b - \sqrt{b^2 - 4ac}}{2a}.$$

When the discriminant, $b^2 - 4ac$, is positive, the roots r_1 and r_2 are real and distinct; if $b^2 - 4ac = 0$, the roots are real and equal; and when $b^2 - 4ac < 0$, the roots are complex conjugate numbers. We consider the case of complex roots in the next section.

DISTINCT REAL ROOTS

If the auxiliary equation has distinct real roots r_1 and r_2, then $e^{r_1 x}$ and $e^{r_2 x}$ are linearly independent solutions to (1). Therefore, a general solution of (1) is

$$y(x) = c_1 e^{r_1 x} + c_2 e^{r_2 x},$$

where c_1 and c_2 are arbitrary constants.

● **EXAMPLE 1**   Find a general solution to

(3) $y'' + 5y' - 6y = 0.$

Solution   The auxiliary equation associated with (3) is

$$r^2 + 5r - 6 = (r - 1)(r + 6) = 0,$$

which has the roots $r_1 = 1$, $r_2 = -6$. Thus $\{e^x, e^{-6x}\}$ is a fundamental solution set for equation (3), and so a general solution is

$$y(x) = c_1 e^x + c_2 e^{-6x}. \quad \bullet$$

● **EXAMPLE 2**   Solve the initial value problem

(4) $y'' + 2y' - y = 0; \qquad y(0) = 0, \qquad y'(0) = -1.$

Solution   Here the auxiliary equation is

$$r^2 + 2r - 1 = 0.$$

Using the quadratic formula, we find that the roots of this equation are

$$r_1 = -1 + \sqrt{2} \quad \text{and} \quad r_2 = -1 - \sqrt{2}.$$

Consequently, a general solution to the differential equation in (4) is

(5) $y(x) = c_1 e^{(-1+\sqrt{2})x} + c_2 e^{(-1-\sqrt{2})x},$

where c_1 and c_2 are arbitrary constants. To find the specific solution that satisfies the initial conditions given in (4), we first differentiate y as given in (5), then plug y and y' into the initial conditions of (4). This gives

$$y(0) = c_1 e^0 + c_2 e^0,$$
$$y'(0) = (-1 + \sqrt{2})c_1 e^0 + (-1 - \sqrt{2})c_2 e^0,$$

or

$$0 = c_1 + c_2,$$
$$-1 = (-1 + \sqrt{2})c_1 + (-1 - \sqrt{2})c_2.$$

Solving this system yields $c_1 = -\sqrt{2}/4$ and $c_2 = \sqrt{2}/4$. Thus,

$$y(x) = -\frac{\sqrt{2}}{4} e^{(-1+\sqrt{2})x} + \frac{\sqrt{2}}{4} e^{(-1-\sqrt{2})x}$$

is the desired solution.   ●

When the roots of the auxiliary equation (2) are equal, that is, $r_1 = r_2 = r$, then, unlike the previous case, we get only one nontrivial solution, namely $y = e^{rx}$. Of course, constant multiples of this function are also solutions, but they are of no help in finding a second **linearly independent** solution. We can resolve this shortcoming by using the reduction of order method discussed in Section 4.4. When we apply this procedure to the case of a repeated root r, we find that a second linearly independent solution is just xe^{rx}; that is, x times the original solution e^{rx}. We leave the verification of this fact to the exercises (see Problem 50).

> **REPEATED ROOT**
>
> If the auxiliary equation has a repeated root r, then two linearly independent solutions to (1) are e^{rx} and xe^{rx}, and a general solution is
>
> $$y(x) = c_1 e^{rx} + c_2 x e^{rx},$$
>
> where c_1 and c_2 are arbitrary constants.

● EXAMPLE 3   Find a general solution to

(6) $y'' + 4y' + 4y = 0.$

Solution   The auxiliary equation for (6) is

$$r^2 + 4r + 4 = (r + 2)^2 = 0.$$

Since $r = -2$ is a double root, two linearly independent solutions are e^{-2x} and xe^{-2x}. These give

$$y(x) = c_1 e^{-2x} + c_2 x e^{-2x}$$

as a general solution.   ●

The method we have described for solving homogeneous linear second order equations with constant coefficients applies to any order (even first order) homogeneous linear equations with constant coefficients. We give a detailed treatment of such higher order equations in Chapter 6. For now, we shall be content to illustrate the method by means of an example. We remark that, in general, a homogeneous linear nth order equation has a fundamental solution set consisting of n linearly independent solutions.

● EXAMPLE 4   Find a general solution to

(7) $y''' + 3y'' - y' - 3y = 0.$

Solution   If we try to find solutions of the form $y = e^{rx}$, then, as with second order equations, we are led to finding roots of the auxiliary equation

(8) $r^3 + 3r^2 - r - 3 = 0.$

We observe that $r = 1$ is a root of the above equation, and dividing the polynomial on the left-hand side of (8) by $r - 1$ leads to the factorization

$$(r - 1)(r^2 + 4r + 3) = (r - 1)(r + 1)(r + 3) = 0.$$

Hence the roots of the auxiliary equation are 1, -1, and -3, and so three solutions of (7) are e^x, e^{-x}, and e^{-3x}. For now we assume that these form a fundamental solution set. (A discussion of linear independence for sets containing more than two functions is given in Section 6.2.) A general solution to (7) is then

(9) $y(x) = c_1 e^x + c_2 e^{-x} + c_3 e^{-3x}.$ ●

In the last example we can actually verify that (9) is a "general solution" by showing, as we did in the proof of Theorem 3 on page 140, that every solution to equation (7) can be written in the form (9) for a suitable choice of constants c_1, c_2, and c_3. For this verification, we need a generalization of the existence and uniqueness theorem to linear third order equations (see Section 6.2).

In the general case of a homogeneous linear nth order equation with constant coefficients, the auxiliary equation is a polynomial equation of order n. The fundamental theorem of algebra[†] states that a polynomial equation of order n has n roots, some of which may be complex. Unfortunately, a result from Galois theory by Abel and Ruffini[††] also states that for arbitrary polynomial equations of order greater than four, there does *not* exist a formula involving radicals for determining the roots of the equation. Thus, when $n > 4$, we will usually have to rely on numerical techniques for approximating the roots (see Problem 49).

Cauchy-Euler Equations

There is a special class of equations with variable coefficients that can be transformed into equations with constant coefficients.

CAUCHY-EULER EQUATIONS

Definition 4. A linear second order equation that can be expressed in the form

$$(10) \qquad ax^2 \frac{d^2y}{dx^2} + bx \frac{dy}{dx} + cy = h(x),$$

where a, b, and c are constants is called a **Cauchy-Euler equation.**[†††]

For example, the differential equation

$$3x^2 y'' - 2xy' + 7y = \sin x$$

is a Cauchy-Euler equation, whereas

$$2y'' - 3xy' + 11y = 3x - 1$$

is *not*, because the coefficient of y'' is 2, which is not a constant times x^2.

To solve a homogeneous Cauchy-Euler equation (i.e., $h(x) \equiv 0$), we make the substitution $x = e^t$, which transforms equation (10) into an equation with constant coefficients. The technique is illustrated in the next example.

[†] See, for example, *Elements of Abstract Algebra*, by Richard A. Dean, John Wiley and Sons, Inc., New York, 1966, Section 4.6.

[††] Ibid., Section 10.5.

[†††] *Historical Footnote*: Although work on this equation was published by Leonhard Euler in 1769 and later by Augustin Cauchy, its solution was known to John Bernoulli prior to 1700. These equations are also called **equidimensional equations.**

● **EXAMPLE 5**   Find a general solution to

(11) $$3x^2\frac{d^2y}{dx^2} + 11x\frac{dy}{dx} - 3y = 0, \qquad x > 0.$$

Solution   We make the substitution $x = e^t$ to transform (11) into an equation in which the new independent variable is t. Since $x = e^t$, it follows by the chain rule that

$$\frac{dy}{dt} = \frac{dy}{dx}\frac{dx}{dt} = \frac{dy}{dx}e^t = x\frac{dy}{dx};$$

and hence

(12) $$x\frac{dy}{dx} = \frac{dy}{dt}.$$

Differentiating (12) with respect to t, we find from the product rule that

$$\frac{d^2y}{dt^2} = \frac{d}{dt}\left(x\frac{dy}{dx}\right) = \frac{dx}{dt}\frac{dy}{dx} + x\frac{d}{dt}\left(\frac{dy}{dx}\right)$$

$$= \frac{dy}{dt} + x\frac{d^2y}{dx^2}\frac{dx}{dt} = \frac{dy}{dt} + x\frac{d^2y}{dx^2}e^t$$

$$= \frac{dy}{dt} + x^2\frac{d^2y}{dx^2}.$$

Hence

(13) $$x^2\frac{d^2y}{dx^2} = \frac{d^2y}{dt^2} - \frac{dy}{dt}.$$

Substituting into (11) the expressions for $x\, dy/dx$ and $x^2\, d^2y/dx^2$ given in (12) and (13) yields

$$3\left(\frac{d^2y}{dt^2} - \frac{dy}{dt}\right) + 11\frac{dy}{dt} - 3y = 0,$$

(14) $$3\frac{d^2y}{dt^2} + 8\frac{dy}{dt} - 3y = 0.$$

The auxiliary equation associated with (14) is

$$3r^2 + 8r - 3 = (3r - 1)(r + 3) = 0.$$

Since the roots of this equation are $\frac{1}{3}$ and -3, the general solution to (14) is

$$y(t) = c_1e^{t/3} + c_2e^{-3t} = c_1(e^t)^{1/3} + c_2(e^t)^{-3}.$$

Expressing y in terms of the original variable x, we find

(15) $$y(x) = c_1x^{1/3} + c_2x^{-3} \quad \text{for } x > 0. \quad ●$$

It turns out that the solution (15) for the special equation (11) is valid for all $x \neq 0$ even though it was derived only for $x > 0$. In general, it can be shown that if $y(x)$ is a solution to a homogeneous Cauchy-Euler equation for $x > 0$, then $\phi(x) := y(-x)$ is a solution for $x < 0$ (see Problem 43).

Using formulas (12) and (13), the reader can verify that the substitution $x = e^t$ transforms the general Cauchy-Euler equation

$$(16) \qquad ax^2\frac{d^2y}{dx^2} + bx\frac{dy}{dx} + cy = h(x), \qquad x > 0$$

into the constant coefficient equation

$$(17) \qquad a\frac{d^2y}{dt^2} + (b - a)\frac{dy}{dt} + cy = h(e^t).$$

A method similar to the one discussed here for second order Cauchy-Euler equations can be derived for higher order Cauchy-Euler equations of the form

$$a_n x^n y^{(n)}(x) + a_{n-1} x^{n-1} y^{(n-1)}(x) + \cdots + a_0 y(x) = h(x),$$

where $a_n, a_{n-1}, \ldots, a_0$ are real constants (see Problems 46–48).

Another method for finding solutions to a homogeneous Cauchy-Euler equation involves guessing a solution of the form $y = x^r$, which leads to an auxiliary equation for r. This method is discussed in Section 8.5. However, there is often an advantage in making the substitution $x = e^t$ when trying to solve a *nonhomogeneous* Cauchy-Euler equation by the methods of undetermined coefficients (see Section 4.8) or Laplace transforms (see Chapter 7).

EXERCISES 4.5

1, 3, 5, 7, 9, 11, 13, 15, 17, 19, 21, 23

In Problems 1 through 12 find a general solution to the given differential equation.

1. $y'' - y' - 2y = 0.$

2. $y'' + 5y' + 6y = 0.$

3. $y'' + 8y' + 16y = 0.$

4. $y'' + 6y' + 9y = 0.$

5. $z'' + z' - z = 0.$

6. $y'' - 5y' + 6y = 0.$

7. $2u'' + 7u' - 4u = 0.$

8. $6y'' + y' - 2y = 0.$

9. $y'' - y' - 11y = 0.$

10. $4y'' - 4y' + y = 0.$

11. $4w'' + 20w' + 25w = 0.$

12. $3y'' + 11y' - 7y = 0.$

In Problems 13 through 20 solve the given initial value problem.

13. $y'' + 2y' - 8y = 0;$ $y(0) = 3,$ $y'(0) = -12.$

14. $y'' + y' = 0;$ $y(0) = 2,$ $y'(0) = 1.$

15. $y'' + 2y' + y = 0;$ $y(0) = 1,$ $y'(0) = -3.$

16. $y'' - 4y' + 3y = 0;$ $y(0) = 1,$ $y'(0) = \frac{1}{3}.$

17. $z'' - 2z' - 2z = 0;$ $z(0) = 0,$ $z'(0) = 3.$

18. $y'' - 6y' + 9y = 0;$ $y(0) = 2,$ $y'(0) = \frac{25}{3}.$

19. $y'' - 4y' - 5y = 0;$ $y(-1) = 3,$ $y'(-1) = 9.$

20. $y'' - 4y' + 4y = 0;$ $y(1) = 1,$ $y'(1) = 1.$

21. First Order Constant Coefficient Equations.

(a) Substituting $y = e^{rx}$, find the auxiliary equation for the first order linear equation

$$ay' + by = 0,$$

where a, b are constants with $a \neq 0$.

(b) Use the result of part (a) to find the general solution.

In Problems 22 through 25 use the method described in Problem 21 to find a general solution to the given equation.

22. $3y' - 7y = 0.$

23. $5y' + 4y = 0.$

24. $3z' + 11z = 0.$

25. $6w' - 13w = 0.$

In Problems 26 through 31 find a general solution to the given third order differential equation.

26. $y''' + y'' - 6y' + 4y = 0$.

27. $y''' - 6y'' - y' + 6y = 0$.

28. $z''' + 2z'' - 4z' - 8z = 0$.

29. $y''' - 7y'' + 7y' + 15y = 0$.

30. $y''' + 3y'' - 4y' - 12y = 0$.

31. $w''' + w'' - 4w' + 2w = 0$.

32. Solve the initial value problem

$$y''' - y' = 0;$$
$$y(0) = 2, \qquad y'(0) = 3, \qquad y''(0) = -1.$$

In Problems 33 through 38 use the method discussed under "Cauchy-Euler equations" to find a general solution to the given equation for $x > 0$.

33. $x^2 y''(x) + 7xy'(x) - 7y(x) = 0$.

34. $x^2 \dfrac{d^2 y}{dx^2} + 2x \dfrac{dy}{dx} - 6y = 0$.

35. $\dfrac{d^2 w}{dx^2} + \dfrac{6}{x} \dfrac{dw}{dx} + \dfrac{4}{x^2} w = 0$.

36. $x^2 \dfrac{d^2 z}{dx^2} + 5x \dfrac{dz}{dx} + 4z = 0$.

37. $9x^2 y''(x) + 15xy'(x) + y(x) = 0$.

38. $x^2 y''(x) - 3xy'(x) + 4y(x) = 0$.

In Problems 39 and 40 use the method discussed under "Cauchy-Euler equations" to solve the given initial value problem.

39. $x^2 y''(x) - 4xy'(x) + 4y(x) = 0$;
$y(1) = -2, \qquad y'(1) = -11$.

40. $x^2 y''(x) + 7xy'(x) + 5y(x) = 0$;
$y(1) = -1, \qquad y'(1) = 13$.

In Problems 41 and 42 use a modification of the method discussed under "Cauchy-Euler equations" to find a general solution to the given equation.

41. $(x-2)^2 y''(x) - 7(x-2)y'(x) + 7y(x) = 0$, $\qquad x > 2$.

42. $(x+1)^2 y''(x) + 10(x+1)y'(x) + 14y(x) = 0$, $\qquad x > -1$.

43. Let $y(x)$ be a solution for $x > 0$ to the *homogeneous* Cauchy-Euler equation $ax^2 y'' + bxy' + cy = 0$. Show that $\phi(x) := y(-x)$ is a solution to the equation for $x < 0$. Is the analogous statement true for the *non-homogeneous* Cauchy-Euler equation?

In Problems 44 and 45 use the result of Problem 43 to find a general solution to the given differential equation for $x < 0$.

44. $x^2 y''(x) + xy'(x) - 2y(x) = 0$.

45. $2x^2 y''(x) + 7xy'(x) + 2y(x) = 0$.

46. Use the substitution $x = e^t$ to show that the third order Cauchy-Euler equation

$$ax^3 y'''(x) + bx^2 y''(x) + cxy'(x) + dy(x) = 0,$$
$$x > 0$$

is equivalent to the constant coefficient equation

$$ay'''(t) + (b - 3a)y''(t)$$
$$+ (2a - b + c)y'(t) + dy(t) = 0.$$

In Problem 47 and 48 use the result of Problem 46 to find a general solution to the given differential equation for $x > 0$.

47. $x^3 y'''(x) - 2x^2 y''(x) + 3xy'(x) - 3y(x) = 0$.

48. $x^3 y'''(x) + x^2 y''(x) - 8xy'(x) - 4y(x) = 0$.

49. Find a general solution to

$$3y''' + 18y'' + 33y' - 19y = 0$$

by using Newton's method or some other numerical procedure to approximate the roots of the auxiliary equation.

50. When $b^2 - 4ac = 0$, the roots of the auxiliary equation (2) are equal; that is, $r_1 = r_2 = r = -b/2a$. Hence $f(x) = e^{(-b/2a)x}$ is a solution to (1). Use the reduction of order formula (6) of Section 4.4 to derive $xe^{(-b/2a)x}$ as a second linearly independent solution to (1).

51. Suppose the auxiliary equation (2) has a repeated root r. Then, on dividing by a, equation (1) can be written in the equivalent form

$$(D - r)^2 y = y'' - 2ry' + r^2 y = 0,$$

where D is the differential operator d/dx.

(a) Show that for any twice differentiable function u,

$$(D - r)^2 [e^{rx} u] = e^{rx} D^2 u,$$

and hence the problem of solving $(D - r)^2 y = 0$ reduces to that of solving $D^2 u = 0$, where $y = e^{rx} u$.

(b) Use the result of part (a) to show that e^{rx} and xe^{rx} are linearly independent solutions of $(D - r)^2 y = 0$.

52. Let $L[y] := ay'' + by' + c$, where $b^2 - 4ac = 0$.

(a) Show that $L[e^{rx}](x) = a(r - r_0)^2 e^{rx}$, where $r_0 = -b/2a$.

(b) Show that $L[(\partial/\partial r)e^{rx}] = (\partial/\partial r)L[e^{rx}]$, and thus

$$L[xe^{rx}](x) = 2a(r - r_0)e^{rx} + ax(r - r_0)^2 e^{rx}.$$

(c) Use the results of parts (a) and (b) to show that $e^{r_0 x}$ and $xe^{r_0 x}$ are linearly independent solutions to (1).

53. The suspension in an automobile can be modeled as a vibrating spring with damping due to the shock absorbers (see Figure 4.3). This leads to the equation

$$mx''(t) + bx'(t) + kx(t) = 0,$$

where m is the mass of the automobile, b is the damp-ing constant of the shocks, k is the spring constant, and $x(t)$ is the vertical displacement of the automobile at time t. In the next section we will find that the solu-tion of a homogeneous linear second order equation with constant coefficients oscillates when the auxiliary equation has complex roots. If the mass of an auto-mobile is 1000 kilograms (kg) and the spring constant is 3000 kg/sec^2, determine the minimum value for the damping constant in kilograms per second that will provide a smooth, vibration-free (theoretically!) ride. If we replace the springs with heavy duty ones having twice the spring constant, how does this minimum value for b change?

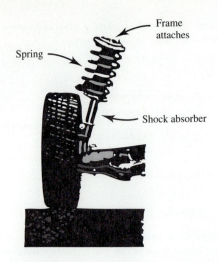

Figure 4.3 Diagram of an automobile suspension

• **4.6** • **AUXILIARY EQUATIONS WITH COMPLEX ROOTS**

The *simple harmonic equation* $y'' + y = 0$, so-called because of its relation to the funda-mental vibration of a musical tone, has as solutions $y_1(x) = \cos x$ and $y_2(x) = \sin x$. Notice that the auxiliary equation associated with the harmonic equation is $r^2 + 1 = 0$ which has imaginary roots $r = \pm i$, where i denotes $\sqrt{-1}$.[†] In the previous section, we expressed the solutions to a linear second order equation with constant coefficients in terms of expo-nential functions. To see how the trigonometric functions arise from complex exponential functions, we use Euler's formula, which is discussed below.

When $b^2 - 4ac < 0$, the roots of the auxiliary equation

(1) $ar^2 + br + c = 0$

associated with the homogeneous equation

(2) $ay'' + by' + cy = 0$

are the complex conjugate numbers

$$r_1 = \alpha + i\beta \quad \text{and} \quad r_2 = \alpha - i\beta \qquad (i = \sqrt{-1}),$$

where α, β are the real numbers

(3) $\alpha = -\dfrac{b}{2a} \quad \text{and} \quad \beta = \dfrac{\sqrt{4ac - b^2}}{2a}.$

As in the previous section, we would like to assert that the functions $e^{r_1 x}$ and $e^{r_2 x}$ are solu-

---

[†] Electrical engineers frequently use the symbol j to denote $\sqrt{-1}$.

tions to the equation (2). This is, in fact, the case, but before we can proceed we need to address some fundamental questions. For example, if $r_1 = \alpha + i\beta$ is a complex number, what do we mean by the expression $e^{(\alpha + i\beta)x}$? If we assume that the law of exponents applies to complex numbers, then

(4) $e^{(\alpha + i\beta)x} = e^{\alpha x + i\beta x} = e^{\alpha x}e^{i\beta x}$.

We now need only clarify the meaning of $e^{i\beta x}$.

For this purpose, let's assume that the Maclaurin series for e^z is the same for complex numbers z as it is for real numbers. Recalling that $i^2 = -1$, then for θ real we have

$$e^{i\theta} = 1 + (i\theta) + \frac{(i\theta)^2}{2!} + \cdots + \frac{(i\theta)^n}{n!} + \cdots$$

$$= 1 + i\theta - \frac{\theta^2}{2!} - \frac{i\theta^3}{3!} + \frac{\theta^4}{4!} + \frac{i\theta^5}{5!} + \cdots$$

$$= \left(1 - \frac{\theta^2}{2!} + \frac{\theta^4}{4!} + \cdots\right) + i\left(\theta - \frac{\theta^3}{3!} + \frac{\theta^5}{5!} + \cdots\right).$$

Recognizing the series expansions for the real and imaginary parts to be Maclaurin series for $\cos\theta$ and $\sin\theta$, respectively, we can simplify the above expansion to

(5) $e^{i\theta} = \cos\theta + i\sin\theta,$

which is known as **Euler's formula.**[†]

When Euler's formula (with $\theta = \beta x$) is used in equation (4), we find

(6) $e^{(\alpha + i\beta)x} = e^{\alpha x}(\cos\beta x + i\sin\beta x),$

which expresses the complex function $e^{(\alpha + i\beta)x}$ in terms of familiar real functions. Having made sense out of $e^{(\alpha + i\beta)x}$, we can now show (see Problem 30) that

(7) $\dfrac{d}{dx}e^{(\alpha + i\beta)x} = (\alpha + i\beta)e^{(\alpha + i\beta)x},$

and with the choices of α and β as given in (3), the complex function $e^{(\alpha + i\beta)x}$ is indeed a solution to equation (2). In general, if $z(x)$ is a complex-valued function of the real variable x, we can write

$$z(x) = u(x) + iv(x),$$

where $u(x)$ and $v(x)$ are real-valued functions. The derivatives of $z(x)$ are then given by

$$\frac{dz}{dx} = \frac{du}{dx} + i\frac{dv}{dx}, \qquad \frac{d^2z}{dx^2} = \frac{d^2u}{dx^2} + i\frac{d^2v}{dx^2}.$$

With the following lemma we will show that the complex-valued solution $e^{(\alpha + i\beta)x}$ gives rise to two linearly independent *real-valued* solutions.

[†] *Historical Footnote*: This formula first appeared in Leonhard Euler's monumental two-volume *Introductio in Analysin Infinitorum* (1748).

> ### REAL SOLUTIONS DERIVED FROM COMPLEX SOLUTIONS
>
> **Lemma 2.**   Let $z(x) = u(x) + iv(x)$ be a solution to equation (2), where a, b, and c are real numbers. Then, the real part $u(x)$ and the imaginary part $v(x)$ are real-valued solutions of (2).

Proof.   By assumption, $az'' + bz' + cz = 0$, and hence

$$a(u'' + iv'') + b(u' + iv') + c(u + iv) = 0,$$
$$(au'' + bu' + cu) + i(av'' + bv' + cv) = 0.$$

But a complex number is zero if and only if its real and imaginary parts are both zero. Thus we must have

$$au'' + bu' + cu = 0 \quad \text{and} \quad av'' + bv' + cv = 0,$$

which means that both $u(x)$ and $v(x)$ are real-valued solutions of (2).   ◄◄◄

When we apply Lemma 2 to the solution

$$e^{(\alpha + i\beta)x} = e^{\alpha x} \cos \beta x + i e^{\alpha x} \sin \beta x,$$

we obtain the following.

> ### COMPLEX CONJUGATE ROOTS
>
> If the auxiliary equation has complex conjugate roots $\alpha \pm i\beta$, then two linearly independent solutions to (2) are
>
> $$e^{\alpha x} \cos \beta x \quad \text{and} \quad e^{\alpha x} \sin \beta x,$$
>
> and a general solution is
>
> $$(8) \qquad y(x) = c_1 e^{\alpha x} \cos \beta x + c_2 e^{\alpha x} \sin \beta x,$$
>
> where c_1 and c_2 are arbitrary constants.

In the preceding discussion we glossed over some important details concerning complex numbers and complex-valued functions. In particular, further analysis is required to justify the use of the law of exponents, Euler's formula, and even the fact that the derivative of e^{rx} is re^{rx} when r is a complex constant.[†] Readers who feel uneasy about our conclusions can easily check the answer by substituting it into (2).

The reader may also be wondering what would have happened if we had worked with the function $e^{(\alpha - i\beta)x}$ instead of $e^{(\alpha + i\beta)x}$. We leave it as an exercise to verify that $e^{(\alpha - i\beta)x}$ gives rise to the same general solution (8).

[†] For a detailed treatment of these topics see, for example, *Fundamentals of Complex Analysis*, Second Edition, by E. B. Saff and A. D. Snider, Prentice-Hall, Inc., Englewood Cliffs, New Jersey, 1993.

● EXAMPLE 1 Find a general solution to

(9) $y'' + 2y' + 4y = 0.$

Solution The auxiliary equation is

$$r^2 + 2r + 4 = 0,$$

which has roots

$$r = \frac{-2 \pm \sqrt{4 - 16}}{2} = \frac{-2 \pm \sqrt{-12}}{2} = -1 \pm i\sqrt{3}.$$

Hence with $\alpha = -1$, $\beta = \sqrt{3}$, a general solution for (9) is

$$y(x) = c_1 e^{-x} \cos(\sqrt{3}\,x) + c_2 e^{-x} \sin(\sqrt{3}\,x). \quad ●$$

When the auxiliary equation has complex conjugate roots, the (real) solutions oscillate between positive and negative values. This type of behavior is observed in vibrating springs.

● EXAMPLE 2 In the study of a vibrating spring with damping, we are led to an initial value problem of the form

(10) $mx''(t) + bx'(t) + kx(t) = 0; \qquad x(0) = x_0, \qquad x'(0) = v_0,$

where m is the mass of the spring system, b is the damping constant, k is the spring constant, x_0 is the initial displacement, v_0 is the initial velocity, and $x(t)$ is the displacement from equilibrium of the spring system at time t (see Figure 4.4). Determine the equation of motion for this spring system when $m = 36$ kg, $b = 12$ kg/sec, $k = 37$ kg/sec^2, $x_0 = 70$ centimeters (cm), and $v_0 = 10$ cm/sec. Also find $x(10)$, the displacement after 10 sec.

Solution The equation of motion is given by $x(t)$, the solution of the initial value problem (10) for the specified values of m, b, k, x_0, and v_0. That is, we seek the solution to

(11) $36x'' + 12x' + 37x = 0; \qquad x(0) = 70, \qquad x'(0) = 10.$

The auxiliary equation for (11) is

$$36r^2 + 12r + 37 = 0,$$

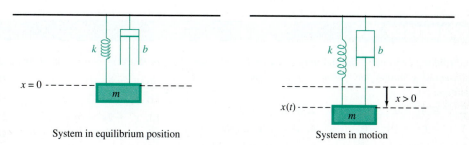

Figure 4.4 Vibrating spring with damping

which has roots

$$r = \frac{-12 \pm \sqrt{144 - 4(36)(37)}}{72} = \frac{-12 \pm 12\sqrt{1 - 37}}{72} = -\frac{1}{6} \pm i.$$

Hence with $\alpha = -\frac{1}{6}$, $\beta = 1$, the displacement $x(t)$ can be expressed in the form

(12) $\qquad x(t) = c_1 e^{-t/6} \cos t + c_2 e^{-t/6} \sin t.$

We can find c_1 and c_2 by substituting $x(t)$ and $x'(t)$ into the initial conditions given in (11). Differentiating (12), we get a formula for $x'(t)$:

$$x'(t) = \left(-\frac{c_1}{6} + c_2 \right) e^{-t/6} \cos t + \left(-c_1 - \frac{c_2}{6} \right) e^{-t/6} \sin t.$$

Substituting into the initial conditions now results in the system

$$c_1 = 70,$$

$$-\frac{c_1}{6} + c_2 = 10.$$

Upon solving, we find $c_1 = 70$ and $c_2 = \frac{65}{3}$. With these values, the equation of motion is

$$x(t) = 70e^{-t/6} \cos t + \frac{65}{3}e^{-t/6} \sin t,$$

and

$$x(10) = 70e^{-5/3} \cos 10 + \frac{65}{3}e^{-5/3} \sin 10$$
$$\approx -13.32 \text{ cm.} \quad \bullet$$

If the constants a, b, and c in (2) are complex numbers, then the roots r_1, r_2 of the auxiliary equation (1) are, in general, complex numbers, but not necessarily complex conjugates of each other. If $r_1 \neq r_2$, then a general solution is

$$y(x) = c_1 e^{r_1 x} + c_2 e^{r_2 x},$$

where c_1 and c_2 are now arbitrary *complex-valued* constants.

We remark that a complex differential equation can be regarded as a system of two real differential equations, since we can always work separately with its real and imaginary parts. Such systems are discussed in Chapters 9 and 10.

EXERCISES 4.6

In Problems 1 through 8 the auxiliary equation for the given differential equation has complex roots. Find a general solution.

1. $y'' + 4y = 0.$

2. $y'' + y = 0.$

3. $z'' - 6z' + 10z = 0.$

4. $y'' - 10y' + 26y = 0.$

5. $w'' + 4w' + 6w = 0.$

6. $y'' - 4y' + 7y = 0.$

7. $4y'' - 4y' + 26y = 0.$

8. $4y'' + 4y' + 6y = 0.$

In Problems 9 through 20 find a general solution.

9. $y'' - 8y' + 7y = 0.$

10. $y'' + 4y' + 8y = 0.$

11. $z'' + 10z' + 25z = 0.$

12. $u'' + 7u = 0.$

13. $y'' + 2y' + 5y = 0.$

14. $y'' - 2y' + 26y = 0.$

15. $y'' + 10y' + 41y = 0.$

16. $y'' - 3y' - 11y = 0.$

17. $y'' - y' + 7y = 0.$

18. $2y'' + 13y' - 7y = 0.$

19. $3v'' + 4v' + 9v = 0.$

20. $9y'' - 12y' + 4y = 0.$

In Problems 21 through 27 solve the given initial value problem.

21. $y'' + 2y' + 2y = 0$; $y(0) = 2$, $y'(0) = 1$.

22. $y'' + 2y' + 17y = 0$; $y(0) = 1$, $y'(0) = -1$.

23. $w'' - 4w' + 2w = 0$; $w(0) = 0$, $w'(0) = 1$.

24. $y'' + 9y = 0$; $y(0) = 1$, $y'(0) = 1$.

25. $y'' - 4y' + 5y = 0$; $y(0) = 1$, $y'(0) = 6$.

26. $y'' - 2y' + y = 0$; $y(0) = 1$, $y'(0) = -2$.

27. $y'' - 2y' + 2y = 0$; $y(\pi) = e^{\pi}$, $y'(\pi) = 0$.

28. To see the effect of changing the parameter b in the initial value problem

$$y'' + by' + 4y = 0; \qquad y(0) = 1, \qquad y'(0) = 0,$$

solve the problem for $b = 5$, 4, and 2 and sketch the solutions.

29. Find a general solution to the following third order equations.

(a) $y''' - y'' + y' + 3y = 0$.

(b) $y''' + 2y'' + 5y' - 26y = 0$.

30. Using the representation for $e^{(\alpha + i\beta)x}$ in (6), verify the differentiation formula (7).

31. Verify, by directly substituting into equation (2), that $e^{\alpha x} \cos \beta x$ and $e^{\alpha x} \sin \beta x$, where α, β are given in (3), are solutions to equation (2).

In Problems 32 through 35 find a general solution to the following Cauchy-Euler equation for $x > 0$ using the substitution $x = e^t$ as discussed in Section 4.5.

32. $y''(x) - \dfrac{1}{x} y'(x) + \dfrac{5}{x^2} y(x) = 0$.

33. $x^2 y''(x) - 3xy'(x) + 6y(x) = 0$.

34. $x^2 y''(x) + 9xy'(x) + 17y(x) = 0$.

35. $x^2 y''(x) + 3xy'(x) + 5y(x) = 0$.

36. Vibrating Spring without Damping. A vibrating spring without damping can be modeled by the initial value problem (10) in Example 2 by taking $b = 0$.

(a) If $m = 10$ kg, $k = 250$ kg/sec^2, $x_0 = 30$ cm, and $v_0 = -10$ cm/sec, find the equation of motion for this undamped vibrating spring.

(b) When the equation of motion is of the form displayed in (8), the motion is said to be **oscillatory** with **frequency** $\beta/2\pi$. Find the frequency of oscillation for the spring system of part (a).

37. Vibrating Spring with Damping. Using the model for a vibrating spring with damping discussed in Example 2:

(a) Find the equation of motion for the vibrating spring with damping if $m = 10$ kg, $b = 60$ kg/sec, $k = 250$ kg/sec^2, $x_0 = 30$ cm, and $v_0 = -10$ cm/sec.

(b) Find the frequency of oscillation for the spring system of part (a). [Hint: See the definition of frequency given in Problem 36(b).]

(c) Comparing the results of Problems 36 and 37, what effect does the damping have on the frequency of oscillation? What other effects does it have on the solution?

38. *RLC* Series Circuit. In the study of an electrical circuit consisting of a resistor, capacitor, inductor, and an electromotive force (see Figure 4.5), we are led to an initial value problem of the form

(13)
$$L\dfrac{dI}{dt} + RI + \dfrac{q}{C} = E(t);$$
$$q(0) = q_0,$$
$$I(0) = I_0,$$

where L is the inductance in henrys, R is the resistance in ohms, C is the capacitance in farads, $E(t)$ is the electromotive force in volts, $q(t)$ is the charge in coulombs on the capacitor at time t, and $I = dq/dt$ is the current in amperes. Find the current at time t if the charge on the capacitor is initially zero, the initial current is zero, $L = 10$ henrys, $R = 20$ ohms, $C = (6260)^{-1}$ farads, and $E(t) = 100$ volts. [Hint: Differentiate both sides of the differential equation in (13) to obtain a homogeneous linear second order equation for $I(t)$. Then use (13) to determine dI/dt at $t = 0$.]

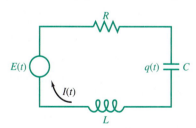

Figure 4.5 *RLC* series circuit

39. Swinging Door. The motion of a swinging door with an adjustment screw that controls the amount of friction on the hinges is governed by the initial value problem

$$I\theta'' + b\theta' + k\theta = 0; \qquad \theta(0) = \theta_0, \qquad \theta'(0) = v_0,$$

where θ is the angle that the door is open, I is the moment of inertia of the door about its hinges, $b > 0$ is a damping constant that varies with the amount of friction on the door, $k > 0$ is the spring constant associated with the swinging door, θ_0 is the initial angle that the door is opened, and v_0 is the initial angular velocity imparted to the door (see Figure 4.6). If I and k are fixed, determine for which values of b the door will *not* continually swing back and forth when closing.

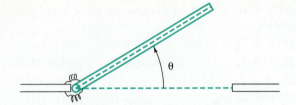

Figure 4.6 Top view of swinging door

• 4.7 • SUPERPOSITION AND NONHOMOGENEOUS EQUATIONS

The differential operator L defined by

(1) $$L[y](x) := y'' + p(x)y' + q(x)y$$

can be viewed as a "black box" with input the function $y(x)$ and output the function on the right-hand side of (1) (see Figure 4.7).

Suppose the input functions $y_1(x)$ and $y_2(x)$ yield, respectively, the output functions $g_1(x)$ and $g_2(x)$; that is,

$$L[y_1](x) = g_1(x), \qquad L[y_2](x) = g_2(x).$$

Then, since L is a linear operator, an input consisting of the linear combination $c_1 y_1(x) + c_2 y_2(x)$ produces the output $c_1 g_1(x) + c_2 g_2(x)$ in the same combination. Stated in terms of solutions to differential equations, we have the following result.

> **SUPERPOSITION PRINCIPLE**
>
> **Theorem 5.** Let y_1 be a solution of the differential equation
>
> $$L[y](x) = g_1(x)$$
>
> and let y_2 be a solution of
>
> $$L[y](x) = g_2(x),$$
>
> where L is a linear differential operator. Then for any constants c_1 and c_2, the function $c_1 y_1 + c_2 y_2$ is a solution to the differential equation
>
> $$L[y](x) = c_1 g_1(x) + c_2 g_2(x).$$

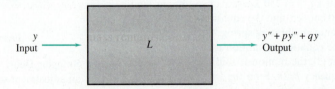

Figure 4.7 Black box diagram for L

● **EXAMPLE 1**   Given that $y_1(x) = -x/3 - \frac{2}{9}$ is a solution to

$$y'' + 2y' - 3y = x,$$

and $y_2(x) = e^{2x}/5$ is a solution to

$$y'' + 2y' - 3y = e^{2x},$$

find a solution to

(2) $y'' + 2y' - 3y = 4x - 5e^{2x}.$

Solution   Let $L[y] := y'' + 2y' - 3y$. We are given that

$$L[y_1](x) = g_1(x) = x \quad \text{and} \quad L[y_2](x) = g_2(x) = e^{2x}.$$

Since we can express

$$4x - 5e^{2x} = 4g_1(x) - 5g_2(x),$$

then, by the superposition principle, the function

$$4y_1(x) - 5y_2(x) = 4\left(\frac{-x}{3} - \frac{2}{9}\right) - 5\left(\frac{e^{2x}}{5}\right)$$

$$= \frac{-4x}{3} - \frac{8}{9} - e^{2x}$$

is a solution to equation (2).   ●

If we combine the superposition principle with the representation theorem for solutions of homogeneous equations, we can prove that *all* solutions of a *non*homogeneous equation can be found if we know just one of its solutions (called a *particular solution*) and a general solution to the corresponding homogeneous equation.

> **REPRESENTATION OF SOLUTIONS (NONHOMOGENEOUS CASE)**
>
> **Theorem 6.**   Let $y_p(x)$ be a particular solution to the nonhomogeneous equation
>
> **(3)** $y'' + p(x)y' + q(x)y = g(x),$
>
> on the interval (a, b) and let $y_1(x)$, $y_2(x)$ be linearly independent solutions on (a, b) of the corresponding homogeneous equation
>
> **(4)** $y'' + p(x)y' + q(x)y = 0.$
>
> Then every solution of (3) on the interval (a, b) can be expressed in the form
>
> **(5)** $y(x) = y_p(x) + c_1 y_1(x) + c_2 y_2(x).$

The linear combination of y_p, y_1, and y_2 in (5), written with arbitrary constants c_1 and c_2, is referred to as a **general solution** to (3).

Proof of Theorem 6.   Let $\phi(x)$ be any solution to (3). Since both $\phi(x)$ and $y_p(x)$ are solutions to (3), the superposition principle states that the difference $\phi(x) - y_p(x)$ is a solution to the homogeneous equation (4). It then follows from Theorem 3 of Section 4.3 that

$$\phi(x) - y_p(x) = c_1 y_1(x) + c_2 y_2(x)$$

for suitable constants c_1 and c_2. Since the last equation is equivalent to (5) (with $\phi(x)$ in place of $y(x)$), the theorem is proved. ◀◀◀

PROCEDURE FOR SOLVING NONHOMOGENEOUS EQUATIONS

To solve $y'' + py' + qy = g$:

(a) Determine a general solution, $c_1 y_1 + c_2 y_2$, of the corresponding homogeneous equation.
(b) Find a particular solution, y_p, of the given nonhomogeneous equation.
(c) A general solution of the given equation is the sum of a particular solution and a general solution to the homogeneous equation; that is,

$$y = y_p + c_1 y_1 + c_2 y_2.$$

● EXAMPLE 2 Given that $y_p(x) = x^2$ is a particular solution to

(6) $$y'' - y = 2 - x^2,$$

find a general solution of (6).

Solution The corresponding homogeneous equation,

$$y'' - y = 0,$$

has the associated auxiliary equation $r^2 - 1 = 0$. Since $r = \pm 1$ are the roots of this equation, a general solution to the homogeneous equation is $c_1 e^x + c_2 e^{-x}$. Combining this with the particular solution $y_p(x) = x^2$ of the nonhomogeneous equation (6), we find that a general solution is

$$y(x) = x^2 + c_1 e^x + c_2 e^{-x}. \quad ●$$

In Sections 4.8, 4.9, and 4.10 we study methods for actually finding a particular solution to a nonhomogeneous linear second order differential equation.

EXERCISES 4.7

In Problems 1 through 10 a nonhomogeneous equation and a particular solution are given. Find a general solution for the equation.

1. $y'' - y = x, \quad y_p(x) = -x.$

2. $y'' + y' = 1, \quad y_p(x) = x.$

3. $\theta'' - \theta' - 2\theta = 1 - 2x, \quad \theta_p(x) = x - 1.$

4. $\dfrac{d^2x}{dt^2} - 4\dfrac{dx}{dt} + 3x = -2e^t, \quad x_p(t) = te^t.$

5. $y'' + 2y' + 4y - 4\cos 2x = 0, \quad y_p(x) = \sin 2x.$

6. $\dfrac{d^2y}{dx^2} + y = 5e^{2x}, \quad y_p(x) = e^{2x}.$

7. $\dfrac{d^2\theta}{dt^2} - \dfrac{d\theta}{dt} + \theta = \sin t, \quad \theta_p(t) = \cos t.$

8. $y'' + 5y' + 6y = 6x^2 + 10x + 2 + 12e^x,$
$y_p(x) = e^x + x^2.$

9. $y'' = 2y' - y + 2e^x, \quad y_p(x) = x^2 e^x.$

10. $y'' = 2y + 2\tan^3 x, \quad y_p(x) = \tan x.$

11. Given that $y_1(x) = \cos x$ is a solution to

$$y'' - y' + y = \sin x,$$

and $y_2(x) = e^{2x}/3$ is a solution to

$$y'' - y' + y = e^{2x},$$

find solutions to the following differential equations:

(a) $y'' - y' + y = 5\sin x$.
(b) $y'' - y' + y = \sin x - 3e^{2x}$.
(c) $y'' - y' + y = 4\sin x + 18e^{2x}$.

12. Let $L[y] := y'' + 2y' + 4y$, $y_1(x) := \frac{1}{4}\sin 2x$, and $y_2(x) := x/4 - \frac{1}{8}$. Verify that $L[y_1](x) = \cos 2x$ and $L[y_2](x) = x$. Now use the superposition principle to find solutions to the following differential equations:

(a) $L[y](x) = x + \cos 2x$.
(b) $L[y](x) = 2x - 3\cos 2x$.
(c) $L[y](x) = 11x - 12\cos 2x$.
(d) $L[y](x) = 0$.

13. Let $L[y] := y'' - 2y$, $y_1(x) := (\tan x)/2$, $y_2(x) := -x/2$, and $y_3(x) := -\frac{1}{2}$. Verify that $L[y_1](x) = \tan^3 x$, $L[y_2](x) = x$, and $L[y_3](x) = 1$. Find a general solution to each of the following differential equations:

(a) $L[y](x) = \tan^3 x - 1$.
(b) $L[y](x) = x - 2$.
(c) $L[y](x) = 2\tan^3 x + 3x - 4$.
(d) $L[y](x) = (x - \tan^3 x)/4$.

14. Let $L[y] := y'' - 4y' + 3y$, $y_1(x) := x$, and $y_2(x) := e^{-x}$. Verify that $L[y_1](x) = 3x - 4$ and $L[y_2](x) = 8e^{-x}$. Find the solution to the given initial value problem.

(a) $L[y](x) = 8e^{-x} + 8 - 6x$;
$y(0) = 2, \quad y'(0) = -2$.
(b) $L[y](x) = 3x - 4 - 8e^{-x}$; $\quad y(0) = 1, \quad y'(0) = 6$.

15. Equivalence of General Solutions. Let y_p and $\hat{y}_p$ be two particular solutions to $y'' + py' + qy = g$ and let $\{y_1, y_2\}$ be a fundamental solution set to the corresponding homogeneous equation. Show that any function of the form $y_p + c_1 y_1 + c_2 y_2$ can be written in the form $\hat{y}_p + k_1 y_1 + k_2 y_2$ for suitable choice of constants k_1, k_2 (and vice-versa). Consequently, it makes no difference which particular solution is used in the expression for the general solution of a nonhomogeneous equation.

16. Third Order Superposition Principle. Let $L[y] := y''' + py'' + qy' + ry$. Show that if $L[y_1] = g_1$ and $L[y_2] = g_2$, then for any constants c_1 and c_2, the function $c_1 y_1 + c_2 y_2$ is a solution of the differential equation $L[y] = c_1 g_1 + c_2 g_2$.

17. Let $L[y] := y''' + 2y'' - 13y' + 10y$, $y_1(x) := x^2$, and $y_2(x) := x^3$. Verify that $L[y_1](x) = 10x^2 - 26x + 4$ and $L[y_2](x) = 10x^3 - 39x^2 + 12x + 6$. Using the third order superposition principle (see Problem 16), find a general solution to each of the following differential equations.

(a) $L[y](x) = -10x^3 + 59x^2 - 64x + 2$.
(b) $L[y](x) = 20x^3 - 68x^2 - 2x + 16$.

18. Reduction of Order. The reduction of order procedure (see Section 4.4) can also be applied to the nonhomogeneous equation

$$y'' + p(x)y' + q(x)y = g(x).$$

(a) Show that the substitution $y = v(x)f(x)$, where $f(x)$ is a known nontrivial solution to the corresponding *homogeneous* equation, reduces the nonhomogeneous equation to the linear first order equation

$$fw' + (2f' + pf)w = g,$$

where $w = v'$. (Such an equation can be solved using the techniques of Section 2.4.)

(b) Use the procedure in part (a) to find a general solution to

$$y'' + x^{-1}y' - 4x^{-2}y = 1 - x^{-3}, \qquad x > 0,$$

given that $f(x) = x^2$ is a solution to the corresponding homogeneous equation.

19. There is another type of reduction of order procedure that can be applied to equations of the form $L[y](x) = g(x)$, provided that L is a differential operator that can be written as the product of first order factors. To illustrate this method consider the equation

(7) $\quad L[y](x) := y'' - y' - 2y = e^x$.

Notice that $L[y] = (D^2 - D - 2)y$ can be written as $L[y] = (D + 1)(D - 2)y$. Let $v = (D - 2)y$ and observe that solving (7) is equivalent to solving the two first order equations

(8) $\quad (D + 1)v = e^x$,

(9) $\quad (D - 2)y = v$,

which, in turn, can be solved using the method dis-discussed in Section 2.4. Carry out this procedure by completing the following steps.

(a) Solve equation (8) for v.
(b) Now substitute the v obtained in part (a) into equation (9) and solve for y. This gives a general solution to (7).

20. Use the method of Problem 19 to find a general solution to

$$y'' + 6y' + 5y = 10x + 5.$$

• 4.8 • METHOD OF UNDETERMINED COEFFICIENTS

In this section we give a simple procedure for finding a particular solution to a nonhomogeneous linear equation with constant coefficients when the nonhomogeneous term $g(x)$ is of a special type. To motivate the procedure, let's first look at a few examples.

● **EXAMPLE 1**   Find a particular solution to

$$L[y](x) := y'' + 3y' + 2y = 3x + 1.$$

Solution   We want to find a function $y_p(x)$ such that $L[y_p](x) = 3x + 1$. Notice that if L is applied to any linear function $y_p = Ax + B$, we obtain the linear function

$$L[y_p](x) = 0 + 3A + 2(Ax + B) = 2Ax + (3A + 2B).$$

Let's therefore try to solve the equation

$$2Ax + (3A + 2B) = 3x + 1$$

for the constants A and B. Since two polynomials are equal when their corresponding coefficients are equal, we set

$$2A = 3,$$
$$3A + 2B = 1.$$

Solving this system gives $A = \frac{3}{2}$ and $B = -\frac{7}{4}$. Thus the function

$$y_p(x) = \frac{3}{2}x - \frac{7}{4}$$

is a particular solution since $L[y_p](x) = 3x + 1$.   ●

Throughout this section, L denotes a linear differential operator with **constant coefficients.** For such operators, Example 1 suggests the following method for finding a particular solution to the equation

$$L[y](x) = p_n(x),$$

where $p_n(x)$ is a polynomial of degree n. Guess a solution of the form

$$y_p(x) = A_n x^n + \cdots + A_1 x + A_0$$

and solve

$$L[y_p](x) = p_n(x),$$

for the coefficients $A_0, A_1, \ldots, A_n$. This procedure involves solving $n + 1$ linear equations in the $n + 1$ unknowns $A_0, \ldots, A_n$ and usually has a solution. The technique is called the **method of undetermined coefficients.** [†]

[†] *Historical Footnote:* This method was developed by Leonhard Euler in his theory of the motion of the moon in 1753.

● **EXAMPLE 2**   Find a particular solution to

$$L[y](x) := y'' + 3y' + 2y = e^{3x}.$$

Solution   Here we seek a function y_p such that $L[y_p](x) = e^{3x}$. If we guess $y_p(x) = Ae^{3x}$, where A is a constant, then

$$L[y_p](x) = 9Ae^{3x} + 3(3Ae^{3x}) + 2(Ae^{3x}) = 20Ae^{3x}.$$

Setting $20Ae^{3x} = e^{3x}$ and solving for A gives $A = \frac{1}{20}$. Hence

$$y_p(x) = \frac{e^{3x}}{20}$$

is a particular solution since $L[y_p](x) = e^{3x}$.   ●

The method of undetermined coefficients will also work for equations of the form

$$L[y](x) = ae^{\alpha x},$$

where a and α are given constants. Here we would guess

$$y_p(x) = Ae^{\alpha x}$$

and solve $L[y_p](x) = ae^{\alpha x}$ for the unknown coefficient A.

● **EXAMPLE 3**   Find a particular solution to

(1) $$L[y](x) := y'' - y' - y = \sin x.$$

Solution   This time we seek a function y_p that satisfies $L[y_p](x) = \sin x$. Our initial action might be to guess $y_p(x) = A \sin x$, in which case we get

$$L[y_p](x) = -2A \sin x - A \cos x.$$

However, since the right-hand side of (1) involves only $\sin x$, this choice of solution would force A (and hence y_p) to be zero, which is absurd!
Instead, to compensate for the $\cos x$ term, let's guess

$$y_p(x) = A \cos x + B \sin x,$$

so that

$$L[y_p](x) = (-2A - B) \cos x + (A - 2B) \sin x.$$

Then setting

$$(-2A - B) \cos x + (A - 2B) \sin x = \sin x$$

and equating the coefficients of $\cos x$ and of $\sin x$ leads to the system of equations

$$-2A - B = 0,$$
$$A - 2B = 1.$$

Solving for A and B yields $A = \frac{1}{5}$ and $B = -\frac{2}{5}$. Hence

$$y_p(x) = \tfrac{1}{5} \cos x - \tfrac{2}{5} \sin x$$

is a particular solution to equation (1).   ●

More generally, for an equation of the form

$$L[y](x) = a\cos\beta x + b\sin\beta x,$$

the method of undetermined coefficients suggests that we guess

$$y_p(x) = A\cos\beta x + B\sin\beta x$$

and solve $L[y_p](x) = a\cos\beta x + b\sin\beta x$ for the unknowns A and B.

● **EXAMPLE 4**   Find a particular solution to

(2) $L[y](x) := y'' + y' = 5.$

Solution   Since the right-hand side is the constant polynomial $p_0(x) \equiv 5$, Example 1 suggests that we guess $y_p(x) = A$ and set $L[y_p](x) = 5$. But then we get

$$L[y_p](x) = 0 \neq 5.$$

This unfortunate circumstance occurs because any constant function is a solution to the corresponding homogeneous equation $L[y](x) = 0$. To circumvent this difficulty, we temporarily abandon the method of undetermined coefficients and observe that we can integrate both sides of equation (2) to obtain the linear first order equation

(3) $y' + y = 5x + c_1,$

where c_1 is an arbitrary constant. Using the method of Section 2.4 for solving linear first order equations, we find that

(4) $y(x) = 5x - 5 + c_1 + c_2 e^{-x}$

is a general solution to equation (3) and hence to equation (2).

Notice that with $c_1 = 5$ and $c_2 = 0$ in (4), we get

(5) $y_p(x) = 5x$

as a particular solution to equation (2). This suggests that in our original attempt to use undetermined coefficients we should have guessed

$$y_p(x) = Ax$$

instead of $y_p = A$. Indeed, with $y_p = Ax$, setting $L[y_p](x) = 5$ immediately gives $A = 5$, and we again obtain (5).   ●

In the preceding example the trial choice for y_p did not work, because it was a solution to the corresponding homogeneous equation. However, when we replaced y_p by the new function $\hat{y}_p = xy_p$, we were able to find a particular solution to the nonhomogeneous equation. Consequently the procedure of undetermined coefficients must be modified as follows.

If any term in the trial expression for y_p is a solution to the corresponding homogeneous equation, then replace y_p by $x^s y_p$, where s is the smallest nonnegative integer such that no term in $x^s y_p$ is a solution to the corresponding homogeneous equation.

● **EXAMPLE 5**   Determine the form of a particular solution to

(6) $y'' - y' - 12y = e^{4x}$.

Solution   The auxiliary equation for the corresponding homogeneous equation

(7) $y'' - y' - 12y = 0$

has the roots $r_1 = 4$ and $r_2 = -3$. A general solution for (7) is therefore

$$y_h(x) = c_1 e^{4x} + c_2 e^{-3x}.$$

Now, since the right-hand side of (6) is e^{4x}, we first guess $y_p = Ae^{4x}$. But, because e^{4x} is a solution to equation (7), we replace this choice by $y_p = Axe^{4x}$. Since xe^{4x} is not a solution to equation (7), there exists a particular solution to (6) of the form

$$y_p(x) = Axe^{4x}. ●$$

As our examples illustrate, when the nonhomogeneous term $g(x)$ is a polynomial, an exponential, a sine, or a cosine function, the function $g(x)$ itself suggests the form of the particular solution $y_p(x)$. In fact, we can expand the list of functions $g(x)$ for which the method of undetermined coefficients can be applied to include products of these functions as well. In Table 4.1 we have listed the various forms for $g(x)$ and the corresponding forms for the particular solution $y_p(x)$. Observe that Type VII includes as special cases all the previous types; we have included the first six for easy reference. In using the table, the reader is reminded to choose s as the smallest nonnegative integer such that no term in y_p is a solution to the corresponding homogeneous equation. For linear second order equations, s will be either 0, 1, or 2. An incorrect choice for s leads to an inconsistent system of equations for the coefficients.

TABLE 4.1 THE FORM OF A PARTICULAR SOLUTION $y_p(x)$ OF $L[y](x) = g(x)$ WHEN $L[y]$ HAS CONSTANT COEFFICIENTS

Type	$g(x)$	$y_p(x)$
(I)	$p_n(x) = a_n x^n + \cdots + a_1 x + a_0$	$x^s P_n(x) = x^s \{A_n x^n + \cdots + A_1 x + A_0\}^{\dagger}$
(II)	$ae^{\alpha x}$	$x^s A e^{\alpha x}$
(III)	$a \cos \beta x + b \sin \beta x$	$x^s \{A \cos \beta x + B \sin \beta x\}$
(IV)	$p_n(x)e^{\alpha x}$	$x^s P_n(x) e^{\alpha x}$
(V)	$p_n(x) \cos \beta x + q_m(x) \sin \beta x,$ where $q_m(x) = b_m x^m + \cdots + b_1 x + b_0$	$x^s \{P_N(x) \cos \beta x + Q_N(x) \sin \beta x\},$ where $Q_N(x) = B_N x^N + \cdots + B_1 x + B_0$ and $N = \max(n, m)$
(VI)	$ae^{\alpha x} \cos \beta x + be^{\alpha x} \sin \beta x$	$x^s \{A e^{\alpha x} \cos \beta x + B e^{\alpha x} \sin \beta x\}$
(VII)	$p_n(x)e^{\alpha x} \cos \beta x + q_m(x)e^{\alpha x} \sin \beta x$	$x^s e^{\alpha x} \{P_N(x) \cos \beta x + Q_N(x) \sin \beta x\},$ where $N = \max(n, m)$

The nonnegative integer s is chosen to be the smallest integer so that no term in the particular solution $y_p(x)$ is a solution to the corresponding homogeneous equation $L[y](x) = 0$.

---

† $P_n(x)$ must include all its terms even if $p_n(x)$ has some terms that are zero.

● EXAMPLE 6   Using Table 4.1, find the form for a particular solution y_p to

(8) $y'' + 2y' - 3y = g(x)$,

where $g(x)$ equals

(a) $7 \cos 3x$,   (b) $5e^{-3x}$,   (c) $x^2 \cos \pi x$,   (d) $2xe^x \sin x - e^x \cos x$,
(e) $x^2 e^x + 3xe^x$,   (f) $\tan x$.

Solution   The auxiliary equation for the corresponding homogeneous equation,

(9) $y'' + 2y' - 3y = 0$,

has roots $r_1 = 1$ and $r_2 = -3$. Thus a general solution for (9) is

$$y_h(x) = c_1 e^x + c_2 e^{-3x}.$$

(a) Referring to Table 4.1, the function $g(x) = 7 \cos 3x$ is of Type III (with $a = 7$, $b = 0$, $\beta = 3$). Hence y_p has the form

(10) $y_p(x) = A \cos 3x + B \sin 3x$,

since no term in (10) is a solution to (9).

(b) The function $g(x) = 5e^{-3x}$ is of Type II, and so y_p has the form $x^s A e^{-3x}$. Here we take $s = 1$ since e^{-3x} is a solution to equation (9) and xe^{-3x} is not. Thus

$$y_p(x) = Axe^{-3x}.$$

(c) The function $g(x) = x^2 \cos \pi x$ is of Type V, with $p_2(x) = x^2$, $q_0(x) \equiv 0$. Since $N = \max\{2, 0\} = 2$,

(11) $y_p(x) = (A_2 x^2 + A_1 x + A_0) \cos \pi x + (B_2 x^2 + B_1 x + B_0) \sin \pi x$,

where we have taken $s = 0$ because no term in (11) is a solution to the homogeneous equation.

(d) The function $g(x) = 2xe^x \sin x - e^x \cos x$ is of Type VII, with $p_0(x) = -1$ and $q_1(x) = 2x$. For these polynomials, $N = \max\{0, 1\} = 1$, so

$$x^s e^x \{(A_1 x + A_0) \cos x + (B_1 x + B_0) \sin x\}$$

is the form for the particular solution. Since none of the terms $e^x x \cos x$, $e^x \cos x$, $e^x x \sin x$, and $e^x \sin x$ is a solution of the homogeneous equation (9), we set $s = 0$ and obtain

$$y_p(x) = e^x \{(A_1 x + A_0) \cos x + (B_1 x + B_0) \sin x\}.$$

(e) Here $g(x) = x^2 e^x + 3xe^x = (x^2 + 3x)e^x$ is of Type IV with $p_2(x) = x^2 + 3x$. Thus

$$y_p(x) = x(A_2 x^2 + A_1 x + A_0)e^x,$$

where we have taken $s = 1$ since the term $A_0 e^x$ in

$$(A_2 x^2 + A_1 x + A_0)e^x$$

is a solution to equation (9).

(f) Unfortunately, $g(x) = \tan x$ is not one of the forms for which the method of undetermined coefficients can be used. In Section 4.10 we discuss a different method
that can handle such nonhomogeneous terms (also see Project A, the Convolution Method).   ●

Let's summarize:

▶ **METHOD OF UNDETERMINED COEFFICIENTS**

To determine a particular solution to $L[y] = g$:

(a) Check that the linear equation has constant coefficients and that the
nonhomogeneous term $g(x)$ is one of the types suitable for the method.
(b) Solve the corresponding homogeneous equation $L[y](x) = 0$.
(c) Based on the form of the nonhomogeneous term $g(x)$, determine the
appropriate form of the particular solution $y_p(x)$. Remember that no term in
the trial expression for $y_p(x)$ may be a solution to the corresponding
homogeneous equation.
(d) Since we want $L[y_p](x) = g(x)$, set the corresponding coefficients from both
sides of this equation equal to each other to form a system of linear equations.
(e) Solve this system of linear equations for the coefficients of $y_p(x)$.

If we combine the method of undetermined coefficients with the superposition principle, we can solve for $y_p(x)$ when $g(x)$ is a finite *sum* of terms like those listed in Table 4.1.

● **EXAMPLE 7**   Find a general solution to the equation

(12) $y'' - y = 2e^{-x} - 4xe^{-x} + 10 \cos 2x.$

Solution   We first solve the associated homogeneous equation $y'' - y = 0$ and obtain as a general
solution $y_h(x) = c_1 e^x + c_2 e^{-x}$.

Notice that the right-hand side of (12) is not one of the forms given in Table 4.1 but
is the sum of two such forms. This suggests that we use the superposition principle and
consider separately the equations

(13) $y'' - y = 2e^{-x} - 4xe^{-x}$

and

(14) $y'' - y = 10 \cos 2x.$

A particular solution to equation (13) has the form $y_q(x) = x^s(A_1 x + A_0)e^{-x}$. Since
$A_0 e^{-x}$ is a solution to the corresponding homogeneous equation, we take $s = 1$. Then

$$y_q(x) = (A_1 x^2 + A_0 x)e^{-x}.$$

Substituting y_q for y in (13) gives

(15) $-4A_1 xe^{-x} + (2A_1 - 2A_0)e^{-x} = -4xe^{-x} + 2e^{-x}.$

When we equate the coefficients of xe^{-x} and e^{-x} on both sides of equation (15), we obtain

the system

$$-4A_1 = -4,$$
$$2A_1 - 2A_0 = 2.$$

Solving, we find $A_1 = 1$ and $A_0 = 0$; thus

$$y_q(x) = x^2 e^{-x}$$

is a particular solution to (13).

Next, consider equation (14). A particular solution to this equation has the form

$$y_r(x) = A\cos 2x + B\sin 2x$$

(here we can take $s = 0$). Substituting y_r for y in (14) gives

(16) $$-5A\cos 2x - 5B\sin 2x = 10\cos 2x,$$

which yields $A = -2$ and $B = 0$. Thus

$$y_r(x) = -2\cos 2x.$$

It follows from the superposition principle that a particular solution to the original equation (12) is given by the sum $y_q + y_r$; that is,

$$y_p(x) = x^2 e^{-x} - 2\cos 2x.$$

Hence a general solution to equation (12), given by the sum of the general solution to the corresponding homogeneous equation and a particular solution, is

$$y(x) = c_1 e^x + c_2 e^{-x} + x^2 e^{-x} - 2\cos 2x. \quad \bullet$$

Success in applying the method of undetermined coefficients lies in the ability to guess the form of a particular solution. As we have seen, when L is a second order linear differential operator with constant coefficients, there is a class of functions $g(x)$ for which the choice of the form of a particular solution to the equation $L[y](x) = g(x)$ is quite easy to determine. This class consists of functions of the form

$$p_n(x)e^{\alpha x}\cos\beta x + q_m(x)e^{\alpha x}\sin\beta x,$$

where $p_n(x)$ and $q_m(x)$ are polynomials. The method works because when we apply the operator L to a function of this form we get back a function of the same type. As you might have guessed, these functions are themselves solutions of some (higher order) homogeneous linear differential equation with constant coefficients. We verify in Chapter 6 that the method of undetermined coefficients applies to such functions.

EXERCISES 4.8

In Problems 1 through 16 find a particular solution to the differential equation.

1. $y'' + 2y = -8.$

2. $y'' + 2y' - y = 10.$

3. $z'' + z = 5e^{2x}.$

4. $x'' - x = 3e^{-2t}.$

5. $2x' + x = 3t^2 + 10t.$

6. $y'' - y' - 2y = -2x^3 - 3x^2 + 8x + 1.$

7. $y'' - y' + 9y = 3\sin 3x.$

8. $y'' + y' + y = 2\cos 2x - 3\sin 2x.$

9. $\dfrac{d^2\theta}{dr^2} - 5\dfrac{d\theta}{dr} + 6\theta = re^r.$

10. $\dfrac{d^2x}{dt^2} + 2\dfrac{dx}{dt} - 8x = te^{-t} + e^{-t}.$

11. $\theta'' - \theta = t\sin t.$

12. $y'' + y = 2^x.$

13. $y'' - 2y' + y = 8e^x.$

14. $w'' - 4w' = x^2 - 1.$

15. $z'' - 6z' + 9z = x^2 + e^x.$

16. $y'' - 4y' - 5y = 2e^{-x}.$

In Problems 17 through 30 find a general solution to the differential equation.

17. $y'' - y = -11x + 1.$

18. $\theta'' + \theta = 7x + 5.$

19. $y'' + y' - 2y = x^2 - 2x + 3.$

20. $y'' - 2y' - 3y = 3x^2 - 5.$

21. $y'' - 3y' + 2y = e^t \sin t.$

22. $y'' + 4y = \sin\theta - \cos\theta.$

23. $y'' + 2y' + 2y = e^{-\theta}\cos\theta.$

24. $y'' + 6y' + 10y = 10x^4 + 24x^3 + 2x^2 - 12x + 18.$

25. $x'' - 4x' + 4x = te^{2t}.$

26. $y'' - 7y' = x^2.$

27. $y'' - 3y = x^2 - e^x.$

28. $y'' + 4y' + 5y = e^{-x} - \sin 2x.$

29. $y'' + y' + y = \cos x - x^2 e^x.$

30. $y'' - 2y' - 35y = 13\sin x - e^{3x} + 1.$

In Problems 31 through 40 find the solution to the initial value problem.

31. $y' - y = 1, \quad y(0) = 0.$

32. $h'' = 6x; \quad h(0) = 3, \quad h'(0) = -1.$

33. $z'' + z = 2e^{-x}; \quad z(0) = 0, \quad z'(0) = 0.$

34. $y'' + 9y = 27; \quad y(0) = 4, \quad y'(0) = 6.$

35. $y'' - y' - 2y = \cos x - \sin 2x;$
$y(0) = -\frac{7}{20}, \quad y'(0) = \frac{1}{5}.$

36. $y'' - y = e^\theta - e^{-\theta} + 2; \quad y(0) = 0, \quad y'(0) = 0.$

37. $y'' - 7y' + 10y = x^2 - 4 + e^x;$
$y(0) = 3, \quad y'(0) = -3.$

38. $y'' + y' - 12y = e^x + e^{2x} - 1;$
$y(0) = 1, \quad y'(0) = 3.$

39. $y'' - y = \sin\theta - e^{2\theta}; \quad y(0) = 1, \quad y'(0) = -1.$

40. $y'' + 2y' + y = x^2 + 1 - e^x; \quad y(0) = 0, \quad y'(0) = 2.$

In Problems 41 through 48 determine the form of a particular solution for the differential equation.

41. $y'' + y = \sin x + x\cos x + 10^x.$

42. $y'' - y = e^{2x} + xe^{2x} + x^2e^{2x}.$

43. $x'' - x' - 2x = e^t\cos t - t^2 + t + 1.$

44. $y'' + 5y' + 6y = \sin x - \cos 2x.$

45. $y'' - 4y' + 5y = e^{5x} + x\sin 3x - \cos 3x.$

46. $y'' - 4y' + 4y = x^2e^{2x} - e^{2x}.$

47. $v'' - v = e^u - 7 + \cos u.$

48. $y'' + 2y' + 2y = x^2 + e^{-x}\cos x + xe^{-x}\sin x.$

In Problems 49 through 56 decide whether the method of undetermined coefficients (possibly together with superposition) can be applied to find a particular solution of the given equation.

49. $y'' + 2y' - y = x^{-1}e^x.$

50. $y'' + 3y' - y = \sec\theta.$

51. $x''(t) - 5x'(t) + 2x(t) = 3^t.$

52. $x''(t) + tx'(t) + x(t) = \sin t.$

53. $xy'' - y' + 2y = \cos x.$

54. $y'' - y' + 2y = x^2 + \sin x.$

55. $u' + 2u = (x^2 + 1)e^x.$

56. $y'' + y' + 2y = \cos^2 x.$

In Problems 57 and 58 find a particular solution to the given third order equation.

57. $y''' - y'' + y = \sin x.$

58. $2y''' + 3y'' + y' - 4y = e^{-x}.$

In Problems 59 through 64 use the method discussed under "Cauchy-Euler equations" in Section 4.5 to find a general solution to the given equation for $x > 0$.

59. $xy''(x) + 3y'(x) - \dfrac{3}{x}y(x) = x^2.$

60. $x^2y''(x) - y(x) = (\ln x)^2 - 1.$

61. $x^4y''(x) - 6x^2y(x) = 1 - 6x^2.$

62. $xy''(x) + y'(x) = 9x^2.$

63. $x^2y''(x) + xy'(x) + y(x) = (\ln x)\sin(\ln x).$

64. $x^2y''(x) + 3xy'(x) + 4y(x) = \cos(4\ln x).$

65. Discontinuous Forcing Term. In certain physical models the nonhomogeneous term, or **forcing term,** $g(x)$ in the equation $L[y](x) = g(x)$ may not be continuous, but may have a jump discontinuity. If this occurs, we can still obtain a reasonable solution using the following procedure. Consider the initial value problem

$$y'' + 2y' + 5y = g(x); \qquad y(0) = 0, \qquad y'(0) = 0,$$

where

$$g(x) := \begin{cases} 10, & 0 \le x \le \dfrac{3\pi}{2}, \\ 0, & x > \dfrac{3\pi}{2}. \end{cases}$$

(a) Find a solution to the initial value problem for $0 \le x \le 3\pi/2$.

(b) Find a general solution for $x > 3\pi/2$.

(c) Now choose the constants in the general solution from part (b) so that the solution from part (a) and the solution from part (b) agree at $x = 3\pi/2$. This gives us a continuous function that satisfies the differential equation except at $x = 3\pi/2$.

Problems 66 through 68 concern justifications for the method of undetermined coefficients.

66. Consider the nonhomogeneous equation

(17) $$ay'' + by' + cy = a_n x^n + \cdots + a_1 x + a_0,$$

where $a_n \ne 0$. We want to determine the form of a particular solution y_p to equation (17).

(a) Assume

(18) $$y_p(x) = A_n x^n + \cdots + A_0,$$

and substitute y_p into equation (17) to obtain, after simplifying,

$$cA_n x^n + (cA_{n-1} + nbA_n)x^{n-1} + \cdots + (2aA_2 + bA_1 + cA_0)$$
$$= a_n x^n + \cdots + a_0.$$

(b) Show that if $c \ne 0$, then no (nonzero) term in (18) is a solution to the homogeneous equation

(19) $$ay'' + by' + cy = 0.$$

(c) Show that if $c \ne 0$, then you can solve for the constants $A_n, A_{n-1}, \ldots, A_0$ in terms of the constants $a_n, a_{n-1}, \ldots, a_0$; hence there exists a solution to equation (17) of the form (18).

67. Consider the nonhomogeneous equation

(20) $$ay'' + by' + cy = e^{\alpha x}(a_n x^n + \cdots + a_0).$$

We want to determine the form of a particular solution y_p to equation (20).

(a) Let $y_p(x) = e^{\alpha x} v(x)$. Substitute for y, y', and y'' in equation (20) and divide by the nonzero factor $e^{\alpha x}$ to obtain

(21) $$av'' + (2a\alpha + b)v' + (a\alpha^2 + b\alpha + c)v$$
$$= a_n x^n + \cdots + a_0.$$

(b) Using the results from Problem 66, show that if $e^{\alpha x}$ is not a solution to the homogeneous equation (19), then equation (20) has a solution of the form

$$y_p(x) = e^{\alpha x}(A_n x^n + \cdots + A_0).$$

68. Annihilator Method. The annihilator method, which is discussed in Chapter 6, can be used to justify the method of undetermined coefficients. To illustrate the annihilator approach, consider the equation

(22) $$(D - 1)(D + 1)y = y'' - y = e^{3x}.$$

(a) Show that applying $(D - 3)$ to both sides of equation (22) gives

(23) $$(D - 3)(D - 1)(D + 1)y$$
$$= y''' - 3y'' - y' + 3y = 0.$$

(Since $(D - 3)e^{3x} \equiv 0$, we say that $(D - 3)$ "annihilates" e^{3x}.)

(b) If y_p satisfies equation (22), it will also satisfy equation (23). Now verify that

(24) $$c_1 e^{3x} + c_2 e^x + c_3 e^{-x}$$

is a general solution to (23) and conclude that there exists a particular solution to (22) of the form (24).

(c) Using the fact that $c_2 e^x + c_3 e^{-x}$ is a solution to the homogeneous equation

$$(D - 1)(D + 1)y = 0,$$

conclude that there must be a particular solution to (22) of the form $y_p(x) = c_1 e^{3x}$.

(d) Use the above procedure to show that there is a particular solution to

$$(D - 1)(D + 1)y = e^x$$

of the form $y_p(x) = Cxe^x$. [Hint: Use the annihilator $(D - 1)$.]

69. Forced Vibrations. A vibrating spring with damping that is under external force can be modeled by

(25) $$mx'' + bx' + kx = g(t),$$

where $m > 0$ is the mass of the spring system, $b > 0$ is the damping constant, $k > 0$ is the spring constant, $g(t)$ is the force on the system at time t, and $x(t)$ is the displacement from the equilibrium of the spring system at time t. Assume $b^2 < 4mk$.

(a) Determine the form of the equation of motion for the spring system when $g(t) = \sin \beta t$ by finding a general solution to equation (25).

(b) Discuss the long-term behavior of this system. [Hint: Consider what happens to the general solution obtained in part (a) as $t \to +\infty$.]

SKIP

• 4.9 • UNDETERMINED COEFFICIENTS USING COMPLEX ARITHMETIC (Optional)

The technique of undetermined coefficients described in the preceding section can be streamlined by using complex arithmetic and properties of the complex exponential function. Although we have previously mentioned complex numbers in the text, here we shall actually be computing with them. Therefore, we begin by briefly reviewing the basic notation and manipulation of complex numbers.

An expression of the form $a + bi$, where a and b are real numbers and $i^2 = -1$, is called a complex number in standard (or Cartesian) form. We refer to a and b as the **real** and **imaginary** parts, respectively, of $a + bi$ and write

$$a = \text{Re}(a + bi), \qquad b = \text{Im}(a + bi).$$

Two complex numbers $a + bi$ and $c + di$ are equal if and only if they have the same real and imaginary parts, i.e., $a = c$ and $b = d$. Thus, any equation involving complex numbers can be interpreted as a pair of real equations.

The operations of addition and subtraction of complex numbers are given by

$$(a + bi) \pm (c + di) = (a \pm c) + (b \pm d)i.$$

In accordance with the distributive law and the proviso that $i^2 = -1$, the multiplication of two complex numbers is given by

(1) $\qquad (a + bi)(c + di) = (ac - bd) + (bc + ad)i.$

To compute the quotient of two complex numbers, we use a technique that is similar to the familiar method of "rationalizing the denominator":

(2) $\qquad \dfrac{a + bi}{c + di} = \dfrac{a + bi}{c + di} \cdot \dfrac{c - di}{c - di} = \dfrac{ac + bd}{c^2 + d^2} + \dfrac{bc - ad}{c^2 + d^2} i,$

provided $c + di \neq 0$. Thus the division of complex numbers is accomplished by multiplying the numerator and denominator by the **complex conjugate** $c - di$ of the denominator $c + di$. For example,

$$\frac{6 + 2i}{-1 + i} = \frac{6 + 2i}{-1 + i} \cdot \frac{-1 - i}{-1 - i} = \frac{-6 + 2 - 2i - 6i}{1 + 1} = -2 - 4i.$$

The complex exponential function $e^{(\alpha + i\beta)x}$, with α, β, and x real, was discussed in Section 4.6, page 160, and we recall the essential formulas:

(3) $\qquad e^{(\alpha + i\beta)x} = e^{\alpha x}(\cos \beta x + i \sin \beta x),$

(4) $\qquad \dfrac{d}{dx} e^{(\alpha + i\beta)x} = (\alpha + i\beta)e^{(\alpha + i\beta)x}.$

From (3) we see that

$$\text{Re } e^{(\alpha + i\beta)x} = e^{\alpha x} \cos \beta x, \qquad \text{Im } e^{(\alpha + i\beta)x} = e^{\alpha x} \sin \beta x,$$

and, more generally,

(5) $\text{Re}\{e^{(\alpha\pm i\beta)x}(a + bi)\} = e^{\alpha x}(a\cos\beta x \mp b\sin\beta x)$,

(6) $\text{Im}\{e^{(\alpha\pm i\beta)x}(a + bi)\} = e^{\alpha x}(b\cos\beta x \pm a\sin\beta x)$.

Formula (4) provides the key to the usefulness of the complex exponential: differentiation of $e^{(\alpha + i\beta)x}$ is equivalent to multiplication by the factor $(\alpha + i\beta)$; in other words, for this function, a calculus problem is reduced to a simple algebraic manipulation.

With this background, we now describe how to simplify the method of undetermined coefficients. Consider a second order equation of the form

(7) $L[y] := ay'' + by' + cy = g$,

where a, b, and c are real numbers and g is of the special form

(8) $g(x) = e^{\alpha x}\{(a_n x^n + \cdots + a_1 x + a_0)\cos\beta x + (b_n x^n + \cdots + b_1 x + b_0)\sin\beta x\}$,

with the a_j's, b_j's, α, and β real numbers. Such a function can always be expressed as the real or imaginary part of a function involving the complex exponential; for example, using equation (5), one can quickly check that

(9) $g(x) = \text{Re}\{G(x)\}$,

where

(10) $G(x) = e^{(\alpha - i\beta)x}[(a_n + ib_n)x^n + \cdots + (a_1 + ib_1)x + (a_0 + ib_0)]$.

Now suppose for the moment that we can find a complex-valued solution Y to the equation

(11) $L[Y] = aY'' + bY' + cY = G$.

Then, since a, b, and c are real numbers, we get a real-valued solution y to (7) by simply taking the real part of Y; that is, $y = \text{Re } Y$ solves (7). (Recall that in Lemma 2, page 162, we proved this fact for homogeneous equations!) Thus we need only focus on finding a solution to (11).

The method of undetermined coefficients asserts that any differential equation of the form

(12) $L[Y] = e^{(\alpha \pm i\beta)x}[(a_n + ib_n)x^n + \cdots + (a_1 + ib_1)x + (a_0 + ib_0)]$

has a solution of the form

(13) $Y_p(x) = x^s e^{(\alpha \pm i\beta)x}[A_n x^n + \cdots + A_1 x + A_0]$,

where $A_n, \ldots, A_0$ are complex constants and s is the smallest nonnegative integer such that no term in (13) is a (complex) solution to the corresponding homogeneous equation $L[Y] = 0$. As in the preceding section, we can solve for the unknown constants A_j by substituting (13) into (12) and equating coefficients of like terms. With these facts in mind, we can (for the small price of using complex arithmetic) dispense with Table 4.1 on page 173 and avoid the unpleasant task of computing derivatives of a function like $e^{3x}\sin(2x)(2 + 3x + x^2)$ which involves both exponential and trigonometric factors.

We illustrate the technique in the following examples.

● **EXAMPLE 1**   Find a particular solution to the equation

(14) $y'' + y = e^{-x}[\cos 2x - 3\sin 2x]$.

Solution   The right-hand side of (14) is of the required form (8) with $\alpha = -1$, $\beta = 2$, $a_0 = 1$, and $b_0 = -3$. From (9) and (10) we have

$$e^{-x}[\cos 2x - 3\sin 2x] = \text{Re}\{e^{(-1-2i)x}(1 - 3i)\},$$

and so the desired solution is the real part of a solution to

(15) $Y'' + Y = e^{(-1-2i)x}(1 - 3i)$.

The homogeneous equation corresponding to (15) has a general solution $Y_h(x) = c_1 \cos x + c_2 \sin x$ or, equivalently,

(16) $Y_h(x) = C_1 e^{ix} + C_2 e^{-ix}$.

(It is important to express this solution in terms of complex exponentials for purposes of comparison with the form of the solution to the nonhomogeneous equation.) Since the right-hand side of (15) is of the form $e^{(\alpha - \beta i)x}(a_0 + b_0 i)$, we look for a solution to (15) of the form

(17) $Y_p(x) = e^{(\alpha - \beta i)x} A = A e^{(-1-2i)x}$

(we take $s = 0$ since $e^{(-1-2i)x}$ is not a solution to the corresponding homogeneous equation; see (16)). Differentiation gives

(18) $Y_p''(x) = A(-1 - 2i)^2 e^{(-1-2i)x} = A(-3 + 4i)e^{(-1-2i)x}$,

and substituting (17) and (18) into (15) yields

$$A e^{(-1-2i)x}[(-3 + 4i) + 1] = e^{(-1-2i)x}(1 - 3i).$$

Thus, equating the coefficients of the exponential term, we get

$$A(-2 + 4i) = 1 - 3i,$$

$$A = \frac{1 - 3i}{-2 + 4i} = \frac{1 - 3i}{-2 + 4i} \cdot \frac{-2 - 4i}{-2 - 4i} = \frac{-14 + 2i}{4 + 16} = -\frac{7}{10} + \frac{1}{10}i,$$

and so a particular solution to (15) is

$$Y_p(x) = \left(-\frac{7}{10} + \frac{1}{10}i\right)e^{(-1-2i)x}.$$

Finally, on taking the real part of Y_p, we obtain

(19) $y_p(x) = \text{Re}\left\{\left(-\frac{7}{10} + \frac{1}{10}i\right)e^{(-1-2i)x}\right\}$

$$= e^{-x}\left(-\frac{7}{10}\cos 2x + \frac{1}{10}\sin 2x\right)$$

as a particular solution to (14).   ●

In using the method of undetermined coefficients with complex arithmetic, one can use the imaginary part in place of the real part, as we illustrate in the next example.

● **EXAMPLE 2** Find a particular solution to the equation

(20) $\quad y'' - 2y' + 10y = xe^x \sin 3x.$

Solution Since $\sin 3x = \text{Im}\{e^{3ix}\}$, we see that

$$xe^x \sin 3x = \text{Im}\{xe^{(1+3i)x}\},$$

and so the desired solution is the imaginary part of a solution to

(21) $\quad Y'' - 2Y' + 10Y = xe^{(1+3i)x}.$

The corresponding homogeneous equation has a general solution (expressed in exponential form) given by

$$Y_h(x) = C_1 e^{(1+3i)x} + C_2 e^{(1-3i)x}.$$

From (13) we know that the form of a solution to (21) is

$$x^s e^{(1+3i)x}[Ax + B],$$

and, since $e^{(1+3i)x}$ is a solution to the homogeneous equation, we take $s = 1$; that is, we set

(22) $\quad Y_p(x) = xe^{(1+3i)x}[Ax + B] = e^{(1+3i)x}[Ax^2 + Bx].$

The reader can verify that differentiation of this expression and substitution into (21) ultimately gives

$$[12iAx + (2A + 6iB)]e^{(1+3i)x} = xe^{(1+3i)x}.$$

Thus $12iA = 1$ and $2A + 6iB = 0$, which gives

$$A = \frac{1}{12i} = -\frac{i}{12}, \qquad B = -\frac{A}{3i} = \frac{1}{36}.$$

Using these values in (22) yields

$$Y_p(x) = e^{(1+3i)x}\left[-\frac{i}{12}x^2 + \frac{1}{36}x\right]$$

as a particular solution to (21). Consequently,

$$y_p(x) = \text{Im}\left\{e^{(1+3i)x}\left[-\frac{i}{12}x^2 + \frac{1}{36}x\right]\right\}$$

$$= e^x\left\{-\frac{1}{12}x^2 \cos 3x + \frac{1}{36}x \sin 3x\right\}$$

is a particular solution to (20). ●

The use of complex arithmetic not only streamlines the computations, but proves very useful in analyzing the response of a linear system to a sinusoidal input. Electrical engineers make good use of this in their study of RLC circuits by introducing the concept of **impedance.** This is illustrated in Problem 27.

EXERCISES 4.9

In Problems 1 through 6 determine the real part and the imaginary part of the given function.

1. $e^{(2-3i)x}(ix + 3 - 2i)$.

2. $e^{-2ix}\{(2-i)x + 3i\}$.

3. $e^{ix}(2-3i)(x^2 + (1-i)x)$.

4. $e^{-(2+i)x}(x^2 + ix)^2$.

5. $e^{-x}(ix + 2)^2 + e^{3ix}$.

6. $\{e^{(1-i)x}(x^2 + i)\}^2$.

In Problems 7 through 13 a real-valued function g is given. Using formulas (8), (9), and (10), determine a function G involving the complex exponential function whose real part equals g.

7. $e^{3x}\cos x - 5e^{3x}\sin x$.

8. $e^{-x}\{\cos 2x + (3 - 5x^2)\sin 2x\}$.

9. $(x^3 - x^2 + 1)\cos 2x$.   10. $e^{2x}\cos(3x - \pi/3)$.

11. $e^{-x}\sin(2x + \pi/4)$.   12. $[(1 + x)e^x \sin x]^2$.

13. $e^{2x}\{(x-1)\cos 4x + (3x - 2)\sin 4x\}$.

14. Show that if a, b, ω, and ϕ are real, then
$$a\cos(\omega x + \phi) + b\sin(\omega x + \phi)$$
$$= \text{Re}\{e^{i(\omega x + \phi)}(a - ib)\}$$
$$= \text{Im}\{ie^{i(\omega x + \phi)}(a - ib)\}.$$

In Problems 15 through 24 find a particular solution to the given equation using the method of undetermined coefficients with complex arithmetic.

15. $y'' + y = e^x(\cos 3x - 2\sin 3x)$.

16. $y'' - 2y' + y = \cos x - \sin x$.

17. $x'' - 2x' + 5x = te^t \sin 2t$.

18. $y'' + y = 10^x \cos 3x$.

19. $z'' - 4z' + 5z = x\sin 3x - \cos 3x$.

20. $y'' + 2y' + 2y = 4xe^{-x}\cos x$.

21. $y'' + 2y' + 2y = e^{-x}\cos x + xe^{-x}\sin x$.

22. $y'' + 4y = 16\theta \cos 2\theta$.

23. $y''' - 2y'' - y = 4\cos(x + 3)$.

24. $y^{(4)} + 3y''' - 2y'' - 6y = \cos(\sqrt{3}x + 1)$.

25. Prove that a particular solution of
$$ay'' + by' + cy = K\cos\omega x,$$

where a, b, c, K, and ω are real, is given by
$$y_p(x) = \text{Re}\left\{\frac{Ke^{i\omega x}}{a(i\omega)^2 + b(i\omega) + c}\right\},$$
if $a(i\omega)^2 + b(i\omega) + c \neq 0$;
$$y_p(x) = \text{Re}\left\{\frac{Kxe^{i\omega x}}{2a(i\omega) + b}\right\},$$
if $a(i\omega)^2 + b(i\omega) + c = 0$.

26. Prove that a particular solution of
$$ay'' + by' + cy = K\sin\omega x,$$
where a, b, c, K, and ω are real, is given by
$$y_p(x) = \text{Im}\left\{\frac{Ke^{i\omega x}}{a(i\omega)^2 + b(i\omega) + c}\right\},$$
if $a(i\omega)^2 + b(i\omega) + c \neq 0$;
$$y_p(x) = \text{Im}\left\{\frac{Kxe^{i\omega x}}{2a(i\omega) + b}\right\},$$
if $a(i\omega)^2 + b(i\omega) + c = 0$.

27. **Impedance.** For an electric circuit consisting of a resistor, inductor, and capacitor connected in series with a power supply, Kirchhoff's laws lead to the equation
$$L\frac{d^2I}{dt^2} + R\frac{dI}{dt} + \frac{1}{C}I = \frac{dE}{dt},$$

where L is the inductance, R is the resistance, C is the capacitance, and $E(t)$ is the electromotive force (voltage) of the power supply. Given that $E(t) = A\sin\omega t$, the *steady state* current I_s in the circuit can be found by applying the method of undetermined coefficients. Engineers find this current by considering a complex voltage $Ae^{i\omega t}$ for which the capacitor behaves like a resistor with resistance
$$R_C = \frac{1}{i\omega C}$$

and an inductor behaves like a resistor with resistance
$$R_L = i\omega L.$$

The effective resistance, or **impedance** Z of the series circuit is just the sum of these complex resistances:
$$Z = R + R_C + R_L.$$

Using undetermined coefficients, show that the steady state current in the circuit is given by the formula
$$I_s(t) = \text{Im}\left(\frac{Ae^{i\omega t}}{Z}\right).$$

● 4.10 ● VARIATION OF PARAMETERS

We have seen that the method of undetermined coefficients is a simple procedure for determining a particular solution when the equation has constant coefficients and the non-homogeneous term is of a special type. Here we present a more general method, called **variation of parameters,**[†] which can be used to determine a particular solution. This method applies even when the coefficients of the differential equation are functions of x, *provided* we know a fundamental solution set for the corresponding homogeneous equation.

Consider the nonhomogeneous linear second order equation

(1) $L[y](x) := y'' + p(x)y' + q(x)y = g(x),$

where the coefficient of y'' is taken to be 1, and let $\{y_1(x), y_2(x)\}$ be a fundamental solution set for the corresponding homogeneous equation

(2) $L[y](x) = 0.$

Then we know that the solutions to the homogeneous equation (2) are given by

(3) $y_h(x) = c_1 y_1(x) + c_2 y_2(x),$

where c_1 and c_2 are constants. To find a particular solution to the nonhomogeneous equation, the idea behind variation of parameters is to replace the constants in (3) by functions of x. That is, we seek a solution of (1) of the form[††]

(4) $y_p(x) = v_1(x)y_1(x) + v_2(x)y_2(x).$

Since we have introduced two unknown functions, $v_1(x)$ and $v_2(x)$, it is reasonable to expect that we will need two equations involving these functions in order to determine them. Naturally, one of these two equations should come from (1). Let's therefore plug $y_p(x)$ given by (4) into (1). To accomplish this, we must first compute $y_p'(x)$ and $y_p''(x)$. From (4) we obtain

(5) $y_p' = (v_1'y_1 + v_2'y_2) + (v_1 y_1' + v_2 y_2').$

To simplify the computation and to avoid second order derivatives for the unknowns v_1, v_2 in the expression for y_p'', let us require

(6) $v_1'y_1 + v_2'y_2 = 0.$

Then (5) becomes

(7) $y_p' = v_1 y_1' + v_2 y_2',$

so

(8) $y_p'' = v_1'y_1' + v_1 y_1'' + v_2'y_2' + v_2 y_2''.$

[†] *Historical Footnote:* The method of variation of parameters was discovered by Joseph Lagrange in 1774.

[††] In Problem 34 in Exercises 2.4 we developed this approach for first order linear equations.

Now, substituting y_p, y'_p, and y''_p, as given in (4), (7), and (8), into (1) we find

(9)
$$g = L[y_p]$$
$$= (v'_1 y'_1 + v_1 y''_1 + v'_2 y'_2 + v_2 y''_2) + p(v_1 y'_1 + v_2 y'_2) + q(v_1 y_1 + v_2 y_2)$$
$$= (v'_1 y'_1 + v'_2 y'_2) + v_1(y''_1 + py'_1 + qy_1) + v_2(y''_2 + py'_2 + qy_2)$$
$$= (v'_1 y'_1 + v'_2 y'_2) + v_1 L[y_1] + v_2 L[y_2].$$

Since y_1 and y_2 are solutions to the homogeneous equation, we have $L[y_1] = L[y_2] = 0$. Thus (9) becomes

(10)
$$v'_1 y'_1 + v'_2 y'_2 = g.$$

If we can find v_1 and v_2 that satisfy both (6) and (10), that is,

know   (11)
$$y_1 v'_1 + y_2 v'_2 = 0,$$
$$y'_1 v'_1 + y'_2 v'_2 = g,$$

then y_p given by (4) will be a particular solution to (1). To determine v_1 and v_2, we first solve the linear system (11) for v'_1 and v'_2. Algebraic manipulation or Cramer's rule immediately gives

$$v'_1(x) = \frac{-g(x)y_2(x)}{W[y_1, y_2](x)}, \qquad v'_2(x) = \frac{g(x)y_1(x)}{W[y_1, y_2](x)},$$

where $W[y_1, y_2](x)$, which occurs in the denominator, is the Wronskian of $y_1(x)$ and $y_2(x)$. Notice that this Wronskian is never zero, because $\{y_1, y_2\}$ is a fundamental solution set. Upon integrating these equations, we finally obtain

(12)
$$v_1(x) = \int \frac{-g(x)y_2(x)}{W[y_1, y_2](x)} \, dx, \qquad v_2(x) = \int \frac{g(x)y_1(x)}{W[y_1, y_2](x)} \, dx.$$

Let's review this procedure:

▶ METHOD OF VARIATION OF PARAMETERS

To determine a particular solution to $y'' + py' + qy = g$:

(a) Find a fundamental solution set $\{y_1(x), y_2(x)\}$ for the corresponding homogeneous equation and take

$$y_p(x) = v_1(x)y_1(x) + v_2(x)y_2(x).$$

(b) Determine $v_1(x)$ and $v_2(x)$ by solving the system in (11) for $v'_1(x)$ and $v'_2(x)$ and integrating. (If the given equation is $a_2(x)y'' + a_1(x)y' + a_0(x)y = b(x)$, then it must be put in the form $y'' + p(x)y' + q(x)y = g(x)$, so that $g(x) = b(x)/a_2(x)$.)

(c) Substitute $v_1(x)$ and $v_2(x)$ into the expression for $y_p(x)$ to obtain a particular solution.

Of course in step (b) one can use the formulas in (12), but the reader is advised not to memorize them.

● **EXAMPLE 1**   Find a general solution on $(-\pi/2, \pi/2)$ to

(13) $\dfrac{d^2y}{dx^2} + y = \tan x.$

Solution   Recall that a fundamental solution set for the homogeneous equation $y'' + y = 0$ is $\{\cos x, \sin x\}$. We now set

(14) $y_p(x) = v_1(x)\cos x + v_2(x)\sin x$

and referring to (11), solve the system

$$(\cos x)v_1'(x) + (\sin x)v_2'(x) = 0,$$
$$(-\sin x)v_1'(x) + (\cos x)v_2'(x) = \tan x$$

for $v_1'(x)$ and $v_2'(x)$. This gives

$$v_1'(x) = -\tan x \sin x,$$
$$v_2'(x) = \tan x \cos x = \sin x.$$

Integrating, we obtain

(15) $v_1(x) = -\displaystyle\int \tan x \sin x\, dx = -\int \dfrac{\sin^2 x}{\cos x}\, dx$

$= -\displaystyle\int \dfrac{1 - \cos^2 x}{\cos x}\, dx = \int (\cos x - \sec x)\, dx$

$= \sin x - \ln(\sec x + \tan x) + C_1,$

(16) $v_2(x) = \displaystyle\int \sin x\, dx = -\cos x + C_2.$

Since we need only one particular solution, we take both C_1 and C_2 to be zero for simplicity. Then, substituting $v_1(x)$ and $v_2(x)$ in (14), we obtain

$$y_p(x) = [\sin x - \ln(\sec x + \tan x)]\cos x - \cos x \sin x,$$

which simplifies to

$$y_p(x) = -(\cos x)\ln(\sec x + \tan x).$$

Recall from Theorem 6 in Section 4.7 that a general solution to a nonhomogeneous equation is given by the sum of a general solution to the homogeneous equation and a particular solution. Consequently, a general solution to equation (13) on the interval $(-\pi/2, \pi/2)$ is

(17) $y(x) = c_1 \cos x + c_2 \sin x - (\cos x)\ln(\sec x + \tan x).$ ●

In the above example, the constants C_1 and C_2 appearing in (15) and (16) were chosen to be zero. If we had retained these arbitrary constants, the ultimate effect would be just to add a solution of the homogeneous equation. Hence, leaving C_1 and C_2 arbitrary and substituting $v_1(x)$ and $v_2(x)$ in (14) leads to the general solution given in (17).

● **EXAMPLE 2**   Find a particular solution on $(-\pi/2, \pi/2)$ to

(18) $$\frac{d^2y}{dx^2} + y = \tan x + 3x - 1.$$

Solution   With $g(x) = \tan x + 3x - 1$, the variation of parameters procedure will lead to a solution of (18). But it is simpler in this case to consider separately the equations

(19) $$\frac{d^2y}{dx^2} + y = \tan x,$$

(20) $$\frac{d^2y}{dx^2} + y = 3x - 1$$

and then use the superposition principle (Theorem 5, page 166).

In Example 1 we found that

$$y_q(x) = -(\cos x)\ln(\sec x + \tan x)$$

is a particular solution for equation (19). For equation (20), the method of undetermined coefficients can be applied. On seeking a solution to (20) of the form $y_r(x) = Ax + B$, we quickly obtain

$$y_r(x) = 3x - 1.$$

Finally, we apply the superposition principle to get

$$\begin{aligned} y_p(x) &= y_q(x) + y_r(x) \\ &= -(\cos x)\ln(\sec x + \tan x) + 3x - 1 \end{aligned}$$

as a particular solution for equation (18).   ●

One important advantage the method of variation of parameters has over the method of undetermined coefficients is its applicability to linear equations whose coefficients are functions of x (see Problems 23–26). The fly in the ointment is that variation of parameters requires that we know a fundamental solution set for the corresponding homogeneous equation. For equations with variable coefficients, this set may be extremely difficult to determine.

EXERCISES 4.10

In Problems 1 through 12 find a general solution to the differential equation using the method of variation of parameters.

1. $y'' + 4y = \tan 2x.$

2. $y'' + y = \sec x.$

3. $2x'' - 2x' - 4x = 2e^{3t}.$

4. $y'' - y = 2x + 4.$

5. $y'' - 2y' + y = x^{-1}e^x.$

6. $y'' + 2y' + y = e^{-x}.$

7. $y'' + 16y = \sec 4\theta.$

8. $y'' + 9y = \sec^2 3x.$

9. $y'' + 4y = \csc^2 2x.$

10. $y'' + 4y' + 4y = e^{-2x}\ln x.$

11. $x^2 z'' + xz' + 9z = -\tan(3\ln x).$

12. $x^2 y'' + 3xy' + y = x^{-1}.$

In Problems 13 through 21 find a general solution to the differential equation.

13. $y'' + y = \tan^2 x.$

14. $y'' + y = \tan x + e^{3x} - 1$.

15. $v'' + 4v = \sec^4 2x$.

16. $y'' + y = \sec^3 \theta$.

17. $y'' + y = 3 \sec x - x^2 + 1$.

18. $y'' + 5y' + 6y = 18x^2$.

19. $\frac{1}{2}y'' + 2y = \tan 2x - \frac{1}{2}e^x$.

20. $y'' - 6y' + 9y = x^{-3}e^{3x}$.

21. $x^2 z'' - xz' + z = x\left(1 + \dfrac{3}{\ln x}\right)$.

22. The **Bessel equation** of order one-half,
$$x^2 y'' + xy' + (x^2 - \tfrac{1}{4})y = 0, \qquad x > 0,$$
has two linearly independent solutions,
$$y_1(x) = x^{-1/2} \cos x, \qquad y_2(x) = x^{-1/2} \sin x.$$
Find a general solution to the nonhomogeneous equation
$$x^2 y'' + xy' + (x^2 - \tfrac{1}{4})y = x^{5/2}, \qquad x > 0.$$

In Problems 23 through 26 find a general solution to the differential equation, given that the functions y_1 and y_2 are linearly independent solutions to the corresponding homogeneous equation for $x > 0$. Remember to put the equation in the same form as equation (1).

23. $xy'' - (x + 1)y' + y = x^2$; $y_1 = e^x$, $y_2 = x + 1$.

24. $x^2 y'' - 4xy' + 6y = x^3 + 1$; $y_1 = x^2$, $y_2 = x^3$.

25. $xy'' + (5x - 1)y' - 5y = x^2 e^{-5x}$; $y_1 = 5x - 1$, $y_2 = e^{-5x}$.

26. $xy'' + (1 - 2x)y' + (x - 1)y = xe^x$; $y_1 = e^x$, $y_2 = e^x \ln x$.

27. Express the solution to the initial value problem
$$y'' - y = \frac{1}{x}; \qquad y(1) = 0, \qquad y'(1) = -2$$
using definite integrals. Using Simpson's rule to approximate the integrals, find an approximation for $y(2)$.

28. Use the method of variation of parameters to show that
$$y(x) = c_1 \cos x + c_2 \sin x + \int_0^x f(s) \sin(x - s)\, ds$$
is a general solution to the differential equation
$$y'' + y = f(x),$$
where $f(x)$ is a continuous function on $(-\infty, \infty)$. [Hint: Use the trigonometric identity $\sin(x - s) = \sin x \cos s - \sin s \cos x$.]

• CHAPTER SUMMARY •

In this chapter we discussed the theory of linear second order equations and presented methods for solving these equations. The important features and techniques are listed below.

Homogeneous Linear Equations

$$y'' + p(x)y' + q(x)y = 0$$

Fundamental Solution Set: $\{y_1, y_2\}$.

Two solutions y_1 and y_2 to the homogeneous equation on the interval I form a fundamental solution set provided that their Wronskian,

$$W[y_1, y_2](x) := y_1(x)y_2'(x) - y_1'(x)y_2(x),$$

is different from zero for some x in I. If y_1 and y_2 are linearly independent solutions on I, then $W[y_1, y_2](x) \neq 0$ on I, and hence y_1 and y_2 form a fundamental solution set.

General Solution to Homogeneous Equation: $c_1 y_1 + c_2 y_2$.

If y_1 and y_2 are linearly independent solutions to the homogeneous equation, then a general solution is

$$y(x) = c_1 y_1(x) + c_2 y_2(x),$$

where c_1 and c_2 are arbitrary constants.

Homogeneous Equation with Constant Coefficients: $ay'' + by' + cy = 0.$

The form of a general solution for a homogeneous equation with constant coefficients depends on the roots

$$r_1 = \frac{-b + \sqrt{b^2 - 4ac}}{2a}, \qquad r_2 = \frac{-b - \sqrt{b^2 - 4ac}}{2a}$$

of the auxiliary equation

$$ar^2 + br + c = 0, \qquad a \neq 0.$$

(a) When $b^2 - 4ac > 0$, the auxiliary equation has two distinct real roots r_1 and r_2, and a general solution is

$$y(x) = c_1 e^{r_1 x} + c_2 e^{r_2 x}.$$

(b) When $b^2 - 4ac = 0$, the auxiliary equation has a repeated real root $r = r_1 = r_2$, and a general solution is

$$y(x) = c_1 e^{rx} + c_2 x e^{rx}.$$

(c) When $b^2 - 4ac < 0$, the auxiliary equation has complex conjugate roots $r = \alpha \pm i\beta$, and a general solution is

$$y(x) = c_1 e^{\alpha x} \cos \beta x + c_2 e^{\alpha x} \sin \beta x.$$

Reduction of Order Formula: $y(x) = v(x)f(x).$

Let $f(x)$ be a nontrivial solution to the homogeneous equation. The substitution $y(x) = v(x)f(x)$ reduces the homogeneous equation to a first order equation in v'. When solved for v' and integrated to obtain v, a second linearly independent solution $y(x)$ is given by the reduction of order formula

$$y(x) = f(x) \int \frac{e^{-\int p(x)\,dx}}{[f(x)]^2}\,dx.$$

Nonhomogeneous Linear Equations

$$y'' + p(x)y' + q(x)y = g(x)$$

General Solution to Nonhomogeneous Equation: $y_p + c_1 y_1 + c_2 y_2.$

If y_p is any particular solution to the nonhomogeneous equation, and y_1 and y_2 are linearly independent solutions to the corresponding homogeneous equation, then a general solution is

$$y(x) = y_p(x) + c_1 y_1(x) + c_2 y_2(x),$$

where c_1 and c_2 are arbitrary constants.

Two methods for finding a particular solution y_p are those of undetermined coefficients and variation of parameters.

Undetermined Coefficients: $\quad g(x) = e^{\alpha x} p_n(x) \begin{Bmatrix} \cos \beta x \\ \sin \beta x \end{Bmatrix}.$

If the right-hand side $g(x)$ of a nonhomogeneous equation with constant coefficients is a polynomial, an exponential of the form $e^{\alpha x}$, a trigonometric function of the form $\cos \beta x$ or $\sin \beta x$, or any product of these special types of functions, then a particular solution of an appropriate form exists. This special form is given in Table 4.1, page 173. The method of undetermined coefficients can be streamlined by using complex arithmetic and properties of the complex exponential function as discussed in Section 4.9.

Variation of Parameters: $\quad y(x) = v_1(x)y_1(x) + v_2(x)y_2(x).$

If y_1 and y_2 are two linearly independent solutions to the corresponding homogeneous equation, then a particular solution to the nonhomogeneous equation is

$$y(x) = v_1(x)y_1(x) + v_2(x)y_2(x),$$

where

$$v_1(x) = \int \frac{-g(x)y_2(x)}{W[y_1, y_2](x)} \, dx, \qquad v_2(x) = \int \frac{g(x)y_1(x)}{W[y_1, y_2](x)} \, dx.$$

Superposition Principle: $\quad L[y] = c_1 g_1 + c_2 g_2.$

If y_1 and y_2 are solutions to the equations

$$y'' + py' + qy = g_1 \quad \text{and} \quad y'' + py' + qy = g_2,$$

respectively, then $c_1 y_1 + c_2 y_2$ is a solution to the equation

$$y'' + py' + qy = c_1 g_1 + c_2 g_2.$$

Cauchy-Euler Equations: $\quad ax^2 \dfrac{d^2 y}{dx^2} + bx \dfrac{dy}{dx} + cy = h(x), \qquad x > 0.$

Let $x = e^t$. This substitution transforms the original equation into the following constant coefficient equation in t:

$$a \frac{d^2 y}{dt^2} + (b - a)\frac{dy}{dt} + cy = h(e^t).$$

REVIEW PROBLEMS

1, 3, 5, 7, 13, 21, 29

In Problems 1 through 28 find a general solution to the given differential equation.

1. $y'' + 8y' - 9y = 0.$

2. $49y'' + 14y' + y = 0.$

3. $4y'' - 4y' + 10y = 0.$

4. $9y'' - 30y' + 25y = 0.$

5. $6y'' - 11y' + 3y = 0.$

6. $y'' + 8y' - 14y = 0.$

7. $36y'' + 24y' + 5y = 0.$

8. $25y'' + 20y' + 4y = 0.$

9. $16z'' - 56z' + 49z = 0.$

10. $u'' + 11u = 0.$

11. $t^2 x''(t) + 5x(t) = 0, \qquad t > 0.$

12. $2y''' - 3y'' - 12y' + 20y = 0.$

13. $y'' + 16y = xe^x.$

14. $v'' - 4v' + 7v = 0.$

15. $3y''' + 10y'' + 9y' + 2y = 0.$

16. $y''' + 3y'' + 5y' + 3y = 0.$

17. $y''' + 10y' - 11y = 0.$

18. $h^{(4)} = 120x.$

19. $4y''' + 8y'' - 11y' + 3y = 0.$

20. $2y'' - y = x \sin x.$

21. $y'' - 3y' + 7y = 7x^2 - e^x.$

22. $y'' - 8y' - 33y = 546 \sin x.$

23. $y'' + 16y = \tan 4\theta.$

24. $10y'' + y' - 3y = x - e^{x/2}.$

25. $4y'' - 12y' + 9y = e^{5x} + e^{3x}.$

26. $y'' + 6y' + 15y = e^{2x} + 75.$

27. $x^2 y'' + 2xy' - 2y = 6x^{-2} + 3x, \qquad x > 0.$

28. $y'' = 5x^{-1} y' - 13x^{-2} y, \qquad x > 0.$

In Problems 29 through 36 find the solution to the initial value problem.

29. $y'' + 4y' + 7y = 0; \qquad y(0) = 1, \qquad y'(0) = -2.$

30. $y'' + 2y' + y = 2 \cos \theta; \qquad y(0) = 3, \qquad y'(0) = 0.$

31. $y'' - 2y' + 10y = 6 \cos 3x - \sin 3x;$
    $y(0) = 2, \qquad y'(0) = -8.$

32. $4y'' - 4y' + 5y = 0; \qquad y(0) = 1, \qquad y'(0) = -\frac{11}{2}.$

33. $y''' - 12y'' + 27y' + 40y = 0;$
    $y(0) = -3, \qquad y'(0) = -6, \qquad y''(0) = -12.$

34. $y'' + 5y' - 14y = 0; \qquad y(0) = 5, \qquad y'(0) = 1.$

35. $y'' + y = \sec \theta; \qquad y(0) = 1, \qquad y'(0) = 2.$

36. $9y'' + 12y' + 4y = 0; \qquad y(0) = -3, \qquad y'(0) = 3.$

In Problems 37 through 39 a differential equation and a nontrivial solution $f(x)$ are given. Find a general solution for the equation.

37. $xy'' + (x - 1)y' - y = 0, \qquad x > 0; \qquad f(x) = e^{-x}.$

38. $(x - 1)y'' - xy' + y = 0, \qquad x > 1; \qquad f(x) = e^x.$

39. $(1 - x^2)w'' - 2xw' + 2w = 0, \qquad -1 < x < 1;$
    $f(x) = x.$

40. Determine whether the statement made is *always true* or *sometimes false*.

    (a) Two differentiable functions are linearly independent on $[a, b]$ if and only if their Wronskian is not zero on $[a, b]$.

    (b) The differential operator

    $$L[y](x) := 2y''(x) + x^2 y'(x) + \frac{1}{x} y(x)$$

    defined for twice differentiable functions on $(0, \infty)$ is linear.

    (c) The initial value problem

    $$xy'' + \frac{5}{x - 3} y' + 2y = \cos x;$$

    $$y(1) = 6, \qquad y'(1) = 23$$

    has a unique solution on the interval $(0, 3)$.

    (d) A general solution of $y'' + y = x$ is given by

    $$y = c_1 \cos x + c_2 \sin x + c_3 x,$$

    where $c_1, c_2,$ and c_3 are arbitrary constants.

TECHNICAL WRITING EXERCISES

1. Compare the two methods—undetermined coefficients and variation of parameters—for determining a particular solution to a nonhomogeneous equation. What are the advantages and disadvantages of each?

2. Consider the differential equation

$$\frac{d^2 y}{dx^2} + 2b \frac{dy}{dx} + y = 0,$$

where b is a constant. Describe how the behavior of solutions to this equation changes as b varies.

3. Consider the differential equation

$$\frac{d^2 y}{dx^2} + cy = 0,$$

where c is a constant. Describe how the behavior of solutions to this equation changes as c varies.

4. For students with a background in linear algebra: Compare the theory for linear second order equations with that for systems of n linear equations in n unknowns whose coefficient matrix has rank $n - 2$. Use the terminology from linear algebra; for example, subspace, basis, dimension, linear transformation, and kernel. Discuss both homogeneous and nonhomogeneous equations.

GROUP PROJECTS FOR CHAPTER 4

▶ A. Convolution Method

The convolution of two functions g and f is the function $g * f$ defined by

$$(g * f)(x) := \int_0^x g(x - v) f(v)\, dv.$$

The aim of this project is to show how convolutions can be used to obtain a particular solution to a nonhomogeneous equation of the form

(1) $\qquad ay'' + by' + cy = f(x),$

where a, b, and c are constants, $a \neq 0$.

(a) Use Leibniz's rule,

$$\frac{d}{dx} \int_a^x h(x, v)\, dv = \int_a^x \frac{\partial h}{\partial x}(x, v)\, dv + h(x, x),$$

to show that

$$(y * f)'(x) = (y' * f)(x) + y(0) f(x),$$

and

$$(y * f)''(x) = (y'' * f)(x) + y'(0) f(x) + y(0) f'(x),$$

assuming y and f are sufficiently differentiable.

(b) Let $y_s(x)$ be the solution to the homogeneous equation $ay'' + by' + cy = 0$ satisfying $y_s(0) = 0$, $y_s'(0) = 1/a$. Show that $y_s * f$ is the particular solution to equation (1) satisfying $y(0) = y'(0) = 0$.

(c) Let $y_k(x)$ be the solution to the homogeneous equation $ay'' + by' + cy = 0$ satisfying $y(0) = y_0$, $y'(0) = y_1$, and let y_s be as defined in part (b). Show that

$$(y_s * f)(x) + y_k(x)$$

is the unique solution to the initial value problem

(2) $\qquad ay'' + by' + cy = f(x); \qquad y(0) = y_0, \qquad y'(0) = y_1.$

(d) Use the result of part (c) to determine the solution to each of the following initial value problems. Carry out all integrations and express your answers in terms of elementary functions.

(i) $y'' + y = \tan x;$ $\qquad y(0) = 1, \qquad y'(0) = 1.$
(ii) $2y'' + y' - y = e^{-x} \sin x;$ $\qquad y(0) = -1, \qquad y'(0) = 1.$
(iii) $y'' - 2y' + y = \sqrt{x} e^x;$ $\qquad y(0) = 1, \qquad y'(0) = 0.$

▶ B. Asymptotic Behavior of Solutions

In the application of linear systems theory to mechanical and electrical problems, one often encounters the equation

(3) $\qquad y'' + py' + qy = f(x),$

where p and q are positive constants with $p^2 < 4q$, and $f(x)$ is a forcing function for the system. In many cases it is important for the design engineer to know that a bounded forcing function gives rise only to bounded solutions. More specifically, how does the behavior of $f(x)$ for large values of x affect the asymptotic behavior of the solution? To answer this question:

(a) Show that the homogeneous equation associated with equation (3) has two linearly independent solutions given by

$$e^{\alpha x} \cos \beta x, \qquad e^{\alpha x} \sin \beta x,$$

where $\alpha = -p/2 < 0$ and $\beta = \frac{1}{2}\sqrt{4q - p^2}$.

(b) Let $f(x)$ be a continuous function defined on the interval $[0, \infty)$. Use the variation of parameters formula to show that any solution to (3) on $[0, \infty)$ can be expressed in the form

(4)
$$y(x) = c_1 e^{\alpha x} \cos \beta x + c_2 e^{\alpha x} \sin \beta x$$

$$-\frac{1}{\beta} e^{\alpha x} \cos \beta x \int_0^x f(t) e^{-\alpha t} \sin \beta t \, dt$$

$$+\frac{1}{\beta} e^{\alpha x} \sin \beta x \int_0^x f(t) e^{-\alpha t} \cos \beta t \, dt.$$

(c) Assuming that f is bounded on $[0, \infty)$, that is, that there exists a constant K such that $|f(t)| \leq K$ for all $t \geq 0$, use the triangle inequality and other properties of the absolute value to show that $y(x)$ given in (4) satisfies

$$|y(x)| \leq \{|c_1| + |c_2|\} e^{\alpha x} + \frac{2K}{|\alpha|\beta}(1 - e^{\alpha x})$$

for all $x > 0$.

(d) In a similar fashion, show that if $f_1(x)$ and $f_2(x)$ are two bounded continuous functions on $[0, \infty)$ such that $|f_1(x) - f_2(x)| \leq \varepsilon$ for all $x > x_0$, and if ϕ_1 is a solution to (3) with $f = f_1$ and ϕ_2 is a solution to (3) with $f = f_2$, then

$$|\phi_1(x) - \phi_2(x)| \leq M e^{\alpha x} + \frac{2\varepsilon}{|\alpha|\beta}(1 - e^{\alpha(x - x_0)})$$

for all $x > x_0$, where M is a constant that depends on ϕ_1 and ϕ_2 but not on x.

(e) Now assume that $f(x) \to F_0$ as $x \to \infty$, where F_0 is a constant. By using the result of part (d), prove that any solution ϕ to (3) must satisfy $\phi(x) \to F_0/q$ as $x \to \infty$. [Hint: Choose $f_1 = f$, $f_2 \equiv F_0$, $\phi_1 = \phi$, $\phi_2 \equiv F_0/q$.]

C. Linearization of Nonlinear Problems

A useful approach to analyzing a nonlinear equation is to study its **linearized equation,** which is obtained by replacing the nonlinear terms by linear approximations. For example, the nonlinear equation

(5)
$$\frac{d^2\theta}{dt^2} + \sin \theta = 0,$$

which governs the motion of a simple pendulum, has

(6)
$$\frac{d^2\theta}{dt^2} + \theta = 0$$

as a linearization for small θ. (The nonlinear term $\sin \theta$ has been replaced by the linear approximation θ.)

Since a general solution to equation (5) involves Jacobi elliptic functions (see Project D), let's try to approximate the solutions. For this purpose we consider two methods: Taylor series and linearization.

(a) Derive the first six terms of the Taylor series about $t = 0$ of the solution to equation (5) with initial conditions $\theta(0) = \pi/12$, $\theta'(0) = 0$. (The Taylor series method is discussed in Project A of Chapter 1 and Section 8.2.)

(b) Solve equation (6) subject to the same initial conditions $\theta(0) = \pi/12$, $\theta'(0) = 0$.

(c) On the same coordinate axes, graph the two approximations found in parts (a) and (b).

(d) Discuss the advantages and disadvantages of the Taylor series method and the linearization method.

(e) Give a linearization for the initial value problem

$$x''(t) + 0.1[1 - x^2(t)]x'(t) + x(t) = 0; \qquad x(0) = 0.5, \qquad x'(0) = 0$$

for x small. Solve this linearized problem to obtain an approximation for the nonlinear problem.

◢ D. Nonlinear Equations Solvable by First Order Techniques

Certain *nonlinear* second order equations, namely those with dependent or independent variables missing, can be solved by reducing them to a pair of *first order equations*. This is accomplished by making the substitution $w = dy/dx$.

(a) To solve an equation of the form $y'' = F(x, y')$ in which the dependent variable y is missing, setting $w = y'$ (so that $w' = y''$) yields the pair of equations

$$w' = F(x, w),$$

$$y' = w.$$

Since $w' = F(x, w)$ is a first order equation, we have available the techniques of Chapter 2 to solve it for $w(x)$. Once $w(x)$ is determined, we integrate it to obtain $y(x)$.

Using this method, solve

$$2xy'' - y' + \frac{1}{y'} = 0, \qquad x > 0.$$

(b) To solve an equation of the form $y'' = F(y, y')$ in which the independent variable x is missing, setting $w = dy/dx$ yields, via the chain rule,

$$\frac{d^2y}{dx^2} = \frac{dw}{dx} = \frac{dw}{dy}\frac{dy}{dx} = w\frac{dw}{dy}.$$

Thus $y'' = F(y, y')$ is equivalent to the pair of equations

(7) $$w\frac{dw}{dy} = F(y, w),$$

(8) $$\frac{dy}{dx} = w.$$

In equation (7), notice that y plays the role of the *independent* variable; hence solving it yields $w(y)$. Then substituting $w(y)$ into (8), we obtain a separable equation that determines $y(x)$.

Using this method, solve

$$2y\frac{d^2y}{dx^2} = 1 + \left(\frac{dy}{dx}\right)^2.$$

(c) **Suspended Cable.**   In the study of a cable suspended between two fixed points (see Figure 4.8), one encounters the initial value problem

$$\frac{d^2y}{dx^2} = \frac{1}{a}\sqrt{1 + \left(\frac{dy}{dx}\right)^2}\,; \qquad y(0) = a, \qquad y'(0) = 0,$$

where $a\ (\neq 0)$ is a constant. Solve this initial value problem for y. The resulting curve is called a **catenary.**

(d) **Newton Equation.**   A mechanical system with one degree of freedom is a conservative system if the total energy remains constant during motion. Such a system can always be expressed (using generalized coordinates) in the form

$$\frac{d^2y}{dx^2} = f(y),$$

where $f(y)$ is a continuous function for all y, which is called a **Newton equation.** Derive the following implicit solution to the Newton equation:

$$\int \left(2\int f(y)\,dy + c_1\right)^{-1/2}\,dy = x + c_2.$$

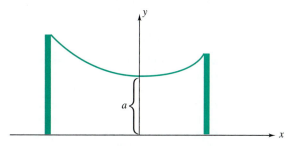

Figure 4.8 Suspended cable

E. Simple Pendulum

In Section 4.1 we discussed the simple pendulum consisting of a mass m suspended by a rod of length l having negligible mass and derived the nonlinear initial value problem

(9) $$\frac{d^2\theta}{dt^2} + \frac{g}{l}\sin\theta = 0; \qquad \theta(0) = \alpha, \qquad \theta'(0) = 0,$$

where g is the acceleration due to gravity and $\theta(t)$ is the angle the rod makes with the vertical at time t (see Figure 4.1). Here it is assumed that the mass is released with zero velocity at an initial angle α, $0 < \alpha < \pi$. We would like to determine the equation of motion for the pendulum and its period of oscillation.

(a) Use equation (9) and the technique discussed in Project D, part (b), to show that

$$\left(\frac{d\theta}{dt}\right)^2 = \frac{2g}{l}(\cos\theta - \cos\alpha)$$

and hence

$$dt = \sqrt{\frac{l}{2g}}\,\frac{d\theta}{\sqrt{\cos\theta - \cos\alpha}}\,.$$

(b) Use the trigonometric identity

$$\cos x = 1 - 2\sin^2(x/2)$$

to express dt by

$$dt = \frac{1}{2}\sqrt{\frac{l}{g}}\frac{d\theta}{\sqrt{\sin^2(\alpha/2) - \sin^2(\theta/2)}}.$$

(c) Make the change of variables

$$\sin(\theta/2) = \sin(\alpha/2)\sin\phi,$$

to rewrite dt in the form

(10) $$dt = \sqrt{\frac{l}{g}}\frac{d\phi}{\sqrt{1 - k^2\sin^2\phi}},$$

where $k := \sin(\alpha/2)$.

(d) The period $P(\alpha)$ of the pendulum is defined to be the time required for the pendulum to swing from one extreme to the other and back, that is, from α to $-\alpha$ back to α. Show that the period of oscillation is given by

(11) $$P(\alpha) = 4\sqrt{\frac{l}{g}}\int_0^{\pi/2}\frac{d\phi}{\sqrt{1 - k^2\sin^2\phi}},$$

where $k := \sin(\alpha/2)$. [Hint: The period is just four times the time it takes the pendulum to go from $\theta = 0$ to $\theta = \alpha$.]

The integral in (11) is called an **elliptic integral of the first kind** and is denoted by $F(k, \pi/2)$. As you might expect, the period of the simple pendulum depends on the length l of the rod and the initial displacement α. In fact, a check of an elliptic integral table will show that the period nearly doubles as the initial displacement increases from $\pi/8$ to $15\pi/16$ (for fixed l). What happens as α approaches π?

(e) To determine the equation of motion of the pendulum, we observe that as t varies from 0 to t, θ varies from α to θ, and ϕ varies from $\pi/2$ to ϕ. Integrating equation (10), show that

(12) $$t + \frac{P(\alpha)}{4} = \sqrt{\frac{l}{g}}\int_0^{\phi}\frac{ds}{\sqrt{1 - k^2\sin^2 s}} = \sqrt{\frac{l}{g}}F(k, \phi).$$

For fixed k, $F(k, \phi)$ has an "inverse," denoted by $\mathrm{sn}(k, u)$, that satisfies $u = F(k, \phi)$ if and only if $\mathrm{sn}(k, u) = \sin\phi$. The function $\mathrm{sn}(k, u)$ is called a **Jacobi elliptic function** and has many properties that are similar to those of the sine function. Using the Jacobi elliptic function $\mathrm{sn}(k, u)$, express the equation of motion for the pendulum in the form

(13) $$\theta = 2\arcsin\left\{k\,\mathrm{sn}\left(k, \sqrt{\frac{g}{l}}\left(t + \frac{P(\alpha)}{4}\right)\right)\right\},$$

where $k := \sin(\alpha/2)$.

F. Phase Plane Diagrams and Periodic Solutions

In Project E a procedure is given for computing the period of a simple pendulum in terms of elliptic integrals. That computation was based on the premise that the simple pendulum problem *has a* periodic solution.

The differential equation that governs the motion of a simple pendulum is

(14) $$\frac{d^2\theta}{dt^2} + \frac{g}{l}\sin\theta = 0,$$

where θ is the angular displacement from the vertical, l is the length of the pendulum, and g is the acceleration due to gravity (see Figure 4.1 on page 129). To simplify the computations let's assume that $l = g$, so that equation (14) becomes

(15) $$\frac{d^2\theta}{dt^2} + \sin\theta = 0.$$

To show that equation (15) has a periodic solution, proceed as follows:

(a) Show that solving (15) is equivalent to solving the system

(16)
$$\frac{d\theta}{dt} = v,$$
$$\frac{dv}{dt} = -\sin\theta.$$

(b) The **phase plane** for system (16) is the plane of the variables $v = d\theta/dt$ and θ. A solution to system (16) traces out a **path** in the θv-phase plane. Show that this path is a solution to the differential equation

(17) $$\frac{dv}{d\theta} = -\frac{\sin\theta}{v}.$$

(c) Solve equation (17).

(d) Show that there are closed paths in the phase plane associated with the simple pendulum equation, and hence that the simple pendulum problem has periodic solutions (see Figure 4.9).

(e) Give a physical interpretation of the motion of a pendulum that corresponds to the three trajectories labeled A, B, and C in Figure 4.9.

(f) The procedure outlined in parts (a)–(d) can be used to show the existence of periodic solutions for other nonlinear equations. In particular, show that the **nonlinear spring equation**

$$\frac{d^2x}{dt^2} + x + \beta x^3 = 0,$$

where $\beta > 0$ is a fixed parameter, has a nontrivial periodic solution.

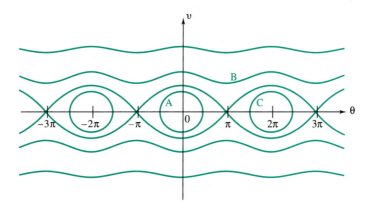

Figure 4.9 Phase plane trajectories for a simple pendulum

CHAPTER 5

Applications and Numerical Methods for Second Order Equations

•5.1• MECHANICAL VIBRATIONS AND SIMPLE HARMONIC MOTION

Each day we encounter many types of mechanical vibrations. The bouncing motion of an automobile due to the bumps and cracks in the pavement, the vibrations of a bridge caused by traffic and wind, and the normal flutter of an airplane wing due to the vibration of the engine and the air rushing past are some common examples. To study mechanical vibrations, we shall start with the simple mechanical system consisting of a coil spring suspended from a rigid support with a mass attached to the end of the spring.

To analyze this spring-mass system, we need to recall two laws of physics: Hooke's law and Newton's second law of motion. Hooke's law states that the spring exerts a restoring force opposite to the direction of elongation of the spring and with a magnitude directly proportional to the amount of elongation.[†] That is, the spring exerts a restoring force $\mathcal{F}$ whose magnitude is ks, where s is the amount of elongation and $k(>0)$ is the **spring constant.**

For example, if a 20-lb weight stretches a spring 6 in., then (in the ft-lb system) Hooke's law gives

(1) $\qquad 20 = |\mathcal{F}| = ks = k(\tfrac{1}{2}).$

Hence, the spring constant is $k = 40$ lb/ft.

In Section 3.4 we discussed Newton's second law and how to apply it to problems in mechanics. Recall that when mass remains constant, this law can be expressed as

(2) $\qquad m\dfrac{d^2x}{dt^2} = ma = F(t, x, dx/dt),$

[†] Hooke's law is a reasonable approximation until the spring is stretched to near its elastic limit.

where x is the position, dx/dt is the velocity, and d^2x/dt^2 is the acceleration of the mass at time t. Here F, the total force acting on the mass, is assumed to depend only on time, position, and velocity.

A first step in the analysis of the spring-mass system is to choose a coordinate axis in which to represent the motion of the mass. For this purpose, we observe that the spring has a certain natural length L when hanging from its support (see Figure 5.1(a)). Attaching a mass m elongates the spring, and when it comes to rest (equilibrium), the spring has been stretched an amount, say l, beyond its natural length (Figure 5.1(b)). Therefore, let's choose a vertical coordinate axis passing through the spring, with the origin at the equilibrium position of the mass. We let x denote the displacement of the mass from its equilibrium position, taking x positive when the mass is below its equilibrium position, as shown in Figure 5.1(c).

We now consider the various forces acting on the mass m.

Gravity.   The force of gravity F_1 is a downward force with magnitude mg, where g is the acceleration due to gravity. Hence

$$F_1 = mg.$$

Restoring Force.   The spring exerts a restoring force F_2 whose magnitude is proportional to the elongation of the spring. Referring to Figure 5.1(c), we see that the spring is stretched $x + l$ units beyond its natural length. Hence the magnitude of F_2 is $k(x + l)$, where k is the spring constant. Since the spring pulls upward (in the negative x direction), we have

$$F_2 = -k(x + l).$$

We should observe that when $x = 0$, that is, when the system is at equilibrium, the force of gravity and the force due to the spring balance each other. Thus, $mg = kl$. We can now express the restoring force as

$$F_2 = -kx - mg.$$

Damping Force.   There is a damping or frictional force F_3 acting on the mass. For example, this force may be air resistance or friction due to a shock absorber. In either case we

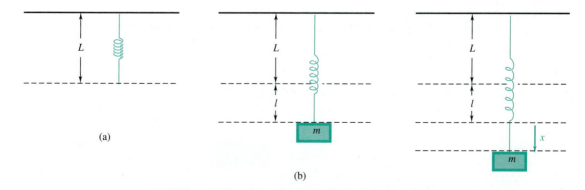

(a)

(b)

(c)

Figure 5.1 Spring **(a)** in natural position, **(b)** in equilibrium, and **(c)** in motion *assume* that the damping force is proportional to the magnitude of the velocity of the mass, but opposite in direction. That is,

$$F_3 = -b\frac{dx}{dt}, \qquad b > 0,$$

where b is the **damping constant** given in units of mass/time (or force-time/length). In certain circumstances a more elaborate formula for the damping force may be valid, but empirical studies show that when the velocity is small, the above expression for F_3 is a reasonable one.

External Forces.   Any external forces acting on the mass (for example, a magnetic force or the forces exerted on a car by bumps in the pavement) will be denoted by $F_4 = f(t)$. For simplicity, we assume that these forces depend only on time and *not* on the location of the mass or its velocity. (We can model the vertical displacements of a car in this way if its horizontal position is known as a function of time.)

The total force F acting on the mass m is the sum of the four forces F_1, F_2, F_3, and F_4:

(3) $\qquad F(t, x, dx/dt) = mg - kx - mg - b\frac{dx}{dt} + f(t).$

Applying Newton's second law to the system gives

$$m\frac{d^2x}{dt^2} = mg - kx - mg - b\frac{dx}{dt} + f(t),$$

which simplifies to

(4) $\qquad m\frac{d^2x}{dt^2} + b\frac{dx}{dt} + kx = f(t).$

When $b = 0$, the system is said to be **undamped;** otherwise it is **damped.** When $f(t) \equiv 0$, the motion is said to be **free;** otherwise the motion is **forced.** A convenient schematic representation for a simple spring-mass system with damping is given in Figure 5.2.

Let's begin with the simple system in which $b = 0$ and $f(t) \equiv 0$, the so-called **undamped, free** case. In this case equation (4) reduces to

(5) $\qquad m\frac{d^2x}{dt^2} + kx = 0,$

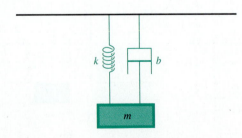

Figure 5.2 Spring-mass system with spring constant k, damping constant b, and mass m

and, when divided by m, becomes

(6) $\qquad \dfrac{d^2x}{dt^2} + \omega^2 x = 0,$

where $\omega = \sqrt{k/m}$. The auxiliary equation associated with (6) is $r^2 + \omega^2 = 0$, which has complex conjugate roots $\pm \omega i$. Hence a general solution to (6) is

(7) $\qquad x(t) = C_1 \cos \omega t + C_2 \sin \omega t.$

We can express $x(t)$ in the more convenient form

(8) $\qquad x(t) = A \sin(\omega t + \phi)$

by letting $C_1 = A \sin \phi$ and $C_2 = A \cos \phi$. That is,

$$A \sin(\omega t + \phi) = A \cos \omega t \sin \phi + A \sin \omega t \cos \phi$$
$$= C_1 \cos \omega t + C_2 \sin \omega t.$$

Solving for A and ϕ in terms of C_1 and C_2, we find

(9) $\qquad A = \sqrt{C_1^2 + C_2^2} \quad \text{and} \quad \tan \phi = \dfrac{C_1}{C_2},$

where the quadrant in which ϕ lies is determined by the signs of C_1 and C_2 since $\sin \phi$ has the same sign as C_1 ($\sin \phi = C_1/A$) and $\cos \phi$ has the same sign as C_2 ($\cos \phi = C_2/A$). For example, if $C_1 > 0$ and $C_2 < 0$, then ϕ is in Quadrant II.

It is evident from (8) that the motion of a mass in an *undamped*, *free* system is simply a sine wave, or what is called **simple harmonic motion.** The constant A is the **amplitude** of the motion and ϕ is the **phase angle.** The motion is periodic with **period** $2\pi/\omega$ and **natural frequency** $\omega/2\pi$, where $\omega = \sqrt{k/m}$. (The constant ω is the **angular frequency** for the sine function in (8).)

Observe that the amplitude and phase angle depend on the constants C_1 and C_2, which, in turn, are determined by the initial position and initial velocity of the mass. However, the period and frequency depend only on k and m and not on the initial conditions (see Figure 5.3).

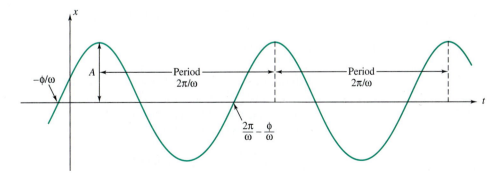

Figure 5.3 Simple harmonic motion of undamped, free vibrations

● **EXAMPLE 1** A mass weighing 4 lb stretches a spring 3 in. after coming to rest at equilibrium. The mass is then pulled down 6 in. below the equilibrium point and given a downward velocity of $\sqrt{2}$ ft/sec. Neglecting any damping or external forces that may be present, determine the equation of motion of the mass along with its amplitude, period, and natural frequency. How long after release does the mass pass through the equilibrium position?

Solution Since we have a case of undamped, free vibration, the equation of motion is given by (7). To find k, we observe that the 4-lb mass stretches the spring 3 in. or $\frac{1}{4}$ ft. Using Hooke's law, we have

$$4 = mg = k(\tfrac{1}{4}),$$

or $k = 16$ lb/ft. Since $g \approx 32$ ft/sec^2, we find $m = \frac{4}{32} = \frac{1}{8}$ slugs, and so

$$\omega = \sqrt{\frac{k}{m}} = \sqrt{\frac{16}{1/8}} = 8\sqrt{2}.$$

Substituting this value for ω into (7) gives

(10) $x(t) = C_1 \cos(8\sqrt{2}\,t) + C_2 \sin(8\sqrt{2}\,t).$

Now we use the initial conditions, $x(0) = 6$ in. $= \frac{1}{2}$ ft and $x'(0) = \sqrt{2}$ ft/sec, to solve for C_1 and C_2 in (10). That is,

$$\tfrac{1}{2} = x(0) = C_1,$$
$$\sqrt{2} = x'(0) = 8\sqrt{2}\,C_2,$$

and so $C_1 = \frac{1}{2}$ and $C_2 = \frac{1}{8}$. Hence the equation of motion of the mass is

(11) $x(t) = \frac{1}{2}\cos(8\sqrt{2}\,t) + \frac{1}{8}\sin(8\sqrt{2}\,t).$

To express $x(t)$ in the alternate form (8), we set

$$A = \sqrt{C_1^2 + C_2^2} = \sqrt{(\tfrac{1}{2})^2 + (\tfrac{1}{8})^2} = \frac{\sqrt{17}}{8},$$

$$\tan \phi = \frac{C_1}{C_2} = \frac{1/2}{1/8} = 4.$$

Since both C_1 and C_2 are positive, ϕ is in the first quadrant, and so $\phi = \arctan 4$. Hence

(12) $x(t) = \dfrac{\sqrt{17}}{8} \sin(8\sqrt{2}\,t + \phi).$

Thus the amplitude A is $\sqrt{17}/8$ and the phase angle ϕ is $\arctan 4 \approx 1.326$. The period is $P = 2\pi/\omega = 2\pi/(8\sqrt{2}) = \sqrt{2}\,\pi/8$, and the natural frequency is $1/P = 8/(\sqrt{2}\,\pi)$.

Finally, to determine when the mass will pass through the equilibrium position, $x = 0$, we must solve the trigonometric equation

(13) $x(t) = \dfrac{\sqrt{17}}{8} \sin(8\sqrt{2}\,t + \phi) = 0$

for t. Equation (13) will be satisfied whenever

(14) $8\sqrt{2}\,t + \phi = n\pi,$

n an integer. Since $0 < \phi < \pi/2$, putting $n = 1, 2, \dots$ in (14) determines the (positive) times t when the mass crosses its equilibrium position. The first such occurrence is when

$$8\sqrt{2}\,t + \phi = \pi;$$

that is,

$$t = \frac{\pi - \phi}{8\sqrt{2}} = \frac{\pi - \arctan 4}{8\sqrt{2}} \approx 0.16 \text{ sec.} \quad \bullet$$

EXERCISES 5.1

In the following problems take $g = 32$ ft/sec² for the U.S. Customary system and $g = 9.8$ m/sec² for the MKS system.

1. A mass of 100 kg is attached to a spring suspended from the ceiling, causing the spring to stretch 20 cm on coming to rest at equilibrium. The mass is then pulled down 5 cm below the equilibrium point and released (zero initial velocity). Neglecting any damping or external forces that may be present, determine the equation of motion of the mass, along with its amplitude, period, and natural frequency. Sketch the graph of this simple harmonic motion.

2. A mass weighing 8 lb is attached to a spring suspended from the ceiling. When the mass comes to rest at equilibrium, the spring has been stretched 6 in. The mass is then pulled down 3 in. below the equilibrium point and given an upward velocity of 0.5 ft/sec. Neglecting any damping or external forces that may be present, determine the equation of motion of the mass, along with its amplitude, period, and natural frequency. Sketch the graph of this simple harmonic motion.

3. A mass weighing 8 lb stretches a spring 2 ft on coming to rest at equilibrium. The mass is then lifted up 6 in. above the equilibrium point and given a downward velocity of 1 ft/sec. Determine the simple harmonic motion of the mass. How fast and in what direction will the mass be moving 10 sec after being released?

4. Show that the period of the simple harmonic motion of a mass hanging from a spring is $2\pi\sqrt{l/g}$, where l denotes the amount (beyond its natural length) that the spring is stretched when the mass is at equilibrium.

5. A mass of 5 kg is attached to a spring hanging from a ceiling, thereby stretching the spring 0.5 m on coming to rest at equilibrium. The mass is then pulled down 0.1 m below the equilibrium point and given an upward velocity of 0.1 m/sec. Determine the equation for the simple harmonic motion of the mass. When will the mass first reach its minimum height after being set in motion?

6. A mass of 5 kg is attached to a spring suspended from a ceiling, thereby stretching the spring 2 m on coming to rest at equilibrium. The mass is then lifted up 1 m above the equilibrium point and given an upward velocity of $\frac{1}{3}$ m/sec. Determine the equation for the simple harmonic motion of the mass. When will the mass first reach its maximum height after being set in motion?

7. A mass of $\frac{1}{2}$ kg is attached to the lower end of a spring, thereby stretching the spring 2 m on coming to rest at equilibrium. The mass is then pulled down $\frac{1}{2}$ m and given an upward velocity of $\frac{1}{2}$ m/sec. Determine the equation for the simple harmonic motion of the mass. How long until the mass first returns to the equilibrium position?

8. A weight hanging on a spring oscillates with a period of 3 sec. After 2 lb is added, the period becomes 4 sec. Assuming that we can neglect any damping or external forces, determine how much weight was originally attached to the spring.

• 5.2 • DAMPED FREE VIBRATIONS

In the previous section we considered vibrations in a rather idyllic setting—no external frictional forces were assumed present—and the result was simple harmonic motion. In most applications, however, there is at least some type of frictional or damping force that plays a significant role. This force may be due to a component in the system, such as a shock absorber in a car, or to the medium that surrounds the system, such as air or some liquid.

In this section we study the effect of a damping force on free vibrations. In particular, setting $f(t) \equiv 0$ in equation (4) of Section 5.1, we consider the motion of a system that is governed by

(1) $\qquad m\dfrac{d^2x}{dt^2} + b\dfrac{dx}{dt} + kx = 0.$

The auxiliary equation associated with (1) is

(2) $\qquad mr^2 + br + k = 0;$

its roots are

(3) $\qquad \dfrac{-b \pm \sqrt{b^2 - 4mk}}{2m} = -\dfrac{b}{2m} \pm \dfrac{1}{2m}\sqrt{b^2 - 4mk}.$

As we found in Chapter 4, the form of the solution to (1) depends on the nature of these roots and, in particular, on the discriminant $b^2 - 4mk$.

Underdamped or Oscillatory Motion ($b^2 < 4mk$)

When $b^2 < 4mk$, the discriminant $b^2 - 4mk$ is negative and there are two complex conjugate roots to the auxiliary equation (2). These roots are $\alpha \pm i\beta$, where

(4) $\qquad \alpha := -\dfrac{b}{2m}, \qquad \beta := \dfrac{1}{2m}\sqrt{4mk - b^2}.$

Hence a general solution to (1) is

(5) $\qquad x(t) = e^{\alpha t}(C_1 \cos \beta t + C_2 \sin \beta t).$

As we did with simple harmonic motion, we can express $x(t)$ in the alternate form

(6) $\qquad x(t) = Ae^{\alpha t}\sin(\beta t + \phi),$

where $A = \sqrt{C_1^2 + C_2^2}$ and $\tan \phi = C_1/C_2$. It is now evident that $x(t)$ is the product of an exponential factor

$\qquad Ae^{\alpha t} = Ae^{-(b/2m)t}$

called the **damping factor,** and a sine factor, $\sin(\beta t + \phi)$, that accounts for the oscillatory motion. Since the sine factor varies between -1 and 1 with period $2\pi/\beta$, the solution $x(t)$ varies between $-Ae^{\alpha t}$ and $Ae^{\alpha t}$ with **quasiperiod** $P = 2\pi/\beta = 4m\pi/\sqrt{4mk - b^2}$ and **quasifrequency** $1/P$. Moreover, since b and m are positive, $\alpha = -b/2m$ is negative, and thus the damping factor tends to zero as $t \to \infty$. A graph of a typical solution $x(t)$ is given in Figure 5.4. The system is called **underdamped** because there is not enough damping present

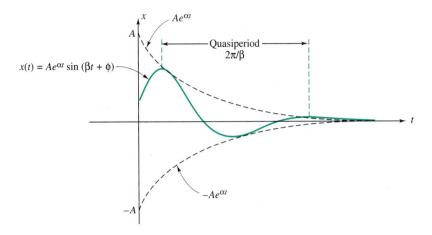

Figure 5.4 Oscillatory motion

(b is too small) to prevent the system from oscillating.

It is easily seen that as $b \to 0$ the damping factor approaches the constant A and the quasifrequency approaches the natural frequency of the simple harmonic motion corresponding to $b = 0$. It is also worth mentioning that the values of t where the graph of $x(t)$ touches the exponential curves $\pm Ae^{\alpha t}$ are *not* the same values of t at which $x(t)$ attains its relative maximum and minimum values (see Problem 11).

Critically Damped Motion ($b^2 = 4mk$)

When $b^2 = 4mk$, the discriminant $b^2 - 4mk$ is zero and the auxiliary equation has the repeated root $-b/2m$. Hence a general solution to (1) is now

(7) $$x(t) = (C_1 + C_2 t)e^{-(b/2m)t}.$$

To understand the motion described by $x(t)$ in (7), we first consider the behavior of $x(t)$ as $t \to \infty$. Using L'Hôpital's rule,

(8) $$\lim_{t \to \infty} x(t) = \lim_{t \to \infty} \frac{C_1 + C_2 t}{e^{(b/2m)t}} = \lim_{t \to \infty} \frac{C_2}{(b/2m)e^{(b/2m)t}} = 0,$$

(recall that $b/2m > 0$). Hence, $x(t)$ dies off to zero as $t \to \infty$. Moreover, since

$$x'(t) = \left(C_2 - \frac{b}{2m}C_1 - \frac{b}{2m}C_2 t \right)e^{-(b/2m)t},$$

we see that the derivative is either identically zero (when $C_1 = C_2 = 0$) or vanishes for at most one point (when the linear factor in parentheses is zero). If the trivial solution is ignored, it follows that $x(t)$ has at most one local maximum or minimum for $t > 0$. Therefore, $x(t)$ *does not oscillate*. This leaves, qualitatively, only three possibilities for the motion of $x(t)$, depending on the initial conditions. These are illustrated in Figure 5.5 on page 206. This special case when $b^2 = 4mk$ is called **critically damped** motion, since if b were any smaller, oscillation would occur.

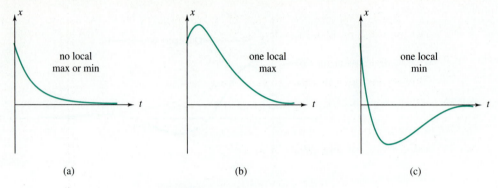

Figure 5.5 Critically damped vibrations

Overdamped Motion ($b^2 > 4mk$)

When $b^2 > 4mk$, the discriminant $b^2 - 4mk$ is positive and there are two distinct real roots to the auxiliary equation (2):

$$(9) \qquad r_1 = -\frac{b}{2m} + \frac{1}{2m}\sqrt{b^2 - 4mk}, \qquad r_2 = -\frac{b}{2m} - \frac{1}{2m}\sqrt{b^2 - 4mk}.$$

Hence a general solution to (1) is

$$(10) \qquad x(t) = C_1 e^{r_1 t} + C_2 e^{r_2 t}.$$

Obviously r_2 is negative. And since $b^2 > b^2 - 4mk$ (that is, $b > \sqrt{b^2 - 4mk}$), it follows that r_1 is also negative. Therefore, as $t \to \infty$, both of the exponentials in (10) decay and $x(t) \to 0$. Moreover, since

$$x'(t) = C_1 r_1 e^{r_1 t} + C_2 r_2 e^{r_2 t} = e^{r_1 t}(C_1 r_1 + C_2 r_2 e^{(r_2 - r_1)t}),$$

it follows that $x'(t) = 0$ only when $C_1 r_1 + C_2 r_2 e^{(r_2 - r_1)t} = 0$. Thus a nontrivial solution $x(t)$ can have at most one local maximum or minimum for $t > 0$. As in the case of critical damping, the motion is *non*oscillatory and must be qualitatively like one of the three sketches in Figure 5.5. This case where $b^2 > 4mk$ is called **overdamped** motion.

● **EXAMPLE 1**   Assume that the motion of a spring-mass system with damping is governed by

$$(11) \qquad \frac{d^2 x}{dt^2} + b\frac{dx}{dt} + 25x = 0; \qquad x(0) = 1, \qquad x'(0) = 0.$$

Find the equation of motion and sketch its graph for the three cases when $b = 8$, 10, and 12.

Solution   The auxiliary equation for (11) is

$$(12) \qquad r^2 + br + 25 = 0,$$

whose roots are

$$(13) \qquad r = -\frac{b}{2} \pm \frac{1}{2}\sqrt{b^2 - 100}.$$

Case 1. When $b = 8$, the roots (13) are $-4 \pm 3i$. This is thus a case of underdamping, and the equation of motion has the form

(14) $x(t) = C_1 e^{-4t} \cos 3t + C_2 e^{-4t} \sin 3t.$

Setting $x(0) = 1$ and $x'(0) = 0$ gives the system

$$C_1 = 1, \qquad -4C_1 + 3C_2 = 0,$$

whose solution is $C_1 = 1$, $C_2 = \frac{4}{3}$. To express $x(t)$ as the product of a damping factor and a sine factor (recall equation (6)), we set

$$A = \sqrt{C_1^2 + C_2^2} = \frac{5}{3}, \qquad \tan \phi = \frac{C_1}{C_2} = \frac{3}{4},$$

where ϕ is a first-quadrant angle since C_1 and C_2 are both positive. Then

(15) $x(t) = \frac{5}{3} e^{-4t} \sin(3t + \phi),$

where $\phi = \arctan \frac{3}{4} \approx 0.64$ (see Figure 5.6(a)).

Case 2. When $b = 10$, there is only one (repeated) root to the auxiliary equation (12), namely $r = -5$. This is a case of critical damping, and the equation of motion has the form

(16) $x(t) = (C_1 + C_2 t)e^{-5t}.$

Setting $x(0) = 1$ and $x'(0) = 0$ now gives

$$C_1 = 1, \qquad C_2 - 5C_1 = 0,$$

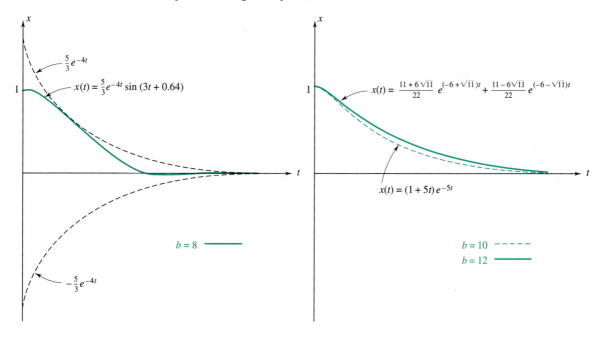

(a)   **Figure 5.6** Solutions for various values of b (b) and so $C_1 = 1$, $C_2 = 5$. Thus

(17) $\qquad x(t) = (1 + 5t)e^{-5t}$.

The graph of $x(t)$ given in (17) is represented by the dashed line in Figure 5.6(b) on page 207. Notice that $x(t)$ is zero only for $t = -\frac{1}{5}$ and hence doesn't cross the t-axis for $t > 0$.

  Case 3. When $b = 12$, the roots to the auxiliary equation are $-6 \pm \sqrt{11}$. This is a case of overdamping, and the equation of motion has the form

(18) $\qquad x(t) = C_1 e^{(-6+\sqrt{11})t} + C_2 e^{(-6-\sqrt{11})t}$.

Setting $x(0) = 1$ and $x'(0) = 0$ gives

$$C_1 + C_2 = 1, \qquad (-6 + \sqrt{11})C_1 + (-6 - \sqrt{11})C_2 = 0,$$

from which we find $C_1 = (11 + 6\sqrt{11})/22$ and $C_2 = (11 - 6\sqrt{11})/22$. Hence

(19) $\qquad x(t) = \dfrac{11 + 6\sqrt{11}}{22}e^{(-6+\sqrt{11})t} + \dfrac{11 - 6\sqrt{11}}{22}e^{(-6-\sqrt{11})t}$

$$\qquad\quad = \dfrac{e^{(-6+\sqrt{11})t}}{22}\{11 + 6\sqrt{11} + (11 - 6\sqrt{11})e^{-2\sqrt{11}t}\}.$$

The graph of this overdamped motion is represented by the solid line in Figure 5.6(b). ●

  It is interesting to observe in Example 1 that when the system is underdamped ($b = 8$), the solution goes to zero like e^{-4t}; when the system is critically damped ($b = 10$), the solution tends to zero roughly like e^{-5t}; and when the system is overdamped ($b = 12$), the solution goes to zero like $e^{(-6+\sqrt{11})t} \approx e^{-2.68t}$. This means that if the system is underdamped, it not only oscillates, but also dies off slower than if it were critically damped. Moreover, if the system is overdamped, it again dies off slower than if it were critically damped. (This agrees with our physical intuition that the damping forces hinder the return to equilibrium.)

● **EXAMPLE 2**   An 8-lb weight is attached to a spring hanging from the ceiling. When the weight comes to rest at equilibrium, the spring has been stretched 2 ft. The damping constant b for the system is 1 lb-sec/ft. If the weight is raised 6 in. above equilibrium and given an initial upward velocity of 1 ft/sec, find the equation of motion for the weight. What is the maximum displacement above equilibrium that the weight will attain?

Solution   Since an 8-lb weight stretches the spring 2 ft, Hooke's law gives $8 = k2$, and so the spring constant is $k = 4$. The 8-lb weight has mass $m = W/g$, where g, the acceleration due to gravity, is 32 ft/sec². Thus $m = \frac{8}{32} = \frac{1}{4}$ slugs. Substituting these values for m, k, and the given value $b = 1$ into equation (1) and using the initial conditions, we obtain the initial value problem

(20) $\qquad \dfrac{1}{4}\dfrac{d^2x}{dt^2} + \dfrac{dx}{dt} + 4x = 0; \qquad x(0) = -\dfrac{1}{2}, \qquad x'(0) = -1;$

the negative signs for the initial conditions reflect the facts that the initial displacement and push are *upward* (see Figure 5.7(a)).

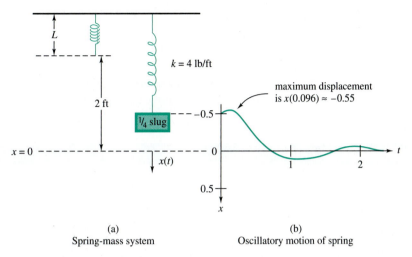

(a)
Spring-mass system

(b)
Oscillatory motion of spring

Figure 5.7 Spring-mass system and graph of motion for Example 2

The reader can verify that the solution to problem (20) is

(21) $x(t) = -\dfrac{1}{2}e^{-2t}\cos(2\sqrt{3}\,t) - \dfrac{1}{\sqrt{3}}e^{-2t}\sin(2\sqrt{3}\,t),$

or

(22) $x(t) = \sqrt{\dfrac{7}{12}}\,e^{-2t}\sin(2\sqrt{3}\,t + \phi),$

where $\tan\phi = \sqrt{3}/2$ and ϕ lies in the third quadrant because $C_1 = -1/2$ and $C_2 = -1/\sqrt{3}$ are both negative. (See Figure 5.7(b) for a sketch of $x(t)$ where we have inverted the x-axis so that the graph corresponds to the motion of the system in Figure 5.7(a)).

To determine the maximum displacement above equilibrium, we must determine the *highest* point on the graph of $x(t)$ in Figure 5.7(b). Since $x(t)$ dies off exponentially, this will occur at the first positive critical point of $x(t)$. Computing $x'(t)$ from (21), setting it equal to zero, and solving gives

$$x'(t) = e^{-2t}\left\{\frac{5}{\sqrt{3}}\sin(2\sqrt{3}\,t) - \cos(2\sqrt{3}\,t)\right\} = 0,$$

$$\frac{5}{\sqrt{3}}\sin(2\sqrt{3}\,t) = \cos(2\sqrt{3}\,t),$$

$$\tan(2\sqrt{3}\,t) = \frac{\sqrt{3}}{5}.$$

Thus the first positive root is

$$t = \frac{1}{2\sqrt{3}} \arctan \frac{\sqrt{3}}{5} \approx 0.096.$$

Substituting this value for t back into equation (21) or (22) gives $x(0.096) \approx -0.55$. Hence the maximum displacement above equilibrium is approximately 0.55 ft. ●

EXERCISES 5.2

In the following problems take $g = 32$ ft/sec^2 for the U.S. Customary system and $g = 9.8$ m/sec^2 for the MKS system.

1. The motion of a spring-mass system with damping is governed by

   $$x''(t) + bx'(t) + 16x(t) = 0; \qquad x(0) = 1,$$
   $$x'(0) = 0.$$

   Find the equation of motion and sketch its graph for $b = 6, 8$, and 10.

2. The motion of a spring-mass system with damping is governed by

   $$x''(t) + bx'(t) + 64x(t) = 0; \qquad x(0) = 1,$$
   $$x'(0) = 0.$$

   Find the equation of motion and sketch its graph for $b = 10, 16$, and 20.

3. The motion of a spring-mass system with damping is governed by

   $$x''(t) + 10x'(t) + kx(t) = 0; \qquad x(0) = 1,$$
   $$x'(0) = 0.$$

   Find the equation of motion and sketch its graph for $k = 20, 25$, and 30.

4. The motion of a spring-mass system with damping is governed by

   $$x''(t) + 4x'(t) + kx(t) = 0; \qquad x(0) = 1,$$
   $$x'(0) = 0.$$

   Find the equation of motion and sketch its graph for $k = 2, 4$, and 6.

5. A 4-lb weight is attached to a spring suspended from the ceiling. When the weight comes to rest at equilibrium, the spring has been stretched 3 in. The damping constant for the system is $\frac{1}{2}$ lb-sec/ft. If the weight is raised 2 in. above equilibrium and given an initial upward velocity of $\frac{1}{2}$ ft/sec, determine the equation of motion of the weight and give its damping factor, quasiperiod, and quasifrequency.

6. A mass of 20 kg stretches a spring 98 cm on coming to rest at equilibrium. The damping constant for the system is 140 N-sec/m. If the mass is pulled 25 cm below equilibrium and given an upward velocity of 1 m/sec, when will it return to its equilibrium position?

7. A mass of 2 kg stretches a spring 49 cm on coming to rest at equilibrium. The damping constant for the system is $8\sqrt{5}$ kg/sec. If the mass is pulled 10 cm below equilibrium and given a downward velocity of 2 m/sec, what is the maximum displacement from equilibrium that it will attain?

8. An 8-lb weight stretches a spring 1 ft on coming to rest at equilibrium. The damping constant for the system is $\frac{1}{4}$ lb-sec/ft. If the weight is raised 1 ft above equilibrium and released, what is the maximum displacement below equilibrium that it will attain?

9. A mass of 1 kg stretches a spring 9.8 cm on coming to rest at equilibrium. The damping constant for the system is 0.2 N-sec/m. If the mass is pushed downward from the equilibrium position with a velocity of 1 m/sec, when will it attain its maximum displacement below equilibrium?

10. An 8-lb weight stretches a spring 1 ft on coming to rest at equilibrium. The damping constant for the system is 2 lb-sec/ft. If the weight is raised 6 in. above equilibrium and given an upward velocity of 2 ft/sec, when will the mass attain its maximum displacement above equilibrium?

11. Show that for the underdamped system of Example 2, the times when the solution curve $x(t)$ in (22) touches the exponential curves $\pm\sqrt{7/12}\,e^{-2t}$ are not the same values of t for which the function $x(t)$ attains its relative extrema.

12. For an underdamped system, verify that as $b \to 0$ the damping factor approaches the constant A and the quasifrequency approaches the natural frequency $\sqrt{k/m}/2\pi$.

•5.3• FORCED VIBRATIONS

We now consider the vibrations of a spring-mass system when an external force is applied. Of particular interest is the response of the system to a *periodic* forcing term. As a paradigm, let's investigate the effect of a cosine forcing function on the system governed by the differential equation

(1) $$m\frac{d^2x}{dt^2} + b\frac{dx}{dt} + kx = F_0\cos\gamma t,$$

where F_0, γ are nonnegative constants and $0 < b^2 < 4mk$.

A solution to (1) has the form $x = x_h + x_p$, where x_p is a particular solution and x_h is a general solution to the corresponding homogeneous equation. We found in equation (6) of Section 5.2 that

(2) $$x_h(t) = Ae^{-(b/2m)t}\sin\left(\frac{\sqrt{4mk - b^2}}{2m}t + \phi\right),$$

where A and ϕ are constants.

To determine x_p we can use the method of undetermined coefficients (Section 4.8). From the form of the nonhomogeneous term, we know that

(3) $$x_p(t) = A_1\cos\gamma t + A_2\sin\gamma t,$$

where A_1, A_2 are constants to be determined. Substituting this expression into equation (1) and simplifying gives

(4) $$[(k - m\gamma^2)A_1 + b\gamma A_2]\cos\gamma t + [(k - m\gamma^2)A_2 - b\gamma A_1]\sin\gamma t = F_0\cos\gamma t.$$

Setting the corresponding coefficients on both sides equal, we have

$$(k - m\gamma^2)A_1 + b\gamma A_2 = F_0,$$
$$-b\gamma A_1 + (k - m\gamma^2)A_2 = 0,$$

and, solving, we obtain

(5) $$A_1 = \frac{F_0(k - m\gamma^2)}{(k - m\gamma^2)^2 + b^2\gamma^2}, \qquad A_2 = \frac{F_0 b\gamma}{(k - m\gamma^2)^2 + b^2\gamma^2}.$$

Hence a particular solution to (1) is

(6) $$x_p(t) = \frac{F_0}{(k - m\gamma^2)^2 + b^2\gamma^2}\{(k - m\gamma^2)\cos\gamma t + b\gamma\sin\gamma t\}.$$

Since the expression in braces can also be written as

$$\sqrt{(k - m\gamma^2)^2 + b^2\gamma^2}\,\sin(\gamma t + \theta),$$

we can express x_p in the alternate form

(7) $$x_p(t) = \frac{F_0}{\sqrt{(k - m\gamma^2)^2 + b^2\gamma^2}}\sin(\gamma t + \theta),$$

where $\tan\theta = A_1/A_2 = (k - m\gamma^2)/(b\gamma)$ and the quadrant in which θ lies is determined by the signs of A_1 and A_2.

Combining equations (2) and (7), we have the following representation of a general solution to (1) in the case $0 < b^2 < 4mk$:

(8) $$x(t) = Ae^{-(b/2m)t}\sin\left(\frac{\sqrt{4mk - b^2}}{2m}t + \phi\right) + \frac{F_0}{\sqrt{(k - m\gamma^2)^2 + b^2\gamma^2}}\sin(\gamma t + \theta).$$

The solution (8) is the sum of two terms. The first term, x_h, represents damped oscillation and depends only on the parameters of the system and the initial conditions. Because of the damping factor $e^{-(b/2m)t}$, this term tends to zero as $t \to \infty$. Consequently, it is referred to as a **transient** solution. The second term, x_p, in (8) is the offspring of the external forcing function $f(t) = F_0\cos\gamma t$. Like the forcing function, x_p is a sinusoid with angular frequency γ; however, x_p is out of phase with $f(t)$ (by the angle $\theta - \pi/2$), and its magnitude is different by the factor

(9) $$\frac{1}{\sqrt{(k - m\gamma^2)^2 + b^2\gamma^2}}.$$

As the transient term dies off, the motion of the spring-mass system becomes essentially that of the second term x_p (see Figure 5.8). Hence this term is called the **steady-state** solution. The factor appearing in (9) is referred to as the **frequency gain** since it represents the change in the magnitude of the sinusoidal input. The frequency gain has units of length/force.

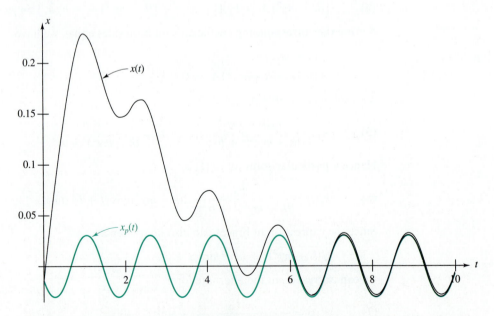

Figure 5.8 Convergence of $x(t)$ to the steady-state solution $x_p(t)$ when $m = 4$, $b = 6$, $k = 3$, $F_0 = 2$, $\gamma = 4$

● **EXAMPLE 1**   A 10-kg mass is attached to a spring hanging from the ceiling. This causes the spring to stretch 2 m on coming to rest at equilibrium. At time $t=0$, an external force $f(t)=20\cos 4t$ N is applied to the system. The damping constant for the system is 3 N-sec/m. Determine the steady-state solution for the system.

Solution   Since the mass m, damping constant b, and forcing term are given, we need only determine the spring constant k to obtain the differential equation that governs the system. The constant acceleration g due to gravity is approximately 9.8 m/sec². Thus the force of gravity on the system is $mg = (10)(9.8) = 98$ N. From Hooke's law, we have $98 = k2$, and so $k = 49$ N/m. Now, substituting into equation (1), we obtain

(10) $$10\frac{d^2x}{dt^2} + 3\frac{dx}{dt} + 49x = 20\cos 4t,$$

where $x(t)$ is the displacement (from equilibrium) of the mass at time t.

To find the steady-state response we must produce a particular solution to (10) that is a sinusoid. We can do this using the method of undetermined coefficients, guessing a solution of the form $A_1\cos 4t + A_2\sin 4t$. But this is precisely how we derived equation (7). Thus we substitute directly into (7) and find

(11) $$x_p(t) = \frac{20}{\sqrt{(49-160)^2 + (9)(16)}}\sin(4t+\theta) \approx (0.18)\sin(4t+\theta),$$

where $\tan\theta = (49-160)/12 \approx -9.25$. Since the numerator, $(49-160)$, is negative and the denominator, 12, positive, θ is a fourth-quadrant angle. Thus

$$\theta \approx \arctan(-9.25) \approx -1.46,$$

and the steady-state solution is given approximately by

(12) $$x_p(t) = (0.18)\sin(4t - 1.46). \quad ●$$

The above example illustrates an important point made earlier: the steady-state response (12) to the sinusoidal forcing function $20\cos 4t$ is a sinusoid of the same frequency but different amplitude. The gain factor (see (9)) in this case is $(0.18)/20 = 0.009$ m/N.

In general, the amplitude of the steady-state solution to equation (1) depends on the angular frequency γ of the forcing function and is given by $A(\gamma) = F_0 M(\gamma)$, where

(13) $$M(\gamma) := \frac{1}{\sqrt{(k - m\gamma^2)^2 + b^2\gamma^2}} = \frac{1/m}{\sqrt{\left(\frac{k}{m} - \gamma^2\right)^2 + \left(\frac{b}{m}\right)^2\gamma^2}}$$

is the frequency gain (see (9)). This formula is valid even when $b^2 \geq 4mk$. For a given system (m, b, and k fixed), it is often of interest to know how this system reacts to sinusoidal inputs of various frequencies (γ is a variable). For this purpose, the graph of the gain $M(\gamma)$, called the **frequency response curve** or **resonance curve** for the system, is enlightening.

In order to sketch the frequency response curve for $\gamma \geq 0$, we observe that when $\gamma = 0$, we have $M(0) = 1/k$. Also note that as $\gamma \to \infty$, the gain $M(\gamma) \to 0$. As a further aid in describing the graph, we compute from (13)

(14) $M'(\gamma) = \dfrac{-\left(\dfrac{2\gamma}{m}\right)\left[\gamma^2 - \left(\dfrac{k}{m} - \dfrac{b^2}{2m^2}\right)\right]}{\left[\left(\dfrac{k}{m} - \gamma^2\right)^2 + \left(\dfrac{b}{m}\right)^2\gamma^2\right]^{3/2}}.$

It follows from (14) that $M'(\gamma) = 0$ if and only if

(15) $\gamma = 0$ or $\gamma = \gamma_r := \sqrt{\dfrac{k}{m} - \dfrac{b^2}{2m^2}}.$

Now when $b^2 > 2mk$, the term inside the radical in (15) is negative and hence $M'(\gamma) = 0$ only when $\gamma = 0$. In this case, as γ increases from zero to infinity, $M(\gamma)$ decreases from $M(0) = 1/k$ to a limit value of zero.

When $b^2 < 2mk$, then γ_r is real and positive, and it is easy to verify that $M(\gamma)$ has a *maximum* at γ_r. Substituting γ_r into (13) gives

(16) $M(\gamma_r) = \dfrac{1}{b\sqrt{\dfrac{k}{m} - \dfrac{b^2}{4m^2}}}.$

The value $\gamma_r/2\pi$ is called the **resonance frequency** for the system, and, when stimulated by an external force at this frequency, the system is said to be **at resonance.**

Observe that since we must have $b^2 < 2mk$ for resonance to occur, a system cannot be at resonance unless it is underdamped ($b^2 < 4mk$).

To illustrate the effect of the damping constant b on the resonance curve, let's consider a system in which $m = k = 1$. In this case the frequency response curves are given by

(17) $M(\gamma) = \dfrac{1}{\sqrt{(1 - \gamma^2)^2 + b^2\gamma^2}},$

and, for $b < \sqrt{2}$, the resonance frequency is $\gamma_r/2\pi = (1/2\pi)\sqrt{1 - b^2}/2$. Figure 5.9 displays the graphs of these frequency response curves for $b = \frac{1}{4}, \frac{1}{2}, 1, \frac{3}{2}$, and 2. Observe that as $b \to 0$, the maximum magnitude of the frequency gain increases and the resonance frequency $\gamma_r/2\pi$ for the damped system approaches $1/2\pi = \sqrt{k/m}/2\pi$, the natural frequency for the undamped system.

To understand what is occurring, let's consider the undamped system ($b = 0$) with forcing term $F_0 \cos \gamma t$. This system is governed by

(18) $m\dfrac{d^2x}{dt^2} + kx = F_0 \cos \gamma t.$

A general solution to (18) is the sum of a particular solution and a general solution to the homogeneous equation. In Section 5.1 we showed that the latter describes simple harmonic motion:

(19) $x_h(t) = A \sin(\omega t + \phi),$ $\omega := \sqrt{k/m}.$

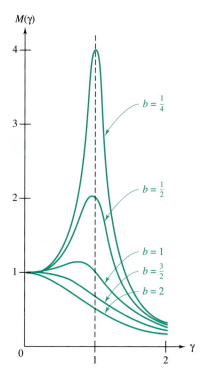

Figure 5.9 Frequency response curves for various values of b

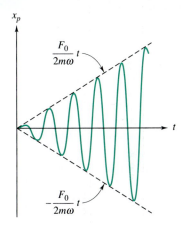

Figure 5.10 Undamped oscillation of the particular solution in (21)

The formula for the particular solution given in (7) is valid for $b = 0$ provided that $\gamma \neq \omega = \sqrt{k/m}$. However, when $b = 0$ and $\gamma = \omega$, then the form we used with undetermined coefficients to derive (7) does not work, because $\cos \omega t$ and $\sin \omega t$ are solutions to the corresponding homogeneous equation. The correct form is

(20) $x_p(t) = A_1 t \cos \omega t + A_2 t \sin \omega t,$

which leads to the solution

(21) $x_p(t) = \dfrac{F_0}{2m\omega} t \sin \omega t.$

(The verification of (21) is left to the reader.) Hence, in the *undamped resonant case* (when $\gamma = \omega$), a general solution to (18) is

(22) $x(t) = A \sin(\omega t + \phi) + \dfrac{F_0}{2m\omega} t \sin \omega t.$

Returning to the question of resonance, observe that the particular solution in (21) oscillates between $\pm (F_0 t)/(2m\omega)$. Hence as $t \rightarrow \infty$, the maximum magnitude of (21) approaches ∞ (see Figure 5.10).

It is obvious from the above discussion that **if the damping constant *b* is very small, the system is subject to large oscillations when the forcing function has a frequency near the resonance frequency for the system.** It is these large vibrations at resonance that concern engineers. Indeed, resonance vibrations have been known to cause airplane wings to snap, bridges to collapse,[†] and (less catastrophic) wine glasses to shatter.

EXERCISES 5.3

In the following problems take $g = 32$ ft/sec² for the U.S. Customary system and $g = 9.8$ m/sec² for the MKS system.

1. An 8-kg mass is attached to a spring hanging from the ceiling. This causes the spring to stretch 1.96 m upon coming to rest at equilibrium. At time $t = 0$, an external force $f(t) = \cos 2t$ N is applied to the system. The damping constant for the system is 3 N-sec/m. Determine the steady-state solution for the system.

2. A mass weighing 32 lb is attached to a spring hanging from the ceiling and comes to rest at its equilibrium position. At time $t=0$, an external force $f(t)=3\cos 4t$ lb is applied to the system. If the spring constant is 5 lb/ft and the damping constant is 2 lb-sec/ft, find the steady-state solution for the system.

3. A mass weighing 8 lb is attached to a spring hanging from the ceiling and comes to rest at its equilibrium position. At time $t = 0$, an external force $f(t) = 2\cos 2t$ lb is applied to the system. If the spring constant is 16 lb/ft and the damping constant is $\frac{1}{4}$ lb-sec/ft, find the equation of motion of the mass. What is the resonance frequency for the system?

4. A 2-kg mass is attached to a spring hanging from the ceiling. This causes the spring to stretch 20 cm upon coming to rest at equilibrium. At time $t = 0$, the mass is displaced 5 cm below the equilibrium position and released. At this same instant, an external force $f(t) = 0.3\cos t$ N is applied to the system. If the damping constant for the system is 5 N-sec/m, determine the

equation of motion for the mass. What is the resonance frequency for the system?

5. Sketch the frequency response curve (13) for the system in which $m = 4$, $k = 1$, $b = 2$.

6. Sketch the frequency response curve (13) for the system in which $m = 2$, $k = 3$, $b = 3$.

7. Determine the equation of motion for an undamped system at resonance governed by

$$\frac{d^2x}{dt^2} + 9x = 2\cos 3t; \qquad x(0) = 1, \qquad x'(0) = 0.$$

Sketch the solution.

8. Determine the equation of motion for an undamped system at resonance governed by

$$\frac{d^2x}{dt^2} + x = 5\cos t; \qquad x(0) = 0, \qquad x'(0) = 1.$$

Sketch the solution.

9. An undamped system is governed by

$$m\frac{d^2x}{dt^2} + kx = F_0\cos\gamma t, \qquad x(0) = x'(0) = 0,$$

where $\gamma \neq \omega := \sqrt{k/m}$.

   (a) Find the equation of motion of the system.
   (b) Use trigonometric identities to show that the solution can be written in the form

$$x(t) = \frac{2F}{m(\omega^2 - \gamma^2)}\sin\left(\frac{\omega + \gamma}{2}\right)t\sin\left(\frac{\omega - \gamma}{2}\right)t.$$

[†] An interesting discussion of one such disaster, involving the Tacoma Narrows bridge in Washington State, can be found in *Differential Equations and Their Applications*, by M. Braun, Springer-Verlag, New York, 1978. Researchers are still trying to understand this disaster; see the recent articles "Large-amplitude periodic oscillations in suspension bridges: some new connections with nonlinear analysis," by A. C. Lazer and P. J. McKenna, *SIAM Review*, Vol. 32, 1990, pp. 537–578 or "Still twisting," by Henry Petroski, *American Scientist*, Vol. 19, 1991, pp. 398–401.

(c) When γ is near ω, then $\omega - \gamma$ is small, while $\omega + \gamma$ is relatively large compared with $\omega - \gamma$. Hence, $x(t)$ can be viewed as the product of a slowly varying sine function, $\sin[(\omega - \gamma)t/2]$, and a rapidly varying sine function, $\sin[(\omega + \gamma)t/2]$. The net effect is a sine function $x(t)$ with frequency $(\omega - \gamma)/4\pi$, whose amplitude is a sine function with frequency $(\omega + \gamma)/4\pi$. This vibration phenomenon is referred to as **beats** and is used in tuning stringed instruments. This same phenomenon in electronics is called **amplitude modulation.** To illustrate this phenomenon, sketch the curve $x(t)$ for $F_0 = 32$, $m = 2$, $\omega = 9$, and $\gamma = 7$.

10. Derive the formula for $x_p(t)$ given in (21).

•5.4• ELEMENTARY ELECTRIC CIRCUITS

In this section we consider the application of differential equations to an elementary electric circuit consisting of an electromotive force (e.g., a battery or generator), resistor, inductor, and capacitor in series. These circuits are called *RLC* **series circuits** and are represented schematically as shown in Figure 5.11.

Two physical principles governing *RLC* series circuits are (I) conservation of charge and (II) conservation of energy. These conservation laws were formulated for electric circuits by G. R. Kirchhoff in 1859 and are called **Kirchhoff's laws.** They state:

(I) The current I passing through each of the elements (resistor, inductor, capacitor, or electromotive force) in the series circuit must be the same.

(II) The algebraic sum of the instantaneous changes in potential (voltage drops) around a closed circuit must be zero.

In order to apply Kirchhoff's laws we obviously need to know the voltage drop across each element of the circuit. These voltage formulas are stated below (the reader can consult an introductory physics text for further details[†]).

(a) According to Ohm's law, the voltage drop E_R across a resistor is proportional to

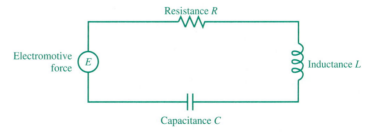

Figure 5.11 Schematic representation of an *RLC* series circuit

---

[†] For example, *Physics*, by Paul A. Tipler, Worth Publishers, Inc., New York, 1983.

the current I passing through the resistor:

$$E_R = RI.$$

The proportionality constant R is called the **resistance**.

(b) It can be shown using Faraday's law and Lenz's law that the voltage drop E_L across an inductor is proportional to the instantaneous rate of change of the current I:

$$E_L = L\frac{dI}{dt}.$$

The proportionality constant L is called the **inductance**.

(c) The voltage drop E_C across a capacitor is proportional to the electric charge q on the capacitor:

$$E_C = \frac{1}{C}q.$$

The proportionality constant $1/C$ is called the **elastance** and C the **capacitance**.

An electromotive force is assumed to *add* voltage or potential energy to the circuit. If we let $E(t)$ denote the voltage supplied to the circuit at time t, then Kirchhoff's conservation of energy law gives

(1) $E_L + E_R + E_C = E(t).$

Substituting into (1) the expressions for E_L, E_R, and E_C gives

(2) $L\dfrac{dI}{dt} + RI + \dfrac{1}{C}q = E(t).$

Now the current is just the instantaneous rate of change in charge; that is, $I = dq/dt$. Therefore, we can express (2) in terms of the single variable q as follows:

(3) $L\dfrac{d^2q}{dt^2} + R\dfrac{dq}{dt} + \dfrac{1}{C}q = E(t).$

In many applications we will be interested in determining the current $I(t)$. If we differentiate (3) with respect to t and substitute I for dq/dt, we obtain

(4) $L\dfrac{d^2I}{dt^2} + R\dfrac{dI}{dt} + \dfrac{1}{C}I = \dfrac{dE}{dt}.$

For an RLC series circuit, one is usually given the initial charge on the capacitor $q(0)$ and the initial current $I(0) = q'(0)$. These constants provide the initial conditions for equation (3). However, for equation (4), we also need to know the initial rate of change in the current, $I'(0)$. This can be obtained from equation (2) by substituting in the initial values $q(0)$, $I(0)$, and $E(0)$.

The common units and symbols used for electric circuits are listed in Table 5.1.

	TABLE 5.1 COMMON UNITS AND SYMBOLS USED WITH ELECTRIC CIRCUITS			

Quantity	Letter Representation	Units	Symbol Representation	
Electromotive force (impressed voltage)	E	volt (V)	—○— ┤├	Generator Battery
Resistance	R	ohm (Ω)	—⌇⌇—	
Inductance	L	henry (H)	—◠◠◠◠—	
Capacitance	C	farad (F)	┤├	
Charge	q	coulomb		
Current	I	ampere		

● **EXAMPLE 1**   An RLC series circuit has an electromotive force given by $E(t) = \sin 100t$ volts, a resistor of 0.02 ohms, an inductor of 0.001 henrys, and a capacitor of 2 farads. If the initial current and the initial charge on the capacitor are both zero, determine the current in the circuit for $t > 0$.

Solution   Here we have $L = 0.001$, $R = 0.02$, $C = 2$, and $E(t) = \sin 100t$. Substituting these into equation (4) for the current gives

(5) $$(0.001)\frac{d^2I}{dt^2} + (0.02)\frac{dI}{dt} + (0.5)I = 100\cos 100t,$$

or, equivalently,

(6) $$\frac{d^2I}{dt^2} + 20\frac{dI}{dt} + 500I = 100,000\cos 100t.$$

The homogeneous equation associated with (6) has the auxiliary equation

$$r^2 + 20r + 500 = (r + 10)^2 + (20)^2 = 0,$$

whose roots are $-10 \pm 20i$. Hence the solution to the homogeneous equation is

(7) $$I_h(t) = C_1 e^{-10t}\cos 20t + C_2 e^{-10t}\sin 20t.$$

To find a particular solution for (6), we can use the method of undetermined coefficients. Setting

$$I_p(t) = A\cos 100t + B\sin 100t$$

and carrying out the procedure in Section 4.8, we ultimately find

(8) $$A = -\frac{95}{9.425}, \qquad B = \frac{20}{9.425}.$$

Hence a particular solution to (6) is given by

(9) $$I_p(t) = -\frac{95}{9.425}\cos 100t + \frac{20}{9.425}\sin 100t.$$

Since $I = I_h + I_p$, we find from (7) and (9) that

(10) $I(t) = e^{-10t}(C_1 \cos 20t + C_2 \sin 20t) - \dfrac{95}{9.425} \cos 100t + \dfrac{20}{9.425} \sin 100t.$

To determine the constants C_1 and C_2, we need the values $I(0)$ and $I'(0)$. We were given $I(0) = q(0) = 0$. To find $I'(0)$, we substitute the values for L, R, and C into equation (2) and equate the two sides at $t = 0$. This gives

$$(0.001)I'(0) + (0.02)I(0) + (0.5)q(0) = \sin 0.$$

Since $I(0) = q(0) = 0$, we find $I'(0) = 0$. Finally, using $I(t)$ in (10) and the initial conditions $I(0) = I'(0) = 0$, we obtain the system

$$I(0) = C_1 - \frac{95}{9.425} = 0,$$

$$I'(0) = -10C_1 + 20C_2 + \frac{2000}{9.425} = 0.$$

Solving this system yields $C_1 = 95/9.425$ and $C_2 = -105/18.85$. Hence the current in the *RLC* series circuit is

(11) $I(t) = e^{-10t}\left(\dfrac{95}{9.425} \cos 20t - \dfrac{105}{18.85} \sin 20t \right)$

$$- \frac{95}{9.425} \cos 100t + \frac{20}{9.425} \sin 100t. \quad \bullet$$

Observe that, as was the case with mechanical vibrations, the current in (11) is made up of two components. The first, I_h, is a **transient current** that tends to zero as $t \to \infty$. The second,

$$I_p(t) = -\frac{95}{9.425} \cos 100t + \frac{20}{9.425} \sin 100t,$$

is a **steady-state current** that remains. If we had chosen to solve for the charge $q(t)$ in Example 1, we would also have found that there is a **transient charge** q_h that dies off and a **steady-state charge** q_p that remains.

We leave it for the reader to verify that the steady-state solutions $q_p(t)$ and $I_p(t)$ that arise from the electromotive force $E(t) = E_0 \sin \gamma t$ are

(12) $q_p(t) = \dfrac{-E_0 \cos(\gamma t + \theta)}{\sqrt{(1/C - L\gamma^2)^2 + \gamma^2 R^2}},$

(13) $I_p(t) = q'_p(t) = \dfrac{E_0 \sin(\gamma t + \theta)}{\sqrt{R^2 + [\gamma L - 1/(\gamma C)]^2}},$

where $\tan \theta = (1/C - L\gamma^2)/(\gamma R)$. The quantity $\sqrt{R^2 + [\gamma L - 1/(\gamma C)]^2}$ is called the **impedance** of the circuit and is a function of the frequency γ of the electromotive force $E(t)$.

It should be obvious to the reader that the differential equations that describe mechanical vibrations and RLC series circuits are essentially the same. And, in fact, there is a natural identification of the parameters m, b, and k for a spring-mass system with the parameters L, R, and C that describe circuits. This is illustrated in Table 5.2. Moreover, the terms *transient, steady-state, overdamped, critically damped, underdamped,* and *resonance frequency* described in the preceding sections apply to RLC series circuits as well.

TABLE 5.2 ANALOGY BETWEEN MECHANICAL AND ELECTRIC SYSTEMS			
Mechanical Spring-Mass System with Damping		**Electric RLC Series Circuit**	
$mx'' + bx' + kx = f(t)$		$Lq'' + Rq' + (1/C)q = E(t)$	
Displacement	x	Charge	q
Velocity	x'	Current	$q' = I$
Mass	m	Inductance	L
Damping constant	b	Resistance	R
Spring constant	k	Elastance	$1/C$
External force	$f(t)$	Electromotive force	$E(t)$

This analogy between a simple mechanical system and an elementary electric circuit extends to large-scale systems and circuits. An interesting consequence of this is the use of analog simulation and, in particular, analog computers to analyze mechanical systems. Due to size, time, and cost constraints, large-scale mechanical systems are modeled by building a corresponding electric system and then measuring the charge $q(t)$ and current $I(t)$. While such analog simulations are important, both large-scale mechanical and electric systems are currently modeled using computer simulation. This involves the numerical solution of the initial value problem governing the system. Still, the analogy between mechanical and electric systems means that basically the same computer software can be used to analyze both systems.

EXERCISES 5.4

1. An RLC series circuit has an electromotive force given by $E(t) = 20$ volts, a resistor of 100 ohms, an inductor of 4 henrys, and a capacitor of 0.01 farads. If the initial current is zero and the initial charge on the capacitor is 4 coulombs, determine the current in the circuit for $t > 0$.

2. An RLC series circuit has an electromotive force given by $E(t) = 40 \cos 2t$ volts, a resistor of 2 ohms, an inductor of $\frac{1}{4}$ henrys, and a capacitor of $\frac{1}{13}$ farads. If the initial current is zero and the initial charge on the capacitor is 3.5 coulombs, determine the charge on the capacitor for $t > 0$.

3. An RLC series circuit has an electromotive force given by $E(t) = 10 \cos 20t$ volts, a resistor of 120 ohms, an in-

ductor of 4 henrys, and a capacitor of $(2200)^{-1}$ farads. Find the steady-state current (solution) for this circuit. What is the resonance frequency of the circuit?

4. An LC series circuit has an electromotive force given by $E(t) = 30 \sin 50t$ volts, an inductor of 2 henrys, and a capacitor of 0.02 farads (but no resistor). What is the current in this circuit for $t > 0$ if at $t = 0$, $I(0) = q(0) = 0$?

5. An RLC series circuit has an electromotive force of the form $E(t) = E_0 \cos \gamma t$ volts, a resistor of 10 ohms, an inductor of 4 henrys, and a capacitor of 0.01 farads. Sketch the frequency response curve for this circuit.

6. Show that when the electromotive force in (3) is of the

form $E(t) = E_0 \sin \gamma t$, then the steady-state solutions q_p and I_p are as given in equations (12) and (13).

7. A spring-mass system with damping consists of a 7-kg mass, a spring with spring constant 3 N/m, a frictional component with damping constant 2 N-sec/m, and an external force given by $f(t) = 10 \cos 10t$ N. Using a 10-ohm resistor, construct an *RLC* series circuit that is the analog of this mechanical system in the sense that the two systems are governed by the same differential equation.

8. A spring-mass system with damping consists of a 16-lb weight, a spring with spring constant 64 lb/ft, a frictional component with damping constant 10 lb-sec/ft, and an external force given by $f(t) = 20 \cos 8t$ lb. Using an inductor of 0.01 henrys, construct an *RLC* series circuit that is the analog of this mechanical system.

9. Because of Euler's formula, $e^{i\theta} = \cos \theta + i \sin \theta$, it is often convenient to treat the electromotive forces $E_0 \cos \gamma t$ and $E_0 \sin \gamma t$ simultaneously, using $E(t) = E_0 e^{i\gamma t}$, where $i = \sqrt{-1}$. In this case equation (3) becomes

$$(14) \qquad L\frac{d^2q}{dt^2} + R\frac{dq}{dt} + \frac{1}{C}q = E_0 e^{i\gamma t}.$$

(a) Show that the steady-state solution to (14) is

$$q_p(t) = \frac{E_0}{1/C - \gamma^2 L + i\gamma R} e^{i\gamma t}.$$

[Hint: Use the method of undetermined coefficients with the guess $q_p = Ae^{i\gamma t}$, where A denotes a complex constant. The technique is discussed in detail in Section 4.9.]

(b) Now show that the steady-state current is

$$I_p(t) = \frac{E_0}{R + i[\gamma L + 1/(\gamma C)]} e^{i\gamma t}.$$

(c) Use the relation $\alpha + i\beta = \sqrt{\alpha^2 + \beta^2}\, e^{i\theta}$, where $\tan \theta = \beta/\alpha$, to show that I_p can be expressed in the form

$$I_p(t) = \frac{E_0}{\sqrt{R^2 + [\gamma L - 1/(\gamma C)]^2}} e^{i(\gamma t + \theta)},$$

where $\tan \theta = (1/C - L\gamma^2)/(\gamma R)$.

(d) Since the imaginary part of $e^{i\gamma t}$ is $\sin \gamma t$, then the imaginary part of the solution to (14) must be the solution to equation (3) for $E(t) = E_0 \sin \gamma t$. Verify that this is also the case for the current by showing that the imaginary part of I_p in part (c) is the same as that given in equation (13).

•5.5• LINEAR DIFFERENCE EQUATIONS

In this section we give a brief discussion of linear difference equations.[†] These equations occur in mathematical models of physical processes and as tools in numerical analysis. For example, a model for population dynamics under immigration involves the equation

$$P_{t+1} - P_t = aP_t + b,$$

where $P_{t+1} - P_t$ is the difference between the population at time $t + 1$ and at time t, the constant a is the difference between the average birth rate and average death rate, and b is the rate at which people immigrate to the country. First order difference equations of this type were studied in Project D in Chapter 3. Another example is the equation

$$y_{n+1} - y_{n-1} = -2hy_n,$$

which arises in the study of the stability of a numerical scheme that approximates the

---

[†] For a more in-depth discussion of difference equations and their applications we refer the reader to *Introduction to Difference Equations* by Samuel Goldberg, Dover Publications, Inc., New York, 1986, or *Differential and Difference Equations* by Louis Brand, John Wiley and Sons, Inc., New York, 1966.

solution to a differential equation. Here y_n is the approximation of the solution at x_n and h is the step size.

As these examples suggest, difference equations can be thought of as discrete analogs of differential equations. In fact, the theory of linear difference equations parallels the theory of linear differential equations as discussed in Chapter 4. Using the results of that chapter as a model both for the statements of the theorems and for their proofs, we can develop a theory and devise methods for solving linear difference equations.

Theory

A **kth-order linear difference equation** is an equation of the form

$$(1) \qquad a_k(n)y_{n+k} + a_{k-1}(n)y_{n+k-1} + \cdots + a_1(n)y_{n+1} + a_0(n)y_n = g_n,$$

$n = 0, 1, 2, \ldots$, where $a_k(n), \ldots, a_0(n)$, and g_n are defined for all nonnegative integers n. By a **solution** to (1) we mean a sequence of numbers $\{y_n\}_{n=0}^{\infty}$ that satisfies (1) for all integers $n \geq 0$. For example, the sequence $\{2^n\}_{n=0}^{\infty}$ is a solution to the equation

$$y_{n+2} + y_{n+1} - 6y_n = 0,$$

since

$$2^{n+2} + 2^{n+1} - (6)2^n = 2^n(4 + 2 - 6) = 0,$$

for $n = 0, 1, 2, \ldots$.

When $a_k(n) \neq 0$ for all integers $n \geq 0$, we can solve equation (1) for y_{n+k}:

$$(2) \qquad y_{n+k} = \frac{-1}{a_k(n)}\{a_{k-1}(n)y_{n+k-1} + \cdots + a_0(n)y_n - g_n\}.$$

If we are given the first k values for $\{y_n\}$, that is, $y_0, y_1, \ldots, y_{k-1}$, then we can use equation (2) with $n = 0$ to solve uniquely for y_k. But knowing y_k then allows us to use equation (2) with $n = 1$ to determine y_{k+1} uniquely. Since we can continue this process indefinitely, we see that there is a unique solution to (1) that satisfies the given initial conditions. We summarize these observations in the following theorem.

EXISTENCE AND UNIQUENESS OF SOLUTION

Theorem 1.   Let $Y_0, Y_1, \ldots, Y_{k-1}$ be given constants and assume that $a_k(n) \neq 0$ for all integers $n \geq 0$. Then, there exists a unique solution $\{y_n\}_{n=0}^{\infty}$ to (1) that satisfies the initial conditions $y_0 = Y_0, y_1 = Y_1, \ldots, y_{k-1} = Y_{k-1}$.

When $g_n = 0$ for $n \geq 0$, equation (1) becomes the **homogeneous** (or **reduced**) equation

$$(3) \qquad a_k(n)y_{n+k} + \cdots + a_0(n)y_n = 0, \qquad n = 0, 1, 2, \ldots.$$

As is the case for homogeneous differential equations, if two sequences $\{y_n\}_{n=0}^{\infty}$ and $\{z_n\}_{n=0}^{\infty}$ are solutions to the homogeneous equation (3), then so is any linear combination of these sequences (see Problem 28). In this context, by a linear combination of sequences $\{y_n\}_{n=0}^{\infty}$ and $\{z_n\}_{n=0}^{\infty}$, we mean a sequence of the form

$$\{c_1 y_n + c_2 z_n\}_{n=0}^{\infty},$$

where c_1 and c_2 are constants.

As you would probably guess, if we can find two "different" solutions $\{y_n\}_{n=0}^{\infty}$ and $\{z_n\}_{n=0}^{\infty}$ to a homogeneous linear *second* order equation

(4) $\qquad a_2(n)y_{n+2} + a_1(n)y_{n+1} + a_0(n)y_n = 0, \qquad n = 0, 1, 2, \ldots,$

then we can express any solution to (4) as a linear combination of these solutions. To show this (and to decide what we should mean by "different"), let $\{w_n\}_{n=0}^{\infty}$ be a solution to (4). If it is possible to choose constants c_1 and c_2 so that

(5) $\qquad \begin{aligned} c_1 y_0 + c_2 z_0 &= w_0, \\ c_1 y_1 + c_2 z_1 &= w_1, \end{aligned}$

then, since $\{w_n\}_{n=0}^{\infty}$ and $\{c_1 y_n + c_2 z_n\}_{n=0}^{\infty}$ are two solutions to (4) that satisfy the same initial conditions (5), the uniqueness conclusion of Theorem 1 gives

$$w_n = c_1 y_n + c_2 z_n, \qquad n = 0, 1, 2, \ldots.$$

A necessary and sufficient condition for (5) to have a unique solution for every w_0, w_1 is that

(6) $\qquad \begin{vmatrix} y_0 & z_0 \\ y_1 & z_1 \end{vmatrix} = y_0 z_1 - y_1 z_0 \neq 0.$

This is equivalent to saying that the first two terms in the sequences are not multiples of each other, that is, $(z_0, z_1) \neq c(y_0, y_1)$.

We now state the representation theorem for solutions to the homogeneous equation that we have just proved.

▶ REPRESENTATION OF SOLUTIONS (HOMOGENEOUS CASE)

Theorem 2. Assume $a_2(n) \neq 0$ for all integers $n \geq 0$, and let $\{y_n\}_{n=0}^{\infty}$ and $\{z_n\}_{n=0}^{\infty}$ be any two solutions to (4) that satisfy the condition[†]

(7) $\qquad y_0 z_1 - y_1 z_0 \neq 0.$

Then, every solution $\{w_n\}_{n=0}^{\infty}$ to (4) can be expressed in the form

(8) $\qquad w_n = c_1 y_n + c_2 z_n, \qquad n = 0, 1, 2, \ldots,$

where c_1 and c_2 are constants.

[†] Condition (7) is similar to the condition we encountered in Chapter 4 for linear second order differential equations. There we introduced the Wronskian function and discussed its connection with linear independence. The analog of the Wronskian function is the **Casoratian** sequence, which is named for the Italian mathematician Felice Casorati (1835–1890).

A solution set $\{\{y_n\}_{n=0}^{\infty}, \{z_n\}_{n=0}^{\infty}\}$ that satisfies condition (7) is called a **fundamental solution set** for (4). The linear combination of $\{y_n\}_{n=0}^{\infty}$ and $\{z_n\}_{n=0}^{\infty}$ in (8), written with arbitrary constants c_1 and c_2, is referred to as a **general solution** to (4).

● EXAMPLE 1   Given that $\{2^{-n}\}_{n=0}^{\infty}$ and $\{(-3)^n\}_{n=0}^{\infty}$ are solutions to

$$(9) \qquad 2y_{n+2} + 5y_{n+1} - 3y_n = 0, \qquad n = 0, 1, 2, \ldots,$$

find the solution to (9) that satisfies $y_0 = 0$ and $y_1 = 1$.

Solution   Since $(2^{-0})(-3) - (2^{-1})(-3)^0 = -3 - \frac{1}{2} \neq 0$, condition (7) is satisfied and so a general solution to (9) is

$$(10) \qquad y_n = c_1 2^{-n} + c_2 (-3)^n, \qquad n = 0, 1, 2, \ldots,$$

where c_1 and c_2 are arbitrary constants. Using the initial conditions to solve for c_1 and c_2 gives the system

$$y_0 = c_1 + c_2 = 0,$$
$$y_1 = (1/2)c_1 - 3c_2 = 1,$$

whose solution is $c_1 = -c_2 = \frac{2}{7}$. Thus the solution we seek is

$$(11) \qquad y_n = \left(\frac{2}{7}\right)2^{-n} - \left(\frac{2}{7}\right)(-3)^n, \qquad n = 0, 1, 2, \ldots.$$

The first 12 terms of the solution are displayed in Table 5.3.   ●

TABLE 5.3	SOME TERMS IN THE SOLUTION TO THE INITIAL VALUE PROBLEM IN EXAMPLE 1				
n	y_n	n	y_n	n	y_n
0	0	4	-23.125	8	-1874.57031
1	1	5	69.4375	9	5623.71485
2	-2.5	6	-208.28125	10	-16871.14258
3	7.75	7	624.85938	11	50613.42872

When $\{g_n\}$ is not the zero sequence, equation (1) is said to be **nonhomogeneous** (or **complete**). Using the superposition principle for linear difference equations (see Problem 29) and the representation theorem for homogeneous equations, we can prove the following representation theorem for nonhomogeneous equations. We leave its proof as an exercise (see Problem 30).

> ▸ **REPRESENTATION OF SOLUTIONS (NONHOMOGENEOUS CASE)**
>
> **Theorem 3.**   Assume $a_2(n) \neq 0$ for all integers $n \geq 0$, and let $\{p_n\}_{n=0}^{\infty}$ be a particular solution to the nonhomogeneous equation
>
> **(12)** $a_2(n)y_{n+2} + a_1(n)y_{n+1} + a_0(n)y_n = g_n, \qquad n = 0, 1, 2, \ldots .$
>
> Then, every solution to (12) can be expressed in the form
>
> **(13)** $w_n = p_n + c_1 y_n + c_2 z_n, \qquad n = 0, 1, 2, \ldots ,$
>
> where $\{\{y_n\}_{n=0}^{\infty}, \{z_n\}_{n=0}^{\infty}\}$ is a fundamental solution set for the corresponding homogeneous equation and c_1 and c_2 are constants.

The linear combination of $\{p_n\}_{n=0}^{\infty}$, $\{y_n\}_{n=0}^{\infty}$, and $\{z_n\}_{n=0}^{\infty}$ in (13) written with arbitrary constants c_1 and c_2 is referred to as a **general solution** to (12).

Constant Coefficient Equations

When a_0, a_1, and a_2 are constants, the equation

(14) $a_2 y_{n+2} + a_1 y_{n+1} + a_0 y_n = g_n, \qquad n = 0, 1, 2, \ldots$

is said to have **constant coefficients.** To find a general solution to (14), we begin with the homogeneous equation

(15) $a_2 y_{n+2} + a_1 y_{n+1} + a_0 y_n = 0, \qquad n = 0, 1, 2, \ldots .$

We assume that both a_0 and a_2 are not zero so that we have a genuine second order difference equation.

Our experience with differential equations suggests that we try to find solutions of the form $y_n = r^n, n = 0, 1, 2, \ldots$, where $r \neq 0$ is a fixed (real or complex) number. Substituting in (15) gives

$$a_2 r^{n+2} + a_1 r^{n+1} + a_0 r^n = r^n(a_2 r^2 + a_1 r + a_0) = 0.$$

Hence, a nontrivial solution to (15) is given by $y_n = r^n, n = 0, 1, 2, \ldots$, if and only if r satisfies the **auxiliary equation**

(16) $a_2 r^2 + a_1 r + a_0 = 0.$

Now the form of a fundamental solution set for (15) depends upon the type of roots to the auxiliary equation. These results are summarized in the following theorem.

> **FORM OF GENERAL SOLUTION TO** $a_2y_{n+2} + a_1y_{n+1} + a_0y_n = 0$

Theorem 4.   Let r_1 and r_2 be the roots of the auxiliary equation (16), where a_2, a_1, a_0 are real constants, $a_2 \neq 0$, $a_1 \neq 0$.

(a) If r_1 and r_2 are distinct real roots, then a general solution to (15) is

$$y_n = c_1 r_1^n + c_2 r_2^n, \qquad n = 0, 1, 2, \ldots .$$

(b) If $r_1 = r_2 = r = -a_1/(2a_2)$ is a repeated root, then a general solution to (15) is

$$y_n = c_1 r^n + c_2 n r^n, \qquad n = 0, 1, 2, \ldots .$$

(c) If r_1 and r_2 are complex conjugate roots and we express r_1 in the polar form $r_1 = \rho e^{i\theta}$, then a general solution to (15) is

$$y_n = c_1 \rho^n \cos n\theta + c_2 \rho^n \sin n\theta, \qquad n = 0, 1, 2, \ldots .$$

Proof.   Parts (a) and (b) are analogous to the results for linear second order differential equations with constant coefficients, and so we omit their proofs. To prove part (c), we observe that if $r_1 = \rho e^{i\theta}$, then $r_1^n = \rho^n e^{in\theta}$. Using Euler's formula, we find

$$r_1^n = \rho^n \{\cos n\theta + i \sin n\theta\}, \qquad n = 0, 1, 2, \ldots .$$

If $\{r_1^n\}_{n=0}^{\infty}$ is a complex-valued solution, then the two real-valued sequences that make up the real and imaginary parts of $\{r_1^n\}_{n=0}^{\infty}$ must also be solutions. Hence, two solutions are

$$\{\rho^n \cos n\theta\}_{n=0}^{\infty} \quad \text{and} \quad \{\rho^n \sin n\theta\}_{n=0}^{\infty}.$$

Here condition (7) is $\rho \sin \theta = \text{Im}(r_1) \neq 0$. But we have assumed that r_1 is complex, and so $\text{Im}(r_1) \neq 0$. Thus, these two real-valued sequences form a fundamental solution set. ◀◀◀

● **EXAMPLE 2**   Find a general solution to

$$y_{n+2} + 6y_{n+1} + 9y_n = 0, \qquad n = 0, 1, 2, \ldots .$$

Solution   The auxiliary equation is

$$r^2 + 6r + 9 = (r + 3)^2 = 0,$$

which has roots $r_1 = r_2 = -3$. Hence, by Theorem 4, part (b), a general solution is

$$y_n = c_1(-3)^n + c_2 n(-3)^n, \qquad n = 0, 1, 2, \ldots . \quad ●$$

● **EXAMPLE 3**   Find a general solution to

$$y_{n+2} + 2y_{n+1} + 5y_n = 0, \qquad n = 0, 1, 2, \ldots .$$

Solution   The auxiliary equation is

$$r^2 + 2r + 5 = (r + 1)^2 + 2^2 = 0,$$

which has roots $r = -1 \pm 2i$. Writing $r_1 = -1 + 2i$ in polar form, we find $\rho = \sqrt{(-1)^2 + 2^2} = \sqrt{5}$ and $\tan\theta = 2/(-1) = -2$. Since θ lies in the second quadrant, we have $\theta = \pi - \arctan 2 \approx 2.03$ radians. Hence, by Theorem 4, part (c), a general solution is

$$y_n = c_1(\sqrt{5})^n \cos n\theta + c_2(\sqrt{5})^n \sin n\theta, \qquad n = 0, 1, 2, \ldots,$$

where $\theta = \pi - \arctan 2$. ●

● EXAMPLE 4 To approximate the solution to the initial value problem

$$\textbf{(17)} \qquad y' = -y, \qquad y(0) = 1,$$

a method based upon the center-difference formula, $y'(x) \approx [y(x+h) - y(x-h)]/2h$, leads to the difference equation

$$\textbf{(18)} \qquad y_{n+1} = y_{n-1} - 2hy_n, \qquad n = 1, 2, 3, \ldots, \qquad y_0 = 1,$$

where $h\ (>0)$ is the step size and y_n is the approximation to y at the point $x_n = nh$, $n = 0, 1, 2, \ldots$. Discuss the stability of this method; that is, compare the behavior (as $n \to \infty$) of the solutions to (18) with the solution to (17).

Solution The exact solution to equation (17) is just $\phi(x) = e^{-x}$. For fixed $h\ (>0)$, if $x_n = nh$, then $\phi(x_n) = e^{-x_n} = e^{-nh}$ and $\phi(x_n) \to 0$ as $n \to \infty$.

Equation (18) is a linear second order difference equation with auxiliary equation

$$r^2 + 2hr - 1 = 0,$$

which has roots

$$r_1 = -h + \sqrt{1 + h^2} \quad \text{and} \quad r_2 = -h - \sqrt{1 + h^2}.$$

Hence, a general solution to (18) is

$$y_n = c_1 r_1^n + c_2 r_2^n.$$

To solve for c_1 and c_2, we must know y_0 and y_1. The initial condition for the initial value problem (17) only provides $y_0 = 1$. Thus the value for y_1 must be determined by some other numerical method which provides an approximation to the solution of (17) at $x = h$. This, however, is not the problem. To understand the difficulty, let's consider what happens to r_1^n and r_2^n as $n \to \infty$.

For any $h > 0$, it is easy to check that $0 < r_1 < 1$. Hence $r_1^n \to 0$ as $n \to \infty$. This is the same behavior as the exact solution $\phi(x) = e^{-x}$. But, $r_2 = -h - \sqrt{1 + h^2} < -1$, and so r_2^n diverges as $n \to \infty$. Consequently, if $c_2 \neq 0$, then $|y_n| \to \infty$.

This particular method for approximating the solution to (17) has given rise to the extraneous "solution" $c_2 r_2^n$ which exhibits larger and larger oscillations. Moreover, since y_1 can only be approximated due to truncation error (or because of round-off), the coefficient c_2, however small, will be nonzero. Hence, at some point the term $c_2 r_2^n$ will dominate the term $c_1 r_1^r$, giving rise to wild oscillations of the form $(-h - \sqrt{1 + h^2})^n$ (see Figure 5.12). Therefore, this method is unstable for any choice of the step size h. ●

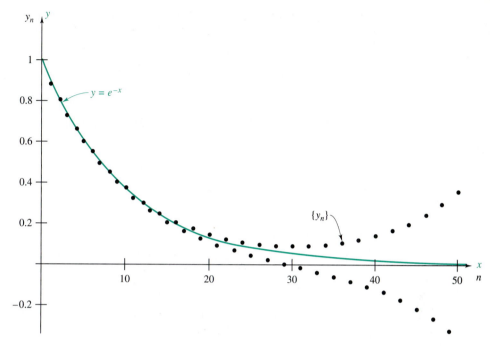

Figure 5.12 Approximation of $y = e^{-x}$ using the solution $\{y_n\}_{n=0}^{\infty}$ to the difference equation (18) with $h = 0.1$ and $y_1 = 0.9$

Undetermined Coefficients

To find a solution to the nonhomogeneous equation (14), we use the analog of the **method of undetermined coefficients** discussed in Section 4.8. We can use Table 4.1 on page 173 with a minor modification. For difference equations it is more convenient to express an exponential function in the form α^n rather than $e^{\alpha t}$. (Recall $\alpha^n = e^{(\ln \alpha)n}$.) The method of undetermined coefficients is illustrated in the following example.

● **EXAMPLE 5**   Find a general solution to

(19) $y_{n+2} - 2y_{n+1} + 2y_n = 3 \sin n$, $n = 0, 1, 2, \ldots$.

Solution   We first solve the associated homogeneous equation

$$y_{n+2} - 2y_{n+1} + 2y_n = 0,$$

which has the auxiliary equation $r^2 - 2r + 2 = (r - 1)^2 + 1^2 = 0$. The roots of the auxiliary equation are $r = 1 \pm i$. Writing $1 + i = \rho e^{i\theta}$, we find $\rho = \sqrt{1^2 + 1^2} = \sqrt{2}$ and $\theta = \arctan 1 = \pi/4$. Hence, a general solution to the homogeneous equation is

(20) $h_n = c_1(\sqrt{2})^n \cos\left(\dfrac{n\pi}{4}\right) + c_2(\sqrt{2})^n \sin\left(\dfrac{n\pi}{4}\right)$, $n = 0, 1, 2, \ldots$.

Since the nonhomogeneous term is $g_n = \sin n$, we seek a particular solution of the form

(21) $p_n = A \cos n + B \sin n$,

where the constants A and B are to be determined. Substituting this expression for p_n into (19) yields

$$A\cos(n + 2) + B\sin(n + 2) - 2(A\cos(n + 1) + B\sin(n + 1))$$
$$+ 2(A\cos n + B\sin n) = 3\sin n, \qquad n = 0, 1, 2, \ldots.$$

Using the addition formulas for the trigonometric functions and simplifying, we have

$$[(2 + \cos 2 - 2\cos 1)A + (\sin 2 - 2\sin 1)B]\cos n$$
$$+ [(2\sin 1 - \sin 2)A + (2 + \cos 2 - 2\cos 1)B]\sin n = 3\sin n.$$

When we set corresponding coefficients equal, we obtain the system

$$(2 + \cos 2 - 2\cos 1)A + (\sin 2 - 2\sin 1)B = 0,$$
$$(2\sin 1 - \sin 2)A + (2 + \cos 2 - 2\cos 1)B = 3,$$

which has the solution

(22)
$$A = \frac{-3(\sin 2 - 2\sin 1)}{(2 + \cos 2 - 2\cos 1)^2 + (\sin 2 - 2\sin 1)^2},$$
$$B = \frac{3(2 + \cos 2 - 2\cos 1)}{(2 + \cos 2 - 2\cos 1)^2 + (\sin 2 - 2\sin 1)^2}.$$

Thus, a general solution to (26) is

$$y_n = h_n + p_n, \qquad n = 0, 1, 2, \ldots,$$

where h_n is given by equation (20) and p_n is given by equation (21) with A and B given in (22). ●

When one of the terms in our initial guess $\{p_n\}_{n=0}^{\infty}$ is a solution to the corresponding homogeneous equation, we must modify our guess by multiplying by n^s where s is the smallest nonnegative integer such that no term in $\{n^s p_n\}_{n=0}^{\infty}$ is a solution to the corresponding homogeneous equation.

EXERCISES 5.5

In Problems 1 through 10 find a general solution to the given homogeneous equation for $n = 0, 1, 2, \ldots$.

1. $y_{n+2} - 6y_{n+1} + 8y_n = 0.$

2. $y_{n+2} - 4y_{n+1} + 3y_n = 0.$

3. $y_{n+2} + 3y_{n+1} - y_n = 0.$

4. $y_{n+2} + 4y_{n+1} + 4y_n = 0.$

5. $y_{n+2} - 6y_{n+1} + 11y_n = 0.$

6. $y_{n+2} + 4y_{n+1} + 9y_n = 0.$

7. $y_{n+2} + 10y_{n+1} + 25y_n = 0.$

8. $4y_{n+2} + 4y_{n+1} - 3y_n = 0.$

9. $y_{n+2} - y_{n+1} + 4y_n = 0.$

10. $y_{n+2} + y_{n+1} - y_n = 0.$

In Problems 11 through 18 find a general solution to the given nonhomogeneous equation for $n = 0, 1, 2, \ldots$.

11. $y_{n+2} + y_n = n^2.$

12. $y_{n+2} + y_{n+1} - 2y_n = n - 1.$

13. $y_{n+2} - y_n = \cos 3n.$

14. $y_{n+2} - 4y_n = \sin 2n.$

15. $y_{n+2} + 6y_{n+1} + 5y_n = 2^n$.

16. $y_{n+2} + 2y_{n+1} - y_n = (-1)^n$.

17. $y_{n+2} - 4y_{n+1} + 4y_n = 2^n$.

18. $y_{n+2} - y_{n+1} - 2y_n = (-1)^n$.

In Problems 19 through 22 find a general solution to the given third order equation for $n = 0, 1, 2, \dots$.

19. $y_{n+3} + y_{n+2} - 4y_{n+1} - 4y_n = 0$.

20. $y_{n+3} - 4y_{n+2} + 5y_{n+1} - 2y_n = 0$.

21. $y_{n+3} - 3y_{n+2} + 4y_{n+1} - 2y_n = 0$.

22. $y_{n+3} + 3y_{n+2} + 3y_{n+1} + y_n = 0$.

In Problems 23 through 27 find the solution to the initial value problem for $n = 0, 1, 2, \dots$.

23. $y_{n+2} + 2y_{n+1} + 5y_n = 0$; $y_0 = 1$, $y_1 = -1$.

24. $y_{n+2} - 5y_{n+1} + 6y_n = 0$; $y_0 = 3$, $y_1 = -2$.

25. $y_{n+2} - 6y_{n+1} + 9y_n = 1$; $y_0 = 0$, $y_1 = 0$.

26. $y_{n+2} - 4y_{n+1} + 3y_n = 3^n$; $y_0 = 1$, $y_1 = 0$.

27. $y_{n+2} - 2y_{n+1} + y_n = n$; $y_0 = 1$, $y_1 = 3$.

28. Prove that if the two sequences $\{y_n\}_{n=0}^\infty$ and $\{z_n\}_{n=0}^\infty$ are solutions to the homogeneous equation (3), then so is any linear combination $\{c_1 y_n + c_2 z_n\}_{n=0}^\infty$.

29. Superposition Principle.   Let $\{y_n\}_{n=0}^\infty$ be a solution to the nonhomogeneous equation (1) with nonhomogeneous term g_n and $\{z_n\}_{n=0}^\infty$ be a solution to (1) with nonhomogeneous term h_n. Show that the sequence $\{c_1 y_n + c_2 z_n\}_{n=0}^\infty$ is a solution to (1) with nonhomogeneous term $c_1 g_n + c_2 h_n$.

30. Using the representation theorem for homogeneous

equations and the superposition principle given in Problem 29, prove Theorem 3, the representation theorem for nonhomogeneous equations.

31. The linear multistep numerical method

$$y_{n+2} - y_{n+1} = \frac{h}{2}(3f_{n+1} - f_n), \qquad h > 0,$$

is used to approximate the solution to

$$y' = f(x, y), \qquad y(0) = y_0,$$

where $x_n := nh$, $f_n := f(x_n, y_n)$. As in Example 4, discuss the stability (as $n \to \infty$) of this method when it is applied to

$$y' = -y, \qquad y(0) = 1.$$

32. National Income Model.   P. A. Samuelson[†] developed a model for national income Y_t in period t under fixed government spending. His model involved the difference equation

$$Y_{t+2} - \alpha(1 + \beta)Y_{t+1} + \alpha\beta Y_t = 1, \qquad t = 0, 1, 2, \dots,$$

where α is a constant called the *marginal propensity to consume* and β is a constant of proportionality called the *relation*, which is a consequence of the so-called acceleration principle of economics.

(a) Solve for the national income in the special case when $\alpha = \frac{1}{2}$, $\beta = 1$, $Y_0 = 2$, and $Y_1 = 3$.

(b) Show that as $t \to \infty$, the national income Y experiences damped oscillations that converge to $1/(1 - \alpha) = 2$. (Samuelson concluded that a fixed level of government spending will result in damped oscillations in the national income that will gradually approach the quantity $1/(1 - \alpha)$ times the level of government spending.)

•5.6• NUMERICAL METHODS FOR SECOND ORDER EQUATIONS

The standard approach for obtaining numerical approximations of the solution to an initial value problem for a second order differential equation is first to convert the second order equation to a system of first order equations. This can be accomplished by the following reduction procedure.

Any second order initial value problem

(1) $y''(t) = F(t, y, y')$; $y(t_0) = y_0$, $y'(t_0) = y_1$,

[†] P. A. Samuelson, "Interactions Between the Multiplier Analysis and the Principle of Acceleration," *Review of Economic Statistics*, 21 (1939), 75–78.

can be converted into a first order system by setting

(2) $x_1(t) := y(t), \qquad x_2(t) := y'(t).$

With this substitution, equation (1) is equivalent to the system

(3) $\begin{aligned} x_1'(t) &= y'(t) = x_2(t), \\ x_2'(t) &= y''(t) = F(t, x_1, x_2), \end{aligned}$

and the initial conditions become

(4) $x_1(t_0) = y(t_0) = y_0, \qquad x_2(t_0) = y'(t_0) = y_1.$

● **EXAMPLE 1**   Convert the initial value problem

(5) $y''(t) + 3y'(t) + 2y(t) = 0; \qquad y(0) = 1, \qquad y'(0) = 1,$

into an initial value problem for a system of first order equations.

Solution   We first express the differential equation in (5) as

$$y''(t) = -3y'(t) - 2y(t).$$

Setting $x_1(t) := y(t)$, $x_2(t) := y'(t)$, we obtain

(6) $\begin{aligned} x_1'(t) &= x_2(t), \\ x_2'(t) &= -3x_2(t) - 2x_1(t). \end{aligned}$

The initial conditions in (5) transform to $x_1(0) = 1$, $x_2(0) = 1$.   ●

The numerical methods for solving initial value problems for systems of first order differential equations are just extensions of techniques for solving a single first order differential equation. To see how these extensions are done, let's recall the classical Runge-Kutta method of order four, which was discussed in Section 3.6.

For the initial value problem

(7) $x' = f(t, x), \qquad x(t_0) = x_0,$

the recursive formulas for the fourth order Runge-Kutta method are

(8) $\begin{aligned} t_{n+1} &= t_n + h, \qquad n = 0, 1, \ldots, \\ x_{n+1} &= x_n + \tfrac{1}{6}(k_1 + 2k_2 + 2k_3 + k_4), \end{aligned}$

where h is the step size and

(9) $\begin{aligned} k_1 &= hf(t_n, x_n), \qquad k_2 = hf\left(t_n + \frac{h}{2}, x_n + \frac{1}{2}k_1\right), \\ k_3 &= hf\left(t_n + \frac{h}{2}, x_n + \frac{1}{2}k_2\right), \qquad k_4 = hf(t_n + h, x_n + k_3). \end{aligned}$

Now suppose that we wish to approximate the solution $x_1(t)$, $x_2(t)$ to the system

$$
\begin{aligned}
x_1'(t) &= f_1(t, x_1, x_2), \\
x_2'(t) &= f_2(t, x_1, x_2),
\end{aligned}
$$
(10)

that satisfies the initial conditions

(11) $x_1(t_0) = a_1$, $x_2(t_0) = a_2$.

Let $x_{n,1}$ and $x_{n,2}$ denote approximations to $x_1(t_n)$ and $x_2(t_n)$, respectively, where $t_n = t_0 + nh$ for $n = 0, 1, 2, \dots$. The recursive formulas for the fourth order Runge-Kutta method for the system (10)–(11) are obtained from the formulas for a single equation given in (8)–(9) by treating each of the quantities x_n, k_1, k_2, k_3, and k_4 as *vectors*. Namely, we set $\mathbf{x}_n :=$ $(x_{n,1}, x_{n,2})$, $\mathbf{k}_1 := (k_{1,1}, k_{1,2})$, $\mathbf{k}_2 := (k_{2,1}, k_{2,2})$, etc. This approach gives the following recursive formulas for the fourth order Runge-Kutta method for system (10)–(11):

(12) $t_{n+1} := t_n + h$, $n = 0, 1, 2, \dots$,

(13) $x_{n+1,1} := x_{n,1} + \frac{1}{6}(k_{1,1} + 2k_{2,1} + 2k_{3,1} + k_{4,1})$,

(14) $x_{n+1,2} := x_{n,2} + \frac{1}{6}(k_{1,2} + 2k_{2,2} + 2k_{3,2} + k_{4,2})$,

where h is the step size and, for $i = 1$ and 2,

$$k_{1,i} := hf_i(t_n, x_{n,1}, x_{n,2}),$$

$$k_{2,i} := hf_i\left(t_n + \frac{h}{2}, x_{n,1} + \frac{1}{2}k_{1,1}, x_{n,2} + \frac{1}{2}k_{1,2} \right),$$

$$k_{3,i} := hf_i\left(t_n + \frac{h}{2}, x_{n,1} + \frac{1}{2}k_{2,1}, x_{n,2} + \frac{1}{2}k_{2,2} \right),$$

$$k_{4,i} := hf_i(t_n + h, x_{n,1} + k_{3,1}, x_{n,2} + k_{3,2}).$$

It is important to note that both $k_{1,1}$ and $k_{1,2}$ must be computed before either $k_{2,1}$ or $k_{2,2}$. Similarly, both $k_{2,1}$ and $k_{2,2}$ are needed to compute $k_{3,1}$ and $k_{3,2}$, etc.

● **EXAMPLE 2**   Use the Runge-Kutta method with step size $h = 0.125$ to find an approximation at $t = 1$ for the solution to the initial value problem

(15) $y''(t) + 3y'(t) + 2y(t) = 0$; $y(0) = 1$, $y'(0) = 1$.

Solution   In Example 1 we showed that setting $x_1 := y$, $x_2 := y'$ reduces the initial value problem in (15) to

$$
\begin{aligned}
x_1' &= x_2; & x_1(0) &= 1, \\
x_2' &= -3x_2 - 2x_1; & x_2(0) &= 1.
\end{aligned}
$$
(16)

Comparing this system with (10)–(11), we see that $f_1(t, x_1, x_2) = x_2$ and $f_2(t, x_1, x_2) = -3x_2 - 2x_1$. With the starting values of $t_0 = 0$, $x_{0,1} = 1$, and $x_{0,2} = 1$, we compute

$$k_{1,1} = hf_1(t_0, x_{0,1}, x_{0,2}) = hx_{0,2} = (0.125)(1) = 0.125,$$
$$k_{1,2} = hf_2(t_0, x_{0,1}, x_{0,2}) = h(-3x_{0,2} - 2x_{0,1})$$
$$= (0.125)[(-3)(1) - 2(1)] = -0.625,$$

$$k_{2,1} = hf_1\left(t_0 + \frac{h}{2}, x_{0,1} + \frac{1}{2}k_{1,1}, x_{0,2} + \frac{1}{2}k_{1,2}\right)$$

$$= h\left(x_{0,2} + \frac{1}{2}k_{1,2}\right) = (0.125)(1 - 0.3125) = 0.08594,$$

$$k_{2,2} = hf_2\left(t_0 + \frac{h}{2}, x_{0,1} + \frac{1}{2}k_{1,1}, x_{0,2} + \frac{1}{2}k_{1,2}\right)$$

$$= -3h\left(x_{0,2} + \frac{1}{2}k_{1,2}\right) - 2h\left(x_{0,1} + \frac{1}{2}k_{1,1}\right)$$

$$= -3(0.125)(1 - 0.3125) - 2(0.125)(1 + 0.0625)$$

$$= -0.52344.$$

Similarly, we find

$$k_{3,1} = hf_1\left(t_0 + \frac{h}{2}, x_{0,1} + \frac{1}{2}k_{2,1}, x_{0,2} + \frac{1}{2}k_{2,2}\right) = 0.09229,$$

$$k_{3,2} = hf_2\left(t_0 + \frac{h}{2}, x_{0,1} + \frac{1}{2}k_{2,1}, x_{0,2} + \frac{1}{2}k_{2,2}\right) = -0.53760,$$

$$k_{4,1} = hf_1(t_0 + h, x_{0,1} + k_{3,1}, x_{0,2} + k_{3,2}) = 0.05780,$$
$$k_{4,2} = hf_2(t_0 + h, x_{0,1} + k_{3,1}, x_{0,2} + k_{3,2}) = -0.44647.$$

Hence from (13) and (14), we compute

$$x_{1,1} = x_{0,1} + \tfrac{1}{6}(k_{1,1} + 2k_{2,1} + 2k_{3,1} + k_{4,1})$$
$$= 1 + \tfrac{1}{6}(0.125 + 2(0.08594) + 2(0.09229) + 0.05780)$$
$$= 1.08987,$$
$$x_{1,2} = x_{0,2} + \tfrac{1}{6}(k_{1,2} + 2k_{2,2} + 2k_{3,2} + k_{4,2})$$
$$= 1 - \tfrac{1}{6}(0.625 + 2(0.52344) + 2(0.53760) + 0.44647)$$
$$= 0.46774.$$

Continuing the algorithm, we compute $x_{i,1}$ and $x_{i,2}$ for $i = 2, 3, \ldots, 8$. These values, along with the values of the actual solution to (15), $y(t) = x_1(t) = 3e^{-t} - 2e^{-2t}$, are given in Table 5.4 (to five decimal places). We note that the $x_{i,2}$ column gives approximations for $y'(t)$ since $x_2(t) = y'(t)$. ●

TABLE 5.4	FOURTH ORDER RUNGE-KUTTA APPROXIMATION OF THE SOLUTION TO (15) USING $h = 0.125$			
t_i	$x_{i,1}$	$x_{i,2}$	$y(t_i) = x_1(t_i)$	$\lvert y(t_i) - x_{i,1} \rvert$
0.000	1.00000	1.00000	1.00000	0
0.125	1.08987	0.46774	1.08989	2×10^{-5}
0.250	1.12332	0.08977	1.12334	2×10^{-5}
0.375	1.11711	-0.17235	1.11713	2×10^{-5}
0.500	1.08381	-0.34802	1.08383	2×10^{-5}
0.625	1.03275	-0.45971	1.03277	2×10^{-5}
0.750	0.97981	-0.52453	0.97984	3×10^{-5}
0.875	0.90302	-0.55544	0.90304	2×10^{-5}
1.000	0.83295	-0.56226	0.83297	2×10^{-5}

Notice that the Runge-Kutta method as described in (12)–(14) applies to any system of two first order equations. For the second order equation (1), with the reduction procedure of setting $x_1 = y$, $x_2 = y'$, we find by comparing system (3) with system (10) that

$$f_1(t, x_1, x_2) = x_2,$$
$$f_2(t, x_1, x_2) = F(t, x_1, x_2).$$

Using these functions, we can easily modify the Runge-Kutta algorithm of Chapter 3 to obtain the Runge-Kutta algorithm for approximating the solution to an initial value problem for a second order equation (see box below). Notice that the new procedure stops when two successive approximations for *both* the function $y = x_1$ and its derivative $y' = x_2$ differ by less than a prescribed tolerance ε, or it stops after a prescribed maximum number of iterations.

RUNGE-KUTTA ALGORITHM FOR SECOND ORDER EQUATIONS

Purpose   To approximate the solution to the initial value problem

$$y'' = F(t, y, y'); \qquad y(t_0) = a_1, \qquad y'(t_0) = a_2$$

at $t = c$.

INPUT t_0, a_1, a_2, c
ε (tolerance)
M (maximum number of iterations)

Step 1   Set $z_1 = a_1$, $z_2 = a_2$
Step 2   For $n = 0$ to M do Steps 3–9
Step 3   Set $h = (c - t_0)2^{-n}$, $t = t_0$, $x_1 = a_1$, $x_2 = a_2$
Step 4   For $j = 1$ to 2^n do Steps 5 and 6

Step 5 Set

$$k_{1,1} = hx_2$$
$$k_{1,2} = hF(t, x_1, x_2)$$
$$k_{2,1} = h\left(x_2 + \frac{1}{2}k_{1,2}\right)$$
$$k_{2,2} = hF\left(t + \frac{h}{2}, x_1 + \frac{1}{2}k_{1,1}, x_2 + \frac{1}{2}k_{1,2}\right)$$
$$k_{3,1} = h\left(x_2 + \frac{1}{2}k_{2,2}\right)$$
$$k_{3,2} = hF\left(t + \frac{h}{2}, x_1 + \frac{1}{2}k_{2,1}, x_2 + \frac{1}{2}k_{2,2}\right)$$
$$k_{4,1} = h(x_2 + k_{3,2})$$
$$k_{4,2} = hF(t + h, x_1 + k_{3,1}, x_2 + k_{3,2})$$

Step 6 Set

$$t = t + h$$
$$x_1 = x_1 + \frac{1}{6}(k_{1,1} + 2k_{2,1} + 2k_{3,1} + k_{4,1})$$
$$x_2 = x_2 + \frac{1}{6}(k_{1,2} + 2k_{2,2} + 2k_{3,2} + k_{4,2})$$

Step 7 Print h, x_1, x_2

Step 8 If $|z_1 - x_1| < \varepsilon$ and $|z_2 - x_2| < \varepsilon$, go to Step 12

Step 9 Set $z_1 = x_1, z_2 = x_2$

Step 10 Print "$y(c)$ is approximately"; x_1; "but may not be within the tolerance"; ε

Step 11 Go to Step 13

Step 12 Print "$y(c)$ is approximately"; x_1; "with tolerance"; ε

Step 13 Stop

OUTPUT Approximations of the solution to the initial value problem at $t = c$, using 2^n steps.

EXAMPLE 3 Use the Runge-Kutta algorithm for second order equations outlined above with a stopping procedure based upon the absolute error and a tolerance of $\varepsilon = 0.001$ to approximate the solution $y(t)$ of the initial value problem

(17) $y'' + ty' + y^3 = 0; \quad y(0) = 1, \quad y'(0) = 0,$

at $t = 2$.

Solution   As input for the algorithm we have $t_0 = 0$, $a_1 = y(0) = 1$, $a_2 = y'(0) = 0$, $c = 2$, $\varepsilon = 0.001$, and $M = 10$ (say). Since $F(t, y, y') = -ty' - y^3$, the four formulas depending on $F(t, y, y')$ in Step 5 of the algorithm are

$$k_{1,2} = h(-tx_2 - x_1^3),$$

$$k_{2,2} = h\left[-\left(t + \frac{h}{2}\right)\left(x_2 + \frac{1}{2}k_{1,2}\right) - \left(x_1 + \frac{1}{2}k_{1,1}\right)^3\right],$$

$$k_{3,2} = h\left[-\left(t + \frac{h}{2}\right)\left(x_2 + \frac{1}{2}k_{2,2}\right) - \left(x_1 + \frac{1}{2}k_{2,1}\right)^3\right],$$

$$k_{4,2} = h[-(t + h)(x_2 + k_{3,2}) - (x_1 + k_{3,1})^3].$$

In Table 5.5 we give the approximations $x_1(2, 2^{-m+1})$ for $y(2)$ and $x_2(2, 2^{-m+1})$ for $y'(2)$ for $m = 0, 1, 2$, and 3. The algorithm stops at $m = 3$ because

$$|x_1(2, 0.5) - x_1(2, 0.25)| = 0.00046 < \varepsilon = 0.001,$$

and

$$|x_2(2, 0.5) - x_2(2, 0.25)| = 0.00040 < \varepsilon = 0.001.$$

Hence $y(2) \approx x_1(2) = 0.34400$ with a tolerance of 0.001.   ●

TABLE 5.5   RUNGE-KUTTA APPROXIMATION FOR $y(2)$ FROM EXAMPLE 3

		Approximations	
m	h	$y(2) \approx x_1(2, h)$	$y'(2) \approx x_2(2, h)$
0	2	0.33333	−0.66667
1	1	0.33417	−0.13764
2	0.5	0.34446	−0.14822
3	0.25	0.34400	−0.14782

$y(2)$ is approximately 0.34400 with a tolerance of $\varepsilon = 0.001$.

The reader may have questioned why we selected a stopping procedure in Step 8 of the algorithm which required two successive approximations for *both* the function $y = x_1$ and its derivative $y' = x_2$ to differ by less than the prescribed tolerance ε. The reason is that if we only require two successive approximations for $y = x_1$ to differ by less than the prescribed tolerance, then significant error can result. For instance, in the previous example $|x_1(2, 2) - x_1(2, 1)| = 0.00084 < 0.001 = \varepsilon$. This yields the approximation $y(2) \approx x_1(2, 1) = 0.33417$. Comparing this with the approximation $y(2) \approx x_1(2, 0.25) = 0.34400$ that we obtained in Example 3 we see that they differ by 0.00983, or by roughly ten times the tolerance. Numerical methods for higher order equations and systems are discussed in Section 9.5.

EXERCISES 5.6

The reader will find it helpful to have access to either a micro-computer or a mainframe.

In Problems 1 through 4 use the Runge-Kutta method with $h = 0.25$ to find an approximation for the solution to the given initial value problem on the specified interval.

1. $y'' + ty' + y = 0$; $y(0) = 1$, $y'(0) = 0$ on $[0, 1]$.

2. $(1 + t^2)y'' + y' - y = 0$;
   $y(0) = 1$, $y'(0) = -1$ on $[0, 1]$.

3. $t^2 y'' + y = t + 2$;
   $y(1) = 1$, $y'(1) = -1$ on $[1, 2]$.

4. $y'' = t^2 - y^2$; $y(0) = 0$, $y'(0) = 1$ on $[0, 1]$.

5. Using the Runge-Kutta method with $h = 0.5$, approximate the solution to the initial value problem
$$3t^2 y'' - 5ty' + 5y = 0; \qquad y(1) = 0, \qquad y'(1) = \tfrac{2}{3}$$
at $t = 8$. Compare this approximation to the actual solution $y(t) = t^{5/3} - t$.

6. Using the Runge-Kutta algorithm with the stopping procedure based on the absolute error and a tolerance of $\varepsilon = 0.01$, approximate the solution to the initial value problem
$$y'' = t^2 + y^2; \qquad y(0) = 1, \qquad y'(0) = 0$$
at $t = 1$.

7. Using the Runge-Kutta algorithm with the stopping procedure based on the relative error and a tolerance of $\varepsilon = 0.02$, approximate the solution to the initial value problem
$$y'' = ty^3; \qquad y(0) = 1, \qquad y'(0) = 1$$
at $t = 1$.

8. In Section 3.5 we discussed Euler's method for approximating the solution to a first order equation. Extend Euler's method to systems and give the recursive formulas in component form.

9. In Section 3.5 we discussed the improved Euler's method for approximating the solution to a first order equation. Extend this method to systems and give the recursive formulas for solving the initial value problem (1).

10. In Project F of Chapter 4 it was shown that the simple pendulum equation
$$\theta''(t) + \sin \theta(t) = 0$$
has periodic solutions when the initial displacement and velocity are small. Show that the period of the solution may depend upon the initial conditions by using the Runge-Kutta method with $h = 0.02$ to approximate the solutions to the simple pendulum problem on $[0, 4]$ for the initial conditions:

   (a) $\theta(0) = 0.1$; $\theta'(0) = 0$.
   (b) $\theta(0) = 0.5$; $\theta'(0) = 0$.
   (c) $\theta(0) = 1.0$; $\theta'(0) = 0$.

[Hint: Approximate the length of time from $t = 0$ to the time of the next maximum.]

11. **Fluid Ejection.** In the design of a sewage treatment plant, the following equation arose:[†]
$$60 - H = (77.7)H'' + (19.42)(H')^2;$$
$$H(0) = H'(0) = 0,$$
where H is the level of the fluid in an ejection chamber and t is the time in seconds. Use the Runge-Kutta method with $h = 0.5$ to approximate $H(t)$ over the interval $[0, 5]$.

12. **Oscillations and Nonlinear Equations.** For the initial value problem
$$x'' + (0.1)(1 - x^2)x' + x = 0;$$
$$x(0) = x_0, \qquad x'(0) = 0,$$
use the Runge-Kutta method with $h = 0.02$ to illustrate that as t increases from 0 to 20 the solution x exhibits damped oscillations with $x_0 = 1$, whereas x exhibits expanding oscillations when $x_0 = 3$.

13. **Nonlinear Spring.** The **Duffing equation**
$$y'' + y + ry^3 = 0,$$
where r is a constant, is a model for the vibrations of a mass attached to a *nonlinear* spring. For this model, does the period of vibration vary as the parameter r is varied? Does the period vary as the initial conditions are varied? [Hint: Use the Runge-Kutta method with $h = 0.1$ to approximate the solutions, for $r = 1$ and 2, with initial conditions $y(0) = a$, $y'(0) = 0$ for $a = 1, 2$, and 3.]

14. **Pendulum with Varying Length.** A pendulum is

[†] See *Numerical Solution of Differential Equations*, by William Milne, Dover Publications, Inc., New York, 1970, p. 82.

formed by a mass m attached to the end of a wire that is 'attached to the ceiling. Assume that the length $l(t)$ of the wire varies with time in some predetermined fashion. If $\theta(t)$ is the angle between the pendulum and the vertical, then the motion of the pendulum is governed by the initial value problem

$$l^2(t)\theta''(t) + 2l(t)l'(t)\theta'(t) + gl(t)\theta(t) = 0;$$

$$\theta(0) = \theta_0, \qquad \theta'(0) = \theta_1,$$

where g is the acceleration due to gravity. Assume that

$$l(t) = l_0 + l_1\cos(\omega t - \phi),$$

where l_1 is much smaller than l_0. (This might be a model for a person on a swing, where the *pumping* action changes the distance from the center of mass of the swing to the point where the swing is attached.) To simplify the computations, take $g = 1$. Using the Runge-Kutta method with $h = 0.1$, study the motion of the pendulum when $\theta_0 = 0.5$, $\theta_1 = 0$, $l_0 = 1$, $l_1 = 0.1$, $\omega = 1$, and $\phi = 0.02$. In particular, does the pendulum ever attain an angle greater in absolute value than the initial angle θ_0? Does the total arc traversed during one-half of a swing ever exceed 1?

GROUP PROJECTS FOR CHAPTER 5

A. Finite-Difference Method for Boundary Value Problems

A classical method for numerically approximating a solution to a boundary value problem for a linear second order differential equation consists of replacing each derivative in the differential equation by a difference quotient that approximates that derivative. To illustrate this procedure, consider the linear boundary value problem

(1) $\qquad y''(x) + p(x)y'(x) + q(x)y(x) = g(x); \qquad a < x < b.$

(2) $\qquad y(a) = \alpha, \qquad y(b) = \beta,$

where p, q, and g are continuous on $[a, b]$ and α, β are given constants.

We begin by partitioning the interval $[a, b]$ into $N + 1$ subintervals of equal length h. That is, let $h := (b - a)/(N + 1)$ and for $k = 0, 1, \ldots, N + 1$ set

$$x_k := a + kh.$$

Then $a = x_0 < x_1 < \cdots < x_N < x_{N+1} = b$.

Assuming that a solution $y(x)$ to (1)–(2) exists, our aim is to approximate $y(x)$ at the interior mesh points x_k, $k = 1, 2, \ldots, N$. At these points we know that

(3) $\qquad y''(x_k) + p(x_k)y'(x_k) + q(x_k)y(x_k) = g(x_k).$

The essence of the finite-difference method is to find difference quotients that closely approximate the derivatives $y'(x_k)$ and $y''(x_k)$ when h is small.

(a) Assuming that y has a Taylor series representation of the form

(4) $\qquad y(x) = y(x_k) + y'(x_k)(x - x_k) + \dfrac{y''(x_k)}{2}(x - x_k)^2 + \dfrac{y'''(x_k)}{3!}(x - x_k)^3 + \cdots$

$$= \sum_{j=0}^{\infty} \frac{y^{(j)}(x_k)}{j!}(x - x_k)^j,$$

derive the following **centered-difference formula for** $y'(x_k)$:

(5) $$y'(x_k) = \frac{1}{2h}[y(x_{k+1}) - y(x_{k-1})] + [\text{terms involving } h^2, h^3, \ldots].$$

[Hint: Work with the Taylor representations for $y(x_{k+1})$ and $y(x_{k-1})$.]

(b) Again, using (4) above, derive the following centered-difference formula for $y''(x_k)$:

(6) $$y''(x_k) = \frac{1}{h^2}[y(x_{k+1}) - 2y(x_k) + y(x_{k-1})] + [\text{terms involving } h^2, h^3, \ldots].$$

(c) If we let w_k denote the approximation (to be computed) for $y(x_k)$, $k = 1, 2, \ldots, N$ and set $w_0 = y(x_0) = \alpha$, $w_{N+1} = y(x_{N+1}) = \beta$, then from (5) and (6) we obtain the approximations

(7) $$y''(x_k) \approx \frac{1}{h^2}(w_{k+1} - 2w_k + w_{k-1}),$$

(8) $$y'(x_k) \approx \frac{1}{2h}(w_{k+1} - w_{k-1}).$$

Using the centered-difference approximations (7) and (8) in place of $y''(x_k)$ and $y'(x_k)$ in (3) yields the system of equations

(9) $$\left(\frac{w_{k+1} - 2w_k + w_{k-1}}{h^2}\right) + p(x_k)\left(\frac{w_{k+1} - w_{k-1}}{2h}\right) + q(x_k)w_k = g(x_k),$$

$k = 1, 2, \ldots, N$. The equations (9) represent N linear equations in the N unknowns $w_1, \ldots, w_N$. Thus, on solving (9), we obtain approximations for $y(x)$ at the interior mesh points x_k, $k = 1, \ldots, N$.

Use this method to approximate the solution to the boundary value problem

(10) $$y''(x) - 4y(x) = 0; \qquad y(0) = 1, \qquad y(1) = e^{-2},$$

taking $h = \frac{1}{4}$ ($N = 3$).

(d) Compare the approximations obtained in part (c) with the actual solution $y = e^{-2x}$.

(e) Let E denote the electrostatic potential between two concentric metal spheres of radii r_1 and r_2 ($r_1 < r_2$), where the potential of the inner sphere is kept constant at 100 volts and the potential of the outer sphere at 0 volts. The potential E will then depend only on the distance r from the center of the spheres and must satisfy Laplace's equation. This leads to the following boundary value problem for E:

(11) $$\frac{d^2E}{dr^2} + \frac{2}{r}\frac{dE}{dr} = 0; \qquad E(r_1) = 100, \qquad E(r_2) = 0,$$

where $r_1 < r < r_2$.

For $r_1 = 1$ and $r_2 = 3$, obtain finite-difference approximations for $E(2)$ using $h = \frac{1}{2}$ ($N = 3$) and $h = \frac{1}{4}$ ($N = 7$).

(f) Compare your approximations in part (e) with the actual solution $E(r) = 150r^{-1} - 50$.

B. The Transmission of Information

A signaling system consists of two signals, one short and one long. A message is transmitted over a channel by encoding it into a sequence of the two signals. Let M_t denote the number of messages of length t, that is, the number of different messages that take precisely t units of time to transmit. Assume it takes 1 unit of time to transmit a short signal and 2 units of time to transmit a long signal.

(a) Show that the number of messages of length t is

(12) $\qquad M_t = M_{t-1} + M_{t-2}.$

[Hint: Determine the number of messages of length t that end in a short and the number that end in a long, then add.]

(b) Find a general solution for equation (12).

(c) Determine the initial conditions M_0 and M_1 and solve for M_t.

(d) C. E. Shannon defines the **capacity** C of a channel to be

$$C := \lim_{t \to \infty} \frac{\log_2 M_t}{t}.$$

Compute the capacity of the channel just discussed.

CHAPTER 6

Higher Order Linear Differential Equations

6.1 INTRODUCTION: COUPLED SPRING-MASS SYSTEM

On a smooth horizontal surface a mass of 2 kg is attached to a vertical surface by a spring with spring constant 4 N/m. Another mass of 1 kg is attached to the first object by a spring with spring constant 2 N/m. The objects are aligned horizontally so that the springs are their natural lengths (Figure 6.1). If both objects are displaced 3 m to the right of their equilibrium positions (Figure 6.2) and then released, what are the equations of motion for the two objects?

Let's assume that the surface is so smooth that we can neglect any friction due to the motion of the objects along the surface. Since there are no outside forces acting on the objects, the only forces we need to consider are those due to the springs themselves. Recall that Hooke's law asserts that the force acting on an object due to a spring has magnitude proportional to the displacement of the spring from its natural length and has direction opposite to its displacement. That is, if the spring is either stretched or compressed, then it tries to return to its natural length.

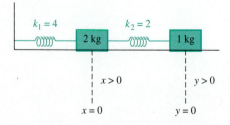

Figure 6.1 Coupled system at equilibrium

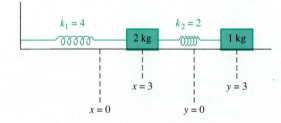

Figure 6.2 Coupled system at initial displacement

Since each mass is free to move, we apply Newton's second law to each object. Let $x(t)$ denote the displacement (to the right) of the 2-kg mass from its equilibrium position, and, similarly, let $y(t)$ denote the corresponding displacement for the 1-kg mass. The 2-kg mass has a force F_1 acting on its left side due to one spring and a force F_2 acting on its right side due to the second spring. Referring to Figure 6.1 and applying Hooke's law, we see that

(1) $F_1 = -k_1 x, \qquad F_2 = +k_2(y - x),$

where $(y - x)$ is the net displacement of the second spring from its natural length. There is only one force acting on the 1-kg mass; namely the force due to the second spring, which is

(2) $F_3 = -k_2(y - x).$

Applying Newton's second law to these objects, we now obtain

$$\begin{cases} m_1 \dfrac{d^2 x}{dt^2} = F_1 + F_2 = -k_1 x + k_2(y - x), \\[2mm] m_2 \dfrac{d^2 y}{dt^2} = F_3 = -k_2(y - x), \end{cases}$$

or

(3)
$$\begin{cases} m_1 \dfrac{d^2 x}{dt^2} + (k_1 + k_2)x - k_2 y = 0, \\[2mm] m_2 \dfrac{d^2 y}{dt^2} + k_2 y - k_2 x = 0. \end{cases}$$

In this problem, we know that $m_1 = 2$, $m_2 = 1$, $k_1 = 4$, and $k_2 = 2$. Substituting these values into system (3) yields

(4) $2 \dfrac{d^2 x}{dt^2} + 6x - 2y = 0,$

(5) $\dfrac{d^2 y}{dt^2} + 2y - 2x = 0.$

One approach to solving for x and y is to convert (4) and (5) into a system of four first order equations in four unknowns. (Such equations can be solved using matrix methods as discussed in Chapter 10.) Here we illustrate another approach called the **elimination method** which reduces our system to a higher order equation in a single unknown function.

For the system (4) and (5), we can obtain a single differential equation in x by solving (4) for y and then substituting the expression for y back into (5). From (4) we get

(6) $y = \dfrac{d^2 x}{dt^2} + 3x,$

and, substituting for y in (5), we find

(7) $\dfrac{d^2}{dt^2}\left[\dfrac{d^2 x}{dt^2} + 3x\right] + 2\left[\dfrac{d^2 x}{dt^2} + 3x\right] - 2x = 0,$

which simplifies to

(8) $$\frac{d^4x}{dt^4} + 5\frac{d^2x}{dt^2} + 4x = 0.$$

Notice that equation (8) is linear with constant coefficients. To solve it, let's proceed as we did with linear second order equations and try to find solutions of the form $x = e^{rt}$. Substituting e^{rt} in equation (8) gives

(9) $$(r^4 + 5r^2 + 4)e^{rt} = 0.$$

Thus we have a solution to (8), provided that r satisfies the auxiliary equation

(10) $$r^4 + 5r^2 + 4 = 0.$$

From the factorization $r^4 + 5r^2 + 4 = (r^2 + 1)(r^2 + 4)$, we see that the roots of the auxiliary equation are the complex numbers $i, -i, 2i, -2i$. Using Euler's formula, it follows that

$$z_1(t) = e^{it} = \cos t + i\sin t$$

and

$$z_2(t) = e^{2it} = \cos 2t + i\sin 2t$$

are complex-valued solutions to equation (8). To obtain real-valued solutions to (8), we take the real and imaginary parts of $z_1(t)$ and $z_2(t)$. Thus four real-valued solutions are

(11) $$x_1(t) = \cos t, \qquad x_2(t) = \sin t,$$
$$x_3(t) = \cos 2t, \qquad x_4(t) = \sin 2t.$$

Because (8) is a linear homogeneous equation, any linear combination of $x_1(t), x_2(t), x_3(t)$, and $x_4(t)$ will also be a solution. Hence a (general) solution to (8) is

(12) $$x(t) = a_1 \cos t + a_2 \sin t + a_3 \cos 2t + a_4 \sin 2t,$$

where a_1, a_2, a_3, and a_4 are arbitrary constants.

To obtain a formula for $y(t)$, we substitute the above representation for $x(t)$ into equation (6):

(13) $$y(t) = \frac{d^2x}{dt^2} + 3x$$
$$= -a_1 \cos t - a_2 \sin t - 4a_3 \cos 2t - 4a_4 \sin 2t$$
$$+ 3a_1 \cos t + 3a_2 \sin t + 3a_3 \cos 2t + 3a_4 \sin 2t,$$

and so

(14) $$y(t) = 2a_1 \cos t + 2a_2 \sin t - a_3 \cos 2t - a_4 \sin 2t.$$

In order to determine the constants a_1, a_2, a_3, and a_4, let's return to the original problem. We were told that the objects were originally displaced 3 m to the right and then released. Hence

(15) $$x(0) = 3, \qquad \frac{dx}{dt}(0) = 0; \qquad y(0) = 3, \qquad \frac{dy}{dt}(0) = 0.$$

On differentiating equations (12) and (14), we find

(16) $\quad \dfrac{dx}{dt} = -a_1 \sin t + a_2 \cos t - 2a_3 \sin 2t + 2a_4 \cos 2t,$

(17) $\quad \dfrac{dy}{dt} = -2a_1 \sin t + 2a_2 \cos t + 2a_3 \sin 2t - 2a_4 \cos 2t.$

Now, if we put $t = 0$ in the formulas for x, dx/dt, y, and dy/dt, the initial conditions (15) give the four equations

$$x(0) = a_1 + a_3 = 3, \qquad \frac{dx}{dt}(0) = a_2 + 2a_4 = 0,$$

$$y(0) = 2a_1 - a_3 = 3, \qquad \frac{dy}{dt}(0) = 2a_2 - 2a_4 = 0.$$

From this system we find $a_1 = 2$, $a_2 = 0$, $a_3 = 1$, and $a_4 = 0$. Hence the equations of motion for the two objects are

(18) $\qquad x(t) = 2\cos t + \cos 2t,$

(19) $\qquad y(t) = 4\cos t - \cos 2t,$

which are depicted in Figure 6.3.

The analysis of the coupled spring-mass system illustrates that many of the techniques developed for second order linear equations have natural extensions to higher order linear equations. This chapter is devoted to exploring these generalizations.

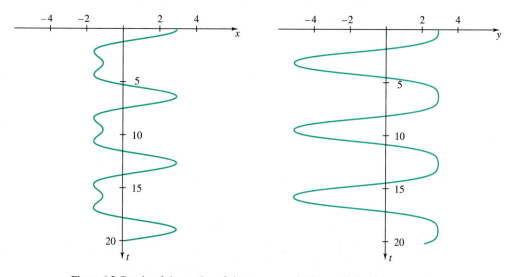

Figure 6.3 Graphs of the motion of the two masses in the coupled spring-mass system

•6.2• BASIC THEORY OF LINEAR DIFFERENTIAL EQUATIONS

In this section we discuss the basic theory of linear higher order differential equations. The material is a generalization of the results we obtained in Chapter 4 for second order equations. In the statements and proofs of these results we will be using concepts usually covered in an elementary linear algebra course; namely linear dependence, determinants, and solving systems of linear equations. These concepts also arise in the matrix approach for solving systems of differential equations, and are discussed in Section 10.2.

Recall that a *linear* differential equation is an equation that can be written in the form

(1) $a_n(x)y^{(n)}(x) + a_{n-1}(x)y^{(n-1)}(x) + \cdots + a_0(x)y(x) = b(x),$

where $a_0(x), a_1(x), \ldots, a_n(x)$, and $b(x)$ depend only on x, not y. When $a_0, a_1, \ldots, a_n$ are constants, we say that equation (1) has **constant coefficients;** otherwise it has **variable coefficients.** If $b(x) \equiv 0$, equation (1) is called **homogeneous;** otherwise it is **nonhomogeneous.**

In developing a basic theory, we assume that $a_0(x), a_1(x), \ldots, a_n(x)$, and $b(x)$ are continuous on an interval I and $a_n(x) \neq 0$ on I. Then, on dividing by $a_n(x)$, we can rewrite (1) in the **standard form**

(2) $y^{(n)}(x) + p_1(x)y^{(n-1)}(x) + \cdots + p_n(x)y(x) = g(x),$

where the functions $p_1(x), \ldots, p_n(x)$, and $g(x)$ are continuous on I.

For a linear higher order differential equation the initial value problem always has a unique solution.

▶ EXISTENCE AND UNIQUENESS

Theorem 1.   Suppose that $p_1(x), \ldots, p_n(x)$, and $g(x)$ are each continuous on an interval (a, b) that contains the point x_0. Then, for any choice of the initial values $y_0, y_1, \ldots, y_{n-1}$, there exists a unique solution $y(x)$ on the whole interval (a, b) to the initial value problem

(3) $y^{(n)}(x) + p_1(x)y^{(n-1)}(x) + \cdots + p_n(x)y(x) = g(x),$

(4) $y(x_0) = y_0, \ y'(x_0) = y_1, \ldots, y^{(n-1)}(x_0) = y_{n-1}.$

The proof of Theorem 1 can be found in Chapter 14.[†]

● EXAMPLE 1   For the initial value problem

(5) $x(x-1)y''' - 3xy'' + 6x^2y' - (\cos x)y = \sqrt{x+5};$

(6) $y(x_0) = 1, \qquad y'(x_0) = 0, \qquad y''(x_0) = 7,$

determine the values of x_0 and the intervals (a, b) containing x_0 for which Theorem 1 guarantees the existence of a unique solution on (a, b).

[†] All references to Chapters 12–14 refer to the expanded text *Fundamentals of Differential Equations and Boundary Value Problems.*

Solution Putting equation (5) in standard form, we find that $p_1(x) = -3/(x-1)$, $p_2(x) = 6x/(x-1)$, $p_3(x) = -(\cos x)/[x(x-1)]$, and $g(x) = \sqrt{x+5}/[x(x-1)]$. Now $p_1(x)$ and $p_2(x)$ are continuous on every interval not containing $x = 1$, while $p_3(x)$ is continuous on every inteval not containing $x = 0$ or $x = 1$. The function $g(x)$ is not defined for $x < -5$, $x = 0$, and $x = 1$, but is continuous on $(-5, 0)$, $(0, 1)$, and $(1, \infty)$. Hence the functions p_1, p_2, p_3, and g are *simultaneously* continuous on the intervals $(-5, 0)$, $(0, 1)$, and $(1, \infty)$. From Theorem 1 it follows that if we choose $x_0 \in (-5, 0)$, then there exists a unique solution to the initial value problem (5)–(6) on the whole interval $(-5, 0)$. Similarly, for $x_0 \in (0, 1)$ there is a unique solution on $(0, 1)$, and for $x_0 \in (1, \infty)$ a unique solution exists on $(1, \infty)$. ●

If we let the left-hand side of equation (3) define the differential operator L,

(7) $$L[y] := \frac{d^n y}{dx^n} + p_1 \frac{d^{n-1} y}{dx^{n-1}} + \cdots + p_n y,$$

then we can express equation (3) in the operator form

(8) $$L[y](x) = g(x).$$

It is essential to keep in mind that L is a *linear* operator: it satisfies

(9) $$L[y_1 + y_2] = L[y_1] + L[y_2],$$

(10) $$L[cy_1] = cL[y_1], \qquad c \text{ any constant}$$

(see Problem 25).

As a consequence of this linearity, if $y_1, \ldots, y_m$ are solutions to the homogeneous equation

(11) $$L[y](x) = 0,$$

then any linear combination of these functions, $C_1 y_1 + \cdots + C_m y_m$, is also a solution to (11). Imagine now that we have found n solutions $y_1, \ldots, y_n$ to the nth order linear equation (11). Is it true that *every* solution to (11) can be represented by

(12) $$C_1 y_1 + \cdots + C_n y_n$$

for appropriate choices of the constants $C_1, \ldots, C_n$? The answer is yes, provided that the solutions $y_1, \ldots, y_n$ satisfy a certain property of independence that we shall now derive.

Let $\phi(x)$ be a solution to (11) on the interval (a, b) and let x_0 be a fixed number in (a, b). If it is possible to choose the constants $C_1, \ldots, C_n$ so that

(13) $$\begin{aligned} C_1 y_1(x_0) &+ \cdots + C_n y_n(x_0) &&= \phi(x_0), \\ C_1 y_1'(x_0) &+ \cdots + C_n y_n'(x_0) &&= \phi'(x_0), \\ &\vdots && \vdots \\ C_1 y_1^{(n-1)}(x_0) &+ \cdots + C_n y_n^{(n-1)}(x_0) &&= \phi^{(n-1)}(x_0), \end{aligned}$$

then, since $\phi(x)$ and $C_1 y_1(x) + \cdots + C_n y_n(x)$ are two solutions satisfying the same initial conditions at x_0, the uniqueness conclusion of Theorem 1 gives

(14) $$\phi(x) = C_1 y_1(x) + \cdots + C_n y_n(x)$$

for all x in (a, b).

The system (13) consists of n linear equations in the n unknowns $C_1, \ldots, C_n$. It has a unique solution for all possible values of $\phi(x_0), \ldots, \phi^{(n-1)}(x_0)$ if and only if the determinant of the coefficients is different from zero; that is, if and only if

(15)
$$\begin{vmatrix} y_1(x_0) & y_2(x_0) & \cdots & y_n(x_0) \\ y_1'(x_0) & y_2'(x_0) & \cdots & y_n'(x_0) \\ \vdots & \vdots & & \vdots \\ y_1^{(n-1)}(x_0) & y_2^{(n-1)}(x_0) & \cdots & y_n^{(n-1)}(x_0) \end{vmatrix} \neq 0.$$

Hence, if $y_1, \ldots, y_n$ are solutions to equation (11) and there is some point x_0 in (a, b) such that (15) holds, then every solution $\phi(x)$ to (11) is a linear combination of $y_1, \ldots, y_n$.

Before putting this observation in the form of a theorem, it is convenient to define the Wronskian of a set of n functions.

> ### WRONSKIAN
>
> **Definition 1.**   Let $f_1, \ldots, f_n$ be any n functions that are $(n-1)$ times differentiable. The function
>
> **(16)** $$W[f_1, \ldots, f_n](x) := \begin{vmatrix} f_1(x) & f_2(x) & \cdots & f_n(x) \\ f_1'(x) & f_2'(x) & \cdots & f_n'(x) \\ \vdots & \vdots & & \vdots \\ f_1^{(n-1)}(x) & f_2^{(n-1)}(x) & \cdots & f_n^{(n-1)}(x) \end{vmatrix}$$
>
> is called the **Wronskian** of $f_1, \ldots, f_n$.

We now state the representation theorem for solutions to homogeneous linear differential equations that we proved above.

> ### REPRESENTATION OF SOLUTIONS (HOMOGENEOUS CASE)
>
> **Theorem 2.**   Let $y_1, \ldots, y_n$ be n solutions on (a, b) of
>
> **(17)** $$y^{(n)}(x) + p_1(x)y^{(n-1)}(x) + \cdots + p_n(x)y(x) = 0,$$
>
> where $p_1, \ldots, p_n$ are continuous on (a, b). If at some point x_0 in (a, b) these solutions satisfy
>
> **(18)** $$W[y_1, \ldots, y_n](x_0) \neq 0,$$
>
> then every solution of (17) on (a, b) can be expressed in the form
>
> **(19)** $$y(x) = C_1 y_1(x) + \cdots + C_n y_n(x),$$
>
> where $C_1, \ldots, C_n$ are constants.

A set of solutions $\{y_1, \ldots, y_n\}$ satisfying (18) for some x_0 in (a, b) is called a **fundamental solution set** for (17) on (a, b). The linear combination of $y_1, \ldots, y_n$ in (19), written with arbitrary constants $C_1, \ldots, C_n$, is referred to as a **general solution** to (17).

● **EXAMPLE 2**   Given that $y_1(x) = x$, $y_2(x) = x^2$, and $y_3(x) = x^{-1}$ are solutions to

(20) $x^3 y''' + x^2 y'' - 2xy' + 2y = 0$, $x > 0$,

find a general solution.

Solution   We first show that $\{y_1, y_2, y_3\}$ is a fundamental solution set for equation (20) on $(0, \infty)$. Since we are told that y_1, y_2, and y_3 satisfy (20) (the reader can easily verify this), we need only consider

$$W[y_1, y_2, y_3](x) = \begin{vmatrix} y_1(x) & y_2(x) & y_3(x) \\ y_1'(x) & y_2'(x) & y_3'(x) \\ y_1''(x) & y_2''(x) & y_3''(x) \end{vmatrix} = \begin{vmatrix} x & x^2 & x^{-1} \\ 1 & 2x & -x^{-2} \\ 0 & 2 & 2x^{-3} \end{vmatrix}.$$

Evaluating the determinant, we find after a little algebra that

$$W[y_1, y_2, y_3](x) = 6x^{-1},$$

which is not zero for $x > 0$. Thus $\{y_1, y_2, y_3\}$ is a fundamental solution set, and hence a general solution is

(21) $y(x) = C_1 x + C_2 x^2 + C_3 x^{-1}$, $x > 0$.   ●

The important condition (18) concerning the nonvanishing of the Wronskian can also be described in terms of the linear independence of the solutions $y_1, \ldots, y_n$.

▶ LINEAR DEPENDENCE OF FUNCTIONS

Definition 2.   The m functions $f_1, \ldots, f_m$ are said to be **linearly dependent on an interval I** if there exist constants $c_1, \ldots, c_m$, not all zero, such that

(22) $c_1 f_1(x) + \cdots + c_m f_m(x) = 0$

for all x in I. If the functions $f_1, \ldots, f_m$ are not linearly dependent on I, they are said to be **linearly independent on I**.

● **EXAMPLE 3**   Show that the functions $f_1(x) = e^x$, $f_2(x) = e^{-2x}$, and $f_3(x) = 3e^x - 2e^{-2x}$ are linearly dependent on $(-\infty, \infty)$.

Solution   Notice that f_3 is a linear combination of f_1 and f_2:

$$f_3(x) = 3e^x - 2e^{-2x} = 3f_1(x) - 2f_2(x).$$

Therefore, we have

(23) $3f_1(x) - 2f_2(x) - f_3(x) = 0$

for all x in $(-\infty, \infty)$. Consequently, f_1, f_2, and f_3 are linearly dependent on $(-\infty, \infty)$.   ●

To prove that functions $f_1, f_2, \ldots, f_m$ are linearly *independent* on (a, b), a convenient approach is the following: *Assume* that equation (22) holds on (a, b) and show that this forces $c_1 = c_2 = \cdots = c_m = 0$.

● **EXAMPLE 4**   Show that the functions $f_1(x) = x$, $f_2(x) = x^2$, and $f_3(x) = 1 - 2x^2$ are linearly independent on $(-\infty, \infty)$.

Solution   Assume that c_1, c_2, and c_3 are constants for which

(24) $c_1 x + c_2 x^2 + c_3(1 - 2x^2) = 0$

holds at every x. If we can prove that (24) implies $c_1 = c_2 = c_3 = 0$, then linear independence follows. Let's set $x = 0, 1$, and -1 in equation (24); these x values are, essentially, "picked out of a hat," but will get the job done. Substituting in (24) gives

(25)
$$c_3 = 0 \quad (x = 0),$$
$$c_1 + c_2 - c_3 = 0 \quad (x = 1),$$
$$-c_1 + c_2 - c_3 = 0 \quad (x = -1).$$

When we solve this system (or compute the determinant of the coefficients), we find that the only possible solution is $c_1 = c_2 = c_3 = 0$. Consequently, the functions f_1, f_2, and f_3 are linearly independent on $(-\infty, \infty)$. ●

The connection between linear independence and fundamental solution sets is stated in the next theorem.

> **LINEAR INDEPENDENCE AND FUNDAMENTAL SOLUTIONS**
>
> **Theorem 3.**   Let $y_1, \ldots, y_n$ be n solutions to $y^{(n)} + p_1 y^{(n-1)} + \cdots + p_n y = 0$ on (a, b). Then $\{y_1, \ldots, y_n\}$ is a fundamental solution set on (a, b) if and only if these functions are linearly independent on (a, b).

Theorem 3 is a generalization of Corollary 1 in Section 4.3. Since its proof is similar, we leave it as an exercise (see Problem 30).

As with linear second order equations, *the Wronskian of solutions is either identically zero or never zero on (a, b).* This fact follows from the proof of Theorem 3 or from **Abel's identity,**

(26) $W[y_1, \ldots, y_n](x) = W[y_1, \ldots, y_n](x_0) \exp\left(-\int_{x_0}^x p_1(t)\, dt\right),$

which holds for any n solutions $y_1, \ldots, y_n$ to $y^{(n)} + p_1 y^{(n-1)} + \cdots + p_n y = 0$ on (a, b). In (26), the point x_0 can be anywhere in (a, b). For a proof of Abel's identity when $n = 3$, see Problem 31.

Notice that Theorem 3 provides a simpler approach to proving that the solutions $y_1(x) = x$, $y_2(x) = x^2$, $y_3(x) = x^{-1}$ in Example 2 form a fundamental solution set. For, without having to compute a determinant, we can see (practically at a glance) that these functions are linearly independent on $(0, \infty)$.

It is useful to keep in mind that the following sets consist of functions that are linearly independent on every open interval (a, b):

$$\{1, x, x^2, \ldots, x^n\},$$

$$\{1, \cos x, \sin x, \cos 2x, \sin 2x, \ldots, \cos nx, \sin nx\},$$

$$\{e^{\alpha_1 x}, e^{\alpha_2 x}, \ldots, e^{\alpha_n x}\}, \qquad \alpha_i\text{'s distinct constants.}$$

(See Problems 27, 28.)

Recall that the **superposition principle** (Chapter 4, page 166) for linear differential operators states that if $L[y_1] = g_1$ and $L[y_2] = g_2$, then the linear combination $C_1 y_1 + C_2 y_2$ satisfies

$$L[C_1 y_1 + C_2 y_2] = C_1 g_1 + C_2 g_2.$$

If we combine this principle with the representation theorem for solutions of the homogeneous equation, we obtain the following representation theorem for nonhomogeneous equations.

REPRESENTATION OF SOLUTIONS (NONHOMOGENEOUS CASE)

Theorem 4. Let $y_p(x)$ be a particular solution to the nonhomogeneous equation

(27) $$y^{(n)}(x) + p_1(x)y^{(n-1)}(x) + \cdots + p_n(x)y(x) = g(x)$$

on the interval (a, b) and let $\{y_1, \ldots, y_n\}$ be a fundamental solution set on (a, b) for the corresponding homogeneous equation

(28) $$y^{(n)}(x) + p_1(x)y^{(n-1)}(x) + \cdots + p_n(x)y(x) = 0.$$

Then every solution of (27) on the interval (a, b) can be expressed in the form

(29) $$y(x) = y_p(x) + C_1 y_1(x) + \cdots + C_n y_n(x).$$

Proof. Let $\phi(x)$ be any solution to (27). Since both $\phi(x)$ and $y_p(x)$ are solutions to (27), the superposition principle states that the difference $\phi(x) - y_p(x)$ is a solution to the homogeneous equation (28). It then follows from Theorem 2 that

$$\phi(x) - y_p(x) = C_1 y_1(x) + \cdots + C_n y_n(x)$$

for suitable constants $C_1, \ldots, C_n$. Since the last equation is equivalent to (29) (with $\phi(x)$ in place of $y(x)$), the theorem is proved. ◀◀◀

The linear combination of $y_p, y_1, \ldots, y_n$ in (29) written with arbitrary constants $C_1, \ldots, C_n$ is, for obvious reasons, referred to as a **general solution** to (27).

● **EXAMPLE 5** Given that $y_p(x) = x^2$ is a particular solution to

(30) $$y''' - 2y'' - y' + 2y = 2x^2 - 2x - 4$$

on the interval $(-\infty, \infty)$ and that $y_1(x) = e^{-x}$, $y_2(x) = e^x$, $y_3(x) = e^{2x}$ are solutions to the corresponding homogeneous equation, find a general solution to (30).

Solution We previously remarked that the functions e^{-x}, e^{x}, e^{2x} are linearly independent because the exponents $-1, 1$, and 2 are distinct. Since each of these functions is a solution to the corresponding homogeneous equation, then $\{e^{-x}, e^{x}, e^{2x}\}$ is a fundamental solution set. It now follows from the representation theorem for nonhomogeneous equations that a general solution is

(31) $$y(x) = x^2 + C_1e^{-x} + C_2e^{x} + C_3e^{2x}. \quad \bullet$$

EXERCISES 6.2

1, 3, 5, 7, 9, 15, 19

In Problems 1 through 6 determine the largest interval (a, b) for which Theorem 1 guarantees the existence of a unique solution on (a, b) to the given initial value problem.

1. $xy''' - 3y' + e^xy = x^2 - 1$;
$y(-2) = 1, \quad y'(-2) = 0, \quad y''(-2) = 2.$

2. $y''' - \sqrt{x}y = \sin x$;
$y(\pi) = 0, \quad y'(\pi) = 11, \quad y''(\pi) = 3.$

3. $y''' - y'' + \sqrt{x-1}\,y = \tan x; \quad y(5) = y'(5) = y''(5) = 1.$

4. $x(x+1)y''' - 3xy' + y = 0$;
$y(-\frac{1}{2}) = 1, \quad y'(-\frac{1}{2}) = y''(-\frac{1}{2}) = 0.$

5. $x\sqrt{x+1}\,y''' - y' + xy = 0$;
$y(\frac{1}{2}) = y'(\frac{1}{2}) = -1, \quad y''(\frac{1}{2}) = 1.$

6. $(x^2 - 1)y''' + e^xy = \ln x$;
$y(\frac{3}{4}) = 1, \quad y'(\frac{3}{4}) = y''(\frac{3}{4}) = 0.$

In Problems 7 through 14 determine whether the given functions are linearly dependent or linearly independent on the specified interval. Also compute their Wronskian.

7. $\{e^{3x}, e^{5x}, e^{-x}\}$ on $(-\infty, \infty)$.

8. $\{x^2, x^2 - 1, 5\}$ on $(-\infty, \infty)$.

9. $\{\sin^2 x, \cos^2 x, 1\}$ on $(-\infty, \infty)$.

10. $\{\sin x, \cos x, \tan x\}$ on $(-\pi/2, \pi/2)$.

11. $\{x^{-1}, x^{1/2}, x\}$ on $(0, \infty)$.

12. $\{\cos 2x, \cos^2 x, \sin^2 x\}$ on $(-\infty, \infty)$.

13. $\{x, x^2, x^3, x^4\}$ on $(-\infty, \infty)$.

14. $\{x, xe^x, 1\}$ on $(-\infty, \infty)$.

In Problems 15 through 18 verify that the given functions form a fundamental solution set for the given differential equation and find a general solution.

15. $y''' + 2y'' - 11y' - 12y = 0; \quad \{e^{3x}, e^{-x}, e^{-4x}\}.$

16. $y''' - y'' + 4y' - 4y = 0; \quad \{e^x, \cos 2x, \sin 2x\}.$

17. $x^3y''' - 3x^2y'' + 6xy' - 6y = 0, \quad x > 0;$
$\{x, x^2, x^3\}.$

18. $y^{(4)} - y = 0; \quad \{e^x, e^{-x}, \cos x, \sin x\}.$

In Problems 19 through 22 a particular solution and a fundamental solution set are given for a nonhomogeneous equation and its corresponding homogeneous equation. (a) Find a general solution to the nonhomogeneous equation. (b) Find the solution that satisfies the specified initial conditions.

19. $y''' + y'' + 3y' - 5y = 2 + 6x - 5x^2$;
$y(0) = -1, \quad y'(0) = 1, \quad y''(0) = -3;$
$y_p = x^2; \quad \{e^x, e^{-x}\cos 2x, e^{-x}\sin 2x\}.$

20. $xy''' - y'' = -2; \quad y(1) = 2, \quad y'(1) = -1, \quad y''(1) = -4;$
$y_p = x^2; \quad \{1, x, x^3\}.$

21. $x^3y''' + xy' - y = 3 - \ln x$;
$y(1) = 3, \quad y'(1) = 3, \quad y''(1) = 0; \quad y_p = \ln x;$
$\{x, x\ln x, x(\ln x)^2\}.$

22. $y^{(4)} + 4y = 5\cos x$;
$y(0) = 2, \quad y'(0) = 1, \quad y''(0) = -1, \quad y'''(0) = -2;$
$y_p = \cos x; \quad \{e^x\cos x, e^x\sin x, e^{-x}\cos x, e^{-x}\sin x\}.$

23. Let $L[y] := y''' + y' + xy$, $y_1(x) := \sin x$, and $y_2(x) := x$. Verify that $L[y_1](x) = x\sin x$ and $L[y_2](x) = x^2 + 1$. Now use the superposition principle to find a solution to the differential equation:

(a) $L[y] = 2x\sin x - x^2 - 1$.
(b) $L[y] = 4x^2 + 4 - 6x\sin x$.

24. Let $L[y] := y''' - xy'' + 4y' - 3xy$, $y_1(x) := \cos 2x$, and $y_2(x) := -\frac{1}{3}$. Verify that $L[y_1](x) = x\cos 2x$ and $L[y_2](x) = x$. Now use the superposition principle to find a solution to the differential equation:

(a) $L[y] = 7x\cos 2x - 3x$.
(b) $L[y] = -6x\cos 2x + 11x$.

25. Prove that L defined in (7) is a linear operator by veri-

fying that properties (9) and (10) hold for any two n-times differentiable functions y_1, y_2 on (a, b).

26. Show that a fundamental solution set always exists for equation (17). [Hint: By Theorem 1 we know that there is a unique solution for each choice of initial conditions. Choose n sets of initial conditions wisely!]

27. Show that the set of functions $\{1, x, x^2, \ldots, x^n\}$, where n is a positive integer, is linearly independent on every open interval (a, b). [Hint: Use the fact that a polynomial of degree at most n has no more than n zeros unless it is identically zero.]

28. The set of functions

$$\{1, \cos x, \sin x, \ldots, \cos nx, \sin nx\},$$

where n is a positive integer, is linearly independent on every interval (a, b). Prove this in the special case $n = 2$ and $(a, b) = (-\infty, \infty)$.

29. (a) Show that if $f_1, \ldots, f_m$ are linearly independent on $(-1, 1)$, then they are linearly independent on $(-\infty, \infty)$.

   (b) Give an example to show that if $f_1, \ldots, f_m$ are linearly independent on $(-\infty, \infty)$, then they need not be linearly independent on $(-1, 1)$.

30. To prove Theorem 3, proceed as follows:

   (a) Let $x_0 \in (a, b)$. Prove that the solutions $y_1, \ldots, y_n$ are linearly independent on (a, b) if and only if the initial vectors $(y_k(x_0), y_k'(x_0), \ldots, y_k^{(n-1)}(x_0))$, $k = 1, 2, \ldots, n$, are linearly dependent.

   (b) Use the result of part (a) and the fact that *the column vectors of a matrix are linearly independent if and only if the determinant of the matrix is not zero* to complete the proof of Theorem 3.

31. To prove Abel's identity for $n = 3$ proceed as follows:

   (a) Let $W(x) := W[y_1, y_2, y_3](x)$. Use the product rule for differentiation to show

$$W'(x) = \begin{vmatrix} y_1' & y_2' & y_3' \\ y_1' & y_2' & y_3' \\ y_1'' & y_2'' & y_3'' \end{vmatrix} + \begin{vmatrix} y_1 & y_2 & y_3 \\ y_1'' & y_2'' & y_3'' \\ y_1'' & y_2'' & y_3'' \end{vmatrix}$$

$$+ \begin{vmatrix} y_1 & y_2 & y_3 \\ y_1' & y_2' & y_3' \\ y_1''' & y_2''' & y_3''' \end{vmatrix}.$$

   (b) Show that the above expression reduces to

$$(32) \qquad W'(x) = \begin{vmatrix} y_1 & y_2 & y_3 \\ y_1' & y_2' & y_3' \\ y_1''' & y_2''' & y_3''' \end{vmatrix}.$$

   (c) Since each y_i satisfies (17), show that

$$(33) \quad y_i^{(3)}(x) = -\sum_{k=1}^{3} p_k(x) y_i^{(3-k)}(x), \quad i = 1, 2, 3.$$

   (d) Substituting the expressions in (33) into (32), show that

$$(34) \qquad W'(x) = -p_1(x) W(x).$$

   (e) Deduce Abel's identity by solving equation (34).

• 6.3 • HOMOGENEOUS LINEAR EQUATIONS WITH CONSTANT COEFFICIENTS

Our goal in this section is to obtain a general solution to an nth order differential equation with constant coefficients. Based on the experience gained with second order equations in Chapter 4, the reader should have little trouble guessing the form of such a solution. However, our interest here is to help the reader understand *why* these techniques work. This is done using an operator approach—a technique that is useful in tackling many other problems in analysis such as solving partial differential equations.

Let's consider the homogeneous linear nth order differential equation

$$(1) \qquad a_n y^{(n)}(x) + a_{n-1} y^{(n-1)}(x) + \cdots + a_1 y'(x) + a_0 y(x) = 0,$$

where $a_n \, (\neq 0), a_{n-1}, \ldots, a_0$ are real constants.[†] Since constant functions are everywhere

---

[†] *Historical Footnote*: In a letter to John Bernoulli dated September 15, 1739, Leonhard Euler claimed to have solved the general case of the homogeneous linear nth order equation with constant coefficients.

continuous, equation (1) has solutions defined for all x in $(-\infty, \infty)$ (recall Theorem 1 in Section 6.2). If we can find n linearly independent solutions to (1) on $(-\infty, \infty)$, say $y_1, \ldots, y_n$, then we can express a general solution to (1) in the form

$$\textbf{(2)} \qquad y(x) = C_1 y_1(x) + \cdots + C_n y_n(x),$$

with $C_1, \ldots, C_n$ as arbitrary constants.

To find these n linearly independent solutions, we shall capitalize on our previous success with second order equations. Namely, experience suggests that we begin by trying a function of the form $y = e^{rx}$.

If we let L be the differential operator defined by the left-hand side of (1), that is,

$$\textbf{(3)} \qquad L[y] := a_n y^{(n)} + a_{n-1} y^{(n-1)} + \cdots + a_1 y' + a_0 y,$$

then we can write (1) in the operator form

$$\textbf{(4)} \qquad L[y](x) = 0.$$

For $y = e^{rx}$, we find

$$\textbf{(5)} \qquad L[e^{rx}](x) = a_n r^n e^{rx} + a_{n-1} r^{n-1} e^{rx} + \cdots + a_0 e^{rx}$$
$$= e^{rx}(a_n r^n + a_{n-1} r^{n-1} + \cdots + a_0) = e^{rx} P(r),$$

where $P(r)$ is the polynomial $a_n r^n + a_{n-1} r^{n-1} + \cdots + a_0$. Thus e^{rx} is a solution to equation (4), provided that r is a root of the **auxiliary equation**

$$\textbf{(6)} \qquad P(r) = a_n r^n + a_{n-1} r^{n-1} + \cdots + a_0 = 0.$$

According to the fundamental theorem of algebra, the auxiliary equation has n roots (counting multiplicities), which may be either real or complex. As we mentioned in Chapter 4, there are no formulas for determining the zeros of an arbitrary polynomial of degree greater than four. However, if we can determine one zero r_1, then we can divide out the factor $r - r_1$ and be left with a polynomial of lower degree. (For convenience we have chosen most of our examples and exercises so that 0, ± 1, or ± 2 are zeros of any polynomial of degree greater than two that we must factor.) When a zero cannot be exactly determined, numerical algorithms such as Newton's method or the quotient-difference algorithm can be used to compute approximate roots of the polynomial equation.[†] Some pocket calculators even have these algorithms built in.

Distinct Real Roots

If the roots $r_1, \ldots, r_n$ of the auxiliary equation (6) are real and distinct, then n solutions to equation (1) are

$$\textbf{(7)} \qquad y_1(x) = e^{r_1 x}, \; y_2(x) = e^{r_2 x}, \ldots, y_n(x) = e^{r_n x}.$$

[†] See, for example, *Applied and Computational Complex Analysis*, by P. Henrici, Wiley-Interscience, New York, 1974, Volume 1, or *Numerical Analysis*, Second Edition, by R. L. Burden, J. D. Faires, and A. C. Reynolds, Prindle, Weber & Schmidt, Boston, 1981.

As stated in the previous section, these functions are linearly independent on $(-\infty, \infty)$—a fact that we shall now officially verify. Let's assume that $c_1, \ldots, c_n$ are constants such that

(8) $$c_1 e^{r_1 x} + \cdots + c_n e^{r_n x} = 0$$

for all x in $(-\infty, \infty)$. Our goal is to prove that $c_1 = c_2 = \cdots = c_n = 0$.

One way to show this is to construct a linear operator L_k that annihilates (maps to zero) everything on the left-hand side of (8) except the kth term. For this purpose, we note that since $r_1, \ldots, r_n$ are the zeros of the auxiliary polynomial $P(r)$, then $P(r)$ can be factored as

(9) $$P(r) = a_n(r - r_1) \cdots (r - r_n).$$

Consequently, the operator $L[y] = a_n y^{(n)} + a_{n-1} y^{(n-1)} + \cdots + a_0 y$ can be expressed as the following composition:[†]

(10) $$L = P(D) = a_n(D - r_1) \cdots (D - r_n).$$

We now construct the polynomial $P_k(r)$ by deleting the factor $(r - r_k)$ from $P(r)$. Then we set $L_k := P_k(D)$; that is,

(11) $$L_k := P_k(D) = a_n(D - r_1) \cdots (D - r_{k-1})(D - r_{k+1}) \cdots (D - r_n).$$

Applying L_k to both sides of (8), we get, via linearity,

(12) $$c_1 L_k[e^{r_1 x}] + \cdots + c_n L_k[e^{r_n x}] = 0.$$

Also, since $L_k = P_k(D)$, we find (just as in equation (5)) that $L_k[e^{rx}](x) = e^{rx} P_k(r)$ for all r. Thus (12) can be written as

$$c_1 e^{r_1 x} P_k(r_1) + \cdots + c_n e^{r_n x} P_k(r_n) = 0,$$

which simplifies to

(13) $$c_k e^{r_k x} P_k(r_k) = 0,$$

because $P_k(r_i) = 0$ for $i \neq k$. Since r_k is not a root of $P_k(r)$, then $P_k(r_k) \neq 0$. It now follows from (13) that $c_k = 0$. But as k is arbitrary, all the constants $c_1, \ldots, c_n$ must be zero. Thus $y_1(x), \ldots, y_n(x)$ as given in (7) are linearly independent.

We have proved that, in the case of n distinct real roots, a general solution to (1) is

(14) $$y(x) = C_1 e^{r_1 x} + \cdots + C_n e^{r_n x},$$

where $C_1, \ldots, C_n$ are arbitrary constants.

● **EXAMPLE 1**   Find a general solution to

(15) $$y''' - 2y'' - 5y' + 6y = 0.$$

Solution   The auxiliary equation is

(16) $$r^3 - 2r^2 - 5r + 6 = 0.$$

[†] *Historical footnote:* The symbolic notation $P(D)$ was introduced by Augustin Cauchy in 1827.

By inspection, we find that $r = 1$ is a root. Then, using polynomial division, we get

$$r^3 - 2r^2 - 5r + 6 = (r - 1)(r^2 - r - 6),$$

which further factors into $(r - 1)(r + 2)(r - 3)$. Hence the roots of equation (16) are $r_1 = 1, r_2 = -2, r_3 = 3$. Since these roots are real and distinct, a general solution to (15) is

$$y(x) = C_1 e^x + C_2 e^{-2x} + C_3 e^{3x}. \quad \bullet$$

Complex Roots

If $\alpha + i\beta$ (α, β real) is a complex root of the auxiliary equation (6), then so is its complex conjugate $\alpha - i\beta$, since the coefficients of $P(r)$ are real valued (see Problem 22). If we accept complex-valued functions as solutions, then both $e^{(\alpha + i\beta)x}$ and $e^{(\alpha - i\beta)x}$ are solutions to (1). Moreover, if there are no repeated roots, then a general solution to (1) is again given by (14). To find two real-valued solutions corresponding to the roots $\alpha \pm i\beta$, we can just take the real and imaginary parts of $e^{(\alpha + i\beta)x}$. That is, since

(17) $e^{(\alpha + i\beta)x} = e^{\alpha x} \cos \beta x + i e^{\alpha x} \sin \beta x,$

then two linearly independent solutions to (1) are

(18) $e^{\alpha x} \cos \beta x, \qquad e^{\alpha x} \sin \beta x.$

In fact, using these solutions in place of $e^{(\alpha + i\beta)x}$ and $e^{(\alpha - i\beta)x}$ in (14) preserves the linear independence of the set of n solutions. Thus, treating each of the conjugate pairs of roots in this manner, we obtain a real-valued general solution to (1).

● **EXAMPLE 2**   Find a general solution to

(19) $y''' + y'' + 3y' - 5y = 0.$

Solution   The auxiliary equation is

(20) $r^3 + r^2 + 3r - 5 = (r - 1)(r^2 + 2r + 5) = 0,$

which has distinct roots $r_1 = 1, r_2 = -1 + 2i, r_3 = -1 - 2i$. Thus a general solution is

(21) $y(x) = C_1 e^x + C_2 e^{-x} \cos 2x + C_3 e^{-x} \sin 2x. \quad \bullet$

Repeated Roots

If r_1 is a root of multiplicity m, then the n solutions given in (7) are not even distinct, let alone linearly independent. Recall that, for a second order equation, when we had a repeated root r_1 to the auxiliary equation, we obtained two linearly independent solutions by taking $e^{r_1 x}$ and $x e^{r_1 x}$. So if r_1 is a root of (6) of multiplicity m, we might expect that m linearly independent solutions are

(22) $e^{r_1 x}, \quad x e^{r_1 x}, \quad x^2 e^{r_1 x}, \quad \dots, \quad x^{m-1} e^{r_1 x}.$

To see that this is the case, observe that if r_1 is a root of multiplicity m, then the auxiliary equation can be written in the form

(23) $a_n(r - r_1)^m(r - r_{m+1})\cdots(r - r_n) = 0,$

$$(r - r_1)^m\tilde{P}(r) = 0,$$

where $\tilde{P}(r) := a_n(r - r_{m+1})\cdots(r - r_n)$ and $\tilde{P}(r_1) \neq 0$. With this notation,

(24) $L[e^{rx}](x) = e^{rx}(r - r_1)^m\tilde{P}(r)$

(see (5)). Setting $r = r_1$ in (24), we again see that $e^{r_1 x}$ is a solution.

To find other solutions, we take the kth partial derivative with respect to r of both sides of (24):

(25) $\dfrac{\partial^k}{\partial r^k} L[e^{rx}](x) = \dfrac{\partial^k}{\partial r^k} [e^{rx}(r - r_1)^m\tilde{P}(r)].$

Carrying out the differentiation on the right-hand side of (25), the resulting expression will still have $(r - r_1)$ as a factor, provided that $k \leq m - 1$. Thus, setting $r = r_1$ in (25) gives

(26) $\dfrac{\partial^k}{\partial r^k} L[e^{rx}](x)\big|_{r=r_1} = 0$ if $k \leq m - 1.$

Now notice that the function e^{rx} has continuous partial derivatives of all orders with respect to r and x. Hence for mixed partial derivatives of e^{rx}, it makes no difference whether the differentiation is done first with respect to x, then with respect to r, or vice versa. Since L involves derivatives with respect to x, this means we can interchange the order of differentiation in (26) to obtain

$$L\left[\dfrac{\partial^k}{\partial r^k}(e^{rx})\big|_{r=r_1}\right](x) = 0.$$

Thus

(27) $\dfrac{\partial^k}{\partial r^k}(e^{rx})\big|_{r=r_1} = x^k e^{r_1 x}$

will be a solution to (1) for $k = 0, 1, \ldots, m - 1$. So m distinct solutions to (1), due to the root $r = r_1$ of multiplicity m, are indeed given by (22). We leave it as an exercise to show that the m functions in (22) are linearly independent on $(-\infty, \infty)$ (see Problem 23).

If $\alpha + i\beta$ is a repeated complex root of multiplicity m, then we can replace the $2m$ complex-valued functions

$$e^{(\alpha+i\beta)x}, \quad xe^{(\alpha+i\beta)x}, \quad \ldots, \quad x^{m-1}e^{(\alpha+i\beta)x},$$
$$e^{(\alpha-i\beta)x}, \quad xe^{(\alpha-i\beta)x}, \quad \ldots, \quad x^{m-1}e^{(\alpha-i\beta)x}$$

by the $2m$ linearly independent real-valued functions

(28)
$$e^{\alpha x}\cos\beta x, \quad xe^{\alpha x}\cos\beta x, \quad \ldots, \quad x^{m-1}e^{\alpha x}\cos\beta x,$$
$$e^{\alpha x}\sin\beta x, \quad xe^{\alpha x}\sin\beta x, \quad \ldots, \quad x^{m-1}e^{\alpha x}\sin\beta x.$$

Using the results of the three cases discussed above, we can obtain a set of n linearly independent solutions that yield a real-valued general solution for (1).

● **EXAMPLE 3**   Find a general solution to

(29) $y^{(4)} - y^{(3)} - 3y'' + 5y' - 2y = 0.$

Solution   The auxiliary equation is

$$r^4 - r^3 - 3r^2 + 5r - 2 = (r - 1)^3(r + 2) = 0,$$

which has roots $r_1 = 1, r_2 = 1, r_3 = 1, r_4 = -2.$ Since the root at 1 has multiplicity 3, a general solution is

(30) $y(x) = C_1 e^x + C_2 x e^x + C_3 x^2 e^x + C_4 e^{-2x}.$ ●

● **EXAMPLE 4**   Find a general solution to

(31) $y^{(4)} - 8y^{(3)} + 26y'' - 40y' + 25y = 0,$

whose auxiliary equation can be factored as

(32) $r^4 - 8r^3 + 26r^2 - 40r + 25 = (r^2 - 4r + 5)^2 = 0.$

Solution   The auxiliary equation (32) has repeated complex roots: $r_1 = 2 + i, r_2 = 2 + i, r_3 = 2 - i,$ and $r_4 = 2 - i.$ Hence a general solution is

$$y(x) = C_1 e^{2x} \cos x + C_2 x e^{2x} \cos x + C_3 e^{2x} \sin x + C_4 x e^{2x} \sin x.$$ ●

When the auxiliary equation is a polynomial of order greater than four, there are no formulas involving radicals for determining the roots of the equation (see discussion at the end of Section 4.5 on page 156). Numerical routines such as Newton's method can be used to approximate both the real *and* complex roots (see Problems 25–27).

EXERCISES 6.3

1, 3, 5, 9

In Problems 1 through 14 find a general solution for the differential equation.

1. $y''' + 2y'' - 8y' = 0.$

2. $y''' - 3y'' - y' + 3y = 0.$

3. $6z''' + 7z'' - z' - 2z = 0.$

4. $y''' + 2y'' - 19y' - 20y = 0.$

5. $y''' + 3y'' + 28y' + 26y = 0.$

6. $y''' - y'' + 2y = 0.$

7. $2y''' - y'' - 10y' - 7y = 0.$

8. $y''' + 5y'' - 13y' + 7y = 0.$

9. $u''' - 9u'' + 27u' - 27u = 0.$

10. $y''' + 3y'' - 4y' - 6y = 0.$

11. $y^{(4)} + 4y''' + 6y'' + 4y' + y = 0.$

12. $y''' + 5y'' + 3y' - 9y = 0.$

13. $y^{(4)} + 4y'' + 4y = 0.$

14. $\theta''' + 5\theta'' + 5\theta' - 11\theta = 0.$

In Problems 15 through 18 find a general solution to the linear homogeneous differential equation with constant coefficients whose auxiliary equation is given.

15. $(r - 1)^2(r + 3)(r^2 + 2r + 5)^2 = 0.$

16. $(r + 1)^2(r - 6)^3(r + 5)(r^2 + 1)(r^2 + 4) = 0.$

17. $(r + 4)(r - 3)(r + 2)^3(r^2 + 4r + 5)^2 r^5 = 0.$

18. $(r - 1)^3(r - 2)(r^2 + r + 1)(r^2 + 6r + 10)^3 = 0.$

In Problems 19 through 21 solve the given initial value problem.

19. $y''' - y'' - 4y' + 4y = 0;$
    $y(0) = -4,$ $y'(0) = -1,$ $y''(0) = -19.$

20. $y''' + 7y'' + 14y' + 8y = 0$;
$$y(0) = 1, \qquad y'(0) = -3, \qquad y''(0) = 13.$$

21. $y''' - 4y'' + 7y' - 6y = 0$;
$$y(0) = 1, \qquad y'(0) = 0, \qquad y''(0) = 0.$$

22. Let $P(r) = a_n r^n + \cdots + a_1 r + a_0$ be a polynomial with real coefficients $a_n, \ldots, a_0$. Prove that if r_1 is a zero of $P(r)$, then so is its complex conjugate $\bar{r}_1$. [Hint: Show that $\overline{P(r)} = P(\bar{r})$, where the bar denotes complex conjugation.]

23. Show that the m functions $e^{rx}, xe^{rx}, \ldots, x^{m-1}e^{rx}$ are linearly independent on $(-\infty, \infty)$. [Hint: Show that these functions are linearly independent if and only if $1, x, \ldots, x^{m-1}$ are linearly independent.]

24. As an alternate proof that $\{e^{r_1 x}, \ldots, e^{r_n x}\}$ is linearly independent when $r_1, \ldots, r_n$ are distinct, proceed as follows:

(a) Show that
$$W[e^{r_1 x}, \ldots, e^{r_n x}](x) = e^{(r_1 + \cdots + r_n)x} V(r_1, \ldots, r_n),$$
where $V(r_1, \ldots, r_n)$ is the determinant of the $n \times n$ matrix whose jth column is $\text{col}[1 \; r_j \; r_j^2 \cdots r_j^{n-1}]$. V is called a **Vandermonde determinant.**

(b) Use the fact that
$$V(r_1, \ldots, r_n) = [r_2 - r_1][(r_3 - r_1)(r_3 - r_2)] \cdots$$
$$[(r_n - r_1) \cdots (r_n - r_{n-1})]$$
$$= \prod_{i < j = 2}^{n} (r_j - r_i)$$
to show that the exponential functions are linearly independent on any interval (a, b).

25. Find a general solution to
$$y^{(4)} + 2y''' - 3y'' - y' + \tfrac{1}{2}y = 0,$$
by using Newton's method or some other numerical procedure to approximate the roots of the auxiliary equation.

26. Find a general solution to $y''' - 3y' - y = 0$ by using Newton's method or some other numerical procedure to approximate the roots of the auxiliary equation.

27. Find a general solution to
$$y^{(4)} + 2y^{(3)} + 4y'' + 3y' + 2y = 0,$$
by using Newton's method to numerically approximate the roots of the auxiliary equation. [Hint: To find complex roots, use the Newton recursion formula $z_{n+1} = z_n - f(z_n)/f'(z_n)$, and start with an appropriate initial guess z_0 in the complex plane.]

28. (a) Show that a general solution for the equation $y^{(n)} - y = 0$ is given by
$$y(x) = A_0 e^x + \sum_{k=1}^{(n-1)/2} e^{\alpha_k x}\{A_k \cos(\beta_k x) + B_k \sin(\beta_k x)\},$$
where $\alpha_k = \cos(2\pi k/n)$ and $\beta_k = \sin(2\pi k/n)$ when n is odd and when n is even by
$$y(x) = A_0 e^x + B_0 e^{-x}$$
$$+ \sum_{k=1}^{(n/2)-1} e^{\alpha_k x}\{A_k \cos(\beta_k x) + B_k \sin(\beta_k x)\}.$$
[Hint: Recall that the nth roots of unity are given by $e^{2\pi k i/n}, k = 0, 1, \ldots, n - 1$.]

(b) Find a general solution for the equation $y^{(n)} - 2y = 0$.

29. Higher Order Cauchy-Euler Equations. A differential equation that can be expressed in the form
$$a_n x^n y^{(n)}(x) + a_{n-1} x^{n-1} y^{(n-1)}(x) + \cdots + a_0 y(x) = 0,$$
where $a_n, a_{n-1}, \ldots, a_0$ are constants, is called a homogeneous **Cauchy-Euler** equation. (The second order case is discussed in Sections 4.5 and 8.5.) Use the substitution $y = x^r$ to help determine a fundamental solution set for the following Cauchy-Euler equations.

(a) $x^3 y''' + x^2 y'' - 2xy' + 2y = 0, \qquad x > 0.$

(b) $x^4 y^{(4)} + 6x^3 y''' + 2x^2 y'' - 4xy' + 4y = 0,$
$x > 0.$

(c) $x^3 y''' - 2x^2 y'' + 13xy' - 13y = 0, \qquad x > 0.$
[Hint: $x^{\alpha + i\beta} = e^{(\alpha + i\beta)\ln x}$
$$= x^\alpha\{\cos(\beta \ln x) + i \sin(\beta \ln x)\}.]$$

30. Let $y(x) = Ce^{rx}$, where $C \neq 0$ and r are real numbers, be a solution to a differential equation. Suppose we cannot determine r exactly, but can only approximate it by $\tilde{r}$. Let $\tilde{y}(x) := Ce^{\tilde{r}x}$ and consider the error $y(x) - \tilde{y}(x)$.

(a) If r and $\tilde{r}$ are positive, $r \neq \tilde{r}$, show that the error grows exponentially large as x approaches ∞.

(b) If r and $\tilde{r}$ are negative, $r \neq \tilde{r}$, show that the error goes to zero exponentially as x approaches ∞.

31. On a smooth horizontal surface a mass of m_1 kg is attached to a vertical surface by a spring with spring constant k_1 N/m. Another mass of m_2 kg is attached to the first object by a spring with spring constant k_2 N/m. The objects are aligned horizontally so that the springs are their natural lengths. As we showed in Section 6.1, this coupled spring-mass system is governed by the system of differential equations

$$(33) \qquad m_1 \frac{d^2x}{dt^2} + (k_1 + k_2)x - k_2 y = 0,$$

$$(34) \qquad m_2 \frac{d^2y}{dt^2} - k_2 x + k_2 y = 0.$$

Let's assume that $m_1 = m_2 = 1$, $k_1 = 3$, and $k_2 = 2$. If both objects are displaced 1 m to the right of their equilibrium positions and then released, determine the equations of motion for the objects as follows:

(a) Show that $x(t)$ satisfies the equation

$$(35) \qquad x^{(4)}(t) + 7x''(t) + 6x(t) = 0.$$

[Hint: Solve equation (33) for y in terms of x and x'' and substitute into equation (34).]

(b) Find a general solution $x(t)$ to (35).

(c) Substitute $x(t)$ back into (33) to obtain a general solution for $y(t)$.

(d) Use the initial conditions to determine the solu-

tions, $x(t)$ and $y(t)$, which are the equations of motion.

32. Suppose the two springs in the coupled spring-mass system discussed in Problem 31 are switched, giving the new data $m_1 = m_2 = 1$, $k_1 = 2$, and $k_2 = 3$. If both objects are now displaced 1 m to the right of their equilibrium positions and then released, determine the equations of motion of the two objects.

33. Vibrating Beam. In studying the transverse vibrations of a beam one encounters the homogeneous equation

$$EI \frac{d^4y}{dx^4} - ky = 0,$$

where $y(x)$ is related to the displacement of the beam at position x, the constant E is Young's modulus, I is the area moment of inertia, and k is a parameter. Assuming E, I, and k are positive constants, find a general solution in terms of sines, cosines, hyperbolic sines, and hyperbolic cosines.

•6.4• UNDETERMINED COEFFICIENTS AND THE ANNIHILATOR METHOD

When a higher order linear equation has constant coefficients, the method of undetermined coefficients discussed in Section 4.8 can still be used to obtain a particular y_p provided the forcing term is a sum of functions of the type given in Table 4.1 on page 173.

● **EXAMPLE 1** Use the method of undetermined coefficients to find a general solution to

$$(1) \qquad y'''(x) - 3y''(x) + 4y(x) = xe^{2x}.$$

Solution We first solve the corresponding homogeneous equation

$$(2) \qquad y'''(x) - 3y''(x) + 4y(x) = 0.$$

The auxiliary equation

$$r^3 - 3r^2 + 4 = (r + 1)(r - 2)^2 = 0$$

has a single root at -1 and a double root at 2, and hence a general solution to (2) is

$$(3) \qquad y_h(x) = C_1 e^{-x} + C_2 e^{2x} + C_3 x e^{2x}.$$

We can find the form for a particular solution y_p to equation (1) by referring to Table 4.1 on page 173. Since $g(x) = xe^{2x}$ is of Type IV, y_p has the form

$$y_p(x) = x^s(A_1 x + A_0)e^{2x}.$$

Since e^{2x} and xe^{2x} are solutions to the corresponding homogeneous equation (2), we take

$s = 2$. That is, y_p has the form

(4) $y_p(x) = (A_1 x^3 + A_0 x^2) e^{2x}$.

Substituting y_p for y in (1) gives

$$[8A_1 x^3 + (36A_1 + 8A_0)x^2 + (36A_1 + 24A_0)x + (6A_1 + 12A_0)]e^{2x}$$
$$- 3[4A_1 x^3 + (12A_1 + 4A_0)x^2 + (6A_1 + 8A_0)x + 2A_0]e^{2x}$$
$$+ 4[A_1 x^3 + A_0 x^2]e^{2x} = xe^{2x},$$

which reduces to

$$[18A_1 x + (6A_1 + 6A_0)]e^{2x} = xe^{2x}.$$

This leads to the system of equations

$$18A_1 = 1, \qquad 6A_1 + 6A_0 = 0,$$

which yields $A_1 = 1/18$ and $A_0 = -1/18$. Thus,

(5) $y_p(x) = \dfrac{1}{18}(x^3 - x^2)e^{2x}$.

A general solution to (1) is $y_h + y_p$, that is,

(6) $y(x) = C_1 e^{-x} + C_2 e^{2x} + C_3 xe^{2x} + \dfrac{1}{18}(x^3 - x^2)e^{2x}$.   ●

For a better understanding of why the method of undetermined coefficients works for linear equations with constant coefficients, we will study a procedure for solving non-homogeneous equations that begins by transforming the nonhomogeneous equation into a homogeneous one. For this purpose, we introduce the concept of an annihilator.

ANNIHILATOR

Definition 3.   A linear differential operator A is said to **annihilate** a function f if

(7) $A[f](x) = 0$,   all x.

That is, A annihilates f if f is a solution to the homogeneous linear differential equation (7) on $(-\infty, \infty)$.

For example, $A = D - 3$ annihilates $f(x) = e^{3x}$ since

$$(D - 3)[e^{3x}] = 3e^{3x} - 3e^{3x} = 0.$$

Also, $A = D^2 - 4D + 20$ is an annihilator of $e^{2x}\sin 4x$ since this function satisfies the equation

$$y'' - 4y' + 20y = 0.$$

From what we learned in the previous section about auxiliary equations with repeated roots, it follows that the differential operator $(D - r)^m$, m a positive integer, annihilates each of the functions

(8) $e^{rx}, xe^{rx}, \ldots, x^{m-1}e^{rx}.$

Moreover, the differential operator $[(D - \alpha)^2 + \beta^2]^m$ annihilates each of the functions

(9)
$$e^{\alpha x}\cos\beta x, \ xe^{\alpha x}\cos\beta x, \ldots, \ x^{m-1}e^{\alpha x}\cos\beta x,$$
$$e^{\alpha x}\sin\beta x, \ xe^{\alpha x}\sin\beta x, \ldots, \ x^{m-1}e^{\alpha x}\sin\beta x,$$

since these are the $2m$ linearly independent solutions to $[(D - \alpha)^2 + \beta^2]^m y = 0$.

In the next example we make use of the fact that linear differential operators with *constant* coefficients commute (see Exercises 4.2, Problem 32).

● **EXAMPLE 2**   Find a differential operator that annihilates

(10) $6xe^{-4x} + 5e^x\sin 2x.$

Solution   Let's consider the two functions whose sum appears in (10). Observe that $(D + 4)^2$ annihilates the function $f_1(x) := 6xe^{-4x}$. Further, $f_2(x) := 5e^x\sin 2x$ is annihilated by the operator $(D - 1)^2 + 4$. Hence the composite operator

$$A := (D + 4)^2[(D - 1)^2 + 4],$$

which is the same as the operator

$$[(D - 1)^2 + 4](D + 4)^2,$$

annihilates both f_1 and f_2. But then, by linearity, A also annihilates the sum $f_1 + f_2$.   ●

We now show how annihilators can be used to determine particular solutions to certain *non*homogeneous equations. Consider the nth order differential equation with constant coefficients

(11) $a_n y^{(n)}(x) + a_{n-1}y^{(n-1)}(x) + \cdots + a_0 y(x) = g(x),$

which can be written in the operator form

(12) $L[y](x) = g(x),$

where

$$L[y] := a_n y^{(n)} + a_{n-1}y^{(n-1)} + \cdots + a_0 y.$$

In this section we restrict our attention to nonhomogeneous terms that can be annihilated by a linear differential operator with *constant* coefficients. Assuming that A is such an annihilator of g, then applying A to both sides of (12) yields

$$A[L[y]](x) = A[g](x) = 0.$$

Therefore, if y is a solution to (12), then y is also a solution to the *homogeneous* linear differential equation

(13) $AL[y](x) = 0$

involving the composition of the operators A and L. Since equation (13) has constant coefficients (why?), we can apply the methods of Section 6.3 to obtain its general solution. Comparing this with a general solution of $L[y] = 0$, it is then possible to determine the *form* of a particular solution to equation (12). This procedure, known as the **annihilator method,** is illustrated in the following examples.

● **EXAMPLE 3**   Find a general solution to

(14) $y''(x) + y(x) = e^{2x} + 1$.

Solution   Since $D - 2$ annihilates e^{2x} and D annihilates 1, then $A := D(D - 2)$ annihilates $e^{2x} + 1$. Therefore, applying A to both sides of (14) yields

$$A[y'' + y] = A[e^{2x} + 1]$$

(15) $D(D - 2)(D^2 + 1)[y] = 0$.

The auxiliary equation associated with (15) is

$$r(r - 2)(r^2 + 1) = 0,$$

which has the roots i, $-i$, 2, 0. Hence a general solution to (15) is

(16) $y(x) = C_1 \cos x + C_2 \sin x + C_3 e^{2x} + C_4$.

Now recall that a general solution to (14) is of the form $y_h + y_p$, where y_p is a particular solution to (14) and y_h is a general solution to

$$y''(x) + y(x) = 0.$$

Since every solution to (14) is also a solution to (15), then $y_h + y_p$ must have the form displayed on the right-hand side of (16). But we recognize that $y_h(x) = C_1 \cos x + C_2 \sin x$, and so there exists a particular solution of the form

(17) $y_p(x) = C_3 e^{2x} + C_4$.

To determine the constants C_3 and C_4, we substitute y_p given by (17) into (14):

$$4C_3 e^{2x} + C_3 e^{2x} + C_4 = e^{2x} + 1,$$
$$5C_3 e^{2x} + C_4 = e^{2x} + 1.$$

Equating the coefficients of e^{2x} and setting the constant terms equal yields $C_3 = \frac{1}{5}$ and $C_4 = 1$. Thus

$$y_p(x) = \tfrac{1}{5} e^{2x} + 1,$$

and so a general solution to (14) is

$$y(x) = C_1 \cos x + C_2 \sin x + \tfrac{1}{5} e^{2x} + 1. \quad ●$$

● **EXAMPLE 4**   Use the annihilator method to determine the form of a particular solution to

(18) $y'' - y = e^{-2x} \sin x$.

Solution   The function $g(x) = e^{-2x} \sin x$ is annihilated by the operator

$$A := (D+2)^2 + 1^2 = D^2 + 4D + 5.$$

If we apply A to both sides of (18), we obtain

$$A[y'' - y] = A[e^{-2x}\sin x]$$

(19) $(D^2 + 4D + 5)(D^2 - 1)[y] = 0.$

Now the auxiliary equation associated with (19) is

$$(r^2 + 4r + 5)(r-1)(r+1) = 0,$$

which has roots $1, -1, -2+i, -2-i$. Hence a general solution to (19) is

(20) $y(x) = C_1 e^x + C_2 e^{-x} + C_3 e^{-2x}\cos x + C_4 e^{-2x}\sin x.$

Since a general solution to the corresponding homogeneous equation $y'' - y = 0$ is $y_h(x) = C_1 e^x + C_2 e^{-x}$, we see that a particular solution for (18) has the form

(21) $y_p(x) = C_3 e^{-2x}\cos x + C_4 e^{-2x}\sin x.$ ●

Of course the forms of the particular solutions derived in Examples 3 and 4 using the annihilator method can be quickly obtained from Table 4.1, page 173, for the method of undetermined coefficients. The essential point is that the annihilator method can be used to *justify* the method of undetermined coefficients; that is, to derive the entries that appear in Table 4.1. This applies not only to second order, but also to higher order linear differential equations with constant coefficients.

The connection between the two methods is even more evident once one realizes that those functions for which the method of undetermined coefficients works are *exactly* those functions that are solutions to higher order linear differential equations with constant coefficients; that is, those functions $g(x)$ that are annihilated by a linear differential operator with constant coefficients. In particular, these methods work for polynomials, exponentials, sines, and cosines, and any product or sum of these functions. (Why?)

We leave it for the reader to show how the annihilator method yields various entries in Table 4.1 (see Problems 34–37). The derivation for the general case, where

$$g(x) = p_n(x)e^{\alpha x}\cos \beta x + q_m(x)e^{\alpha x}\sin \beta x,$$

p_n, q_m polynomials, is discussed in Project A at the end of this chapter.

EXERCISES 6.4

In Problems 1 through 4 use the method of undetermined coefficients (Table 4.1, inside back cover) to determine the form of a particular solution for the given equation.

1. $y''' - 2y'' - 5y' + 6y = e^x + x^2.$

2. $y''' + y'' - 5y' + 3y = e^{-x} + \sin x.$

3. $y''' + 3y'' - 4y = e^{-2x}.$

4. $y''' + y'' - 2y = xe^x + 1.$

In Problems 5 through 10 find a general solution to the given equation.

5. $y''' - 2y'' - 5y' + 6y = e^x + x^2.$

6. $y''' + y'' - 5y' + 3y = e^{-x} + \sin x$.

7. $y''' + 3y'' - 4y = e^{-2x}$.

8. $y''' + y'' - 2y = xe^x + 1$.

9. $y''' - 3y'' + 3y' - y = e^x$.

10. $y''' + 4y'' + y' - 26y = e^{-3x}\sin 2x + x$.

In Problems 11 through 20 find a differential operator that annihilates the given function.

11. $x^4 - x^2 + 11$.

12. $3x^2 - 6x + 1$.

13. e^{-7x}.

14. e^{5x}.

15. $e^{2x} - 6e^x$.

16. $x^2 - e^x$.

17. $x^2 e^{-x}\sin 2x$.

18. $xe^{3x}\cos 5x$.

19. $xe^{-2x} + xe^{-5x}\sin 3x$.

20. $x^2 e^x - x\sin 4x + x^3$.

In Problems 21 through 30 use the annihilator method to determine the form of a particular solution for the given equation.

21. $u'' - 5u' + 6u = \cos 2x + 1$.

22. $y'' + 6y' + 8y = e^{3x} - \sin x$.

23. $y'' - 5y' + 6y = e^{3x} - x^2$.

24. $\theta'' - \theta = xe^x$.

25. $y'' - 6y' + 9y = \sin 2x + x$.

26. $y'' + 2y' + y = x^2 - x + 1$.

27. $y'' + 2y' + 2y = e^{-x}\cos x + x^2$.

28. $y'' - 6y' + 10y = e^{3x} - x$.

29. $z''' - 2z'' + z' = x - e^x$.

30. $y''' + 2y'' - y' - 2y = e^x - 1$.

In Problems 31 through 33 solve the given initial value problem.

31. $y''' + 2y'' - 9y' - 18y = -18x^2 - 18x + 22$;
$y(0) = -2$, $y'(0) = -8$, $y''(0) = -12$.

32. $y''' - 2y'' + 5y' = -24e^{3x}$;
$y(0) = 4$, $y'(0) = -1$, $y''(0) = -5$.

33. $y''' - 2y'' - 3y' + 10y$
$\quad = 34xe^{-2x} - 16e^{-2x} - 10x^2 + 6x + 34$;
$y(0) = 3$, $y'(0) = 0$, $y''(0) = 0$.

34. Use the annihilator method to show that if $a_0 \ne 0$ in equation (11) and $g(x)$ has the form

(22) $\qquad g(x) = b_m x^m + b_{m-1}x^{m-1} + \cdots + b_1 x + b_0$,

then

$\qquad y_p(x) = B_m x^m + B_{m-1}x^{m-1} + \cdots + B_1 x + B_0$

is the form of a particular solution to equation (11).

35. Use the annihilator method to show that if $a_0 = 0$ and $a_1 \ne 0$ in (11) and $g(x)$ has the form given in (22), then equation (11) has a particular solution of the form

$\qquad y_p(x) = x\{B_m x^m + B_{m-1}x^{m-1} + \cdots + B_1 x + B_0\}$.

36. Use the annihilator method to show that if $g(x)$ in (11) has the form $g(x) = Be^{\alpha x}$, then equation (11) has a particular solution of the form $y_p(x) = x^s Be^{\alpha x}$, where s is chosen to be the smallest nonnegative integer such that $x^s e^{\alpha x}$ is not a solution to the corresponding homogeneous equation.

37. Use the annihilator method to show that if $g(x)$ in (11) has the form

$\qquad g(x) = a\cos\beta x + b\sin\beta x$,

then equation (11) has a particular solution of the form

(23) $\qquad y_p(x) = x^s\{A\cos\beta x + B\sin\beta x\}$,

where s is chosen to be the smallest nonnegative integer such that $x^s\cos\beta x$ and $x^s\sin\beta x$ are not solutions to the corresponding homogeneous equation.

38. Referring to the coupled spring-mass system discussed in Section 6.1, suppose that an external force $E(t) = 37\cos 3t$ is applied to the second object of mass 1 kg. The displacement functions $x(t)$ and $y(t)$ now satisfy the system

(24) $\qquad 2x''(t) + 6x(t) - 2y(t) = 0$,

(25) $\qquad y''(t) + 2y(t) - 2x(t) = 37\cos 3t$.

(a) Show that $x(t)$ satisfies the equation

(26) $\qquad x^{(4)}(t) + 5x''(t) + 4x(t) = 37\cos 3t$.

[Hint: Solve equation (24) for y in terms of x and x'' and substitute into equation (25).]

(b) Find a general solution $x(t)$ to equation (26).

(c) Substitute $x(t)$ back into (24) to obtain a general solution for $y(t)$.

(d) If both masses are displaced 2 m to the right of their equilibrium positions and then released, find the displacement functions $x(t)$ and $y(t)$.

39. Suppose the displacement functions $x(t)$ and $y(t)$ for a coupled spring-mass system (similar to the one discussed in Problem 38) satisfy the initial value problem

$\qquad x''(t) + 5x(t) - 2y(t) = 0$; $x(0) = x'(0) = 0$,

$\qquad y''(t) + 2y(t) - 2x(t) = 3\sin 2t$;

$\qquad y(0) = 1$, $y'(0) = 0$.

Solve for $x(t)$ and $y(t)$.

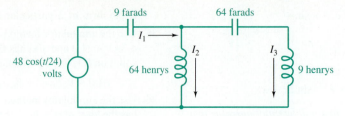

Figure 6.4 An electric network

40. The currents in the electric network in Figure 6.4 satisfy the system

$$\frac{1}{9}I_1 + 64I_2'' = -2\sin\frac{t}{24},$$

$$\frac{1}{64}I_3 + 9I_3'' - 64I_2'' = 0,$$

$$I_1 = I_2 + I_3,$$

where $I_1, I_2,$ and I_3 are the currents through the different branches of the network. By completing the following steps, determine the currents if initially $I_1(0) = I_2(0) = I_3(0) = 0,$ $I_1'(0) = \frac{73}{12},$ $I_2'(0) = \frac{3}{4},$ and $I_3'(0) = \frac{16}{3}.$

(a) By eliminating I_1 and I_3 from the system, show that I_2 satisfies

$$9^2(64)^2 I_2^{(4)} + (82)(64)I_2'' + I_2 = 0.$$

(b) Find a general solution for the equation in part (a). (This gives I_2 with four arbitrary constants.)

(c) Substitute I_2 into the first equation of the system to determine a general expression for I_1. Then use the third equation of the system to find a general expression for I_3.

(d) Use the initial conditions to determine $I_1, I_2,$ and I_3.

•6.5• METHOD OF VARIATION OF PARAMETERS

In the previous section we discussed the method of undetermined coefficients and the annihilator method. These methods work only for linear equations with constant coefficients *and* when the nonhomogeneous term is a solution to some homogeneous linear equation with constant coefficients. In this section we show how the method of **variation of parameters** discussed in Section 4.10 generalizes to higher order linear equations with variable coefficients.

Our goal is to find a particular solution to

(1) $L[y](x) = g(x),$

where $L[y] := y^{(n)} + p_1 y^{(n-1)} + \cdots + p_n y$ and the coefficient functions $p_1, \ldots, p_n$, as well as g, are continuous on (a, b). The method to be described requires that we already know a fundamental solution set $\{y_1, \ldots, y_n\}$ for the corresponding homogeneous equation

(2) $L[y](x) = 0.$

A general solution to (2) is then

(3) $y_h(x) = C_1 y_1(x) + \cdots + C_n y_n(x),$

where $C_1, \ldots, C_n$ are arbitrary constants. In the method of variation of parameters, we assume that there exists a particular solution to (1) of the form

(4) $y_p(x) = v_1(x)y_1(x) + \cdots + v_n(x)y_n(x)$

and try to determine the functions $v_1, \ldots, v_n$.

Since there are n unknown functions, we will need n conditions (equations) to determine them. These conditions are obtained as follows. Differentiating y_p in (4) gives

(5) $y_p' = (v_1 y_1' + \cdots + v_n y_n') + (v_1' y_1 + \cdots + v_n' y_n).$

To prevent second and higher order derivatives of the unknowns $v_1, \ldots, v_n$ from entering our later computations, we impose the condition

$$v_1' y_1 + \cdots + v_n' y_n = 0.$$

In a like manner, on computing $y_p'', y_p''', \ldots, y_p^{(n-1)}$, we impose $(n-2)$ additional conditions involving $v_1', \ldots, v_n'$; namely

$$v_1' y_1' + \cdots + v_n' y_n' = 0, \ldots, v_1' y_1^{(n-2)} + \cdots + v_n' y_n^{(n-2)} = 0.$$

Finally, the nth condition that we impose is that y_p satisfy the given equation (1). Using the previous conditions and the fact that $y_1, \ldots, y_n$ are solutions to the homogeneous equation, then $L[y_p] = g$ reduces to

(6) $v_1' y_1^{(n-1)} + \cdots + v_n' y_n^{(n-1)} = g,$

(see Problem 12). We therefore seek n functions $v_1', \ldots, v_n'$ that satisfy the system

(7)
$$
\begin{aligned}
y_1 v_1' \quad + \cdots + \ y_n v_n' \quad &= 0, \\
\vdots \qquad\qquad \vdots \quad \vdots\ \vdots \quad & \\
y_1^{(n-2)} v_1' + \cdots + y_n^{(n-2)} v_n' &= 0, \\
y_1^{(n-1)} v_1' + \cdots + y_n^{(n-1)} v_n' &= g.
\end{aligned}
$$

A sufficient condition for the existence of a solution to system (7) for x in (a, b) is that the determinant of the matrix made up of the coefficients of $v_1', \ldots, v_n'$ be different from zero for all x in (a, b). But this determinant is just the Wronskian:

(8)
$$
\begin{vmatrix}
y_1 & \cdots & y_n \\
\vdots & & \vdots \\
y_1^{(n-2)} & \cdots & y_n^{(n-2)} \\
y_1^{(n-1)} & \cdots & y_n^{(n-1)}
\end{vmatrix} = W[y_1, \ldots, y_n](x),
$$

which is never zero on (a, b), because $\{y_1, \ldots, y_n\}$ is a fundamental solution set. Solving (7) via Cramer's rule, we find

(9) $v_k'(x) = \dfrac{g(x) W_k(x)}{W[y_1, \ldots, y_n](x)}, \qquad k = 1, \ldots, n,$

where $W_k(x)$ is the determinant of the matrix obtained from the Wronskian $W[y_1, \ldots, y_n](x)$ by replacing the kth column by $\mathrm{col}[0, \ldots, 0, 1]$. Using a cofactor expansion about this column, we can express $W_k(x)$ in terms of an $(n-1)$st order Wronskian:

(10) $W_k(x) = (-1)^{n-k} W[y_1, \ldots, y_{k-1}, y_{k+1}, \ldots, y_n](x), \qquad k = 1, \ldots, n.$

Integrating $v_k'(x)$ in (9) gives

(11) $\qquad v_k(x) = \int \dfrac{g(x)W_k(x)}{W[y_1,\ldots,y_n](x)}\,dx, \qquad k = 1,\ldots,n.$

Finally, substituting the v_k's back into (4), we obtain the particular solution

(12) $\qquad y_p(x) = \displaystyle\sum_{k=1}^{n} y_k(x) \int \dfrac{g(x)W_k(x)}{W[y_1,\ldots,y_n](x)}\,dx.$

Although equation (12) gives a neat formula for a particular solution to (1), its implementation requires one to evaluate $n + 1$ determinants and then perform n integrations. This may entail several tedious computations. However, the method works even in cases when the technique of undetermined coefficients does not apply, provided, of course, that we know a fundamental solution set.

● **EXAMPLE 1**   Find a general solution to the Cauchy-Euler equation

(13) $\qquad x^3 y''' + x^2 y'' - 2xy' + 2y = x^3 \sin x, \qquad x > 0,$

given that $\{x, x^{-1}, x^2\}$ is a fundamental solution set to the corresponding homogeneous equation.

Solution   An important first step is to divide (13) by x^3 to obtain the standard form

(14) $\qquad y''' + \dfrac{1}{x}y'' - \dfrac{2}{x^2}y' + \dfrac{2}{x^3}y = \sin x, \qquad x > 0,$

from which we see that $g(x) = \sin x$. Since $\{x, x^{-1}, x^2\}$ is a fundamental solution set, we can obtain a particular solution of the form

(15) $\qquad y_p(x) = v_1(x)x + v_2(x)x^{-1} + v_3(x)x^2.$

To use formula (12), we must first evaluate the four determinants:

$$W[x, x^{-1}, x^2](x) = \begin{vmatrix} x & x^{-1} & x^2 \\ 1 & -x^{-2} & 2x \\ 0 & 2x^{-3} & 2 \end{vmatrix} = -6x^{-1},$$

$$W_1(x) = (-1)^{(3-1)}W[x^{-1}, x^2](x) = (-1)^2 \begin{vmatrix} x^{-1} & x^2 \\ -x^{-2} & 2x \end{vmatrix} = 3,$$

$$W_2(x) = (-1)^{(3-2)} \begin{vmatrix} x & x^2 \\ 1 & 2x \end{vmatrix} = -x^2,$$

$$W_3(x) = (-1)^{(3-3)} \begin{vmatrix} x & x^{-1} \\ 1 & -x^{-2} \end{vmatrix} = -2x^{-1}.$$

Substituting the above expressions into (12), we find

$$y_p(x) = x \int \frac{(\sin x)3}{-6x^{-1}} \, dx + x^{-1} \int \frac{(\sin x)(-x^2)}{-6x^{-1}} \, dx + x^2 \int \frac{(\sin x)(-2x^{-1})}{-6x^{-1}} \, dx$$

$$= x \int -\tfrac{1}{2} x \sin x \, dx + x^{-1} \int \tfrac{1}{6} x^3 \sin x \, dx + x^2 \int \tfrac{1}{3} \sin x \, dx,$$

which simplifies to

(16) $y_p(x) = \cos x - x^{-1} \sin x + C_1 x + C_2 x^{-1} + C_3 x^2,$

where $C_1, C_2,$ and C_3 denote the constants of integration. Since $\{x, x^{-1}, x^2\}$ is a fundamental solution set for the homogeneous equation, we can take $C_1, C_2,$ and C_3 to be arbitrary constants; the right-hand side of (16) then gives the desired general solution. ●

In the preceding example the fundamental solution set $\{x, x^{-1}, x^2\}$ can be derived by substituting $y = x^r$ into the homogeneous equation corresponding to (13) (see Problem 27, Exercises 6.3). However, in dealing with other equations that have variable coefficients, the determination of a fundamental set may be extremely difficult. In Chapter 8 we tackle this problem using power series methods.

EXERCISES 6.5

In Problems 1 through 6 use the method of variation of parameters to determine a particular solution to the given equation.

1. $y''' - 3y'' + 4y = e^{2x}.$

2. $y''' - 2y'' + y' = x.$

3. $z''' + 3z'' - 4z = e^{2x}.$

4. $y''' - 3y'' + 3y' - y = e^x.$

5. $y''' + y' = \tan x,$ $0 < x < \pi/2.$

6. $y''' + y' = \sec \theta \tan \theta,$ $0 < \theta < \pi/2.$

7. Find a general solution to the Cauchy-Euler equation

$$x^3 y''' - 3x^2 y'' + 6xy' - 6y = x^{-1}, \qquad x > 0,$$

. given that $\{x, x^2, x^3\}$ is a fundamental solution set for the corresponding homogeneous equation.

8. Find a general solution to the Cauchy-Euler equation

$$x^3 y''' - 2x^2 y'' + 3xy' - 3y = x^2, \qquad x > 0,$$

given that $\{x, x \ln x, x^3\}$ is a fundamental solution set for the corresponding homogeneous equation.

9. Given that $\{e^x, e^{-x}, e^{2x}\}$ is a fundamental solution set for the homogeneous equation corresponding to the

equation

$$y''' - 2y'' - y' + 2y = g(x),$$

determine a formula involving integrals for a particular solution.

10. Given that $\{x, x^{-1}, x^4\}$ is a fundamental solution set for the homogeneous equation corresponding to the equation

$$x^3 y''' - x^2 y'' - 4xy' + 4y = g(x), \qquad x > 0,$$

determine a formula involving integrals for a particular solution.

11. Find a general solution to the Cauchy-Euler equation

$$x^3 y''' - 3xy' + 3y = x^4 \cos x, \qquad x > 0.$$

12. Derive the system (7) in the special case when $n = 3$. [Hint: To determine the last equation, require that $L[y_p] = g$ and use the fact that $y_1, y_2,$ and y_3 satisfy the corresponding homogeneous equation.]

13. Show that

$$W_k(x) = (-1)^{(n-k)} W[y_1, \dots, y_{k-1}, y_{k+1}, \dots, y_n](x).$$

14. **Deflection of a Beam under Axial Force.**   A uniform beam under a load and subject to a constant axial force is governed by the differential equation

$$y^{(4)}(x) - k^2 y''(x) = q(x), \qquad 0 < x < L,$$

where $y(x)$ is the deflection of the beam, L is the length of the beam, k^2 is proportional to the axial force, and $q(x)$ is proportional to the load (see Figure 6.5).

(a) Show that a general solution can be written in the form

$$y(x) = C_1 + C_2 x + C_3 e^{kx} + C_4 e^{-kx}$$

$$+ \frac{1}{k^3} \int q(x)(kx - 1)\, dx - \frac{x}{k^2} \int q(x)\, dx$$

$$+ \frac{e^{kx}}{2k^3} \int q(x)e^{-kx}\, dx - \frac{e^{-kx}}{2k^3} \int q(x)e^{kx}\, dx.$$

(b) Show that the general solution in part (a) can be rewritten in the form

$$y(x) = c_1 + c_2 x + c_3 e^{kx} + c_4 e^{-kx}$$

$$+ \int_0^x q(s)G(s, x)\, ds,$$

where

$$G(s, x) := \frac{s}{k^2} - \frac{1}{k^3} - \frac{x}{k^2} + \frac{e^{kx}e^{-ks}}{2k^3} - \frac{e^{-kx}e^{ks}}{2k^3}.$$

(c) Let $q(x) \equiv 1$. First compute the general solution

using the formula in part (a), then using the formula in part (b). Compare these two general solutions with the general solution

$$y(x) = B_1 + B_2 x + B_3 e^{kx} + B_4 e^{-kx} - \frac{1}{2k^2} x^2,$$

which one would obtain using the method of undetermined coefficients.

(d) What are some advantages of the formula in part (b)?

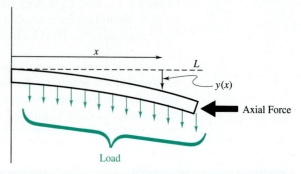

x

L

$y(x)$

Axial Force

Load

Figure 6.5 Deformation of a beam under axial force and load

• CHAPTER SUMMARY •

The theory and techniques for solving an nth order linear differential equation

$$(1) \qquad y^{(n)} + p_1(x)y^{(n-1)} + \cdots + p_n(x)y = g(x)$$

are natural extensions of the development for second order equations given in Chapter 4. Assuming that $p_1, \ldots, p_n$, and g are continuous functions on an open interval I, there is a unique solution to (1) on I that satisfies the n initial conditions: $y(x_0) = y_0$, $y'(x_0) = y_1, \ldots, y^{(n-1)}(x_0) = y_{n-1}$, where $x_0 \in I$.

For the corresponding homogeneous equation

$$(2) \qquad y^{(n)} + p_1(x)y^{(n-1)} + \cdots + p_n(x)y = 0,$$

there exists a set of n **linearly independent** solutions $\{y_1, \ldots, y_n\}$ on I. Such functions are said to form a **fundamental solution set,** and every solution to (2) can be written as a linear combination of these functions:

$$y(x) = C_1 y_1(x) + C_2 y_2(x) + \cdots + C_n y_n(x).$$

The linear independence of solutions to (2) is equivalent to the nonvanishing on I of the **Wronskian**

$$W[y_1,\ldots,y_n](x) := \det \begin{bmatrix} y_1(x) & \cdots & y_n(x) \\ y_1'(x) & \cdots & y_n'(x) \\ \vdots & & \vdots \\ y_1^{(n-1)}(x) & \cdots & y_n^{(n-1)}(x) \end{bmatrix}.$$

When equation (2) has (real) constant coefficients, so that it is of the form

(3) $\qquad a_n y^{(n)} + a_{n-1} y^{(n-1)} + \cdots + a_0 y = 0, \qquad a_n \neq 0,$

then the problem of determining a fundamental solution set can be reduced to the algebraic problem of solving the **auxiliary equation**

(4) $\qquad a_n r^n + a_{n-1} r^{n-1} + \cdots + a_0 = 0.$

If the n roots of (4), say $r_1, r_2, \ldots, r_n$, are all distinct, then

(5) $\qquad \{e^{r_1 x}, e^{r_2 x}, \ldots, e^{r_n x}\}$

is a fundamental solution set for (3). If some real root, say r_1, occurs with multiplicity m (e.g., $r_1 = r_2 = \cdots = r_m$), then m of the functions in (5) are replaced by

$$e^{r_1 x}, xe^{r_1 x}, \ldots, x^{m-1} e^{r_1 x}.$$

When a complex root $\alpha + i\beta$ to (4) occurs with multiplicity m, then so does its conjugate, and $2m$ members of the set (5) are replaced by the real-valued functions

$$e^{\alpha x} \sin \beta x, \, xe^{\alpha x} \sin \beta x, \ldots, x^{m-1} e^{\alpha x} \sin \beta x,$$
$$e^{\alpha x} \cos \beta x, \, xe^{\alpha x} \cos \beta x, \ldots, x^{m-1} e^{\alpha x} \cos \beta x.$$

A general solution to the nonhomogeneous equation (1) can be written as

$$y(x) = y_p(x) + y_h(x),$$

where y_p is some particular solution to (1) and y_h is a general solution to the corresponding homogeneous equation. Two useful techniques for finding particular solutions are the **annihilator method** (undetermined coefficients) and the method of **variation of parameters.**

The annihilator method applies to equations of the form

(6) $\qquad L[y] = g(x),$

where L is a linear differential operator with constant coefficients and the forcing term $g(x)$ is a polynomial, exponential, sine, or cosine, or a linear combination of products of these. Such a function $g(x)$ is annihilated (mapped to zero) by a linear differential operator A that also has constant coefficients. Every solution to the nonhomogeneous equation (6) is then a solution to the homogeneous equation $AL[y] = 0$, and, by comparing the solutions of the latter equation with a general solution to $L[y] = 0$, we can obtain the *form* of a particular solution to (6). These forms have previously been compiled in Table 4.1, page 173, for the method of undetermined coefficients.

The method of variation of parameters is more general in that it applies to arbitrary equations of the form (1). The idea is: Starting with a fundamental solution set $\{y_1, \ldots, y_n\}$ for (2), determine functions $v_1, \ldots, v_n$ such that

(7) $y_p(x) = v_1(x)y_1(x) + \cdots + v_n(x)y_n(x)$

satisfies (1). This method leads to the formula

(8) $$y_p(x) = \sum_{k=1}^{n} y_k(x) \int \frac{g(x)W_k(x)}{W[y_1, \ldots, y_n](x)} dx,$$

where

$$W_k(x) = (-1)^{n-k} W[y_1, \ldots, y_{k-1}, y_{k+1}, \ldots, y_n](x), \qquad k = 1, \ldots, n.$$

REVIEW PROBLEMS

1. Determine the intervals for which Theorem 1 on page 246 guarantees the existence of a solution in that interval.

. **(a)** $y^{(4)} - (\ln x)y'' + xy' + 2y = \cos 3x$.
   (b) $(x^2 - 1)y''' + (\sin x)y'' + \sqrt{x+4}\, y' + e^x y = x^2 + 3$.

2. Determine whether the given functions are linearly dependent or linearly independent on the interval $(0, \infty)$.

   (a) $\{e^{2x}, x^2 e^{2x}, e^{-x}\}$.
   (b) $\{e^x \sin 2x, xe^x \sin 2x, e^x, xe^x\}$.
   (c) $\{2e^{2x} - e^x, e^{2x} + 1, e^{2x} - 3, e^x + 1\}$.

3. Show that the set of functions $\{\sin x, x \sin x, x^2 \sin x, x^3 \sin x\}$ is linearly independent on $(-\infty, \infty)$.

4. Find a general solution for the given differential equation.

   (a) $y^{(4)} + 2y''' - 4y'' - 2y' + 3y = 0$.
   (b) $y''' + 3y'' - 5y' + y = 0$.
   (c) $y^{(5)} - y^{(4)} + 2y''' - 2y'' + y' - y = 0$.
   (d) $y''' - 2y'' - y' + 2y = e^x + x$.

5. Find a general solution for the homogeneous linear differential equation with constant coefficients whose auxiliary equation is

   (a) $(r + 5)^2(r - 2)^3(r^2 + 1)^2 = 0$.
   (b) $r^4(r - 1)^2(r^2 + 2r + 4)^2 = 0$.

6. Given that $y_p = \sin(x^2)$ is a particular solution to
   $$y^{(4)} + y = (16x^4 - 11)\sin(x^2) - 48x^2 \cos(x^2)$$
   on $(0, \infty)$, find a general solution.

7. Find a differential operator that annihilates the given function.

   (a) $x^2 - 2x + 5$.   **(b)** $e^{3x} + x - 1$.
   (c) $x \sin 2x$.   **(d)** $x^2 e^{-2x} \cos 3x$.
   (e) $x^2 - 2x + xe^{-x} + \sin 2x - \cos 3x$.

8. Use the annihilator method to determine the form of a particular solution for the given equation.

   (a) $y'' + 6y' + 5y = e^{-x} + x^2 - 1$.
   (b) $y''' + 2y'' - 19y' - 20y = xe^{-x}$.
   (c) $y^{(4)} + 6y'' + 9y = x^2 - \sin 3x$.
   (d) $y''' - y'' + 2y = x \sin x$.

9. Find a general solution to the Cauchy-Euler equation
   $$x^3 y''' - 2x^2 y'' - 5xy' + 5y = x^{-2}, \qquad x > 0,$$
   given that $\{x, x^5, x^{-1}\}$ is a fundamental solution set to the corresponding homogeneous equation.

10. Find a general solution to the given Cauchy-Euler equation.

    (a) $4x^3 y''' + 8x^2 y'' - xy' + y = 0$, $x > 0$.
    (b) $x^3 y''' + 2x^2 y'' + 2xy' + 4y = 0$, $x > 0$.

TECHNICAL WRITING EXERCISES

1. Describe the differences and similarities between second order and higher order linear differential equations. Include in your comparisons both theoretical results and the methods of solution; for example, what complications arise in solving higher order equations that are not present for the second order case?

2. Explain the relationship between the method of undetermined coefficients and the annihilator method. What difficulties would you encounter in applying the annihilator method if the linear equation did not have constant coefficients?

3. Section 5.5 contains a discussion of second order linear difference equations. Extend that discussion to nth order linear difference equations using the results for nth order differential equations as a guide.

4. For students with a background in linear algebra: Compare the theory for kth order linear differential equations with that for systems of n linear equations in n unknowns whose coefficient matrix has rank $n - k$. Use the terminology from linear algebra; for example, subspaces, basis, dimension, linear transformation, and kernel. Discuss both homogeneous and nonhomogeneous equations.

GROUP PROJECTS FOR CHAPTER 6

A. Justifying the Method of Undetermined Coefficients

The annihilator method discussed in Section 6.4 can be used to derive the entries in Table 4.1, page 173, for the method of undetermined coefficients. To show this, it suffices to work with type VII functions—that is, functions of the form

(1) $g(x) = p_n(x)e^{\alpha x} \cos \beta x + q_m(x)e^{\alpha x} \sin \beta x,$

where p_n and q_m are polynomials of degrees n and m, respectively—since the other types listed in Table 4.1 are just special cases of (1).

Consider the nonhomogeneous equation

(2) $L[y](x) = g(x),$

where L is the linear operator

(3) $L[y] := a_n y^{(n)} + a_{n-1} y^{(n-1)} + \cdots + a_0 y,$

with $a_n, a_{n-1}, \ldots, a_0$ constants, and $g(x)$ as given in equation (1). Let $N := \max(n, m)$.

(a) Show that

$$A := [(D - \alpha)^2 + \beta^2]^{N+1}$$

is an annihilator for g.

(b) Show that the auxiliary equation associated with $AL[y] = 0$ is of the form

(4) $a_n[(r - \alpha)^2 + \beta^2]^{s+N+1}(r - r_{2s+1})\cdots(r - r_n) = 0,$

where $s\ (\geq 0)$ is the multiplicity of $\alpha \pm i\beta$ as roots of the auxiliary equation associated with $L[y] = 0$, and $r_{2s+1}, \ldots, r_n$ are the remaining roots of this equation.

(c) Find a general solution for $AL[y] = 0$ and compare it with a general solution for $L[y] = 0$ to verify that equation (2) has a particular solution of the form

$$y_p(x) = x^s e^{\alpha x}\{P_N(x)\cos \beta x + Q_N(x)\sin \beta x\},$$

where P_N and Q_N are polynomials of degree N.

B. Transverse Vibrations of a Beam

In applying elasticity theory to study the transverse vibrations of a beam one encounters the equation

$$EIy^{(4)}(x) - \gamma\lambda y(x) = 0,$$

where $y(x)$ is related to the displacement of the beam at position x, the constant E is Young's modulus, I is the area moment of inertia which we assume is constant, γ is the constant mass per unit length of the beam, and λ is a positive parameter to be determined. We can simplify the equation by letting $r^4 := \gamma\lambda/EI$, that is, we consider

(5) $$y^{(4)}(x) - r^4 y(x) = 0.$$

When the beam is clamped at each end, we seek a solution to (5) that satisfies the boundary conditions

(6) $$y(0) = y'(0) = 0 \quad \text{and} \quad y(L) = y'(L) = 0,$$

where L is the length of the beam. The problem is to determine those values of r for which equation (5) has a nontrivial solution ($y(x) \not\equiv 0$) that satisfies (6). To do this proceed as follows:

(a) Represent the general solution to (5) in terms of sines, cosines, hyperbolic sines, and hyperbolic cosines.
(b) Substitute the general solution obtained in part (a) into the equations in (6) to obtain four equations in four unknowns.
(c) Determine those values of r for which the system of equations in part (b) has nontrivial solutions.
(d) For each value of r found in part (c), find the corresponding nontrivial solutions.

CHAPTER 7

Laplace Transforms

7.1 INTRODUCTION: A SIMPLE ELECTRICAL CIRCUIT

At time $t = 0$ a 10-volt battery is connected to the electrical circuit shown in Figure 7.1. Two seconds later ($t = 2$) the poles on the battery are reversed so that the potential is now -10 volts. Another three seconds later ($t = 5$) the circuit remains closed but the battery is by-passed. Determine the current $I(t)$ in the circuit at time $t > 0$.

To analyze the circuit we recall **Kirchhoff's laws,** which were discussed in Section 5.4.

1. **Conservation of Charge.** The current I passing through each element in the circuit must be the same.
2. **Conservation of Energy.** The algebraic sum of the instantaneous changes in potential around a closed circuit must be zero.

We further recall that the drop in potential at a resistor is RI and at an inductor is $L\,dI/dt$.

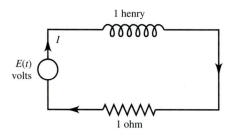

Figure 7.1 An electrical circuit

For the circuit of Figure 7.1 on the previous page, we have $R = 1$ ohm, $L = 1$ henry, and an electromotive force $E(t)$ given by

$$E(t) = \begin{cases} 10 \text{ volts}, & 0 < t < 2, \\ -10 \text{ volts}, & 2 < t < 5, \\ 0 \text{ volts}, & 5 < t. \end{cases}$$

Thus, on applying Kirchhoff's second law to the circuit, we obtain

(1) $$\frac{dI}{dt} + I = E(t), \qquad I(0) = 0.$$

This initial value problem can be solved using the techniques of earlier chapters, but since $E(t)$ is discontinuous, it would require solving *three* separate initial value problems—the first on the interval $[0, 2]$, a second on $[2, 5]$, and a third on $[5, \infty)$. Our purpose here is to illustrate a new approach that involves the use of Laplace transforms. As we shall see, this method offers some advantages over the previous techniques. In particular, it is ideal for solving initial value problems for linear equations with constant coefficients when the forcing term is discontinuous.

The idea behind Laplace transforms is to "simplify the problem" in much the same spirit as the substitution of $y = e^{rt}$ helped to simplify the problem of solving a linear differential equation with constant coefficients. The **Laplace transform** of a function $f(t)$ defined on $[0, \infty)$ is given by[†]

$$\mathcal{L}\{f\}(s) := \int_0^\infty e^{-st} f(t)\, dt.$$

Thus the transform takes a function of t and produces a function of s. This is done by multiplying $f(t)$ by e^{-st} and then integrating with respect to t from 0 to ∞. A listing of the Laplace transforms of some commonly occurring functions appears in the table on the inside of the back cover of this text.

Because of the properties of the exponential function, there is a simple algebraic relation between the Laplace transform of a function $f(t)$ and that of its derivative $f'(t)$; namely,

(2) $$\mathcal{L}\{f'\}(s) = s\mathcal{L}\{f\}(s) - f(0).$$

This formula can be verified using integration by parts:

$$\mathcal{L}\{f'\}(s) = \int_0^\infty e^{-st} f'(t)\, dt = f(t)e^{-st}\Big|_{t=0}^{t=\infty} + s\int_0^\infty e^{-st} f(t)\, dt$$

$$= -f(0) + s\mathcal{L}\{f\}(s),$$

assuming that $f(t)e^{-st} \to 0$ as $t \to \infty$.

Thanks to formula (2) (and its extension to higher order derivatives), the Laplace transform can be used to reduce certain differential equations to algebraic equations. Let's now use it to solve the initial value problem given in (1).

[†] *Historical Footnote*: The Laplace transform was first introduced by Pierre Laplace in 1779 in his research on probability. G. Doetsch helped develop the use of Laplace transforms to solve differential equations. His work in the 1930s served to justify the operational calculus procedures earlier used by Oliver Heaviside.

If we take the Laplace transform of both sides of the differential equation in (1)—that is, if we multiply equation (1) by e^{-st} and integrate with respect to t—we get

$$\mathscr{L}\{I'\}(s) + \mathscr{L}\{I\}(s) = \mathscr{L}\{E(t)\}(s),$$

where

$$\mathscr{L}\{E(t)\}(s) = \int_0^2 10e^{-st}\, dt + \int_2^5 (-10)e^{-st}\, dt$$

$$= \frac{10}{s} - \frac{20}{s}e^{-2s} + \frac{10}{s}e^{-5s}$$

$$= \frac{1}{s}(10 - 20e^{-2s} + 10e^{-5s}).$$

Next, from formula (2) and the condition $I(0)=0$, we see that $\mathscr{L}\{I'\}(s)=s\mathscr{L}\{I\}(s)$. Therefore,

$$s\mathscr{L}\{I\}(s) + \mathscr{L}\{I\}(s) = \frac{1}{s}(10 - 20e^{-2s} + 10e^{-5s}),$$

which is a simple algebraic equation for the function $\mathscr{L}\{I\}(s)$. Solving for $\mathscr{L}\{I\}(s)$ gives

$$\mathscr{L}\{I\}(s) = \frac{1}{(s+1)s}(10 - 20e^{-2s} + 10e^{-5s}).$$

The next step is to recover the function $I(t)$ from its Laplace transform. For this purpose it is convenient to use partial fractions and write

$$\mathscr{L}\{I\}(s) = \left(\frac{1}{s} - \frac{1}{s+1}\right)(10 - 20e^{-2s} + 10e^{-5s})$$

(3) $$\mathscr{L}\{I\}(s) = 10\left(\frac{1}{s} - \frac{1}{s+1}\right) - 20\left(\frac{1}{s} - \frac{1}{s+1}\right)e^{-2s} + 10\left(\frac{1}{s} - \frac{1}{s+1}\right)e^{-5s}.$$

Now we are in a position to use the table of Laplace transforms, on the inside back cover, to look up the three functions whose Laplace transforms appear on the right-hand side of (3). From entry **14** of the table, we see that

$$\mathscr{L}\{1\}(s) = \frac{1}{s}, \qquad \mathscr{L}\{e^{-t}\}(s) = \frac{1}{s+1},$$

and so, using the linearity of the Laplace transform, we get

(4) $$\mathscr{L}\{1 - e^{-t}\}(s) = \frac{1}{s} - \frac{1}{s+1}.$$

Next from (4) and entry **10** of the table it follows that

(5) $$\mathscr{L}\{(1 - e^{-(t-2)})u(t-2)\}(s) = e^{-2s}\left(\frac{1}{s} - \frac{1}{s+1}\right),$$

(6) $$\mathscr{L}\{(1 - e^{-(t-5)})u(t-5)\}(s) = e^{-5s}\left(\frac{1}{s} - \frac{1}{s+1}\right),$$

where $u(t - a)$ is the unit step function defined by

$$u(t - a) := \begin{cases} 0, & t < a, \\ 1, & t > a. \end{cases}$$

Combining (4), (5), and (6), we see from (3) that

$$I(t) = 10(1 - e^{-t}) - 20(1 - e^{-(t-2)})u(t - 2) + 10(1 - e^{-(t-5)})u(t - 5).$$

Using the definition of the unit step function, we can rewrite $I(t)$ in the equivalent form

$$\text{(7)} \qquad I(t) = \begin{cases} 10(1 - e^{-t}), & 0 < t < 2, \\ -10 + (20e^2 - 10)e^{-t}, & 2 < t < 5, \\ (20e^2 - 10 - 10e^5)e^{-t}, & 5 < t. \end{cases}$$

Thus $I(t)$ as given in equation (7) is the desired solution to (1).

The above example illustrates how easily the Laplace transform method handles initial value problems and discontinuous forcing functions. We first transform an initial value problem into an algebraic equation. After solving the algebraic equation for the unknown, we "untransform" to get the solution to the initial value problem. The method also works well for *impulse* forcing functions; that is, for functions that are zero except on a very short interval where they are relatively large.

In the next three sections we discuss in detail the definition and properties of the Laplace transform. Section 7.5 explains how Laplace transforms are used to solve initial value problems. The remaining sections describe additional properties of the transform that are useful in applications.

•7.2• DEFINITION OF THE LAPLACE TRANSFORM

In earlier chapters we studied differential operators. These operators took a function and mapped or transformed it (via differentiation) into another function. We now present an important transformation, denoted by $\mathcal{L}$, which is an integral operator.

> ### LAPLACE TRANSFORM
>
> **Definition 1.** Let $f(t)$ be a function on $[0, \infty)$. The **Laplace transform** of f is the function F defined by the integral
>
> $$\text{(1)} \qquad F(s) := \int_0^\infty e^{-st} f(t) \, dt.$$
>
> The domain of $F(s)$ is all the values of s for which the integral in (1) exists.[†] The Laplace transform of f is denoted by both F and $\mathcal{L}\{f\}$.

[†] We treat s as real valued, but in certain applications s may be a complex variable. For a detailed treatment of complex-valued Laplace transforms see *Complex Variables and the Laplace Transform for Engineers*, by Wilbur R. LePage, Dover Publications, New York, 1980, or *Basic Complex Analysis*, by J. E. Marsden, W. H. Freeman and Company, San Francisco, 1973.

Notice that the integral in (1) is an **improper** integral that is defined by

$$\int_0^\infty e^{-st}f(t)\,dt := \lim_{N\to\infty} \int_0^N e^{-st}f(t)\,dt$$

whenever the limit exists.

● **EXAMPLE 1** Determine the Laplace transform of the constant function $f(t) = 1, t \geq 0$.

Solution Using the definition of the transform, we compute

$$F(s) = \int_0^\infty e^{-st} \cdot 1\,dt = \lim_{N\to\infty} \int_0^N e^{-st}\,dt$$

$$= \lim_{N\to\infty} \frac{-e^{-st}}{s}\Big|_{t=0}^{t=N} = \lim_{N\to\infty}\left[\frac{1}{s} - \frac{e^{-sN}}{s}\right].$$

Since $e^{-sN} \to 0$ when $s > 0$ is fixed and $N \to \infty$, we get

$$F(s) = \frac{1}{s} \quad \text{for} \quad s > 0.$$

When $s \leq 0$, the integral $\int_0^\infty e^{-st}\,dt$ diverges (Why?). Hence $F(s) = 1/s$, with the domain of $F(s)$ being all $s > 0$. ●

● **EXAMPLE 2** Determine the Laplace transform of $f(t) = e^{at}$, where a is a constant.

Solution Using the definition of the transform,

$$F(s) = \int_0^\infty e^{-st}e^{at}\,dt = \int_0^\infty e^{-(s-a)t}\,dt$$

$$= \lim_{N\to\infty} \int_0^N e^{-(s-a)t}\,dt = \lim_{N\to\infty} \frac{-e^{-(s-a)t}}{s-a}\Big|_0^N$$

$$= \lim_{N\to\infty}\left[\frac{1}{s-a} - \frac{e^{-(s-a)N}}{s-a}\right]$$

$$= \frac{1}{s-a} \quad \text{for} \quad s > a.$$

Again, if $s \leq a$ the integral diverges, and hence the domain of $F(s)$ is all $s > a$. ●

It is comforting to note from Example 2 that the transform of the constant function $f(t) = 1 = e^{0t}$ is $1/(s-0) = 1/s$, which agrees with the solution in Example 1.

● **EXAMPLE 3** Find $\mathcal{L}\{\sin bt\}$, where b is a nonzero constant.

Solution We need to compute

$$\mathcal{L}\{\sin bt\}(s) = \int_0^\infty e^{-st}\sin bt\,dt = \lim_{N\to\infty} \int_0^N e^{-st}\sin bt\,dt.$$

Using integration by parts twice, we ultimately find (see Problem 32)

$$\mathscr{L}\{\sin bt\}(s) = \lim_{N\to\infty}\left[-\frac{e^{-st}}{s^2+b^2}(s\sin bt + b\cos bt)\Big|_0^N\right]$$

$$= \lim_{N\to\infty}\left[\frac{b}{s^2+b^2} - \frac{e^{-sN}}{s^2+b^2}(s\sin bN + b\cos bN)\right]$$

$$= \frac{b}{s^2+b^2} \quad \text{for} \quad s > 0. \quad \bullet$$

● **EXAMPLE 4**   Determine the Laplace transform of

$$f(t) = \begin{cases} 2, & 0 < t < 5, \\ 0, & 5 < t < 10, \\ e^{4t}, & 10 < t. \end{cases}$$

Solution   Since $f(t)$ is defined by a different formula on different intervals, we begin by breaking up the integral in (1) into three separate parts.[†] Thus

$$F(s) = \int_0^\infty e^{-st}f(t)\,dt$$

$$= \int_0^5 e^{-st}\cdot 2\,dt + \int_5^{10} e^{-st}\cdot 0\,dt + \int_{10}^\infty e^{-st}e^{4t}\,dt$$

$$= 2\int_0^5 e^{-st}\,dt + \lim_{N\to\infty}\int_{10}^N e^{-(s-4)t}\,dt$$

$$= \frac{2}{s} - \frac{2e^{-5s}}{s} + \lim_{N\to\infty}\left[\frac{e^{-10(s-4)}}{s-4} - \frac{e^{-(s-4)N}}{s-4}\right]$$

$$= \frac{2}{s} - \frac{2e^{-5s}}{s} + \frac{e^{-10(s-4)}}{s-4} \quad \text{for} \quad s > 4. \quad \bullet$$

Notice that the function $f(t)$ of Example 4 has jump discontinuities at $t = 5$ and $t = 10$. These values are reflected in the exponential terms e^{-5s} and e^{-10s} that appear in the formula for $F(s)$. We shall make this connection more precise when we discuss the unit step function in Section 7.6.

An important property of the Laplace transform is its **linearity.** That is, the Laplace transform $\mathscr{L}$ is a linear operator.

[†] Notice that $f(t)$ is not defined at the points $t = 0, 5,$ and 10. Nevertheless, the integral in (1) is still meaningful and is unaffected by function values at finitely many points.

> ### LINEARITY OF THE TRANSFORM
>
> **Theorem 1.**   Let f_1 and f_2 be functions whose Laplace transforms exist for $s > \alpha$, and let c be a constant. Then
>
> **(2)** $\mathcal{L}\{f_1 + f_2\} = \mathcal{L}\{f_1\} + \mathcal{L}\{f_2\},$
> **(3)** $\mathcal{L}\{cf_1\} = c\mathcal{L}\{f_1\}.$

Proof.   Using the linearity properties of integration, we have for $s > \alpha$

$$\mathcal{L}\{f_1 + f_2\}(s) = \int_0^\infty e^{-st}[f_1(t) + f_2(t)]\,dt$$

$$= \int_0^\infty e^{-st}f_1(t)\,dt + \int_0^\infty e^{-st}f_2(t)\,dt$$

$$= \mathcal{L}\{f_1\}(s) + \mathcal{L}\{f_2\}(s).$$

Hence equation (2) is satisfied. In a similar fashion we see that

$$\mathcal{L}\{cf_1\}(s) = \int_0^\infty e^{-st}[cf_1(t)]\,dt = c\int_0^\infty e^{-st}f_1(t)\,dt$$

$$= c\mathcal{L}\{f_1\}(s). \quad \blacktriangleleft\blacktriangleleft\blacktriangleleft$$

● **EXAMPLE 5**   Determine $\mathcal{L}\{11 + 5e^{4t} - 6\sin 2t\}$.

Solution   From the linearity property, we know that the Laplace transform of the sum of any finite number of functions is the sum of their Laplace transforms. Thus

$$\mathcal{L}\{11 + 5e^{4t} - 6\sin 2t\} = \mathcal{L}\{11\} + \mathcal{L}\{5e^{4t}\} + \mathcal{L}\{-6\sin 2t\}$$
$$= 11\mathcal{L}\{1\} + 5\mathcal{L}\{e^{4t}\} - 6\mathcal{L}\{\sin 2t\}.$$

In Examples 1, 2, and 3 we determined that

$$\mathcal{L}\{1\}(s) = \frac{1}{s}, \qquad \mathcal{L}\{e^{4t}\}(s) = \frac{1}{s - 4}, \qquad \mathcal{L}\{\sin 2t\}(s) = \frac{2}{s^2 + 2^2}.$$

Using these results, we find

$$\mathcal{L}\{11 + 5e^{4t} - 6\sin 2t\}(s) = 11\left(\frac{1}{s}\right) + 5\left(\frac{1}{s - 4}\right) - 6\left(\frac{2}{s^2 + 4}\right)$$

$$= \frac{11}{s} + \frac{5}{s - 4} - \frac{12}{s^2 + 4}.$$

Since $\mathcal{L}\{1\}$, $\mathcal{L}\{e^{4t}\}$, and $\mathcal{L}\{\sin 2t\}$ are all defined for $s > 4$, so is $\mathcal{L}\{11 + 5e^{4t} - 6\sin 2t\}$.

●

There are functions for which the improper integral in (1) fails to converge for any value of s. For example, this is the case for the function $f(t) = 1/t$, which grows too fast near zero. Likewise, no Laplace transform exists for the function $f(t) = e^{t^2}$, which increases too rapidly as $t \to \infty$. Fortunately, the set of functions for which the Laplace transform *is* defined includes many of the functions that arise in applications involving linear differential equations. We now discuss some properties that will (collectively) ensure the existence of the Laplace transform.

A function $f(t)$ on $[a, b]$ is said to have a **jump discontinuity** at $t_0 \in (a, b)$ if $f(t)$ is discontinuous at t_0 and the one-sided limits

$$\lim_{t \to t_0^-} f(t) \quad \text{and} \quad \lim_{t \to t_0^+} f(t)$$

exist as finite numbers. If the discontinuity occurs at an endpoint, $t_0 = a$ (or b), a jump discontinuity occurs if the one-sided limit of $f(t)$ as $t \to a^+$ $(t \to b^-)$ exists as a finite number. We can now define:

PIECEWISE CONTINUITY

Definition 2. A function $f(t)$ is said to be **piecewise continuous on a finite interval** $[a, b]$ if $f(t)$ is continuous at every point in $[a, b]$ except possibly for a finite number of points at which $f(t)$ has a jump discontinuity.

A function $f(t)$ is said to be **piecewise continuous on $[0, \infty)$** if $f(t)$ is piecewise continuous on $[0, N]$ for all $N > 0$.

EXAMPLE 6 Show that

$$f(t) = \begin{cases} t, & 0 < t < 1, \\ 2, & 1 < t < 2, \\ (t-2)^2, & 2 \le t \le 3 \end{cases}$$

is piecewise continuous on $[0, 3]$.

Solution From the graph of $f(t)$ sketched in Figure 7.2, we see that $f(t)$ is continuous on the intervals $(0, 1)$, $(1, 2)$, and $(2, 3]$. Moreover, at the points of discontinuity, $t = 0, 1$, and 2, the function has jump discontinuities since the one-sided limits exist as finite numbers. In particular, at $x = 1$, the left-hand limit is 1 and the right-hand limit is 2. Therefore, $f(t)$ is piecewise continuous on $[0, 3]$. ●

The reader should observe that the function $f(t)$ of Example 4 is piecewise continuous on $[0, \infty)$ because it is piecewise continuous on every finite interval of the form $[0, N]$, with $N > 0$. In contrast, the function $f(t) = 1/t$ is not piecewise continuous on any interval containing the origin since it has an "infinite jump" at the origin (see Figure 7.3).

A function that is piecewise continuous on a *finite* interval is necessarily integrable over that interval. However, piecewise continuity on $[0, \infty)$ is not enough to guarantee the existence (as a finite number) of the improper integral over $[0, \infty)$; we also need to consider the growth of the integrand for large t. Roughly speaking, we shall show that the Laplace transform of a piecewise continuous function will exist, provided that the function does not grow "faster than an exponential."

> ### EXPONENTIAL ORDER α
>
> **Definition 3.**   A function $f(t)$ is said to be of **exponential order** α if there exist positive constants T and M such that
>
> **(4)** $|f(t)| \le Me^{\alpha t}, \quad \text{for all } t \ge T.$

For example, $f(t) = e^{5t} \sin 2t$ is of exponential order $\alpha = 5$ since

$$|e^{5t} \sin 2t| \le e^{5t},$$

and hence (4) holds with $M = 1$ and T any positive constant.

We use the phrase $f(t)$ *is of exponential order* to mean that for *some* value of α, the function $f(t)$ satisfies the conditions of Definition 3; that is, $f(t)$ grows no faster than a function of the form $Me^{\alpha t}$. The function e^{t^2} is *not* of exponential order. To see this, observe that

$$\lim_{t \to \infty} \frac{e^{t^2}}{e^{\alpha t}} = \lim_{t \to \infty} e^{t(t-\alpha)} = +\infty$$

for any α. Consequently, e^{t^2} grows faster than $e^{\alpha t}$ for every choice of α.

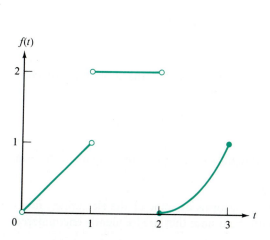

Figure 7.2 Graph of $f(t)$ in Example 6

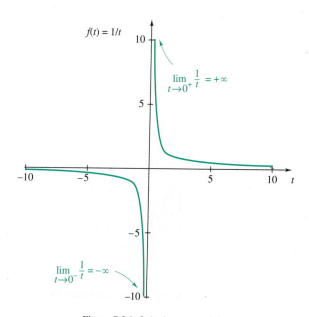

Figure 7.3 Infinite jump at origin

The usual functions encountered in solving linear differential equations with constant coefficients (polynomials, exponentials, sines, and cosines) are both piecewise continuous and of exponential order. As we now show, the Laplace transforms of such functions exist for large enough values of s.

> ▶ **CONDITIONS FOR EXISTENCE OF THE TRANSFORM**
>
> **Theorem 2.** If $f(t)$ is piecewise continuous on $[0, \infty)$ and of exponential order α, then $\mathscr{L}\{f\}(s)$ exists for $s > \alpha$.

Proof. We need to show that the integral

$$\int_0^\infty e^{-st} f(t)\, dt$$

converges for $s > \alpha$. We begin by breaking up this integral into two separate integrals:

$$\text{(5)} \qquad \int_0^T e^{-st} f(t)\, dt + \int_T^\infty e^{-st} f(t)\, dt,$$

where T is chosen so that inequality (4) holds. The first integral in (5) exists because $f(t)$ and hence $e^{-st} f(t)$ are piecewise continuous on the interval $[0, T]$ for any fixed s. To see that the second integral in (5) converges, we use the **comparison test for improper integrals.**

Since $f(t)$ is of exponential order α, we have for $t \geq T$

$$|f(t)| \leq M e^{\alpha t},$$

and hence

$$|e^{-st} f(t)| = e^{-st} |f(t)| \leq M e^{-(s-\alpha)t},$$

for all $t \geq T$. Now for $s > \alpha$,

$$\int_T^\infty M e^{-(s-\alpha)t}\, dt = M \int_T^\infty e^{-(s-\alpha)t}\, dt = \frac{M e^{-(s-\alpha)T}}{s-\alpha} < \infty.$$

Since $|e^{-st} f(t)| \leq M e^{-(s-\alpha)t}$ for $t \geq T$ and the improper integral of the larger function converges for $s > \alpha$, then, by the comparison test, the integral

$$\int_T^\infty e^{-st} f(t)\, dt$$

converges for $s > \alpha$. Finally, since the two integrals in (5) exist, the Laplace transform $\mathscr{L}\{f\}(s)$ exists for $s > \alpha$. ◀◀◀

In Table 7.1 we have listed the Laplace transforms of some of the elementary functions. The reader should become familiar with these since they are frequently encountered

in solving linear differential equations with constant coefficients. The entries in Table 7.1 can be derived from the definition of the Laplace transform. A more elaborate table of transforms is given on the inside back cover.

TABLE 7.1   BRIEF TABLE OF LAPLACE TRANSFORMS

$f(t)$	$F(s) = \mathcal{L}\{f\}(s)$
1	$\dfrac{1}{s}, \quad s > 0$
e^{at}	$\dfrac{1}{s-a}, \quad s > a$
$t^n, \quad n = 1, 2, \ldots$	$\dfrac{n!}{s^{n+1}}, \quad s > 0$
$e^{at}t^n, \quad n = 1, 2, \ldots$	$\dfrac{n!}{(s-a)^{n+1}}, \quad s > a$
$\sin bt$	$\dfrac{b}{s^2 + b^2}, \quad s > 0$
$\cos bt$	$\dfrac{s}{s^2 + b^2}, \quad s > 0$
$e^{at}\sin bt$	$\dfrac{b}{(s-a)^2 + b^2}, \quad s > a$
$e^{at}\cos bt$	$\dfrac{s-a}{(s-a)^2 + b^2}, \quad s > a$

EXERCISES 7.2

In Problems 1 through 12 use Definition 1 to determine the Laplace transform of the given function.

1. t.

2. t^2.

3. e^{6t}.

4. te^{3t}.

5. $\cos 2t$.

6. $\cos bt, \quad b$ a constant.

7. $e^{2t}\cos 3t$.

8. $e^{-t}\sin 2t$.

9. $f(t) = \begin{cases} 0, & 0 < t < 2, \\ t, & 2 < t. \end{cases}$

10. $f(t) = \begin{cases} 1-t, & 0 < t < 1, \\ 0, & 1 < t. \end{cases}$

11. $f(t) = \begin{cases} \sin t, & 0 < t < \pi, \\ 0, & \pi < t. \end{cases}$

12. $f(t) = \begin{cases} e^{2t}, & 0 < t < 3, \\ 1, & 3 < t. \end{cases}$

In Problems 13 through 20 use the Laplace transform table and the linearity of the Laplace transform to determine the following transforms.

13. $\mathcal{L}\{6e^{-3t} - t^2 + 2t - 8\}$.

14. $\mathcal{L}\{5 - e^{2t} + 6t^2\}$.

15. $\mathscr{L}\{t^3 - te^t + e^{4t}\cos t\}$.

16. $\mathscr{L}\{t^2 - 3t - 2e^{-t}\sin 3t\}$.

17. $\mathscr{L}\{e^{3t}\sin 6t - t^3 + e^t\}$.

18. $\mathscr{L}\{t^4 - t^2 - t + \sin\sqrt{2}t\}$.

19. $\mathscr{L}\{t^4 e^{5t} - e^t\cos\sqrt{7}t\}$.

20. $\mathscr{L}\{e^{-2t}\cos\sqrt{3}t - t^2 e^{-2t}\}$.

In Problems 21 through 28 determine whether $f(t)$ is continuous, piecewise continuous, or neither on $[0,10]$ and sketch the graph of $f(t)$.

21. $f(t) = \begin{cases} 1, & 0 \le t \le 1, \\ (t-2)^2, & 1 < t \le 10. \end{cases}$

22. $f(t) = \begin{cases} 0, & 0 \le t < 2, \\ t, & 2 \le t \le 10. \end{cases}$

23. $f(t) = \begin{cases} 1, & 0 \le t < 1, \\ t - 1, & 1 < t < 3, \\ t^2 - 4, & 3 < t \le 10. \end{cases}$

24. $f(t) = \dfrac{t^2 - 3t + 2}{t^2 - 4}$.

25. $f(t) = \dfrac{t^2 - t - 20}{t^2 + 7t + 10}$.

26. $f(t) = \dfrac{t}{t^2 - 1}$.

27. $f(t) = \begin{cases} 1/t, & 0 < t < 1, \\ 1, & 1 \le t \le 2, \\ 1 - t, & 2 < t \le 10. \end{cases}$

28. $f(t) = \begin{cases} \dfrac{\sin t}{t}, & t \ne 0, \\ 1, & t = 0. \end{cases}$

29. Which of the following functions are of exponential order?

(a) $t^3\sin t$. (b) $100e^{49t}$.

(c) e^{t^3}. (d) $t\ln t$.

(e) $\cosh(t^2)$. (f) $\dfrac{1}{t^2+1}$.

(g) $\sin(t^2) + t^4 e^{6t}$. (h) $3 - e^{t^2} + \cos 4t$.

(i) $\exp\{t^2/(t+1)\}$. (j) $\sin(e^{t^2}) + e^{\sin t}$.

30. For the transforms $F(s)$ in Table 7.1, what can be said about $\lim_{s\to\infty} F(s)$?

31. Using Euler's formula (page 161) and the algebraic properties of complex numbers, several of the entries of Table 7.1 can be derived from a single formula, namely,

(6) $\qquad \mathscr{L}\{e^{(a+ib)t}\}(s) = \dfrac{s - a + ib}{(s-a)^2 + b^2}, \qquad s > a.$

(a) Use the definition of Laplace transform to show that

$$\mathscr{L}\{e^{(a+ib)t}\}(s) = \frac{1}{s - (a+ib)}, \qquad s > a.$$

(b) Deduce (6) from part (a) by showing that

$$\frac{1}{s - (a+ib)} = \frac{s - a + ib}{(s-a)^2 + b^2}.$$

(c) By equating the real and imaginary parts in formula (6), deduce the last two entries in Table 7.1.

32. Prove that for fixed $s > 0$ we have

$$\lim_{N\to\infty} e^{-sN}(s\sin bN + b\cos bN) = 0.$$

33. Prove that if f is piecewise continuous on $[a,b]$ and g is continuous on $[a,b]$, then the product fg is piecewise continuous on $[a,b]$.

•7.3• PROPERTIES OF THE LAPLACE TRANSFORM

In the previous section we defined the Laplace transform of a function $f(t)$ as

$$\mathscr{L}\{f\}(s) := \int_0^\infty e^{-st}f(t)\,dt.$$

Using this definition to get an explicit expression for $\mathscr{L}\{f\}$ requires the evaluation of the improper integral—frequently a tedious task! We have already seen how the linearity

property of the transform can help relieve this burden. In this section we discuss some further properties of the Laplace transform that simplify its computation. These new properties will also enable us to use the Laplace transform to solve initial value problems.

TRANSLATION PROPERTY OF TRANSFORM

Theorem 3.   If the Laplace transform $\mathscr{L}\{f\}(s) = F(s)$ exists for $s > \alpha$, then

(1) $\mathscr{L}\{e^{at}f(t)\}(s) = F(s - a)$

know

for $s > \alpha + a$.

Proof.   We simply compute

$$\mathscr{L}\{e^{at}f(t)\}(s) = \int_0^\infty e^{-st}e^{at}f(t)\,dt$$

$$= \int_0^\infty e^{-(s-a)t}f(t)\,dt$$

$$= F(s - a). \quad \blacktriangleleft\blacktriangleleft\blacktriangleleft$$

Theorem 3 illustrates the effect that multiplying a function $f(t)$ by e^{at} has on the Laplace transform.

● **EXAMPLE 1**   Determine the Laplace transform of $e^{at}\sin bt$.

Solution   In Example 3 in Section 7.2 we found that

$$\mathscr{L}\{\sin bt\}(s) = F(s) = \frac{b}{s^2 + b^2}.$$

everywhere there is an s substitute (s-a)

Thus, by the translation property of $F(s)$, we have

$$\mathscr{L}\{e^{at}\sin bt\}(s) = F(s - a) = \frac{b}{(s - a)^2 + b^2}. \quad ●$$

LAPLACE TRANSFORM OF THE DERIVATIVE

Theorem 4.   Let $f(t)$ be continuous on $[0, \infty)$ and $f'(t)$ be piecewise continuous on $[0, \infty)$, with both of exponential order α. Then, for $s > \alpha$,

(2) $\mathscr{L}\{f'\}(s) = s\mathscr{L}\{f\}(s) - f(0).$

Proof.   Since $\mathscr{L}\{f'\}$ exists, we can use integration by parts (with $u = e^{-st}$ and $dv = $

$f'(t)\,dt$) to obtain

(3) $$\mathscr{L}\{f'\}(s) = \int_0^\infty e^{-st}f'(t)\,dt = \lim_{N\to\infty} \int_0^N e^{-st}f'(t)\,dt$$

$$= \lim_{N\to\infty}\left[e^{-st}f(t)\Big|_0^N + s\int_0^N e^{-st}f(t)\,dt \right]$$

$$= \lim_{N\to\infty} e^{-sN}f(N) - f(0) + s\lim_{N\to\infty} \int_0^N e^{-st}f(t)\,dt$$

$$= \lim_{N\to\infty} e^{-sN}f(N) - f(0) + s\mathscr{L}\{f\}(s).$$

To evaluate $\lim_{N\to\infty} e^{-sN}f(N)$, we observe that since $f(t)$ is of exponential order α, there exists a constant M such that for N large,

$$|e^{-sN}f(N)| \le e^{-sN}Me^{\alpha N} = Me^{-(s-\alpha)N}.$$

Hence, for $s > \alpha$,

$$\lim_{N\to\infty} |e^{-sN}f(N)| \le \lim_{N\to\infty} Me^{-(s-\alpha)N} = 0,$$

and so

$$\lim_{N\to\infty} e^{-sN}f(N) = 0$$

for $s > \alpha$. Equation (3) now reduces to

$$\mathscr{L}\{f'\}(s) = s\mathscr{L}\{f\}(s) - f(0). \quad \blacktriangleleft\blacktriangleleft\blacktriangleleft$$

Using induction, we can extend the last theorem to higher order derivatives of $f(t)$. For example,

$$\mathscr{L}\{f''\}(s) = s\mathscr{L}\{f'\}(s) - f'(0)$$
$$= s\{s\mathscr{L}\{f\}(s) - f(0)\} - f'(0),$$

which simplifies to

(4) $$\mathscr{L}\{f''\}(s) = s^2\mathscr{L}\{f\}(s) - sf(0) - f'(0).$$

In general we obtain the following result.

LAPLACE TRANSFORM OF HIGHER ORDER DERIVATIVES

Theorem 5.   Let $f(t), f'(t), \ldots, f^{(n-1)}(t)$ be continuous on $[0,\infty)$ and $f^{(n)}(t)$ be piecewise continuous on $[0,\infty)$, with all these functions of exponential order α. Then, for $s > \alpha$,

(5) $$\mathscr{L}\{f^{(n)}\}(s) = s^n\mathscr{L}\{f\}(s) - s^{n-1}f(0) - s^{n-2}f'(0) - \cdots - f^{(n-1)}(0).$$

The last two theorems shed light on the reason why the Laplace transform is such a useful tool in solving initial value problems. Roughly speaking, they tell us that by using the Laplace transform we can replace "differentiation with respect to t" with "multiplication by s," thereby converting a differential equation into an algebraic one. This idea is explored in Section 7.5. For now, we show how Theorem 4 can be helpful in computing a Laplace transform.

● **EXAMPLE 2**   Using the fact that

$$\mathscr{L}\{\sin bt\}(s) = \frac{b}{s^2 + b^2},$$

determine $\mathscr{L}\{\cos bt\}$.

Solution   Let $f(t) = \sin bt$. Then $f(0) = 0$ and $f'(t) = b \cos bt$. Substituting into equation (2), we have

$$\mathscr{L}\{f'\}(s) = s\mathscr{L}\{f\}(s) - f(0),$$
$$\mathscr{L}\{b \cos bt\}(s) = s\mathscr{L}\{\sin bt\}(s) - 0,$$
$$b\mathscr{L}\{\cos bt\}(s) = \frac{sb}{s^2 + b^2}.$$

Dividing by b gives

$$\mathscr{L}\{\cos bt\}(s) = \frac{s}{s^2 + b^2}. \quad ●$$

Another question arises concerning the Laplace transform: If $F(s)$ is the Laplace transform of $f(t)$, is $F'(s)$ also a Laplace transform of some function of t? The answer is yes:

$$F'(s) = \mathscr{L}\{-tf(t)\}(s).$$

In fact, the following more general assertion holds.

> ### DERIVATIVES OF THE LAPLACE TRANSFORM
>
> **Theorem 6.**   Let $F(s) = \mathscr{L}\{f\}(s)$ and assume that $f(t)$ is piecewise continuous on $[0, \infty)$ and of exponential order α. Then, for $s > \alpha$,
>
> (6) $\mathscr{L}\{t^n f(t)\}(s) = (-1)^n \dfrac{d^n F}{ds^n}(s).$

Proof.   Consider the identity

$$\frac{dF}{ds}(s) = \frac{d}{ds} \int_0^\infty e^{-st} f(t) \, dt.$$

Because of the assumptions on $f(t)$, we can apply a theorem from advanced calculus (sometimes called **Leibniz's rule**) to interchange the order of integration and differentiation:

$$\frac{dF}{ds}(s) = \int_0^\infty \frac{d}{ds}(e^{-st})f(t)\,dt$$

$$= -\int_0^\infty e^{-st}tf(t)\,dt = -\mathscr{L}\{tf(t)\}(s).$$

Thus

$$\mathscr{L}\{tf(t)\}(s) = (-1)\frac{dF}{ds}(s).$$

The general result (6) now follows by induction on n. ◄◄◄

A consequence of the above theorem is that if $f(t)$ is piecewise continuous and of exponential order, then its transform $F(s)$ has derivatives of all orders.

● **EXAMPLE 3** Determine $\mathscr{L}\{t\sin bt\}$.

Solution We already know that

$$\mathscr{L}\{\sin bt\}(s) = F(s) = \frac{b}{s^2 + b^2}.$$

Differentiating $F(s)$, we obtain

$$\frac{dF}{ds}(s) = \frac{-2bs}{(s^2 + b^2)^2}.$$

Hence, using formula (6), we have

$$\mathscr{L}\{t\sin bt\}(s) = -\frac{dF}{ds}(s) = \frac{2bs}{(s^2 + b^2)^2}. \quad ●$$

For easy reference, we have listed in Table 7.2 some of the basic properties of the Laplace transform derived in this section.

TABLE 7.2   PROPERTIES OF LAPLACE TRANSFORMS
1. $\mathscr{L}\{f + g\} = \mathscr{L}\{f\} + \mathscr{L}\{g\}$.
2. $\mathscr{L}\{cf\} = c\mathscr{L}\{f\}$ for any constant c.
3. $\mathscr{L}\{e^{at}f(t)\}(s) = \mathscr{L}\{f\}(s - a)$.
4. $\mathscr{L}\{f'\}(s) = s\mathscr{L}\{f\}(s) - f(0)$.
5. $\mathscr{L}\{f''\}(s) = s^2\mathscr{L}\{f\}(s) - sf(0) - f'(0)$.
6. $\mathscr{L}\{f^{(n)}\}(s) = s^n\mathscr{L}\{f\}(s) - s^{n-1}f(0) - s^{n-2}f'(0) - \cdots - f^{(n-1)}(0)$.
7. $\mathscr{L}\{t^n f(t)\}(s) = (-1)^n \dfrac{d^n}{ds^n}(\mathscr{L}\{f\}(s))$.

EXERCISES 7.3

In Problems 1 through 20 determine the Laplace transform of the given function using Table 7.1, page 285, and the properties of the transform given in Table 7.2. [*Hint*: *In Problems 12 through 20 use an appropriate trigonometric identity.*]

1. $t^2 + e^t \sin 2t$.

2. $3t^2 - e^{2t}$.

3. $e^{-t}\cos 3t + e^{6t} - 1$.

4. $3t^4 - 2t^2 + 1$.

5. $2t^2e^{-t} - t + \cos 4t$.

6. $e^{-2t}\sin 2t + e^{3t}t^2$.

7. $(t-1)^4$.

8. $(1 + e^{-t})^2$.

9. $e^{-t}t \sin 2t$.

10. $te^{2t}\cos 5t$.

11. $\cosh bt$.

12. $\sin 3t \cos 3t$.

13. $\sin^2 t$.

14. $e^{7t}\sin^2 t$.

15. $\cos^3 t$.

16. $t \sin^2 t$.

17. $\sin 2t \sin 5t$.

18. $\cos nt \cos mt, \quad m \neq n$.

19. $\cos nt \sin mt, \quad m \neq n$.

20. $t \sin 2t \sin 5t$.

21. Given that $\mathscr{L}\{\cos bt\}(s) = s/(s^2 + b^2)$, use the translation property to compute $\mathscr{L}\{e^{at}\cos bt\}$.

22. Starting with the transform $\mathscr{L}\{1\}(s) = 1/s$, use formula (6) for the derivatives of the Laplace transform to show that $\mathscr{L}\{t^n\}(s) = n!/s^{n+1}$.

23. Use Theorem 4 on the Laplace transform of the derivative of $f(t)$ to determine $\mathscr{L}\{\cos^2 t\}$. Verify your answer by using the half-angle formula to derive $\mathscr{L}\{\cos^2 t\}$.

24. Show that $\mathscr{L}\{e^{at}t^n\}(s) = n!/(s-a)^{n+1}$ in two ways:

(a) using the translation property for $F(s)$ and

(b) using formula (6) for the derivatives of the Laplace transform.

25. Use formula (6) to help determine:

(a) $\mathscr{L}\{t \cos bt\}$. **(b)** $\mathscr{L}\{t^2 \cos bt\}$.

26. Let $f(t)$ be piecewise continuous on $[0, \infty)$ and of exponential order.

(a) Show that there exist constants K and α such that

$$|f(t)| \leq Ke^{\alpha t} \quad \text{for all } t \geq 0.$$

(b) By using the definition of the transform and estimating the integral with the help of part (a), prove that

$$\lim_{s\to\infty} \mathscr{L}\{f\}(s) = 0.$$

27. Let $f(t)$ be piecewise continuous on $[0, \infty)$ and of exponential order α and assume that $\lim_{t\to 0^+} \dfrac{f(t)}{t}$ exists. Show that

$$\mathscr{L}\left\{\frac{f(t)}{t}\right\}(s) = \int_s^\infty F(u)\,du,$$

where $F(s) = \mathscr{L}\{f\}(s)$. [Hint: Use formula (6) for the derivative of the Laplace transform and the result of Problem 26.]

28. The **initial-value theorem** for Laplace transforms states that if $f(t)$ is continuous on $[0, \infty)$, $f'(t)$ is piecewise continuous on $[0, \infty)$, and both are of exponential order α, then

$$\lim_{s\to\infty} sF(s) = f(0),$$

where $F(s) = \mathscr{L}\{f\}(s)$. Verify this formula for $f(t) = e^{at}\sin bt$ and $f(t) = e^{at}\cos bt$.

29. The **transfer function** of a linear system is defined as the ratio of the Laplace transform of the output function $y(t)$ to the Laplace transform of the input function $g(t)$, assuming that all initial conditions are zero. If a linear system is governed by the differential equation

$$y''(t) + 6y'(t) + 10y(t) = g(t), \qquad t > 0,$$

use the linearity property of the Laplace transform and Theorem 5 on the Laplace transform of higher order derivatives to determine the transfer function $H(s) = Y(s)/G(s)$ for this system.

30. Find the transfer function, as defined in Problem 29, for the linear system governed by

$$y''(t) + 5y'(t) + 6y(t) = g(t), \qquad t > 0.$$

7.4 INVERSE LAPLACE TRANSFORM

In Section 7.2 we defined the Laplace transform as an integral operator that maps a function $f(t)$ into a function $F(s)$. In this section we consider the problem of finding the function $f(t)$ when we are given the transform $F(s)$. That is, we seek an **inverse mapping** for the Laplace transform.

To see the usefulness of such an inverse, let's consider the simple initial value problem

(1) $y'' - y = -t;$ $y(0) = 0,$ $y'(0) = 1.$

If we take the transform of both sides of equation (1) and use the linearity property of the transform, we find

$$\mathscr{L}\{y''\}(s) - Y(s) = -\frac{1}{s^2},$$

where $Y(s) := \mathscr{L}\{y\}(s)$. Since we know the initial values of the solution $y(t)$, we can use Theorem 5 on the Laplace transform of higher order derivatives to express

$$\mathscr{L}\{y''\}(s) = s^2 Y(s) - sy(0) - y'(0) = s^2 Y(s) - 1.$$

Substituting for $\mathscr{L}\{y''\}(s)$ yields

$$s^2 Y(s) - 1 - Y(s) = -\frac{1}{s^2}.$$

Solving this algebraic equation for $Y(s)$ gives

$$Y(s) = \frac{1 - \left(\dfrac{1}{s^2}\right)}{s^2 - 1} = \frac{s^2 - 1}{s^2(s^2 - 1)} = \frac{1}{s^2}.$$

We now recall that $\mathscr{L}\{t\}(s) = 1/s^2$, and since $Y(s) = \mathscr{L}\{y\}(s)$, we have

$$\mathscr{L}\{y\}(s) = 1/s^2 = \mathscr{L}\{t\}(s).$$

It therefore seems reasonable to conclude that $y(t) = t$ is the solution to the initial value problem (1). A quick check confirms this!

Notice that, in the above procedure, a crucial step is to determine $y(t)$ from its Laplace transform $Y(s) = 1/s^2$. As we noted, $y(t) = t$ is such a function, but it is *not* the only function whose Laplace function is $1/s^2$. For example, the transform of

$$g(t) := \begin{cases} t, & t \neq 6, \\ 0, & t = 6 \end{cases}$$

is also $1/s^2$. This is because the transform is an integral, and integrals are not affected by changing a function's values at isolated points. The significant difference between $y(t)$ and $g(t)$ as far as we are concerned is that $y(t)$ is continuous on $[0, \infty)$ while $g(t)$ is not. Naturally, we prefer to work with continuous functions since solutions to differential equations are continuous. Fortunately, it can be shown that if two different functions have the same Laplace transform, at most one of them can be continuous.[†] With this in mind we give the following definition.

[†] For this result and further properties of the Laplace transform and its inverse we refer the reader to *Operational Mathematics*, Third Edition, by R. V. Churchill, McGraw-Hill, New York, 1972.

> ### ▸ INVERSE LAPLACE TRANSFORM
>
> **Definition 4.** By the **inverse Laplace transform** of the function $F(s)$ we mean the unique function $f(t)$ that is continuous on $[0, \infty)$ and satisfies
>
> **(2)** $\qquad \mathscr{L}\{f\}(s) = F(s)$.
>
> We denote f by $\mathscr{L}^{-1}\{F\}$. In case all functions that satisfy (2) are discontinuous on $[0, \infty)$, we select a piecewise continuous function that satisfies (2) to be $\mathscr{L}^{-1}\{F\}$.

Naturally the Laplace transform tables will be a great help in determining the inverse Laplace transform of a given function $F(s)$.

● **EXAMPLE 1** Determine $\mathscr{L}^{-1}\{F\}$, where

$$\text{(a)} \quad F(s) = \frac{2}{s^3}. \qquad \text{(b)} \quad F(s) = \frac{3}{s^2 + 9}. \qquad \text{(c)} \quad F(s) = \frac{s - 1}{s^2 - 2s + 5}.$$

Solution To compute $\mathscr{L}^{-1}\{F\}$, we refer to the Laplace transform table on page 285.

$$\text{(a)} \quad \mathscr{L}^{-1}\left\{\frac{2}{s^3}\right\}(t) = \mathscr{L}^{-1}\left\{\frac{2!}{s^3}\right\}(t) = t^2.$$

$$\text{(b)} \quad \mathscr{L}^{-1}\left\{\frac{3}{s^2 + 9}\right\}(t) = \mathscr{L}^{-1}\left\{\frac{3}{s^2 + 3^2}\right\}(t) = \sin 3t.$$

$$\text{(c)} \quad \mathscr{L}^{-1}\left\{\frac{s - 1}{s^2 - 2s + 5}\right\}(t) = \mathscr{L}^{-1}\left\{\frac{s - 1}{(s - 1)^2 + 2^2}\right\}(t) = e^t \cos 2t.$$

In part (c) we used the technique of completing the square to rewrite the denominator in a form that we could find in the table. ●

In practice, we do not always encounter a transform $F(s)$ that exactly corresponds to an entry in the second column of the Laplace transform table. To handle more complicated functions $F(s)$, we use properties of $\mathscr{L}^{-1}$, just as we used properties of $\mathscr{L}$. One such tool is the linearity of the inverse Laplace transform—a property that is inherited from the linearity of the operator $\mathscr{L}$.

> ### ▸ LINEARITY OF THE INVERSE TRANSFORM
>
> **Theorem 7.** Assume that $\mathscr{L}^{-1}\{F_1\}$ and $\mathscr{L}^{-1}\{F_2\}$ exist and are continuous on $[0, \infty)$, and let c be any constant. Then
>
> **(3)** $\qquad \mathscr{L}^{-1}\{F_1 + F_2\} = \mathscr{L}^{-1}\{F_1\} + \mathscr{L}^{-1}\{F_2\}$,
>
> **(4)** $\qquad \mathscr{L}^{-1}\{cF_1\} = c\mathscr{L}^{-1}\{F_1\}$.

The proof of Theorem 7 is outlined in Problem 37. We illustrate the usefulness of this theorem in the next example.

● **EXAMPLE 2**   Determine $\mathcal{L}^{-1}\left\{\dfrac{5}{s-6} - \dfrac{6s}{s^2+9} + \dfrac{3}{2s^2+8s+10}\right\}$.

Solution   We begin by using the linearity property. Thus

$$\mathcal{L}^{-1}\left\{\frac{5}{s-6} - \frac{6s}{s^2+9} + \frac{3}{2(s^2+4s+5)}\right\}$$

$$= 5\mathcal{L}^{-1}\left\{\frac{1}{s-6}\right\} - 6\mathcal{L}^{-1}\left\{\frac{s}{s^2+9}\right\} + \frac{3}{2}\mathcal{L}^{-1}\left\{\frac{1}{s^2+4s+5}\right\}.$$

Referring to the Laplace transform tables, we see that

$$\mathcal{L}^{-1}\left\{\frac{1}{s-6}\right\}(t) = e^{6t} \quad\text{and}\quad \mathcal{L}^{-1}\left\{\frac{s}{s^2+3^2}\right\}(t) = \cos 3t.$$

This gives us the first two terms. To determine $\mathcal{L}^{-1}\{1/(s^2+4s+5)\}$, we complete the square of the denominator to obtain $s^2+4s+5 = (s+2)^2+1$. We now recognize from the tables that

$$\mathcal{L}^{-1}\left\{\frac{1}{(s+2)^2+1^2}\right\}(t) = e^{-2t}\sin t.$$

Hence

$$\mathcal{L}^{-1}\left\{\frac{5}{s-6} - \frac{6s}{s^2+9} + \frac{3}{2s^2+8s+10}\right\}(t) = 5e^{6t} - 6\cos 3t + \frac{3e^{-2t}}{2}\sin t. \quad ●$$

● **EXAMPLE 3**   Determine $\mathcal{L}^{-1}\left\{\dfrac{5}{(s+2)^4}\right\}$.

Solution   The $(s+2)^4$ in the denominator suggests that we work with the formula

$$\mathcal{L}^{-1}\left\{\frac{n!}{(s-a)^{n+1}}\right\}(t) = e^{at}t^n.$$

Here we have $a = -2$ and $n = 3$, and so $\mathcal{L}^{-1}\{6/(s+2)^4\}(t) = e^{-2t}t^3$. Using the linearity property, we find

$$\mathcal{L}^{-1}\left\{\frac{5}{(s+2)^4}\right\}(t) = \frac{5}{6}\mathcal{L}^{-1}\left\{\frac{3!}{(s+2)^4}\right\}(t) = \frac{5}{6}e^{-2t}t^3. \quad ●$$

● **EXAMPLE 4**   Determine $\mathcal{L}^{-1}\left\{\dfrac{3s+2}{s^2+2s+10}\right\}$.

Solution   By completing the square, the quadratic term in the denominator can be written

$$s^2+2s+10 = s^2+2s+1+9 = (s+1)^2+3^2.$$

The form of $F(s)$ now suggests that we use one or both of the formulas

$$\mathscr{L}^{-1}\left\{\frac{s-a}{(s-a)^2+b^2}\right\}(t) = e^{at}\cos bt,$$

$$\mathscr{L}^{-1}\left\{\frac{b}{(s-a)^2+b^2}\right\}(t) = e^{at}\sin bt.$$

In this case, $a = -1$ and $b = 3$. The next step is to express

(5) $$\frac{3s+2}{s^2+2s+10} = A\frac{s+1}{(s+1)^2+3^2} + B\frac{3}{(s+1)^2+3^2},$$

where A, B are constants to be determined. Multiplying both sides of (5) by $s^2 + 2s + 10$ leaves

$$3s + 2 = A(s+1) + 3B = As + (A + 3B),$$

which is an identity between two polynomials in s. Equating the coefficients of like terms gives

$$A = 3, \qquad A + 3B = 2,$$

and so $A = 3$ and $B = -\frac{1}{3}$. Finally, from (5) and the linearity property, we find

$$\mathscr{L}^{-1}\left\{\frac{3s+2}{s^2+2s+10}\right\}(t) = 3\mathscr{L}^{-1}\left\{\frac{s+1}{(s+1)^2+3^2}\right\}(t) - \frac{1}{3}\mathscr{L}^{-1}\left\{\frac{3}{(s+1)^2+3^2}\right\}(t)$$

$$= 3e^{-t}\cos 3t - \frac{1}{3}e^{-t}\sin 3t. \quad \bullet$$

Given the choice of finding the inverse Laplace transform of

$$F_1(s) = \frac{7s^2+10s-1}{s^3+3s^2-s-3}$$

or of

$$F_2(s) = \frac{2}{s-1} + \frac{1}{s+1} + \frac{4}{s+3},$$

which would you select? No doubt $F_2(s)$ is the easier one. Actually, the two functions $F_1(s)$ and $F_2(s)$ are identical. This can be checked by combining the simple fractions that form $F_2(s)$. Thus, if we are faced with the problem of computing $\mathscr{L}^{-1}$ of a rational function such as $F_1(s)$, we will first express it, as we did $F_2(s)$, as a sum of simple rational functions. This is accomplished by the **method of partial fractions.**

We briefly review this method. Recall from calculus that a rational function of the form $P(s)/Q(s)$, where $P(s)$ and $Q(s)$ are polynomials with the degree of P less than the degree of Q, has a partial fraction expansion whose form is based on the linear and quadratic factors of $Q(s)$. (We assume the coefficients of the polynomials to be real numbers.)

There are three cases to consider:

1. Nonrepeated linear factors.
2. Repeated linear factors.
3. Quadratic factors.

1. Nonrepeated Linear Factors

If $Q(s)$ can be factored into a product of distinct linear factors,

$$Q(s) = (s - r_1)(s - r_2) \cdots (s - r_n),$$

where the r_i's are all distinct real numbers, then the partial fraction expansion has the form

$$\frac{P(s)}{Q(s)} = \frac{A_1}{s - r_1} + \frac{A_2}{s - r_2} + \cdots + \frac{A_n}{s - r_n},$$

where the A_i's are real numbers. There are various ways of determining the constants $A_1, \ldots, A_n$. In the next example we demonstrate two such methods.

● **EXAMPLE 5**   Determine $\mathscr{L}^{-1}\{F\}$, where

$$F(s) = \frac{7s - 1}{(s + 1)(s + 2)(s - 3)}.$$

Solution   We begin by finding the partial fraction expansion for $F(s)$. Since the denominator consists of three distinct linear factors, the expansion has the form

(6)
$$\frac{7s - 1}{(s + 1)(s + 2)(s - 3)} = \frac{A}{s + 1} + \frac{B}{s + 2} + \frac{C}{s - 3},$$

where A, B, and C are real numbers to be determined.

One procedure that works for all partial fraction expansions is to first multiply the expansion equation by the denominator of the given rational function. This leaves us with two identical polynomials. Equating the coefficients of s^k leads to a system of linear equations that we can solve to determine the unknown constants. In this example we multiply (6) by $(s + 1)(s + 2)(s - 3)$ and find

(7) $7s - 1 = A(s + 2)(s - 3) + B(s + 1)(s - 3) + C(s + 1)(s + 2),$

which reduces to

$$7s - 1 = (A + B + C)s^2 + (-A - 2B + 3C)s + (-6A - 3B + 2C).$$

Equating the coefficients of s^2, s, and 1 gives the system of linear equations

$$A + B + C = 0,$$
$$-A - 2B + 3C = 7,$$
$$-6A - 3B + 2C = -1.$$

Solving this system yields $A = 2$, $B = -3$, and $C = 1$. Hence

(8)
$$\frac{7s - 1}{(s + 1)(s + 2)(s - 3)} = \frac{2}{s + 1} - \frac{3}{s + 2} + \frac{1}{s - 3}.$$

An alternative method for finding the constants A, B, and C from (7) is to choose three values for s and substitute them into (7) to obtain three linear equations in the three unknowns. If we are careful in our choice of the values for s, the system is easy to solve. In this case, let's take $s = -1$, -2, and 3, which are the zeros of $Q(s)$. Putting $s = -1$ in (7) gives

$$-7 - 1 = A(1)(-4) + B(0) + C(0),$$
$$-8 = -4A.$$

Hence $A = 2$. Next, setting $s = -2$ gives

$$-14 - 1 = A(0) + B(-1)(-5) + C(0),$$
$$-15 = 5B,$$

and so $B = -3$. Finally, letting $s = 3$, we similarly find that $C = 1$. In the case of non-repeated linear factors the alternative method is easier to use.

Now that we have obtained the partial fraction expansion (8), we use linearity to compute

$$\mathcal{L}^{-1}\left\{\frac{7s - 1}{(s + 1)(s + 2)(s - 3)}\right\}(t) = \mathcal{L}^{-1}\left\{\frac{2}{s + 1} - \frac{3}{s + 2} + \frac{1}{s - 3}\right\}(t)$$

$$= 2\mathcal{L}^{-1}\left\{\frac{1}{s + 1}\right\}(t) - 3\mathcal{L}^{-1}\left\{\frac{1}{s + 2}\right\}(t)$$

$$+ \mathcal{L}^{-1}\left\{\frac{1}{s - 3}\right\}(t)$$

$$= 2e^{-t} - 3e^{-2t} + e^{3t}. \quad \bullet$$

2.   Repeated Linear Factors

Let $s - r$ be a factor of $Q(s)$ and suppose $(s - r)^m$ is the highest power of $s - r$ that divides $Q(s)$. Then the portion of the partial fraction expansion of $P(s)/Q(s)$ that corresponds to the term $(s - r)^m$ is

$$\frac{A_1}{s - r} + \frac{A_2}{(s - r)^2} + \cdots + \frac{A_m}{(s - r)^m},$$

where the A_i's are real numbers.

● **EXAMPLE 6**   Determine $\mathcal{L}^{-1}\left\{\dfrac{s^2 + 9s + 2}{(s - 1)^2(s + 3)}\right\}$.

Solution   Since $s - 1$ is a repeated linear factor with multiplicity two and $s + 3$ is a nonrepeated linear factor, the partial fraction expansion has the form

$$\frac{s^2 + 9s + 2}{(s - 1)^2(s + 3)} = \frac{A}{s - 1} + \frac{B}{(s - 1)^2} + \frac{C}{s + 3}.$$

We begin by multiplying both sides by $(s - 1)^2(s + 3)$ to obtain

(9) $s^2 + 9s + 2 = A(s - 1)(s + 3) + B(s + 3) + C(s - 1)^2.$

Now observe that when we set $s = 1$ (or $s = -3$), two terms on the right-hand side of (9) vanish, leaving a linear equation that we can solve for B (or C). Setting $s = 1$ in (9) gives

$$1 + 9 + 2 = A(0) + 4B + C(0),$$
$$12 = 4B,$$

and hence $B = 3$. Similarly, setting $s = -3$ in (9) gives

$$9 - 27 + 2 = A(0) + B(0) + 16C,$$
$$-16 = 16C.$$

Thus $C = -1$. Finally, to find A, we pick a different value for s, say $s = 0$. Then, since $B = 3$ and $C = -1$, plugging $s = 0$ into (9) yields

$$2 = -3A + 3B + C = -3A + 9 - 1,$$

so that $A = 2$. Hence

(10) $$\frac{s^2 + 9s + 2}{(s - 1)^2(s + 3)} = \frac{2}{s - 1} + \frac{3}{(s - 1)^2} - \frac{1}{s + 3}.$$

We could also have determined the constants A, B, and C by first rewriting equation (9) in the form

$$s^2 + 9s + 2 = (A + C)s^2 + (2A + B - 2C)s + (-3A + 3B + C).$$

Then, equating the corresponding coefficients of s^2, s, and 1 and solving the resulting system, we again find $A = 2$, $B = 3$, and $C = -1$.

Now that we have derived the partial fraction expansion (10) for the given rational function, we can determine its inverse Laplace transform:

$$\mathscr{L}^{-1}\left\{\frac{s^2 + 9s + 2}{(s - 1)^2(s + 3)}\right\}(t) = \mathscr{L}^{-1}\left\{\frac{2}{s - 1} + \frac{3}{(s - 1)^2} - \frac{1}{s + 3}\right\}(t)$$

$$= 2\mathscr{L}^{-1}\left\{\frac{1}{s - 1}\right\}(t) + 3\mathscr{L}^{-1}\left\{\frac{1}{(s - 1)^2}\right\}(t)$$

$$- \mathscr{L}^{-1}\left\{\frac{1}{s + 3}\right\}(t)$$

$$= 2e^t + 3te^t - e^{-3t}. \quad \bullet$$

3. Quadratic Factors

Let $(s - \alpha)^2 + \beta^2$ be a quadratic factor of $Q(s)$ that cannot be reduced to linear factors with real coefficients. Suppose that m is the highest power of $(s - \alpha)^2 + \beta^2$ that divides $Q(s)$. Then the portion of the partial fraction expansion that corresponds to $(s - \alpha)^2 + \beta^2$ is

$$\frac{C_1 s + D_1}{(s - \alpha)^2 + \beta^2} + \frac{C_2 s + D_2}{[(s - \alpha)^2 + \beta^2]^2} + \cdots + \frac{C_m s + D_m}{[(s - \alpha)^2 + \beta^2]^m}.$$

As we saw in Example 4, it is more convenient to express $C_i s + D_i$ in the form $A_i(s - \alpha) + \beta B_i$ when we look up the inverse Laplace transforms. So let's agree to write this portion of the partial fraction expansion in the equivalent form

$$\frac{A_1(s - \alpha) + \beta B_1}{(s - \alpha)^2 + \beta^2} + \frac{A_2(s - \alpha) + \beta B_2}{[(s - \alpha)^2 + \beta^2]^2} + \cdots + \frac{A_m(s - \alpha) + \beta B_m}{[(s - \alpha)^2 + \beta^2]^m}.$$

● **EXAMPLE 7**   Determine $\mathcal{L}^{-1}\left\{\dfrac{2s^2 + 10s}{(s^2 - 2s + 5)(s + 1)}\right\}$.

Solution   We first observe that the quadratic factor $s^2 - 2s + 5$ is irreducible (check the sign of the discriminant in the quadratic formula). Next we write the quadratic in the form $(s - \alpha)^2 + \beta^2$ by completing the square:

$$s^2 - 2s + 5 = (s - 1)^2 + 2^2.$$

Since $s^2 - 2s + 5$ and $s + 1$ are nonrepeated factors, the partial fraction expansion has the form

$$\frac{2s^2 + 10s}{(s^2 - 2s + 5)(s + 1)} = \frac{A(s - 1) + 2B}{(s - 1)^2 + 2^2} + \frac{C}{s + 1}.$$

When we multiply both sides by the common denominator, we obtain

(11) $2s^2 + 10s = [A(s - 1) + 2B](s + 1) + C(s^2 - 2s + 5).$

In equation (11), let's put $s = -1$, 1, and 0. With $s = -1$ we find

$$2 - 10 = [A(-2) + 2B](0) + C(8),$$
$$-8 = 8C,$$

and hence $C = -1$. With $s = 1$ in (11), we obtain

$$2 + 10 = [A(0) + 2B](2) + C(4),$$

and since $C = -1$, the last equation becomes $12 = 4B - 4$. Thus $B = 4$. Finally, setting $s = 0$ in (11) and using $C = -1$ and $B = 4$ gives

$$0 = [A(-1) + 2B](1) + C(5),$$
$$0 = -A + 8 - 5,$$
$$A = 3.$$

Hence $A = 3$, $B = 4$, and $C = -1$, so that

$$\frac{2s^2 + 10s}{(s^2 - 2s + 5)(s + 1)} = \frac{3(s - 1) + 2(4)}{(s - 1)^2 + 2^2} - \frac{1}{s + 1}.$$

With this partial fraction expansion in hand, we can immediately determine the inverse Laplace transform:

$$\begin{aligned}\mathscr{L}^{-1}\left\{\frac{2s^2 + 10s}{(s^2 - 2s + 5)(s + 1)}\right\}(t) &= \mathscr{L}^{-1}\left\{\frac{3(s - 1) + 2(4)}{(s - 1)^2 + 2^2} - \frac{1}{s + 1}\right\}(t) \\ &= 3\mathscr{L}^{-1}\left\{\frac{s - 1}{(s - 1)^2 + 2^2}\right\}(t) \\ &\quad + 4\mathscr{L}^{-1}\left\{\frac{2}{(s - 1)^2 + 2^2}\right\}(t) - \mathscr{L}^{-1}\left\{\frac{1}{s + 1}\right\}(t) \\ &= 3e^t \cos 2t + 4e^t \sin 2t - e^{-t}.\end{aligned}$$

In Section 7.7 we discuss a different method (involving convolutions) for computing inverse transforms, which does not require partial fraction decompositions. Moreover, the convolution method is convenient in the case of a rational function with a repeated quadratic factor in the denominator. Other helpful tools are described in Problems 33–36 and 38–40.

EXERCISES 7.4

In Problems 1 through 10 determine the inverse Laplace transform of the given function.

1. $\dfrac{6}{(s - 1)^4}$.
2. $\dfrac{2}{s^2 + 4}$.
3. $\dfrac{s + 1}{s^2 + 2s + 10}$.
4. $\dfrac{4}{s^2 + 9}$.
5. $\dfrac{1}{s^2 + 4s + 8}$.
6. $\dfrac{3}{(2s + 5)^3}$.
7. $\dfrac{2s + 16}{s^2 + 4s + 13}$.
8. $\dfrac{1}{s^5}$.
9. $\dfrac{3s - 15}{2s^2 - 4s + 10}$.
10. $\dfrac{s - 1}{2s^2 + s + 6}$.

In Problems 11 through 20 determine the partial fraction expansion for the given rational function.

11. $\dfrac{s^2 - 26s - 47}{(s - 1)(s + 2)(s + 5)}$.
12. $\dfrac{-s - 7}{(s + 1)(s - 2)}$.
13. $\dfrac{-2s^2 - 3s - 2}{s(s + 1)^2}$.
14. $\dfrac{-8s^2 - 5s + 9}{(s + 1)(s^2 - 3s + 2)}$.
15. $\dfrac{8s - 2s^2 - 14}{(s + 1)(s^2 - 2s + 5)}$.
16. $\dfrac{-5s - 36}{(s + 2)(s^2 + 9)}$.
17. $\dfrac{3s + 5}{s(s^2 + s - 6)}$.
18. $\dfrac{3s^2 + 5s + 3}{s^4 + s^3}$.
19. $\dfrac{1}{(s - 3)(s^2 + 2s + 2)}$.
20. $\dfrac{s}{(s - 1)(s^2 - 1)}$.

In Problems 21 through 30 determine $\mathscr{L}^{-1}\{F\}$.

21. $F(s) = \dfrac{6s^2 - 13s + 2}{s(s - 1)(s - 6)}$.
22. $F(s) = \dfrac{s + 11}{(s - 1)(s + 3)}$.
23. $F(s) = \dfrac{5s^2 + 34s + 53}{(s + 3)^2(s + 1)}$.
24. $F(s) = \dfrac{7s^2 - 41s + 84}{(s - 1)(s^2 - 4s + 13)}$.
25. $F(s) = \dfrac{7s^2 + 23s + 30}{(s - 2)(s^2 + 2s + 5)}$.

26. $F(s) = \dfrac{7s^3 - 2s^2 - 3s + 6}{s^3(s-2)}$.

27. $s^2 F(s) - 4F(s) = \dfrac{5}{s+1}$.

28. $s^2 F(s) + sF(s) - 6F(s) = \dfrac{s^2 + 4}{s^2 + s}$.

29. $sF(s) + 2F(s) = \dfrac{10s^2 + 12s + 14}{s^2 - 2s + 2}$.

30. $sF(s) - F(s) = \dfrac{2s+5}{s^2 + 2s + 1}$.

31. Determine the Laplace transform of each of the following functions.

(a) $f_1(t) = \begin{cases} 0, & t = 2, \\ t, & t \neq 2. \end{cases}$

(b) $f_2(t) = \begin{cases} 5, & t = 1, \\ 2, & t = 6, \\ t, & t \neq 1, 6. \end{cases}$

(c) $f_3(t) = t$.
Which of the preceding functions is the inverse Laplace transform of $1/s^2$?

32. Determine the Laplace transform of each of the following functions.

(a) $f_1(t) = \begin{cases} t, & t = 1, 2, 3, \ldots, \\ e^t, & t \neq 1, 2, 3, \ldots. \end{cases}$

(b) $f_2(t) = \begin{cases} e^t, & t \neq 5, 8, \\ 6, & t = 5, \\ 0, & t = 8. \end{cases}$

(c) $f_3(t) = e^t$.
Which of the preceding functions is the inverse Laplace transform of $1/(s-1)$?

Theorem 6 in Section 7.3 can be expressed in terms of the inverse Laplace transform as

$$\mathscr{L}^{-1}\left\{\dfrac{d^n F}{ds^n}\right\}(t) = (-t)^n f(t),$$

where $f = \mathscr{L}^{-1}\{F\}$. Use this equation in Problems 33 through 36 to compute $\mathscr{L}^{-1}\{F\}$.

33. $F(s) = \ln\left(\dfrac{s+2}{s-5}\right)$.

34. $F(s) = \ln\left(\dfrac{s-4}{s-3}\right)$.

35. $F(s) = \ln\left(\dfrac{s^2 + 9}{s^2 + 1}\right)$.

36. $F(s) = \arctan(1/s)$.

37. Prove Theorem 7 on the linearity of the inverse transform. [Hint: Show that the right-hand side of equation (3) is a continuous function on $[0, \infty)$ whose Laplace transform is $F_1(s) + F_2(s)$.]

38. Residue Computation. Let $P(s)/Q(s)$ be a rational function with $\deg P < \deg Q$ and suppose that $s - r$ is a nonrepeated linear factor of $Q(s)$. Prove that the portion of the partial fraction expansion of $P(s)/Q(s)$ corresponding to $s - r$ is

$$\dfrac{A}{s-r},$$

where A (called the **residue**) is given by the formula

$$A = \lim_{s \to r} \dfrac{(s-r)P(s)}{Q(s)}.$$

39. Use the residue computation formula derived in Problem 38 to determine quickly the partial fraction expansion for

$$F(s) = \dfrac{2s+1}{s(s-1)(s+2)}.$$

40. Heaviside's Expansion Formula. [†] Let $P(s)$ and $Q(s)$ be polynomials with the degree of $P(s)$ less than the degree of $Q(s)$. Let

$$Q(s) = (s - r_1)(s - r_2)\cdots(s - r_n),$$

where the r_i's are distinct real numbers. Show that

$$\mathscr{L}^{-1}\left\{\dfrac{P}{Q}\right\}(t) = \sum_{i=1}^{n} \dfrac{P(r_i)}{Q'(r_i)} e^{r_i t}.$$

41. Use Heaviside's expansion formula derived in Problem 40 to determine the inverse Laplace transform of

$$F(s) = \dfrac{3s^2 - 16s + 5}{(s+1)(s-3)(s-2)}.$$

42. Complex Residues. Let $P(s)/Q(s)$ be a rational function with $\deg P < \deg Q$ and suppose $(s - \alpha)^2 + \beta^2$ is a nonrepeated quadratic factor of Q. (That is, $\alpha \pm i\beta$ are complex conjugate zeros of Q.) Prove that the portion of the partial fraction expansion of $P(s)/Q(s)$ corresponding to $(s - \alpha)^2 + \beta^2$ is

$$\dfrac{A(s - \alpha) + \beta B}{(s - \alpha)^2 + \beta^2},$$

[†] *Historical Footnote:* This formula played an important role in the "operational solution" to ordinary differential equations developed by Oliver Heaviside in the 1890s.

where the **complex residue** $\beta B + i\beta A$ is given by the formula

$$\beta B + i\beta A = \lim_{s \to \alpha + i\beta} \frac{[(s-\alpha)^2 + \beta^2]P(s)}{Q(s)}.$$

(Thus, we can determine B and A by taking the real and imaginary parts of the limit and dividing them by β.)

43. Use the residue formulas derived in Problems 38 and 42 to determine the partial fraction expansion for

$$F(s) = \frac{6s^2 + 28}{(s^2 - 2s + 5)(s + 2)}.$$

7.5 • SOLVING INITIAL VALUE PROBLEMS

Our goal is to show how Laplace transforms can be used to solve initial value problems for linear differential equations. Recall that we have already studied ways of solving such initial value problems in Chapter 4. These previous methods required that we first find a *general solution* of the differential equation and then use the initial conditions to determine the desired solution. As we shall see, the method of Laplace transforms leads to the solution of the initial value problem *without* first finding a general solution.

Other advantages to the transform method are worth noting. For example, the technique can easily handle equations involving discontinuous forcing functions as illustrated in Section 7.1. Furthermore, the method can be used for certain linear differential equations with variable coefficients, a special class of integral equations, systems of differential equations, and partial differential equations.

► METHOD OF LAPLACE TRANSFORMS

To solve an initial value problem:

(a) Take the Laplace transform of both sides of the equation.
(b) Use the properties of the Laplace transform and the initial conditions to obtain an equation for the Laplace transform of the solution and then solve this equation for the transform.
(c) Determine the inverse Laplace transform of the solution by looking it up in a table or by using a suitable method (such as partial fractions) in combination with the table.

In step (a) we are tacitly assuming that the solution is piecewise continuous on $[0, \infty)$ and of exponential order. Once we have obtained the inverse Laplace transform in step (c), we can verify that these tacit assumptions are satisfied.

● **EXAMPLE 1**   Solve the initial value problem

(1) $y'' - 2y' + 5y = -8e^{-t}; \qquad y(0) = 2, \qquad y'(0) = 12.$

Solution   The differential equation in (1) is an identity between two functions of t. Hence equality holds for the Laplace transforms of these functions:

$$\mathscr{L}\{y'' - 2y' + 5y\} = \mathscr{L}\{-8e^{-t}\}.$$

Using the linearity property of $\mathscr{L}$ and the previously computed transform of the exponential function, we can write

(2) $\mathscr{L}\{y''\}(s) - 2\mathscr{L}\{y'\}(s) + 5\mathscr{L}\{y\}(s) = \dfrac{-8}{s+1}.$

Now let $Y(s) := \mathscr{L}\{y\}(s)$. From the formulas for the Laplace transform of higher order derivatives (see Section 7.3) and the initial conditions in (1), we find

$$\mathscr{L}\{y'\}(s) = sY(s) - y(0) = sY(s) - 2,$$
$$\mathscr{L}\{y''\}(s) = s^2Y(s) - sy(0) - y'(0) = s^2Y(s) - 2s - 12.$$

Substituting these expressions into (2) and solving for $Y(s)$ yields

$$[s^2Y(s) - 2s - 12] - 2[sY(s) - 2] + 5Y(s) = \dfrac{-8}{s+1},$$

$$(s^2 - 2s + 5)Y(s) = 2s + 8 - \dfrac{8}{s+1},$$

$$(s^2 - 2s + 5)Y(s) = \dfrac{2s^2 + 10s}{s+1},$$

$$Y(s) = \dfrac{2s^2 + 10s}{(s^2 - 2s + 5)(s+1)}.$$

Our remaining task is to compute the inverse transform of the rational function $Y(s)$. This was done in Example 7 of Section 7.4, where, using a partial fraction expansion, we found

(3) $y(t) = 3e^t \cos 2t + 4e^t \sin 2t - e^{-t},$

which is the solution to the initial value problem (1).   ●

As a quick check on the accuracy of our computations, the reader is advised to verify that the computed solution satisfies the given initial conditions.

The reader is probably questioning the wisdom of using the Laplace transform method to solve an initial value problem that can be easily handled by the methods discussed in Chapter 4. The objective of the first few examples in this section is simply to make the reader familiar with the Laplace transform procedure. We will see in Example 4 and in later sections that the method is applicable to problems that cannot be readily handled by the techniques discussed in the previous chapters.

● **EXAMPLE 2**   Solve the initial value problem

(4) $y'' + 4y' - 5y = te^t;\qquad y(0) = 1,\qquad y'(0) = 0.$

Solution   Let $Y(s) := \mathscr{L}\{y\}(s)$. Taking the Laplace transform of both sides of the differential equation in (4) gives

(5) $\mathscr{L}\{y''\}(s) + 4\mathscr{L}\{y'\}(s) - 5Y(s) = \dfrac{1}{(s-1)^2}.$

Using the initial conditions, we can express $\mathscr{L}\{y'\}(s)$ and $\mathscr{L}\{y''\}(s)$ in terms of $Y(s)$. That is,

$$\mathscr{L}\{y'\}(s) = sY(s) - y(0) = sY(s) - 1,$$
$$\mathscr{L}\{y''\}(s) = s^2Y(s) - sy(0) - y'(0) = s^2Y(s) - s.$$

Substituting back into (5) and solving for $Y(s)$ gives

$$[s^2Y(s) - s] + 4[sY(s) - 1] - 5Y(s) = \frac{1}{(s-1)^2},$$

$$(s^2 + 4s - 5)Y(s) = s + 4 + \frac{1}{(s-1)^2},$$

$$(s+5)(s-1)Y(s) = \frac{s^3 + 2s^2 - 7s + 5}{(s-1)^2},$$

$$Y(s) = \frac{s^3 + 2s^2 - 7s + 5}{(s+5)(s-1)^3}.$$

The partial fraction expansion for $Y(s)$ has the form

(6)
$$\frac{s^3 + 2s^2 - 7s + 5}{(s+5)(s-1)^3} = \frac{A}{s+5} + \frac{B}{s-1} + \frac{C}{(s-1)^2} + \frac{D}{(s-1)^3}.$$

Solving for the numerators, we ultimately obtain $A = \frac{35}{216}$, $B = \frac{181}{216}$, $C = -\frac{1}{36}$, and $D = \frac{1}{6}$. Substituting these values into (6) gives

$$Y(s) = \frac{35}{216}\left(\frac{1}{s+5}\right) + \frac{181}{216}\left(\frac{1}{s-1}\right) - \frac{1}{36}\left(\frac{1}{(s-1)^2}\right) + \frac{1}{12}\left(\frac{2}{(s-1)^3}\right),$$

where we have written $D = \frac{1}{6} = (\frac{1}{12})2$ to facilitate the final step of taking the inverse transform. From the tables, we now obtain

(7)
$$y(t) = \frac{35}{216}e^{-5t} + \frac{181}{216}e^{t} - \frac{1}{36}te^{t} + \frac{1}{12}t^2e^{t}$$

as the solution to the initial value problem (4). ●

● **EXAMPLE 3** Solve the initial value problem

(8)
$$w''(t) - 2w'(t) + 5w(t) = -8e^{\pi - t}; \qquad w(\pi) = 2, \qquad w'(\pi) = 12.$$

Solution To use the method of Laplace transforms, we first move the initial conditions to $t = 0$. This can be done by setting $y(t) := w(t + \pi)$. Then

$$y'(t) = w'(t + \pi), \qquad y''(t) = w''(t + \pi).$$

Replacing t by $t + \pi$ in the differential equation in (8), we have

(9)
$$w''(t + \pi) - 2w'(t + \pi) + 5w(t + \pi) = -8e^{\pi - (t + \pi)} = -8e^{-t}.$$

Substituting $y(t) = w(t + \pi)$ in (9), the initial value problem in (8) becomes

$$y''(t) - 2y'(t) + 5y(t) = -8e^{-t}; \qquad y(0) = 2, \qquad y'(0) = 12.$$

Since the initial conditions are now given at the origin, the Laplace transform method is applicable. In fact, we carried out the procedure in Example 1, where we found

(10) $y(t) = 3e^t \cos 2t + 4e^t \sin 2t - e^{-t}.$

Since $w(t + \pi) = y(t)$, then $w(t) = y(t - \pi)$. Hence, replacing t by $t - \pi$ in (10) gives

$$w(t) = y(t - \pi) = 3e^{t-\pi} \cos 2(t - \pi) + 4e^{t-\pi} \sin 2(t - \pi) - e^{-(t-\pi)}$$
$$= 3e^{t-\pi} \cos 2t + 4e^{t-\pi} \sin 2t - e^{\pi-t}. \quad \bullet$$

Thus far we have applied the Laplace transform method only to linear equations with constant coefficients. Yet several important equations in mathematical physics involve linear equations whose coefficients are polynomials in t. To solve such equations using Laplace transforms, we shall apply Theorem 6, page 289, where we proved that

(11) $\mathcal{L}\{t^n f(t)\}(s) = (-1)^n \dfrac{d^n F}{ds^n}(s).$

If we let $n = 1$ and $f(t) = y'(t)$, we find

$$\mathcal{L}\{ty'(t)\}(s) = -\frac{d}{ds} \mathcal{L}\{y'\}(s)$$

$$= -\frac{d}{ds}[sY(s) - y(0)] = -sY'(s) - Y(s).$$

Similarly, with $n = 1$ and $f(t) = y''(t)$, we obtain from (11)

$$\mathcal{L}\{ty''(t)\}(s) = -\frac{d}{ds} \mathcal{L}\{y''\}(s)$$

$$= -\frac{d}{ds}[s^2 Y(s) - sy(0) - y'(0)]$$

$$= -s^2 Y'(s) - 2s Y(s) + y(0).$$

Thus, we see that for a linear differential equation in $y(t)$ whose coefficients are polynomials in t, the method of Laplace transforms will convert the given equation into a linear differential equation in $Y(s)$ whose coefficients are polynomials in s. Moreover, if the coefficients of the given equation are linear polynomials in t, then (regardless of the order of the given equation) the differential equation for $Y(s)$ is just a linear *first order* equation. Since we know how to solve this first order equation, the only serious obstacle we may encounter is obtaining the inverse Laplace transform of $Y(s)$. This problem may be insurmountable since the solution $y(t)$ may *not* have a Laplace transform.

In illustrating the technique, we make use of the following fact. If $f(t)$ is piecewise continuous on $[0, \infty)$ and of exponential order, then

(12) $\lim\limits_{s \to \infty} \mathcal{L}\{f\}(s) = 0.$

(The reader may have already guessed this from the entries in Table 7.1, page 285.) An outline of the proof of (12) is given in Exercises 7.3, Problem 26.

● **EXAMPLE 4**   Solve the initial value problem

(13) $y'' + 2ty' - 4y = 1,$ $y(0) = y'(0) = 0.$

Solution   Let $Y(s) = \mathscr{L}\{y\}(s)$ and take the Laplace transform of both sides of the equation in (13):

(14) $\mathscr{L}\{y''\}(s) + 2\mathscr{L}\{ty'(t)\}(s) - 4Y(s) = \dfrac{1}{s}.$

Using the initial conditions, we find

$$\mathscr{L}\{y''\}(s) = s^2 Y(s) - sy(0) - y'(0) = s^2 Y(s)$$

and

$$\mathscr{L}\{ty'(t)\}(s) = -\frac{d}{ds}\mathscr{L}\{y'\}(s)$$

$$= -\frac{d}{ds}[sY(s) - y(0)] = -sY'(s) - Y(s).$$

Substituting these expressions into (14) gives

$$s^2 Y(s) + 2[-sY'(s) - Y(s)] - 4Y(s) = \frac{1}{s},$$

$$-2sY'(s) + (s^2 - 6)Y(s) = \frac{1}{s},$$

(15) $Y'(s) + \left(\dfrac{3}{s} - \dfrac{s}{2}\right)Y(s) = \dfrac{-1}{2s^2}.$

Equation (15) is a linear first order equation and has the integrating factor

$$\mu(s) = e^{\int (3/s - s/2)\,ds} = e^{\ln s^3 - s^2/4} = s^3 e^{-s^2/4}$$

(see Section 2.4). Multiplying (15) by $\mu(s)$, we obtain

$$\frac{d}{ds}\{\mu(s)Y(s)\} = \frac{d}{ds}\{s^3 e^{-s^2/4}Y(s)\} = -\frac{s}{2}e^{-s^2/4}.$$

Integrating and solving for $Y(s)$ yields

$$s^3 e^{-s^2/4}Y(s) = -\int \frac{s}{2}e^{-s^2/4}\,ds = e^{-s^2/4} + C,$$

(16) $Y(s) = \dfrac{1}{s^3} + C\dfrac{e^{s^2/4}}{s^3}.$

Now if $Y(s)$ is the Laplace transform of a piecewise continuous function of exponential order, then it follows from equation (12) that

$$\lim_{s \to \infty} Y(s) = 0.$$

For this to occur, the constant C in equation (16) must be zero. Hence $Y(s) = 1/s^3$, and taking the inverse transform gives $y(t) = t^2/2$. We can easily verify that $y(t) = t^2/2$ is the solution to the given initial value problem by substituting it into (13).   ●

We end this section with an application from **control theory.** Let's consider a servo-mechanism that models an automatic pilot. Such a mechanism applies a torque to the steering control shaft so that a plane or boat will follow a prescribed course. If we let $y(t)$ be the true direction (angle) of the craft at time t and $g(t)$ be the desired direction at time t, then

$$e(t) := y(t) - g(t)$$

denotes the **error** or **deviation** between the desired direction and the true direction.

Let's assume that the servomechanism can measure the error $e(t)$ and feed back to the steering shaft a component of torque that is proportional to $e(t)$ but opposite in sign (see Figure 7.4). Newton's second law, expressed in terms of torques, states that

(moment of inertia) × (angular acceleration) = total torque.

For the servomechanism described, this becomes

(17) $Iy''(t) = -ke(t),$

where I is the moment of inertia of the steering shaft and k is a positive proportionality constant.

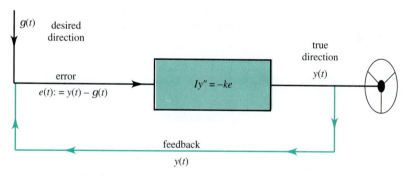

Figure 7.4 Servomechanism with feedback

● **EXAMPLE 5**   Determine the error $e(t)$ for the automatic pilot if the steering shaft is initially at rest in the zero direction and the desired direction is given by $g(t) = at$, where a is a constant.

Solution   Based on the discussion leading to equation (17), a model for the mechanism is given by the initial value problem

(18) $Iy''(t) = -ke(t);$ $y(0) = 0,$ $y'(0) = 0,$

where $e(t) = y(t) - g(t) = y(t) - at$. We begin by taking the Laplace transform of both sides of (18):

$$I\mathscr{L}\{y''\}(s) = -k\mathscr{L}\{e\}(s),$$

$$I[s^2Y(s) - sy(0) - y'(0)] = -kE(s),$$

(19)
$$s^2IY(s) = -kE(s),$$

where $Y(s) = \mathscr{L}\{y\}(s)$ and $E(s) = \mathscr{L}\{e\}(s)$. Since

$$E(s) = \mathscr{L}\{y(t) - at\}(s) = Y(s) - \mathscr{L}\{at\}(s) = Y(s) - as^{-2},$$

we find from (19) that

$$s^2IE(s) + aI = -kE(s).$$

Solving this equation for $E(s)$ gives

$$E(s) = -\frac{aI}{s^2I + k} = \frac{-a}{\sqrt{k/I}}\frac{\sqrt{k/I}}{s^2 + k/I}.$$

Hence, on taking the inverse Laplace transform, we obtain the error

(20)
$$e(t) = -\frac{a}{\sqrt{k/I}}\sin\sqrt{k/I}\,t. \quad \bullet$$

As we can see from equation (20), the automatic pilot will oscillate back and forth about the desired course, always "oversteering" by the factor $a/\sqrt{k/I}$. Clearly we can make the error small by making k large relative to I, but then the term $\sqrt{k/I}$ becomes large, causing the error to oscillate more rapidly. As with vibrations, the oscillations or oversteering can be controlled by introducing a damping torque proportional to $e'(t)$ but opposite in sign (see Problem 40).

EXERCISES 7.5

1, 3, 5, 7, 9, 11, 13, 15, 17

In Problems 1 through 14 solve the given initial value problem using the method of Laplace transforms.

1. $y'' - 2y' + 5y = 0;$ $y(0) = 2,$ $y'(0) = 4.$

2. $y'' - y' - 2y = 0;$ $y(0) = -2,$ $y'(0) = 5.$

3. $y'' + 6y' + 9y = 0;$ $y(0) = -1,$ $y'(0) = 6.$

4. $y'' + 6y' + 5y = 12e^t;$ $y(0) = -1,$ $y'(0) = 7.$

5. $w'' + w = t^2 + 2;$ $w(0) = 1,$ $w'(0) = -1.$

6. $y'' - 4y' + 5y = 4e^{3t};$ $y(0) = 2,$ $y'(0) = 7.$

7. $y'' - 7y' + 10y = 9\cos t + 7\sin t;$
    $y(0) = 5,$ $y'(0) = -4.$

8. $y'' + 4y = 4t^2 - 4t + 10;$ $y(0) = 0,$ $y'(0) = 3.$

9. $z'' + 5z' - 6z = 21e^t;$ $z(0) = -1,$ $z'(0) = 9.$

10. $y'' - 4y = 4t - 8e^{-2t};$ $y(0) = 0,$ $y'(0) = 5.$

11. $y'' - y = t - 2;$ $y(2) = 3,$ $y'(2) = 0.$

12. $w'' - 2w' + w = 6t - 2;$ $w(-1) = 3,$ $w'(-1) = 7.$

13. $y'' - y' - 2y = -8\cos t - 2\sin t;$
    $y(\pi/2) = 1,$ $y'(\pi/2) = 0.$

14. $y'' + y = t;$ $y(\pi) = 0,$ $y'(\pi) = 0.$

In Problems 15 through 24 solve for $Y(s)$, the Laplace transform of the solution $y(t)$ to the given initial value problem.

15. $y'' - 3y' + 2y = \cos t;$ $y(0) = 0,$ $y'(0) = -1.$

16. $y'' + 6y = t^2 - 1;$ $y(0) = 0,$ $y'(0) = -1.$

17. $y'' + y' - y = t^3;$ $y(0) = 1,$ $y'(0) = 0.$

18. $y'' - 2y' - y = e^{2t} - e^t;$ $y(0) = 1,$ $y'(0) = 3.$

19. $y'' + 5y' - y = e^t - 1;$ $y(0) = 1,$ $y'(0) = 1.$

20. $y'' + 3y' = t^3;$ $y(0) = 0,$ $y'(0) = 0.$

21. $y'' - 2y' + y = \cos t - \sin t;$ $y(0) = 1,$ $y'(0) = 3.$

22. $y'' - 6y' + 5y = te^t;$ $y(0) = 2,$ $y'(0) = -1.$

23. $y'' + 4y = g(t);$ $y(0) = -1,$ $y'(0) = 0,$ where

$$g(t) = \begin{cases} t, & t < 2, \\ 5, & t > 2. \end{cases}$$

24. $y'' - y = g(t);$ $y(0) = 1,$ $y'(0) = 2,$ where

$$g(t) = \begin{cases} 1, & t < 3, \\ t, & t > 3. \end{cases}$$

In Problems 25 through 28 solve the given third order initial value problem for y(t) using the method of Laplace transforms.

25. $y''' - y'' + y' - y = 0;$
   $y(0) = 1,$ $y'(0) = 1,$ $y''(0) = 3.$

26. $y''' + 4y'' + y' - 6y = -12;$
   $y(0) = 1,$ $y'(0) = 4,$ $y''(0) = -2.$

27. $y''' + 3y'' + 3y' + y = 0;$
   $y(0) = -4,$ $y'(0) = 4,$ $y''(0) = -2.$

28. $y''' + y'' + 3y' - 5y = 16e^{-t};$
   $y(0) = 0,$ $y'(0) = 2,$ $y''(0) = -4.$

In Problems 29 through 32 use the method of Laplace transforms to find a general solution to the given differential equation by assuming y(0) = a and y'(0) = b, where a and b are arbitrary constants.

29. $y'' - 4y' + 3y = 0.$

30. $y'' + 6y' + 5y = t.$

31. $y'' + 2y' + 2y = 5.$

32. $y'' - 5y' + 6y = -6te^{2t}.$

33. Use Theorem 6 in Section 7.3 to show that
$$\mathscr{L}\{t^2 y'(t)\}(s) = sY''(s) + 2Y'(s),$$

where $Y(s) = \mathscr{L}\{y\}(s).$

34. Use Theorem 6 in Section 7.3 to show that
$$\mathscr{L}\{t^2 y''(t)\}(s) = s^2 Y''(s) + 4sY'(s) + 2Y(s),$$
where $Y(s) = \mathscr{L}\{y\}(s).$

In Problems 35 through 38 find solutions to the given initial value problem.

35. $y'' + 3ty' - 6y = 1;$ $y(0) = 0,$ $y'(0) = 0.$

36. $ty'' - ty' + y = 2;$ $y(0) = 2,$ $y'(0) = -1.$

37. $ty'' - 2y' + ty = 0;$ $y(0) = 1,$ $y'(0) = 0.$
   [Hint: $\mathscr{L}^{-1}\{1/(s^2 + 1)^2\}(t) = (\sin t - t\cos t)/2.$]

38. $y'' + ty' - y = 0;$ $y(0) = 0,$ $y'(0) = 3.$

39. Determine the error $e(t)$ for the automatic pilot in Example 5 if the shaft is initially at rest in the zero direction and the desired direction is $g(t) = a$, where a is a constant.

40. In Example 5, assume that in order to control oscillations, a component of torque proportional to $e'(t)$, but opposite in sign, is also fed back to the steering shaft. Show that equation (17) is now replaced by

$$Iy''(t) = -ke(t) - \mu e'(t),$$

where μ is a positive constant. Determine the error $e(t)$ for the automatic pilot with mild damping (i.e., $\mu < 2\sqrt{Ik}$) if the steering shaft is initially at rest in the zero direction and the desired direction is given by $g(t) = a$, where a is a constant.

41. In Problem 40, determine the error $e(t)$ when the desired direction is given by $g(t) = at$, where a is a constant.

●7.6● **LAPLACE TRANSFORMS AND SPECIAL FUNCTIONS**

In this section we study special functions that often arise when the method of Laplace transforms is applied to physical problems. Of particular interest are methods for handling functions with jump discontinuities. Jump discontinuities occur naturally in physical problems such as electric circuits with on/off switches. To handle such behavior, O. Heaviside introduced the following step function.

> ### UNIT STEP FUNCTION
>
> **Definition 5.**   The **unit step function** $u(t)$ is defined by
>
> (1) $u(t) := \begin{cases} 0, & t < 0, \\ 1, & 0 < t. \end{cases}$

By shifting the argument of $u(t)$, the jump can be moved to a different location. That is,

(2) $u(t - a) = \begin{cases} 0, & t - a < 0, \\ 1, & 0 < t - a \end{cases} = \begin{cases} 0, & t < a, \\ 1, & a < t \end{cases}$

has its jump at $t = a$. By multiplying by a constant M, the height of the jump can also be modified (see Figure 7.5):

$$Mu(t - a) = \begin{cases} 0, & t < a, \\ M, & a < t. \end{cases}$$

Many discontinuous functions can be expressed in terms of unit step functions. For example,

$$f(t) := \begin{cases} \sin t, & t < \pi, \\ t, & \pi < t \end{cases}$$

can be expressed as

$$f(t) = \sin t + \begin{cases} 0, & t < \pi, \\ t - \sin t, & \pi < t \end{cases}$$

$$= \sin t + (t - \sin t)u(t - \pi).$$

The Laplace transform of $u(t - a)$ with $a \geq 0$ is

(3) $\mathscr{L}\{u(t - a)\}(s) = \dfrac{e^{-as}}{s},$

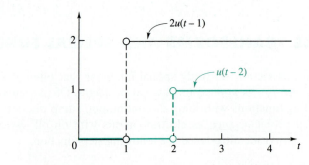

Figure 7.5 Two step functions expressed using the unit step function

since, for $s > 0$,

$$\mathscr{L}\{u(t-a)\}(s) = \int_0^\infty e^{-st}u(t-a)\,dt = \int_a^\infty e^{-st}\,dt$$

$$= \lim_{N\to\infty} \frac{-e^{-st}}{s}\bigg|_a^N = \frac{e^{-as}}{s}.$$

Observe that $\mathscr{L}\{u(t)\} = \mathscr{L}\{1\}$ since $u(t) = 1$ for $t > 0$.

● **EXAMPLE 1**   Find the Laplace transform of

$$(4) \qquad f(t) := \begin{cases} 3, & t < 2, \\ -1, & 2 < t < 5, \\ 7, & 5 < t. \end{cases}$$

Solution   We could use the definition of the Laplace transform to compute $\mathscr{L}\{f\}$, but instead let's express $f(t)$ in terms of unit step functions and then use formula (3).

The function $f(t)$ equals 3 until t reaches 2, at which point $f(t)$ jumps to -1. We can express this jump by $3 - 4u(t-2)$ since $u(t-2)$ is zero until t reaches 2, after which it has the value 1. At $t = 5$ the function $f(t)$ jumps from -1 to 7. This can be expressed by adding on the term $8u(t-5)$. Hence

$$f(t) = 3 - 4u(t-2) + 8u(t-5).$$

Finally, taking the Laplace transform and using formula (3), we find

$$\mathscr{L}\{f\}(s) = 3\mathscr{L}\{1\}(s) - 4\mathscr{L}\{u(t-2)\}(s) + 8\mathscr{L}\{u(t-5)\}(s)$$

$$= \frac{3}{s} - \frac{4e^{-2s}}{s} + \frac{8e^{-5s}}{s}. \quad ●$$

The translation property of $F(s)$ discussed in Section 7.3 described the effect on the Laplace transform of multiplying a function by e^{at}. The next theorem illustrates an analogous effect of multiplying the Laplace transform of a function by e^{-as}.

SHIFTING PROPERTY

Theorem 8.   Let $F(s) = \mathscr{L}\{f\}(s)$ exist for $s > \alpha \geq 0$. If a is a positive constant, then

$$(5) \qquad \mathscr{L}\{f(t-a)u(t-a)\}(s) = e^{-as}F(s),$$

and, if $f(t)$ is continuous on $[0, \infty)$, then[†]

$$(6) \qquad \mathscr{L}^{-1}\{e^{-as}F(s)\}(t) = f(t-a)u(t-a).$$

[†] Although $f(t)$ is continuous on $[0, \infty)$, the function $f(t-a)u(t-a)$ may have a discontinuity at $t = a$. In such cases, it can be shown that there is no continuous function on $[0, \infty)$ whose Laplace transform is $e^{-as}F(s)$.

Proof. By the definition of the Laplace transform, we have

(7) $\qquad \mathcal{L}\{f(t-a)u(t-a)\}(s) = \int_0^\infty e^{-st}f(t-a)u(t-a)\,dt$

$\qquad\qquad\qquad\qquad\quad = \int_a^\infty e^{-st}f(t-a)\,dt,$

where, in the last equation, we used the fact that $u(t-a)$ is zero for $t < a$ and equals 1 for $t > a$. Now let $v = t - a$. Then we have $dv = dt$, and equation (7) becomes

$$\mathcal{L}\{f(t-a)u(t-a)\}(s) = \int_0^\infty e^{-as}e^{-sv}f(v)\,dv$$

$$= e^{-as}\int_0^\infty e^{-sv}f(v)\,dv = e^{-as}F(s). \quad \blacktriangleleft\blacktriangleleft\blacktriangleleft$$

Notice that formula (5) includes as a special case the formula for $\mathcal{L}\{u(t-a)\}$; indeed, if we take $f(t) \equiv 1$, then $F(s) = 1/s$, and (5) becomes $\mathcal{L}\{u(t-a)\}(s) = e^{-as}/s$.

In practice it is more common to be faced with the problem of computing the transform of a function expressed as $g(t)u(t-a)$ rather than $f(t-a)u(t-a)$. To compute $\mathcal{L}\{g(t)u(t-a)\}$, we simply identify $g(t)$ with $f(t-a)$, so that $f(t) = g(t+a)$. Equation (5) then gives

(8) $\qquad \mathcal{L}\{g(t)u(t-a)\}(s) = e^{-as}\mathcal{L}\{g(t+a)\}(s).$

● **EXAMPLE 2**   Determine the Laplace transform of $t^2 u(t-1)$.

Solution   To apply the shifting property in equation (8), we observe that $g(t) = t^2$ and $a = 1$. Hence

$$g(t+a) = g(t+1) = (t+1)^2 = t^2 + 2t + 1.$$

Now the Laplace transform of $g(t+a)$ is

$$\mathcal{L}\{g(t+a)\}(s) = \frac{2}{s^3} + \frac{2}{s^2} + \frac{1}{s}.$$

Hence, by formula (8), we have

$$\mathcal{L}\{t^2 u(t-1)\}(s) = e^{-s}\left\{\frac{2}{s^3} + \frac{2}{s^2} + \frac{1}{s}\right\}. \quad ●$$

● **EXAMPLE 3**   Determine $\mathcal{L}\{(\cos t)u(t-\pi)\}$.

Solution   Here $g(t) = \cos t$ and $a = \pi$. Hence

$$g(t+a) = g(t+\pi) = \cos(t+\pi) = -\cos t,$$

and so the Laplace transform of $g(t+a)$ is

$$\mathcal{L}\{g(t+a)\}(s) = -\mathcal{L}\{\cos t\}(s) = -\frac{s}{s^2+1}.$$

Thus, from formula (8), we get

$$\mathcal{L}\{(\cos t)u(t-\pi)\}(s) = -e^{-\pi s}\frac{s}{s^2+1}. \quad \bullet$$

In Examples 2 and 3 we could also have computed the Laplace transform directly from the definition. In dealing with inverse transforms, however, we do not have a simple alternative formula† upon which to rely, and so formula (6) is especially useful whenever the transform has e^{-as} as a factor.

● EXAMPLE 4   Determine $\mathcal{L}^{-1}\left\{\dfrac{e^{-2s}}{s^2}\right\}$.

Solution   To use shifting property (6), we first express e^{-2s}/s^2 as the product $e^{-as}F(s)$. For this purpose, we put $e^{-as} = e^{-2s}$ and $F(s) = 1/s^2$. Thus $a = 2$ and

$$f(t) = \mathcal{L}^{-1}\left\{\frac{1}{s^2}\right\}(t) = t.$$

It now follows from the shifting property that

$$\mathcal{L}^{-1}\left\{\frac{e^{-2s}}{s^2}\right\}(t) = f(t-2)u(t-2) = (t-2)u(t-2). \quad \bullet$$

As illustrated by the next example, step functions arise in the modeling of on/off switches, changes in polarity, etc.

● EXAMPLE 5   The current I in an LC series circuit is governed by the initial value problem

(9) $I''(t) + 4I(t) = g(t);$ $I(0) = 0,$ $I'(0) = 0,$

where

$$g(t) := \begin{cases} 1, & 0 < t < 1, \\ -1, & 1 < t < 2, \\ 0, & 2 < t. \end{cases}$$

Determine the current as a function of time t.

Solution   Let $J(s) := \mathcal{L}\{I\}(s)$. Then we have $\mathcal{L}\{I''\}(s) = s^2J(s)$. Expressing

$$g(t) = u(t) - 2u(t-1) + u(t-2),$$

---

† Under certain conditions the inverse transform is given by the contour integral

$$\mathcal{L}^{-1}\{F\}(t) = \frac{1}{2\pi i}\int_{\alpha-i\infty}^{\alpha+i\infty} e^{st}F(s)\,ds.$$

See, for example, *Complex Variables and the Laplace Transform for Engineers*, by Wilbur R. LePage, Dover Publications, New York, 1980.

we find that

$$\mathscr{L}\{g\}(s) = \frac{1}{s} - \frac{2e^{-s}}{s} + \frac{e^{-2s}}{s}.$$

Thus, when we take the Laplace transform of both sides of (9), we obtain

$$\mathscr{L}\{I''\}(s) + 4\mathscr{L}\{I\}(s) = \mathscr{L}\{g\}(s),$$

$$s^2 J(s) + 4J(s) = \frac{1}{s} - \frac{2e^{-s}}{s} + \frac{e^{-2s}}{s},$$

$$J(s) = \frac{1}{s(s^2+4)} - \frac{2e^{-s}}{s(s^2+4)} + \frac{e^{-2s}}{s(s^2+4)}.$$

To find $I = \mathscr{L}^{-1}\{J\}$, we first observe that

$$J(s) = F(s) - 2e^{-s}F(s) + e^{-2s}F(s),$$

where

$$F(s) := \frac{1}{s(s^2+4)} = \frac{1}{4}\left(\frac{1}{s}\right) - \frac{1}{4}\left(\frac{s}{s^2+4}\right).$$

Computing the inverse transform of $F(s)$ gives

$$f(t) := \mathscr{L}^{-1}\{F\}(t) = \tfrac{1}{4} - \tfrac{1}{4}\cos 2t.$$

Hence, via the shifting property (6), we find

$$
\begin{aligned}
I(t) &= \mathscr{L}^{-1}\{F(s) - 2e^{-s}F(s) + e^{-2s}F(s)\}(t) \\
&= f(t) - 2f(t-1)u(t-1) + f(t-2)u(t-2) \\
&= (\tfrac{1}{4} - \tfrac{1}{4}\cos 2t) - [\tfrac{1}{2} - \tfrac{1}{2}\cos 2(t-1)]u(t-1) \\
&\quad + [\tfrac{1}{4} - \tfrac{1}{4}\cos 2(t-2)]u(t-2). \quad \bullet
\end{aligned}
$$

Periodic functions are another class of functions that occur frequently in applications.

PERIODIC FUNCTION

Definition 6.   A function $f(t)$ is said to be **periodic of period T** if

$$f(t+T) = f(t)$$

for all t in the domain of f.

As we know, the sine and cosine functions are periodic with period 2π, and the tangent function is periodic with period π.[†] To specify a periodic function, it is sufficient to give

---

[†] A function that has period T will also have period $2T$, $3T$, etc. For example, the sine function has periods $2\pi, 4\pi, 6\pi$, etc. Some authors refer to the smallest period as the **fundamental period** or just the period of the function.

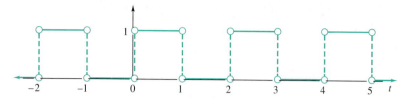

Figure 7.6 Graph of square wave function $f(t)$

its values over one period. For example, the square wave function in Figure 7.6 can be expressed as

(10) $f(t) := \begin{cases} 1, & 0 < t < 1, \\ 0, & 1 < t < 2 \end{cases}$ and $f(t)$ has period 2.

● **EXAMPLE 6**   Determine $\mathcal{L}\{f\}$, where $f(t)$ is the square wave function in (10).

Solution   Since $f(t)$ has period 2, it will prove convenient to write

$$\mathcal{L}\{f\}(s) = \int_0^\infty e^{-st} f(t)\, dt$$

$$= \int_0^2 e^{-st} f(t)\, dt + \int_2^4 e^{-st} f(t)\, dt + \int_4^6 e^{-st} f(t)\, dt + \cdots .$$

If we put $v = t$ in the first integral, $v = t - 2$ in the second integral, $v = t - 4$ in the third integral, and so on, we get

$$\mathcal{L}\{f\}(s) = \int_0^2 e^{-sv} f(v)\, dv + \int_0^2 e^{-s(v+2)} f(v+2)\, dv$$

$$+ \int_0^2 e^{-s(v+4)} f(v+4)\, dv + \cdots .$$

Since $f(v) = f(v+2) = f(v+4) = \cdots$, we find

$$\mathcal{L}\{f\}(s) = \int_0^2 e^{-sv} f(v)\, dv + e^{-2s}\int_0^2 e^{-sv} f(v)\, dv + e^{-4s}\int_0^2 e^{-sv} f(v)\, dv + \cdots$$

$$= \{1 + e^{-2s} + e^{-4s} + \cdots\} \int_0^2 e^{-sv} f(v)\, dv.$$

Now the series in braces is a **geometric series** with ratio $r = e^{-2s}$, which sums to $1/(1 - r)$ or $1/(1 - e^{-2s})$ for $s > 0$. Hence

(11) $\mathcal{L}\{f\}(s) = \dfrac{\displaystyle\int_0^2 e^{-sv} f(v)\, dv}{1 - e^{-2s}}.$

Evaluating the integral in equation (11), we obtain (see (10))

$$\int_0^2 e^{-sv} f(v)\, dv = \int_0^1 e^{-sv}\, dv = \frac{1}{s} - \frac{e^{-s}}{s} = \frac{1 - e^{-s}}{s}.$$

Thus,

(12) $$\mathscr{L}\{f\}(s) = \frac{1 - e^{-s}}{s(1 - e^{-2s})} = \frac{1}{s(1 + e^{-s})}. \quad \bullet$$

With minor modifications the derivation of formula (11) gives the following more general result.

> ### TRANSFORM OF PERIODIC FUNCTION
>
> **Theorem 9.**   If f has period T and is piecewise continuous on $[0, T]$, then
>
> (13) $$\mathscr{L}\{f\}(s) = \frac{\displaystyle\int_0^T e^{-st} f(t)\, dt}{1 - e^{-sT}}.$$

Returning to Example 6, we could have chosen to express the square wave in terms of unit step functions and then computed its Laplace transform. That is, $f(t)$ in (10) can be written in the form

$$f(t) = u(t) - u(t - 1) + u(t - 2) - u(t - 3) + \cdots,$$

for $t > 0$. Using the linearity of the Laplace transform, we find

$$\mathscr{L}\{f\}(s) = \mathscr{L}\{u(t)\}(s) - \mathscr{L}\{u(t - 1)\}(s) + \mathscr{L}\{u(t - 2)\}(s) + \cdots$$

$$= \frac{1}{s} - \frac{e^{-s}}{s} + \frac{e^{-2s}}{s} - \frac{e^{-3s}}{s} + \cdots$$

$$= \frac{1}{s}\{1 - e^{-s} + (e^{-s})^2 - (e^{-s})^3 + \cdots\}$$

$$= \frac{1}{s}\left\{\frac{1}{1 + e^{-s}}\right\} = \frac{1}{s(1 + e^{-s})}.$$

This agrees with equation (12).

In the preceding computation we determined the Laplace transform of $f(t)$ by writing $f(t)$ as a series of functions with known Laplace transforms. This approach can also be used for functions that have a power series expansion since we know that

$$\mathscr{L}\{t^n\}(s) = \frac{n!}{s^{n+1}}, \qquad n = 0, 1, 2, \dots.$$

● **EXAMPLE 7**   Determine $\mathcal{L}\{f\}$, where

$$f(t) := \begin{cases} \dfrac{\sin t}{t}, & t \neq 0, \\ 1, & t = 0. \end{cases}$$

Solution   We begin by expressing $f(t)$ in a Taylor series about $t = 0$. Since

$$\sin t = t - \frac{t^3}{3!} + \frac{t^5}{5!} - \frac{t^7}{7!} + \cdots,$$

then dividing by t we obtain

$$f(t) = \frac{\sin t}{t} = 1 - \frac{t^2}{3!} + \frac{t^4}{5!} - \frac{t^6}{7!} + \cdots$$

for $t > 0$. This representation also holds at $t = 0$ since

$$\lim_{t \to 0} f(t) = \lim_{t \to 0} \frac{\sin t}{t} = 1.$$

Observe that $f(t)$ is continuous on $[0, \infty)$ and of exponential order, and hence its Laplace transform exists for all s large. Because of the linearity of the Laplace transform, we would expect that

$$\mathcal{L}\{f\}(s) = \mathcal{L}\{1\}(s) - \frac{1}{3!}\mathcal{L}\{t^2\}(s) + \frac{1}{5!}\mathcal{L}\{t^4\}(s) + \cdots$$

$$= \frac{1}{s} - \frac{2!}{3! s^3} + \frac{4!}{5! s^5} - \frac{6!}{7! s^7} + \cdots$$

$$= \frac{1}{s} - \frac{1}{3s^3} + \frac{1}{5s^5} - \frac{1}{7s^7} + \cdots.$$

Indeed, using tools from analysis it can be verified that this series representation is valid for all $s > 1$. Moreover, one can show that the series converges to the function $\arctan(1/s)$ (see Problem 54). Thus,

(14) $\mathcal{L}\left\{\dfrac{\sin t}{t}\right\}(s) = \arctan \dfrac{1}{s}.$ ●

A similar procedure involving the series expansion for $F(s)$ in powers of $1/s$ can be used to compute $f(t) = \mathcal{L}^{-1}\{F\}(t)$ (see Problems 55–57).

We have previously shown, for every nonnegative integer n, that $\mathcal{L}\{t^n\}(s) = n!/s^{n+1}$. But what if the power of t is *not* an integer—is this formula still valid? To answer this question we need to extend the idea of "factorial." This is accomplished by the gamma function.[†]

[†] *Historical Footnote:* The gamma function was introduced by Leonhard Euler.

> ### GAMMA FUNCTION
>
> **Definition 7.**   The **gamma function** $\Gamma(t)$ is defined by
>
> $$(15) \qquad \Gamma(t) := \int_0^\infty e^{-u} u^{t-1}\, du, \qquad t > 0.$$

It can be shown that the integral in (15) converges for $t > 0$. A useful property of the gamma function is the recursive relation

$$(16) \qquad \Gamma(t+1) = t\Gamma(t).$$

This identity follows from the definition (15) after performing an integration by parts:

$$
\begin{aligned}
\Gamma(t+1) &= \int_0^\infty e^{-u} u^t\, du = \lim_{N\to\infty} \int_0^N e^{-u} u^t\, du \\
&= \lim_{N\to\infty} \left\{ -e^{-u} u^t \Big|_0^N + \int_0^N t e^{-u} u^{t-1}\, du \right\} \\
&= \lim_{N\to\infty} -e^{-N} N^t + t \lim_{N\to\infty} \int_0^N e^{-u} u^{t-1}\, du \\
&= 0 + t\Gamma(t) = t\Gamma(t).
\end{aligned}
$$

When t is a positive integer, say $t = n$, then the recursive relation (16) can be repeatedly applied to obtain

$$
\begin{aligned}
\Gamma(n+1) &= n\Gamma(n) = n(n-1)\Gamma(n-1) = \cdots \\
&= n(n-1)(n-2)\cdots 2\Gamma(1).
\end{aligned}
$$

It follows from the definition (15) that $\Gamma(1) = 1$, and so we find

$$\Gamma(n+1) = n!.$$

Thus the gamma function extends the notion of factorial!

As an application of the gamma function, let's return to the problem of determining the Laplace transform of an arbitrary power of t. We now verify that the formula

$$(17) \qquad \mathscr{L}\{t^r\}(s) = \frac{\Gamma(r+1)}{s^{r+1}}$$

holds for every constant $r > -1$.

By definition,

$$\mathscr{L}\{t^r\}(s) = \int_0^\infty e^{-st} t^r\, dt.$$

Let's make the substitution $u = st$. Then $du = s\,dt$, and we find

$$\mathscr{L}\{t^r\}(s) = \int_0^\infty e^{-u}\left(\frac{u}{s}\right)^r\left(\frac{1}{s}\right)du$$

$$= \frac{1}{s^{r+1}}\int_0^\infty e^{-u}u^r\,du = \frac{\Gamma(r+1)}{s^{r+1}}.$$

Notice that when $r = n$ is a nonnegative integer, then $\Gamma(n+1) = n!$, and so formula (17) reduces to the familiar formula for $\mathscr{L}\{t^n\}$.

EXERCISES 7.6

In Problems 1 through 4 sketch the graph of the given function and determine its Laplace transform.

1. $(t-1)^2u(t-1)$.

2. $u(t-1) - u(t-4)$.

3. $t^2u(t-2)$.

4. $tu(t-1)$.

In Problems 5 through 8 express the given function using unit step functions and compute its Laplace transform.

5. $g(t) = \begin{cases} 0, & 0 < t < 1, \\ 2, & 1 < t < 2, \\ 1, & 2 < t < 3, \\ 3, & 3 < t. \end{cases}$

6. $g(t) = \begin{cases} 0, & 0 < t < 2, \\ t+1, & 2 < t. \end{cases}$

7.

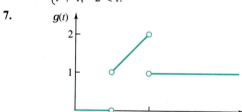

Figure 7.7 Function in Problem 7

8.

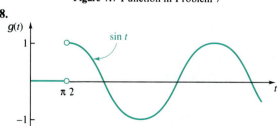

Figure 7.8 Function in Problem 8

In Problems 9 through 16 determine the inverse Laplace transform of the given function.

9. $\dfrac{e^{-2s}}{s-1}$.

10. $\dfrac{e^{-3s}}{s^2}$.

11. $\dfrac{e^{-2s} - 3e^{-4s}}{s+2}$.

12. $\dfrac{e^{-3s}}{s^2+9}$.

13. $\dfrac{se^{-3s}}{s^2+4s+5}$.

14. $\dfrac{e^{-s}}{s^2+4}$.

15. $\dfrac{e^{-3s}(s-5)}{(s+1)(s+2)}$.

16. $\dfrac{e^{-s}(3s^2-s+2)}{(s-1)(s^2+1)}$.

17. The current $I(t)$ in an RLC series circuit is governed by the initial value problem

$$I''(t) + 2I'(t) + 2I(t) = g(t);$$
$$I(0) = 10, \qquad I'(0) = 0,$$

where

$$g(t) := \begin{cases} 20, & 0 < t < 3\pi, \\ 0, & 3\pi < t < 4\pi, \\ 20, & 4\pi < t. \end{cases}$$

Determine the current as a function of time t.

18. The current $I(t)$ in an LC series circuit is governed by the initial value problem

$$I''(t) + 4I(t) = g(t); \qquad I(0) = 1, \qquad I'(0) = 3,$$

where

$$g(t) := \begin{cases} 3\sin t, & 0 \le t \le 2\pi, \\ 0, & 2\pi < t. \end{cases}$$

Determine the current as a function of time t.

In Problems 19 through 22 determine $\mathscr{L}\{f\}$, where $f(t)$ is periodic with given period. Also graph $f(t)$.

19. $f(t) = t$, $0 < t < 2$, and $f(t)$ has period 2.

20. $f(t) = e^t$, $0 < t < 1$, and $f(t)$ has period 1.

21. $f(t) = \begin{cases} e^{-t}, & 0 < t < 1, \\ 1, & 1 < t < 2, \end{cases}$ and $f(t)$ has period 2.

22. $f(t) = \begin{cases} t, & 0 < t < 1, \\ 1 - t, & 1 < t < 2, \end{cases}$ and $f(t)$ has period 2.

In Problems 23 through 26 determine $\mathscr{L}\{f\}$, where the periodic function is described by its graph.

23. $f(t)$

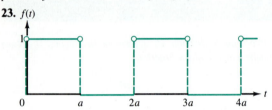

Figure 7.9 Square wave

24. $f(t)$

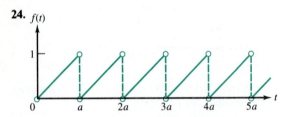

Figure 7.10 Sawtooth wave

25. $f(t)$

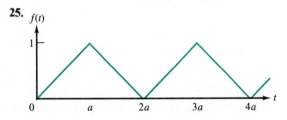

Figure 7.11 Triangular wave

26. $f(t)$

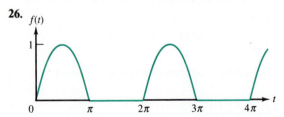

Figure 7.12 Half-rectified sine wave

In Problems 27 through 30 solve the given initial value problem using the method of Laplace transforms. Sketch the graph of the solution.

27. $y'' + y = u(t - 3)$; $\quad y(0) = 0$, $\quad y'(0) = 1$.

28. $w'' + w = u(t - 2) - u(t - 4)$; $\quad w(0) = 1$, $\quad w'(0) = 0$.

29. $y'' + y = t - (t - 4)u(t - 2)$; $\quad y(0) = 0$, $\quad y'(0) = 1$.

30. $y'' + y = 3\sin 2t - 3(\sin 2t)u(t - 2\pi)$;
$y(0) = 1$, $\quad y'(0) = -2$.

In Problems 31 through 40 solve the given initial value problem using the method of Laplace transforms.

31. $y'' + 2y' + 2y = u(t - 2\pi) - u(t - 4\pi)$;
$y(0) = 1$, $\quad y'(0) = 1$.

32. $y'' + 4y' + 4y = u(t - \pi) - u(t - 2\pi)$;
$y(0) = 0$, $\quad y'(0) = 0$.

33. $z'' + 3z' + 2z = e^{-3t}u(t - 2)$;
$z(0) = 2$, $\quad z'(0) = -3$.

34. $y'' + 5y' + 6y = tu(t - 2)$; $\quad y(0) = 0$, $\quad y'(0) = 1$.

35. $y'' - y = 3u(t - 1) - 2u(t - 2)$;
$y(0) = 0$, $\quad y'(0) = 2$.

36. $y'' - y = u(t - 1) - u(t - 2) + u(t - 3) - u(t - 4)$;
$y(0) = 2$, $\quad y'(0) = 0$.

37. $y'' + 4y = g(t)$; $\quad y(0) = 1$, $\quad y'(0) = 3$,
where $g(t) = \begin{cases} \sin t, & 0 \le t \le 2\pi, \\ 0, & 2\pi < t. \end{cases}$

38. $y'' + 2y' + 10y = g(t)$; $\quad y(0) = -1$, $\quad y'(0) = 0$,
where $g(t) = \begin{cases} 10, & 0 \le t \le 10, \\ 20, & 10 < t < 20, \\ 0, & 20 < t. \end{cases}$

39. $y'' + 5y' + 6y = g(t)$; $\quad y(0) = 0$, $\quad y'(0) = 2$,
where $g(t) = \begin{cases} 0, & 0 \le t < 1, \\ t, & 1 < t < 5, \\ 1, & 5 < t. \end{cases}$

40. $y'' + 3y' + 2y = g(t)$; $\quad y(0) = 2$, $\quad y'(0) = -1$,
where $g(t) = \begin{cases} e^{-t}, & 0 \le t < 3, \\ 1, & 3 < t. \end{cases}$

41. Show that if $\mathscr{L}\{g\}(s) = [(s + \alpha)(1 - e^{-Ts})]^{-1}$, where $T > 0$ is fixed, then

(18) $g(t) = e^{-\alpha t} + e^{-\alpha(t - T)}u(t - T) + e^{-\alpha(t - 2T)}u(t - 2T)$
$+ e^{-\alpha(t - 3T)}u(t - 3T) + \cdots$.

[Hint: Use the fact that $1 + x + x^2 + \cdots = 1/(1 - x)$.]

42. The function $g(t)$ in (18) can be expressed in a more convenient form as follows:

(a) Show that for each $n = 0, 1, 2, \ldots$,

$$g(t) = e^{-\alpha t}\left[\frac{e^{(n+1)\alpha T} - 1}{e^{\alpha T} - 1}\right] \quad \text{for} \quad nT < t < (n+1)T.$$

[Hint: Use the fact that $1 + x + x^2 + \cdots + x^n = (x^{n+1} - 1)/(x - 1)$.]

(b) Let $v = t - (n + 1)T$. Show that when $nT < t < (n + 1)T$, then $-T < v < 0$ and

(19) $$g(t) = \frac{e^{-\alpha v}}{e^{\alpha T} - 1} - \frac{e^{-\alpha t}}{e^{\alpha T} - 1}.$$

(c) Use the facts that the first term in (19) is periodic with period T and the second term is independent of n to sketch the graph of $g(t)$ in (19) for $\alpha = 1$ and $T = 2$.

43. Show that if $\mathcal{L}\{g\}(s) = \beta[(s^2 + \beta^2)(1 - e^{-Ts})]^{-1}$, then

$$g(t) = \sin \beta t + [\sin \beta(t - T)]u(t - T)$$
$$+ [\sin \beta(t - 2T)]u(t - 2T)$$
$$+ [\sin \beta(t - 3T)]u(t - 3T) + \cdots.$$

44. Use the result of Problem 43 to show that

$$\mathcal{L}^{-1}\left\{\frac{1}{(s^2 + 1)(1 - e^{-\pi s})}\right\}(t) = g(t),$$

where $g(t)$ is periodic with period 2π and

$$g(t) := \begin{cases} \sin t, & 0 \le t \le \pi, \\ 0, & \pi \le t \le 2\pi. \end{cases}$$

In Problems 45 and 46 use the method of Laplace transforms and the results of Problems 41 and 42 to solve the initial value problem

$$y'' + 3y' + 2y = f(t); \quad y(0) = 0, \quad y'(0) = 0,$$

where $f(t)$ is the periodic function defined in the stated problem.

45. Problem 20. **46.** Problem 23 with $a = 1$.

In Problems 47 through 50 find a Taylor series for $f(t)$ about $t = 0$. Assuming that the Laplace transform of $f(t)$ can be computed term by term, find an expansion for $\mathcal{L}\{f\}(s)$ in powers of $1/s$. If possible, sum the series.

47. $f(t) = e^t$. **48.** $f(t) = \sin t$.

49. $f(t) = \dfrac{1 - \cos t}{t}$. **50.** $f(t) = e^{-t^2}$.

51. Using the recursive relation (16) and the fact that $\Gamma(\frac{1}{2}) = \sqrt{\pi}$, determine

(a) $\mathcal{L}\{t^{-1/2}\}$. (b) $\mathcal{L}\{t^{7/2}\}$.

52. Use the recursive relation (16) and the fact that $\Gamma(\frac{1}{2}) = \sqrt{\pi}$ to show that

$$\mathcal{L}^{-1}\{s^{-(n+1/2)}\}(t) = \frac{2^n t^{n-1/2}}{1 \cdot 3 \cdot 5 \cdots (2n - 1)\sqrt{\pi}},$$

where n is a positive integer.

53. Use the method discussed in Example 6 to show that if $f(t)$ has period T and is piecewise continuous on $[0, T)$, then equation (13) holds.

54. By replacing s by $1/s$ in the Maclaurin series expansion for arctan s, show that

$$\arctan\frac{1}{s} = \frac{1}{s} - \frac{1}{3s^3} + \frac{1}{5s^5} - \frac{1}{7s^7} + \cdots.$$

55. Find an expansion for $e^{-1/s}$ in powers of $1/s$. Use the expansion for $e^{-1/s}$ to obtain an expansion for $s^{-1/2}e^{-1/s}$ in terms of $1/s^{n+1/2}$. Assuming that the inverse Laplace transform can be computed term by term, show that

$$\mathcal{L}^{-1}\{s^{-1/2}e^{-1/s}\}(t) = \frac{1}{\sqrt{\pi t}}\cos 2\sqrt{t}.$$

[Hint: Use the result of Problem 52.]

56. Use the procedure discussed in Problem 55 to show that

$$\mathcal{L}^{-1}\{s^{-3/2}e^{-1/s}\}(t) = \frac{1}{\sqrt{\pi}}\sin 2\sqrt{t}.$$

57. Find an expansion for $\ln[1 + (1/s^2)]$ in powers of $1/s$. Assuming that the inverse Laplace transform can be computed term by term, show that

$$\mathcal{L}^{-1}\left\{\ln\left(1 + \frac{1}{s^2}\right)\right\}(t) = \frac{2}{t}(1 - \cos t).$$

58. The **unit gate function** $G_a(t)$ is defined by

$$G_a(t) := \begin{cases} 0, & t < 0, \\ 1, & 0 < t < a, \\ 0, & a < t. \end{cases}$$

(a) Show that $G_a(t) = u(t) - u(t - a)$.

(b) Verify that $\mathcal{L}\{G_a\}(s) = (1/s)(1 - e^{-as})$.

(c) Show that $\mathcal{L}\{G_a(t - b)\}(s) = (1/s)(e^{-bs} - e^{-(a+b)s})$.

In Problems 59 and 60 use the results of Problem 58 and the method of Laplace transforms to solve the given initial value problem.

59. $y'' - y = G_3(t - 1);$ $y(0) = 0,$ $y'(0) = 2.$

60. $y'' - y = G_4(t - 3);$ $y(0) = 1,$ $y'(0) = -1.$

• 7.7 • CONVOLUTION (Optional)

Consider the initial value problem

(1) $y'' + y = g(t);$ $y(0) = 0,$ $y'(0) = 0.$

If we let $Y(s) = \mathcal{L}\{y\}(s)$ and $G(s) = \mathcal{L}\{g\}(s)$, then taking the Laplace transform of both sides of (1) yields

$$s^2 Y(s) + Y(s) = G(s),$$

and hence

(2) $Y(s) = \left(\dfrac{1}{s^2 + 1}\right) G(s).$

That is, the Laplace transform of the solution to (1) is the product of the Laplace transform of $\sin t$ and the Laplace transform of the forcing term $g(t)$. What we would now like to have is a simple formula for $y(t)$ in terms of $\sin t$ and $g(t)$. Just as the integral of a product is not the product of the integrals, $y(t)$ is not the product of $\sin t$ and $g(t)$. However, we can express $y(t)$ as the "convolution" of $\sin t$ and $g(t)$.

> ### ◢ CONVOLUTION
>
> **Definition 8.**   Let $f(t)$ and $g(t)$ be piecewise continuous on $[0, \infty)$. The **convolution** of $f(t)$ and $g(t)$, denoted $f * g$, is defined by
>
> **(3)** $(f * g)(t) := \displaystyle\int_0^t f(t - v)g(v)\,dv.$

For example, the convolution of t and t^2 is

$$t * t^2 = \int_0^t (t - v)v^2\,dv = \int_0^t (tv^2 - v^3)\,dv$$

$$= \left(\frac{tv^3}{3} - \frac{v^4}{4}\right)\Bigg|_0^t = \frac{t^4}{3} - \frac{t^4}{4} = \frac{t^4}{12}.$$

Convolution is certainly different from ordinary multiplication. For example, $1 * 1 = t \neq 1$ and in general $1 * f \neq f$. However, convolution does satisfy some of the same properties as multiplication.

> ### ◢ PROPERTIES OF CONVOLUTION
>
> **Theorem 10.**   Let $f(t)$, $g(t)$, and $h(t)$ be piecewise continuous on $[0, \infty)$. Then
>
> **(4)** $f * g = g * f,$
> **(5)** $f * (g + h) = (f * g) + (f * h),$
> **(6)** $(f * g) * h = f * (g * h),$
> **(7)** $f * 0 = 0.$

Proof.   To prove equation (4), we begin with the definition

$$(f * g)(t) := \int_0^t f(t - v)g(v)\,dv.$$

Using the change of variables $w = t - v$, we have

$$(f * g)(t) = \int_t^0 f(w)g(t - w)(-dw) = \int_0^t g(t - w)f(w)\,dw = (g * f)(t),$$

which proves (4). The proofs of equations (5) and (6) are left to the exercises (see Problems 33 and 34). Equation (7) is obvious since $f(t - v) \cdot 0 \equiv 0$.   ◄◄◄

Returning to our original goal, we now prove that if $Y(s)$ is the product of the Laplace transforms $F(s)$ and $G(s)$, then $y(t)$ is the convolution $(f * g)(t)$.

.CONVOLUTION THEOREM

Theorem 11.   Let $f(t)$ and $g(t)$ be piecewise continuous on $[0, \infty)$ and of exponential order α, and set $F(s) = \mathscr{L}\{f\}(s)$ and $G(s) = \mathscr{L}\{g\}(s)$. Then

(8) $\mathscr{L}\{f * g\}(s) = F(s)G(s),$

or, equivalently,

(9) $\mathscr{L}^{-1}\{F(s)G(s)\}(t) = (f * g)(t).$

Proof.   Starting with the left-hand side of (8), we use the definition of convolution to write for $s > \alpha$

$$\mathscr{L}\{f * g\}(s) = \int_0^\infty e^{-st}\left(\int_0^t f(t - v)g(v)\,dv\right)dt.$$

To simplify the evaluation of this iterated integral, we introduce the unit step function $u(t - v)$ and write

$$\mathscr{L}\{f * g\}(s) = \int_0^\infty e^{-st}\left(\int_0^\infty u(t - v)f(t - v)g(v)\,dv\right)dt,$$

where we have used the fact that $u(t - v) = 0$ if $v > t$. Reversing the order of integration[†] gives

(10) $\mathscr{L}\{f * g\}(s) = \int_0^\infty g(v)\left(\int_0^\infty e^{-st}u(t - v)f(t - v)\,dt\right)dv.$

Recall from the shifting property in Section 7.6 that the integral in parentheses in equation (10) equals $e^{-sv}F(s)$. Hence

$$\mathscr{L}\{f * g\}(s) = \int_0^\infty g(v)e^{-sv}F(s)\,dv = F(s)\int_0^\infty e^{-sv}g(v)\,dv = F(s)G(s).$$

This proves formula (8).   ◄◄◄

[†] This is permitted since, for each $s > \alpha$, the absolute value of the integrand is integrable on $(0, \infty) \times (0, \infty)$.

For the initial value problem (1), recall that we found

$$Y(s) = \left(\frac{1}{s^2 + 1}\right) G(s) = \mathscr{L}\{\sin t\}(s)\mathscr{L}\{g\}(s).$$

It now follows from the convolution theorem that

$$y(t) = \sin t * g(t) = \int_0^t \sin(t - v)g(v)\,dv.$$

Thus we have obtained an integral representation for the solution to the initial value problem (1) for any forcing function $g(t)$ that is piecewise continuous on $[0, \infty)$ and of exponential order.

● **EXAMPLE 1**   Use the convolution theorem to solve the initial value problem

(11) $y'' - y = g(t);$ $y(0) = 1,$ $y'(0) = 1,$

where $g(t)$ is piecewise continuous on $[0, \infty)$ and of exponential order.

Solution   Let $Y(s) = \mathscr{L}\{y\}(s)$ and $G(s) = \mathscr{L}\{g\}(s)$. Taking the Laplace transform of both sides of the differential equation in (11) and using the initial conditions gives

$$s^2 Y(s) - s - 1 - Y(s) = G(s).$$

Solving for $Y(s)$, we have

$$Y(s) = \frac{s + 1}{s^2 - 1} + \left(\frac{1}{s^2 - 1}\right) G(s) = \frac{1}{s - 1} + \left(\frac{1}{s^2 - 1}\right) G(s).$$

Hence

$$y(t) = \mathscr{L}^{-1}\left\{\frac{1}{s - 1}\right\}(t) + \mathscr{L}^{-1}\left\{\frac{1}{s^2 - 1} G(s)\right\}(t)$$

$$= e^t + \mathscr{L}^{-1}\left\{\frac{1}{s^2 - 1} G(s)\right\}(t).$$

Referring to the table of Laplace transforms on the inside back cover, we find

$$\mathscr{L}\{\sinh t\}(s) = \frac{1}{s^2 - 1},$$

so we can now express

$$\mathscr{L}^{-1}\left\{\frac{1}{s^2 - 1} G(s)\right\}(t) = \sinh t * g(t).$$

Thus

$$y(t) = e^t + \int_0^t \sinh(t - v)g(v)\,dv$$

is the solution to the initial value problem (11).   ●

● **EXAMPLE 2**   Use the convolution theorem to find $\mathscr{L}^{-1}\{1/(s^2 + 1)^2\}$.

Solution   Write

$$\frac{1}{(s^2 + 1)^2} = \left(\frac{1}{s^2 + 1}\right)\left(\frac{1}{s^2 + 1}\right).$$

Since $\mathscr{L}\{\sin t\}(s) = 1/(s^2 + 1)$, it follows from the convolution theorem that

$$\mathscr{L}^{-1}\left\{\frac{1}{(s^2 + 1)^2}\right\}(t) = \sin t * \sin t = \int_0^t \sin(t - v)\sin v\, dv$$

$$= \frac{1}{2}\int_0^t [\cos(2v - t) - \cos t]\, dv^\dagger$$

$$= \frac{1}{2}\left[\frac{\sin(2v - t)}{2}\right]_0^t - \frac{1}{2}t\cos t$$

$$= \frac{1}{2}\left[\frac{\sin t}{2} - \frac{\sin(-t)}{2}\right] - \frac{1}{2}t\cos t$$

$$= \frac{\sin t - t\cos t}{2}. \qquad ●$$

As the preceding example attests, the convolution theorem is useful in determining the inverse transforms of rational functions of s. In fact, it provides an alternative to the method of partial fractions. For example,

$$\mathscr{L}^{-1}\left\{\frac{1}{(s - a)(s - b)}\right\}(t) = \mathscr{L}^{-1}\left\{\left(\frac{1}{s - a}\right)\left(\frac{1}{s - b}\right)\right\}(t) = e^{at} * e^{bt},$$

and all that remains in finding the inverse is to compute the convolution $e^{at} * e^{bt}$.

In the early 1900s, V. Volterra introduced **integro-differential** equations in his study of population growth. These equations enabled him to take into account "hereditary influences." In certain cases, these equations involved a convolution. As the next example shows, the convolution theorem helps to solve such integro-differential equations.

● **EXAMPLE 3**   Solve the integro-differential equation

(12) $\displaystyle y'(t) = 1 - \int_0^t y(t - v)e^{-2v}\, dv, \qquad y(0) = 1.$

Solution   Equation (12) can be written as

(13) $y'(t) = 1 - y(t) * e^{-2t}.$

---

† Here we used the identity $\sin \alpha \sin \beta = \frac{1}{2}[\cos(\beta - \alpha) - \cos(\beta + \alpha)]$.

Let $Y(s) = \mathscr{L}\{y\}(s)$. Taking the Laplace transform of (13) (with the help of the convolution theorem) and solving for $Y(s)$, we obtain

$$sY(s) - 1 = \frac{1}{s} - Y(s)\left(\frac{1}{s+2}\right)$$

$$sY(s) + \left(\frac{1}{s+2}\right)Y(s) = 1 + \frac{1}{s}$$

$$\left(\frac{s^2 + 2s + 1}{s+2}\right)Y(s) = \frac{s+1}{s}$$

$$Y(s) = \frac{(s+1)(s+2)}{s(s+1)^2} = \frac{s+2}{s(s+1)}$$

$$Y(s) = \frac{2}{s} - \frac{1}{s+1}.$$

Hence $y(t) = 2 - e^{-t}$. ●

The **transfer function** $H(s)$ of a linear system is defined as the ratio of the Laplace transform of the output function $y(t)$ to the Laplace transform of the input function $g(t)$, assuming that all initial conditions are zero. That is, $H(s) = Y(s)/G(s)$. If the linear system is governed by the differential equation

(14) $ay'' + by' + cy = g(t), \qquad t > 0,$

where a, b, and c are constants, we can compute the transfer function as follows. Take the Laplace transform of both sides of (14) to get

$$as^2 Y(s) - asy(0) - ay'(0) + bsY(s) - by(0) + cY(s) = G(s).$$

Since the initial conditions are assumed to be zero, the equation reduces to

$$(as^2 + bs + c)Y(s) = G(s).$$

Thus the transfer function for equation (14) is

(15) $H(s) = \dfrac{Y(s)}{G(s)} = \dfrac{1}{as^2 + bs + c}.$

The function $h(t) := \mathscr{L}^{-1}\{H\}(t)$ is called the **impulse response function** for the system because, physically speaking, it describes the solution when a spring-mass system is struck by a hammer (see Section 7.8). We can also characterize $h(t)$ as the unique solution to the homogeneous problem

(16) $ah'' + bh' + ch = 0; \qquad h(0) = 0, \qquad h'(0) = 1/a.$

Indeed, observe that taking the Laplace transform of the equation in (16) gives

(17) $a[s^2 H(s) - sh(0) - h'(0)] + b[sH(s) - h(0)] + cH(s) = 0.$

Substituting in $h(0) = 0$ and $h'(0) = 1/a$ and solving for $H(s)$ yields

$$H(s) = \frac{1}{as^2 + bs + c},$$

which is the same as the formula for the transfer function given in equation (15).

One nice feature of the impulse response function h is that it can help us describe the solution to the *general* initial value problem

(18) $ay'' + by' + cy = g(t);$ $y(0) = y_0,$ $y'(0) = y_1.$

From the discussion of equation (14), we see that the convolution $h * g$ is the solution to (18) in the special case when the initial conditions are zero (i.e., $y_0 = y_1 = 0$). To deal with nonzero initial conditions, let y_k denote the solution to the corresponding *homogeneous* initial value problem; that is, y_k solves

(19) $ay'' + by' + cy = 0;$ $y(0) = y_0,$ $y'(0) = y_1.$

Then, the desired solution to the general initial value problem (18) must be $h * g + y_k$. Indeed, it follows from the superposition principle (see Theorem 6 in Section 4.7) that since $h * g$ is a solution to equation (14) and y_k is a solution to the corresponding homogeneous equation, then $h * g + y_k$ is a solution to equation (14). Moreover,

$$(h * g)(0) + y_k(0) = 0 + y_0 = y_0,$$
$$(h * g)'(0) + y'_k(0) = 0 + y_1 = y_1.$$

We summarize these observations in the following theorem.

IMPULSE RESPONSE FUNCTION

Theorem 12.   Let I be an interval containing the origin. The unique solution to the initial value problem

$$ay'' + by' + cy = g;   y(0) = y_0,   y'(0) = y_1,$$

where a, b, and c are constants and g is continuous on I, is given by

(20) $(h * g)(t) + y_k(t),$

where h is the impulse response function for the system and y_k is the unique solution to (19).

A proof of Theorem 12 that does not involve Laplace transforms is outlined in Project A in Chapter 4.

In the next example we use Theorem 12 to find a formula for the solution to an initial value problem.

● EXAMPLE 4 A linear system is governed by the differential equation

(21) $\quad y'' + 2y' + 5y = g(t); \qquad y(0) = 2, \qquad y'(0) = -2.$

Find the transfer function for the system, the impulse response function, and give a formula for the solution.

Solution According to formula (15), the transfer function for (21) is

$$H(s) = \frac{1}{as^2 + bs + c} = \frac{1}{s^2 + 2s + 5} = \frac{1}{(s+1)^2 + 2^2}.$$

The inverse Laplace transform of $H(s)$ is the impulse response function

$$h(t) = \mathcal{L}^{-1}\{H\}(t) = \frac{1}{2}\mathcal{L}^{-1}\left\{\frac{2}{(s+1)^2 + 2^2}\right\}(t)$$

$$= \frac{1}{2}e^{-t}\sin 2t. $$

To solve the initial value problem, we need the solution to the corresponding homogeneous problem. The auxiliary equation for the homogeneous equation is $r^2 + 2r + 5 = 0$, which has roots $r = -1 \pm 2i$. Thus, a general solution is $C_1 e^{-t}\cos 2t + C_2 e^{-t}\sin 2t$. Choosing C_1 and C_2 so that the initial conditions in (21) are satisfied, we obtain $y_k(t) = 2e^{-t}\cos 2t$. Hence, a formula for the solution to the initial value problem (21) is

$$(h * g)(t) + y_k(t) = \frac{1}{2}\int_0^t e^{-(t-v)}\sin 2(t-v)g(v)\,dv + 2e^{-t}\cos 2t. \quad ●$$

EXERCISES 7.7

In Problems 1 through 4 use the convolution theorem to obtain a formula for the solution to the given initial value problem, where g(t) is piecewise continuous on $[0, \infty)$ and of exponential order.

1. $y'' - 2y' + y = g(t); \qquad y(0) = -1, \qquad y'(0) = 1.$

2. $y'' + 9y = g(t); \qquad y(0) = 1, \qquad y'(0) = 0.$

3. $y'' + 4y' + 5y = g(t); \qquad y(0) = 1, \qquad y'(0) = 1.$

4. $y'' + y = g(t); \qquad y(0) = 0, \qquad y'(0) = 1.$

In Problems 5 through 12 use the convolution theorem to find the inverse Laplace transform of the given function.

5. $\dfrac{1}{s(s^2 + 1)}.$

6. $\dfrac{1}{(s+1)(s+2)}.$

7. $\dfrac{14}{(s+2)(s-5)}.$

8. $\dfrac{1}{(s^2+4)^2}.$

9. $\dfrac{s}{(s^2+1)^2}.$

10. $\dfrac{1}{s^3(s^2+1)}.$

11. $\dfrac{s}{(s-1)(s+2)}.$ $\left[\text{Hint: } \dfrac{s}{s-1} = 1 + \dfrac{1}{s-1}.\right]$

12. $\dfrac{s+1}{(s^2+1)^2}.$

13. Find the Laplace transform of

$$f(t) := \int_0^t (t-v)e^{3v}\,dv.$$

14. Find the Laplace transform of

$$f(t) := \int_0^t \sin(t-v)e^v\,dv.$$

In Problems 15 through 22 solve the given integral equation or integro-differential equation for y(t).

15. $y(t) + 3\displaystyle\int_0^t y(v)\sin(t-v)\,dv = t.$

16. $y(t) + \displaystyle\int_0^t e^{t-v}y(v)\,dv = \sin t.$

17. $y(t) + \int_0^t (t-v)y(v)\,dv = 1.$

18. $y(t) + \int_0^t (t-v)y(v)\,dv = t^2.$

19. $y(t) + \int_0^t (t-v)^2 y(v)\,dv = t^3 + 3.$

20. $y'(t) + \int_0^t (t-v)y(v)\,dv = t,$ $y(0) = 0.$

21. $y'(t) + y(t) - \int_0^t y(v)\sin(t-v)\,dv = -\sin t,$ $y(0) = 1.$

22. $y'(t) - 2\int_0^t e^{t-v}y(v)\,dv = t,$ $y(0) = 2.$

In Problems 23 through 28 a linear system is governed by the given initial value problem. Find the transfer function H(s) for the system, the impulse response function h(t), and give a formula for the solution to the initial value problem.

23. $y'' + 9y = g(t);$ $y(0) = 2,$ $y'(0) = -3.$

24. $y'' - 9y = g(t);$ $y(0) = 2,$ $y'(0) = 0.$

25. $y'' - y' - 6y = g(t);$ $y(0) = 1,$ $y'(0) = 8.$

26. $y'' + 2y' - 15y = g(t);$ $y(0) = 0,$ $y'(0) = 8.$

27. $y'' - 2y' + 5y = g(t);$ $y(0) = 0,$ $y'(0) = 2.$

28. $y'' - 4y' + 5y = g(t);$ $y(0) = 0,$ $y'(0) = 1.$

In Problems 29 and 30 the current I(t) in an RLC circuit with an electromotive force E(t) is governed by the initial value problem

$$LI''(t) + RI'(t) + \frac{1}{C}I(t) = e(t),\quad I(0) = a,\quad I'(0) = b,$$

where $e(t) = E'(t)$ (see Figure 7.13). Given the constants R, L, C, a, and b find a formula for the solution I(t) in terms of e(t).

29. $R = 20$ ohms, $L = 5$ henrys, $C = 0.005$ farads, $a = -1$ amps, $b = 8$ amps/sec.

30. $R = 80$ ohms, $L = 10$ henrys, $C = 1/410$ farads, $a = 2$ amps, $b = -8$ amps/sec.

31. Use the convolution theorem and Laplace transforms to compute $1 * 1 * 1.$

32. Use the convolution theorem and Laplace transforms to compute $1 * t * t^2.$

33. Prove property (5) in Theorem 10.

34. Prove property (6) in Theorem 10.

35. Use the convolution theorem to show that

$$\mathcal{L}^{-1}\left\{\frac{F(s)}{s}\right\}(t) = \int_0^t f(v)\,dv,$$

where $F(s) = \mathcal{L}\{f\}(s).$

36. Using Theorem 5 in Section 7.3 and the convolution theorem, show that

$$\int_0^t \int_0^v f(z)\,dz\,dv = \mathcal{L}^{-1}\left\{\frac{F(s)}{s^2}\right\}(t)$$

$$= t\int_0^t f(v)\,dv - \int_0^t vf(v)\,dv,$$

where $F(s) = \mathcal{L}\{f\}(s).$

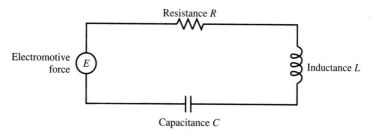

Figure 7.13 Schematic representation of an *RLC* series circuit

•7.8 • IMPULSES AND THE DIRAC DELTA FUNCTION (Optional)

In mechanical systems, electric circuits, bending of beams, and other applications, one encounters functions that have a very large value over a very short interval. For example, the strike of a hammer exerts a relatively large force over a relatively short time, and a heavy weight concentrated at a spot on a suspended beam exerts a large force over a very small section of the beam. In order to deal with violent forces of short duration, physicists and engineers use the delta function introduced by Paul A. M. Dirac.

DIRAC DELTA FUNCTION

Definition 9. The **Dirac delta function** $\delta(t)$ is characterized by the following two properties:

$$(1) \qquad \delta(t) = \begin{cases} 0, & t \neq 0, \\ \text{``infinite,''} & t = 0, \end{cases}$$

and

$$(2) \qquad \int_{-\infty}^{\infty} f(t)\delta(t)\,dt = f(0)$$

for any function $f(t)$ that is continuous on an open interval containing $t = 0$.

By shifting the argument of $\delta(t)$, we have $\delta(t - a) = 0, t \neq a$, and

$$(3) \qquad \int_{-\infty}^{\infty} f(t)\delta(t - a)\,dt = f(a)$$

for any function $f(t)$ that is continuous on an interval containing $t = a$.

It is obvious that $\delta(t - a)$ is not a function in the usual sense; instead it is an example of what is called a **generalized function** or a **distribution**. In spite of this shortcoming, the Dirac delta function was successfully used for several years to solve various physics and engineering problems before Laurent Schwartz mathematically justified its use!

A heuristic argument for the existence of the Dirac delta function can be made by considering the impulse of a force over a short interval. If a force $\mathscr{F}(t)$ is applied from time t_0 to time t_1, then the **impulse** due to $\mathscr{F}$ is the integral

$$\text{Impulse} := \int_{t_0}^{t_1} \mathscr{F}(t)\,dt.$$

By Newton's second law we see that

$$(4) \qquad \int_{t_0}^{t_1} \mathscr{F}(t)\,dt = \int_{t_0}^{t_1} m\frac{dv}{dt}\,dt = mv(t_1) - mv(t_0),$$

where m denotes mass and v denotes velocity. Since mv represents the momentum, we can interpret equation (4) as saying: **the impulse equals the change in momentum.**

When a hammer strikes an object, it transfers momentum to the object. This change in momentum takes place over a very short period of time, say $[t_0, t_1]$. If we let $\mathscr{F}_1(t)$ represent the force due to the hammer, then the *area* under the curve $\mathscr{F}_1(t)$ is the impulse or change in momentum (see Figure 7.14). If, as is illustrated in Figure 7.15, the same change in momentum takes place over shorter and shorter time intervals, say $[t_0, t_2]$ or $[t_0, t_3]$, then the average force must get greater and greater in order for the impulses (the areas under the curves $\mathscr{F}_n$) to remain the same. In fact, if the forces $\mathscr{F}_n$ having the same impulse act, respectively, over the intervals $[t_0, t_n]$, where $t_n \to t_0$ as $n \to \infty$, then $\mathscr{F}_n$ approaches a function that is zero for $t \neq t_0$ but has an infinite value for $t = t_0$. Moreover, the areas under the $\mathscr{F}_n$'s have a common value. Normalizing this value to be 1 gives

$$\int_{-\infty}^{\infty} \mathscr{F}_n(t)\, dt = 1 \qquad \text{for all } n.$$

When $t_0 = 0$, we derive from the limiting properties of the $\mathscr{F}_n$'s a "function" δ that satisfies property (1) and the integral condition

(5) $\qquad \int_{-\infty}^{\infty} \delta(t)\, dt = 1.$

Notice that (5) is a special case of property (2) that is obtained by taking $f(t) \equiv 1$. It is interesting to note that (5) actually implies the general property (2) (see Problem 33).

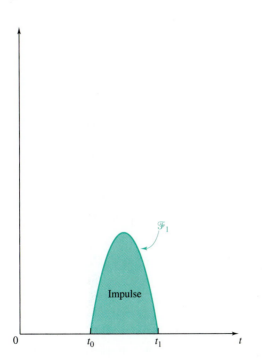

Figure 7.14 Force due to a blow from a hammer

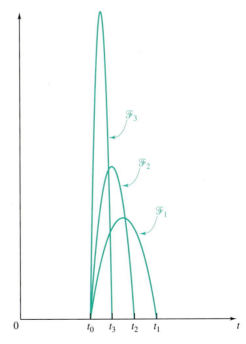

Figure 7.15 Forces with the same impulse

The Laplace transform of the Dirac delta function can be quickly derived from property (3). Since $\delta(t - a) = 0$ for $t \neq a$, setting $f(t) = e^{-st}$ in (3), we find for $a \geq 0$

$$\int_0^\infty e^{-st}\delta(t - a)\, dt = \int_{-\infty}^\infty e^{-st}\delta(t - a)\, dt = e^{-as}.$$

Thus, for $a \geq 0$,

(6) $\mathscr{L}\{\delta(t - a)\}(s) = e^{-as}.$

An interesting connection exists between the unit step function and the Dirac delta function. Observe that as a consequence of equation (5) and the fact that $\delta(x - a)$ is zero for $x < a$ and for $x > a$, we have

(7) $\displaystyle\int_{-\infty}^t \delta(x - a)\, dx = \begin{cases} 0, & t < a, \\ 1, & t > a \end{cases}$

$$= u(t - a).$$

If we formally differentiate both sides of (7) with respect to t (in the spirit of the fundamental theorem of calculus), we find

$$\delta(t - a) = u'(t - a).$$

Thus it appears that the Dirac delta function is the derivative of the unit step function. That is, in fact, the case if we consider "differentiation" in a more general sense.[†]

The Dirac delta function is used in modeling mechanical vibration problems involving an impulse. For example, a spring-mass system at rest that is struck by a hammer exerting an impulse on the mass, might be governed by the *symbolic* initial value problem

(8) $x'' + x = \delta(t);$ $x(0) = 0,$ $x'(0) = 0,$

where $x(t)$ denotes the displacement from equilibrium at time t. We refer to this as a symbolic problem because, while the left-hand side of equation (8) represents an ordinary function, the right-hand side does *not*. Consequently, it is not clear what we mean by a solution to problem (8). Since $\delta(t)$ is zero everywhere, except at $t = 0$, one might be tempted to treat (8) as a homogeneous equation with zero initial conditions. But, the solution to the latter is zero everywhere, which certainly does not describe the motion of the spring after being struck by the hammer.

To define what is meant by a solution to (8), recall that $\delta(t)$ is thought of as the limit of forces $\mathscr{F}_n(t)$ having unit impulse and acting over shorter and shorter intervals. If we let $y_n(t)$ be the solution to the initial value problem

(9) $y_n'' + y_n = \mathscr{F}_n(t);$ $y_n(0) = 0,$ $y_n'(0) = 0,$

where δ is replaced by $\mathscr{F}_n$, then we can think of the solution $x(t)$ to (8) as the limit of the solutions $y_n(t)$.

[†] See *Distributions, Complex Variables, and Fourier Transforms*, by H. J. Bremermann, Addison-Wesley, Reading, Massachusetts, 1966.

For example, let

$$\mathcal{F}_n(t) := n - nu(t - 1/n) = \begin{cases} n, & 0 < t < 1/n, \\ 0, & \text{otherwise.} \end{cases}$$

Taking the Laplace transform of equation (9), we find

$$(s^2 + 1)Y_n(s) = \frac{n}{s}(1 - e^{-s/n}),$$

and so

$$Y_n(s) = \frac{n}{s(s^2 + 1)} - e^{-s/n}\frac{n}{s(s^2 + 1)}.$$

Now

$$\frac{n}{s(s^2 + 1)} = \frac{n}{s} - \frac{ns}{s^2 + 1} = \mathscr{L}\{n - n\cos t\}(s).$$

Hence,

(10) $y_n(t) = n - n\cos t - (n - n\cos(t - 1/n))u(t - 1/n)$

$$= \begin{cases} n - n\cos t, & 0 < t < 1/n, \\ n\cos(t - 1/n) - n\cos t, & 1/n < t. \end{cases}$$

Fix $t > 0$. Then for n large enough, we have $1/n < t$. Thus,

$$\lim_{n\to\infty} y_n(t) = \lim_{n\to\infty}[n\cos(t - 1/n) - n\cos t]$$

$$= -\lim_{n\to\infty}\frac{\cos(t - 1/n) - \cos t}{-1/n}$$

$$= -\lim_{h\to 0}\frac{\cos(t + h) - \cos t}{h}, \quad \text{(where } h = -1/n),$$

$$= -\frac{d}{dt}(\cos t) = \sin t.$$

Also, for $t = 0$, $\lim_{n\to\infty} y_n(0) = 0 = \sin 0$. Therefore,

$$\lim_{n\to\infty} y_n(t) = \sin t.$$

Hence, the solution to the symbolic initial value problem (8) is $x(t) = \sin t$.

Fortunately, we do not have to go through the tedious process of solving for each y_n in order to find the solution x of the symbolic problem. It turns out that the Laplace transform method when applied directly to (8) yields the derived solution $x(t)$. Indeed, simply taking the Laplace transform of both sides of (8), we obtain from (6) (with $a = 0$)

$$(s^2 + 1)X(s) = 1,$$

$$X(s) = \frac{1}{s^2 + 1},$$

which gives

$$x(t) = \mathscr{L}^{-1}\left\{\frac{1}{s^2 + 1}\right\}(t) = \sin t.$$

A peculiarity of using the Dirac delta function is that the solution $x(t) = \sin t$ of the symbolic initial value problem (8) does not satisfy both initial conditions, that is, $x'(0) = 1 \neq 0$. This may be disturbing, but since $\delta(t)$ is "infinite" at $t = 0$, we might expect something unusual to occur.

In the next example the Dirac delta function is used in modeling a mechanical vibration problem.

● **EXAMPLE 1**   A mass attached to a spring is released from rest 1 m below the equilibrium position for the spring-mass system and begins to vibrate. After π seconds the mass is struck by a hammer exerting an impulse on the mass. The system is governed by the symbolic initial value problem

(11) $$\frac{d^2x}{dt^2} + 9x = 3\delta(t - \pi); \qquad x(0) = 1, \qquad \frac{dx}{dt}(0) = 0,$$

where $x(t)$ denotes the displacement from equilibrium at time t. Determine $x(t)$.

Solution   Let $X(s) = \mathscr{L}\{x\}(s)$. Since

$$\mathscr{L}\{x''\}(s) = s^2X(s) - s \quad \text{and} \quad \mathscr{L}\{\delta(t - \pi)\}(s) = e^{-\pi s},$$

taking the Laplace transform of both sides of (11) and solving for $X(s)$ yields

$$s^2X(s) - s + 9X(s) = 3e^{-\pi s}$$

$$X(s) = \frac{s}{s^2 + 9} + e^{-\pi s}\frac{3}{s^2 + 9}$$

$$= \mathscr{L}\{\cos 3t\}(s) + e^{-\pi s}\mathscr{L}\{\sin 3t\}(s).$$

Using the shifting property (cf. Section 7.6) to determine the inverse Laplace transform of $X(s)$, we find

$$x(t) = \cos 3t + \sin 3(t - \pi)u(t - \pi)$$

$$= \begin{cases} \cos 3t, & t < \pi, \\ \cos 3t - \sin 3t, & \pi < t \end{cases}$$

$$= \begin{cases} \cos 3t, & t < \pi, \\ \sqrt{2}\cos\left(3t + \frac{\pi}{4}\right), & \pi < t. \end{cases}$$

The graph of $x(t)$ is given in color in Figure 7.16. For comparison, the dashed curve depicts the displacement of an undisturbed vibrating spring.   ●

In Section 7.7 we defined the **impulse response function** for

(12) $$ay'' + by' + cy = g(t)$$

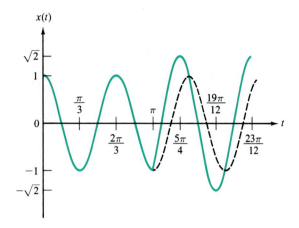

Figure 7.16 Displacement of a vibrating spring that is struck by a hammer at $t = \pi$

as the function $h(t) := \mathscr{L}^{-1}\{H\}(t)$, where $H(s)$ is the **transfer function.** Recall that $H(s)$ is the ratio

$$H(s) := \frac{Y(s)}{G(s)},$$

where $Y(s)$ is the Laplace transform of the solution to (12) with zero initial conditions and $G(s)$ is the Laplace transform of $g(t)$. It is important to note that $H(s)$, and hence $h(t)$, does not depend on the choice of the function $g(t)$ in (12) (see equation (15) in Section 7.7). However, it is useful to think of the impulse response function as the solution of the symbolic initial value problem

$$(13) \qquad ay'' + by' + cy = \delta(t); \qquad y(0) = 0, \qquad y'(0) = 0.$$

Indeed, with $g(t) = \delta(t)$, we have $G(s) = 1$, and hence $H(s) = Y(s)$. Consequently $h(t) = y(t)$. So we see that the function $h(t)$ is the response to the impulse $\delta(t)$ for a mechanical system governed by the symbolic initial value problem (13).

EXERCISES 7.8

In Problems 1 through 6 evaluate the given integral.

1. $\displaystyle\int_{-\infty}^{\infty} (t^2 - 1)\delta(t)\,dt.$

2. $\displaystyle\int_{-\infty}^{\infty} e^{3t}\delta(t)\,dt.$

3. $\displaystyle\int_{-\infty}^{\infty} (\sin 3t)\delta\left(t - \frac{\pi}{2}\right)dt.$

4. $\displaystyle\int_{-\infty}^{\infty} e^{-2t}\delta(t + 1)\,dt.$

5. $\displaystyle\int_{0}^{\infty} e^{-2t}\delta(t - 1)\,dt.$

6. $\displaystyle\int_{-1}^{1} (\cos 2t)\delta(t)\,dt.$

In Problems 7 through 12 determine the Laplace transform of the given generalized function.

7. $\delta(t - 1) - \delta(t - 3).$ **8.** $3\delta(t - 1).$

9. $t\delta(t - 1).$ **10.** $t^3\delta(t - 3).$

11. $\delta(t - \pi)\sin t.$ **12.** $e^t\delta(t - 3).$

In Problems 13 through 20 solve the given symbolic initial value problem.

13. $w'' + w = \delta(t - \pi);$ $w(0) = 0,$ $w'(0) = 0.$

14. $y'' + 2y' + 2y = \delta(t - \pi);$ $y(0) = 1,$ $y'(0) = 1.$

15. $y'' + 2y' - 3y = \delta(t - 1) - \delta(t - 2);$
  $y(0) = 2,$ $y'(0) = -2.$

16. $y'' - 2y' - 3y = 2\delta(t - 1) - \delta(t - 3);$
  $y(0) = 2,$ $y'(0) = 2.$

17. $y'' - y = 4\delta(t - 2) + t^2;$ $y(0) = 0,$ $y'(0) = 2.$

18. $y'' - y' - 2y = 3\delta(t - 1) + e^t;$ $y(0) = 0,$ $y'(0) = 3.$

19. $w'' + 6w' + 5w = e^t\delta(t - 1);$ $w(0) = 0,$ $w'(0) = 4.$

20. $y'' + 5y' + 6y = e^{-t}\delta(t - 2);$
  $y(0) = 2,$ $y'(0) = -5.$

In Problems 21 through 24 solve the given symbolic initial value problem and sketch a graph of the solution.

21. $y'' + y = \delta(t - 2\pi);$ $y(0) = 0,$ $y'(0) = 1.$

22. $y'' + y = \delta(t - \pi/2);$ $y(0) = 0,$ $y'(0) = 1.$

23. $y'' + y = -\delta(t - \pi) + \delta(t - 2\pi);$
  $y(0) = 0,$ $y'(0) = 1.$

24. $y'' + y = \delta(t - \pi) - \delta(t - 2\pi);$ $y(0) = 0,$ $y'(0) = 1.$

In Problems 25 through 28 find the impulse response function h(t) by using the fact that h(t) is the solution to the symbolic initial value problem with $g(t) = \delta(t)$ and zero initial conditions.

25. $y'' + 4y' + 8y = g(t).$ **26.** $y'' - 6y' + 13y = g(t).$

27. $y'' - 2y' + 5y = g(t).$ **28.** $y'' - y = g(t).$

29. A mass attached to a spring is released from rest 1 m below the equilibrium position for the spring-mass system and begins to vibrate. After $\pi/2$ seconds, the mass is struck by a hammer exerting an impulse on the mass. The system is governed by the symbolic initial value problem

$$\frac{d^2x}{dt^2} + 9x = -3\delta\left(t - \frac{\pi}{2}\right);$$

$$x(0) = 1,$$

$$\frac{dx}{dt}(0) = 0,$$

where $x(t)$ denotes the displacement from equilibrium at time t. What happens to the mass after it is struck?

30. You have probably heard that soldiers are told not to march in cadence when crossing a bridge. By solving

the symbolic initial value problem

$$y'' + y = \sum_{k=1}^{\infty} \delta(t - 2k\pi); \qquad y(0) = 0, \qquad y'(0) = 0,$$

explain why soldiers are so instructed. [Hint: See Section 5.3.]

31. A linear system is said to be **stable** if its impulse response function $h(t)$ remains bounded as $t \to \infty$. If the linear system is governed by

$$ay'' + by' + cy = g(t),$$

where b and c are not both zero, show that the system is stable if and only if the real parts of the roots to

$$ar^2 + br + c = 0$$

are less than or equal to zero.

32. A linear system is said to be **asymptotically stable** if its impulse response function satisfies $h(t) \to 0$ as $t \to \infty$. If the linear system is governed by

$$ay'' + by' + cy = g(t),$$

show that the system is asymptotically stable if and only if the real parts of the roots to

$$ar^2 + br + c = 0$$

are strictly less than zero.

33. The Dirac delta function may also be characterized by the properties

$$\delta(t) = \begin{cases} 0, & t \neq 0, \\ \text{"infinite,"} & t = 0 \end{cases} \quad \text{and}$$

$$\int_{-\infty}^{\infty} \delta(t)\,dt = 1.$$

Formally using the mean value theorem for definite integrals, verify that if $f(t)$ is continuous, then the above properties imply

$$\int_{-\infty}^{\infty} f(t)\delta(t)\,dt = f(0).$$

34. Formally using integration by parts, show that

$$\int_{-\infty}^{\infty} f(t)\delta'(t)\,dt = -f'(0).$$

Also show that, in general,

$$\int_{-\infty}^{\infty} f(t)\delta^{(n)}(t)\,dt = (-1)^n f^{(n)}(0).$$

35. Figure 7.17 shows a beam of length 2λ that is imbedded in a support on the left and free on the right. The vertical deflection of the beam a distance x from the support is denoted by $y(x)$. If the beam has a concentrated load L acting on it in the center of the beam, then the deflection must satisfy the symbolic boundary value problem

$$EIy^{(4)}(x) = L\delta(x - \lambda);$$

$$y(0) = y'(0) = y''(2\lambda) = y'''(2\lambda) = 0,$$

where E, the modulus of elasticity, and I, a moment of inertia, are constants. Find a formula for the displacement $y(x)$ in terms of the constants $\lambda, L, E,$ and I. [Hint: Let $y''(0) = A$ and $y'''(0) = B$. First solve the fourth order symbolic initial value problem and then use the conditions $y''(2\lambda) = y'''(2\lambda) = 0$ to determinne A and B.]

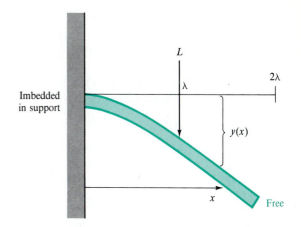

Figure 7.17 Beam imbedded in a support under a concentrated load at $x = \lambda$

● CHAPTER SUMMARY ●

The use of the Laplace transform helps to simplify the process of solving initial value problems for certain differential and integral equations, especially when a discontinuous forcing function is involved. The Laplace transform $\mathscr{L}\{f\}$ of a function $f(t)$ is defined by

$$\mathscr{L}\{f\}(s) := \int_0^\infty e^{-st}f(t)\,dt,$$

for all values of s for which the improper integral exists. If $f(t)$ is piecewise continuous on $[0, \infty)$ and of exponential order α (that is, $|f(t)|$ grows no faster than a constant times $e^{\alpha t}$ as $t \to \infty$), then $\mathscr{L}\{f\}(s)$ exists for all $s > \alpha$.

The Laplace transform can be interpreted as an integral operator that maps a function $f(t)$ to a function $F(s)$. The transforms of commonly occurring functions appear in Table 7.1, page 285, and on the inside back cover. The use of these tables is enhanced by several important properties of the operator $\mathscr{L}$.

Linearity:　$\mathscr{L}\{af + bg\} = a\mathscr{L}\{f\} + b\mathscr{L}\{g\}.$

Translation Property:　$\mathscr{L}\{e^{at}f(t)\}(s) = F(s - a)$, where $F = \mathscr{L}\{f\}.$

Shifting Property:　$\mathscr{L}\{g(t)u(t - a)\}(s) = e^{-as}\mathscr{L}\{g(t + a)\}(s)$, where $u(t - a)$ is the step function that equals 1 for $t > a$ and 0 for $t < a$. If $f(t)$ is continuous, then

$$\mathscr{L}^{-1}\{e^{-as}F(s)\}(t) = f(t - a)u(t - a),$$

where $f = \mathscr{L}^{-1}\{F\}.$

Convolution Property:　$\mathscr{L}\{f * g\} = \mathscr{L}\{f\}\mathscr{L}\{g\}$, where $f * g$ denotes the convolution function

$$(f * g)(t) := \int_0^t f(t - v)g(v)\,dv.$$

One reason for the usefulness of the Laplace transform lies in the simple formula for the transform of the derivative f':

(1) $\mathcal{L}\{f'\}(s) = sF(s) - f(0),$ where $F = \mathcal{L}\{f\}.$

This formula shows that, by using the Laplace transform, "differentiation with respect to t" can be replaced by the simple operation of "multiplication by s." The extension of (1) to higher order derivatives is

(2) $\mathcal{L}\{f^{(n)}\}(s) = s^n F(s) - s^{n-1}f(0) - s^{n-2}f'(0) - \cdots - f^{(n-1)}(0).$

To solve an initial value problem of the form

(3) $ay'' + by' + cy = f(t);$ $y(0) = \alpha,$ $y'(0) = \beta$

via the Laplace transform method, one takes the transform of both sides of the differential equation in (3). Using the linearity of $\mathcal{L}$ and formula (2) leads to an equation involving the Laplace transform $Y(s)$ of the (unknown) solution $y(t)$. The next step is to solve this simpler equation for $Y(s)$. Finally, one computes the *inverse* Laplace transform of $Y(s)$ to obtain the desired solution. This last step of finding $\mathcal{L}^{-1}\{Y\}$ is often the most difficult— sometimes requiring a partial fractions decomposition, a judicious use of the properties of the transform, an excursion through the Laplace transform tables, or the evaluation of a contour integral in the complex plane.

For the special problem in (3), where a, b, and c are constants, the differential equation is transformed to a simple *algebraic* equation for $Y(s)$. Another nice feature of this latter equation is that it incorporates the initial conditions. When the coefficients of the equation in (3) depend on t, the following formula may be helpful in taking the transform:

$$\mathcal{L}\{t^n f(t)\}(s) = (-1)^n \frac{d^n F}{ds^n}(s), \quad \text{where} \quad F = \mathcal{L}\{f\}.$$

If the forcing function $f(t)$ in equation (3) has jump discontinuities, it is often convenient to write $f(t)$ in terms of unit step functions $u(t-a)$ before proceeding with the Laplace transform method. The transform of a periodic forcing function $f(t)$ with period T is given by

$$\mathcal{L}\{f\}(s) = \frac{\displaystyle\int_0^T e^{-st}f(t)\,dt}{1 - e^{-sT}}.$$

The Dirac delta function $\delta(t)$ is useful in modeling a system that is excited by a large force applied over a short time interval. It is not a function in the usual sense, but can be roughly interpreted as the derivative of a unit step function. The transform of $\delta(t - a)$ is

$$\mathcal{L}\{\delta(t - a)\}(s) = e^{-as}, \quad a \geq 0.$$

REVIEW PROBLEMS

1, 3, 7, 11, 13, 19, 21

In Problems 1 and 2 use the definition of the Laplace transform to determine $\mathcal{L}\{f\}$.

1. $f(t) = \begin{cases} 3, & 0 \le t \le 2, \\ 6 - t, & 2 < t. \end{cases}$

2. $f(t) = \begin{cases} e^{-t}, & 0 \le t \le 5, \\ -1, & 5 < t. \end{cases}$

In Problems 3 through 10 determine the Laplace transform of the given function.

3. $t^2 e^{-9t}$.

4. $e^{3t} \sin 4t$.

5. $e^{2t} - t^3 + t^2 - \sin 5t$.

6. $7e^{2t} \cos 3t - 2e^{7t} \sin 5t$.

7. $t \cos 6t$.

8. $(t + 3)^2 - (e^t + 3)^2$.

9. $t^2 u(t - 4)$.

10. $f(t) = \cos t$, $-\pi/2 \le t \le \pi/2$, and $f(t)$ has period π.

In Problems 11 through 17 determine the inverse Laplace transform of the given function.

11. $\dfrac{7}{(s + 3)^3}$.

12. $\dfrac{2s - 1}{s^2 - 4s + 6}$.

13. $\dfrac{4s^2 + 13s + 19}{(s - 1)(s^2 + 4s + 13)}$.

14. $\dfrac{s^2 + 16s + 9}{(s + 1)(s + 3)(s - 2)}$.

15. $\dfrac{2s^2 + 3s - 1}{(s + 1)^2(s + 2)}$.

16. $\dfrac{1}{(s^2 + 9)^2}$.

17. $\dfrac{e^{-2s}(4s + 2)}{(s - 1)(s + 2)}$.

18. Find the Taylor series for $f(t) = e^{-t^2}$ about $t = 0$, and, assuming that the Laplace transform of $f(t)$ can be computed term by term, find an expansion for $\mathcal{L}\{f\}(s)$ in powers of $1/s$.

In Problems 19 through 24 solve the given initial value problem for $y(t)$ using the method of Laplace transforms.

19. $y'' - 7y' + 10y = 0$; $y(0) = 0$, $y'(0) = -3$.

20. $y'' + 6y' + 9y = 0$; $y(0) = -3$, $y'(0) = 10$.

21. $y'' + 2y' + 2y = t^2 + 4t$; $y(0) = 0$, $y'(0) = -1$.

22. $y'' + 9y = 10e^{2t}$; $y(0) = -1$, $y'(0) = 5$.

23. $y'' + 3y' + 4y = u(t - 1)$; $y(0) = 0$, $y'(0) = 1$.

24. $y'' - 4y' + 4y = t^2 e^t$; $y(0) = 0$, $y'(0) = 0$.

In Problems 25 and 26 find solutions to the given initial value problem.

25. $ty'' + 2(t - 1)y' - 2y = 0$; $y(0) = 0$, $y'(0) = 0$.

26. $ty'' + 2(t - 1)y' + (t - 2)y = 0$;
$y(0) = 1$, $y'(0) = -1$.

In Problems 27 and 28 solve the given equation for $y(t)$.

27. $y(t) + \displaystyle\int_0^t (t - v)y(v)\,dv = e^{-3t}$.

28. $y'(t) - 2\displaystyle\int_0^t y(v)\sin(t - v)\,dv = 1$; $y(0) = -1$.

29. A linear system is governed by
$$y'' - 5y' + 6y = g(t).$$
Find the transfer function and the impulse response function.

30. Solve the symbolic initial value problem
$$y'' + 4y = \delta\left(t - \frac{\pi}{2}\right); \qquad y(0) = 0, \qquad y'(0) = 1.$$

TECHNICAL WRITING EXERCISES

1. Compare the use of Laplace transforms in solving linear differential equations with constant coefficients with the use of logarithms in solving algebraic equations of the form $x^r = a$.

2. Explain why the method of Laplace transforms works so well for linear differential equations with constant coefficients *and* integro-differential equations involving a convolution.

3. Discuss several examples of initial value problems where the method of Laplace transforms *cannot* be applied.

4. A linear system is said to be **asymptotically stable** if its impulse response function $h(t) \to 0$ as $t \to \infty$. Assume $H(s)$, the Laplace transform of $h(t)$, is a rational function in reduced form with the degree of its numerator less than the degree of its denominator. Explain in detail how the asymptotic stability of the linear system can be characterized in terms of the zeros of the denominator of $H(s)$. Give examples.

5. Can the method of Laplace transforms be modified to solve *difference* equations (see Section 5.5)? If so, give the precise definition of the transform and describe some of its properties.

GROUP PROJECTS FOR CHAPTER 7

A. Duhamel's Formulas

For a linear system governed by the equation

$$(1) \qquad ay'' + by' + cy = g(t),$$

where a, b, and c are real constants, the function

$$(2) \qquad H(s) := \frac{\mathscr{L}\{y\}(s)}{\mathscr{L}\{g\}(s)} = \frac{\mathscr{L}\{\text{output}\}}{\mathscr{L}\{\text{input}\}},$$

where all initial conditions are taken to be zero, is called the **transfer function** for the system. (As mentioned in Section 7.7, the transfer function $H(s)$ depends only on the constants a, b, c of the system; it is not affected by the choice of g.) If the input function $g(t)$ is the unit step function $u(t)$, then equation (2) yields

$$\mathscr{L}\{y\}(s) = \mathscr{L}\{u\}(s)H(s) = \frac{H(s)}{s}.$$

The solution (output function) in this special case is called the **indicial admittance** and is denoted by $A(t)$. Hence $\mathscr{L}\{A\}(s) = H(s)/s$.

It is possible to express the response $y(t)$ of the system to a general input function $g(t)$ in terms of $g(t)$ and the indicial admittance $A(t)$. To derive these relations, proceed as follows:

(a) Show that

$$(3) \qquad \mathscr{L}\{y\}(s) = s\mathscr{L}\{A\}(s)\mathscr{L}\{g\}(s).$$

(b) Now apply the convolution theorem to (3) and show that

$$(4) \qquad y(t) = \frac{d}{dt}\left[\int_0^t A(t-v)g(v)\,dv\right] = \frac{d}{dt}\left[\int_0^t A(v)g(t-v)\,dv\right].$$

(c) To perform the differentiation indicated in (4), one can use **Leibniz's rule**:

$$\frac{d}{dt}\left[\int_{a(t)}^{b(t)} f(v,t)\,dv\right] = \int_{a(t)}^{b(t)} \frac{\partial f}{\partial t}(v,t)\,dv + f(b(t),t)\frac{db}{dt}(t) - f(a(t),t)\frac{da}{dt}(t),$$

where f and $\partial f/\partial t$ are assumed continuous in v and t, and $a(t)$, $b(t)$ are differentiable functions of t. Applying this rule to (4), derive the formulas

(5) $$y(t) = \int_0^t A'(t - v)g(v)\, dv,$$

(6) $$y(t) = \int_0^t A(v)g'(t - v)\, dv + A(t)g(0).$$

[Hint: Recall that the initial conditions on A are $A(0) = A'(0) = 0$.]

(d) In equations (5) and (6), make the change of variables $w = t - v$ and show that

(7) $$y(t) = \int_0^t A'(w)g(t - w)\, dw,$$

(8) $$y(t) = \int_0^t A(t - w)g'(w)\, dw + A(t)g(0).$$

Equations (5)–(8) are referred to as **Duhamel's formulas,** in honor of the French mathematician J. M. C. Duhamel. These formulas are helpful in determining the response of the system to a general input $g(t)$, since the indicial admittance of the system can be determined experimentally by measuring the response of the system to a unit step function.

(e) The impulse response function $h(t)$ is defined as $h(t) := \mathscr{L}^{-1}\{H\}(t)$, where $H(s)$ is the transfer function. Show that $h(t) = A'(t)$, so that equations (5) and (7) can be written in the form

(9) $$y(t) = \int_0^t h(t - v)g(v)\, dv = \int_0^t h(v)g(t - v)\, dv.$$

We remark that the indicial admittance is the response of the system to a unit step function, and the impulse response function is the response to the unit impulse or delta function (see Section 7.8). But, the delta function is the derivative (in a generalized sense) of the unit step function. Therefore, the fact that $h(t) = A'(t)$ is not really surprising.

B. Frequency Response Modeling

Frequency response modeling of a linear system is based on the premise that the dynamics of a linear system can be recovered from a knowledge of how the system responds to sinusoidal inputs. (This will be made mathematically precise in Theorem 13.) In other words, to determine (or identify) a linear system, all one has to do is observe how the system reacts to sinusoidal inputs.

Let's assume that we have a linear system governed by

(10) $$y'' + py' + qy = g(t),$$

where p and q are real constants. The function $g(t)$ is called the **forcing function** or **input function.** When $g(t)$ is a sinusoid, the particular solution to (10) obtained by the method of undetermined coefficients is the **steady-state solution** or **output function** $y_{ss}(t)$ corresponding to $g(t)$. We can think of a linear system as a compartment or block into which goes an input function g and out of which comes the output function y_{ss} (see Figure 7.18 on page 342). To **identify** a linear system means to determine the coefficients p and q in equation (10).

It will be convenient for us to work with complex variables. A complex number z is usually expressed in the form $z = \alpha + i\beta$, with α, β real numbers and i denoting $\sqrt{-1}$. We can also express z in

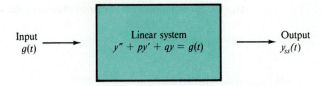

Figure 7.18 Block diagram depicting a linear system

polar form, $z = re^{i\theta}$, where $r^2 = \alpha^2 + \beta^2$ and $\tan\theta = \beta/\alpha$. Here $r\ (\geq 0)$ is called the **magnitude** and θ the **phase angle** of z.

The following theorem gives the relationship between the linear system and its response to sinusoidal inputs in terms of the **transfer function** $H(s)$ (see Project A, equation (2)).

STEADY-STATE SOLUTIONS TO SINUSOIDAL INPUTS

Theorem 13. Let $H(s)$ be the transfer function for equation (10). If $H(s)$ is finite at $s = i\omega$, with ω real, then the steady-state solution to (10) for $g(t) = e^{i\omega t}$ is

(11) $\qquad y_{ss}(t) = H(i\omega)e^{i\omega t} = H(i\omega)\{\cos\omega t + i\sin\omega t\}.$

(a) Prove Theorem 13. [Hint: Guess $y_{ss}(t) = Ae^{i\omega t}$ and show that $A = H(i\omega)$.]

(b) Use Theorem 13 to show that if $g(t) = \sin\omega t$, then the steady-state solution to (10) is $M(\omega)\sin(\omega t + N(\omega))$, where $H(i\omega) = M(\omega)e^{iN(\omega)}$ is the polar form for $H(i\omega)$.

(c) Solve for $M(\omega)$ and $N(\omega)$ in terms of p and q.

(d) Experimental results for modeling done by frequency response methods are usually presented in **frequency response**[†] or **Bode plots**. There are two types of Bode plots. The first is of the log of the magnitude $M(\omega)$ of $H(i\omega)$ versus the angular frequency ω using a log scale for ω. The second is a plot of the phase angle or argument $N(\omega)$ of $H(i\omega)$ versus the angular frequency using a log scale for ω. The Bode plots for the transfer function $H(s) = (1 + 0.2s + s^2)^{-1}$ are given in Figure 7.19.

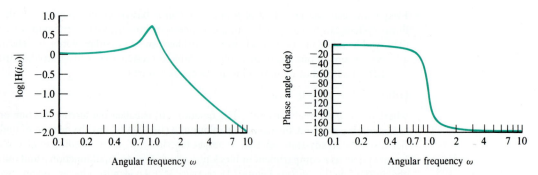

Figure 7.19 Bode plots for $H(i\omega) = [1 + 0.2(i\omega) + (i\omega)^2]^{-1}$

[†] Frequency response curves are also discussed in Section 5.3.

Sketch the Bode plots of the linear system governed by equation (10) with $p = 0.5$ and $q = 1.0$. Use $\omega = 0.3, 0.6, 0.9, 1.2$, and 1.5 for the plot of $M(\omega)$ and $\omega = 0.5, 0.8, 1, 2,$ and 5 for the plot of $N(\omega)$.

(e) Assume we know that $q = 1$. When we input a sine wave with $\omega = 2$, the system settles into a steady-state sinusoidal output with magnitude $M(2) = 0.325$. Find p and thus identify the linear system.

(f) Suppose that a sine wave input with $\omega = 2$ produces a steady-state sinusoidal output with magnitude $M(2) = 0.447$ and that when $\omega = 4$, then $M(4) = 0.085$. Find p and q and thus identify the system.

We remark that in most applications there are some inaccuracies in the measurement of the magnitudes and frequencies. To compensate for these errors, sinusoids with several different frequencies are used as input. A least squares approximation for p and q is then found. For a discussion of frequency response modeling as a mathematical modeling tool see the chapter by W. F. Powers, "Modeling Linear Systems by Frequency Response Methods," in *Differential Equations Models*, by M. Braun, C. Coleman, and D. Drew (eds.), Springer-Verlag, New York, 1983, Chapter 9. Additional examples may be found in *Schaum's Outline on Feedback and Control Systems*, by J. J. DiStefano, A. R. Stubberud, and I. J. Williams, McGraw-Hill, New York, 1967, Chapter 15.

• CHAPTER 8

Series Solution of Differential Equations

8.1 INTRODUCTION: A PROBLEM IN ASTROPHYSICS

In the early 1900s a German astrophysicist, Robert Emden, encountered the following problem in his study of the thermal behavior of a spherical gaseous cloud.

Determine the first point on the positive x-axis where the solution to the initial value problem

$$(1) \qquad xy''(x) + 2y'(x) + xy(x) = 0; \qquad y(0) = 1, \qquad y'(0) = 0$$

is zero.

The variables x and y in (1) are proportional to, respectively, the distance to the center of the cloud and the gravitational potential of the gas. The initial conditions correspond to the fact that $y(x)$, the potential, has been normalized to be 1 at the center of the cloud and that, at the center of the cloud, the gravitational forces cancel and the net force is zero. Astrophysicists and astronomers use equation (1) to approximate the density and internal temperatures of certain stars and nebula.†

The differential equation in (1) appears nice enough since it is a linear second order equation; however, there are two difficulties. First, while (1) is linear, it has variable coefficients, and hence the methods discussed in Chapter 4 do not apply. Second, we are given

† Equation (1) is called the **Lane-Emden equation of index $n = 1$** and is discussed in detail in *An Introduction to the Study of Stellar Structure*, by S. Chandrasekhar, Dover Publications, Inc., New York, 1958, Chapter IV.

initial conditions at $x = 0$, which is precisely where the coefficient of $y''(x)$ is zero. Consequently, even the theoretical results of Chapter 4 do not apply to this problem. While the prospects for finding an explicit solution to the initial value problem appear dim, the physical situation that is modeled suggests that the problem *has* a solution and the solution should be "nice" (analytic). Having faith in the model, let's assume that the solution has a power series expansion denoted by[†]

$$(2) \qquad y(x) = a_0 + a_1 x + a_2 x^2 + \cdots = \sum_{n=0}^{\infty} a_n x^n.$$

If we differentiate $y(x)$ as given in (2), term by term, we obtain

$$y'(x) = a_1 + 2a_2 x + 3a_3 x^2 + \cdots = \sum_{n=1}^{\infty} n a_n x^{n-1},$$

$$y''(x) = 2a_2 + 6a_3 x + 12a_4 x^2 + \cdots = \sum_{n=2}^{\infty} n(n-1) a_n x^{n-2}.$$

Substituting these series representations for y, y', and y'' in equation (1), we get

$$x(2a_2 + 6a_3 x + 12a_4 x^2 + \cdots + (n+1)n a_{n+1} x^{n-1} + \cdots)$$
$$+ 2(a_1 + 2a_2 x + 3a_3 x^2 + 4a_4 x^3 + \cdots + (n+1)a_{n+1} x^n + \cdots)$$
$$+ x(a_0 + a_1 x + a_2 x^2 + \cdots + a_{n-1} x^{n-1} + \cdots) = 0.$$

Grouping like terms gives

$$2a_1 + (6a_2 + a_0)x + (12a_3 + a_1)x^2 + (20a_4 + a_2)x^3$$
$$+ \cdots + ([(n+1)n + 2(n+1)]a_{n+1} + a_{n-1})x^n + \cdots = 0.$$

Now, in order for the power series to sum identically to zero, each of the coefficients must be zero. This means

$$2a_1 = 0, \qquad 6a_2 + a_0 = 0, \qquad 12a_3 + a_1 = 0,$$

and, in general,

$$(n+2)(n+1)a_{n+1} + a_{n-1} = 0.$$

Solving, we find

$$a_1 = 0, \qquad a_2 = -\tfrac{1}{6}a_0, \qquad a_3 = -\tfrac{1}{12}a_1 = 0,$$

and, in general,

$$(3) \qquad a_{n+1} = \frac{-a_{n-1}}{(n+2)(n+1)}.$$

Since $a_1 = 0$, it follows from (3) that a_3, a_5, a_7, etc., are zero. For the even coefficients, a_2,

---

[†] As discussed in Section 8.6 the point $x = 0$ is a *regular singular point* for equation (1) and according to the method of Frobenius one would typically try a solution of the more general form $y(x) = x^r(a_0 + a_1 x + a_2 x^2 + \cdots)$.

a_4, etc., we see from repeated applications of (3) that

$$a_{2k} = \frac{-a_{2k-2}}{(2k+1)(2k)} = \frac{-1}{(2k+1)(2k)} \cdot \frac{-a_{2k-4}}{(2k-1)(2k-2)} = \cdots$$

$$= \frac{(-1)^k a_0}{(2k+1)(2k)\cdots(3)(2)} = \frac{(-1)^k a_0}{(2k+1)!}.$$

Knowing the coefficients of the power series for $y(x)$, we obtain

$$(4) \qquad y(x) = a_0\left(1 - \frac{1}{6}x^2 + \frac{1}{120}x^4 + \cdots + \frac{(-1)^k x^{2k}}{(2k+1)!} + \cdots\right)$$

as a solution to the differential equation.

Next, we substitute in the initial conditions $y(0) = 1$, $y'(0) = 0$. From (4) we see that $y(0) = a_0$ and $y'(0) = 0$. Hence $a_0 = 1$. (Notice our good fortune: Equation (4) has only one parameter, a_0, yet we are able to satisfy *both* initial conditions.) Letting $a_0 = 1$ in (4), we find that the solution to the initial value problem (1) is

$$(5) \qquad y(x) = 1 - \frac{1}{6}x^2 + \frac{1}{120}x^4 - \frac{1}{5040}x^6 + \cdots + \frac{(-1)^k x^{2k}}{(2k+1)!} + \cdots.$$

Now recall that we are interested in the first (positive) zero of $y(x)$. To locate this zero, it would be desirable to have a closed-form expression for $y(x)$. Again good fortune is with us: The expansion in (5) is a recognizable one. It is the Taylor series for $(\sin x)/x$. Hence $y(x) = (\sin x)/x$, and the first positive zero of $y(x)$ occurs when $\sin x = 0$; that is, when $x = \pi$.

In this chapter we study methods for determining series expansions for solutions to differential equations with variable coefficients. In particular, we determine the form of the series expansion, a recurrence relation for determining the coefficients a_n, and the interval of convergence of the expansion.

• 8.2 • POWER SERIES, ANALYTIC FUNCTIONS, AND THE TAYLOR SERIES METHOD

The differential equations studied in earlier chapters often possessed solutions expressible in terms of elementary functions such as polynomials, exponentials, sines, and cosines. However, many important equations arise that do not have solutions that can be so conveniently expressed. In previous chapters, when we encountered such an equation we either expressed the solution as an integral (see Problem 27 in Exercises 2.2) or tried to approximate it numerically (see Sections 3.6 and 3.7). In this chapter, our goal is to obtain

---

[†] This section contains review material from calculus and a discussion of the Taylor series method briefly introduced in Project A in Chapter 1.

representations for solutions as power series. We begin with a brief review of the basic definitions and properties of power series and real analytic functions.

Power Series

A **power series** about the point x_0 is an expression of the form

(1) $$\sum_{n=0}^{\infty} a_n(x - x_0)^n = a_0 + a_1(x - x_0) + a_2(x - x_0)^2 + \cdots,$$

where x is a variable and the a_n's are constants. We say that (1) **converges** at the point $x = c$ if the infinite series (of real numbers) $\sum_{n=0}^{\infty} a_n(c - x_0)^n$ converges; that is, the limit of the partial sums,

$$\lim_{N \to \infty} \sum_{n=0}^{N} a_n(c - x_0)^n,$$

exists (as a finite number). If this limit does not exist, the power series is said to **diverge** at $x = c$. Observe that (1) converges at $x = x_0$ since

$$\sum_{n=0}^{\infty} a_n(x_0 - x_0)^n = a_0 + 0 + 0 + \cdots = a_0.$$

But what about convergence for other values of x? As stated in Theorem 1, a power series of the form (1) converges for all values of x in some "interval" centered at x_0 and diverges for x outside this interval. Moreover, at the interior points of this interval, the power series converges **absolutely** in the sense that $\sum_{n=0}^{\infty} |a_n(x - x_0)^n|$ converges. (Recall that absolute convergence of a series implies (ordinary) convergence of the series.)

> ### RADIUS OF CONVERGENCE
>
> **Theorem 1.**   For each power series of the form (1), there is a number ρ $(0 \le \rho \le \infty)$, called the **radius of convergence** of the power series, such that (1) converges absolutely for $|x - x_0| < \rho$ and diverges for $|x - x_0| > \rho$. (See Figure 8.1.)
>   If the series (1) converges for all values of x, then $\rho = \infty$. When the series (1) converges only at x_0, then $\rho = 0$.

Notice that Theorem 1 settles the question of convergence except at the endpoints $x_0 \pm \rho$. Thus these two points require separate analysis. To determine the radius of convergence ρ, one method that is often easy to apply is the ratio test.

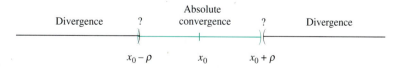

Figure 8.1 Interval of convergence

> ### RATIO TEST
>
> **Theorem 2.**   If
>
> $$\lim_{n \to \infty} \left| \frac{a_{n+1}}{a_n} \right| = L,$$
>
> where $0 \leq L \leq \infty$, then the radius of convergence of the power series $\sum_{n=0}^{\infty} a_n (x - x_0)^n$ is $\rho = 1/L$, with $\rho = \infty$ if $L = 0$ and $\rho = 0$ if $L = \infty$.

Remark.   We caution the reader that if the ratio $|a_{n+1}/a_n|$ does not have a limit, then methods other than the ratio test (e.g., root test) must be used to determine ρ. In particular, if infinitely many of the a_n's are zero, then the ratio test cannot be directly applied.

● **EXAMPLE 1**   Determine the convergence set of

(2) $$\sum_{n=0}^{\infty} \frac{(-2)^n}{n+1} (x - 3)^n.$$

Solution   Since $a_n = (-2)^n/(n+1)$, we have

$$\lim_{n \to \infty} \left| \frac{a_{n+1}}{a_n} \right| = \lim_{n \to \infty} \left| \frac{(-2)^{n+1}(n+1)}{(-2)^n(n+2)} \right|$$

$$= \lim_{n \to \infty} \frac{2(n+1)}{(n+2)} = 2 = L.$$

By the ratio test, the radius of convergence is $\rho = \frac{1}{2}$. Hence the series (2) converges absolutely for $|x - 3| < \frac{1}{2}$ and diverges when $|x - 3| > \frac{1}{2}$. It remains only to determine what happens when $|x - 3| = \frac{1}{2}$, that is, when $x = \frac{5}{2}$ and $x = \frac{7}{2}$.

Setting $x = \frac{5}{2}$, the series (2) becomes the harmonic series $\sum_{n=0}^{\infty}(n+1)^{-1}$, which is known to diverge. When $x = \frac{7}{2}$ the series (2) becomes an **alternating** harmonic series, which is known to converge. Thus the power series converges for each x in the half-open interval $(\frac{5}{2}, \frac{7}{2}]$; outside this interval it diverges.   ●

For each value of x for which the power series $\sum_{n=0}^{\infty} a_n(x - x_0)^n$ converges, we get a number that is the sum of the series. It is appropriate to denote this sum by $f(x)$ since its value depends on the choice of x. Thus we write

$$f(x) = \sum_{n=0}^{\infty} a_n(x - x_0)^n,$$

for all numbers x in the convergence interval. For example, the **geometric series** $\sum_{n=0}^{\infty} x^n$ has the radius of convergence $\rho = 1$ and the sum function $f(x) = 1/(1 - x)$; that is,

(3) $$\frac{1}{1 - x} = 1 + x + x^2 + \cdots = \sum_{n=0}^{\infty} x^n \quad \text{for} \quad -1 < x < 1.$$

Given two power series

(4) $$f(x) = \sum_{n=0}^{\infty} a_n(x - x_0)^n, \qquad g(x) = \sum_{n=0}^{\infty} b_n(x - x_0)^n,$$

with nonzero radii of convergence, we wish to find power series representations for the sum, product, and quotient of the functions $f(x)$ and $g(x)$. The sum is simply obtained by termwise addition:

$$f(x) + g(x) = \sum_{n=0}^{\infty} (a_n + b_n)(x - x_0)^n$$

for all x in the common interval of convergence of the power series in (4). The power series representation for the product $f(x)g(x)$ is a bit more complicated. To provide motivation for the formula, we treat the power series for $f(x)$ and $g(x)$ as "long polynomials," apply the distributive law, and group the terms in powers of $(x - x_0)$:

$$[a_0 + a_1(x - x_0) + a_2(x - x_0)^2 + \cdots] \cdot [b_0 + b_1(x - x_0) + b_2(x - x_0)^2 + \cdots]$$
$$= a_0 b_0 + (a_0 b_1 + a_1 b_0)(x - x_0) + (a_0 b_2 + a_1 b_1 + a_2 b_0)(x - x_0)^2 + \cdots .$$

The general formula for the product is

(5) $$f(x)g(x) = \sum_{n=0}^{\infty} c_n (x - x_0)^n,$$

where

(6) $$c_n := \sum_{k=0}^{n} a_k b_{n-k}.$$

The power series in (5) is called the **Cauchy product,** and it will converge for all x in the common *open* interval of convergence for the power series of f and g.[†]

The quotient $f(x)/g(x)$ will also have a power series expansion about x_0 provided that $g(x_0) \neq 0$. However, the radius of convergence for this quotient series may be smaller than that for $f(x)$ or $g(x)$. Unfortunately, there is no nice formula for obtaining the coefficients in the power series for $f(x)/g(x)$. However, we can use the Cauchy product to divide power series indirectly (see Problem 13). The quotient series can also be obtained by formally carrying out polynomial long division (see Problem 14).

The next theorem explains, in part, why power series are so useful.

DIFFERENTIATION AND INTEGRATION OF POWER SERIES

Theorem 3.   If the series $f(x) = \sum_{n=0}^{\infty} a_n(x - x_0)^n$ has a positive radius of convergence ρ, then f is differentiable in the interval $|x - x_0| < \rho$ and termwise differentiation gives the power series for the derivative:

$$f'(x) = \sum_{n=1}^{\infty} na_n(x - x_0)^{n-1} \quad \text{for} \quad |x - x_0| < \rho.$$

Furthermore, termwise integration gives the power series for the integral of f:

$$\int f(x)\, dx = \sum_{n=0}^{\infty} \frac{a_n}{n+1}(x - x_0)^{n+1} + C \quad \text{for} \quad |x - x_0| < \rho.$$

[†] Actually, it may happen that the radius of convergence of the power series for $f(x)g(x)$ or $f(x) + g(x)$ is larger than that for the power series of f or g.

● **EXAMPLE 2**   Starting with the geometric series (3) for $1/(1 - x)$, find a power series for each of the following functions:

$$\text{(a) } \frac{1}{1 + x^2}, \quad \text{(b) } \frac{1}{(1 - x)^2}, \quad \text{(c) } \arctan x.$$

Solution   (a) Replacing x by $-x^2$ in (3) immediately gives

$$(7) \qquad \frac{1}{1 + x^2} = 1 - x^2 + x^4 - x^6 + \cdots + (-1)^n x^{2n} + \cdots.$$

(b) Notice that $1/(1 - x)^2$ is the derivative of the function $f(x) = 1/(1 - x)$. Hence on differentiating (3) term by term, we obtain

$$(8) \qquad f'(x) = \frac{1}{(1 - x)^2} = 1 + 2x + 3x^2 + 4x^3 + \cdots + nx^{n-1} + \cdots.$$

(c) Since

$$\arctan x = \int_0^x \frac{1}{1 + t^2}\, dt,$$

we can integrate the series in (7) termwise to obtain the series for $\arctan x$. Thus

$$\int_0^x \frac{1}{1 + t^2}\, dt = \int_0^x \{1 - t^2 + t^4 - t^6 + \cdots + (-1)^n t^{2n} + \cdots\}\, dt$$

$$(9) \qquad \arctan x = x - \frac{1}{3}x^3 + \frac{1}{5}x^5 - \frac{1}{7}x^7 + \cdots + \frac{(-1)^n x^{2n+1}}{2n + 1} + \cdots. \quad ●$$

It is important to keep in mind that since the geometric series (3) has the (open) interval of convergence $(-1, 1)$, the representations (7), (8), and (9) are at least valid in this interval. (Actually, the series (9) for $\arctan x$ converges for all $|x| \leq 1$.)

The index of summation in a power series is a dummy index just like the variable of integration in a definite integral. Hence

$$\sum_{n=0}^{\infty} a_n (x - x_0)^n = \sum_{k=0}^{\infty} a_k (x - x_0)^k = \sum_{i=0}^{\infty} a_i (x - x_0)^i.$$

Just as there are times when we want to change the variable of integration, there are situations (and we will encounter many in this chapter) when it is desirable to change or shift the index of summation.

● **EXAMPLE 3**   Express the series

$$\sum_{n=2}^{\infty} n(n - 1)a_n x^{n-2},$$

using the index k, where $k = n - 2$.

Solution   Since $k = n - 2$, we have $n = k + 2$. When $n = 2$, then $k = 0$. Hence, substituting into the given series, we find

$$\sum_{n=2}^{\infty} n(n-1)a_n x^{n-2} = \sum_{k=0}^{\infty} (k+2)(k+1)a_{k+2} x^k. \quad \bullet$$

Analytic Functions

Not all functions are expressible as power series. Those distinguished functions that can be so represented are called **analytic.**

> ### ANALYTIC FUNCTION
>
> **Definition 1.**   A function f is said to be **analytic at** x_0 if, in an open interval about x_0, this function is the sum of a power series $\sum_{n=0}^{\infty} a_n(x - x_0)^n$ that has a positive radius of convergence.

For example, a polynomial function $b_0 + b_1 x + \cdots + b_n x^n$ is analytic at every x_0 since we can always rewrite it in the form $a_0 + a_1(x - x_0) + \cdots + a_n(x - x_0)^n$. A rational function $P(x)/Q(x)$, where $P(x)$ and $Q(x)$ are polynomials without a common factor, is an analytic function except at those x_0 for which $Q(x_0) = 0$. As the reader may recall from calculus, the elementary functions e^x, $\sin x$, and $\cos x$ are analytic for all x, while $\ln x$ is analytic for $x > 0$. Indeed, we have the familiar representations

$$(10) \qquad e^x = 1 + x + \frac{x^2}{2!} + \frac{x^3}{3!} + \cdots \qquad\qquad = \sum_{n=0}^{\infty} \frac{x^n}{n!},$$

$$(11) \qquad \sin x = x - \frac{x^3}{3!} + \frac{x^5}{5!} - \cdots \qquad\qquad = \sum_{n=0}^{\infty} \frac{(-1)^n}{(2n+1)!} x^{2n+1},$$

$$(12) \qquad \cos x = 1 - \frac{x^2}{2!} + \frac{x^4}{4!} - \cdots \qquad\qquad = \sum_{n=0}^{\infty} \frac{(-1)^n}{(2n)!} x^{2n},$$

$$(13) \qquad \ln x = (x-1) - \frac{1}{2}(x-1)^2 + \frac{1}{3}(x-1)^3 - \cdots \qquad = \sum_{n=1}^{\infty} \frac{(-1)^{n-1}}{n}(x-1)^n,$$

where (10), (11), and (12) are valid for all x, whereas (13) is valid for x in the half-open interval $(0, 2]$. In (13) the expansion is about $x_0 = 1$. However, a power series representation for $\ln x$ can be derived about any $x_0 > 0$.

From Theorem 3 on the differentiation of power series, we see that a function f analytic at x_0 is differentiable in a neighborhood of x_0. Moreover, because f' has a power series representation in this neighborhood, it too is analytic at x_0. Repeating this argument, we see that f'', $f^{(3)}$, etc., exist and are analytic at x_0. Consequently, if a function does not have derivatives of all orders at x_0, then it cannot be analytic at x_0. The function $f(x) = |x - 1|$ is not analytic at $x_0 = 1$ because $f'(1)$ does not exist and $f(x) = x^{8/3}$ is not analytic at $x_0 = 0$ because $f'''(0)$ does not exist.

A formula for the coefficients in the power series of an analytic function is given in the next theorem.

TAYLOR AND MACLAURIN SERIES

Theorem 4.   If f is analytic at x_0, then the representation

$$f(x) = \sum_{n=0}^{\infty} \frac{f^{(n)}(x_0)}{n!}(x - x_0)^n$$

$$= f(x_0) + f'(x_0)(x - x_0) + \frac{f''(x_0)}{2!}(x - x_0)^2 + \cdots,$$

holds in some open interval centered at x_0.

This series is called the **Taylor series** for f about x_0. When $x_0 = 0$, it is also referred to as the **Maclaurin series** for f.

To determine the Taylor series for an analytic function f, a direct but sometimes tedious approach is to compute the successive derivatives of f and evaluate them at x_0. For example, the series in (10), (11), (12), and (13) can be derived in this manner.

Power series expansions also have a uniqueness property; namely, if the equation

$$\sum_{n=0}^{\infty} a_n(x - x_0)^n = \sum_{n=0}^{\infty} b_n(x - x_0)^n$$

holds in some *open* interval about x_0, then $a_n = b_n$ for $n = 0, 1, 2, \ldots$. Hence, if "by hook or by crook" we can produce a power series expansion for an analytic function, then this power series must be its Taylor series. For example, the expansion for arctan x given in (9) of Example 2 must be its Maclaurin expansion.

The reader will also find it helpful to keep in mind that if f and g are analytic at x_0, then so are $f + g$, cf, fg, and f/g, provided $g(x_0) \neq 0$. These facts follow from the algebraic properties of power series discussed earlier.

Taylor Series Method

We can use the Taylor series representation to obtain a series solution to an initial value problem. Assuming that we know the problem has an analytic solution, say by some existence theorem, then we can use the initial conditions and the differential equation itself to compute the successive derivatives of the solution at the initial point. Substituting these values into the formula for the Taylor series, we then obtain a representation of the solution as a power series about the initial point x_0. This procedure, called the **Taylor series method**, is illustrated in the next example.

● **EXAMPLE 4**   Determine the first few terms of a power series solution for the initial value problem

$$(14) \qquad y' = \frac{1}{x + y + 1}, \qquad y(0) = 0.$$

Solution   Let's assume that the solution to this initial value problem is analytic at $x = 0$. Then we can write

(15) $$y(x) = \sum_{n=0}^{\infty} \frac{y^{(n)}(0)}{n!} x^n.$$

We already know that $y(0) = 0$, and substituting $x = 0$ and $y = 0$ into equation (14) we find that $y'(0) = 1$. To determine $y''(0)$ we differentiate both sides of the equation in (14) with respect to x, getting an expression for $y''(x)$ in terms of x, $y(x)$, and $y'(x)$. That is,

(16) $$y''(x) = (-1)[x + y(x) + 1]^{-2}[1 + y'(x)].$$

Substituting $x = 0$, $y(0) = 0$, and $y'(0) = 1$ in (16), we obtain

$$y''(0) = (-1)(1)^{-2}(1 + 1) = -2.$$

Similarly, differentiating (16) and substituting, we obtain

$$y^{(3)}(x) = 2[x + y(x) + 1]^{-3}[1 + y'(x)]^2 - (x + y(x) + 1)^{-2}y''(x),$$
$$y^{(3)}(0) = 2(1)^{-3}(1 + 1)^2 - (1)^{-2}(-2) = 10.$$

Repeating this procedure, we can determine $y^{(n)}(0)$ for any value of n, although the amount of work needed to compute $y^{(n)}(0)$ increases as n increases. Finally, substituting into (15), we find that the first few terms of the Maclaurin series of the solution are

$$y(x) = x - x^2 + \tfrac{5}{3}x^3 + \cdots. \quad \bullet$$

The Taylor series method applies to nonlinear as well as linear equations. This is in contrast to the methods in the remainder of this chapter, which are restricted to linear equations.

One disadvantage of the Taylor series method is that by computing finitely many terms of the Taylor expansion, there is no way of knowing the radius of convergence of the series. Fortunately, when the differential equation is linear, there are existence theorems that give a minimum value for this radius (see Theorem 5 in Section 8.4).

EXERCISES 8.2

1, 3, 5, 7, 9, 15, 17, 19, 27

In Problems 1 through 6 determine the convergence set of the given power series.

1. $\displaystyle\sum_{n=0}^{\infty} \frac{2^{-n}}{n+1}(x-1)^n.$

2. $\displaystyle\sum_{n=0}^{\infty} \frac{3^n}{n!}x^n.$

3. $\displaystyle\sum_{n=0}^{\infty} \frac{n^2}{2^n}(x+2)^n.$

4. $\displaystyle\sum_{n=1}^{\infty} \frac{4}{n^2+2n}(x-3)^n.$

5. $\displaystyle\sum_{n=1}^{\infty} \frac{3}{n^3}(x-2)^n.$

6. $\displaystyle\sum_{n=0}^{\infty} \frac{(n+2)!}{n!}(x+2)^n.$

In Problems 7 and 8 find the power series expansion for

$f(x) + g(x)$, *given the expansions for* $f(x)$ *and* $g(x)$.

7. $\displaystyle f(x) = \sum_{n=0}^{\infty} \frac{1}{n+1}x^n, \qquad g(x) = \sum_{n=1}^{\infty} 2^{-n}x^{n-1}.$

8. $\displaystyle f(x) = \sum_{n=3}^{\infty} \frac{2^n}{n!}(x-1)^{n-3}, \qquad g(x) = \sum_{n=1}^{\infty} \frac{n^2}{2^n}(x-1)^{n-1}.$

In Problems 9 through 12 find the first three nonzero terms in the power series expansion for the product $f(x)g(x)$.

9. $\displaystyle f(x) = e^x = \sum_{n=0}^{\infty} \frac{1}{n!}x^n,$

$\displaystyle g(x) = \sin x = \sum_{k=0}^{\infty} \frac{(-1)^k}{(2k+1)!}x^{2k+1}.$

10. $f(x) = \sin x = \sum_{k=0}^{\infty} \frac{(-1)^k}{(2k+1)!} x^{2k+1}$,

$g(x) = \cos x = \sum_{k=0}^{\infty} \frac{(-1)^k}{(2k)!} x^{2k}$.

11. $f(x) = e^{-x} = \sum_{n=0}^{\infty} \frac{(-1)^n}{n!} x^n$,

$g(x) = (1+x)^{-1} = \sum_{n=0}^{\infty} (-1)^n x^n$.

12. $f(x) = e^x = \sum_{n=0}^{\infty} \frac{1}{n!} x^n$, $g(x) = e^{-x} = \sum_{n=0}^{\infty} \frac{(-1)^n}{n!} x^n$.

13. Find the first few terms of the power series for the quotient

$$q(x) = \left(\sum_{n=0}^{\infty} \frac{1}{2^n} x^n \right) \Big/ \left(\sum_{n=0}^{\infty} \frac{1}{n!} x^n \right)$$

by completing the following:

(a) Let $q(x) = \sum_{n=0}^{\infty} a_n x^n$, where the coefficients a_n are to be determined. Show that $\sum_{n=0}^{\infty} x^n/2^n$ is the Cauchy product of $q(x)$ and $\sum_{n=0}^{\infty} x^n/n!$.

(b) Use formula (6) of the Cauchy product to deduce the equations

$$\frac{1}{2^0} = a_0, \qquad \frac{1}{2} = a_0 + a_1, \qquad \frac{1}{2^2} = \frac{a_0}{2} + a_1 + a_2,$$

$$\frac{1}{2^3} = \frac{a_0}{6} + \frac{a_1}{2} + a_2 + a_3, \dots.$$

(c) Solve the equations in part (b) to determine the constants a_0, a_1, a_2, a_3.

14. To find the first few terms in the power series for the quotient $q(x)$ in Problem 13 treat the power series in the the numerator and denominator as "long polynomials" and carry out long division. That is, perform

$$1 + x + \tfrac{1}{2}x^2 + \cdots \overline{) 1 + \tfrac{1}{2}x + \tfrac{1}{4}x^2 + \cdots}.$$

In Problems 15 through 18 find a power series expansion for $f'(x)$, given the expansion for $f(x)$.

15. $f(x) = (1+x)^{-1} = \sum_{n=0}^{\infty} (-1)^n x^n$.

16. $f(x) = \sin x = \sum_{k=0}^{\infty} \frac{(-1)^k}{(2k+1)!} x^{2k+1}$.

17. $f(x) = \sum_{n=0}^{\infty} a_n x^n$.

18. $f(x) = \sum_{n=1}^{\infty} n a_n x^{n-1}$.

In Problems 19 and 20 find a power series expansion for $g(x) := \int_0^x f(t)\,dt$ given the expansion for $f(x)$.

19. $f(x) = (1+x)^{-1} = \sum_{n=0}^{\infty} (-1)^n x^n$.

20. $f(x) = \frac{\sin x}{x} = \sum_{k=0}^{\infty} \frac{(-1)^k}{(2k+1)!} x^{2k}$.

In Problems 21 through 26 express the given power series using the new index k, where the relationship between k and n is given.

21. $\sum_{n=1}^{\infty} n a_n x^{n-1}$, $k = n - 1$.

22. $\sum_{n=2}^{\infty} n(n-1) a_n x^{n+2}$, $k = n + 2$.

23. $\sum_{n=0}^{\infty} a_n x^{n+1}$, $k = n + 1$.

24. $\sum_{n=1}^{\infty} a_n x^n$, $k = n - 1$.

25. $\sum_{n=2}^{\infty} n(n-1) a_n x^{n+1}$, $k = n + 1$.

26. $\sum_{n=1}^{\infty} n a_n x^{n+1}$, $k = n + 1$.

In Problems 27 through 32 determine the Taylor series about the point x_0 for the given functions and values of x_0.

27. $f(x) = \cos x$, $x_0 = \pi$.

28. $f(x) = x^{-1}$, $x_0 = 1$.

29. $f(x) = \dfrac{1+x}{1-x}$, $x_0 = 0$.

30. $f(x) = \ln(1+x)$, $x_0 = 0$.

31. $f(x) = x^3 + 3x - 4$, $x_0 = 1$.

32. $f(x) = \sqrt{x}$, $x_0 = 1$.

33. The Taylor series for $f(x) = \ln x$ about $x_0 = 1$ given in equation (13) can also be obtained as follows.

(a) Starting with the expansion $1/(1-s) = \sum_{n=0}^{\infty} s^n$ and observing that

$$\frac{1}{x} = \frac{1}{1 + (x-1)},$$

obtain the Taylor series for $1/x$ about $x_0 = 1$.

(b) Since $\ln x = \int_1^x 1/t\,dt$, use the result of part (a) and termwise integration to obtain the Taylor series for $f(x) = \ln x$ about $x_0 = 1$.

34. Let $f(x)$ and $g(x)$ be analytic at x_0. Determine whether the following statements are always true or sometimes false.

(a) $3f(x) + g(x)$ is analytic at x_0.
(b) $f(x)/g(x)$ is analytic at x_0.
(c) $f'(x)$ is analytic at x_0.
(d) $[f(x)]^3 - \int_{x_0}^{x} g(t)\,dt$ is analytic at x_0.
(e) $\sin[f(x)]$ is analytic at x_0.

In Problems 35 through 42 use the Taylor series method to determine the first three nonzero terms of a series solution for the given initial value problem.

35. $y' = x^2 + y^2;$ $y(0) = 1.$

36. $y' = y^2;$ $y(0) = 1.$

37. $y' = \sin y + e^x;$ $y(0) = 0.$

38. $y' = \sin(x + y);$ $y(0) = 0.$

39. $x'' + tx = 0;$ $x(0) = 1,$ $x'(0) = 0.$

40. $y'' + y = 0;$ $y(0) = 0,$ $y'(0) = 1.$

41. $y''(\theta) + y(\theta)^3 = \sin\theta;$ $y(0) = 0,$ $y'(0) = 0.$

42. $y'' + \sin y = 0;$ $y(0) = 1,$ $y'(0) = 0.$

43. Let

$$f(x) = \begin{cases} e^{-1/x^2}, & x \neq 0, \\ 0, & x = 0. \end{cases}$$

Show that $f^{(n)}(0) = 0$ for $n = 0, 1, 2, \ldots$, and hence that the Maclaurin series for $f(x)$ is $0 + 0 + 0 + \cdots$, which converges for all x, but is equal to $f(x)$ only when $x = 0$.

44. Van der Pol Equation.   In the study of the vacuum tube, the following equation is encountered:

$$y'' + (0.1)(y^2 - 1)y' + y = 0.$$

Use the Taylor series method to find the first three nonzero terms of the series solution for the initial values $y(0) = 1$, $y'(0) = 0$.

45. Duffing's Equation.   In the study of a nonlinear spring with periodic forcing the following equation arises:

$$y'' + ky + ry^3 = A\cos\omega t.$$

Let $k = r = A = 1$ and $\omega = 10$. Use the Taylor series method to find the first three nonzero terms of the series solution for the initial values $y(0) = 0$, $y'(0) = 1$.

46. Soft *vs* Hard Springs.   For Duffing's equation given in Problem 45, the behavior of the solutions changes as r changes sign. When $r > 0$, the restoring force $ky + ry^3$ becomes stronger than for the linear spring ($r = 0$). Such a spring is called **hard.** When $r < 0$, the restoring force becomes weaker than the linear spring and the spring is called **soft.** Pendulums act like soft springs.

(a) Redo Problem 45 with $r = -1$. Notice that for the initial conditions $y(0) = 0$, $y'(0) = 1$ the soft and hard springs appear to respond in the same way for t small.
(b) Keeping $k = A = 1$ and $\omega = 10$, change the initial conditions to $y(0) = 1$ and $y'(0) = 0$. Now redo Problem 45 with $r = \pm 1$.
(c) Based upon the results of part (b), is there a difference between the behavior of soft and hard springs for t small? Describe.

•8.3• POWER SERIES SOLUTIONS TO LINEAR DIFFERENTIAL EQUATIONS

In this section we demonstrate a method for obtaining a power series solution to a linear differential equation with polynomial coefficients. This method is easier to use than the Taylor series method discussed in Section 8.2 and sometimes gives a nice expression for the general term in the power series expansion. Knowing the form of the general term also allows us to test for the radius of convergence of the power series.

We begin by writing the linear differential equation

(1) $a_2(x)y'' + a_1(x)y' + a_0(x)y = 0$

in the standard form

(2) $y'' + p(x)y' + q(x)y = 0,$

where $p(x) := a_1(x)/a_2(x)$ and $q(x) := a_0(x)/a_2(x)$.

> ### ▶ ORDINARY AND SINGULAR POINTS
>
> **Definition 2.**   A point x_0 is called an **ordinary point** of equation (1) if both $p = a_1/a_2$ and $q = a_0/a_2$ are analytic at x_0. If x_0 is not an ordinary point, it is called a **singular point** of the equation.

● **EXAMPLE 1**   Determine all the singular points of

$$xy'' + x(1 - x)^{-1}y' + (\sin x)y = 0.$$

Solution   Dividing the equation by x, we find that

$$p(x) = \frac{x}{x(1 - x)}, \qquad q(x) = \frac{\sin x}{x}.$$

The singular points are those points where $p(x)$ *or* $q(x)$ fails to be analytic. Observe that $p(x)$ and $q(x)$ are the ratios of functions that are everywhere analytic. Hence $p(x)$ and $q(x)$ are analytic except, *perhaps*, when their denominators are zero. For $p(x)$ this occurs at $x = 0$ and $x = 1$. But since we can cancel an x in the numerator and denominator of $p(x)$, that is

$$p(x) = \frac{x}{x(1 - x)} = \frac{1}{1 - x},$$

we see that $p(x)$ is actually analytic at $x = 0$.[†] Therefore $p(x)$ is analytic except at $x = 1$. For $q(x)$, the denominator is zero at $x = 0$. Just as with $p(x)$, this zero is removable, since $q(x)$ has the power series expansion

$$q(x) = \frac{\sin x}{x} = \frac{x - \dfrac{x^3}{3!} + \dfrac{x^5}{5!} - \cdots}{x} = 1 - \frac{x^2}{3!} + \frac{x^4}{5!} - \cdots.$$

Thus $q(x)$ is everywhere analytic. Consequently the only singular point of the given equation is $x = 1$.   ●

At an ordinary point x_0 of equation (1) (or (2)), the coefficient functions $p(x)$ and $q(x)$ are analytic, and hence we might expect that the solutions to these equations inherit this property. From the discussion in Section 4.3 on linear second order equations, the continuity of p and q in a neighborhood of x_0 is sufficient to imply that equation (2) has two linearly independent solutions defined in that neighborhood. But analytic functions are not merely continuous—they possess derivatives of all orders in a neighborhood of x_0. Thus we can differentiate equation (2) to show that $y^{(3)}$ exists and, by a "bootstrap" argument, prove that solutions to (2) must likewise possess derivatives of all orders. Although we cannot conclude by this reasoning that the solutions enjoy the stronger property of

---

[†] Such points are called **removable singularities.** In this chapter we assume in such cases that the function has been defined (or redefined) so that it is analytic at the point.

analyticity, this is nonetheless the case (see Theorem 5 in Section 8.4). Hence, in a neighborhood of an ordinary point x_0, the solutions to (1) (or (2)) can be expressed as a power series about x_0.

To illustrate the power series method about an ordinary point, let's look at a simple *first order* linear differential equation.

● EXAMPLE 2   Find a power series solution about $x = 0$ to

$$(3) \qquad y' + 2xy = 0.$$

Solution   The coefficient of y is the polynomial $2x$ which is analytic everywhere, and so $x = 0$ is an ordinary point[†] of equation (3). Thus we expect to find a power series solution of the form

$$(4) \qquad y(x) = a_0 + a_1 x + a_2 x^2 + \cdots = \sum_{n=0}^{\infty} a_n x^n.$$

Our task is to determine the coefficients a_n.

For this purpose we need the expansion for $y'(x)$ that is given by termwise differentiation of (4):

$$y'(x) = 0 + a_1 + 2a_2 x + 3a_3 x^2 + \cdots = \sum_{n=1}^{\infty} n a_n x^{n-1}.$$

We now substitute the series expansions for y and y' into (3) and obtain

$$\sum_{n=1}^{\infty} n a_n x^{n-1} + 2x \sum_{n=0}^{\infty} a_n x^n = 0,$$

which simplifies to

$$(5) \qquad \sum_{n=1}^{\infty} n a_n x^{n-1} + \sum_{n=0}^{\infty} 2a_n x^{n+1} = 0.$$

To add the two power series in (5), we add the coefficients of like powers of x. If we write out the first few terms of these summations and add, we get

$$(a_1 + 2a_2 x + 3a_3 x^2 + 4a_4 x^3 + \cdots) + (2a_0 x + 2a_1 x^2 + 2a_2 x^3 + \cdots) = 0,$$
$$(6) \qquad a_1 + (2a_2 + 2a_0)x + (3a_3 + 2a_1)x^2 + (4a_4 + 2a_2)x^3 + \cdots = 0.$$

In order for the power series on the left-hand side of equation (6) to be identically zero, we must have all the coefficients equal to zero. Thus

$$a_1 = 0, \qquad 2a_2 + 2a_0 = 0,$$
$$3a_3 + 2a_1 = 0, \qquad 4a_4 + 2a_2 = 0, \qquad \text{etc.}$$

Solving the preceding system, we find

$$a_1 = 0, \qquad a_2 = -a_0, \qquad a_3 = -\tfrac{2}{3}a_1 = 0,$$
$$a_4 = -\tfrac{1}{2}a_2 = -\tfrac{1}{2}(-a_0) = \tfrac{1}{2}a_0.$$

[†] By an ordinary point of a first order equation $y' + q(x)y = 0$ we mean a point where $q(x)$ is analytic.

Hence the power series for the solution takes the form

(7) $y(x) = a_0 - a_0 x^2 + \frac{1}{2}a_0 x^4 + \cdots$.

While the first few terms displayed in (7) are useful, we would much prefer to have a formula for the *general term* in the power series expansion for the solution. To achieve this goal, let's return to equation (5). This time, instead of just writing out a few terms, let's shift the indices in the two power series so that they sum over the same powers of x, say x^k. To do this, we shift the index in the first summation in (5) by setting $k = n - 1$. Then $n = k + 1$ and $k = 0$ when $n = 1$. Hence the first summation in (5) becomes

(8) $\displaystyle\sum_{n=1}^{\infty} n a_n x^{n-1} = \sum_{k=0}^{\infty} (k + 1)a_{k+1} x^k$.

In the second summation of (5) we put $k = n + 1$ so that $n = k - 1$ and $k = 1$ when $n = 0$. This gives

(9) $\displaystyle\sum_{n=0}^{\infty} 2a_n x^{n+1} = \sum_{k=1}^{\infty} 2a_{k-1} x^k$.

Substituting (8) and (9) into (5) yields

(10) $\displaystyle\sum_{k=0}^{\infty} (k + 1)a_{k+1} x^k + \sum_{k=1}^{\infty} 2a_{k-1} x^k = 0$.

Since the first summation in (10) begins at $k = 0$ and the second at $k = 1$, we break up the first into

$$\sum_{k=0}^{\infty} (k + 1)a_{k+1} x^k = a_1 + \sum_{k=1}^{\infty} (k + 1)a_{k+1} x^k.$$

Then (10) becomes

(11) $a_1 + \displaystyle\sum_{k=1}^{\infty} [(k + 1)a_{k+1} + 2a_{k-1}]x^k = 0$.

When we set all the coefficients in (11) equal to zero, we find

$$a_1 = 0,$$

and, for all $k \geq 1$,

(12) $(k + 1)a_{k+1} + 2a_{k-1} = 0$.

Equation (12) is a **recurrence relation** that we can use to determine the coefficient a_{k+1} in terms of a_{k-1}, that is,

$$a_{k+1} = -\frac{2}{k + 1}a_{k-1}.$$

Setting $k = 1, 2, \ldots, 8$, and using the fact that $a_1 = 0$, we find

$$a_2 = -\frac{2}{2}a_0 = -a_0 \quad (k = 1), \qquad a_3 = -\frac{2}{3}a_1 = 0 \quad (k = 2),$$

$$a_4 = -\frac{2}{4}a_2 = \frac{1}{2}a_0 \quad (k = 3), \qquad a_5 = -\frac{2}{5}a_3 = 0 \quad (k = 4),$$

$$a_6 = -\frac{2}{6}a_4 = -\frac{1}{3!}a_0 \quad (k = 5), \qquad a_7 = -\frac{2}{7}a_5 = 0 \quad (k = 6),$$

$$a_8 = -\frac{2}{8}a_6 = \frac{1}{4!}a_0 \quad (k = 7), \qquad a_9 = -\frac{2}{9}a_7 = 0 \quad (k = 8).$$

After a moment's reflection, we realize that

$$a_{2n} = \frac{(-1)^n}{n!}a_0, \qquad n = 1, 2, \ldots,$$

$$a_{2n+1} = 0, \qquad n = 0, 1, 2, \ldots.$$

Substituting back into the expression (4), we obtain the power series solution

(13) $\quad y(x) = a_0 - a_0 x^2 + \frac{1}{2!}a_0 x^4 + \cdots = a_0 \sum_{n=0}^{\infty} \frac{(-1)^n}{n!} x^{2n}.$

Since a_0 is left undetermined, it serves as an arbitrary constant, and hence (13) gives a general solution to equation (3).   ●

Using the ratio test, it can be verified that the power series in (13) has radius of convergence $\rho = \infty$. Moreover, (13) is reminiscent of the expansion for the exponential function; the reader should check that it converges to

$$y(x) = a_0 e^{-x^2}.$$

This general solution to the simple equation (3) can also be obtained by the method of separation of variables.

In the next example we use the power series method to obtain a general solution to a linear second order differential equation.

● **EXAMPLE 3**   Find a general solution to

(14) $\quad 2y'' + xy' + y = 0$

in the form of a power series about the ordinary point $x = 0$.

Solution   Writing

(15) $\quad y(x) = a_0 + a_1 x + a_2 x^2 + \cdots = \sum_{n=0}^{\infty} a_n x^n,$

we differentiate termwise to obtain

$$y'(x) = a_1 + 2a_2x + 3a_3x^2 + \cdots = \sum_{n=1}^{\infty} na_nx^{n-1},$$

$$y''(x) = 2a_2 + 6a_3x + 12a_4x^2 + \cdots = \sum_{n=2}^{\infty} n(n-1)a_nx^{n-2}.$$

Substituting these power series into equation (14), we find

(16) $$\sum_{n=2}^{\infty} 2n(n-1)a_nx^{n-2} + \sum_{n=1}^{\infty} na_nx^n + \sum_{n=0}^{\infty} a_nx^n = 0.$$

To simplify the addition of the three summations in (16), let's shift the indices so that the general term in each is a constant times x^k. For the first summation we substitute $k = n - 2$ and get

$$\sum_{n=2}^{\infty} 2n(n-1)a_nx^{n-2} = \sum_{k=0}^{\infty} 2(k+2)(k+1)a_{k+2}x^k.$$

In the second and third summations we can take $k = n$. With these changes of indices equation (16) becomes

$$\sum_{k=0}^{\infty} 2(k+2)(k+1)a_{k+2}x^k + \sum_{k=1}^{\infty} ka_kx^k + \sum_{k=0}^{\infty} a_kx^k = 0.$$

Next, we separate the x^0 terms from the others and then combine the like powers of x in the three summations to get

$$4a_2 + a_0 + \sum_{k=1}^{\infty} [2(k+2)(k+1)a_{k+2} + ka_k + a_k]x^k = 0.$$

Setting the coefficients of this power series equal to zero yields

(17) $$4a_2 + a_0 = 0$$

and the recurrence relation

(18) $$2(k+2)(k+1)a_{k+2} + (k+1)a_k = 0, \qquad k \geq 1.$$

We can now use (17) and (18) to determine all the coefficients a_k of the solution in terms of a_0 and a_1. Solving (18) for a_{k+2} gives

(19) $$a_{k+2} = \frac{-1}{2(k+2)}a_k, \qquad k \geq 1.$$

Thus

$$a_2 = \frac{-1}{2^2} a_0,$$

$$a_3 = \frac{-1}{2 \cdot 3} a_1 \qquad\qquad (k = 1),$$

$$a_4 = \frac{-1}{2 \cdot 4} a_2 = \frac{1}{2^2 \cdot 2 \cdot 4} a_0 \qquad (k = 2),$$

$$a_5 = \frac{-1}{2 \cdot 5} a_3 = \frac{1}{2^2 \cdot 3 \cdot 5} a_1 \qquad (k = 3),$$

$$a_6 = \frac{-1}{2 \cdot 6} a_4 = \frac{-1}{2^3 \cdot 2 \cdot 4 \cdot 6} a_0 = \frac{-1}{2^6 \cdot 3!} a_0 \qquad (k = 4),$$

$$a_7 = \frac{-1}{2 \cdot 7} a_5 = \frac{-1}{2^3 \cdot 3 \cdot 5 \cdot 7} a_1 \qquad\qquad (k = 5),$$

$$a_8 = \frac{-1}{2 \cdot 8} a_6 = \frac{1}{2^4 \cdot 2 \cdot 4 \cdot 6 \cdot 8} a_0 = \frac{1}{2^8 \cdot 4!} a_0 \qquad (k = 6).$$

The pattern for the coefficients is now apparent. With a_0 and a_1 taken as arbitrary constants, we find

$$a_{2n} = \frac{(-1)^n}{2^{2n} n!} a_0, \qquad n \geq 1,$$

and

$$a_{2n+1} = \frac{(-1)^n}{2^n [1 \cdot 3 \cdot 5 \cdots (2n + 1)]} a_1, \qquad n \geq 1.$$

From this, two linearly independent solutions emerge; namely

(20) $$y_1(x) = \sum_{n=0}^{\infty} \frac{(-1)^n}{2^{2n} n!} x^{2n},$$

(21) $$y_2(x) = \sum_{n=0}^{\infty} \frac{(-1)^n}{2^n [1 \cdot 3 \cdot 5 \cdots (2n + 1)]} x^{2n+1}.$$

Hence a general solution to (14) is $a_0 y_1(x) + a_1 y_2(x)$. ●

The method illustrated in Example 3 can also be used to solve initial value problems. Suppose that we are given the values of $y(0)$ and $y'(0)$; then, from equation (15) we see that $a_0 = y(0)$ and $a_1 = y'(0)$. Knowing these two coefficients leads to a unique power series solution for the initial value problem.

The recurrence relation (18) in Example 3 involved just two of the coefficients, a_{k+2}

and a_k, and we were fortunate in being able to deduce from this relation the general form for the coefficient a_n. However, many cases arise that lead to more complicated two-term or even to many-term recurrence relations. When this occurs, it may be impossible to determine the general form for the coefficients a_n. In the next example we consider an equation that gives rise to a three-term recurrence relation.

● **EXAMPLE 4**   Find the first few terms in a power series expansion about $x = 0$ for a general solution to

(22) $(1 + x^2)y'' - y' + y = 0.$

Solution   Since $p(x) = -(1 + x^2)^{-1}$ and $q(x) = (1 + x^2)^{-1}$ are analytic at $x = 0$, then $x = 0$ is an ordinary point for equation (22). Hence we can express its general solution in the form

$$y(x) = \sum_{n=0}^{\infty} a_n x^n.$$

Substituting this expansion into (22) yields

$$(1 + x^2) \sum_{n=2}^{\infty} n(n-1)a_n x^{n-2} - \sum_{n=1}^{\infty} na_n x^{n-1} + \sum_{n=0}^{\infty} a_n x^n = 0,$$

(23) $$\sum_{n=2}^{\infty} n(n-1)a_n x^{n-2} + \sum_{n=2}^{\infty} n(n-1)a_n x^n - \sum_{n=1}^{\infty} na_n x^{n-1} + \sum_{n=0}^{\infty} a_n x^n = 0.$$

To sum over like powers x^k, we put $k = n - 2$ in the first summation, $k = n - 1$ in the third, and $k = n$ in the second and fourth summations of (23). This gives

$$\sum_{k=0}^{\infty} (k+2)(k+1)a_{k+2} x^k + \sum_{k=2}^{\infty} k(k-1)a_k x^k - \sum_{k=0}^{\infty} (k+1)a_{k+1} x^k + \sum_{k=0}^{\infty} a_k x^k = 0.$$

Separating the terms corresponding to $k = 0$ and $k = 1$ and combining the rest under one summation, we have

$$(2a_2 - a_1 + a_0) + (6a_3 - 2a_2 + a_1)x$$

$$+ \sum_{k=2}^{\infty} [(k+2)(k+1)a_{k+2} - (k+1)a_{k+1} + (k(k-1)+1)a_k]x^k = 0.$$

Setting the coefficients equal to zero gives

(24) $2a_2 - a_1 + a_0 = 0,$

(25) $6a_3 - 2a_2 + a_1 = 0,$

and the recurrence relation

(26) $(k+2)(k+1)a_{k+2} - (k+1)a_{k+1} + (k^2 - k + 1)a_k = 0,$ $k \geq 2.$

We can solve (24) for a_2 in terms of a_0 and a_1:

$$a_2 = \frac{a_1 - a_0}{2}.$$

Now that we have a_2, we can use (25) to express a_3 in terms of a_0 and a_1:

$$a_3 = \frac{2a_2 - a_1}{6} = \frac{(a_1 - a_0) - a_1}{6} = \frac{-a_0}{6}.$$

Solving the recurrence relation (26) for a_{k+2}, we obtain

$$(27) \qquad a_{k+2} = \frac{(k+1)a_{k+1} - (k^2 - k + 1)a_k}{(k+2)(k+1)}, \qquad k \geq 2.$$

For $k = 2, 3,$ and 4 this gives

$$a_4 = \frac{3a_3 - 3a_2}{4 \cdot 3} = \frac{a_3 - a_2}{4}$$

$$= \frac{\dfrac{-a_0}{6} - \left(\dfrac{a_1 - a_0}{2}\right)}{4} = \frac{2a_0 - 3a_1}{24} \qquad (k = 2),$$

$$a_5 = \frac{4a_4 - 7a_3}{5 \cdot 4} = \frac{3a_0 - a_1}{40} \qquad (k = 3),$$

$$a_6 = \frac{5a_5 - 13a_4}{6 \cdot 5} = \frac{36a_1 - 17a_0}{720} \qquad (k = 4).$$

We can now express a general solution in terms up to order 6, using a_0 and a_1 as the arbitrary constants. Thus

$$(28) \qquad y(x) = a_0 + a_1 x + \left(\frac{a_1 - a_0}{2}\right)x^2 - \frac{a_0}{6}x^3$$

$$+ \left(\frac{2a_0 - 3a_1}{24}\right)x^4 + \left(\frac{3a_0 - a_1}{40}\right)x^5 + \left(\frac{36a_1 - 17a_0}{720}\right)x^6 + \cdots$$

$$= a_0\left(1 - \tfrac{1}{2}x^2 - \tfrac{1}{6}x^3 + \tfrac{1}{12}x^4 + \tfrac{3}{40}x^5 - \tfrac{17}{720}x^6 + \cdots\right)$$

$$+ a_1\left(x + \tfrac{1}{2}x^2 - \tfrac{1}{8}x^4 - \tfrac{1}{40}x^5 + \tfrac{1}{20}x^6 + \cdots\right). \quad \bullet$$

Given specific values for a_0 and a_1, will the partial sums of the power series representation (28) yield useful approximations to the solution when $x = 0.5$? What about when $x = 2.3$ or $x = 7.8$? The answers to these questions certainly depend on the radius of convergence of the power series in (28). But since we were not able to determine a general form for the coefficients a_n in this example, we cannot use the ratio test (or other methods such as the root test, integral test, or comparison test) to compute the radius ρ. In the next section we remedy this situation by giving a simple procedure that determines a lower bound for the radius of convergence of power series solutions.

EXERCISES 8.3

In Problems 1 through 10 determine all the singular points of the given differential equation.

1. $(x + 1)y'' - x^2y' + 3y = 0$.

2. $x^2y'' + 3y' - xy = 0$.

3. $(\theta^2 - 2)y'' + 2y' + (\sin \theta)y = 0$.

4. $(x^2 + x)y'' + 3y' - 6xy = 0$.

5. $(t^2 - t - 2)x'' + (t + 1)x' - (t - 2)x = 0$.

6. $(x^2 - 1)y'' + (1 - x)y' + (x^2 - 2x + 1)y = 0$.

7. $(\sin x)y'' + (\cos x)y = 0$.

8. $e^x y'' - (x^2 - 1)y' + 2xy = 0$.

9. $(\sin \theta)y'' - (\ln \theta)y = 0$.

10. $[\ln(x - 1)]y'' + (\sin 2x)y' - e^x y = 0$.

In Problems 11 through 18 find at least the first four nonzero terms in a power series expansion about $x = 0$ for a general solution to the given differential equation.

11. $y' + (x + 2)y = 0$.   **12.** $y' - y = 0$.

13. $z'' - x^2z = 0$.   **14.** $(x^2 + 1)y'' + y = 0$.

15. $y'' + (x - 1)y' + y = 0$.

16. $y'' - 2y' + y = 0$.

17. $w'' - x^2w' + w = 0$.

18. $(2x - 3)y'' - xy' + y = 0$.

In Problems 19 through 24 find a power series expansion about $x = 0$ for a general solution to the given differential equation. Your answer should include a general formula for the coefficients.

19. $y' - 2xy = 0$.   **20.** $y'' + y = 0$.

21. $y'' - xy' + 4y = 0$.   **22.** $y'' - xy = 0$.

23. $z'' - x^2z' - xz = 0$.

24. $(x^2 + 1)y'' - xy' + y = 0$.

In Problems 25 through 28 find at least the first four non-zero terms in a power series expansion about $x = 0$ for the solution to the given initial value problem.

25. $w'' + 3xw' - w = 0$; $w(0) = 2$, $w'(0) = 0$.

26. $(x^2 - x + 1)y'' - y' - y = 0$; $y(0) = 0$, $y'(0) = 1$.

27. $(x + 1)y'' - y = 0$; $y(0) = 0$, $y'(0) = 1$.

28. $y'' + (x - 2)y' - y = 0$; $y(0) = -1$, $y'(0) = 0$.

In Problems 29 through 31 use the first few terms of the power series expansion to find a cubic polynomial approximation for the solution to the given initial value problem.

29. $y'' + y' - xy = 0$; $y(0) = 1$, $y'(0) = -2$.

30. $y'' - 4xy' + 5y = 0$; $y(0) = -1$, $y'(0) = 1$.

31. $(x^2 + 2)y'' + 2xy' + 3y = 0$; $y(0) = 1$, $y'(0) = 2$.

32. Consider the initial value problem

$$y'' - 2xy' - 2y = 0; \quad y(0) = a_0, \quad y'(0) = a_1,$$

where a_0 and a_1 are constants.

     (a) Show that if $a_0 = 0$, then the solution will be an odd function (that is, $y(-x) = -y(x)$ for all x). What happens when $a_1 = 0$?

     (b) Show that if a_0 and a_1 are positive, then the solution is increasing on $(0, \infty)$.

     (c) Show that if a_0 is negative and a_1 is positive, then the solution is increasing on $(-\infty, 0)$.

     (d) What conditions on a_0 and a_1 would guarantee that the solution is increasing on $(-\infty, \infty)$?

33. Use the ratio test to show that the radius of convergence of the series in equation (13) is infinite. [Hint: First consider the series $\sum_{n=0}^{\infty}(-1)^n x^n/n!$.]

34. **Emden's Equation.** A classical nonlinear equation that occurs in the study of the thermal behavior of a spherical cloud is **Emden's equation**

$$y'' + \frac{2}{x}y' + y^n = 0,$$

with initial conditions $y(0) = 1$, $y'(0) = 0$. (See Section 8.1 for the case when $n = 1$.) Even though $x = 0$ is *not* an ordinary point for this equation (which is nonlinear for $n \neq 1$), it turns out that there does exist a solution analytic at $x = 0$. Assuming n is a positive integer, show that the first few terms in a power series solution are

$$y = 1 - \frac{x^2}{3!} + n\frac{x^4}{5!} + \cdots .$$

[Hint: Substitute $y = 1 + c_2x^2 + c_3x^3 + c_4x^4 + c_5x^5 + \cdots$ into the equation and carefully compute the first few terms in y^n.]

35. Variable Resistor. In Section 5.4 we showed that the charge q on the capacitor in a simple RLC circuit is governed by the equation

$$Lq''(t) + Rq'(t) + \frac{1}{C}q(t) = E(t),$$

where L is the inductance, R the resistance, C the capacitance, and E the electromotive force. Since the resistance of a resistor increases with temperature, let's assume that the resistor is heated so that the resistance at time t is $R(t) = 1 + t/10$ ohms (see Figure 8.2). If $L = 0.1$ henrys, $C = 2$ farads, $E(t) \equiv 0$, $q(0) = 10$ coulombs, and $q'(0) = 0$ amps, find at least the first four nonzero terms in a power series expansion about $t = 0$ for the charge on the capacitor.

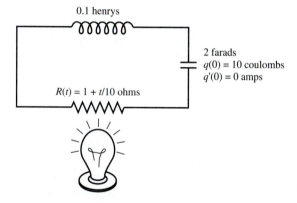

Figure 8.2 An RLC circuit whose resistor is being heated

36. Variable Spring Constant. As a spring is heated, its spring "constant" decreases. Suppose the spring is heated so that the spring "constant" at time t is $k(t) = 6 - t$ N/m (see Figure 8.3). If the unforced spring-mass system has mass $m = 2$ kg and a damping constant $b = 1$ N-sec/m with initial conditions $x(0) = 3$ m and $x'(0) = 0$ m/sec, then the displacement $x(t)$ is governed by the initial value problem

$$2x''(t) + x'(t) + (6 - t)x(t) = 0;$$

$$x(0) = 3, \qquad x'(0) = 0.$$

Find at least the first four nonzero terms in a power series expansion about $t = 0$ for the displacement.

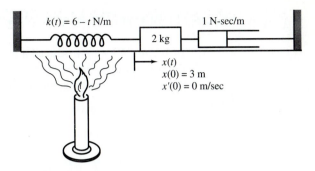

Figure 8.3 A spring-mass system whose spring is being heated

37. Variable Mass. A spring, a mass, and a dashpot are attached in a straight line on a horizontal frictionless surface as shown in Figure 8.4. The mass consists of a container filled with water. However, the water is leaking out so that at time t the mass is $m(t) = 40 - t$ kg.

(a) If the spring constant is $k = 10$ N/m and the damping constant is $b = 11$ N-sec/m, derive the governing equation

$$(40 - t)x''(t) + 10x'(t) + 10x(t) = 0,$$

where $x(t)$ is the displacement of the container from its equilibrium position. [Hint: Refer to Sections 5.1–5.3 for a discussion of vibration; however, in this case Newton's second law has the form $(d/dt)(m(t)x'(t)) = $ force.]

(b) The system is set in motion by displacing the container 0.5 m to the right and releasing it. Find at least the first four nonzero terms in a power series expansion about $t = 0$ for the displacement.

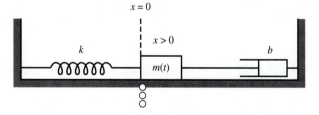

Figure 8.4 Spring-mass system with variable mass

•8.4• EQUATIONS WITH ANALYTIC COEFFICIENTS (Optional)

In Section 8.3 we introduced a method for obtaining a power series solution about an ordinary point. In this section we continue the discussion of this procedure. We begin by stating a basic existence theorem for the equation

(1) $y''(x) + p(x)y'(x) + q(x)y(x) = 0,$

which justifies the power series method.

EXISTENCE OF ANALYTIC SOLUTIONS

Theorem 5.   Suppose that x_0 is an ordinary point for equation (1). Then (1) has two linearly independent analytic solutions of the form

(2) $$y(x) = \sum_{n=0}^{\infty} a_n(x - x_0)^n.$$

Moreover, the radius of convergence of any power series solution of the form given by (2) is at least as large as the distance from x_0 to the nearest singular point (real or complex-valued) of equation (1).

The key element in the proof of Theorem 5 is the construction of a convergent geometric series that dominates the series expansion (2) of a solution to equation (1). The convergence of the series in (2) then follows by the comparison test. The details of the proof can be found in more advanced books on differential equations.[†]

As we saw in Section 8.3, the power series method gives us a general solution in the same form as (2), with a_0 and a_1 as arbitrary constants. The two linearly independent solutions referred to in Theorem 5 can be obtained by taking $a_0 = 1$, $a_1 = 0$ for the first and $a_0 = 0$, $a_1 = 1$ for the second. Thus we can extend Theorem 5 by saying that *equation (1) has a general solution of the form* (2) *with a_0 and a_1 as the arbitrary constants.*

The second part of Theorem 5 gives a simple way of determining a minimum value for the radius of convergence of the power series. We need only find the singular points of equation (1), then determine the distance between the ordinary point x_0 and the nearest singular point.

● **EXAMPLE 1**   Find a minimum value for the radius of convergence of a power series solution about $x = 0$ to

(3) $2y'' + xy' + y = 0.$

Solution   For this equation, $p(x) = x/2$ and $q(x) = \frac{1}{2}$; both of these functions are analytic for all real or complex values of x. Since equation (3) has no singular points, the distance between the ordinary point $x = 0$ and the nearest singular point is infinite. Hence the radius of convergence is infinite. ●

[†] See, for example, *Intermediate Differential Equations*, Second Edition, by Earl D. Rainville, Macmillan, New York, 1964, Chapter 3, Section 22.

The next example helps to answer the questions posed at the end of the last section.

● **EXAMPLE 2**   Find a minimum value for the radius of convergence of a power series solution about $x = 0$ to

(4) $(1 + x^2)y'' - y' + y = 0.$

Solution   Here $p(x) = -1/(1 + x^2)$, $q(x) = 1/(1 + x^2)$, and so the singular points of equation (4) occur when $1 + x^2 = 0$; that is, when $x = \pm\sqrt{-1} = \pm i$. Since the only singular points of equation (4) are the complex numbers $\pm i$, we see that $x = 0$ is an ordinary point. Moreover, the distance[†] from 0 to either $\pm i$ is 1. Thus the radius of convergence of a power series solution about $x = 0$ is at least 1.   ●

In equation (28) of Section 8.3, we found the first few terms of a power series solution to equation (4). Since we now know that this series has radius of convergence at least 1, the partial sums of this series will converge to the solution for $|x| < 1$. However, when $|x| \geq 1$, we have no basis upon which to decide whether we can use the series to approximate the solution.

Power series expansions about $x_0 = 0$ are somewhat easier to manipulate than expansions about nonzero points. As the next example shows, a simple shift in variable enables us always to expand about the origin.

● **EXAMPLE 3**   Find the first few terms in a power series expansion about $x = 1$ for a general solution to

(5) $2y'' + xy' + y = 0.$

Also determine the radius of convergence of the series.

Solution   As seen in Example 1, there are no singular points for equation (5). Thus $x = 1$ is an ordinary point, and, as a consequence of Theorem 5, equation (5) has a general solution of the form

(6) $y(x) = \displaystyle\sum_{n=0}^{\infty} a_n(x - 1)^n.$

Moreover, the radius of convergence of the series in (6) must be infinite.

We can simplify the computation of the coefficients a_n by shifting the center of the expansion (6) from $x_0 = 1$ to $t_0 = 0$. This is accomplished by the substitution $t = x - 1$. Since $y(x) = y(t + 1)$, we find via the chain rule

$$\frac{dy}{dx} = \frac{dy}{dt}, \qquad \frac{d^2y}{dx^2} = \frac{d^2y}{dt^2},$$

and hence equation (5) is changed into

(7) $2\dfrac{d^2y}{dt^2} + (t + 1)\dfrac{dy}{dt} + y = 0.$

[†] Recall that the distance between the two complex numbers $z = a + bi$ and $w = c + di$ is given by $\sqrt{(a - c)^2 + (b - d)^2}$.

We now seek a general solution of the form

(8) $y(t) = \sum\limits_{n=0}^{\infty} a_n t^n$,

where the a_n's in equations (6) and (8) are the same. Proceeding as usual, we substitute the power series for $y(t)$ into (7), derive a recurrence relation for the coefficients, and ultimately find that

$$y(t) = a_0\{1 - \tfrac{1}{4}t^2 + \tfrac{1}{24}t^3 + \cdots\} + a_1\{t - \tfrac{1}{4}t^2 - \tfrac{1}{8}t^3 + \cdots\}$$

(the details are left for the reader). Thus

(9) $y(x) = a_0\{1 - \tfrac{1}{4}(x-1)^2 + \tfrac{1}{24}(x-1)^3 + \cdots\}$
$$+ a_1\{(x-1) - \tfrac{1}{4}(x-1)^2 - \tfrac{1}{8}(x-1)^3 + \cdots\}. \quad \bullet$$

When the coefficients of a linear equation are not polynomials in x, but are analytic functions, we can still find analytic solutions by essentially the same method.

● **EXAMPLE 4**   Find a power series expansion for the solution to

(10) $y''(x) + e^x y'(x) + (1 + x^2)y(x) = 0;$ $y(0) = 1,$ $y'(0) = 0.$

Solution   Here $p(x) = e^x$ and $q(x) = 1 + x^2$, both of which are analytic for all x. Thus by Theorem 5, the initial value problem (10) has a power series solution

(11) $y(x) = \sum\limits_{n=0}^{\infty} a_n x^n$

that converges for all x ($\rho = \infty$). To find the first few terms of this series, we first expand $p(x) = e^x$ in its Maclaurin series:

$$e^x = 1 + x + \frac{x^2}{2!} + \frac{x^3}{3!} + \cdots.$$

Substituting the expansions for $y(x)$, $y'(x)$, $y''(x)$, and e^x into (10) gives

(12) $\sum\limits_{n=2}^{\infty} n(n-1)a_n x^{n-2} + \left(1 + x + \dfrac{x^2}{2} + \dfrac{x^3}{6} + \dfrac{x^4}{24} + \cdots\right)\sum\limits_{n=1}^{\infty} na_n x^{n-1}$

$$+ (1 + x^2)\sum\limits_{n=0}^{\infty} a_n x^n = 0.$$

Because of the computational difficulties due to the appearance of the product of the power series for e^x and $y'(x)$, we concern ourselves with just those terms up to order 4. Writing out (12) and keeping track of all such terms, we find

(13)

$$(2a_2 + 6a_3 x + 12a_4 x^2 + 20a_5 x^3 + 30a_6 x^4 + \cdots)$$
$$\left.\begin{array}{l} + (a_1 + 2a_2 x + 3a_3 x^2 + 4a_4 x^3 + 5a_5 x^4 + \cdots) \\ + (a_1 x + 2a_2 x^2 + 3a_3 x^3 + 4a_4 x^4 + \cdots) \\ + (\tfrac{1}{2}a_1 x^2 + a_2 x^3 + \tfrac{3}{2}a_3 x^4 + \cdots) \\ + (\tfrac{1}{6}a_1 x^3 + \tfrac{1}{3}a_2 x^4 + \cdots) \\ + (\tfrac{1}{24}a_1 x^4 + \cdots) \end{array}\right\}$$
$$+ (a_0 + a_1 x + a_2 x^2 + a_3 x^3 + a_4 x^4 + \cdots)$$
$$+ (a_0 x^2 + a_1 x^3 + a_2 x^4 + \cdots) = 0.$$

$$1 \cdot \sum n a_n x^{n-1}$$
$$x \cdot \sum n a_n x^{n-1}$$
$$\tfrac{1}{2} x^2 \cdot \sum n a_n x^{n-1}$$
$$\vdots$$
$$1 \cdot \sum a_n x^n$$
$$x^2 \cdot \sum a_n x^n$$

Grouping the like powers of x in equation (13) (the x^2 terms are shown in color) and then setting the coefficients equal to zero yields the system of equations

$$2a_2 + a_1 + a_0 = 0,$$
$$6a_3 + 2a_2 + 2a_1 = 0,$$
$$12a_4 + 3a_3 + 3a_2 + \tfrac{1}{2}a_1 + a_0 = 0,$$
$$20a_5 + 4a_4 + 4a_3 + a_2 + \tfrac{7}{6}a_1 = 0,$$
$$30a_6 + 5a_5 + 5a_4 + \tfrac{3}{2}a_3 + \tfrac{4}{3}a_2 + \tfrac{1}{24}a_1 = 0.$$

The initial conditions in (10) imply that $y(0) = a_0 = 1$ and $y'(0) = a_1 = 0$. Using these values for a_0 and a_1, we can solve the above system first for a_2, then a_3, and so on:

$$2a_2 + 0 + 1 = 0 \Rightarrow a_2 = -\tfrac{1}{2},$$
$$6a_3 - 1 + 0 = 0 \Rightarrow a_3 = \tfrac{1}{6},$$
$$12a_4 + \tfrac{1}{2} - \tfrac{3}{2} + 0 + 1 = 0 \Rightarrow a_4 = 0,$$
$$20a_5 + 0 + \tfrac{2}{3} - \tfrac{1}{2} + 0 = 0 \Rightarrow a_5 = -\tfrac{1}{120},$$
$$30a_6 - \tfrac{1}{24} + 0 + \tfrac{1}{4} - \tfrac{2}{3} + 0 = 0 \Rightarrow a_6 = \tfrac{11}{720}.$$

Thus the solution to the initial value problem in (10) is

(14) $y(x) = 1 - \tfrac{1}{2}x^2 + \tfrac{1}{6}x^3 - \tfrac{1}{120}x^5 + \tfrac{11}{720}x^6 + \cdots.$ ●

Thus far we have used the power series method only for homogeneous equations. But the same method applies, with obvious modifications, to nonhomogeneous equations of the form

(15) $y''(x) + p(x)y'(x) + q(x)y(x) = g(x),$

provided the forcing term $g(x)$ and the coefficient functions are analytic at x_0. For example, to find a power series about $x = 0$ for a general solution to

(16) $y''(x) - xy'(x) - y(x) = \sin x,$

we use the substitution $y(x) = \sum a_n x^n$ to obtain a power series expansion for the left-hand side of (16). We then equate the coefficients of this series with the corresponding coefficients of the Maclaurin expansion for $\sin x$:

$$\sin x = \sum_{n=0}^{\infty} \frac{(-1)^n}{(2n+1)!} x^{2n+1}.$$

Carrying out the details (see Problem 20), we ultimately find that an expansion for a general solution to (16) is

(17) $y(x) = a_0 y_1(x) + a_1 y_2(x) + y_p(x),$

where

(18) $y_1(x) = 1 + \frac{1}{2}x^2 + \frac{1}{8}x^4 + \frac{1}{48}x^6 + \cdots,$

(19) $y_2(x) = x + \frac{1}{3}x^3 + \frac{1}{15}x^5 + \frac{1}{105}x^7 + \cdots$

are the solutions to the homogeneous equation associated with equation (16), and

(20) $y_p(x) = \frac{1}{6}x^3 + \frac{1}{40}x^5 + \frac{19}{5040}x^7 + \cdots$

is a particular solution to equation (16).

EXERCISES 8.4

In Problems 1 through 6 find a minimum value for the radius of convergence of a power series solution about x_0.

1. $(x+1)y'' - 3xy' + 2y = 0,$ $x_0 = 1.$

2. $y'' - xy' - 3y = 0,$ $x_0 = 2.$

3. $(1+x^2)y'' - 3y = 0,$ $x_0 = 1.$

4. $(x^2 - 5x + 6)y'' - 3xy' - y = 0,$ $x_0 = 0.$

5. $y'' - (\tan x)y' + y = 0,$ $x_0 = 0.$

6. $(1+x^3)y'' - xy' + 3x^2 y = 0,$ $x_0 = 1.$

In Problems 7 through 12 find at least the first four nonzero terms in a power series expansion about x_0 for a general solution to the given differential equation with the given value for x_0.

7. $y' + 2(x-1)y = 0,$ $x_0 = 1.$

8. $y' - 2xy = 0,$ $x_0 = -1.$

9. $(x^2 - 2x)y'' + 2y = 0,$ $x_0 = 1.$

10. $x^2 y'' - xy' + 2y = 0,$ $x_0 = 2.$

11. $x^2 y'' - y' + y = 0,$ $x_0 = 2.$

12. $y'' + (3x-1)y' - y = 0,$ $x_0 = -1.$

In Problems 13 through 19 find at least the first four non-zero terms in a power series expansion of the solution to the given initial value problem.

13. $x' + (\sin t)x = 0,$ $x(0) = 1.$

14. $y' - e^x y = 0,$ $y(0) = 1.$

15. $(x^2 + 1)y'' - e^x y' + y = 0,$ $y(0) = 1,$ $y'(0) = 1.$

16. $y'' + ty' + e^t y = 0,$ $y(0) = 0,$ $y'(0) = -1.$

17. $y'' - (\sin x)y = 0,$ $y(\pi) = 1,$ $y'(\pi) = 0.$

18. $y'' - (\cos x)y' - y = 0,$ $y(\pi/2) = 1,$ $y'(\pi/2) = 1.$

19. $y'' - e^{2x} y' + (\cos x)y = 0,$ $y(0) = -1,$ $y'(0) = 1.$

20. To derive the general solution given by equations (17)–(20) for the nonhomogeneous equation (16), complete the following steps:

(a) Substitute $y(x) = \sum_{n=0}^{\infty} a_n x^n$ and the Maclaurin series for $\sin x$ into equation (16) to obtain

$$(2a_2 - a_0) + \sum_{k=1}^{\infty} [(k+2)(k+1)a_{k+2} - (k+1)a_k]x^k$$

$$= \sum_{n=0}^{\infty} \frac{(-1)^n}{(2n+1)!} x^{2n+1}.$$

(b) Equate the coefficients of like powers of x on both sides of the equation in part (a) and thereby deduce the equations

$$a_2 = \frac{a_0}{2}, \qquad a_3 = \frac{1}{6} + \frac{a_1}{3}, \qquad a_4 = \frac{a_0}{8},$$

$$a_5 = \frac{1}{40} + \frac{a_1}{15}, \qquad a_6 = \frac{a_0}{48}, \qquad a_7 = \frac{19}{5040} + \frac{a_1}{105}.$$

(c) Show that the relations in part (b) yield the general solution to (16) given in equations (17)–(20).

In Problems 21 through 28 use the procedure illustrated in Problem 20 to find at least the first four nonzero terms in a power series expansion about $x = 0$ of a general solution to the given differential equation.

21. $y' - xy = \sin x$.

22. $w' + xw = e^x$.

23. $z'' + xz' + z = x^2 + 2x + 1$.

24. $y'' - 2xy' + 3y = x^2$.

25. $(1 + x^2)y'' - xy' + y = e^{-x}$.

26. $y'' - xy' + 2y = \cos x$.

27. $(1 - x^2)y'' - y' + y = \tan x$.

28. $y'' - (\sin x)y = \cos x$.

29. The equation
$$(1 - x^2)y'' - 2xy' + n(n + 1)y = 0,$$
where n is a nonnegative integer, is called **Legendre's equation.** This equation occurs in applications of differential equations to physics and engineering.

(a) Find a power series expansion about $x = 0$ for a solution to Legendre's equation.

(b) Show that for n a nonnegative integer, there exists an nth degree polynomial that is a solution to Legendre's equation. These polynomials, up to a constant multiple, are called **Legendre polynomials.**

(c) Determine the first three Legendre polynomials, up to a constant multiple.

30. Aging Spring. As a spring ages, its spring "constant" decreases in value. One such model for a spring-mass system with an aging spring is
$$mx''(t) + bx'(t) + ke^{-\eta t}x(t) = 0,$$
where m is the mass, b the damping constant, k and η positive constants, and $x(t)$ is the displacement of the spring from its equilibrium position. Let $m = 1$ kg, $b = 2$ N-sec/m, $k = 1$ N/m, and $\eta = 1$ (sec)$^{-1}$. The system is set in motion by displacing the mass one meter from its equilibrium position and then releasing it ($x(0) = 1$, $x'(0) = 0$). Find at least the first four nonzero terms in a power series expansion about $t = 0$ for the displacement.

31. Aging Spring without Damping. In the spring-mass system for an aging spring discussed in Problem 30, assume that there is no damping (i.e. $b = 0$), $m = 1$, and $k = 1$. To see the effect of aging, consider η as a positive parameter.

(a) Redo Problem 30 with $b = 0$ and η arbitrary but fixed.

(b) Set $\eta = 0$ in the expansion obtained in part (a). Does this expansion agree with the expansion for the solution to the problem with $\eta = 0$? [Hint: When $\eta = 0$, the solution is $x(t) = \cos t$.]

•8.5• CAUCHY-EULER EQUATIONS REVISITED (**Optional**)

In the previous sections we considered methods for obtaining power series solutions about an ordinary point for a linear second order equation. However, in certain instances we may want a series expansion about a *singular point* of the equation. (This was the case for the Lane-Emden equation discussed in Section 8.1.) To motivate a procedure for finding such expansions, we return to the class of **Cauchy-Euler equations.**[†] In Chapter 4 we solved these equations by making the change of variables $x = e^t$, which transforms a Cauchy-Euler equation into an equation with constant coefficients. However, it is more instructive for our study of series expansions about singular points to work directly in the variable x.

[†] These equations are also called **equidimensional equations.**

Recall that a second order homogeneous Cauchy-Euler equation has the form

(1) $\qquad ax^2y''(x) + bxy'(x) + cy(x) = 0, \qquad x > 0,$

where $a\ (\neq 0)$, b, and c are (real) constants. Since here $p(x) = b/ax$ and $q(x) = c/ax^2$, it follows that $x = 0$ is a singular point for (1).

As we found in Chapter 4, equation (1) has solutions of the form $y = x^r$. To determine the values for r, we can proceed as follows. Let L be the differential operator defined by the left-hand side of equation (1), that is,

(2) $\qquad L[y](x) := ax^2y''(x) + bxy'(x) + cy(x),$

and set

(3) $\qquad w(r, x) := x^r.$

When we substitute $w(r, x)$ for $y(x)$ in (2), we find

$$L[w](x) = ax^2r(r - 1)x^{r-2} + bxrx^{r-1} + cx^r$$
$$= \{ar^2 + (b - a)r + c\}x^r.$$

From this we see that $w = x^r$ is a solution to (1) if and only if r satisfies

(4) $\qquad ar^2 + (b - a)r + c = 0.$

Equation (4) is referred to as the **auxiliary** or **indicial equation** for (1).

When the indicial equation has two distinct roots, we have

$$L[w](x) = a(r - r_1)(r - r_2)x^r,$$

from which it follows that equation (1) has the two linearly independent solutions

(5) $\qquad y_1(x) = w(r_1, x) = x^{r_1}, \qquad x > 0,$

(6) $\qquad y_2(x) = w(r_2, x) = x^{r_2}, \qquad x > 0.$

When r_1 and r_2 are complex conjugates, $\alpha \pm i\beta$, we can use Euler's formula to express

$$x^{\alpha + i\beta} = e^{(\alpha + i\beta)\ln x} = e^{\alpha \ln x}\cos(\beta \ln x) + ie^{\alpha \ln x}\sin(\beta \ln x)$$
$$= x^\alpha \cos(\beta \ln x) + ix^\alpha \sin(\beta \ln x).$$

Since the real and imaginary parts of $x^{\alpha + i\beta}$ must also be solutions to (1), we can replace (5) and (6) by the two linearly independent real-valued solutions

$$y_1(x) = x^\alpha \cos(\beta \ln x), \qquad y_2(x) = x^\alpha \sin(\beta \ln x).$$

If the indicial equation (4) has a repeated real root r_0, then it turns out that x^{r_0} and $x^{r_0}\ln x$ are two linearly independent solutions. This can be deduced using the reduction of order approach of Section 4.4. However, it is more instructive for later applications to see how these two linearly independent solutions can be obtained via the operator approach. If r_0 is a repeated root, then

(7) $\qquad L[w](x) = a(r - r_0)^2 x^r.$

Setting $r = r_0$ gives the solution

(8) $y_1(x) = w(r_0, x) = x^{r_0}, \qquad x > 0.$

To find a second linearly independent solution, we make the following observation. Since the right-hand side of (7) has the factor $(r - r_0)^2$, then taking the partial derivative of (7) with respect to r and setting $r = r_0$, we get *zero*. That is,

(9) $\dfrac{\partial}{\partial r}\{L[w](x)\}\big|_{r=r_0} = \{a(r - r_0)^2 x^r \ln x + 2a(r - r_0)x^r\}\big|_{r=r_0} = 0.$

While it may not appear that we have made any progress toward finding a second solution, a closer look at the expression on the left-hand side of (9) will soon vindicate our efforts.

First note that $w(r, x) = x^r$ has continuous partial derivatives of all orders with respect to both r and x. Hence the mixed partial derivatives are equal:

$$\dfrac{\partial^3 w}{\partial r\, \partial x^2} = \dfrac{\partial^3 w}{\partial x^2\, \partial r}, \qquad \dfrac{\partial^2 w}{\partial r\, \partial x} = \dfrac{\partial^2 w}{\partial x\, \partial r}.$$

Consequently, for the differential operator L, we have

$$\dfrac{\partial}{\partial r} L[w](x) = \dfrac{\partial}{\partial r}\left\{ax^2 \dfrac{\partial^2 w}{\partial x^2} + bx \dfrac{\partial w}{\partial x} + cw\right\}$$

$$= ax^2 \dfrac{\partial^3 w}{\partial r\, \partial x^2} + bx \dfrac{\partial^2 w}{\partial r\, \partial x} + c\dfrac{\partial w}{\partial r}$$

$$= ax^2 \dfrac{\partial^3 w}{\partial x^2\, \partial r} + bx \dfrac{\partial^2 w}{\partial x\, \partial r} + c\dfrac{\partial w}{\partial r}$$

$$= L\left[\dfrac{\partial w}{\partial r}\right](x);$$

that is, the operators $\partial/\partial r$ and L commute. With this fact, (9) can be written as

$$L\left[\dfrac{\partial w}{\partial r}\right]\bigg|_{r=r_0} = 0.$$

Thus for the case of a repeated root r_0, a second linearly independent solution to (1) is

(10) $y_2(x) = \dfrac{\partial w}{\partial r}(r_0, x) = \dfrac{\partial}{\partial r}(x^r)\big|_{r=r_0} = x^{r_0} \ln x, \qquad x > 0.$

● **EXAMPLE 1**   Find a general solution to

(11) $4x^2 y''(x) + y(x) = 0, \qquad x > 0.$

Solution   Let $w(r, x) = x^r$ and let L denote the left-hand side of (11). A short calculation gives

$$L[w](x) = (4r^2 - 4r + 1)x^r.$$

Solving the indicial equation

$$4r^2 - 4r + 1 = (2r - 1)^2 = 4(r - \tfrac{1}{2})^2 = 0$$

yields the repeated root $r_0 = \frac{1}{2}$. Thus a general solution to (11) is obtained from equations (8) and (10) by setting $r_0 = \frac{1}{2}$. That is,

$$y(x) = C_1\sqrt{x} + C_2\sqrt{x}\ln x, \qquad x > 0. \quad \bullet$$

In Section 8.7 we discuss the problem of finding a second linearly independent series solution to certain differential equations. As we shall see, operator methods similar to those described in this section will lead to the desired second solution.

EXERCISES 8.5

In Problems 1 through 10 use the substitution $y = x^r$ to find a general solution to the given equation for $x > 0$.

1. $x^2y''(x) + 6xy'(x) + 6y(x) = 0.$

2. $2x^2y''(x) + 13xy'(x) + 15y(x) = 0.$

3. $x^2y''(x) - xy'(x) + 17y(x) = 0.$

4. $x^2y''(x) + 2xy'(x) - 3y(x) = 0.$

5. $\dfrac{d^2y}{dx^2} = \dfrac{5}{x}\dfrac{dy}{dx} - \dfrac{13}{x^2}y.$

6. $\dfrac{d^2y}{dx^2} = \dfrac{1}{x}\dfrac{dy}{dx} - \dfrac{4}{x^2}y.$

7. $x^3y'''(x) + 4x^2y''(x) + 10xy'(x) - 10y(x) = 0.$

8. $x^3y'''(x) + 4x^2y''(x) + xy'(x) = 0.$

9. $x^3y'''(x) + 3x^2y''(x) + 5xy'(x) - 5y(x) = 0.$

10. $x^3y'''(x) + 9x^2y''(x) + 19xy'(x) + 8y(x) = 0.$

In Problems 11 and 12 use a substitution of the form $y = (x - c)^r$ to find a general solution to the given equation for $x > c$.

11. $2(x - 3)^2y''(x) + 5(x - 3)y'(x) - 2y(x) = 0.$

12. $4(x + 2)^2y''(x) + 5y(x) = 0.$

In Problems 13 and 14 use variation of parameters to find a general solution to the given equation for $x > 0$.

13. $x^2y''(x) - 2xy'(x) + 2y(x) = x^{-1/2}.$

14. $x^2y''(x) + 2xy'(x) - 2y(x) = 6x^{-2} + 3x.$

In Problems 15 through 17 solve the given initial value problem.

15. $t^2x''(t) - 12x(t) = 0; \qquad x(1) = 3, \qquad x'(1) = 5.$

16. $x^2y''(x) + 5xy'(x) + 4y(x) = 0; \qquad y(1) = 3, \qquad y'(1) = 7.$

17. $x^3y'''(x) + 6x^2y''(x) + 29xy'(x) - 29y(x) = 0;$
$y(1) = 2, \qquad y'(1) = -3, \qquad y''(1) = 19.$

18. When r_0 is a repeated root of the auxiliary equation $ar^2 + br + c = 0$, then $y_1(t) = e^{r_0t}$ is a solution to the equation $ay'' + by' + cy = 0$, where a, b, and c are constants. Use a derivation similar to the one given in this section for the case when the indicial equation has a repeated root to show that a second linearly independent solution is $y_2(t) = te^{r_0t}$.

19. Let $L[y](x) := x^3y'''(x) + xy'(x) - y(x)$.
  (a) Show that $L[x^r](x) = (r - 1)^3x^r$.
  (b) Using an extension of the argument given in this section for the case when the indicial equation has a double root, show that $L[y] = 0$ has the general solution

$$y(x) = C_1x + C_2x\ln x + C_3x(\ln x)^2.$$

•8.6• METHOD OF FROBENIUS

In the previous section we showed that a homogeneous Cauchy-Euler equation has a solution of the form $y(x) = x^r$, $x > 0$, where r is a certain constant. Cauchy-Euler equations have, of course, a very special form with only one singular point (at $x = 0$). In this section we show how the theory for Cauchy-Euler equations generalizes to other equations that have a special type of singularity.

To motivate the procedure, let's rewrite the Cauchy-Euler equation,

(1) $ax^2y''(x) + bxy'(x) + cy(x) = 0, \qquad x > 0,$

in the standard form

(2) $y''(x) + p(x)y'(x) + q(x)y(x) = 0,$ $x > 0,$

where

$$p(x) = \frac{p_0}{x}, \qquad q(x) = \frac{q_0}{x^2},$$

and p_0, q_0 are the constants b/a and c/a, respectively. When we substitute $w(r, x) = x^r$ into equation (2), we get

$$[r(r-1) + p_0 r + q_0]x^{r-2} = 0,$$

which yields the indicial equation

(3) $r(r-1) + p_0 r + q_0 = 0.$

Thus, if r_1 is a root of (3), then $w(r_1, x) = x^{r_1}$ is a solution to equations (1) and (2).

Let's now assume, more generally, that (2) is an equation for which $xp(x)$ and $x^2q(x)$, instead of being constants, are *analytic functions*. That is, in some open interval about $x = 0$,

(4) $xp(x) = p_0 + p_1 x + p_2 x^2 + \cdots = \sum_{n=0}^{\infty} p_n x^n,$

(5) $x^2 q(x) = q_0 + q_1 x + q_2 x^2 + \cdots = \sum_{n=0}^{\infty} q_n x^n.$

It follows from (4) and (5) that

(6) $\lim_{x \to 0} xp(x) = p_0$ and $\lim_{x \to 0} x^2 q(x) = q_0,$

and hence, for x near 0 we have $xp(x) \approx p_0$ and $x^2 q(x) \approx q_0$. Therefore, it is reasonable to expect that the solutions to (2) will behave (for x near 0) like the solutions to the Cauchy-Euler equation

$$x^2 y'' + p_0 xy' + q_0 y = 0.$$

When $p(x)$ and $q(x)$ satisfy (4) and (5), we say that the singular point at $x = 0$ is regular. More generally we state:

▶ REGULAR SINGULAR POINT

Definition 3.   A singular point x_0 of

(7) $y''(x) + p(x)y'(x) + q(x)y(x) = 0$

is said to be a **regular singular point** if both $(x - x_0)p(x)$ and $(x - x_0)^2 q(x)$ are analytic at x_0.[†] Otherwise x_0 is called an **irregular singular point**.

[†] In the terminology of complex variables, p has a pole of order at most 1, and q has a pole of order at most 2 at x_0.

● **EXAMPLE 1**   Classify the singular points of the equation

(8) $(x^2 - 1)^2 y''(x) + (x + 1)y'(x) - y(x) = 0.$

Solution   Here

$$p(x) = \frac{x + 1}{(x^2 - 1)^2} = \frac{1}{(x + 1)(x - 1)^2},$$

$$q(x) = \frac{-1}{(x^2 - 1)^2} = \frac{-1}{(x + 1)^2(x - 1)^2},$$

from which we see that ± 1 are the singular points of (8). For the singularity at 1, we have

$$(x - 1)p(x) = \frac{1}{(x + 1)(x - 1)},$$

which is not analytic at $x = 1$. Therefore $x = 1$ is an irregular singular point.
For the singularity at -1, we have

$$(x + 1)p(x) = \frac{1}{(x - 1)^2}, \qquad (x + 1)^2 q(x) = \frac{-1}{(x - 1)^2},$$

both of which are analytic at $x = -1$. Hence $x = -1$ is a regular singular point.   ●

Let's assume that $x = 0$ is a regular singular point for equation (7), so that $p(x)$ and $q(x)$ satisfy (4) and (5); that is,

(9) $p(x) = \displaystyle\sum_{n=0}^{\infty} p_n x^{n-1}, \qquad q(x) = \sum_{n=0}^{\infty} q_n x^{n-2}.$

The idea of the mathematician Frobenius was that, since Cauchy-Euler equations have solutions of the form x^r, then for the regular singular point $x = 0$ there should be solutions to (7) of the form x^r *times an analytic function.*[†] Hence we seek solutions to (7) of the form

(10) $w(r, x) = x^r \displaystyle\sum_{n=0}^{\infty} a_n x^n = \sum_{n=0}^{\infty} a_n x^{n+r}, \qquad x > 0.$

Without loss of generality, we assume that a_0 is an arbitrary nonzero constant, and so we are left with determining r and the coefficients a_n, $n \geq 1$. Differentiating $w(r, x)$ with respect to x, we have

(11) $w'(r, x) = \displaystyle\sum_{n=0}^{\infty} (n + r)a_n x^{n+r-1},$

(12) $w''(r, x) = \displaystyle\sum_{n=0}^{\infty} (n + r)(n + r - 1)a_n x^{n+r-2}.$

If we substitute the above expansions for $w(r, x)$, $w'(r, x)$, $w''(r, x)$, $p(x)$, and $q(x)$ into (7), we

---

[†] *Historical Footnote*: George Frobenius (1848–1917) developed this method in 1873. He is also known for his research on group theory.

obtain

(13)
$$\sum_{n=0}^{\infty} (n+r)(n+r-1)a_n x^{n+r-2} + \left(\sum_{n=0}^{\infty} p_n x^{n-1} \right)\left(\sum_{n=0}^{\infty} (n+r)a_n x^{n+r-1} \right)$$
$$+ \left(\sum_{n=0}^{\infty} q_n x^{n-2} \right)\left(\sum_{n=0}^{\infty} a_n x^{n+r} \right) = 0.$$

Now we use the Cauchy product to perform the series multiplications and then group like powers of x, starting with the lowest power, x^{r-2}. This gives

(14) $[r(r-1) + p_0 r + q_0]a_0 x^{r-2}$
$$+ [(r+1)ra_1 + (r+1)p_0 a_1 + p_1 r a_0 + q_0 a_1 + q_1 a_0]x^{r-1} + \cdots = 0.$$

In order for the expansion on the left-hand side of equation (14) to sum to zero, each coefficient must be zero. Considering the first term, x^{r-2}, we find

(15) $[r(r-1) + p_0 r + q_0]a_0 = 0.$

Since we have assumed that $a_0 \neq 0$, the quantity in brackets must be zero. This gives the indicial equation that is the same as the one we derived for Cauchy-Euler equations.

INDICIAL EQUATION

Definition 4.   If x_0 is a regular singular point of $y'' + py' + qy = 0$, then the **indicial equation** for this point is

(16) $r(r-1) + p_0 r + q_0 = 0,$

where

$$p_0 := \lim_{x \to x_0} (x - x_0)p(x), \qquad q_0 := \lim_{x \to x_0} (x - x_0)^2 q(x).$$

The roots of the indicial equation are called the **exponents (indices)** of the singularity x_0.

● **EXAMPLE 2**   Find the indicial equation and the exponents at the singularity $x = -1$ of

(17) $(x^2 - 1)^2 y''(x) + (x+1)y'(x) - y(x) = 0.$

Solution   In Example 1 we showed that $x = -1$ is a regular singular point. Since $p(x) = (x+1)^{-1}(x-1)^{-2}$ and $q(x) = -(x+1)^{-2}(x-1)^{-2}$, we find

$$p_0 = \lim_{x \to -1} (x+1)p(x) = \lim_{x \to -1} (x-1)^{-2} = \tfrac{1}{4},$$

$$q_0 = \lim_{x \to -1} (x+1)^2 q(x) = \lim_{x \to -1} [-(x-1)^{-2}] = -\tfrac{1}{4}.$$

Substituting these values for p_0 and q_0 into (16), we obtain the indicial equation

(18) $r(r-1) + \tfrac{1}{4}r - \tfrac{1}{4} = 0.$

Multiplying by 4 and factoring gives $(4r + 1)(r - 1) = 0$. Hence $r = 1$, $-\frac{1}{4}$ are the exponents. ●

As we have seen, we can use the indicial equation to determine those values of r for which the coefficient of x^{r-2} in (14) is zero. If we set the coefficient of x^{r-1} in (14) equal to zero, we have

(19) $[(r + 1)r + (r + 1)p_0 + q_0]a_1 + (p_1 r + q_1)a_0 = 0.$

Since a_0 is arbitrary and we know the p_i's, q_i's, and r, we can solve equation (19) for a_1 provided that the coefficient of a_1 in (19) is not zero. This will be the case if we take r to be the larger of the two roots of the indicial equation (see Problem 43). Similarly, when we set the coefficient of x^r equal to zero, we can solve for a_2 in terms of the p_i's, q_i's, r, a_0, and a_1. Continuing in this manner, we can recursively solve for the a_n's. The procedure is illustrated in the following example.

● **EXAMPLE 3**   Find a series expansion about the regular singular point $x = 0$ for a solution to

(20) $(x + 2)x^2 y'' - xy' + (1 + x)y = 0,$ $x > 0.$

Solution   Here $p(x) = -x^{-1}(x + 2)^{-1}$ and $q(x) = x^{-2}(x + 2)^{-1}(1 + x)$, and so

$$p_0 = \lim_{x \to 0} xp(x) = \lim_{x \to 0} [-(x + 2)^{-1}] = -\tfrac{1}{2},$$

$$q_0 = \lim_{x \to 0} x^2 q(x) = \lim_{x \to 0} (x + 2)^{-1}(1 + x) = \tfrac{1}{2}.$$

Since $x = 0$ is a regular singular point, we seek a solution to (20) of the form

(21) $w(r, x) = x^r \sum_{n=0}^{\infty} a_n x^n = \sum_{n=0}^{\infty} a_n x^{n+r}.$

By the previous discussion, r must satisfy the indicial equation (16). Substituting for p_0 and q_0 in (16), we obtain

$$r(r - 1) - \tfrac{1}{2}r + \tfrac{1}{2} = 0,$$

which simplifies to $2r^2 - 3r + 1 = (2r - 1)(r - 1) = 0$. Thus $r = 1$ and $r = \frac{1}{2}$ are the roots of the indicial equation associated with $x = 0$.

Let's use the larger root $r = 1$ and solve for a_1, a_2, etc., to obtain the solution $w(1, x)$. We can simplify the computations by substituting $w(r, x)$ directly into equation (20), where the coefficients are polynomials in x, rather than dividing by $(x + 2)x^2$ and having to work with the rational functions $p(x)$ and $q(x)$. Inserting $w(r, x)$ in (20) and recalling the formulas for $w'(r, x)$ and $w''(r, x)$ in (11) and (12) gives (with $r = 1$)

(22) $(x + 2)x^2 \sum_{n=0}^{\infty} (n + 1)na_n x^{n-1} - x \sum_{n=0}^{\infty} (n + 1)a_n x^n$

$$+ (1 + x) \sum_{n=0}^{\infty} a_n x^{n+1} = 0,$$

which we can write as

(23) $\displaystyle\sum_{n=0}^{\infty} (n + 1)na_n x^{n+2} + \sum_{n=0}^{\infty} 2(n + 1)na_n x^{n+1} - \sum_{n=0}^{\infty} (n + 1)a_n x^{n+1}$

$\displaystyle + \sum_{n=0}^{\infty} a_n x^{n+1} + \sum_{n=0}^{\infty} a_n x^{n+2} = 0.$

Next we shift the indices so that each summation in (23) is over x^k. With $k = n + 2$ in the first and last summations and $k = n + 1$ in the rest, (23) becomes

(24) $\displaystyle\sum_{k=2}^{\infty} [(k - 1)(k - 2) + 1]a_{k-2} x^k + \sum_{k=1}^{\infty} [2k(k - 1) - k + 1]a_{k-1} x^k = 0.$

Separating off the $k = 1$ term and combining the rest under one summation yields

(25) $\displaystyle [2(1)(0) - 1 + 1]a_0 x + \sum_{k=2}^{\infty} [(k^2 - 3k + 3)a_{k-2} + (2k - 1)(k - 1)a_{k-1}]x^k = 0.$

Notice that the coefficient of x in (25) is zero. This is because $r = 1$ is a root of the indicial equation, which is the equation we obtained by setting the coefficient of the lowest power of x equal to zero.

We can now determine the a_k's in terms of a_0 by setting the coefficients of x^k in equation (25) equal to zero for $k = 2, 3$, etc. This gives the recurrence relation

(26) $(k^2 - 3k + 3)a_{k-2} + (2k - 1)(k - 1)a_{k-1} = 0,$

or, equivalently,

(27) $\displaystyle a_{k-1} = -\frac{k^2 - 3k + 3}{(2k - 1)(k - 1)} a_{k-2}, \qquad k \geq 2.$

Setting $k = 2, 3$, and 4 in (27), we find

$\begin{aligned} a_1 &= -\tfrac{1}{3}a_0 && (k = 2), \\ a_2 &= -\tfrac{3}{10}a_1 = \tfrac{1}{10}a_0 && (k = 3), \\ a_3 &= -\tfrac{1}{3}a_2 = -\tfrac{1}{30}a_0 && (k = 4). \end{aligned}$

Substituting these values for r, a_1, a_2, and a_3 into (21) gives

(28) $w(1, x) = a_0 x^1 (1 - \tfrac{1}{3}x + \tfrac{1}{10}x^2 - \tfrac{1}{30}x^3 + \cdots),$

where a_0 is arbitrary. In particular, for $a_0 = 1$, we get the solution

$y_1(x) = x - \tfrac{1}{3}x^2 + \tfrac{1}{10}x^3 - \tfrac{1}{30}x^4 + \cdots, \qquad x > 0.$ ●

To find a second linearly independent solution to equation (20), we could try setting $r = \tfrac{1}{2}$ and solving for $a_1, a_2, \ldots$, to obtain a solution $w(\tfrac{1}{2}, x)$ (see Problem 44). In this particular case, the approach would work. However, if we encounter an indicial equation that has a repeated root, then the method of Frobenius would yield just one solution (apart from constant multiples). To find the desired second solution we must use another technique, such as the reduction of order procedure discussed in Section 4.4. We tackle the problem of finding a second linearly independent solution in the next section.

The method of Frobenius can be summarized as follows:

> **► METHOD OF FROBENIUS**
>
> To derive a series solution about the singular point x_0 of
>
> **(29)** $\qquad a_2(x)y''(x) + a_1(x)y'(x) + a_0(x)y(x) = 0, \qquad x > x_0:$
>
> **(a)** Set $p(x) := a_1(x)/a_2(x)$, $q(x) := a_0(x)/a_2(x)$. If both $(x - x_0)p(x)$ and $(x - x_0)^2 q(x)$ are analytic at x_0, then x_0 is a regular singular point and the remaining steps apply.
> **(b)** Let
>
> **(30)** $\qquad w(r, x) = (x - x_0)^r \displaystyle\sum_{n=0}^{\infty} a_n(x - x_0)^n = \sum_{n=0}^{\infty} a_n(x - x_0)^{n+r},$
>
> and, using termwise differentiation, substitute $w(r, x)$ into equation (29) to obtain an equation of the form
>
> $\qquad A_0(x - x_0)^{r+J} + A_1(x - x_0)^{r+J+1} + \cdots = 0.$
>
> **(c)** Set the coefficients $A_0, A_1, A_2, \dots$ equal to zero. (Notice that the equation $A_0 = 0$ is just a constant multiple of the indicial equation $r(r - 1) + p_0 r + q_0 = 0$.)
> **(d)** Use the system of equations
>
> $\qquad A_0 = 0, \qquad A_1 = 0, \dots, \qquad A_k = 0,$
>
> to find a recurrence relation involving a_k and $a_0, a_1, \dots, a_{k-1}$.
> **(e)** Take $r = r_1$, the larger root of the indicial equation, and use the relation obtained in step (d) to recursively determine $a_1, a_2, \dots$ in terms of a_0 and r_1.
> **(f)** A series expansion of a solution to (29) is
>
> **(31)** $\qquad w(r_1, x) = (x - x_0)^{r_1} \displaystyle\sum_{n=0}^{\infty} a_n(x - x_0)^n, \qquad x > x_0,$
>
> where a_0 is arbitrary and the a_n's are defined in terms of a_0 and r_1.

One important question that remains concerns the radius of convergence of the power series that appears in (31). The following theorem contains an answer.[†]

[†] For a proof of this theorem see *Ordinary Differential Equations*, by E. L. Ince, Dover Publications, Inc., New York, 1956, Chapter XVI.

> ### FROBENIUS'S THEOREM
>
> **Theorem 6.**   If x_0 is a regular singular point of equation (29), then there exists at least one series solution of the form (30), where $r = r_1$ is the larger root of the associated indicial equation. Moreover, this series converges for all x such that $0 < x - x_0 < R$, where R is the distance from x_0 to the nearest other singular point (real or complex) of (29).

For simplicity, in the examples that follow we consider only series expansions about the regular singular point $x = 0$, and only those equations for which the associated indicial equation has real roots.

The following three examples not only illustrate the method of Frobenius, but are important examples to which we refer in later sections.

● **EXAMPLE 4**   Find a series solution about the regular singular point $x = 0$ of

(32) $$x^2 y''(x) - xy'(x) + (1 - x)y(x) = 0, \qquad x > 0.$$

Solution   Here $p(x) = -x^{-1}$ and $q(x) = (1 - x)x^{-2}$. It is easy to check that $x = 0$ is a regular singular point of (32), so we compute

$$p_0 = \lim_{x \to 0} xp(x) = \lim_{x \to 0} -1 = -1,$$

$$q_0 = \lim_{x \to 0} x^2 q(x) = \lim_{x \to 0}(1 - x) = 1.$$

Then the indicial equation is

$$r(r - 1) - r + 1 = r^2 - 2r + 1 = (r - 1)^2 = 0,$$

which has the roots $r_1 = r_2 = 1$.

Next we substitute

(33) $$w(r, x) = x^r \sum_{n=0}^{\infty} a_n x^n = \sum_{n=0}^{\infty} a_n x^{n+r}$$

into (32) and obtain

(34) $$x^2 \sum_{n=0}^{\infty} (n + r)(n + r - 1)a_n x^{n+r-2} - x \sum_{n=0}^{\infty} (n + r)a_n x^{n+r-1}$$

$$+ (1 - x) \sum_{n=0}^{\infty} a_n x^{n+r} = 0,$$

which we write as

(35) $$\sum_{n=0}^{\infty} (n + r)(n + r - 1)a_n x^{n+r} - \sum_{n=0}^{\infty} (n + r)a_n x^{n+r}.$$

$$+ \sum_{n=0}^{\infty} a_n x^{n+r} - \sum_{n=0}^{\infty} a_n x^{n+r+1} = 0.$$

Shifting the indices so that each summation in (35) is over x^{k+r}, we take $k = n + 1$ in the last summation and $k = n$ in the rest. This gives

(36) $\displaystyle\sum_{k=0}^{\infty} [(k + r)(k + r - 1) - (k + r) + 1] a_k x^{k+r} - \sum_{k=1}^{\infty} a_{k-1} x^{k+r} = 0.$

Singling out the term corresponding to $k = 0$ and combining the rest under one summation yields

(37) $[r(r - 1) - r + 1] a_0 x^r$

$\displaystyle + \sum_{k=1}^{\infty} \{[(k + r)(k + r - 1) - (k + r) + 1] a_k - a_{k-1}\} x^{k+r} = 0.$

When we set the coefficients equal to zero, we obtain

(38) $[r(r - 1) - r + 1] a_0 = 0,$

and, for $k \geq 1$, the recurrence relation

(39) $[(k + r)^2 - 2(k + r) + 1] a_k - a_{k-1} = 0,$

which reduces to

(40) $(k + r - 1)^2 a_k - a_{k-1} = 0.$

Relation (40) can be used to solve for a_k in terms of a_{k-1}:

(41) $a_k = \dfrac{1}{(k + r - 1)^2} a_{k-1}, \qquad k \geq 1.$

Setting $r = r_1 = 1$ in (38) gives (as expected) $0 \cdot a_0 = 0$, and in (41) gives

(42) $a_k = \dfrac{1}{k^2} a_{k-1}, \qquad k \geq 1.$

For $k = 1, 2,$ and 3, we now find

$$a_1 = \frac{1}{1^2} a_0 = a_0 \qquad\qquad (k = 1),$$

$$a_2 = \frac{1}{2^2} a_1 = \frac{1}{(2 \cdot 1)^2} a_0 = \frac{1}{4} a_0 \qquad (k = 2),$$

$$a_3 = \frac{1}{3^2} a_2 = \frac{1}{(3 \cdot 2 \cdot 1)^2} a_0 = \frac{1}{36} a_0 \qquad (k = 3).$$

In general, we have

(43) $a_k = \dfrac{1}{(k!)^2} a_0.$

Hence equation (32) has a series solution given by

(44) $\quad w(1, x) = a_0 x \{1 + x + \frac{1}{4}x^2 + \frac{1}{36}x^3 + \cdots\}$

$$= a_0 x \sum_{k=0}^{\infty} \frac{1}{(k!)^2} x^k, \qquad x > 0. \quad \bullet$$

Since $x = 0$ is the only singular point for equation (32), it follows from Frobenius's theorem or directly by the ratio test that the series solution (44) converges for all $x > 0$.

In the next two examples, we only outline the method; the reader is invited to furnish the intermediate steps.

● EXAMPLE 5   Find a series solution about the regular singular point $x = 0$ of

(45) $\quad xy''(x) + 4y'(x) - xy(x) = 0, \qquad x > 0.$

Solution   Since $p(x) = 4/x$ and $q(x) = -1$, we see that $x = 0$ is indeed a regular singular point and

$$p_0 = \lim_{x \to 0} xp(x) = 4, \qquad q_0 = \lim_{x \to 0} x^2 q(x) = 0.$$

The indicial equation is

$$r(r - 1) + 4r = r^2 + 3r = r(r + 3) = 0,$$

with roots $r_1 = 0$ and $r_2 = -3$.

Now substitute

(46) $\quad w(r, x) = x^r \sum_{n=0}^{\infty} a_n x^n = \sum_{n=0}^{\infty} a_n x^{n+r}$

into (45). After a little algebra and a shift in indices, we get

(47) $\quad [r(r - 1) + 4r]a_0 x^{r-1} + [(r + 1)r + 4(r + 1)]a_1 x^r$

$$+ \sum_{k=1}^{\infty} [(k + r + 1)(k + r + 4)a_{k+1} - a_{k-1}]x^{k+r} = 0.$$

Next we set the coefficients equal to zero and find

(48) $\qquad [r(r - 1) + 4r]a_0 = 0,$

(49) $\quad [(r + 1)r + 4(r + 1)]a_1 = 0,$

and, for $k \geq 1$, the recurrence relation

(50) $\quad (k + r + 1)(k + r + 4)a_{k+1} - a_{k-1} = 0.$

For $r = r_1 = 0$, equation (48) becomes $0 \cdot a_0 = 0$ and (49) becomes $4 \cdot a_1 = 0$. Hence, while a_0 is arbitrary, a_1 must be zero. Setting $r = r_1 = 0$ in (50), we find

(51) $\quad a_{k+1} = \dfrac{1}{(k + 1)(k + 4)} a_{k-1}, \qquad k \geq 1,$

from which it follows (after a few experimental computations) that $a_{2k+1} = 0$ for

$k = 0, 1, \ldots,$ and

(52) $\quad a_{2k} = \dfrac{1}{[2 \cdot 4 \cdots (2k)][5 \cdot 7 \cdots (2k + 3)]} \, a_0$

$\qquad\qquad = \dfrac{1}{2^k k! [5 \cdot 7 \cdots (2k + 3)]} \, a_0, \qquad k \geq 1.$

Hence equation (45) has a series solution

(53) $\quad w(0, x) = a_0 \left\{ 1 + \displaystyle\sum_{k=1}^{\infty} \dfrac{1}{2^k k! [5 \cdot 7 \cdots (2k + 3)]} x^{2k} \right\}, \qquad x > 0. \quad \bullet$

If in Example 5 we had worked with the root $r = r_2 = -3$, then we would actually have obtained *two* linearly independent solutions (see Problem 45).

● **EXAMPLE 6**   Find a series solution about the regular singular point $x = 0$ of

(54) $\quad xy''(x) + 3y'(x) - xy(x) = 0, \qquad x > 0.$

Solution   Since $p(x) = 3/x$ and $q(x) = -1$, we see that $x = 0$ is a regular singular point. Moreover,

$$p_0 = \lim_{x \to 0} x p(x) = 3, \qquad q_0 = \lim_{x \to 0} x^2 q(x) = 0,$$

and so the indicial equation is

(55) $\quad r(r - 1) + 3r = r^2 + 2r = r(r + 2) = 0,$

with roots $r_1 = 0$ and $r_2 = -2$.

Substituting

(56) $\quad w(r, x) = x^r \displaystyle\sum_{n=0}^{\infty} a_n x^n = \sum_{n=0}^{\infty} a_n x^{n+r}$

into (54) ultimately gives

(57) $\quad [r(r - 1) + 3r] a_0 x^{r-1} + [(r + 1)r + 3(r + 1)] a_1 x^r$

$\qquad\qquad + \displaystyle\sum_{k=1}^{\infty} [(k + r + 1)(k + r + 3) a_{k+1} - a_{k-1}] x^{k+r} = 0.$

Setting the coefficients equal to zero, we have

(58) $\qquad\quad [r(r - 1) + 3r] a_0 = 0,$

(59) $\quad\; [(r + 1)r + 3(r + 1)] a_1 = 0,$

and, for $k \geq 1$, the recurrence relation

(60) $\quad (k + r + 1)(k + r + 3) a_{k+1} - a_{k-1} = 0.$

With $r = r_1 = 0$, these equations lead to the following formulas: $a_{2k+1} = 0, \; k = 0, 1, \ldots,$ and

(61) $\quad a_{2k} = \dfrac{1}{[2 \cdot 4 \cdots (2k)][4 \cdot 6 \cdots (2k + 2)]} \, a_0 = \dfrac{1}{2^{2k} k! (k + 1)!} \, a_0, \qquad k \geq 0.$

Hence equation (54) has a series solution

(62) $w(0, x) = a_0 \sum\limits_{k=0}^{\infty} \dfrac{1}{2^{2k}k!(k+1)!} x^{2k}, \qquad x > 0.$ ●

Unlike Example 5, if we work with the second root $r = r_2 = -2$ in Example 6, then we do *not* obtain a second linearity independent solution (see Problem 46).

In the preceding examples we were able to use the method of Frobenius to find a series solution valid to the right ($x > 0$) of the regular singular point $x = 0$. For $x < 0$, we can use the change of variables $x = -t$ and then solve the resulting equation for $t > 0$.

The method of Frobenius also applies to higher order linear equations (see Problems 35–38).

EXERCISES 8.6

In Problems 1 through 10 classify each singular point (real or complex) of the given equation as regular or irregular.

1. $(x^2 - 1)y'' + xy' + 3y = 0.$
2. $x^2 y'' + 8xy' - 3xy = 0.$
3. $(x^2 + 1)z'' + 7x^2 z' - 3xz = 0.$
4. $x^2 y'' - 5xy' + 7y = 0.$
5. $(x^2 - 1)^2 y'' - (x - 1)y' + 3y = 0.$
6. $(x^2 - 4)y'' + (x + 2)y' + 3y = 0.$
7. $(t^2 - t - 2)^2 x'' + (t^2 - 4)x' - tx = 0.$
8. $(x^2 - x)y'' + xy' + 7y = 0.$
9. $(x^2 + 2x - 8)^2 y'' + (3x + 12)y' - x^2 y = 0.$
10. $x^3(x - 1)y'' + (x^2 - 3x)(\sin x)y' - xy = 0.$

In Problems 11 through 18 find the indicial equation and the exponents for the specified singularity of the given differential equation.

11. $x^2 y'' - 2xy' - 10y = 0,$ at $x = 0.$
12. $x^2 y'' + 4xy' + 2y = 0,$ at $x = 0.$
13. $(x^2 - x - 2)^2 z'' + (x^2 - 4)z' - 6xz = 0,$ at $x = 2.$
14. $(x^2 - 4)y'' + (x + 2)y' + 3y = 0,$ at $x = -2.$
15. $\theta^3 y'' + \theta(\sin \theta)y' - (\tan \theta)y = 0,$ at $\theta = 0.$
16. $(x^2 - 1)^2 y'' - (x - 1)y' - 3y = 0,$ at $x = 1.$
17. $(x - 1)^2 y'' + (x^2 - 1)y' - 12y = 0,$ at $x = 1.$
18. $4x(\sin x)y'' - 3y = 0,$ at $x = 0.$

In Problems 19 through 24 use the method of Frobenius to find at least the first four nonzero terms in the series expansion about $x = 0$ for a solution to the given equation for $x > 0$.

19. $9x^2 y'' + 9x^2 y' + 2y = 0.$
20. $2x(x - 1)y'' + 3(x - 1)y' - y = 0.$
21. $x^2 y'' + xy' + x^2 y = 0.$
22. $xy'' + y' - 4y = 0.$
23. $x^2 z'' + (x^2 + x)z' - z = 0.$
24. $3xy'' + (2 - x)y' - y = 0.$

In Problems 25 through 30 use the method of Frobenius to find a general formula for the coefficient a_n in a series expansion about $x = 0$ for a solution to the given equation for $x > 0$.

25. $4x^2 y'' + 2x^2 y' - (x + 3)y = 0.$
26. $x^2 y'' + (x^2 - x)y' + y = 0.$
27. $xw'' - w' - xw = 0.$
28. $3x^2 y'' + 8xy' + (x - 2)y = 0.$
29. $xy'' + (x - 1)y' - 2y = 0.$
30. $x(x + 1)y'' + (x + 5)y' - 4y = 0.$

In Problems 31 through 34 first determine a recurrence formula for the coefficients in the (Frobenius) series expansion of the solution about $x = 0$. Use this recurrence formula to determine if there exists a solution to the differential equation that is decreasing for $x > 0$.

31. $xy'' + (1 - x)y' - y = 0.$
32. $x^2 y'' - x(1 + x)y' + y = 0.$
33. $3xy'' + 2(1 - x)y' - 4y = 0.$
34. $xy'' + (x + 2)y' - y = 0.$

In Problems 35 through 38 use the method of Frobenius to find at least the first four nonzero terms in the series expan-

sion about $x = 0$ for a solution to the given linear third order equation for $x > 0$.

35. $6x^3y''' + 13x^2y'' + (x + x^2)y' + xy = 0$.

36. $6x^3y''' + 11x^2y'' - 2xy' - (x - 2)y = 0$.

37. $6x^3y''' + 13x^2y'' - (x^2 + 3x)y' - xy = 0$.

38. $6x^3y''' + (13x^2 - x^3)y'' + xy' - xy = 0$.

In Problems 39 and 40 try to use the method of Frobenius to find a series expansion about the irregular singular point $x = 0$ for a solution to the given differential equation. If the method works, give at least the first four nonzero terms in the expansion. If the method does not work, explain what went wrong.

39. $x^2y'' + (3x - 1)y' + y = 0$.

40. $x^2y'' + y' - 2y = 0$.

In certain applications it is desirable to have an expansion about the point at infinity. To obtain such an expansion, we use the change of variables $z = 1/x$ and expand about $z = 0$. In Problems 41 and 42 show that infinity is a regular singular point of the given differential equation by showing that $z = 0$ is a regular singular point for the transformed equation in z. Also find at least the first four nonzero terms in the series expansion about infinity of a solution to the original equation in x.

41. $x^3y'' - x^2y' - y = 0$.

42. $18(x - 4)^2(x - 6)y'' + 9x(x - 4)y' - 32y = 0$.

43. Show that if r_1 and r_2 are roots of the indicial equation (16), with r_1 the larger root ($\mathrm{Re}\, r_1 \geq \mathrm{Re}\, r_2$), then the coefficient of a_1 in equation (19) is not zero when $r = r_1$.

44. To obtain a second linearly independent solution to equation (20):

(a) Substitute $w(r, x)$ given in (21) into (20) and conclude that the coefficients a_k, $k \geq 1$ must satisfy the recurrence relation

$$(k + r - 1)(2k + 2r - 1)a_k$$
$$+ [(k + r - 1)(k + r - 2) + 1]a_{k-1} = 0.$$

(b) Use the recurrence relation with $r = \frac{1}{2}$ to derive the second series solution

$$w(\tfrac{1}{2}, x) = a_0(x^{1/2} - \tfrac{3}{4}x^{3/2} + \tfrac{7}{32}x^{7/2} - \tfrac{133}{1920}x^{9/2} + \cdots).$$

(c) Use the recurrence relation with $r = 1$ to obtain $w(1, x)$ in (28).

45. In Example 5, show that if we choose $r = r_2 = -3$, then we obtain *two* linearly independent solutions to equation (45). [Hint: a_0 and a_3 are arbitrary constants.]

46. In Example 6, show that if we choose $r = r_2 = -2$, then we obtain a solution that is a constant multiple of the solution given in (62). [Hint: $a_0 = a_1 = 0$ and a_2 is arbitrary.]

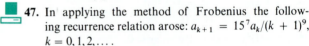 **47.** In applying the method of Frobenius the following recurrence relation arose: $a_{k+1} = 15^7 a_k/(k+1)^9$, $k = 0, 1, 2, \ldots$.

(a) Show that the coefficients are given by the formula $a_k = 15^{7k}a_0/(k!)^9$, $k = 0, 1, 2, \ldots$.

(b) Use the formula obtained in part (a) with $a_0 = 1$ to compute $a_5, a_{10}, a_{15}, a_{20}$, and a_{25}. What goes wrong?

(c) Now use the recurrence relation to compute a_k for $k = 1, 2, 3, \ldots, 25$, assuming $a_0 = 1$.

(d) What advantage does the recurrence relation have over the formula?

• 8.7 • FINDING A SECOND LINEARLY INDEPENDENT SOLUTION

In the previous section we showed that if $x = 0$ is a regular singular point of

(1) $\qquad y''(x) + p(x)y'(x) + q(x)y(x) = 0, \qquad x > 0,$

then the method of Frobenius can be used to find a series solution valid for x near zero. The first step in the method is to find the roots r_1 and r_2 ($\mathrm{Re}\, r_1 \geq \mathrm{Re}\, r_2$) of the associated indicial equation

(2) $\qquad r(r - 1) + p_0 r + q_0 = 0.$

Then, utilizing the larger root r_1, equation (1) has a series solution of the form

(3) $\qquad w(r_1, x) = x^{r_1} \sum_{n=0}^{\infty} a_n x^n = \sum_{n=0}^{\infty} a_n x^{n+r_1},$

where $a_0 \neq 0$. In order to find a second linearly independent solution, our first inclination is to set $r = r_2$ and seek a solution of the form

(4) $$w(r_2, x) = x^{r_2} \sum_{n=0}^{\infty} a_n x^n = \sum_{n=0}^{\infty} a_n x^{n+r_2}.$$

This procedure will work *provided that $r_1 - r_2$ is not an integer.* However, when $r_1 - r_2$ is an integer, the Frobenius method with $r = r_2$ may just lead to the same solution that we obtained using the root r_1. (This is obviously true when $r_1 = r_2$.)

● **EXAMPLE 1**   Find the first few terms in the series expansion about the regular singular point $x = 0$ for a general solution to

(5) $$(x + 2)x^2 y''(x) - xy'(x) + (1 + x)y(x) = 0, \qquad x > 0.$$

Solution   In Example 3 of Section 8.6 we used the method of Frobenius to find a series solution for (5). In the process we determined that $p_0 = -\frac{1}{2}$, $q_0 = \frac{1}{2}$, and that the indicial equation has roots $r_1 = 1$, $r_2 = \frac{1}{2}$. Since these roots do not differ by an integer ($r_1 - r_2 = \frac{1}{2}$), the method of Frobenius will give two linearly independent solutions of the form

(6) $$w(r, x) := x^r \sum_{n=0}^{\infty} a_n x^n = \sum_{n=0}^{\infty} a_n x^{n+r}.$$

In Problem 44 of Exercises 8.6 the reader was requested to show that substituting $w(r, x)$ into (5) leads to the recurrence relation

(7) $$(k + r - 1)(2k + 2r - 1)a_k + [(k + r - 1)(k + r - 2) + 1]a_{k-1} = 0, \qquad k \geq 1.$$

With $r = r_1 = 1$ and $a_0 = 1$ we find

(8) $$y_1(x) = x - \tfrac{1}{3}x^2 + \tfrac{1}{10}x^3 - \tfrac{1}{30}x^4 + \cdots.$$

as obtained in the previous section. Moreover, taking $r = r_2 = \frac{1}{2}$ and $a_0 = 1$ in (7) leads to the second solution,

(9) $$y_2(x) = x^{1/2} - \tfrac{3}{4}x^{3/2} + \tfrac{7}{32}x^{7/2} - \tfrac{133}{1920}x^{9/2} + \cdots.$$

Consequently, a general solution to equation (5) is

(10) $$y(x) = c_1 y_1(x) + c_2 y_2(x), \qquad x > 0,$$

where $y_1(x)$ and $y_2(x)$ are the series solutions given in equations (8) and (9).   ●

When the indicial equation has repeated roots, $r_1 = r_2$, substituting $r = r_2$ just gives us back the first solution and gets us nowhere. However, in such a case, we can use the reduction of order formula discussed in Section 4.4 to obtain a second linearly independent solution.

Recall that if $f(x)$ is a nontrivial solution to equation (1), then a second linearly independent solution is given by the reduction of order formula

(11) $$y(x) = f(x) \int \frac{e^{-\int p(x)\,dx}}{[f(x)]^2} \, dx.$$

In the next example we use this formula to hunt for a second series solution.

● **EXAMPLE 2**   Find the first few terms in the series expansion about the regular singular point $x = 0$ for a general solution to

(12) $x^2 y''(x) - xy'(x) + (1 - x)y(x) = 0, \qquad x > 0.$

Solution   In Example 4 of Section 8.6 we used the method of Frobenius to obtain a series solution to equation (12). In the process we found the indicial equation to be $r^2 - 2r + 1 = 0$, which has roots $r_1 = r_2 = 1$. Working with $r_1 = 1$, we derived the series solution

(13) $y_1(x) = x + x^2 + \frac{1}{4}x^3 + \frac{1}{36}x^4 + \frac{1}{576}x^5 + \cdots = \sum_{k=0}^{\infty} \frac{1}{(k!)^2} x^{k+1}$

(see equation (44) in Section 8.6 with $a_0 = 1$).

Since $p(x) = -x^{-1}$ for equation (12), we can substitute $f(x) = y_1(x)$ and $p(x) = -x^{-1}$ into the reduction of order formula (11) to obtain

(14) $y_2(x) = y_1(x) \int \frac{e^{\int x^{-1} dx}}{[y_1(x)]^2} dx = y_1(x) \int \frac{e^{\ln x}}{[y_1(x)]^2} dx$

$= y_1(x) \int \frac{x}{[x + x^2 + \frac{1}{4}x^3 + \frac{1}{36}x^4 + \frac{1}{576}x^5 + \cdots]^2} dx.$

Using the Cauchy product to square the series in the denominator and canceling out an x gives

(15) $y_2(x) = y_1(x) \int \frac{x}{[x^2 + 2x^3 + \frac{3}{2}x^4 + \frac{5}{9}x^5 + \frac{35}{288}x^6 + \cdots]} dx$

$= y_1(x) \int \left\{ \frac{1}{x + 2x^2 + \frac{3}{2}x^3 + \frac{5}{9}x^4 + \frac{35}{288}x^5 + \cdots} \right\} dx.$

Next, we can use long division to compute the power series for the quotient in the braces, obtaining

(16) $y_2(x) = y_1(x) \int \left\{ \frac{1}{x} - 2 + \frac{5}{2}x - \frac{23}{9}x^2 + \cdots \right\} dx.$

Integrating term by term, we then find

(17) $y_2(x) = y_1(x)\{\ln x - 2x + \frac{5}{4}x^2 - \frac{23}{27}x^3 + \cdots\}$

$= y_1(x)\ln x + y_1(x)(-2x + \frac{5}{4}x^2 - \frac{23}{27}x^3 + \cdots).$

Now we substitute the series for $y_1(x)$ given in (13) and use the Cauchy product to obtain

(18) $y_2(x) = y_1(x)\ln x + (x + x^2 + \frac{1}{4}x^3 + \cdots)(-2x + \frac{5}{4}x^2 - \frac{23}{27}x^3 + \cdots)$

$= y_1(x)\ln x + (-2x^2 - \frac{3}{4}x^3 - \frac{11}{108}x^4 + \cdots).$

Hence a general solution to equation (12) is

(19) $y(x) = c_1 y_1(x) + c_2 y_2(x), \qquad x > 0,$

where c_1 and c_2 are arbitrary constants and $y_1(x)$ and $y_2(x)$ are given by (13) and (18).   ●

The previous example illustrates the drawback to using the reduction of order formula to find a second linearly independent solution: We must square a series and then compute its reciprocal. As a result, it is very difficult to determine the general term in the series expansion for $y_2(x)$.

In Example 2 the roots of the indicial equation are equal, and we found a second linearly independent solution (18) that involves the first solution $y_1(x)$ multiplied by $\ln x$. This should not be too surprising if we recall the analogous situation for a Cauchy-Euler equation having an indicial equation with repeated roots. As we shall soon see, if an indicial equation has roots that differ by an integer, then the expansion of the second solution may also involve the term $y_1(x)\ln x$. In the following theorem we give the *form* of two linearly independent solutions for the three cases where the roots of the indicial equation (a) do not differ by an integer, (b) are equal, or (c) differ by a nonzero integer.

▶ FORM OF SECOND LINEARLY INDEPENDENT SOLUTION

Theorem 7.   Let x_0 be a regular singular point for $y'' + py' + qy = 0$ and let r_1 and r_2 be the roots of the associated indicial equation, where $r_1 \geq r_2$ $(\mathrm{Re}\, r_1 \geq \mathrm{Re}\, r_2)$.

(a) If $r_1 - r_2$ is not an integer, then there exist two linearly independent solutions of the form

$$(20) \qquad y_1(x) = \sum_{n=0}^{\infty} a_n(x - x_0)^{n+r_1}, \qquad a_0 \neq 0,$$

$$(21) \qquad y_2(x) = \sum_{n=0}^{\infty} b_n(x - x_0)^{n+r_2}, \qquad b_0 \neq 0.$$

(b) If $r_1 = r_2$, then there exist two linearly independent solutions of the form

$$(22) \qquad y_1(x) = \sum_{n=0}^{\infty} a_n(x - x_0)^{n+r_1}, \qquad a_0 \neq 0,$$

$$(23) \qquad y_2(x) = y_1(x)\ln(x - x_0) + \sum_{n=1}^{\infty} b_n(x - x_0)^{n+r_1}.$$

(c) If $r_1 - r_2$ is a positive integer, then there exist two linearly independent solutions of the form

$$(24) \qquad y_1(x) = \sum_{n=0}^{\infty} a_n(x - x_0)^{n+r_1}, \qquad a_0 \neq 0,$$

$$(25) \qquad y_2(x) = Cy_1(x)\ln(x - x_0) + \sum_{n=0}^{\infty} b_n(x - x_0)^{n+r_2}, \qquad b_0 \neq 0,$$

where C is a constant that could be zero.

In each case of the theorem, $y_1(x)$ is just the series solution obtained by the method of Frobenius, with $r = r_1$. When $r_1 - r_2$ is not an integer, the method of Frobenius yields a second linearly independent solution by taking $r = r_2$. In Section 8.8 we derive the formulas stated in Theorem 7 for cases (b) and (c). For now, let's see how knowing the form of the second solution enables us to obtain it. Again, for simplicity, we consider only indicial equations having real roots.

● **EXAMPLE 3**   Find the first few terms in the series expansion about the regular singular point $x = 0$ for two linearly independent solutions to

(26) $x^2 y''(x) - xy'(x) + (1 - x)y(x) = 0, \qquad x > 0.$

Solution   This is the same equation we considered in Example 2, where we used the reduction of order formula to find a second linearly independent solution. This time let's try to find a second solution using Theorem 7. Recall that $r_1 = r_2 = 1$ and so $y_2(x)$ has the form given in (23) (with $x_0 = 0$); that is,

(27) $y_2(x) = y_1(x) \ln x + \sum_{n=1}^{\infty} b_n x^{n+1},$

where (cf. (13))

(28) $y_1(x) = \sum_{k=0}^{\infty} \frac{1}{(k!)^2} x^{k+1}.$

Our goal is to determine the coefficients b_n by substituting $y_2(x)$ directly into equation (26). We begin by differentiating $y_2(x)$ in (27) to obtain

$$y_2'(x) = y_1'(x) \ln x + x^{-1} y_1(x) + \sum_{n=1}^{\infty} (n+1) b_n x^n,$$

$$y_2''(x) = y_1''(x) \ln x - x^{-2} y_1(x) + 2x^{-1} y_1'(x) + \sum_{n=1}^{\infty} n(n+1) b_n x^{n-1}.$$

Substituting $y_2(x)$ into (26) yields

(29) $x^2 \left\{ y_1''(x) \ln x - x^{-2} y_1(x) + 2x^{-1} y_1'(x) + \sum_{n=1}^{\infty} n(n+1) b_n x^{n-1} \right\}$

$$- x \left\{ y_1'(x) \ln x + x^{-1} y_1(x) + \sum_{n=1}^{\infty} (n+1) b_n x^n \right\}$$

$$+ (1-x) \left\{ y_1(x) \ln x + \sum_{n=1}^{\infty} b_n x^{n+1} \right\} = 0,$$

which simplifies to

(30) $\{ x^2 y_1''(x) - xy_1'(x) + (1-x)y_1(x) \} \ln x - 2y_1(x) + 2xy_1'(x)$

$$+ \sum_{n=1}^{\infty} n(n+1) b_n x^{n+1} - \sum_{n=1}^{\infty} (n+1) b_n x^{n+1} + \sum_{n=1}^{\infty} b_n x^{n+1} - \sum_{n=1}^{\infty} b_n x^{n+2} = 0.$$

Notice that the factor in front of $\ln x$ is just the left-hand side of equation (26) with $y = y_1$. Since y_1 is a solution to (26), this factor is zero. With this observation and a shift in the indices of summation, equation (30) can be rewritten as

(31) $$2xy_1'(x) - 2y_1(x) + b_1x^2 + \sum_{k=2}^{\infty}(k^2b_k - b_{k-1})x^{k+1} = 0.$$

Before we can set the coefficients equal to zero, we must substitute back in the series expansions for $y_1(x)$ and $y_1'(x)$. From (28), we see that $y_1'(x) = \sum_{k=0}^{\infty}(k+1)x^k/(k!)^2$, and inserting this series together with (28) into (31) we find

(32) $$\sum_{k=0}^{\infty}\frac{2(k+1)-2}{(k!)^2}x^{k+1} + b_1x^2 + \sum_{k=2}^{\infty}(k^2b_k - b_{k-1})x^{k+1} = 0.$$

Listing separately the $k = 0$ and $k = 1$ terms and combining the remaining terms gives

(33) $$(2 + b_1)x^2 + \sum_{k=2}^{\infty}\left[\frac{2k}{(k!)^2} + k^2 b_k - b_{k-1}\right]x^{k+1} = 0.$$

Next, we set the coefficients in (33) equal to zero. From the x^2 term we have $2 + b_1 = 0$, and so $b_1 = -2$. From the x^{k+1} term we obtain

$$\frac{2k}{(k!)^2} + k^2b_k - b_{k-1} = 0,$$

or

(34) $$b_k = \frac{1}{k^2}\left[b_{k-1} - \frac{2k}{(k!)^2}\right], \qquad k \geq 2.$$

Taking $k = 2$ and 3, we compute

(35) $$b_2 = \frac{1}{2^2}[b_1 - 1] = \frac{-3}{4}, \qquad b_3 = \frac{1}{9}\left[-\frac{3}{4} - \frac{6}{36}\right] = \frac{-11}{108}.$$

Hence, a second linearly independent solution to (26) is

(36) $$y_2(x) = y_1(x)\ln x - 2x^2 - \frac{3}{4}x^3 - \frac{11}{108}x^4 + \cdots. \quad \bullet$$

If we compare the different methods used in Examples 2 and 3 for obtaining a second linearly independent solution, we find that the technique of Example 3 has an advantage: It yields a recurrence relation, equation (34), that can be used to solve recursively for the coefficients b_k.

In the next two examples we consider the case when the difference between the roots of the indicial equation is a positive integer. In Example 4, it turns out that the constant C in formula (25) must be taken to be zero (i.e., no $\ln x$ term is present), while in Example 5, this constant is nonzero (i.e., a $\ln x$ term is present). Since the solutions to these examples require several intermediate computations, we do not display all the details. Rather, we encourage the reader to take an active part by bridging the gaps.

● **EXAMPLE 4**   Find the first few terms in the series expansion about the regular singular point $x = 0$ for a general solution to

(37) $xy''(x) + 4y'(x) - xy(x) = 0,$ $x > 0.$

Solution   In Example 5 of Section 8.6 we used the method of Frobenius to find a series expansion about $x = 0$ for a solution to equation (37). There we found the indicial equation to be $r^2 + 3r = 0$, which has roots $r_1 = 0$ and $r_2 = -3$. Working with $r_1 = 0$, we obtained the series solution

(38) $y_1(x) = 1 + \frac{1}{10}x^2 + \frac{1}{280}x^4 + \cdots$

(see equation (53) in Section 8.6, with $a_0 = 1$).

Since $r_1 - r_2 = 3$ is a positive integer, it follows from Theorem 7 that equation (37) has a second linearly independent solution of the form

(39) $y_2(x) = Cy_1(x)\ln x + \sum_{n=0}^{\infty} b_n x^{n-3}.$

When we substitute this expression for y_2 into equation (37), we obtain

(40) $x\left\{ Cy_1''(x)\ln x + 2Cx^{-1}y_1'(x) - Cx^{-2}y_1(x) + \sum_{n=0}^{\infty}(n-3)(n-4)b_n x^{n-5} \right\}$

$+ 4\left\{ Cy_1'(x)\ln x + Cx^{-1}y_1(x) + \sum_{n=0}^{\infty}(n-3)b_n x^{n-4} \right\}$

$- x\left\{ Cy_1(x)\ln x + \sum_{n=0}^{\infty} b_n x^{n-3} \right\} = 0,$

which simplifies to

(41) $\{xy_1''(x) + 4y_1'(x) - xy_1(x)\}C\ln x + 3Cx^{-1}y_1(x) + 2Cy_1'(x)$

$+ \sum_{n=0}^{\infty}(n-3)(n-4)b_n x^{n-4} + \sum_{n=0}^{\infty} 4(n-3)b_n x^{n-4} - \sum_{n=0}^{\infty} b_n x^{n-2} = 0.$

The factor in braces is zero because $y_1(x)$ satisfies equation (37). Combining the summations and simplifying, equation (41) becomes

(42) $3Cx^{-1}y_1(x) + 2Cy_1'(x) - 2b_1 x^{-3} + \sum_{k=2}^{\infty}[k(k-3)b_k - b_{k-2}]x^{k-4} = 0.$

Substituting in the series for $y_1(x)$ and writing out the first few terms of the summation in (42), we obtain

(43) $-2b_1 x^{-3} + (-2b_2 - b_0)x^{-2} + (3C - b_1)x^{-1} + (4b_4 - b_2)$

$+ (\frac{7}{10}C + 10b_5 - b_3)x + (18b_6 - b_4)x^2 + (\frac{11}{280}C + 28b_7 - b_5)x^3$

$+ \cdots = 0.$

Next, we set the coefficients equal to zero:

$$-2b_1 = 0 \Rightarrow b_1 = 0, \qquad\qquad -2b_2 - b_0 = 0 \Rightarrow b_2 = -\tfrac{1}{2}b_0,$$

$$3C - b_1 = 0 \Rightarrow C = \tfrac{1}{3}b_1 = 0, \qquad 4b_4 - b_2 = 0 \Rightarrow b_4 = \tfrac{1}{4}b_2 = -\tfrac{1}{8}b_0,$$

$$\tfrac{7}{10}C + 10b_5 - b_3 = 0 \Rightarrow b_5 = \frac{b_3 - \tfrac{7}{10}C}{10} = \tfrac{1}{10}b_3,$$

$$18b_6 - b_4 = 0 \Rightarrow b_6 = \tfrac{1}{18}b_4 = -\tfrac{1}{144}b_0,$$

$$\tfrac{11}{280}C + 28b_7 - b_5 = 0 \Rightarrow b_7 = \frac{b_5 - \tfrac{11}{280}C}{28} = \tfrac{1}{280}b_3.$$

Substituting the above values for C and the b_n's back into equation (39) gives

(44)
$$y_2(x) = b_0\{x^{-3} - \tfrac{1}{2}x^{-1} - \tfrac{1}{8}x - \tfrac{1}{144}x^3 + \cdots\}$$
$$+ b_3\{1 + \tfrac{1}{10}x^2 + \tfrac{1}{280}x^4 + \cdots\},$$

where b_0 and b_3 are arbitrary constants. Observe that the expression in braces following b_3 is just the series expansion for $y_1(x)$ given in equation (38). Hence, in order to obtain a second linearly independent solution, we must choose b_0 to be nonzero. Taking $b_0 = 1$ and $b_3 = 0$ gives

(45)
$$y_2(x) = x^{-3} - \tfrac{1}{2}x^{-1} - \tfrac{1}{8}x - \tfrac{1}{144}x^3 + \cdots.$$

Therefore, a general solution to equation (37) is

$$y(x) = c_1 y_1(x) + c_2 y_2(x), \qquad x > 0,$$

where $y_1(x)$ and $y_2(x)$ are given in (38) and (45). (Notice that the right-hand side of (44) is also a general solution to (37), with b_0 and b_3 as the two arbitrary constants.) ●

● **EXAMPLE 5**   Find the first few terms in the series expansion about the regular singular point $x = 0$ for two linearly independent solutions to

(46)
$$xy''(x) + 3y'(x) - xy(x) = 0, \qquad x > 0.$$

Solution   In Example 6 of Section 8.6 we used the method of Frobenius to find a series expansion about $x = 0$ for a solution to equation (46). The indicial equation turned out to be $r^2 + 2r = 0$, which has roots $r_1 = 0$ and $r_2 = -2$. Using $r_1 = 0$, we obtained the series solution

(47)
$$y_1(x) = 1 + \tfrac{1}{8}x^2 + \tfrac{1}{192}x^4 + \tfrac{1}{9216}x^6 + \cdots$$

(see equation (62) in Section 8.6 and put $a_0 = 1$).

Since $r_1 - r_2 = 2$ is a positive integer, it follows from Theorem 7 that equation (46) has a second linearly independent solution of the form

(48)
$$y_2(x) = Cy_1(x)\ln x + \sum_{n=0}^{\infty} b_n x^{n-2}.$$

Plugging the expansion for $y_2(x)$ into equation (46) and simplifying yields

$$(49) \quad \{xy_1''(x) + 3y_1'(x) - xy_1(x)\} C \ln x + 2Cx^{-1}y_1(x) + 2Cy_1'(x)$$

$$+ \sum_{n=0}^{\infty} (n-2)(n-3)b_n x^{n-3} + \sum_{n=0}^{\infty} 3(n-2)b_n x^{n-3} - \sum_{n=0}^{\infty} b_n x^{n-1} = 0.$$

Again the factor in braces is zero, because $y_1(x)$ is a solution to equation (46). If we combine the summations and simplify, equation (49) becomes

$$(50) \quad 2Cx^{-1}y_1(x) + 2Cy_1'(x) - b_1 x^{-2} + \sum_{k=2}^{\infty} [k(k-2)b_k - b_{k-2}]x^{k-3} = 0.$$

Substituting in the series expansions for $y_1(x)$ and $y_1'(x)$ and writing out the first few terms of the summation in (50) leads to

$$(51) \quad -b_1 x^{-2} + (2C - b_0)x^{-1} + (3b_3 - b_1) + (\tfrac{3}{4}C + 8b_4 - b_2)x$$
$$+ (15b_5 - b_3)x^2 + (\tfrac{5}{96}C + 24b_6 - b_4)x^3 + \cdots = 0.$$

When we set the coefficients in (51) equal to zero it turns out that we are free to choose C and b_2 as arbitrary constants:

$$-b_1 = 0 \qquad\qquad \Rightarrow b_1 = 0,$$
$$2C - b_0 = 0 \qquad\quad \Rightarrow b_0 = 2C \quad (C \text{ arbitrary}),$$
$$3b_3 - b_1 = 0 \qquad\quad \Rightarrow b_3 = \tfrac{1}{3}b_1 = 0,$$
$$8b_4 - b_2 + \tfrac{3}{4}C = 0 \quad \Rightarrow b_4 = \frac{b_2 - \tfrac{3}{4}C}{8} = \tfrac{1}{8}b_2 - \tfrac{3}{32}C \quad (b_2 \text{ arbitrary}),$$
$$15b_5 - b_3 = 0 \qquad\quad \Rightarrow b_5 = \tfrac{1}{15}b_3 = 0,$$
$$24b_6 - b_4 + \tfrac{5}{96}C = 0 \Rightarrow b_6 = \frac{b_4 - \tfrac{5}{96}C}{24} = \tfrac{1}{192}b_2 - \tfrac{7}{1152}C.$$

Substituting these values for C and the b_n's back into (48), we obtain the solution

$$(52) \quad y_2(x) = C\{y_1(x)\ln x + 2x^{-2} - \tfrac{3}{32}x^2 - \tfrac{7}{1152}x^4 + \cdots\}$$
$$+ b_2\{1 + \tfrac{1}{8}x^2 + \tfrac{1}{192}x^4 + \cdots\},$$

where C and b_2 are arbitrary constants. Since the second series is b_2 times the series expansion for $y_1(x)$, we obtain a second linearly independent solution by choosing $C = 1$ and $b_2 = 0$. Thus a second linearly independent solution is

$$(53) \quad y_2(x) = y_1(x)\ln x + 2x^{-2} - \tfrac{3}{32}x^2 - \tfrac{7}{1152}x^4 + \cdots. \quad \bullet$$

EXERCISES 8.7

In Problems 1 through 14 find at least the first three non-zero terms in the series expansion about $x = 0$ for a general solution to the given equation for $x > 0$. (These are the same equations as in Problems 19 through 32 of Exercises 8.6.)

1. $9x^2y'' + 9x^2y' + 2y = 0.$

2. $2x(x-1)y'' + 3(x-1)y' - y = 0.$
3. $x^2y'' + xy' + x^2y = 0.$
4. $xy'' + y' - 4y = 0.$
5. $x^2z'' + (x^2 + x)z' - z = 0.$

6. $3xy'' + (2 - x)y' - y = 0.$

7. $4x^2y'' + 2x^2y' - (x + 3)y = 0.$

8. $x^2y'' + (x^2 - x)y' + y = 0.$

9. $xw'' - w' - xw = 0.$

10. $3x^2y'' + 8xy' + (x - 2)y = 0.$

11. $xy'' + (x - 1)y' - 2y = 0.$

12. $x(x + 1)y'' + (x + 5)y' - 4y = 0.$

13. $xy'' + (1 - x)y' - y = 0.$

14. $x^2y'' - x(1 + x)y' + y = 0.$

In Problems 15 and 16 determine whether the given equation has a solution that is bounded near the origin, all solutions are bounded near the origin, or none of the solutions is bounded near the origin. (These are the same equations as in Problems 33 and 34 of Exercises 8.6.)

15. $3xy'' + 2(1 - x)y' - 4y = 0.$

16. $xy'' + (x + 2)y' - y = 0.$

In Problems 17 through 20 find at least the first three nonzero terms in the series expansion about $x = 0$ for a general solution to the given linear third order equation for $x > 0$. (These are the same equations as in Problems 35 through 38 in Exercises 8.6.)

17. $6x^3y''' + 13x^2y'' + (x + x^2)y' + xy = 0.$

18. $6x^3y''' + 11x^2y'' - 2xy' - (x - 2)y = 0.$

19. $6x^3y''' + 13x^2y'' - (x^2 + 3x)y' - xy = 0.$

20. $6x^3y''' + (13x^2 - x^3)y'' + xy' - xy = 0.$

21. **Buckling Columns.**   In the study of the buckling of a column whose cross section varies one encounters the equation

(54) $x^n y''(x) + \alpha^2 y(x) = 0, \qquad x > 0,$

where x is related to the height above the ground and y is the deflection away from the vertical. The positive constant α depends on the rigidity of the column, its moment of inertia at the top, and the load. The positive integer n depends upon the type of column. For example, when the column is a truncated cone (see Figure 8.5(a)), we have $n = 4$.

(a) Use the substitution $x = t^{-1}$ to reduce (54) with $n = 4$ to the form

$$\frac{d^2y}{dt^2} + \frac{2}{t}\frac{dy}{dt} + \alpha^2 y = 0, \qquad t > 0.$$

(b) Find at least the first six nonzero terms in the series expansion about $t = 0$ for a general solution to the equation obtained in part (a).

(c) Use the result of part (b) to give an expansion about $x = 0$ for a general solution to (54).

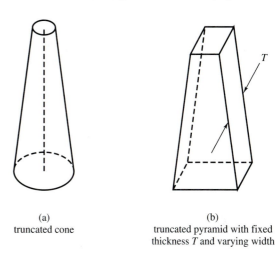

(a)
truncated cone

(b)
truncated pyramid with fixed
thickness T and varying width

Figure 8.5 Buckling columns

22. In Problem 21, consider a column with a rectangular cross section with two sides constant and the other two changing linearly (see Figure 8.5(b)). In this case $n = 1$. Find at least the first four nonzero terms in the series expansion about $x = 0$ for a general solution to equation (54) when $n = 1$.

23. Use the method of Frobenius and the reduction of order formula, equation (11), to find at least the first three nonzero terms in the series expansion about the irregular singular point $x = 0$ for a general solution to the differential equation

$$x^2y'' + y' - 2y = 0.$$

24. The equation

$$xy''(x) + (1 - x)y'(x) + ny(x) = 0,$$

where n is a nonnegative integer, is called **Laguerre's differential equation.** Show that for each n this equation has a polynomial solution of degree n. These polynomials are denoted by $L_n(x)$ and are called **Laguerre polynomials.** The first few Laguerre polynomials are

$$L_0(x) = 1, \qquad L_1(x) = -x + 1,$$
$$L_2(x) = x^2 - 4x + 2.$$

25. Use the results of Problem 24 and the reduction of order formula, equation (11), to obtain the first few terms in a series expansion about $x = 0$ for a general solution for $x > 0$ to Laguerre's differential equation for $n = 0$ and 1.

•8.8• MORE ON THE CASES WHEN ROOTS DIFFER BY AN INTEGER (Optional)

In this section we discuss another procedure for obtaining a second linearly independent solution when the indicial equation has equal roots or the roots differ by a positive integer. As a consequence of the approach, we will verify the formulas given in parts (b) and (c) of Theorem 7 in the previous section.

Suppose that $x = 0$ is a regular singular point for

(1) $$y''(x) + p(x)y'(x) + q(x)y(x) = 0$$

and that the indicial equation has equal roots, $r_1 = r_2$. As usual, let's try to find a solution to equation (1) of the form

(2) $$w(r, x) = x^r \sum_{n=0}^{\infty} a_n x^n = \sum_{n=0}^{\infty} a_n x^{n+r},$$

where a_0 is an arbitrary nonzero constant. Proceeding as in the method of Frobenius, we substitute the series $p(x) = \sum_{n=0}^{\infty} p_n x^{n-1}$, $q(x) = \sum_{n=0}^{\infty} q_n x^{n-2}$ and the expansions for $w(r, x)$, $w'(r, x)$, and $w''(r, x)$ into equation (1). After some algebra we arrive at the expansion

(3) $$[r(r-1) + p_0 r + q_0]a_0 x^{r-2}$$
$$+ [(r+1)ra_1 + p_0(r+1)a_1 + p_1 ra_0 + q_0 a_1 + q_1 a_0]x^{r-1} + \cdots = 0$$

(see equations (9)–(14) in Section 8.6). Until now our next step has been to set the coefficient of x^{r-2} equal to zero, which has the effect of forcing r to be a specific number, namely a root of the indicial equation. As we shall see, there is an advantage to treating r as a *variable*, and so, for the moment, we ignore the x^{r-2} term and set the remaining coefficients in (3) equal to zero. From the x^{r-1} term we get

(4) $$(r+1)ra_1 + p_0(r+1)a_1 + p_1 ra_0 + q_0 a_1 + q_1 a_0 = 0.$$

To simplify the computations, we take $a_0 = 1$ and then solve for a_1 in (4):

(5) $$a_1 = -\frac{p_1 r + q_1}{(r+1)r + p_0(r+1) + q_0}.$$

For a given differential equation, the quantities p_0, p_1, q_0, and q_1 are known, so (5) can be regarded as a formula for a_1 in terms of the variable r. That is, a_1 is a function of r, and we will denote it by $a_1(r)$. In a similar fashion, we set the successive coefficients in (3) equal to zero and obtain a_n in terms of r. Again we denote this functional dependence by $a_n(r)$.

Since we have chosen the $a_n(r)$'s so that all the coefficients in (3) are equal to zero except for the coefficient of x^{r-2}, substituting $w(r, x)$ into the left-hand side of equation (1) yields

(6) $$w''(r, x) + p(x)w'(r, x) + q(x)w(r, x) = [r(r-1) + p_0 r + q_0]x^{r-2}.$$

If we let L denote the differential operator defined by the left-hand side of (1), then we can rewrite (6) as

(7) $\quad L[w(r, x)](x) = [r(r - 1) + p_0 r + q_0] x^{r-2}.$

Now since we are assuming that the indicial equation has roots $r_1 = r_2$, then $r(r - 1) + p_0 r + q_0 = (r - r_1)^2$. Hence equation (7) has the form

(8) $\quad L[w(r, x)](x) = (r - r_1)^2 x^{r-2}.$

Setting $r = r_1$ gives $L[w(r_1, x)](x) = 0$, and therefore, as we know from the method of Frobenius, $w(r_1, x)$ is a solution to equation (1).

To find a second linearly independent solution, let's revive the procedure that effectively dealt with repeated roots associated with a Cauchy-Euler equation (see Section 8.5). Namely, we take the partial derivative with respect to r of both sides of equation (8). This gives

(9) $\quad \dfrac{\partial}{\partial r} L[w(r, x)](x) = 2(r - r_1) x^{r-2} + (r - r_1)^2 x^{r-2} \ln x.$

Setting $r = r_1$, we find

(10) $\quad \dfrac{\partial}{\partial r} L[w(r_1, x)](x) = 0.$

Assuming that $w(r, x)$ has continuous third order partial derivatives with respect to both r and x, we can then interchange the order of differentiation and obtain

(11) $\quad L\left[\dfrac{\partial w}{\partial r}(r_1, x) \right](x) = 0.$

Hence $\partial w / \partial r$ evaluated at $r = r_1$ is a second solution to equation (1). Let's now find its series expansion.

Since

(12) $\quad w(r, x) = x^r \displaystyle\sum_{n=0}^{\infty} a_n(r) x^n,$

then

(13) $\quad \dfrac{\partial w}{\partial r}(r, x) = x^r \ln x \displaystyle\sum_{n=0}^{\infty} a_n(r) x^n + x^r \displaystyle\sum_{n=0}^{\infty} a_n'(r) x^n.$

Here $a_n'(r)$ is the derivative of $a_n(r)$ with respect to r. Moreover, since we have taken $a_0(r) \equiv 1$, then $a_0'(r) \equiv 0$. Putting $r = r_1$ in (13), we then get

$$\dfrac{\partial w}{\partial r}(r_1, x) = \left\{ x^{r_1} \sum_{n=0}^{\infty} a_n(r_1) x^n \right\} \ln x + x^{r_1} \sum_{n=1}^{\infty} a_n'(r_1) x^n,$$

which simplifies to

(14) $\quad \dfrac{\partial w}{\partial r}(r_1, x) = w(r_1, x) \ln x + \displaystyle\sum_{n=1}^{\infty} a_n'(r_1) x^{n+r_1}.$

Notice that with $y_1(x) = w(r_1, x)$ and $b_n = a'_n(r_1)$, the right-hand side of (14) is the same as the formula for the second linearly independent solution claimed in Theorem 7, equation (23).

● **EXAMPLE 1** Find series expansions about the regular singular point $x = 0$ for two linearly independent solutions to

(15) $\qquad x^2 y''(x) - xy'(x) + (1 - x)y(x) = 0, \qquad x > 0.$

Solution We considered this same equation in Example 4 of Section 8.6 and in Examples 2 and 3 of Section 8.7. The indicial equation is $r^2 - 2r + 1 = 0$, which has roots $r_1 = r_2 = 1$. We now use our new procedure to find two linearly independent solutions.

Let

(16) $\qquad w(r, x) = x^r \displaystyle\sum_{n=0}^{\infty} a_n x^n = \sum_{n=0}^{\infty} a_n x^{n+r}.$

Taking the derivatives of $w(r, x)$ with respect to x, substituting them into (15), and setting the coefficients equal to zero yields the recurrence relation

(17) $\qquad a_k = \dfrac{1}{(r + k - 1)^2} a_{k-1}, \qquad k \geq 1,$

(see equation (41) in Section 8.6). Taking $a_0 = 1$ and solving equation (17) recursively for the a_k's in terms of r, we find

$$a_1 = \frac{1}{r^2} a_0 = \frac{1}{r^2}, \qquad a_2 = \frac{1}{(r + 1)^2} a_1 = \frac{1}{[(r + 1)(r)]^2},$$

$$a_3 = \frac{1}{(r + 2)^2} a_2 = \frac{1}{[(r + 2)(r + 1)r]^2},$$

and, in general,

(18) $\qquad a_k(r) = \dfrac{1}{[(r + k - 1)(r + k - 2)\cdots r]^2}, \qquad k \geq 1.$

Setting $r = r_1 = 1$ in equation (18) gives

(19) $\qquad a_k(1) = \dfrac{1}{[k(k - 1)\cdots 2 \cdot 1]^2} = \dfrac{1}{(k!)^2}, \qquad k \geq 1.$

Hence one solution is

(20) $\qquad y_1(x) = w(1, x) = \displaystyle\sum_{n=0}^{\infty} a_n(1) x^{n+1} = \sum_{n=0}^{\infty} \frac{1}{(n!)^2} x^{n+1}.$

Here we used the fact that $a_0 = 1 = (0!)^{-2}$.

To get a second linearly independent solution to (15), we must differentiate $a_k(r)$ in equation (18) with respect to r. This is easily done using logarithmic differentiation. We begin by taking the logarithm of both sides of (18), then implicitly differentiate with respect to r. That is,

(21) $\ln a_k(r) = \ln[(r + k - 1)(r + k - 2)\cdots r]^{-2}$
$$= -2\ln(r + k - 1) - 2\ln(r + k - 2) - \cdots - 2\ln r,$$

and, differentiating with respect to r, we get

$$\frac{a_k'(r)}{a_k(r)} = -\frac{2}{r + k - 1} - \frac{2}{r + k - 2} - \cdots - \frac{2}{r},$$

which yields

$$a_k'(r) = -2a_k(r)\left\{\frac{1}{r + k - 1} + \frac{1}{r + k - 2} + \cdots + \frac{1}{r}\right\}.$$

Setting $r = 1$ and using formula (19) for $a_k(1)$, we obtain

(22) $a_k'(1) = \dfrac{-2}{(k!)^2}\left\{\dfrac{1}{k} + \dfrac{1}{k - 1} + \cdots + \dfrac{1}{1}\right\}, \qquad k \geq 1.$

We now insert these values into formula (14), which we have shown gives a second linearly independent solution:

(23) $y_2(x) = \dfrac{\partial w}{\partial r}(1, x) = w(1, x)\ln x + \displaystyle\sum_{n=1}^{\infty} a_n'(1)x^{n+1}$

$$= w(1, x)\ln x + \sum_{n=1}^{\infty} \frac{-2}{(n!)^2}\left\{\frac{1}{n} + \frac{1}{n - 1} + \cdots + \frac{1}{1}\right\}x^{n+1}.$$

Thus two linearly independent solutions to equation (15) are given by the series for $y_1(x)$ in (20) and the series for $y_2(x)$ in (23).   ●

The previous procedure can be modified to accommodate the case when the roots of the indicial equation differ by a positive integer. For this purpose, the new ingredient is to treat a_0 also as a function of r. Proceeding as in the case of a repeated root, we must now include a_0 in equation (7):

(24) $L[w(r, x)](x) = [r(r - 1) + p_0 r + q_0]a_0 x^{r-2}$
$$= (r - r_1)(r - r_2)a_0 x^{r-2},$$

where r_1, r_2 are the roots of the indicial equation. As we shall soon see, it is convenient to put

(25) $a_0 = r - r_2,$

so that equation (24) becomes

(26) $L[w(r, x)](x) = (r - r_1)(r - r_2)^2 x^{r-2}.$

We now exploit the fact that the right-hand side of (26) has the repeated factor

$(r - r_2)^2$. Taking the partial derivative with respect to r of both sides of equation (26) gives

(27) $\dfrac{\partial}{\partial r} L[w(r, x)](x) = (r - r_2)^2 x^{r-2} + 2(r - r_1)(r - r_2) x^{r-2}$

$$+ (r - r_1)(r - r_2)^2 x^{r-2} \ln x.$$

Assuming, as before, that $w(r, x)$ has continuous third order partial derivatives with respect to r and x, then

(28) $\dfrac{\partial}{\partial r} L[w(r, x)](x) = L\left[\dfrac{\partial w}{\partial r}(r, x)\right](x).$

Hence on setting $r = r_2$ in (27), we see that

(29) $L\left[\dfrac{\partial w}{\partial r}(r_2, x)\right](x) = 0.$

Thus the partial derivative $\partial w/\partial r$ evaluated at $r = r_2$ is a solution to equation (1). To write out the series expansion for this solution, we first express $w(r, x)$ in the form

$$w(r, x) = x^r \sum_{n=0}^{\infty} a_n(r) x^n,$$

where we have indicated the fact that all the coefficients a_n (including a_0) are functions of r. Then

(30) $\dfrac{\partial w}{\partial r}(r, x) = x^r \ln x \sum_{n=0}^{\infty} a_n(r) x^n + x^r \sum_{n=0}^{\infty} a_n'(r) x^n$

$$= w(r, x) \ln x + x^r \sum_{n=0}^{\infty} a_n'(r) x^n,$$

where $a_n'(r)$ is the derivative of $a_n(r)$ with respect to r. Setting $r = r_2$, we obtain the solution

(31) $\dfrac{\partial w}{\partial r}(r_2, x) = w(r_2, x) \ln x + \sum_{n=0}^{\infty} a_n'(r_2) x^{n+r_2},$

where $a_0 = r - r_2$.

It is clear that if the $a_n(r)$'s are defined for $r = r_2$ and $w(r_2, x) \not\equiv 0$, then $w(r_2, x)$ and $(\partial w/\partial r)(r_2, x)$ are linearly independent solutions because of the factor $\ln x$. What is not so obvious, but nevertheless true, is that $a_R(r)$ where $R = r_1 - r_2$, may fail to be defined for $r = r_2$, so that the preceding method cannot be employed. Fortunately, in this situation two linearly independent solutions can be obtained from $w(r_2, x)$ by treating a_0 and a_R as arbitrary constants (see Problem 9).

The preceding analysis suggests the following procedure when the roots of the indicial equation for (1) differ by a positive integer J. First determine $w(r, x)$, where the coefficients $a_n(r)$ depend on r. If substituting $r = r_2$ leaves a_0 *and* a_J arbitrary, then $w(r_2, x)$ is a general solution, with a_0 and a_J the arbitrary constants. If this substitution leaves a_J arbitrary but $a_i \equiv 0$ for $i = 0, 1, \ldots, J - 1$, then $w(r_2, x) = w(r_1, x)$ is one solution and the second solution is given by $(\partial w / \partial r)(r_2, x)$ (see Problem 10).†

The preceding methods are particularly useful in determining the form of the general term in the series expansion of a second linearly independent solution when the recurrence relation is a two-term relation. With suitable modifications these methods can also be used for higher order equations.

EXERCISES 8.8

In Problems 1 through 8 use the method discussed in this section to obtain the series expansion about the regular singular point $x = 0$ for two linearly independent solutions to the given equation whose indicial equation has repeated roots.

1. $xy'' + y' - 4y = 0$.

2. $xy'' + (1 - x)y' - y = 0$.

3. $x^2 y'' + (x^2 - x)y' + y = 0$.

4. $x^2 y'' + xy' + x^2 y = 0$.

5. $x^2 y'' + 3x(1 + x)y' + y = 0$.

6. $x^2 y'' - xy' + (1 - x^3)y = 0$.

7. $xy'' + y' - 3y = 0$.

8. $x^2 y'' + (3x - x^2)y' + y = 0$.

9. To find a series expansion for a general solution to
$$xy''(x) + 4y'(x) - xy(x) = 0, \qquad x > 0:$$
   (a) Show that the coefficients of $w(r, x)$ satisfy the recurrence relation
$$(k + r + 1)(k + r + 4)a_{k+1} - a_{k-1} = 0, \qquad k \geq 1.$$
   (b) Setting $r = r_2$, show that not only is a_0 arbitrary, but so is a_3.
   (c) Use the recurrence relation to obtain a series expansion for a general solution with a_0 and a_3 as the arbitrary constants. Compare your answer with Example 4 in Section 8.7.

10. To find a series expansion for a general solution to
$$xy''(x) + 3y'(x) - xy(x) = 0, \qquad x > 0:$$
   (a) Show that the coefficients of $w(r, x)$ satisfy the recurrence relation
$$(k + r + 1)(k + r + 3)a_{k+1} - a_{k-1} = 0, \qquad k \geq 1.$$
   (b) Show that $a_1 \equiv 0$ and hence
$$a_{2n+1}(r) \equiv 0, \qquad n = 0, 1, 2, \ldots.$$
   (c) Set $r = -2$ and use the recurrence relation to show

that $a_0(-2)$ must be zero.
   (d) Let $a_0(r) = r + 2$ so that $a_0(-2) = 0$. Show that $a_2(r) = (r + 4)^{-1}$ and, for $n \geq 2$,
$$a_{2n}(r) = \frac{1}{(r + 2n + 2)[(r + 2n)^2 \cdots (r + 6)^2(r + 4)^2]}.$$
   (e) Find the series solution $w(-2, x)$.
   (f) Let $a_0(r) = r + 2$ and show that $a_0'(r) = 1$, $a_2'(r) = -(r + 4)^{-2}$, and, for $n \geq 2$,
$$a_{2n}'(r) = \frac{(-1)}{(r + 2n + 2)[(r + 2n)^2 \cdots (r + 4)^2]}$$
$$\times \left\{ \frac{1}{r + 2n + 2} + \frac{2}{r + 2n} + \cdots + \frac{2}{r + 4} \right\}.$$
   [*Hint: Use logarithmic differentiation.*]
   (g) Find the series solution $(\partial w / \partial r)(-2, x)$. Compare your answers with Example 5 in Section 8.7.

In Problems 11 through 16 use the methods discussed in this section to obtain the series expansion about the regular singular point $x = 0$ for two linearly independent solutions for $x > 0$ to the given equation whose roots of the indicial equation differ by a positive integer.

11. $x(1 - x)y'' - 3y' + 2y = 0$.

12. $4x^2 y'' + 2x^2 y' - (x + 3)y = 0$.

13. $xy'' + (x + 2)y' - y = 0$.

14. $xy'' - y' - xy = 0$.

15. $xy'' + (x - 1)y' - 2y = 0$.

16. $xy'' + xy' + y = 0$.

In Problems 17 through 20 use the method discussed in this section to obtain the series expansion about the regular singular point $x = 0$ for three linearly independent solutions to the given linear third order equation for $x > 0$. [Hint: If the

---

† Further details are given in *Intermediate Differential Equations*, Second Edition, by Earl D. Rainville, Macmillan, New York, 1964, Sections 30–32.

indicial equation has a root of multiplicity three, it is helpful to consider $\partial^2 w/\partial r^2$.]

17. $2x^3y''' + (5x^2 - x^3)y'' + xy' - xy = 0.$

18. $3x^3y''' + 5x^2y'' + (x^2 - x)y' + xy = 0.$

19. $x^2y''' + 3xy'' + y' - y = 0.$

20. $x^3y''' + (x^3 + 6x^2)y'' - (x^2 - 7x)y' + (x + 1)y = 0.$

21. In Problem 10, let $a_0 = r - r_1 = r$ (instead of $r - r_2$), then solve for $a_n(r)$. Show that if you now set $r = r_1 = 0$, then you obtain $y_3(x) = w(0, x) \equiv 0$, and

$$y_4(x) = \frac{\partial w}{\partial r}(0, x) = 0 \cdot \ln x + \sum_{n=0}^{\infty} b_n x^n.$$

Hence you get only one linearly independent solution.

• 8.9 •   SPECIAL FUNCTIONS (Optional)

In advanced work in applied mathematics, engineering, and physics a few special second order equations arise with amazing frequency. These equations have been extensively studied, and volumes have been written on the properties of their solutions. Three of these equations are the hypergeometric equation, Bessel's equation, and Legendre's equation. The solutions to these and other equations that occur in applications are often referred to as **special functions.** We shall briefly consider the three equations mentioned above. For a more detailed study of special functions we refer the reader to *Basic Hypergeometric Series*, by G. Gasper and M. Rahman, Cambridge University Press, Cambridge, 1990; *Special Functions*, by E. D. Rainville, Macmillan, New York, 1960; and *Higher Transcendental Functions*, by A. Erdelyi (ed.), McGraw-Hill, New York, 1953, 3 volumes.

Hypergeometric Equation

The linear second order differential equation

(1) $x(1 - x)y'' + [\gamma - (\alpha + \beta + 1)x]y' - \alpha\beta y = 0,$

where α, β, and γ are fixed parameters, is called the **hypergeometric equation.** This equation has singular points at $x=0$ and 1, both of which are regular. Thus a series expansion about $x=0$ for a solution to (1) obtained by the method of Frobenius will converge at least for $0 < x < 1$ (see Theorem 6, page 381). To find this expansion, observe that the indicial equation associated with $x = 0$ is

$$r(r - 1) + \gamma r = r(r - (1 - \gamma)) = 0,$$

which has roots 0 and $1 - \gamma$. If γ is not an integer, then we can use the root $r = 0$ to obtain a solution to (1) of the form

(2) $y_1(x) = \sum_{n=0}^{\infty} a_n x^n.$

Substituting $y_1(x)$ given in (2) into (1), shifting indices, and simplifying ultimately leads to the equation

(3) $\sum_{n=1}^{\infty} [n(n + \gamma - 1)a_n - (n + \alpha - 1)(n + \beta - 1)a_{n-1}]x^{n-1} = 0.$

Setting the series coefficients equal to zero yields the recurrence relation

(4) $n(n + \gamma - 1)a_n - (n + \alpha - 1)(n + \beta - 1)a_{n-1} = 0, \qquad n \geq 1.$

Since $n \geq 1$ and γ is not an integer, there is no fear of dividing by zero when we rewrite (4) as

(5) $$a_n = \frac{(n + \alpha - 1)(n + \beta - 1)}{n(n + \gamma - 1)} a_{n-1}.$$

Solving recursively for a_n, we obtain

(6) $$a_n = \frac{\alpha(\alpha + 1)\cdots(\alpha + n - 1)\beta(\beta + 1)\cdots(\beta + n - 1)}{n!\gamma(\gamma + 1)\cdots(\gamma + n - 1)} a_0, \qquad n \geq 1.$$

If we employ the **factorial function** $(t)_n$, which is defined for nonnegative integers n by

(7) $$(t)_n := t(t + 1)(t + 2)\cdots(t + n - 1), \qquad n \geq 1,$$
$$(t)_0 := 1, \qquad t \neq 0,$$

then we can express a_n more compactly as

(8) $$a_n = \frac{(\alpha)_n(\beta)_n}{n!(\gamma)_n} a_0, \qquad n \geq 1.$$

If we take $a_0 = 1$ and then substitute the expression for a_n in (8) into (2), we obtain the following solution to the hypergeometric equation:

(9) $$y_1(x) = 1 + \sum_{n=1}^{\infty} \frac{(\alpha)_n(\beta)_n}{n!(\gamma)_n} x^n.$$

The solution given in (9) is called a **Gaussian hypergeometric function** and is denoted by the symbol $F(\alpha, \beta; \gamma; x)$.[†] That is,

(10) $$F(\alpha, \beta; \gamma; x) := 1 + \sum_{n=1}^{\infty} \frac{(\alpha)_n(\beta)_n}{n!(\gamma)_n} x^n.$$

Hypergeometric functions are generalizations of the geometric series. To see this, observe that for any constant β that is not zero or a negative integer,

$$F(1, \beta; \beta; x) = 1 + x + x^2 + x^3 + \cdots.$$

It is interesting to note that many other familiar functions can be expressed in terms of the hypergeometric function. For example,

(11) $$F(\alpha, \beta; \beta; x) = (1 - x)^{-\alpha},$$
(12) $$F(1, 1; 2; x) = -x^{-1}\ln(1 - x),$$

(13) $$F\left(\frac{1}{2}, 1; \frac{3}{2}; x^2\right) = \frac{1}{2}x^{-1}\ln\left(\frac{1 + x}{1 - x}\right),$$

(14) $$F\left(\frac{1}{2}, 1; \frac{3}{2}; -x^2\right) = x^{-1}\arctan x.$$

[†] *Historical Footnote*: A detailed study of this function was done by Carl Friedrich Gauss in 1813. The mathematical historian E. T. Bell refers to Gauss as the Prince of Mathematicians.

We leave the verification of these formulas for the reader.

To obtain a second linearly independent solution to (1) when γ is not an integer, we use the other root, $1-\gamma$, of the indicial equation and seek a solution of the form

$$\text{(15)} \qquad y_2(x) = \sum_{n=0}^{\infty} b_n x^{n+1-\gamma}.$$

Substituting $y_2(x)$ into equation (1) and solving for b_n, we eventually arrive at

$$\text{(16)} \qquad y_2(x) = x^{1-\gamma} + \sum_{n=1}^{\infty} \frac{(\alpha+1-\gamma)_n(\beta+1-\gamma)_n}{n!(2-\gamma)_n} x^{n+1-\gamma}.$$

Factoring out $x^{1-\gamma}$, we see that the second solution $y_2(x)$ can be expressed in terms of a hypergeometric function. That is,

$$\text{(17)} \qquad y_2(x) = x^{1-\gamma}F(\alpha+1-\gamma, \beta+1-\gamma; 2-\gamma; x).$$

When γ is an integer, one of the formulas given in (9) or (16) (corresponding to the larger root, 0 or $1-\gamma$) still gives a solution. We then use the techniques of this chapter to obtain a second linearly independent solution, which may or may not involve a logarithmic term. We omit a discussion of these solutions.

In many books the hypergeometric function is expressed in terms of the gamma function $\Gamma(x)$ instead of the factorial function. Recall that in Section 7.6 we defined

$$\text{(18)} \qquad \Gamma(x) := \int_0^{\infty} e^{-u}u^{x-1}\,du, \qquad x > 0$$

and showed that

$$\text{(19)} \qquad \Gamma(x+1) = x\Gamma(x), \qquad x > 0.$$

It follows from repeated use of relation (19) that

$$\text{(20)} \qquad (t)_n = \frac{\Gamma(t+n)}{\Gamma(t)}$$

for $t > 0$ and n any nonnegative integer. Using relation (20), we can express the hypergeometric function as

$$\text{(21)} \qquad F(\alpha, \beta; \gamma; x) = \frac{\Gamma(\gamma)}{\Gamma(\alpha)\Gamma(\beta)} \sum_{n=0}^{\infty} \frac{\Gamma(\alpha+n)\Gamma(\beta+n)}{n!\Gamma(\gamma+n)} x^n.$$

Bessel's Equation

The linear second order differential equation

$$\text{(22)} \qquad x^2y'' + xy' + (x^2 - \nu^2)y = 0,$$

where $\nu \geq 0$ is a fixed parameter, is called **Bessel's equation of order ν.** This equation has a regular singular point at $x = 0$ and no other singular points in the complex plane. Hence a series solution for (22) obtained by the method of Frobenius will converge for $0 < x < \infty$. The indicial equation for (22) is

$$r(r-1) + r - \nu^2 = (r-\nu)(r+\nu) = 0,$$

which has roots $r_1 = v$ and $r_2 = -v$. If v is not an integer, then the method of Frobenius yields two linearly independent solutions, given by

(23) $y_1(x) = a_0 \sum_{n=0}^{\infty} \frac{(-1)^n}{2^{2n}n!(1+v)_n} x^{2n+v},$

(24) $y_2(x) = b_0 \sum_{n=0}^{\infty} \frac{(-1)^n}{2^{2n}n!(1-v)_n} x^{2n-v}.$

If in (23) we take

$$a_0 = \frac{1}{2^v \Gamma(1+v)},$$

then it follows from relation (20) that the function

(25) $J_v(x) := \sum_{n=0}^{\infty} \frac{(-1)^n}{n!\Gamma(1+v+n)} \left(\frac{x}{2}\right)^{2n+v}$

is a solution to (22). We call $J_v(x)$ the **Bessel function of the first kind of order v.**[†] Similarly, taking

$$b_0 = \frac{1}{2^{-v}\Gamma(1-v)}$$

in equation (24) gives the solution

(26) $J_{-v}(x) := \sum_{n=0}^{\infty} \frac{(-1)^n}{n!\Gamma(1-v+n)} \left(\frac{x}{2}\right)^{2n-v},$

which is the Bessel function of the first kind of order $-v$. When $r_1 - r_2 = 2v$ is not an integer, we know by Theorem 7 of Section 8.7 that $J_v(x)$ and $J_{-v}(x)$ are linearly independent. Moreover, it can be shown that if v is not an integer, even though $2v$ is, then $J_v(x)$ and $J_{-v}(x)$ are still linearly independent.

What happens in the remaining case when v is a nonnegative integer, say $v = m$? While $J_m(x)$ is still a solution, the function $J_{-m}(x)$ is not even properly defined, because formula (26) will involve the gamma function evaluated at a nonpositive integer. From a more in-depth study of the gamma function, it turns out that $1/\Gamma(k) = 0$, for $k = 0, -1, -2, \dots$. Hence (26) becomes

(27) $J_{-m}(x) = \sum_{n=0}^{\infty} \frac{(-1)^n}{n!\Gamma(1-m+n)} \left(\frac{x}{2}\right)^{2n-m} = \sum_{n=m}^{\infty} \frac{(-1)^n}{n!\Gamma(1-m+n)} \left(\frac{x}{2}\right)^{2n-m}.$

Comparing (27) with the formula (25) for $J_m(x)$, we see (after a shift in index) that

(28) $J_{-m}(x) = (-1)^m J_m(x),$

[†] *Historical Footnote*: Frederic Wilhelm Bessel (1784–1846) started his career in commercial navigation and later became an astronomer. In 1817, Bessel introduced the functions $J_v(x)$ in his study of planetary orbits.

which means that $J_{-m}(x)$ and $J_m(x)$ are linearly *dependent*. To resolve the problem of finding a second linearly independent solution in the case when v is a nonnegative integer, we can use the method discussed in Section 8.8 (see Problem 27). There is, however, another approach to this problem, which we now describe.

For v not an integer, we can take linear combinations of $J_v(x)$ and $J_{-v}(x)$ to obtain other solutions to (22). In particular, let

(29) $$Y_v(x) := \frac{\cos(v\pi)J_v(x) - J_{-v}(x)}{\sin(v\pi)}, \qquad v \text{ not an integer,}$$

for $x > 0$. The function $Y_v(x)$ is called the **Bessel function of the second kind of order v,** and, as can be verified, $J_v(x)$ and $Y_v(x)$ are linearly independent. Notice that when v is an integer, the denominator in (29) is zero; but, by formula (28), so is the numerator! Hence it is reasonable to hope that, in a limiting sense, formula (29) is still meaningful. In fact, using l'Hôpital's rule, it is possible to show that, for m a nonnegative integer, the function defined by

(30) $$Y_m(x) := \lim_{v \to m} \frac{\cos(v\pi)J_v(x) - J_{-v}(x)}{\sin(v\pi)}$$

for $x > 0$, is a solution to (22) with $v = m$. Furthermore, $J_m(x)$ and $Y_m(x)$ are linearly independent. We again call $Y_m(x)$ the Bessel function of the second kind of order m. In the literature the function $Y_m(x)$ is also denoted by $N_m(x)$ and referred to as the **Neumann function** or **Weber function.**

Figure 8.6 shows the graphs of $J_0(x)$ and $J_1(x)$, and Figure 8.7 the graphs of $Y_0(x)$ and $Y_1(x)$. Notice that the curves for $J_0(x)$ and $J_1(x)$ behave like damped sine waves and have zeros that interlace (see Problem 28). The fact that the Bessel functions J_n have an infinite number of zeros will be helpful in certain applications (see Project A in Chapter 11).

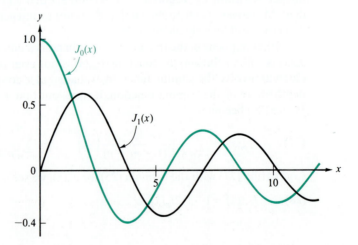

Figure 8.6 Graphs of the Bessel functions $J_0(x)$ and $J_1(x)$

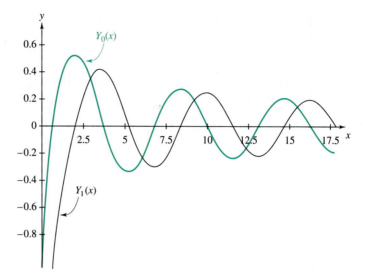

Figure 8.7 Graphs of the Bessel functions $Y_0(x)$ and $Y_1(x)$

There are several useful recurrence relations involving Bessel functions. For example,

(31) $\qquad \dfrac{d}{dx}[x^v J_v(x)] = x^v J_{v-1}(x),$

(32) $\qquad \dfrac{d}{dx}[x^{-v} J_v(x)] = -x^{-v} J_{v+1}(x),$

(33) $\qquad J_{v+1}(x) = \dfrac{2v}{x} J_v(x) - J_{v-1}(x),$

(34) $\qquad J_{v+1}(x) = J_{v-1}(x) - 2J'_v(x).$

Furthermore, analogous equations hold for Bessel functions of the second kind.

To illustrate the techniques involved in proving the recurrence relations, let's verify relation (31). We begin by substituting series (25) for $J_v(x)$ into the left-hand side of (31). Differentiating, we get

(35) $\qquad \dfrac{d}{dx}[x^v J_v(x)] = \dfrac{d}{dx}\left\{ x^v \sum_{n=0}^{\infty} \dfrac{(-1)^n}{n!\Gamma(1+v+n)}\left(\dfrac{x}{2}\right)^{2n+v} \right\}$

$\qquad\qquad\qquad\qquad = \dfrac{d}{dx}\left\{ \sum_{n=0}^{\infty} \dfrac{(-1)^n x^{2n+2v}}{n!\Gamma(1+v+n)2^{2n+v}} \right\}$

$\qquad\qquad\qquad\qquad = \sum_{n=0}^{\infty} \dfrac{(-1)^n(2n+2v)x^{2n+2v-1}}{n!\Gamma(1+v+n)2^{2n+v}}.$

Since $\Gamma(1+v+n) = (v+n)\Gamma(v+n)$, we have from (35)

$\qquad \dfrac{d}{dx}[x^v J_v(x)] = \sum_{n=0}^{\infty} \dfrac{(-1)^n 2x^{2n+2v-1}}{n!\Gamma(v+n)2^{2n+v}},$

and factoring out an x^ν gives

$$\frac{d}{dx}[x^\nu J_\nu(x)] = x^\nu \sum_{n=0}^\infty \frac{(-1)^n}{n!\,\Gamma(1+(\nu-1)+n)}\left(\frac{x}{2}\right)^{2n+\nu-1} = x^\nu J_{\nu-1}(x),$$

as claimed in equation (31). We leave the verifications of the remaining relations as exercises (see Problems 22–24).

Legendre's Equation

The linear second order differential equation

(36) $(1-x^2)y'' - 2xy' + n(n+1)y = 0,$

where n is a fixed nonnegative integer, is called **Legendre's equation.** [†] This equation has a regular singular point at 1, and hence a series solution for (36) about $x = 1$ may be obtained by the method of Frobenius. By setting $z = x - 1$, equation (36) is transformed into

(37) $z(z+2)\dfrac{d^2y}{dz^2} + 2(z+1)\dfrac{dy}{dz} - n(n+1)y = 0.$

The indicial equation for (37) at $z = 0$ is

$$r(r-1) + r = r^2 = 0,$$

which has roots $r_1 = r_2 = 0$. Upon substituting

$$y(z) = \sum_{k=0}^\infty a_k z^k$$

into (37) and proceeding as usual, we arrive at the solution

(38) $y_1(x) = 1 + \displaystyle\sum_{k=1}^\infty \frac{(-n)_k(n+1)_k}{k!\,(1)_k}\left(\frac{1-x}{2}\right)^k,$

where we have expressed y_1 in terms of the original variable x and have taken $a_0 = 1$. We have written the solution in the above form because it is now obvious from (38) that

(39) $y_1(x) = F\left(-n,\, n+1;\, 1;\, \dfrac{1-x}{2}\right),$

where F is the Gaussian hypergeometric function defined in (10).

For n a nonnegative integer, the factor

$$(-n)_k = (-n)(-n+1)(-n+2)\cdots(-n+k-1)$$

will be zero for $k \geq n+1$. Hence the solution given in (38) and (39) is a *polynomial* of degree n. Moreover, $y_1(1) = 1$. These polynomial solutions of equation (36) are called the **Legendre polynomials** or **spherical polynomials** and are traditionally denoted by $P_n(x)$. That is,

(40) $P_n(x) := 1 + \displaystyle\sum_{k=1}^n \frac{(-n)_k(n+1)_k}{k!\,(1)_k}\left(\frac{1-x}{2}\right)^k.$

[†] *Historical Footnote:* Solutions to this equation were obtained by Adrien Marie Legendre (1752–1833) in 1785 and are referred to as Legendre functions.

If we expand about $x = 0$, then $P_n(x)$ takes the form

$$(41) \qquad P_n(x) = 2^{-n} \sum_{m=0}^{[n/2]} \frac{(-1)^m (2n - 2m)!}{(n - m)! m! (n - 2m)!} x^{n-2m},$$

where $[n/2]$ is the greatest integer less than or equal to $n/2$ (see Problem 34). The first three Legendre polynomials are

$$P_0(x) = 1, \qquad P_1(x) = x, \qquad P_2(x) = \tfrac{3}{2} x^2 - \tfrac{1}{2}.$$

The Legendre polynomials satisfy the **orthogonality condition**

$$(42) \qquad \int_{-1}^{1} P_m(x) P_n(x) \, dx = 0 \quad \text{for} \quad n \neq m.$$

To see this we first rewrite equation (36) in what is called the **selfadjoint form:**

$$(43) \qquad [(1 - x^2) y']' + n(n + 1) y = 0.$$

Since $P_n(x)$ and $P_m(x)$ satisfy (43) with parameters n and m respectively, we have

$$(44) \qquad [(1 - x^2) P_n'(x)]' + n(n + 1) P_n(x) = 0,$$
$$(45) \qquad [(1 - x^2) P_m'(x)]' + m(m + 1) P_m(x) = 0.$$

Multiplying (44) by $P_m(x)$ and (45) by $P_n(x)$ and then subtracting, we find

$$P_m(x)[(1 - x^2) P_n'(x)]' - P_n(x)[(1 - x^2) P_m'(x)]'$$
$$+ [n(n + 1) - m(m + 1)] P_m(x) P_n(x) = 0,$$

which can be rewritten in the form

$$(46) \qquad (n^2 - m^2 + n - m) P_m(x) P_n(x) = P_n(x)[(1 - x^2) P_m'(x)]' - P_m(x)[(1 - x^2) P_n'(x)]'.$$

It is a straightforward calculation to show that the right-hand side of (46) is just

$$\{(1 - x^2)[P_n(x) P_m'(x) - P_n'(x) P_m(x)]\}'.$$

Using this fact and the identity $n^2 - m^2 + n - m = (n - m)(n + m + 1)$, equation (46) becomes

$$(47) \qquad (n - m)(n + m + 1) P_m(x) P_n(x) = \{(1 - x^2)[P_n(x) P_m'(x) - P_n'(x) P_m(x)]\}'.$$

Integrating both sides of (47) from $x = -1$ to $x = 1$ yields

$$(48) \qquad (n - m)(n + m + 1) \int_{-1}^{1} P_m(x) P_n(x) \, dx$$

$$= \int_{-1}^{1} \{(1 - x^2)[P_n(x) P_m'(x) - P_n'(x) P_m(x)]\}' \, dx$$

$$= \{(1 - x^2)[P_n(x) P_m'(x) - P_n'(x) P_m(x)]\} \Big|_{-1}^{1}$$

$$= 0,$$

because $1 - x^2 = 0$ for $x = \pm 1$. Since n and m are nonnegative integers with $n \neq m$, then $(n - m)(n + m + 1) \neq 0$, and so equation (42) follows from (48).

Legendre polynomials also satisfy the **recurrence formula**

(49) $(n+1)P_{n+1}(x) = (2n+1)xP_n(x) - nP_{n-1}(x)$

and **Rodrigues's formula**

(50) $P_n(x) = \dfrac{1}{2^n n!}\dfrac{d^n}{dx^n}\{(x^2-1)^n\}$

(see Problems 32 and 33).

The Legendre polynomials are generated by the function $(1-2xz+z^2)^{-1/2}$ in the sense that

(51) $(1-2xz+z^2)^{-1/2} = \sum_{n=0}^{\infty} P_n(x)z^n, \qquad |z|<1, \qquad |x|<1.$

That is, if we expand $(1-2xz+z^2)^{-1/2}$ in a Taylor series about $z=0$, treating x as a fixed parameter, then the coefficients of z^n are the Legendre polynomials $P_n(x)$. The function $(1-2xz+z^2)^{-1/2}$ is called a **generating function** for $P_n(x)$ and can be derived from the recurrence formula (49) (see Problem 35).

The Legendre polynomials are an example of a class of special functions called **classical orthogonal polynomials.**† The latter includes, for example, the **Jacobi polynomials,** $P_n^{(\alpha,\beta)}(x)$; the **Gegenbauer** or **ultraspherical polynomials,** $C_n^{\lambda}(x)$; **Chebyshev** (Tchebichef) **polynomials,** $T_n(x)$ and $U_n(x)$; **Laguerre polynomials,** $L_n^{\alpha}(x)$; and the **Hermite polynomials,** $H_n(x)$. The properties of the classical orthogonal polynomials can be found in the books mentioned earlier in this section or in the *Handbook of Mathematical Functions with Formulas, Graphs, and Mathematical Tables*, by M. Abramowitz and I. A. Stegun (eds.), Dover, New York, 1965, Chapter 22.

EXERCISES 8.9

In Problems 1 through 4 express a general solution to the given equation using Gaussian hypergeometric functions.

1. $x(1-x)y'' + (\frac{1}{2}-4x)y' - 2y = 0.$
2. $3x(1-x)y'' + (1-27x)y' - 45y = 0.$
3. $2x(1-x)y'' + (1-6x)y' - 2y = 0.$
4. $2x(1-x)y'' + (3-10x)y' - 6y = 0.$

In Problems 5 through 8 verify the following formulas by expanding each function in a power series about $x = 0$.

5. $F(1,1;2;x) = -x^{-1}\ln(1-x).$
6. $F(\alpha,\beta;\beta;x) = (1-x)^{-\alpha}.$
7. $F(\frac{1}{2},1;\frac{3}{2};x^2) = \frac{1}{2}x^{-1}\ln\left(\dfrac{1+x}{1-x}\right).$
8. $F(\frac{1}{2},1;\frac{3}{2};-x^2) = x^{-1}\arctan x.$

In Problems 9 and 10 use one of the methods discussed in Section 8.7 to obtain two linearly independent solutions to the given hypergeometric equation.

9. $x(1-x)y'' + (1-3x)y' - y = 0.$
10. $x(1-x)y'' + (2-2x)y' - \frac{1}{4}y = 0.$
11. Show that the **confluent hypergeometric equation**

$$xy'' + (\gamma - x)y' - \alpha y = 0,$$

† Here orthogonality is used in the more general sense that $\int_a^b p_n(x)p_m(x)w(x)\,dx = 0$ for $n \neq m$, where $w(x)$ is a weight function on the interval (a, b).

where α and γ are fixed parameters and γ is not an integer, has two linearly independent solutions

$$y_1(x) = {}_1F_1(\alpha; \gamma; x) := 1 + \sum_{n=1}^{\infty} \frac{(\alpha)_n}{n!(\gamma)_n} x^n$$

and

$$y_2(x) = x^{1-\gamma}{}_1F_1(\alpha + 1 - \gamma; 2 - \gamma; x).$$

12. Use the property of the gamma function given in (19) to derive relation (20).

In Problems 13 through 18 express a general solution to the given equation using Bessel functions of either the first or second kind.

13. $4x^2y'' + 4xy' + (4x^2 - 1)y = 0.$

14. $9x^2y'' + 9xy' + (9x^2 - 16)y = 0.$

15. $x^2y'' + xy' + (x^2 - 1)y = 0.$

16. $x^2y'' + xy' + x^2y = 0.$

17. $9t^2x'' + 9tx' + (9t^2 - 4)x = 0.$

18. $x^2z'' + xz' + (x^2 - 16)z = 0.$

In Problems 19 and 20 a Bessel equation is given. For the appropriate choice of ν, the Bessel function $J_\nu(x)$ is one solution. Use one of the methods discussed in Section 8.7 to obtain a second linearly independent solution.

19. $x^2y'' + xy' + (x^2 - 1)y = 0.$

20. $x^2y'' + xy' + (x^2 - 4)y = 0.$

21. Show that $x^\nu J_\nu(x)$ satisfies the equation

$$xy'' + (1 - 2\nu)y' + xy = 0, \qquad x > 0,$$

and use this result to find a solution for the equation

$$xy'' - 2y' + xy = 0, \qquad x > 0.$$

In Problems 22 through 24 derive the indicated recurrence formulas.

22. Formula (32).

23. Formula (33).

24. Formula (34).

25. Show that $J_{1/2}(x) = (2/\pi x)^{1/2} \sin x$ and

$$J_{-1/2}(x) = (2/\pi x)^{1/2} \cos x.$$

26. The Bessel functions of order $\nu = n + \frac{1}{2}$, n any integer, are called the **spherical Bessel functions.** Use relation (33) and the results of Problem 25 to show that the spherical Bessel functions can be represented in terms of $\sin x$, $\cos x$, and powers of x. Demonstrate this by determining a closed form for $J_{-3/2}(x)$ and $J_{5/2}(x)$.

27. Use the method discussed in Section 8.8 to determine a second linearly independent solution to Bessel's equation of order zero in terms of the Bessel function $J_0(x)$.

28. Show that between two consecutive positive roots (zeros) of $J_1(x)$ there is a root of $J_0(x)$. This interlacing property of the roots of Bessel functions is illustrated in Figure 8.6. [Hint: Use relation (31) and Rolle's theorem from calculus.]

29. Use formula (41) to determine the first five Legendre polynomials.

30. Show that the Legendre polynomials of even degree are even functions of x, while those of odd degree are odd functions.

31. **(a)** Show that the orthogonality condition (42) for Legendre polynomials implies that

$$\int_{-1}^{1} P_n(x)q(x)\,dx = 0$$

for *any* polynomial $q(x)$ of degree at most $n - 1$. [Hint: The polynomials $P_0, P_1, \ldots, P_{n-1}$ are linearly independent and hence span the space of all polynomials of degree at most $n - 1$. Thus $q(x) = a_0P_0(x) + a_1P_1(x) + \cdots + a_{n-1}P_{n-1}(x)$ for suitable constants a_k.]

(b) Prove that if $Q_n(x)$ is a polynomial of degree n such that

$$\int_{-1}^{1} Q_n(x)P_k(x)\,dx = 0 \quad \text{for} \quad k = 0, 1, \ldots, n - 1,$$

then $Q_n(x) = cP_n(x)$ for some constant c. [Hint: Select c so that the coefficient of x^n for $Q_n(x) - cP_n(x)$ is zero. Then, since $P_0, \ldots, P_{n-1}$ is a basis,

$$Q_n(x) - cP_n(x) = a_0P_0(x) + \cdots + a_{n-1}P_{n-1}(x).$$

Multiply the last equation by $P_k(x)$ $(0 \le k \le n - 1)$ and integrate from $x = -1$ to $x = 1$ to show that each a_k is zero.]

32. Deduce the recurrence formula (49) for Legendre polynomials by completing the following steps.

(a) Show that the function $Q_{n-1}(x) := (n+1)P_{n+1}(x) - (2n + 1)xP_n(x)$ is a polynomial of degree $n - 1$. [Hint: Compute the coefficient of the x^{n+1} term using the representation (40). The coefficient of x^n is also zero because $P_{n+1}(x)$ and $xP_n(x)$ are both odd or both even functions, a consequence of Problem 30.]

(b) Using the result of Problem 31(a), show that

$$\int_{-1}^{1} Q_{n-1}(x)P_k(x)\,dx = 0 \quad \text{for} \quad k = 0, 1, \ldots, n-2.$$

(c) From Problem 31(b) conclude that $Q_{n-1}(x) = cP_{n-1}(x)$ and, by taking $x = 1$, show that $c = -n$. [Hint: Recall that $P_m(1) = 1$ for all m.] From the definition of $Q_{n-1}(x)$ in part (a), the recurrence formula now follows.

33. To prove Rodrigues's formula (50) for Legendre polynomials, complete the following steps.

(a) Let $v_n := \dfrac{d^n}{dx^n}\{(x^2 - 1)^n\}$ and show that $v_n(x)$ is a polynomial of degree n with the coefficient of x^n equal to $(2n)!/n!$.

(b) Use integration by parts n times to show that, for any polynomial $q(x)$ of degree less than n,

$$\int_{-1}^{1} v_n(x)q(x)\,dx = 0.$$

[Hint: For example, when $n = 2$,

$$\int_{-1}^{1} \frac{d^2}{dx^2}\{(x^2 - 1)^2\}q(x)\,dx$$

$$= q(x)\frac{d}{dx}\{(x^2 - 1)^2\}\Big|_{-1}^{1} - \{q'(x)(x^2 - 1)\}\Big|_{-1}^{1}$$

$$+ \int_{-1}^{1} q''(x)(x^2 - 1)^2\,dx.$$

Since $n = 2$, the degree of $q(x)$ is at most 1, and so $q''(x) \equiv 0$. Thus

$$\int_{-1}^{1} \frac{d^2}{dx^2}\{(x^2 - 1)^2\}q(x)\,dx = 0.]$$

(c) Use the result of Problem 31(b) to conclude that $P_n(x) = cv_n(x)$ and show that $c = 1/2^n n!$ by comparing the coefficients of x^n in $P_n(x)$ and $v_n(x)$.

34. Use Rodrigues's formula (50) to obtain the representation (41) for the Legendre polynomials $P_n(x)$. Hint: From the binomial formula,

$$P_n(x) = \frac{1}{2^n n!} \frac{d^n}{dx^n}\{(x^2 - 1)^n\}$$

$$= \frac{1}{2^n n!} \frac{d^n}{dx^n}\left\{\sum_{m=0}^{n} \frac{n!(-1)^m}{(n-m)!m!}x^{2n-2m}\right\}.$$

35. The generating function in (51) for Legendre polynomials can be derived from the recurrence formula (49) as follows. Let x be fixed and set $f(z) := \sum_{n=0}^{\infty} P_n(x)z^n$. The goal is to determine an explicit formula for $f(z)$.

(a) Show that multiplying each term in the recurrence formula (49) by z^n and summing the terms from

$n = 1$ to ∞ leads to the differential equation

$$\frac{df}{dz} = \frac{x - z}{1 - 2xz + z^2}f.$$

Hint:

$$\sum_{n=1}^{\infty}(n+1)P_{n+1}(x)z^n = \sum_{n=0}^{\infty}(n+1)P_{n+1}(x)z^n - P_1(x)$$

$$= \frac{df}{dz} - x.$$

(b) Solve the differential equation derived in part (a) and use the initial condition $f(0) = P_0(x) \equiv 1$ to obtain $f(z) = (1 - 2xz + z^2)^{-1/2}$.

36. Find a series solution about $x = 0$ for the equation

$$(1 - x^2)y'' - 2xy' + 2y = 0$$

by first finding a polynomial solution and then using the reduction of order formula given in equation (11) in Section 8.7, page 387, or refer to the inside back cover.

37. The **Hermite polynomials** $H_n(x)$ are polynomial solutions to Hermite's equation

$$y'' - 2xy' + 2ny = 0.$$

The Hermite polynomials are generated by

$$e^{2tx - t^2} = \sum_{n=0}^{\infty} \frac{H_n(x)}{n!}t^n.$$

Use this equation to determine the first four Hermite polynomials.

38. The **Chebyshev** (Tchebichef) **polynomials** $T_n(x)$ are polynomial solutions to Chebyshev's equation

$$(1 - x^2)y'' - xy' + n^2 y = 0.$$

The Chebyshev polynomials satisfy the recurrence relation

$$T_{n+1}(x) = 2xT_n(x) - T_{n-1}(x),$$

with $T_0(x) = 1$ and $T_1(x) = x$. Use this recurrence relation to determine the next three Chebyshev polynomials.

39. The **Laguerre polynomials** $L_n(x)$ are polynomial solutions to Laguerre's equation

$$xy'' + (1 - x)y' + ny = 0.$$

The Laguerre polynomials satisfy Rodrigues's formula,

$$L_n(x) = \frac{e^x}{n!} \frac{d^n}{dx^n}(x^n e^{-x}).$$

Use this formula to determine the first four Laguerre polynomials.

40. Reduction to Bessel's Equation. The class of equations of the form

$$(52) \qquad y''(x) + cx^n y(x) = 0, \qquad x > 0,$$

where c and n are positive constants, can be solved by transforming the equation into Bessel's equation.

(a) First, use the substitution $y = x^{1/2} z$ to transform (52) into an equation involving x and z.

(b) Second, use the substitution

$$s = \frac{2\sqrt{c}}{n+2} x^{n/2+1},$$

to transform the equation obtained in part (a) into the Bessel equation

$$s^2 \frac{d^2 z}{ds^2} + s \frac{dz}{ds} + \left(s^2 - \frac{1}{(n+2)^2} \right) z = 0, \qquad s > 0.$$

(c) A general solution to the equation in part (b) can be given in terms of Bessel functions of the first and second kind. Substituting back in for s and z, obtain a general solution for equation (52).

● CHAPTER SUMMARY ●

Power series solutions to differential equations are useful alternatives when explicit solutions involving elementary functions cannot be found.

Power Series

Every power series $\sum_{n=0}^{\infty} a_n (x - x_0)^n$ has a **radius of convergence** ρ, $0 \le \rho \le \infty$, such that the series converges absolutely for $|x - x_0| < \rho$ and diverges when $|x - x_0| > \rho$. By the ratio test,

$$\frac{1}{\rho} = \lim_{n \to \infty} \frac{|a_{n+1}|}{|a_n|},$$

provided that this limit exists as an extended real number. A function $f(x)$ that is the sum of a power series in some open interval about x_0 is said to be **analytic at x_0**. If f is analytic at x_0, its power series representation about x_0 is the **Taylor series**

$$f(x) = \sum_{n=0}^{\infty} \frac{f^{(n)}(x_0)}{n!} (x - x_0)^n.$$

Taylor Series Method

If it is known that the solution $y(x)$ to the initial value problem

$$y' = F(x, y), \qquad y(x_0) = y_0$$

is analytic at the initial point x_0, then the first few terms of its Taylor expansion about x_0 can be computed by successively differentiating the equation:

$$y(x_0) = y_0, \qquad y'(x_0) = F(x_0, y_0),$$

$$y''(x_0) = \frac{\partial F}{\partial x}(x_0, y_0) + \frac{\partial F}{\partial y}(x_0, y_0) F(x_0, y_0), \qquad \text{etc.}$$

However, this Taylor series method involves increasingly tedious computations.

Power Series Method for an Ordinary Point

In the case of a linear equation of the form

(1) $y'' + p(x)y' + q(x)y = 0,$

where p and q are analytic at x_0, the point x_0 is called an **ordinary point,** and the equation has a pair of linearly independent solutions expressible as power series about x_0. The radii of convergence of these series solutions are at least as large as the distance from x_0 to the nearest singularity (real or complex) of the equation. To find power series solutions to (1), we substitute $y(x) = \sum_{n=0}^{\infty} a_n(x - x_0)^n$ into (1), group like terms, and set the coefficients of the resulting power series equal to zero. This leads to a recurrence relation for the coefficients a_n, which, in some cases, may even yield a general formula for the a_n. The same method applies to the nonhomogeneous version of (1), provided that the forcing function is also analytic at x_0.

Regular Singular Points

If, in equation (1), either p or q fails to be analytic at x_0, then x_0 is a **singular point** of (1). If x_0 is a singular point for which $(x - x_0)p(x)$ and $(x - x_0)^2q(x)$ are both analytic at x_0, then x_0 is a **regular singular point.** The Cauchy-Euler equation,

(2) $ax^2\dfrac{d^2y}{dx^2} + bx\dfrac{dy}{dx} + cy = 0,$ $x > 0,$

has a regular singular point at $x = 0$, and a general solution to (2) can be obtained by substituting $y = x^r$ and examining the roots of the resulting indicial equation $ar^2 + (b - a)r + c = 0.$

Method of Frobenius

For an equation of the form (1) with a regular singular point at x_0, a series solution can be found by the **method of Frobenius.** This is obtained by substituting

$$w(r, x) = (x - x_0)^r \sum_{n=0}^{\infty} a_n(x - x_0)^n$$

into (1), finding a recurrence relation for the coefficients, and choosing $r = r_1$, the larger root of the **indicial equation**

(3) $r(r - 1) + p_0 r + q_0 = 0,$

where $p_0 := \lim_{x \to x_0}(x - x_0)p(x)$, $q_0 := \lim_{x \to x_0}(x - x_0)^2q(x).$

Finding a Second Linearly Independent Solution

If the two roots r_1, r_2 of the indicial equation (3) do not differ by an integer, then a second linearly independent solution to (1) can be found by taking $r = r_2$ in the method of Frobenius. However, if $r_1 = r_2$ or $r_1 - r_2$ is a positive integer, then discovering a second solution requires a different approach. This may be a reduction of order procedure, the utilization of Theorem 7 which gives the *forms* of the solutions, or an operator approach in which, in the method of Frobenius, the coefficients a_n are treated as functions of r.

Special Functions

Some special functions in physics and engineering that arise from series solutions to linear second order equations with polynomial coefficients are Gaussian hypergeometric functions, $F(\alpha, \beta; \gamma; x)$; Bessel functions $J_\nu(x)$; and orthogonal polynomials such as those of Legendre, Chebyshev, Laguerre, and Hermite.

REVIEW PROBLEMS

1. Use the Taylor series method to determine the first four nonzero terms of a series solution for the given initial value problem.

(a) $y' = xy - y^2$; $\quad y(0) = 1$.

(b) $z'' - x^3 z' + xz^2 = 0$; $\quad z(0) = -1$, $\quad z'(0) = 1$.

2. Determine all the singular points of the given equation and classify them as regular or irregular.

(a) $(x^2 - 4)^2 y'' + (x - 4)y' + xy = 0$.

(b) $(\sin x)y'' + y = 0$.

3. Find at least the first four nonzero terms in a power series expansion about $x = 0$ for a general solution to the given equation.

(a) $y'' + x^2 y' - 2y = 0$.

(b) $y'' + e^{-x} y' - y = 0$.

4. Find a general formula for the coefficient a_n in a power series expansion about $x = 0$ for a general solution to the given equation.

(a) $(1 - x^2)y'' + xy' + 3y = 0$.

(b) $(x^2 - 2)y'' + 3y = 0$.

5. Find at least the first four nonzero terms in a power series expansion about $x = 2$ for a general solution to

$$w'' + (x - 2)w' - w = 0.$$

6. Use the substitution $y = x^r$ to find a general solution to the given equation for $x > 0$.

(a) $2x^2 y''(x) + 5xy'(x) - 12y(x) = 0$.

(b) $x^3 y'''(x) + 3x^2 y''(x) - 2xy'(x) - 2y(x) = 0$.

7. Use the method of Frobenius to find at least the first four nonzero terms in the series expansion about $x = 0$ for a solution to the given equation for $x > 0$.

(a) $x^2 y'' - 5xy' + (9 - x)y = 0$.

(b) $x^2 y'' + (x^2 + 2x)y' - 2y = 0$.

8. Find the indicial equation and its roots and state (but do not compute) the form of the series expansion about $x = 0$ (as in Theorem 7 on page 389) for two linearly independent solutions of the given equation for $x > 0$.

(a) $x^2 y'' + (\sin x)y' - 4y = 0$.

(b) $2xy'' + 5y' + xy = 0$.

(c) $(x \sin x)y'' + xy' + (\tan x)y = 0$.

9. Find at least the first three nonzero terms in the series expansion about $x = 0$ for a general solution to the given equation for $x > 0$.

(a) $x^2 y'' - x(1 + x)y' + y = 0$.

(b) $xy'' + y' - 2y = 0$.

(c) $2xy'' + 6y' + y = 0$.

(d) $x^2 y'' + (x - 2)y = 0$.

10. Express a general solution to the given equation using Gaussian hypergeometric functions or Bessel functions.

(a) $x(1 - x)y'' + (\frac{1}{2} - 6x)y' - 6y = 0$.

(b) $9\theta^2 y'' + 9\theta y' + (9\theta^2 - 1)y = 0$.

TECHNICAL WRITING EXERCISES

1. Knowing that a general solution to a nonhomogeneous linear second order equation can be expressed as a particular solution plus a general solution to the corresponding homogeneous equation, what can you say about the form of a general power series solution to the nonhomogeneous equation about an ordinary point?

2. Discuss advantages and disadvantages of power series solutions over numerical solutions.

3. Explain why it is not very helpful to know that $y(x) = \sum_{n=0}^{\infty} a_n x^n$ when you are solving nonlinear equations with variable coefficients.

4. Give an overview of the method of Frobenius that can be understood by a typical calculus student.

GROUP PROJECTS FOR CHAPTER 8

▶ **A. Spherically Symmetric Solutions to Schrödinger's Equation for the Hydrogen Atom**

In quantum mechanics one is interested in determining the wave function and energy states of an atom. These are determined from Schrödinger's equation. In the case of the hydrogen atom, it is possible to find wave functions ψ that are functions only of r, the distance from the proton to the electron. Such functions are called **spherically symmetric** and satisfy the simpler equation

(1)
$$\frac{1}{r}\frac{d^2}{dr^2}(r\psi) = \frac{-8m\pi^2}{h^2}\left(E + \frac{e_0^2}{r}\right)\psi,$$

where e_0^2, m, and h are constants, and E, also a constant, represents the energy of the atom, which we assume here to be negative.

(a) Show that with the substitutions

$$r = \frac{h^2}{4\pi^2 m e_0^2}\rho, \qquad E = \frac{2\pi^2 m e_0^4}{h^2}\varepsilon,$$

equation (1) reduces to

$$\frac{d^2(\rho\psi)}{d\rho^2} = -\left(\varepsilon + \frac{2}{\rho}\right)\rho\psi.$$

(b) If $f := \rho\psi$, then the preceding equation becomes

(2)
$$\frac{d^2 f}{d\rho^2} = -\left(\varepsilon + \frac{2}{\rho}\right)f.$$

Show that the substitution $f(\rho) = e^{-\alpha\rho}g(\rho)$, where α is a positive constant, transforms (2) into

(3)
$$\frac{d^2 g}{d\rho^2} - 2\alpha\frac{dg}{d\rho} + \left(\frac{2}{\rho} + \varepsilon + \alpha^2\right)g = 0.$$

(c) If we choose $\alpha^2 = -\varepsilon$ (ε negative), then (3) becomes

(4)
$$\frac{d^2 g}{d\rho^2} - 2\alpha\frac{dg}{d\rho} + \frac{2}{\rho}g = 0.$$

Show that a power series solution $g(\rho) = \sum_{k=1}^{\infty} a_k \rho^k$ (starting with $k = 1$) for (4) must have coefficients a_k that satisfy the recurrence relation

(5)
$$a_{k+1} = \frac{2(\alpha k - 1)}{k(k+1)}a_k, \qquad k \geq 1.$$

(d) Now for $a_1 = 1$ and k very large, $a_{k+1} \approx (2\alpha/k)a_k$ and so $a_{k+1} \approx (2\alpha)^k/k!$, which are the coefficients for $\rho e^{2\alpha\rho}$. Hence g acts like $\rho e^{2\alpha\rho}$, and so $f(\rho) = e^{-\alpha\rho}g(\rho)$ is like $\rho e^{\alpha\rho}$. Going back further we then see that $\psi \approx e^{\alpha\rho}$. Therefore, when $r = h^2\rho/4\pi^2 m e_0^2$ is large, so is ψ. Roughly speaking, $\psi^2(r)$ is proportional to the probability of finding an electron a distance r from the proton. Thus the above argument would imply that the electron in a hydrogen atom is more likely to be found at a very large distance from the proton! Since this makes no sense physically, we ask: Do there exist positive values for α for which ψ remains bounded as r becomes large?

Show that when $\alpha = 1/n$, $n = 1, 2, 3, \ldots$, then $g(\rho)$ is a polynomial of degree n, and argue that ψ is therefore bounded.

(e) Let E_n and $\psi_n(\rho)$ denote, respectively, the energy state and wave function corresponding to $\alpha = 1/n$. Find E_n (in terms of the constants e_0^2, m, and h) and $\psi_n(\rho)$ for $n = 1, 2,$ and 3.

B. Airy's Equation

In aerodynamics one encounters the following initial value problem for **Airy's equation:**

$$y'' + xy = 0; \qquad y(0) = 1, \qquad y'(0) = 0.$$

(a) Find the first ten terms in a power series expansion about $x = 0$ for the solution, and graph this polynomial for $-10 \leq x \leq 10$.

(b) Using the Runge-Kutta method (see Section 5.6) with $h = 0.05$, approximate the solution on the interval $[0, 10]$, i.e., at the points 0.05, 0.1, 0.15, etc.

(c) Using the Runge-Kutta method with $h = 0.05$, approximate the solution on the interval $[-10, 0]$. [Hint: With the change of variables $z = -x$, it suffices to approximate the solution to $y'' - zy = 0$; $y(0) = 1$, $y'(0) = 0$, on the interval $[0, 10]$.]

(d) Using your knowledge of constant coefficient equations as a basis for guessing the behavior of the solutions to Airy's equation, decide whether the power series approximation obtained in part (a) or the numerical approximation obtained in parts (b) and (c) better describes the true behavior of the solution on the interval $[-10, 10]$.

C. Buckling of a Tower

A tower is constructed of four angle beams connected by diagonals (see Figure 8.8). The deflection curve $y(x)$ for the tower is governed by the equation

(6) $\qquad x^2 \dfrac{d^2 y}{dx^2} + \alpha^2 y = 0, \qquad a < x < a + L,$

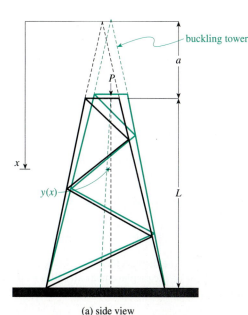

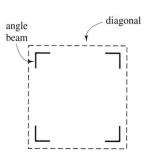

(a) side view   (b) cross section

Figure 8.8 Buckling tower

where x is related to the height above the ground, y is the deflection from the vertical of the curve passing through the center of the buckled tower, α is a positive constant, L is the height of the tower, and a is the length "cut" from the tower. Here $\alpha^2 = Pa^2/EI_1$ where P is the load, E is the modulus of elasticity, and I_1 is the moment of inertia at the top of the tower. We want to determine the critical load P_c at which buckling begins.

(a) Solve equation (6). [Hint: Equation (6) is a Cauchy-Euler equation.]
(b) Since $y = 0$ when $x = a$, show that

$$y = Ax^{1/2} \sin[\beta \ln(x/a)],$$

where $\beta := \sqrt{\alpha^2 - 1/4}$ and A is an arbitrary constant.
(c) Since the bottom of the tower is fixed in place, we must have $dy/dx = 0$ at $x = a + L$. Show that this gives

$$A\left\{ \tan\left[\beta \ln\left(\frac{a+L}{a}\right) \right] + 2\beta \right\} = 0.$$

(d) Use the result of part (c) to argue that *no* buckling takes place if $0 < \beta < \beta_c$, where β_c is the smallest positive real number that makes the expression in braces zero.
(e) The value of the load corresponding to β_c is called the **critical load** P_c. Solve for P_c in terms of β_c, a, E, and I_1.
(f) Find the critical load P_c if $a = 10$, $L = 40$, and $EI_1 = 1000$.

D. Aging Spring and Bessel Functions

In Problems 30 and 31 in Exercises 8.4, page 371, we discussed a model for a spring-mass system with an aging spring. Without damping, the displacement $x(t)$ at time t is governed by the equation

(7) $mx''(t) + ke^{-\eta t}x(t) = 0,$

where m, k, and η are positive constants. The general solution to this equation can be expressed using Bessel functions.

(a) The coefficient of x suggests a change of variables of the form $s = \alpha e^{\beta t}$. Show that (7) transforms into

(8) $\beta^2 s^2 \dfrac{d^2 x}{ds^2} + \beta^2 s \dfrac{dx}{ds} + \dfrac{k}{m}\left(\dfrac{s}{\alpha}\right)^{-\eta/\beta} x = 0.$

(b) Show that choosing α and β to satisfy

$$\frac{-\eta}{\beta} = 2 \quad \text{and} \quad \frac{k}{m\beta^2 \alpha^{-\eta/\beta}} = 1,$$

transforms (8) into the Bessel equation of order zero in s and x.
(c) Using the result of part (b), show that a general solution to (7) is given by

$$x(t) = c_1 J_0\left(\frac{2}{\eta}\sqrt{k/m}\,e^{-\eta t/2}\right) + c_2 Y_0\left(\frac{2}{\eta}\sqrt{k/m}\,e^{-\eta t/2}\right),$$

where J_0 and Y_0 are the Bessel functions of order zero of the first and second kind, respectively.
(d) Discuss the behavior of the displacement $x(t)$ for c_2 positive, negative, and zero.
(e) Compare the behavior of the displacement $x(t)$ for η a small positive number and η a large positive number.

CHAPTER 9

Systems of Differential Equations and Their Applications

Remark: *Chapter 9 uses the elimination method to solve systems of differential equations. The reader with a background in matrix algebra may first want to study Chapter 10, which has a more complete description of linear systems. The reader can then return to Chapter 9 to study applications in Section 9.4, numerical methods in Section 9.5, phase plane analysis in Section 9.6, or dynamical systems and chaos in Section 9.7.*

9.1 INTRODUCTION: ANALYSIS OF AN ELECTRIC NETWORK

At time $t = 0$ the charge on the capacitor in the electric network shown in Figure 9.1 is 2 coulombs, while the current through the capacitor is zero. Determine the charge on the capacitor and the currents in the various branches of the network at any time $t > 0$.

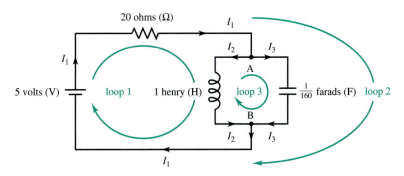

Figure 9.1 Schematic of an electric network

In Section 5.4 we derived a mathematical model for a simple RLC series circuit. There we used the following fundamental conservation laws, referred to as **Kirchhoff's laws:**

1. In a given branch of a network the current I passing through each of the elements must be the same.
2. The algebraic sum of the instantaneous changes in potential (voltage drops) around a closed circuit (loop) must be zero.

For electric networks consisting of more than one circuit, an additional conservation law, also due to Kirchhoff, is required:

3. In an electric network, the algebraic sum of the currents flowing to a junction point must be zero.

To determine the charge and currents in the electric network in Figure 9.1 on page 419, we begin by observing that the network consists of three closed circuits: loop 1 through the battery, resistor, and inductor; loop 2 through the battery, resistor, and capacitor; and loop 3 containing the capacitor and inductor. Taking advantage of the first conservation law, we denote the current passing through the battery and the resistor by I_1, the current through the inductor by I_2, and the current through the capacitor by I_3. For consistency of notation, we denote the charge on the capacitor by q_3; hence, $I_3 = dq_3/dt$.

As discussed in Section 5.4, the voltage drop at a resistor is RI, at an inductor $L\,dI/dt$, and at a capacitor q/C. So, applying Kirchhoff's second law to the electric network in Figure 9.1, we find for loop 1,

(1) $$\frac{dI_2}{dt} + 20I_1 = 5;$$

$$\underset{\text{(inductor)}}{} \quad \underset{\text{(resistor)}}{} \quad \underset{\text{(battery)}}{}$$

for loop 2,

(2) $$20I_1 + 160q_3 = 5;$$

$$\underset{\text{(resistor)}}{} \quad \underset{\text{(capacitor)}}{} \quad \underset{\text{(battery)}}{}$$

and for loop 3,

(3) $$-\frac{dI_2}{dt} + 160q_3 = 0.$$

$$\underset{\text{(inductor)}}{} \quad \underset{\text{(capacitor)}}{}$$

(The minus sign in (3) arises from taking a clockwise path around loop 3, so that the current passing through the inductor is $-I_2$.) Notice that these three equations are not independent: we can obtain equation (3) by subtracting (1) from (2). Hence, we have only two equations from which to determine the three unknowns I_1, I_2, and q_3. If we now apply the third conservation law to the two junction points in the network, we find at point A that $I_1 - I_2 - I_3 = 0$, and at point B that $I_2 + I_3 - I_1 = 0$. In both cases we get $I_1 = I_2 + I_3$, where $I_3 = dq_3/dt$. Substituting for I_1 in (2) now gives

(4) $$20I_2 + 20\frac{dq_3}{dt} + 160q_3 = 5.$$

From equations (3) and (4), we can eliminate the function I_2 and its derivatives and obtain a single second order equation for q_3. This can be accomplished if we multiply equation (3) by 20, differentiate equation (4) with respect to t, and then add the resulting equations. This yields

(5) $$20\frac{d^2 q_3}{dt^2} + 160\frac{dq_3}{dt} + 3200q_3 = 0.$$

To obtain the initial conditions for (5), recall that at time $t = 0$ the charge on the capacitor is 2 coulombs and the current is zero. Hence

(6) $$q_3(0) = 2, \quad \frac{dq_3}{dt}(0) = 0.$$

We can now solve the initial value problem (5)–(6) using the techniques of Chapter 4 or 7. Ultimately we find

$$q_3(t) = 2e^{-4t}\cos 12t + \frac{2}{3}e^{-4t}\sin 12t,$$

$$I_3(t) = \frac{dq_3}{dt}(t) = -\frac{80}{3}e^{-4t}\sin 12t.$$

Next, to determine I_2, we substitute these expressions into (4) and obtain

$$I_2(t) = \frac{1}{4} - \frac{dq_3}{dt}(t) - 8q_3(t)$$

$$= \frac{1}{4} - 16e^{-4t}\cos 12t + \frac{64}{3}e^{-4t}\sin 12t.$$

Finally, from $I_1 = I_2 + I_3$, we obtain

$$I_1(t) = \frac{1}{4} - 16e^{-4t}\cos 12t - \frac{16}{3}e^{-4t}\sin 12t.$$

In retrospect, we see that applying Kirchhoff's laws to the network of Figure 9.1 leads to the system consisting of equations (3) and (4). In this chapter we describe two techniques for solving similar systems. The first, which we successfully used to solve the network problem, is called the **elimination method** and involves transforming a system into a single higher order equation (as in (5)). The second method involves the use of Laplace transforms to convert the linear system of differential equations with initial conditions into a linear system of algebraic equations. A third technique, involving matrix algebra, is developed in Chapter 10.

In Section 9.4 we study the applications to multiple compartment systems and to mechanical systems with multiple degrees of freedom. In Section 9.5 we present a numerical method for approximating the solution to a system of first order differential equations with initial conditions. In Section 9.6, we introduce a qualitative method called **phase plane analysis,** which applies to a class of nonlinear systems. Finally, in Section 9.7 we define Poincaré maps and discuss the phenomena of **strange attractors** and **chaos.**

•9.2• ELIMINATION METHOD FOR LINEAR SYSTEMS

Until now we have concentrated on solving differential equations that involve a single dependent variable. In this chapter and the next we are interested in finding a solution to a *system* of differential equations that involve two or more dependent variables. Here we restrict our study to a system of linear differential equations, or what is called a **linear system.**

The general form for a linear system of two first order differential equations is

$$
\begin{aligned}
a_1(t)x'(t) + a_2(t)x(t) + b_1(t)y'(t) + b_2(t)y(t) &= f_1(t), \\
c_1(t)x'(t) + c_2(t)x(t) + d_1(t)y'(t) + d_2(t)y(t) &= f_2(t),
\end{aligned}
$$

(1)

where the coefficients $a_1, a_2, b_1, b_2, c_1, c_2, d_1, d_2$, and the nonhomogeneous terms f_1 and f_2 are given functions of t. A **solution** to system (1) is a pair of functions $x(t)$, $y(t)$ that satisfy (1) on some interval I.

A fundamental procedure for solving linear systems with constant coefficients is based on the technique of elimination that is used to solve a system of algebraic equations. This method is performed more easily if we express the linear system in operator notation, where $D := d/dt$ (see Section 4.2). We demonstrate the **elimination method** in the following example for the 2×2 case.

● **EXAMPLE 1** Solve the system

$$
\begin{aligned}
x'(t) &= 3x(t) - 4y(t), \\
y'(t) &= 4x(t) - 7y(t).
\end{aligned}
$$

(2)

Solution First we write the system using the operator notation:

$$
\begin{aligned}
(D - 3)[x] + 4y &= 0, \\
-4x + (D + 7)[y] &= 0.
\end{aligned}
$$

(3)

We can eliminate x from this system by adding 4 times the first equation to $(D - 3)$ applied to the second equation; this gives

$$
(16 + (D - 3)(D + 7))[y] = 4 \cdot 0 + (D - 3)[0] = 0,
$$

which simplifies to

(4) $(D^2 + 4D - 5)[y] = 0.$

Now equation (4) is just a second order linear equation in y with constant coefficients that has the general solution

(5) $y(t) = C_1 e^{-5t} + C_2 e^t.$

To find $x(t)$, we have two options. One is to return to system (3) and eliminate y. This is accomplished by "multiplying" the first equation in (3) by $(D + 7)$ and the second equation by -4 and then adding to obtain

$$
(D^2 + 4D - 5)[x] = 0.
$$

Coincidentally, this equation is the same as equation (4) except that here the unknown function is $x(t)$. Hence

(6) $x(t) = K_1 e^{-5t} + K_2 e^t,$

where we have taken K_1 and K_2 to be the arbitrary constants, which are not necessarily the same as C_1 and C_2 used in equation (5).

It is reasonable to expect that system (2) will involve only two arbitrary constants since it consists of two first order equations. To determine the relationships among the four constants C_1, C_2, K_1, and K_2, we substitute the expressions for $x(t)$ and $y(t)$ given in (5) and (6) into one of the equations in (2), say the first. This yields

$$-5K_1 e^{-5t} + K_2 e^t = 3K_1 e^{-5t} + 3K_2 e^t - 4C_1 e^{-5t} - 4C_2 e^t,$$

which simplifies to

$$(4C_1 - 8K_1)e^{-5t} + (4C_2 - 2K_2)e^t = 0.$$

Since e^t and e^{-5t} are linearly independent functions on any interval, this last equation holds for all t only if

$$4C_1 - 8K_1 = 0 \quad \text{and} \quad 4C_2 - 2K_2 = 0.$$

Therefore, $K_1 = C_1/2$ and $K_2 = 2C_2$.

A solution to system (2) is then given by the pair

(7) $x(t) = \tfrac{1}{2}C_1 e^{-5t} + 2C_2 e^t, \qquad y(t) = C_1 e^{-5t} + C_2 e^t.$

As you might expect, the pair in (7) is a **general solution** to (2) in the sense that *any* solution to (2) can be expressed in this fashion (see Section 10.3).

A simpler method for determining $x(t)$ once $y(t)$ is known is to use the system to obtain an equation for $x(t)$ in terms of $y(t)$ and $y'(t)$. In this example, solving the second equation in (2) for $x(t)$ gives

$$x(t) = \tfrac{1}{4}y'(t) + \tfrac{7}{4}y(t).$$

Substituting $y(t)$ as given in (5) yields

$$x(t) = \tfrac{1}{4}[-5C_1 e^{-5t} + C_2 e^t] + \tfrac{7}{4}[C_1 e^{-5t} + C_2 e^t]$$
$$= \tfrac{1}{2}C_1 e^{-5t} + 2C_2 e^t,$$

which agrees with (7).   ●

The above procedure works, more generally, for any linear system of two equations and two unknowns with *constant coefficients* regardless of the order of the equations. For example, if we let L_1, L_2, L_3, and L_4 denote linear differential operators with constant coefficients (i.e., polynomials in D), then the method can be applied to the linear system

$$L_1[x] + L_2[y] = f_1,$$
$$L_3[x] + L_4[y] = f_2.$$

Since the system has constant coefficients, the operators commute (e.g., $L_2 L_4 = L_4 L_2$), and we can eliminate variables in the usual algebraic fashion. Eliminating the variable y gives

(8) $(L_1 L_4 - L_2 L_3)[x] = g_1,$

where $g_1 := L_4[f_1] - L_2[f_2]$. Similarly, eliminating the variable x yields

(9) $(L_1L_4 - L_2L_3)[y] = g_2$,

where $g_2 := L_1[f_2] - L_3[f_1]$. Now if $L_1L_4 - L_2L_3$ is a differential operator of order n, then a general solution for (8) contains n arbitrary constants, and a general solution for (9) also contains n arbitrary constants; thus a total of $2n$ constants arise. However, as we saw in Example 1, there are only n of these that are independent for the system; the remaining constants can be expressed in terms of these.[†] The pair of general solutions to (8) and (9) written in terms of the n independent constants is called a **general solution for the system.**

If it turns out that $L_1L_4 - L_2L_3$ is the zero operator, the system is said to be **degenerate.** As with the anomalous problem of solving for the points of intersection of two parallel or coincident lines, a degenerate system may have no solutions, or if it does possess solutions, they may involve any number of arbitrary constants (see Problems 23 and 24).

> ◀ **ELIMINATION PROCEDURE FOR 2 x 2 SYSTEMS**
>
> To find a general solution for the system
>
> $$L_1[x] + L_2[y] = f_1,$$
> $$L_3[x] + L_4[y] = f_2,$$
>
> where $L_1, L_2, L_3,$ and L_4 are polynomials in $D = d/dt$:
>
> **(a)** Make sure that the system is written in operator form.
> **(b)** Eliminate one of the variables, say y, and solve the resulting equation for $x(t)$. If the system is degenerate, stop! A separate analysis is required to determine whether there are no solutions or infinitely many linearly independent ones.
> **(c)** (*Short cut*) If possible, use the system to derive an equation that involves $y(t)$ but not its derivatives. (Otherwise, go to step d). Substitute the expression for $x(t)$ into this equation to get a formula for $y(t)$. The expressions for $x(t)$, $y(t)$ give the desired general solution.
> **(d)** Eliminate x from the system and solve for $y(t)$. (Solving for $y(t)$ gives more constants, in fact, twice as many as needed.)
> **(e)** Remove the extra constants by substituting the expressions for $x(t)$ and $y(t)$ into one or both of the equations in the system. Write the expressions for $x(t)$ and $y(t)$ in terms of the remaining constants.

● **EXAMPLE 2**   Find a general solution for

(10) $$x''(t) + y'(t) - x(t) + y(t) = -1,$$
$$x'(t) + y'(t) - x(t) = t^2.$$

[†] For a proof of this fact see *Ordinary Differential Equations*, by M. Tenebaum and H. Pollard, Dover, New York, 1985, Chapter 7.

Solution   We begin by expressing the system in operator notation:

$$(D^2 - 1)[x] + (D + 1)[y] = -1,$$

(11)

$$(D - 1)[x] + D[y] = t^2.$$

Here $L_1 := D^2 - 1$, $L_2 := D + 1$, $L_3 := D - 1$, and $L_4 := D$.
    Eliminating y gives (cf. (8)):

$$((D^2 - 1)D - (D + 1)(D - 1))[x] = D[-1] - (D + 1)[t^2],$$

which reduces to

$$(D - 1)((D + 1)D - (D + 1))[x] = -2t - t^2,$$

(12)

$$(D - 1)^2(D + 1)[x] = -2t - t^2.$$

Since $(D - 1)^2(D + 1)$ is third order, there will be three arbitrary constants in a general solution to system (10).

The homogeneous equation associated with (12) has the auxiliary equation $(r - 1)^2(r + 1) = 0$ with roots $r = 1, 1, -1$. Hence a general solution for this homogeneous equation is

$$x_h(t) = C_1 e^t + C_2 t e^t + C_3 e^{-t}.$$

To find a particular solution to (12) we use the method of undetermined coefficients with $x_p(t) = At^2 + Bt + C$. Substituting into (12) and solving for A, B, and C yields (after a little algebra)

$$x_p(t) = -t^2 - 4t - 6.$$

Thus a general solution to equation (12) is

(13) $$x(t) = x_h(t) + x_p(t) = C_1 e^t + C_2 t e^t + C_3 e^{-t} - t^2 - 4t - 6.$$

To find $y(t)$, we take the short cut described in step (c) of the elimination procedure box. Subtracting the second equation in (11) from the first, we find

$$(D^2 - D)[x] + y = -1 - t^2,$$

so that

$$y = (D - D^2)[x] - 1 - t^2.$$

Inserting the expression for $x(t)$, given in (13), we obtain

$$y(t) = C_1 e^t + C_2(t e^t + e^t) - C_3 e^{-t} - 2t - 4$$
$$-[C_1 e^t + C_2(t e^t + 2e^t) + C_3 e^{-t} - 2] - 1 - t^2,$$

(14) $$y(t) = -C_2 e^t - 2C_3 e^{-t} - t^2 - 2t - 3.$$

The formulas for $x(t)$ in (13) and $y(t)$ in (14) give the desired general solution to (10).

●

The elimination method also applies to linear systems with three or more equations and unknowns; however, the process becomes more cumbersome as the number of equations and unknowns increases. In the next example we illustrate the technique for a 3×3 system.

● **EXAMPLE 3**   Find a general solution to

$$x'(t) = x(t) + 2y(t) - z(t),$$
(15) $$y'(t) = x(t) + z(t),$$
$$z'(t) = 4x(t) - 4y(t) + 5z(t).$$

Solution   We begin by expressing the system in operator notation:

$$(D - 1)[x] - \quad 2y + \quad\quad z = 0,$$
(16) $$-x + D[y] - \quad\quad z = 0,$$
$$-4x + \quad 4y + (D - 5)[z] = 0,$$

Eliminating z from the first two equations and then from the second two equations yields

(17) $$(D - 2)[x] + \quad\quad (D - 2)[y] = 0,$$
$$-(D - 1)[x] + (D - 1)(D - 4)[y] = 0.$$

On eliminating x from this 2×2 system, we obtain

$$(D - 1)(D - 2)(D - 3)[y] = 0,$$

which has the general solution

(18) $$y(t) = C_1 e^t + C_2 e^{2t} + C_3 e^{3t}.$$

Taking the short cut approach, we add the two equations in (17) to get an expression for x in terms of y and its derivatives:

$$x = (D^2 - 4D + 2)[y] = y'' - 4y' + 2y.$$

When we substitute the expression (18) for $y(t)$ into this equation, we find

(19) $$x(t) = -C_1 e^t - 2C_2 e^{2t} - C_3 e^{3t}.$$

Finally, using the second equation in (15) to solve for $z(t)$, we get

$$z(t) = y'(t) - x(t),$$

and substituting in for $y(t)$ and $x(t)$ yields

(20) $$z(t) = 2C_1 e^t + 4C_2 e^{2t} + 4C_3 e^{3t}.$$

The expressions for $x(t)$ in (19), $y(t)$ in (18), and $z(t)$ in (20) give a general solution with C_1, C_2, and C_3 as arbitrary constants.   ●

Conversion of Higher Order Equations to First Order Systems

In carrying out the elimination method, we derived from the system a higher order differential equation (involving one dependent variable). In the *converse* direction, it is possible (and often useful) to rewrite an mth order differential equation as a system of m **first order** equations. For instance, most numerical procedures for approximating the solution to an initial value problem for a higher order differential equation require that the problem first be restated as a system of first order equations (see Section 9.5). Furthermore, as we will see in Chapter 10, the powerful machinery of linear algebra can be easily applied in such a setting.

A first order system that is expressed as

$$
\begin{aligned}
x_1'(t) &= f_1(t, x_1, x_2, \dots, x_m), \\
x_2'(t) &= f_2(t, x_1, x_2, \dots, x_m), \\
&\ \ \vdots \\
x_m'(t) &= f_m(t, x_1, x_2, \dots, x_m),
\end{aligned}
$$

is said to be in **normal form.** Any mth order differential equation

$$(21) \qquad y^{(m)}(t) = f(t, y, y', \dots, y^{(m-1)})$$

can be converted into a first order system in normal form by setting

$$x_1(t) := y(t), \quad x_2(t) := y'(t), \quad \dots, \quad x_m(t) := y^{(m-1)}(t).$$

With this substitution, we obtain

$$
(22) \qquad
\begin{aligned}
x_1'(t) &= y'(t) = x_2(t), \\
x_2'(t) &= y''(t) = x_3(t), \\
&\ \ \vdots \\
x_{m-1}'(t) &= y^{(m-1)}(t) = x_m(t), \\
x_m'(t) &= y^{(m)}(t) = f(t, x_1, x_2, \dots, x_m).
\end{aligned}
$$

If equation (21) has initial conditions $y(t_0) = a_1, y'(t_0) = a_2, \dots, y^{(m-1)}(t_0) = a_m$, then system (22) has initial conditions $x_1(t_0) = a_1, x_2(t_0) = a_2, \dots, x_m(t_0) = a_m$.

● **EXAMPLE 4**   Convert the initial value problem

$$(23) \qquad y''(t) + 3y'(t) + 2y(t) = 0; \qquad y(0) = 1, \qquad y'(0) = 3$$

into an initial value problem for a system in normal form.

Solution   We first express the differential equation in (23) as

$$y''(t) = -3y'(t) - 2y(t).$$

Setting $x_1(t) := y(t)$ and $x_2(t) := y'(t)$, we obtain

$$
\begin{aligned}
x_1'(t) &= x_2(t), \\
x_2'(t) &= -3x_2(t) - 2x_1(t).
\end{aligned}
$$

The initial conditions transform to $x_1(0) = 1, x_2(0) = 3$.   ●

EXERCISES 9.2

In Problems 1 through 20 use the elimination method to find a general solution for the given linear system, where x', y', z' denote differentiation with respect to t.

1. $x' + y' = -2y,$
    $y' = x - 2y.$

2. $x' = 3y,$
    $y' = 2x - y.$

3. $x' + 2y = 0,$
    $x' - y' = 0.$

4. $x' = x - y,$
    $y' = y - 4x.$

5. $x' + y' - x = 5,$
    $x' + y' + y = 1.$

6. $x' = 3x - 2y + \sin t,$
    $y' = 4x - y - \cos t.$

7. $(D + 1)[u] - (D + 1)[v] = e^t,$
    $(D - 1)[u] + (2D + 1)[v] = 5.$

8. $(D - 3)[x] + (D - 1)[y] = t,$
    $(D + 1)[x] + (D + 4)[y] = 1.$

9. $x' + y' + 2x = 0,$
    $x' + y' - x - y = \sin t.$

10. $2x' + y' - x - y = e^{-t},$
     $x' + y' + 2x + y = e^t.$

11. $(D^2 - 1)[u] + 5v = e^t,$
     $2u + (D^2 + 2)[v] = 0.$

12. $D^2[u] + D[v] = 2,$
     $4u + D[v] = 6.$

13. $\dfrac{dx}{dt} = x - 4y,$

     $\dfrac{dy}{dt} = x + y.$

14. $\dfrac{dx}{dt} + y = t^2,$

     $-x + \dfrac{dy}{dt} = 1.$

15. $\dfrac{dw}{dt} = 5w + 2z + 5t,$

     $\dfrac{dz}{dt} = 3w + 4z + 17t.$

16. $\dfrac{dx}{dt} + x + \dfrac{dy}{dt} = e^{4t},$

     $2x + \dfrac{d^2y}{dt^2} = 0.$

17. $x'' + 5x - 4y = 0,$
     $-x + y'' + 2y = 0.$

18. $y'' + x' + x = \sin t,$
     $y' - y + x = 0.$

19. $x' = 3x + y - z,$
     $y' = x + 2y - z,$
     $z' = 3x + 3y - z.$

20. $x' = y + z,$
     $y' = x + z,$
     $z' = x + y.$

In Problems 21 and 22 describe how solutions to the given system behave as $t \to +\infty$ for each (fixed) value of the parameter λ between $-\infty$ and $+\infty$.

21. $\dfrac{dx}{dt} = -x + \lambda y,$

     $\dfrac{dy}{dt} = x - y.$

22. $\dfrac{dx}{dt} = x + \lambda y,$

     $\dfrac{dy}{dt} = x - y.$

In Problems 23 and 24 show that the given linear system is degenerate. By attempting to solve the system, determine

whether it has no solutions or infinitely many linearly independent solutions.

23. $(D - 1)[x] + (D - 1)[y] = -3e^{-2t},$
     $(D + 2)[x] + (D + 2)[y] = 3e^t.$

24. $D[x] + (D + 1)[y] = e^t,$
     $D[x] + (D + 1)[y] = 0.$

In Problems 25 through 28 convert the given initial value problem into an initial value problem for a system in normal form.

25. $y''(t) + ty'(t) - 3y(t) = t^2;$
     $y(0) = 3, \qquad y'(0) = -6.$

26. $y''(t) = \cos(t - y) + y^2(t); \qquad y(0) = 1, \qquad y'(0) = 0.$

27. $y^{(4)}(t) - y^{(3)}(t) + 7y(t) = \cos t;$
     $y(0) = y'(0) = 1, \qquad y''(0) = 0, \qquad y^{(3)}(0) = 2.$

28. $y^{(6)}(t) = [y'(t)]^3 - \sin(y(t)) + e^{2t};$
     $y(0) = \cdots = y^{(5)}(0) = 0.$

29. Taylor Series Method. The Taylor series method discussed in Section 8.2 can be applied to systems. Assuming that the solutions $x(t)$, $y(t)$ to the initial value problem

$$\frac{dx}{dt} = xy - 1, \qquad x(0) = 1,$$

$$\frac{dy}{dt} = -x - y, \qquad y(0) = 2,$$

have Taylor series expansions about $t = 0$, determine the first few terms in the expansion by successively differentiating the equations in order to obtain the derivatives of $x(t)$ and $y(t)$ evaluated at $t = 0$.

30. Power Series Method. The power series method of Section 8.3 can be applied to systems. Assuming that the solutions $x(t)$, $y(t)$ to the system

$$\frac{dx}{dt} = ty,$$

$$\frac{dy}{dt} = tx - y,$$

have power series expansions about $t = 0$, determine the first few terms in these series by substituting them into the system and equating like terms. (A general solution will involve two arbitrary constants.)

In Problems 31 through 34 find a system of differential equations and initial conditions for the currents in the net-

works given in the schematic diagram, assuming that all initial currents are zero. Solve for the currents in each branch of the network (see Section 9.1 for a discussion of electric networks).

31.

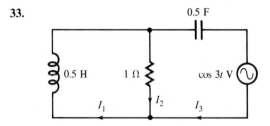

Figure 9.2 *RLC* network for Problem 31

32.

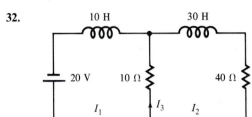

Figure 9.3 *RL* network for Problem 32

33.

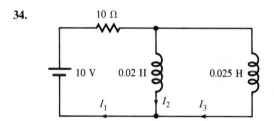

Figure 9.4 *RLC* network for Problem 33

34.

Figure 9.5 *RL* network for Problem 34

• 9.3 • SOLVING LINEAR SYSTEMS WITH LAPLACE TRANSFORMS (Optional)

In Chapter 7 we used the Laplace transform to reduce an initial value problem for a linear differential equation with constant coefficients to a linear algebraic equation, where the unknown was the transform of the solution. We then solved for the transform, took the inverse Laplace transform, and obtained the solution to the initial value problem. In a similar manner we can use the Laplace transform to reduce certain systems of linear differential equations with initial conditions to a system of linear algebraic equations, where again the unknowns are the transforms of the functions that make up the solution. Solving for these unknowns and taking their inverse Laplace transforms, we can then obtain the solution to the initial value problem for the system.

● EXAMPLE 1   Solve the initial value problem

(1)
$$x'(t) - 2y(t) = 4t, \qquad x(0) = 4,$$
$$y'(t) + 2y(t) - 4x(t) = -4t - 2, \qquad y(0) = -5.$$

Solution   Taking the Laplace transform of both sides of the differential equations gives (see the

Laplace transform table on the inside back cover)

$$\mathscr{L}\{x'\}(s) - 2\mathscr{L}\{y\}(s) = \frac{4}{s^2},$$

(2)

$$\mathscr{L}\{y'\}(s) + 2\mathscr{L}\{y\}(s) - 4\mathscr{L}\{x\}(s) = -\frac{4}{s^2} - \frac{2}{s}.$$

Let $X(s) := \mathscr{L}\{x\}(s)$ and $Y(s) := \mathscr{L}\{y\}(s)$. Then, using the initial conditions, we can express $\mathscr{L}\{x'\}(s)$ in terms of $X(s)$, and $\mathscr{L}\{y'\}(s)$ in terms of $Y(s)$. Namely,

$$\mathscr{L}\{x'\}(s) = sX(s) - x(0) = sX(s) - 4,$$
$$\mathscr{L}\{y'\}(s) = sY(s) - y(0) = sY(s) + 5.$$

Substituting these expressions into system (2) and simplifying, we find

$$sX(s) - 2Y(s) = \frac{4s^2 + 4}{s^2},$$

(3)

$$-4X(s) + (s + 2)Y(s) = -\frac{5s^2 + 2s + 4}{s^2}.$$

To eliminate $Y(s)$ from the system, we multiply the first equation by $(s + 2)$ and the second by 2 and then add to obtain

$$[s(s + 2) - 8]X(s) = \frac{(s + 2)(4s^2 + 4)}{s^2} - \frac{10s^2 + 4s + 8}{s^2}.$$

This simplifies to

$$X(s) = \frac{4s - 2}{(s + 4)(s - 2)}.$$

To compute the inverse transform, we first write $X(s)$ in the partial fraction form

$$X(s) = \frac{3}{s + 4} + \frac{1}{s - 2}.$$

Hence, from the Laplace transform table on the inside back cover, we find that

(4) $x(t) = 3e^{-4t} + e^{2t}.$

To determine $y(t)$ we could solve system (3) for $Y(s)$ and then compute its inverse Laplace transform. However, it is easier just to solve the first equation in system (1) for $y(t)$ in terms of $x(t)$. Thus

$$y(t) = \tfrac{1}{2}x'(t) - 2t.$$

Substituting $x(t)$ from equation (4), we find that

(5) $y(t) = -6e^{-4t} + e^{2t} - 2t.$

The solution to the initial value problem (1) consists of the pair of functions $x(t)$, $y(t)$ given by equations (4) and (5).   ●

EXERCISES 9.3

In Problems 1 through 14 use the method of Laplace transforms to solve the given initial value problem. Here x', y', etc., denotes differentiation with respect to t, and so does the symbol D.

1. $x' = 3x - 2y$; $x(0) = 1$,
   $y' = 3y - 2x$; $y(0) = 1$.

2. $x' = x - y$; $x(0) = -1$,
   $y' = 2x + 4y$; $y(0) = 0$.

3. $z' + 2w = 0$; $z(0) = -1$,
   $z' - w' = 0$; $w(0) = 2$.

4. $x' - 3x + 2y = \sin t$; $x(0) = 0$,
   $4x - y' - y = \cos t$; $y(0) = 0$.

5. $x' = y + \sin t$; $x(0) = 2$,
   $y' = x + 2\cos t$; $y(0) = 0$.

6. $x' - x - y = 1$; $x(0) = 0$,
   $-x + y' - y = 0$; $y(0) = -\frac{5}{2}$.

7. $(D - 4)[x] + 6y = 9e^{-3t}$; $x(0) = -9$,
   $x - (D - 1)[y] = 5e^{-3t}$; $y(0) = 4$.

8. $D[x] + y = 0$; $x(0) = \frac{7}{4}$,
   $4x + D[y] = 3$; $y(0) = 4$.

9. $x'' + 2y' = -x$; $x(0) = 2$, $x'(0) = -7$,
   $-3x'' + 2y'' = 3x - 4y$; $y(0) = 4$, $y'(0) = -9$.

10. $x'' + y = 1$; $x(0) = 1$, $x'(0) = 1$,
    $x + y'' = -1$; $y(0) = 1$, $y'(0) = -1$.

11. $x' + y = 1 - u(t - 2)$; $x(0) = 0$,
    $x + y' = 0$; $y(0) = 0$.

12. $x' + y = x$; $x(0) = 0$,
    $2x' + y'' = u(t - 3)$; $y(0) = 1$, $y'(0) = -1$.

13. $x' - y' = (\sin t)u(t - \pi)$; $x(0) = 1$,
    $x + y' = 0$; $y(0) = 1$.

14. $x'' = y + u(t - 1)$; $x(0) = 1$, $x'(0) = 0$,
    $y'' = x + 1 - u(t - 1)$; $y(0) = 0$, $y'(0) = 0$.

15. Use the method of Laplace transforms to solve
$$x' = 3x + y - 2z;   x(0) = -6,$$
$$y' = -x + 2y + z;   y(0) = 2,$$
$$z' = 4x + y - 3z;   z(0) = -12.$$

16. Use the method of Laplace transforms to solve
$$x'' + y' = 2;   x(0) = 3,   x'(0) = 0,$$
$$4x + y' = 6;   y(1) = 4.$$
[Hint: Let $y(0) = c$ and then solve for c.]

In Problems 17 and 18 find a system of differential equations and initial conditions for the currents in the networks given by the schematic diagrams, assuming that the initial currents are all zero. Solve for the currents in each branch of the network (see Section 9.1 for a discussion of electric networks).

17.

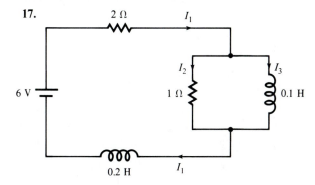

Figure 9.6 *RL* network for Problem 17

18.

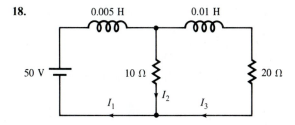

Figure 9.7 *RL* network for Problem 18

•9.4• SOME APPLICATIONS OF LINEAR SYSTEMS

In Section 9.1 we saw how linear systems arise in a mathematical model for an electric network. Such systems also occur in the study of mechanical systems when there is more than one degree of freedom (see Section 6.1) and in compartmental analysis models involving two or more compartments. In this section we demonstrate three applications of linear systems. The first involves a coupled spring-mass system (see Sections 5.1–5.3 for a discussion of single spring-mass systems).

● **EXAMPLE 1**   Three springs, with spring constants k_1, k_2, k_3, and two masses m_1, m_2 are attached in a straight line on a horizontal frictionless surface with the ends of the outside springs fixed (see Figure 9.8). The system is set in motion by holding the mass m_1 at its equilibrium position and pulling the mass m_2 to the right of its equilibrium position a distance α and then releasing both masses. Determine the equations of motion for the two masses in the special case where $m_1 = m_2 = m$ and $k_1 = k_2 = k_3 = k$.

Solution   Let x_1 and x_2 represent the displacements of the masses m_1 and m_2 to the right of their respective equilibrium positions. Since we can neglect friction and there are no other external forces acting on the masses, the only forces we need consider are those due to the springs.

The mass m_1 has a force F_1 acting on its left side due to the left spring and a force F_2 acting on its right side due to the middle spring. Applying Hooke's law, we see that

$$F_1 = -k_1 x_1 \quad \text{and} \quad F_2 = +k_2(x_2 - x_1),$$

where $(x_2 - x_1)$ is the net displacement of the middle spring from its initial or equilibrium length. The mass m_2 has a force F_3 acting on its left side due to the middle spring and a force F_4 acting on its right side due to the right spring. Again, using Hooke's law, we have

$$F_3 = -k_2(x_2 - x_1) \quad \text{and} \quad F_4 = -k_3 x_2.$$

Applying Newton's second law to each of the masses gives

$$m_1 x_1'' = F_1 + F_2 = -k_1 x_1 + k_2(x_2 - x_1),$$
$$m_2 x_2'' = F_3 + F_4 = -k_2(x_2 - x_1) - k_3 x_2,$$

or

(1) $\qquad (m_1 D^2 + k_1 + k_2)[x_1] - k_2 x_2 = 0,$

(2) $\qquad -k_2 x_1 + (m_2 D^2 + k_2 + k_3)[x_2] = 0.$

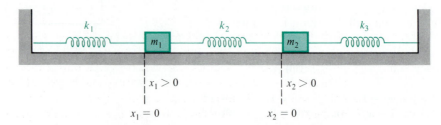

Figure 9.8 Coupled spring-mass system with fixed ends

The initial conditions are

(3) $x_1(0) = 0,$ $x_1'(0) = 0,$ $x_2(0) = \alpha,$ $x_2'(0) = 0.$

To solve the initial value problem (1)–(3), we will use the elimination method of Section 9.2. Eliminating x_2 gives

(4) $((m_2 D^2 + k_2 + k_3)(m_1 D^2 + k_1 + k_2) - k_2^2)[x_1] = 0.$

For $m_1 = m_2 = m$ and $k_1 = k_2 = k_3 = k$, equation (4) reduces to

(5) $((mD^2 + 2k)^2 - k^2)[x_1] = 0.$

This has the auxiliary equation

(6) $(mr^2 + 2k)^2 - k^2 = (mr^2 + k)(mr^2 + 3k) = 0,$

with roots $\pm i\sqrt{k/m}, \pm i\sqrt{3k/m}$. Setting $\omega := \sqrt{k/m}$, we get the following general solution to (5):

(7) $x_1(t) = C_1 \cos \omega t + C_2 \sin \omega t + C_3 \cos \sqrt{3}\omega t + C_4 \sin \sqrt{3}\omega t.$

To obtain $x_2(t)$, we solve for $x_2(t)$ in (1) and substitute $x_1(t)$ as given in (7). Upon simplifying, we get

(8) $x_2(t) = C_1 \cos \omega t + C_2 \sin \omega t - C_3 \cos \sqrt{3}\omega t - C_4 \sin \sqrt{3}\omega t.$

Using the initial conditions in (3) to determine the constants, we find (after a little algebra)

$$C_1 = \frac{\alpha}{2}, \qquad C_2 = 0, \qquad C_3 = -\frac{\alpha}{2}, \qquad C_4 = 0.$$

Thus the equations of motion for the two masses are

(9)

$$x_1(t) = \frac{\alpha}{2}(\cos \omega t - \cos \sqrt{3}\omega t),$$

$$x_2(t) = \frac{\alpha}{2}(\cos \omega t + \cos \sqrt{3}\omega t).$$

These functions are illustrated in Figure 9.9 on page 434 for $\alpha = 2$ and $\omega = 1$.   ●

As in the case of a single spring-mass system (see Section 5.1), the frequencies obtained by solving the auxiliary equation (6),

$$f_1 := \frac{\omega}{2\pi} = \frac{1}{2\pi}\sqrt{\frac{k}{m}} \quad \text{and} \quad f_2 := \frac{\sqrt{3}\omega}{2\pi} = \frac{1}{2\pi}\sqrt{\frac{3k}{m}},$$

are called the **normal** or **natural frequencies** of the system (ω and $\sqrt{3}\omega$ are the **angular frequencies** of the system). It is evident from equation (4) that for most choices of the constants m_1, m_2, k_1, k_2, and k_3, the system would have had two normal frequencies. For more complex systems involving more springs and degrees of freedom, there are many normal frequencies. These frequencies are of great interest to mechanical and civil engineers involved in the stress analysis of such structures as buildings and bridges.

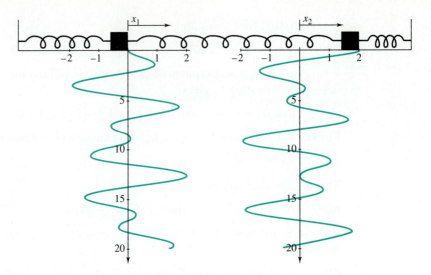

Figure 9.9 Graphs of the solutions to the coupled spring-mass system in Example 1 with $\alpha = 2$ and $\omega = 1$

Corresponding to the two normal frequencies for the coupled system discussed in Example 1, there are two normal or natural modes of vibration. These modes correspond to the special cases when the system is vibrating at only one frequency. For example, when $C_1 = C_2 = 0$ in (7) and (8), both x_1 and x_2 are oscillating, with a normal frequency $\sqrt{3}\,\omega/2\pi$. In this case the normal mode is $x_1(t) = -x_2(t)$, so that the masses vibrate as mirror images of each other (see Figure 9.10(a)). When $C_3 = C_4 = 0$, we get the normal frequency $\omega/2\pi$. Here the normal mode is $x_1(t) = x_2(t)$, and the masses move with the same relative positions (see Figure 9.10(b)).

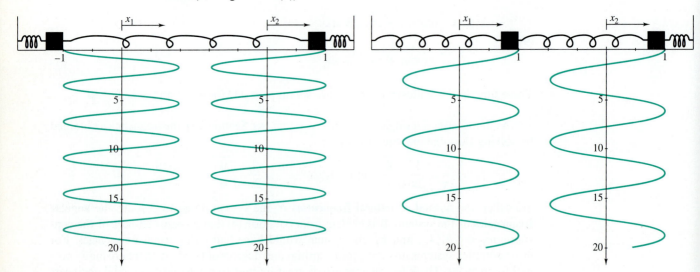

Figure 9.10 Natural modes for the coupled spring-mass system in Example 1 with $\omega = 1$

The next example is a mixing problem involving two interconnected tanks. (See Section 3.2 for a discussion of mixing problems with one tank.)

● EXAMPLE 2   Two large tanks, each holding 100 L of liquid, are interconnected by pipes, with the liquid flowing from tank A into tank B at a rate of 3 L/min and from B into A at a rate of 1 L/min (see Figure 9.11). The liquid inside each tank is kept well stirred. A brine solution with a concentration of 2 kg/L of salt flows into tank A at a rate of 6 L/min. The (diluted) solution flows out of the system from tank A at 4 L/min and from tank B at 2 L/min. If, initially, tank A contains only water and tank B contains 200 kg of salt, determine the mass of salt in each tank at time $t \geq 0$.

Solution   We approach this interconnected tank problem as a compartmental analysis problem involving two compartments.

Let $x(t)$ denote the mass of salt in tank A at time t and $y(t)$ the mass of salt in tank B. Since liquid flows into tank A at 7 L/min and flows out at the same rate, the volume of liquid in tank A remains constant. Similarly, the volume of liquid in tank B remains constant. Let's now determine the rate of change in the amount of salt in tank A. Salt enters tank A from the outside at a rate of $(6)(2) = 12$ kg/min and from tank B at a rate of $y/100$ kg/min. Salt leaves tank A to enter tank B at $3x/100$ kg/min and leaves the system at $4x/100$ kg/min. Using the compartmental analysis model,

$$\frac{dx}{dt} = \text{INPUT RATE} - \text{OUTPUT RATE},$$

the rate of change in the mass of salt in tank A is

(10) $\dfrac{dx}{dt} = 12 + \dfrac{1}{100}y - \dfrac{3}{100}x - \dfrac{4}{100}x.$

In a similar fashion, we can show that the rate of change in the mass of salt in tank B satisfies

(11) $\dfrac{dy}{dt} = \dfrac{3}{100}x - \dfrac{3}{100}y.$

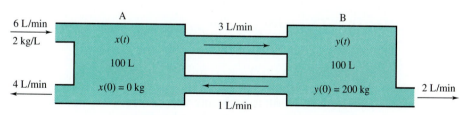

Figure 9.11 Mixing problem for interconnected tanks

We can rewrite (10) and (11) as the system

$$(D + 0.07)[x] - 0.01y = 12,$$
$$-0.03x + (D + 0.03)[y] = 0.$$

Eliminating x from this system gives

$$(-0.0003 + (D + 0.07)(D + 0.03))[y] = 0.36,$$

which simplifies to

(12) $(D^2 + 0.1D + 0.0018)[y] = 0.36.$

The auxiliary equation for the homogeneous equation associated with (12) is $r^2 + 0.1r + 0.0018 = 0$, which has roots

(13) $r_1 = \dfrac{-5 - \sqrt{7}}{100} \approx -0.0765, \qquad r_2 = \dfrac{-5 + \sqrt{7}}{100} \approx -0.0235.$

To determine a particular solution to (12), we just substitute $y_p = a$ into (12) and solve for the constant a. This gives $y_p = 200$. Therefore, a general solution to (12) is

(14) $y(t) = C_1 e^{r_1 t} + C_2 e^{r_2 t} + 200,$

where r_1 and r_2 are given in (13).

To determine $x(t)$, we solve for x in (11) to obtain

$$x = \tfrac{100}{3}(D + 0.03)[y],$$

and then insert the expression for $y(t)$ given in (14). This yields

(15) $x(t) = \dfrac{-2 - \sqrt{7}}{3} C_1 e^{r_1 t} + \dfrac{-2 + \sqrt{7}}{3} C_2 e^{r_2 t} + 200.$

Finally, to determine the constants C_1 and C_2, we use the initial conditions $x(0) = 0$ and $y(0) = 200$. Substituting x and y given in (14) and (15) into these initial conditions and solving yields $C_1 = -C_2 = 300/\sqrt{7} \approx 113$. Hence the masses of salt at time t in tanks A and B are, respectively.

(16) $x(t) = -\left(100 + \dfrac{200}{\sqrt{7}}\right)e^{r_1 t} - \left(100 - \dfrac{200}{\sqrt{7}}\right)e^{r_2 t} + 200 \text{ kg},$

(17) $y(t) = \dfrac{300}{\sqrt{7}} e^{r_1 t} - \dfrac{300}{\sqrt{7}} e^{r_2 t} + 200 \text{ kg.}$ ●

An important observation that we can make here is that since r_1 and r_2 are negative, the amount of salt in each tank approaches 200 kg as $t \to \infty$. Hence the concentration of salt in each tank approaches 2 kg/L, which is the concentration of salt in the brine entering tank A.

The next example concerns a heating problem for a building with two zones (rooms) that utilizes the analysis of Section 3.3.

● **EXAMPLE 3**   A building consists of two zones A and B (see Figure 9.12). Only zone A is heated by a furnace, which generates 80,000 Btu per hour. The heat capacity of zone A is $\frac{1}{4}°$F per thousand Btu. The time constants for heat transfer are: between zone A and the outside, 4 hr; between the unheated zone B and the outside, 5 hr; and between the two zones, 2 hr. If the outside temperature stays at 0°F, how cold can it get in the unheated zone B?

Solution   To answer the question we use a two-compartment model along with Newton's law of cooling (see Section 3.3).

Let $x(t)$ and $y(t)$ denote the temperatures in zones A and B, respectively. The rate at which the temperature in zone A changes depends on the addition of heat from the furnace and any temperature loss (or gain) from the outside or the other zone. The rate at which the furnace affects temperature is just the number of thousands of Btu per hour times the heat capacity; that is, $(80)(\frac{1}{4})°$F or 20°F per hour.

Recall that Newton's law of cooling states that the rate of change in temperature due to a difference in temperature between two regions is proportional to that difference. Between zone A and the outside this is $k_1(0 - x) = -k_1 x$, and between zones A and B it is $k_2(y - x)$. Hence, for room A, the rate of change in temperature is

(18) $x'(t) = 20 - k_1 x(t) + k_2(y(t) - x(t))$.

Recall from the discussion in Section 3.3 that the time constant associated with Newton's law of cooling is just $1/K$, where K is the proportionality constant for heat transfer. Therefore, $1/k_1 = 4$ and $1/k_2 = 2$. Solving for k_1 and k_2 and substituting into (18) gives

(19) $x'(t) = 20 - \frac{1}{4}x(t) + \frac{1}{2}(y(t) - x(t))$.

A similar argument for zone B yields

(20) $y'(t) = \frac{1}{2}(x(t) - y(t)) - \frac{1}{5}y(t)$.

We can rewrite (19) and (20) as the system

$$(D + \tfrac{3}{4})[x] - \tfrac{1}{2}y = 20,$$
$$-\tfrac{1}{2}x + (D + \tfrac{7}{10})[y] = 0.$$

Since we are interested only in the temperature in zone B, we eliminate x from the above system and obtain

$$(-\tfrac{1}{4} + (D + \tfrac{3}{4})(D + \tfrac{7}{10}))[y] = 10,$$

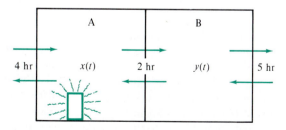

Figure 9.12 Two-zone building with one zone heated

which simplifies to

$$(D^2 + \tfrac{58}{40}D + \tfrac{11}{40})[y] = 10.$$

From standard methods we find that a general solution is

(21) $y(t) = c_1 e^{r_1 t} + c_2 e^{r_2 t} + \tfrac{400}{11},$

where

$$r_1 = \frac{-58 - \sqrt{1604}}{80} \approx -1.23, \qquad r_2 = \frac{-58 + \sqrt{1604}}{80} \approx -0.22.$$

We assume that the initial temperatures of the zones are sufficiently high that the temperature in zone B steadily decreases. (For example, assume $x(0) = y(0) = 70°$F.) Since r_1 and r_2 are negative, the two exponential terms in (21) go to zero as $t \to \infty$, and hence $y(t)$ decreases to $\tfrac{400}{11} \approx 36.4°$F. Consequently, the temperature in zone B can never get below $36.4°$F. ●

EXERCISES 9.4

1. Two springs and two masses are attached in a straight line on a horizontal frictionless surface as illustrated in Figure 9.13. The system is set in motion by holding the mass m_1 at its equilibrium position and pulling the mass m_2 to the right of its equilibrium position a distance 1 m and then releasing both masses. Determine the equations of motion for the two masses if $m_1 = 2$ kg, $m_2 = 1$ kg, $k_1 = 2$ N/m, and $k_2 = 4$ N/m.

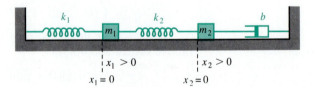

Figure 9.14 Coupled spring-mass system with three degrees of freedom

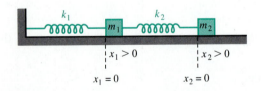

Figure 9.13 Coupled spring-mass system with one end free

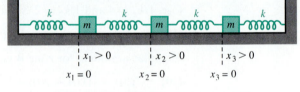

Figure 9.15 Coupled spring-mass system with one end damped

2. Determine the equations of motion for the two masses described in Problem 1 if $m_1 = 1$ kg, $m_2 = 1$ kg, $k_1 = 3$ N/m, and $k_2 = 2$ N/m.

3. Four springs with the same spring constant and three equal masses are attached in a straight line on a horizontal frictionless surface as illustrated in Figure 9.14. Determine the normal frequencies for the system and describe the three normal modes of vibration.

4. Two springs, two masses, and a dashpot are attached in a straight line on a horizontal frictionless surface as shown in Figure 9.15. Derive the system of differential equations for the displacements x_1 and x_2.

5. Two springs, two masses, and a dashpot are attached in a straight line on a horizontal frictionless surface as shown in Figure 9.16. The system is set in motion by holding the mass m_2 at its equilibrium position and pushing the mass m_1 to the left of its equilibrium position a distance 2 m and then releasing both masses. Determine the equations of motion for the two masses if $m_1 = m_2 = 1$ kg, $k_1 = k_2 = 1$ N/m, and $b = 1$ N-sec/m.

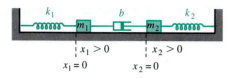

Figure 9.16 Coupled spring-mass system with damping between the masses

6. A double pendulum swinging in a vertical plane under the influence of gravity (see Figure 9.17) satisfies the system

$$(m_1 + m_2)l_1^2\theta_1'' + m_2l_1l_2\theta_2'' + (m_1 + m_2)l_1g\theta_1 = 0,$$
$$m_2l_2^2\theta_2'' + m_2l_1l_2\theta_1'' + m_2l_2g\theta_2 = 0,$$

where θ_1 and θ_2 are small angles. Solve the system when $m_1 = 3$ kg, $m_2 = 2$ kg, $l_1 = l_2 = 5$ m, $\theta_1(0) = \pi/6$, $\theta_2(0) = \theta_1'(0) = \theta_2'(0) = 0$.

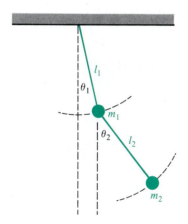

Figure 9.17 Double pendulum

7. The motion of a pair of identical pendulums coupled by a spring is modeled by the system

$$mx_1'' = -\frac{mg}{l}x_1 - k(x_1 - x_2),$$

$$mx_2'' = -\frac{mg}{l}x_2 + k(x_1 - x_2),$$

for small displacements (see Figure 9.18). Determine the two normal frequencies for the system.

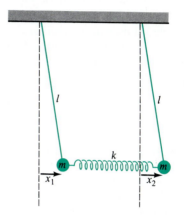

Figure 9.18 Coupled pendulums

8. In Example 2, 3 L/min of liquid flowed from tank A into tank B and 1 L/min from B into A. Determine the mass of salt in each tank at time $t \geq 0$ if, instead, 5 L/min flows from A into B and 3 L/min flows from B into A, with all other data the same.

9. Two large tanks, each holding 50 L of liquid, are interconnected by a pipe with liquid flowing from tank A into tank B at a rate of 5 L/min (see Figure 9.19). The liquid inside each tank is kept well stirred. A salt brine with concentration 3 kg/L of salt flows into tank A at a rate of 5 L/min. The solution flows out of the system from tank B at 5 L/min. If, initially, tank A contains 50 kg of salt and tank B contains 100 kg, determine the mass of salt in each tank at time $t \geq 0$.

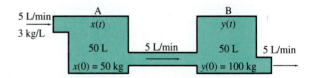

Figure 9.19 Mixing problem for the flow through two tanks

10. In Example 2, assume now that no solution flows out of the system from tank B, only 1 L/min flows from A into B, and only 4 L/min of brine flows into the system at tank A; other data being the same. Determine the mass of salt in each tank at time $t \geq 0$.

11. In Example 3, if a small furnace that generates 1000 Btu/hr is placed in zone B, determine the coldest it could get in zone B if zone B has a heat capacity of 2°F per thousand Btu.

12. In Example 3, if the insulation between zones A and B were better—say if the time constant for heat transfer between zones A and B were 3 hr—then how cold could it get in the unheated zone B?

13. A house, for cooling purposes, consists of two zones: the attic area zone A and the living area zone B (see Figure 9.20). The living area is cooled by a 2-ton air conditioning unit that removes 24,000 Btu/hr. The heat capacity of zone B is $\frac{1}{2}$°F per thousand Btu. The time constant for heat transfer between zone A and the outside is 2 hr, between zone B and the outside is 4 hr, and between the two zones is 4 hr. If the outside temperature stays at 100°F, how warm can it get in the attic zone A?

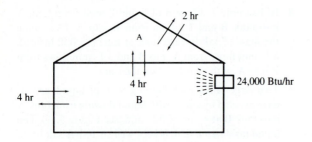

Figure 9.20 Air-conditioned house with attic

14. Feedback System with Pooling Delay. Many physical and biological systems involve time delays. A pure time delay has the output the same as the input, but is shifted in time. A more common type of delay is *pooling delay*. An example of such a feedback system is shown in Figure 9.21. Here the level of fluid in tank B determines the rate at which fluid enters tank A. Suppose this rate is given by $R_1(t) = \alpha[V - V_2(t)]$, where α and V are positive constants and $V_2(t)$ is the volume of fluid in tank B at time t.

(a) If the outflow rate R_3 from tank B is constant and the flow rate R_2 from tank A into B is $R_2(t) = KV_1(t)$, where K is a positive constant and $V_1(t)$ is the volume of fluid in tank A at time t, then show that this feedback system is governed by the system

$$\frac{dV_1}{dt} = \alpha(V - V_2(t)) - KV_1(t),$$

$$\frac{dV_2}{dt} = KV_1(t) - R_3 .$$

(b) Find a general solution for the system in part (a) when $\alpha = 5(\text{min})^{-1}$, $V = 20$ L, $K = 2(\text{min})^{-1}$, and $R_3 = 10$ L/min.

(c) Using the general solution obtained in part (b), what can be said about the volume of fluid in each of the tanks as $t \to \infty$?

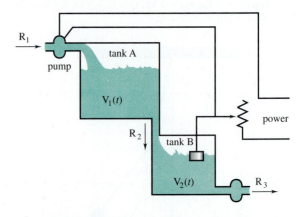

Figure 9.21 Feedback system with pooling delay

• 9.5 • NUMERICAL METHODS FOR HIGHER ORDER EQUATIONS AND SYSTEMS

In Section 5.6 we obtained a numerical method for approximating the solution to an initial value problem for a second order differential equation by converting the second order equation to a system of first order equations. A similar approach works for higher order equations. Using the reduction procedure discussed at the end of Section 9.2, a higher order equation can always be converted to a system of first order equations in normal form. We must now extend the numerical technique for solving an initial value problem for a single first order equation to a technique for a system of equations in normal form. To see how these extentions are done, let's recall the classical Runge-Kutta method of order four, which we discussed in Section 3.6.

For the initial value problem

$$x' = f(t, x), \qquad x(t_0) = x_0,$$

the recursive formulas for the fourth order Runge-Kutta method are

(1)
$$t_{n+1} = t_n + h, \qquad n = 0, 1, \ldots,$$
$$x_{n+1} = x_n + \frac{1}{6}(k_1 + 2k_2 + 2k_3 + k_4),$$

where h is the step size and

(2)
$$k_1 = hf(t_n, x_n), \qquad k_2 = hf\left(t_n + \frac{h}{2}, x_n + \frac{1}{2}k\right),$$
$$k_3 = hf\left(t_n + \frac{h}{2}, x_n + \frac{1}{2}k_2\right), \qquad k_4 = hf(t_n + h, x_n + k_3).$$

Let $x_1(t), \ldots, x_m(t)$ be the components of the solution to the system of m first order equations

$$x_1'(t) = f_1(t, x_1, x_2, \ldots, x_m),$$
$$x_2'(t) = f_2(t, x_1, x_2, \ldots, x_m),$$
$$\vdots$$
$$x_m'(t) = f_m(t, x_1, x_2, \ldots, x_m)$$

that satisfy the initial conditions

$$x_1(t_0) = a_1, \quad x_2(t_0) = a_2, \quad \ldots, \quad x_m(t_0) = a_m.$$

Let $x_{n,i}$ denote an approximation to $x_i(t_n)$ for $i = 1, 2, \ldots, m$ and $t_n = t_0 + nh$ for $n = 0, 1, \ldots$. The recursive formulas for the fourth order Runge-Kutta method for systems are obtained from the formulas for a single equation given in (1)–(2) by treating the quantities x_n, k_1, k_2, k_3, and k_4 as vectors; that is $\mathbf{x}_n = (x_{n,1}, \ldots, x_{n,m})$, $\mathbf{k}_1 = (k_{1,1}, \ldots, k_{1,m})$, $\mathbf{k}_2 = (k_{2,1}, \ldots, k_{2,m})$, and so on. With this approach, the equations in (1)–(2) become vector equations. Writing these equations in component form gives the following recursive formulas for the fourth order Runge-Kutta method for systems:

(3)
$$t_{n+1} := t_n + h, \qquad n = 0, 1, 2, \ldots$$

(4)
$$x_{n+1,i} := x_{n,i} + \frac{1}{6}(k_{1,i} + 2k_{2,i} + 2k_{3,i} + k_{4,i}),$$

$$i = 1, 2, \ldots, m,$$

where h is the step size and

$$k_{1,i} := hf_i(t_n, x_{n,1}, x_{n,2}, \ldots, x_{n,m}),$$
$$i = 1, 2, \ldots, m;$$

$$k_{2,i} := hf_i\left(t_n + \frac{h}{2}, x_{n,1} + \frac{1}{2}k_{1,1}, x_{n,2} + \frac{1}{2}k_{1,2}, \ldots, x_{n,m} + \frac{1}{2}k_{1,m}\right),$$
$$i = 1, 2, \ldots, m;$$

$$k_{3,i} := hf_i\left(t_n + \frac{h}{2}, x_{n,1} + \frac{1}{2}k_{2,1}, x_{n,2} + \frac{1}{2}k_{2,2}, \ldots, x_{n,m} + \frac{1}{2}k_{2,m}\right),$$
$$i = 1, 2, \ldots, m;$$

$$k_{4,i} := hf_i(t_n + h, x_{n,1} + k_{3,1}, x_{n,2} + k_{3,2}, \ldots, x_{n,m} + k_{3,m}),$$
$$i = 1, 2, \ldots, m.$$

It is important to note that each $k_{2,i}$ depends on all the $k_{1,i}$'s and hence all the $k_{1,i}$'s must be computed before any of the $k_{2,i}$'s. Similarly, each $k_{2,i}$ must be computed before the $k_{3,i}$'s and each $k_{3,i}$ before any $k_{4,i}$.

The Runge-Kutta algorithm of Chapter 3 for a single first order equation can be easily modified to handle systems (see page 113). For systems the procedure stops when each component of two successive approximations differs by less than a prescribed tolerance ε, or it stops after a prescribed maximum number of iterations.

▶ FOURTH ORDER RUNGE-KUTTA ALGORITHM FOR SYSTEMS

Purpose To approximate the solution to the initial value problem

$$x_i' = f_i(t, x_1, \ldots, x_m);$$
$$x_i(t_0) = a_i, \qquad i = 1, 2, \ldots, m$$

at $t = c$.

INPUT $m, t_0, a_1, \ldots, a_m, c$
ε (tolerance)
M (maximum number of iterations)

Step 1 Set $z_i = a_i, \qquad i = 1, 2, \ldots, m$

Step 2 For $n = 0$ to M do Steps 3–9

Step 3 Set

$$h = (c - t_0)2^{-n}, \quad t = t_0, \quad x_1 = a_1, \quad \ldots, \quad x_m = a_m$$

Step 4 For $j = 1$ to 2^n do Steps 5 and 6

Step 5   Set

$$k_{1,i} = hf_i(t, x_1, \ldots, x_m),$$
$$i = 1, \ldots, m;$$

$$k_{2,i} = hf_i\left(t + \frac{h}{2}, x_1 + \frac{1}{2}k_{1,1}, \ldots, x_m + \frac{1}{2}k_{1,m}\right),$$
$$i = 1, \ldots, m;$$

$$k_{3,i} = hf_i\left(t + \frac{h}{2}, x_1 + \frac{1}{2}k_{2,1}, \ldots, x_m + \frac{1}{2}k_{2,m}\right),$$
$$i = 1, \ldots, m;$$

$$k_{4,i} = hf_i(t + h, x_1 + k_{3,1}, \ldots, x_m + k_{3,m}),$$
$$i = 1, \ldots, m$$

Step 6   Set

$$t = t + h;$$

$$x_i = x_i + \frac{1}{6}(k_{1,i} + 2k_{2,i} + 2k_{3,i} + k_{4,i}),$$
$$i = 1, \ldots, m$$

Step 7   Print $t, x_1, x_2, \ldots, x_m$

Step 8   If $|z_i - x_i| < \varepsilon$ for $i = 1, \ldots, m$, go to Step 12

Step 9   Set $z_i = x_i,$ $i = 1, \ldots, m$

Step 10   Print "$x_i(c)$ is approximately,"; x_i (for $i = 1, \ldots, m$); "but may not be within the tolerance"; ε

Step 11   Go to Step 13

Step 12   Print "$x_i(c)$ is approximately,"; x_i (for $i = 1, \ldots, m$); "with tolerance"; ε

Step 13   Stop

OUTPUT   Approximations of the solution to the initial value problem at $t = c$, using 2^n steps.

An Application to Population Dynamics

A mathematical model for the population dynamics of competing species, one a predator with population $x_2(t)$ and the other its prey with population $x_1(t)$, was developed independently in the early 1900s by A. J. Lotka and V. Volterra. They assume that there is plenty of food available for the prey to eat, so the birth rate of the prey should follow the Malthusian or exponential law (see Section 3.2); that is, the birth rate of the prey is Ax_1, where A is a positive constant. The death rate of the prey depends on the number of interactions between the predators and the prey. This is modeled by the expression Bx_1x_2, where B is a positive constant. Therefore, the rate of change in the population of the prey per unit time is $dx_1/dt = Ax_1 - Bx_1x_2$. Assuming that the predators depend entirely on the prey for their food, it is argued that the birth rate of the predators depends on the number of interactions with the prey; that is, the birth rate of predators is Dx_1x_2, where D is a positive constant. The death rate of the predators is assumed to be Cx_2, because without food the population would die off at a rate proportional to the population present. Hence the rate of change in the population of predators per unit time is $dx_2/dt = -Cx_2 + Dx_1x_2$. Combining these two equations, we obtain the Volterra-Lotka system for the population dynamics of two competing species:

$$x_1' = Ax_1 - Bx_1x_2,$$
$$x_2' = -Cx_2 + Dx_1x_2.$$

Such systems are in general not explicitly solvable. In the following example, we obtain an approximate solution for such a system.

● EXAMPLE 1   Use the fourth order Runge-Kutta algorithm for systems to approximate the solution of the initial value problem

(5)
$$x_1' = 2x_1 - 2x_1x_2; \qquad x_1(0) = 1,$$
$$x_2' = -x_2 + x_1x_2; \qquad x_2(0) = 3,$$

at $t = 1$ with a tolerance of 0.0001.

Solution   Here $f_1(t, x_1, x_2) = 2x_1 - 2x_1x_2$ and $f_2(t, x_1, x_2) = x_1x_2 - x_2$. With the starting values $t_0 = 0, x_{0,1} = 1$, and $x_{0,2} = 3$, we proceed with the algorithm to compute $x_1(1; 1)$ and $x_2(1; 1)$, the approximations to $x_1(1), x_2(1)$ using $h = 1$. We find

$$k_{1,1} = hf_1(t_0, x_{0,1}, x_{0,2}) = h(2x_{0,1} - 2x_{0,1}x_{0,2}) = 2(1) - 2(1)(3) = -4,$$
$$k_{1,2} = hf_2(t_0, x_{0,1}, x_{0,2}) = h(x_{0,1}x_{0,2} - x_{0,2}) = (1)(3) - 3 = 0,$$

$$k_{2,1} = hf_1\left(t_0 + \frac{h}{2}, x_{0,1} + \frac{1}{2}k_{1,1}, x_{0,2} + \frac{1}{2}k_{1,2}\right)$$

$$= h\left[2\left(x_{0,1} + \frac{1}{2}k_{1,1}\right) - 2\left(x_{0,1} + \frac{1}{2}k_{1,1}\right)\left(x_{0,2} + \frac{1}{2}k_{1,2}\right)\right]$$

$$= 2\left(1 + \frac{1}{2}(-4)\right) - 2\left(1 + \frac{1}{2}(-4)\right)\left(3 + \frac{1}{2}(0)\right)$$

$$= -2 + 2(3) = 4,$$

$$k_{2,2} = hf_2\left(t_0 + \frac{h}{2}, x_{0,1} + \frac{1}{2}k_{1,1}, x_{0,2} + \frac{1}{2}k_{1,2}\right)$$

$$= h\left[\left(x_{0,1} + \frac{1}{2}k_{1,1}\right)\left(x_{0,2} + \frac{1}{2}k_{1,2}\right) - \left(x_{0,2} + \frac{1}{2}k_{1,2}\right)\right]$$

$$= \left[1 + \frac{1}{2}(-4)\right]\left[3 + \frac{1}{2}(0)\right] - \left[3 + \frac{1}{2}(0)\right]$$

$$= (-1)(3) - 3 = -6.$$

Similarly, we find

$$k_{3,1} = hf_1\left(t_0 + \frac{h}{2}, x_{0,1} + \frac{1}{2}k_{2,1}, x_{0,2} + \frac{1}{2}k_{2,2}\right) = 6,$$

$$k_{3,2} = hf_2\left(t_0 + \frac{h}{2}, x_{0,1} + \frac{1}{2}k_{2,1}, x_{0,2} + \frac{1}{2}k_{2,2}\right) = 0,$$

$$k_{4,1} = hf_1(t_0 + h, x_{0,1} + k_{3,1}, x_{0,2} + k_{3,2}) = -28,$$
$$k_{4,2} = hf_2(t_0 + h, x_{0,1} + k_{3,1}, x_{0,2} + k_{3,2}) = 18.$$

Hence from (4) we compute

$$x_{1,1} = x_{0,1} + \frac{1}{6}(k_{1,1} + 2k_{2,1} + 2k_{3,1} + k_{4,1})$$

$$= 1 + \frac{1}{6}[-4 + 8 + 12 - 28] = -1,$$

$$x_{1,2} = x_{0,2} + \frac{1}{6}(k_{1,2} + 2k_{2,2} + 2k_{3,2} + k_{4,2})$$

$$= 3 + \frac{1}{6}(0 - 12 + 0 + 18) = 4.$$

Repeating the algorithm with $h = \frac{1}{2}$, we obtain the approximations $x_1(1; 2^{-1})$ and $x_2(1; 2^{-1})$ for $x_1(1)$ and $x_2(1)$. In Table 9.1 we list the approximations $x_1(1; 2^{-n})$ and $x_2(1; 2^{-n})$ for $x_1(1)$ and $x_2(1)$, using step size $h = 2^{-n}$ for $n = 0, 1, 2, 3$, and 4. We stopped at $n = 4$ since both

$$|x_1(1; 2^{-3}) - x_1(1; 2^{-4})| = 0.00006 < 0.0001$$

and

$$|x_2(1; 2^{-3}) - x_2(1; 2^{-4})| = 0.00001 < 0.0001.$$

Hence $x_1(1) \approx 0.07735$ and $x_2(1) \approx 1.46445$, with tolerance 0.0001.   ●

TABLE 9.1	APPROXIMATIONS OF THE SOLUTION TO SYSTEM (5) IN EXAMPLE 1		
n	h	$x_1(1; h)$	$x_2(1; h)$
0	1.0	-1.0	4.0
1	0.5	0.14662	1.47356
2	0.25	0.07885	1.46469
3	0.125	0.07741	1.46446
4	0.0625	0.07735	1.46445

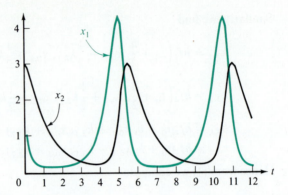

Figure 9.22 Graphs of the components of an approximate solution to the Volterra-Lotka system (5)

To get a better feel for the solution to system (5), we have graphed in Figure 9.22 an approximation of the solution, using linear interpolation to connect the points. This approximation was obtained using step size $h = 0.125$ and the fourth order Runge-Kutta method for systems. From the graph it appears that the components x_1 and x_2 are periodic in the variable t. In the next section we study the qualitative behavior of a system of two first order equations using a technique called **phase plane analysis.** This technique is also used in Project C to study general Volterra-Lotka equations and, in particular, to show that these equations have periodic solutions.

EXERCISES 9.5

The reader will find it helpful to have a microcomputer available or access to a mainframe.

In Problems 1 through 8 use fourth order Runge-Kutta for systems with $h = 0.25$ to find an approximation for the solution to the given initial value problem at the times $t = 0.25$, $t = 0.5, t = 0.75,$ and $t = 1.0$.

1. $x' = 3x - y,$ $x(0) = 0,$
   $y' = x - 2y,$ $y(0) = 2.$

2. $x' = -x^2 + y^2 + 1,$ $x(0) = 0,$
   $y' = -y,$ $y(0) = 1.$

3. $x' = \cos(x + y),$ $x(0) = 1,$
   $y' = (x^2 + t)^{-1} + t,$ $y(0) = 0.5.$

4. $x' = e^{-x^2} + y,$ $x(0) = 1,$
   $y' = x^2 - y + t,$ $y(0) = 1.$

5. $y''' - ty'' + ty' - 3y = 0;$
   $y(0) = 1,$ $y'(0) = -1,$ $y''(0) = 0.$

6. $y^{(4)} + 3y''' + 2y' + (\sin t)y = 0;$
   $y(0) = 0,$ $y'(0) = 1,$ $y''(0) = 0,$ $y'''(0) = 0.$

7. $y^{(4)} - t^2y' + e^t y = t + 1;$
   $y(0) = -1,$ $y'(0) = 1,$ $y''(0) = 1,$ $y'''(0) = -1.$

8. $y''' - \sin y = t;$
   $y(0) = 1,$ $y'(0) = 0,$ $y''(0) = 0.$

9. Using fourth order Runge-Kutta for systems with $h = 0.125$, approximate the solution to the initial value problem

$$x' = 2x - y,   x(0) = 0,$$
$$y' = 3x + 6y,   y(0) = -2,$$

at $t = 1$. Compare this approximation to the actual solution

$$x(t) = e^{5t} - e^{3t},   y(t) = e^{3t} - 3e^{5t}.$$

10. Using the Runge-Kutta algorithm for systems with the stopping procedure based on the relative error and a tolerance of $\varepsilon = 0.02$, approximate the solution to the initial value problem

$$x' = 6x - 3y, \qquad x(0) = -1,$$
$$y' = 2x + y, \qquad y(0) = 0,$$

at $t = 1$.

11. Using the Runge-Kutta algorithm for systems with the stopping procedure based on the absolute error and a tolerance of $\varepsilon = 0.001$, approximate the solution to the initial value problem

$$\frac{du}{dx} = 3u - 4v, \qquad u(0) = 1,$$

$$\frac{dv}{dx} = 2u - 3v, \qquad v(0) = 1,$$

at $x = 1$.

12. Using the Runge-Kutta algorithm for systems with the stopping procedure based on the absolute error and a tolerance of $\varepsilon = 0.0001$, approximate the solution to the initial value problem

$$y''' - y'' - y = t^2;$$
$$y(0) = 0, \qquad y'(0) = 1, \qquad y''(0) = 0,$$

at $t = 1$.

13. Using the Runge-Kutta algorithm for systems with the stopping procedure based on the relative error and a tolerance of $\varepsilon = 0.01$, approximate the solution to the initial value problem

$$y''' + y'' + y^2 = t;$$
$$y(0) = 1, \qquad y'(0) = 0, \qquad y''(0) = 1,$$

at $t = 1$.

14. In Section 3.5, we discussed the improved Euler's method for approximating the solution to a first order equation. Extend the improved Euler's method to systems and give the recursive formulas in component form.

15. Use fourth order Runge-Kutta for systems with $h = 0.1$ to approximate the solution to the initial value problem

$$x' = yz, \qquad x(0) = 0,$$
$$y' = -xz, \qquad y(0) = 1,$$
$$z' = -xy/2, \qquad z(0) = 1,$$

at $t = 1$.

16. **Combat Model.** A mathematical model for conventional versus guerrilla combat is given by the system

$$x_1' = -(0.1)x_1 x_2, \qquad x_1(0) = 10,$$
$$x_2' = -x_1, \qquad x_2(0) = 15,$$

where x_1 and x_2 are the strengths of guerrilla and conventional troops, respectively, and 0.1 and 1 are the *combat effectiveness coefficients.* Who will win the conflict, the conventional troops or the guerrillas? [Hint: Use the Runge-Kutta method for systems with $h = 0.1$ to approximate the solutions.]

17. **Generalized Blasius Equation.** H. Blasius, in his study of laminar flow of a fluid, encountered an equation of the form

$$y''' + yy'' = (y')^2 - 1.$$

Use the Runge-Kutta method for systems with $h = 0.1$ to approximate the solution that satisfies the initial conditions $y(0) = 0$, $y'(0) = 0$, and $y''(0) = 1.32824$. Sketch this solution on the interval $[0, 2]$.

18. **Lunar Orbit.** The motion of a moon moving in a planar orbit about a planet is governed by the equations

$$x'' = -G\frac{mx}{r^3}, \qquad y'' = -G\frac{my}{r^3},$$

where $r := (x^2 + y^2)^{1/2}$, G is the gravitational constant, and m is the mass of the moon. Assume that $Gm = 1$. When $x(0) = 1$, $x'(0) = y(0) = 0$, and $y'(0) = 1$, the motion is a circular orbit of radius 1 and period 2π.

(a) Setting $x_1 = x$, $x_2 = x'$, $x_3 = y$, $x_4 = y'$, express the governing equations as a first order system in normal form.

(b) Using $h = 2\pi/100 \approx 0.0628318$, compute one orbit of this moon (i.e., let $n = 100$). Do your approximations agree with the fact that the orbit is a circle of radius 1?

19. **Predator-Prey Model.** The Volterra-Lotka predator-prey model predicts some rather interesting behavior that is evident in certain biological systems. For example, if you fix the initial population of prey but increase the initial population of predators, then the population cycle for the prey becomes more severe in the sense that there is a long period of time with a reduced population of prey followed by a short period when the population of prey is very large. To demonstrate this behavior, use the fourth order Runge-Kutta method for systems

with $h = 0.5$ to approximate the populations of prey x and of predators y over the period $[0, 5]$ that satisfy the Volterra-Lotka system

$$x' = x(3 - y),$$
$$y' = y(x - 3)$$

under each of the following initial conditions:

(a) $x(0) = 2$, $y(0) = 4$.
(b) $x(0) = 2$, $y(0) = 5$.
(c) $x(0) = 2$, $y(0) = 7$.

20. **Spring Pendulum.**   Let a mass be attached to one end of a spring with spring constant k and the other end attached to the ceiling. Let l_0 be the natural length of the spring and let $l(t)$ be its length at time t. If $\theta(t)$ is the angle between the pendulum and the vertical, then the motion of the spring pendulum is governed by the system

$$l''(t) - l(t)\theta'(t) - g\cos\theta(t) + \frac{k}{m}(l - l_0) = 0,$$

$$l^2(t)\theta''(t) + 2l(t)l'(t)\theta'(t) + gl(t)\sin\theta(t) = 0.$$

Assume $g = 1$, $k = m = 1$, and $l_0 = 4$. When the system is at rest, $l = l_0 + mg/k = 5$.

(a) Describe the motion of the pendulum when $l(0) = 5.5$, $l'(0) = 0$, $\theta(0) = 0$, and $\theta'(0) = 0$. [Hint: Use fourth order Runge-Kutta for systems with $h = 0.1$ to approximate the solution on the interval $[0, 10]$.]
(b) When the pendulum is both stretched and given an angular displacement, the motion of the pendulum is more complicated. Using Runge-Kutta for sys-

tems with $h = 0.1$ to approximate the solution, sketch the graph of the length l and the angular displacement θ on the interval $[0, 10]$ if $l(0) = 5.5$, $l'(0) = 0$, $\theta(0) = 0.5$, and $\theta'(0) = 0$.

21. **Competing Species.**   Let $p_i(t)$ denote, respectively, the populations of three competing species S_i, $i = 1, 2, 3$. Suppose that these species have the same growth rates and that the maximum population that the habitat can support is the same for each species. (We assume it to be one unit.) Moreover, suppose that the competitive advantage that S_1 has over S_2 is the same as that of S_2 over S_3 and S_3 over S_1. This situation is modeled by the system

$$p_1' = p_1(1 - p_1 - ap_2 - bp_3),$$
$$p_2' = p_2(1 - bp_1 - p_2 - ap_3).$$
$$p_3' = p_3(1 - ap_1 - bp_2 - p_3),$$

where a and b are positive constants. To demonstrate the population dynamics of this system when $a = b = 0.5$, use the fourth order Runge-Kutta method for systems with $h = 0.1$ to approximate the populations p_i over the time interval $[0, 10]$ under each of the following initial conditions.

(a) $p_1(0) = 1.0$, $p_2(0) = 0.1$, $p_3(0) = 0.1$.
(b) $p_1(0) = 0.1$, $p_2(0) = 1.0$, $p_3(0) = 0.1$.
(c) $p_1(0) = 0.1$, $p_2(0) = 0.1$, $p_3(0) = 1.0$.

Based upon the results of parts (a)–(c), what do you think will happen to these populations as $t \to \infty$? When $a = 0.5$ and $b = 2.0$, the behavior of these populations is entirely different (see Project E).

•9.6• NONLINEAR AUTONOMOUS SYSTEMS

In this section we study systems of two first order equations of the form

(1)
$$\frac{dx}{dt} = f(x, y),$$

$$\frac{dy}{dt} = g(x, y),$$

where f and g are real-valued functions that do not depend explicitly on t. Such systems are called **autonomous.** An example of an autonomous system is the Volterra-Lotka system

$$\frac{dx}{dt} = Ax - Bxy,$$

$$\frac{dy}{dt} = -Cy + Dxy$$

discussed in Section 9.5. As with all autonomous systems, the parameters involved (here, A, B, C, and D) are *time independent*.

A **solution** to (1) on an interval I is a pair of functions $x(t)$, $y(t)$ that satisfy (1) for all t in I. The set of points $\{(x(t), y(t)); t \in I\}$ in the xy-plane is called a **trajectory,**[†] and the xy-plane is referred to as the **phase plane.**

If the pair $x(t)$, $y(t)$ is a solution to (1), then so is the pair $x(t + c)$, $y(t + c)$, for any constant c, because (thinking of t as time) the system (1) is not time dependent. To be more precise, let $X(t) := x(t + c)$ and $Y(t) := y(t + c)$. Then, by the chain rule,

$$\frac{dX}{dt}(t) = \frac{dx}{dt}(t + c) = f(x(t + c), y(t + c)) = f(X(t), Y(t)),$$

$$\frac{dY}{dt}(t) = \frac{dy}{dt}(t + c) = g(x(t + c), y(t + c)) = g(X(t), Y(t)),$$

which proves that $X(t)$, $Y(t)$ is also a solution to (1).

For nonlinear autonomous systems, obtaining explicit solutions $x(t)$, $y(t)$ may be an extraordinarily difficult task. However, in certain cases, it is possible to obtain the trajectories for the system that give us valuable qualitative information about the solution. Here we give a brief introduction to the study of the qualitative behavior of autonomous systems.[††] A more thorough discussion can be found in *Ordinary Differential Equations*, Third Edition, by G. Birkhoff and G. C. Rota, John Wiley and Sons, Inc., New York, 1978, Chapter 5.

To illustrate the idea of trajectories, we start with a simple example of two uncoupled linear equations.

● **EXAMPLE 1** Find solutions to the autonomous system

$$(2) \qquad \frac{dx}{dt} = x, \qquad \frac{dy}{dt} = 2y$$

and sketch their trajectories.

Solution Solving separately each equation in (2), we quickly find that $x(t) = c_1 e^t$, $y(t) = c_2 e^{2t}$ are solutions. Expressing y in terms of x, we obtain the trajectories

$$(3) \qquad y = c_2(e^t)^2 = c_2(x/c_1)^2 = kx^2,$$

[†] Trajectories are also called **solution curves, paths,** or **orbits.**

[††] *Historical Footnote*: The study of the qualitative behavior of autonomous systems was begun by the great French mathematician **J. Henri Poincaré** (1854–1912) in 1881 in his work in celestial mechanics. His work was continued by **I. Bendixson** in 1901 in his study of closed trajectories.

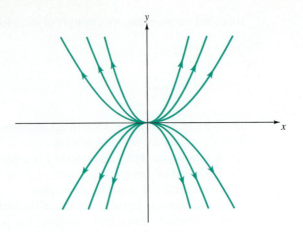

Figure 9.23 Phase plane diagram for $x' = x$, $y' = 2y$

where $k = c_2/c_1^2$ is an arbitrary constant. These parabolic trajectories are sketched in the phase plane diagram, Figure 9.23.

The arrows on the trajectories in Figure 9.23 indicate the direction or **flow** of a particle as it travels along the trajectory. In equation (3) we see that as x moves away from zero, y increases if k is positive (this is the case for the trajectories above the x-axis) and decreases if k is negative (the trajectories below the x-axis). Hence, the flow is always away from the origin. ●

In the preceding example, the origin is the trajectory of the constant solution $x \equiv 0$, $y \equiv 0$. Such points and solutions play an important role in the study of autonomous systems.

CRITICAL POINTS AND EQUILIBRIUM SOLUTIONS

> **Definition 1.**   A point (x_0, y_0), where $f(x_0, y_0) = 0$ and $g(x_0, y_0) = 0$, is called a **critical point** of the system $dx/dt = f(x, y)$, $dy/dt = g(x, y)$, and the corresponding constant solution $x(t) \equiv x_0$, $y(t) \equiv y_0$ is called an **equilibrium solution.** The set of all critical points is called the **critical point set.**

● **EXAMPLE 2**   Find the critical points and corresponding equilibrium solutions of

$$\frac{dx}{dt} = -y(y - 2),$$

(4)

$$\frac{dy}{dt} = (x - 2)(y - 2).$$

Solution   To find the critical points we solve the system

$$-y(y - 2) = 0, \qquad (x - 2)(y - 2) = 0.$$

One family of solutions to this system is given by $y = 2$ with x arbitrary; that is, the line

$y = 2$. If $y \neq 2$, then the system simplifies to $-y = 0$, and $x - 2 = 0$, which has the solution $x = 2$, $y = 0$. Hence the critical point set consists of the isolated point $(2, 0)$ and the horizontal line $y = 2$. The corresponding equilibrium solutions are $x(t) \equiv 2$, $y(t) \equiv 0$, and the family $x(t) \equiv c$, $y(t) \equiv 2$, where c is an arbitrary constant. ●

Although we succeeded in finding the equilibrium solutions to the nonlinear system (4) of Example 2, determining other explicit solutions to this system is beyond our present capabilities. However, we can determine the trajectories for the system using techniques for solving first order equations discussed in Chapter 2.

● **EXAMPLE 3**   Determine the trajectories of system (4).

Solution   Unlike Example 1, we are not able to determine the trajectories by first computing the solutions. But we do know that when $dx/dt \neq 0$, the chain rule yields

(5) $$\frac{dy}{dx} = \frac{dy/dt}{dx/dt} = \frac{(x - 2)(y - 2)}{-y(y - 2)} = -\frac{x - 2}{y}.$$

In (5) we have a first order differential equation whose solutions in x and y contain the trajectories of system (4). Solving (5) by separation of variables, we obtain

$$\frac{y^2}{2} = \int y \, dy = -\int (x - 2) \, dx = -\frac{(x - 2)^2}{2} + c,$$

which is just the family of concentric circles

(6) $$y^2 + (x - 2)^2 = k,$$

with centers located at $(2, 0)$. Thus each trajectory of (4) lies on a circle given in (6). In Figure 9.24 we have sketched these circles and the critical points of (4).

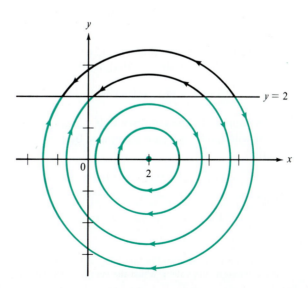

Figure 9.24 Phase plane diagram for $x' = -y(y - 2)$, $y' = (x - 2)(y - 2)$

Now that we know that the trajectories lie on the circles in (6), what remains is to analyze the flow along each trajectory. For this purpose we use the given equation $dx/dt = -y(y-2)$. If $y > 2$, we see that $dx/dt < 0$. Hence x is decreasing when $y > 2$. This means that the flow is from right to left along the arc of a circle that lies above the line $y = 2$. For $0 < y < 2$, we have $dx/dt > 0$, and so in this region the flow is from left to right. Furthermore, for $y < 0$, we have $dx/dt < 0$, and again the flow is from right to left.

We now observe in Figure 9.24 on page 451 that there are four types of trajectories associated with system (4): (a) those that begin above the line $y = 2$ and follow the arc of a circle counter-clockwise back to that line; (b) those that begin below the line $y = 2$ and follow the arc of a circle clockwise back to that line; (c) those that continually move clockwise around a circle centered at $(2, 0)$ with radius less than 2 (i.e., do not intersect the line $y = 2$); and finally, (d) the critical points $(2, 0)$ and $y = 2$, x arbitrary. ●

The curves given in equation (6) are called the **integral curves** of (4). They are the level curves of the function $V(x, y) := (x-2)^2 + y^2$, which is called a **first integral** of (4).[†] As we observed earlier, a trajectory must lie on an integral curve.

In Example 3 we noted that the nonperiodic trajectories (a) and (b) begin and end at a critical point, namely at a point on the line $y = 2$. This fact is a consequence of the following theorem:

> ### LIMIT POINTS ARE CRITICAL POINTS
>
> **Theorem 1.** Let the pair $x(t)$, $y(t)$ be a solution on $[0, \infty)$ to the autonomous system $dx/dt = f(x, y)$, $dy/dt = g(x, y)$, where f and g are continuous in the plane. If the limits
> $$x_\omega := \lim_{t \to \infty} x(t) \quad \text{and} \quad y_\omega := \lim_{t \to \infty} y(t)$$
> exist and are finite, then the limit point (x_ω, y_ω) is a critical point for the system.

Proof. To show that (x_ω, y_ω) is a critical point, we must prove that $f(x_\omega, y_\omega) = g(x_\omega, y_\omega) = 0$. First observe that, by the continuity of f,
$$\lim_{t \to \infty} x'(t) = \lim_{t \to \infty} f(x(t), y(t)) = f(x_\omega, y_\omega).$$

Let's suppose to the contrary that $f(x_\omega, y_\omega) \neq 0$; say, $w := f(x_\omega, y_\omega) > 0$ (a similar argument will handle the case when $w < 0$). For any $\varepsilon > 0$, we can choose N sufficiently large so that
$$|f(x_\omega, y_\omega) - f(x(t), y(t))| < \varepsilon \quad \text{for all } t \geq N.$$

Taking $\varepsilon = w/2$, we have $|w - f(x(t), y(t))| < w/2$ and so $x'(t) = f(x(t), y(t)) > w/2$. But if $x'(t) > w/2$ for all $t \geq N$, then $x(t)$ must lie above some line with positive slope $w/2$. Hence

---

[†] In classical mechanics the first integral V is associated with the total energy of the system, and hence a level curve is a curve along which the energy is constant. In fluid mechanics V is called the **stream function**.

$\lim_{t \to \infty} x(t) = +\infty$, which contradicts the assumption that this limit is finite. Therefore, $f(x_\omega, y_\omega) = 0$. A similar argument also shows that $g(x_\omega, y_\omega) = 0$. Hence the limit point (x_ω, y_ω) is a critical point. ◀◀◀

The proof of Theorem 1 can be easily modified to show that a limit point (x_α, y_α), where

$$x_\alpha := \lim_{t \to -\infty} x(t), \qquad y_\alpha := \lim_{t \to -\infty} y(t),$$

is also a critical point. As a consequence of Theorem 1, we observe that *the only points in the phase plane where a nonperiodic trajectory can begin $(t \to -\infty)$ or end $(t \to +\infty)$ are the critical points.* For example, in Figure 9.24, the trajectories that lie above the line $y = 2$ begin at a critical point on the line $y = 2$ and end at a critical point on the line.

A sketch of the critical point set for a system, along with representative integral curves and their trajectories with arrows indicating the flow, is called a **phase plane diagram.** The use of these diagrams to obtain qualitative information about the solutions of the system is referred to as **phase plane analysis.** In many cases the integral curves or the trajectories must be obtained using numerical techniques, and several software packages are available for this purpose. Two such packages that run on the IBM-PC or its clones are PHASE-PLANE by Brad Ermentrout available from Brooks/Cole Publishing Company, Pacific Groves, California and PHASER by Hüseyin Koçak available from Springer-Verlag, New York. The package MACMATH by John H. Hubbard and Beverly H. West also available from Springer-Verlag, runs on the Macintosh.

PHASE PLANE ANALYSIS

To sketch a phase plane diagram for the system $dx/dt = f(x, y)$, $dy/dt = g(x, y)$:

(a) Find the critical point set consisting of the solutions to the system $f(x, y) = 0$, $g(x, y) = 0$. Sketch this set in the xy-plane.

(b) Solve the first order equation $dy/dx = g(x, y)/f(x, y)$. Use the solution to sketch some representative integral curves for the system. (Numerical techniques may be needed for this purpose.)

(c) Determine the direction of the flow along the integral curves using either of the equations $dx/dt = f(x, y)$ or $dy/dt = g(x, y)$. This gives the trajectories.

It can be shown that system (1) has a unique solution satisfying $x(t_0) = x_0$, $y(t_0) = y_0$, provided that f and g have continuous partial derivatives in the xy-plane. This is helpful in sketching the trajectories of a system because it means that two trajectories cannot intersect (in finite time).

In the next example we will use the software PHASEPLANE to construct a phase plane diagram.

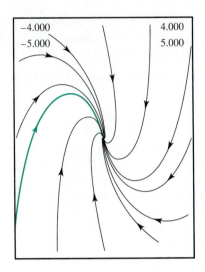

Figure 9.25 Phase plane diagram for system (7) constructed using the software PHASEPLANE

● **EXAMPLE 4**   Use a software package to construct a phase plane diagram about the critical point $(0,0)$ for the linear system

$$\frac{dx}{dt} = -2x + y,$$

(7)

$$\frac{dy}{dt} = -5x - 4y.$$

Solution   With the aid of the software package PHASEPLANE, we have graphed in Figure 9.25 several solutions to system (7) starting at various initial locations within the window $-4 \le x \le 4$, $-5 \le y \le 5$. The trajectories all appear to converge to the origin. In Chapter 13[†] we shall see that the origin is an example of an *asymptotically stable spiral point.* Here we use the zoom feature of PHASEPLANE to observe the spiraling of a trajectory about the origin. In the series of graphs in Figure 9.26, we have zoomed in on the trajectory (in color in Figure 9.25) that passes through the point $(-3.875000, -3.111111)$. Observe how the trajectory spirals about the origin as the window about the origin gets smaller $\{-0.089 \le x \le 0.125$, $-0.111 \le y \le 0.222\}$ and smaller $\{-0.002 \le x \le 0.002$, $-0.004 \le y \le 0.004\}$ and smaller {"$-0.000 \le x \le 0.000$, $-0.000 \le y \le 0.000$"}.   ●

Autonomous systems arise in the study of the motion of a particle governed by Newton's second law, $md^2x/dt^2 = F(x, dx/dt)$, when the forcing function F does not depend explicitly on t. In the next example we use a phase plane analysis to study a nonlinear spring-mass system.

_____

[†] All references to Chapters 12–14 refer to the expanded text *Fundamentals of Differential Equations and Boundary Value Problems.*

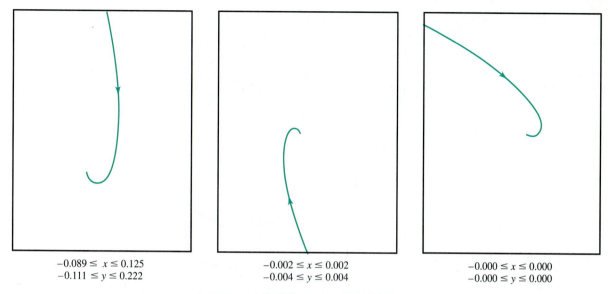

$$-0.089 \le x \le 0.125$$
$$-0.111 \le y \le 0.222$$

$$-0.002 \le x \le 0.002$$
$$-0.004 \le y \le 0.004$$

$$-0.000 \le x \le 0.000$$
$$-0.000 \le y \le 0.000$$

Figure 9.26 Spiraling of a trajectory about the origin

● **EXAMPLE 5**   A mass is attached to a spring and encounters friction as it moves along a surface (see Figure 9.27). The motion is governed by

$$(8) \qquad m\frac{d^2x}{dt^2} + \mu\,\text{sign}\left(\frac{dx}{dt}\right) + kx = 0,$$

where m is the mass, μ a positive constant depending on the mass of the object and the coefficient of friction between the mass and the surface, k is the spring constant, and sign(x') is the function that is $+1$ when $x' > 0$ and -1 when $x' < 0$. Describe the motion of the mass for the case where $m = \mu = k = 1$.

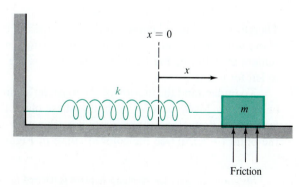

$x = 0$

x

k

m

Friction

Figure 9.27 Spring-mass system with friction

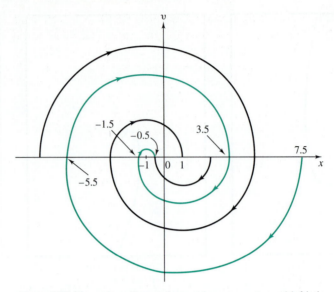

Figure 9.28 Phase plane diagram for a spring-mass system with friction

Solution   To study the motion governed by (8), we convert it to a system by introducing the velocity $v := x'$ as the second dependent variable.[†] Then we get

$$\text{(9)} \qquad \begin{aligned} x' &= v, \\ v' &= x'' = -x \mp 1, \qquad x' \gtrless 0, \end{aligned}$$

where we take $\mp$ to be *minus* for $x' > 0$ and plus for $x' < 0$.

The critical points of (9) in the xv-plane are $(1, 0)$, $(-1, 0)$. (Technically, we have not defined the system for $v = x' = 0$ since sign(x') is defined only for $x' \neq 0$. We shall return to this problem later.)

To find the integral curves of (9), we solve

$$\frac{dv}{dx} = \frac{dv/dt}{dx/dt} = \frac{-x \mp 1}{v}, \qquad v \gtrless 0.$$

Separating variables and integrating gives

$$v^2 + (x \pm 1)^2 = c, \qquad v \gtrless 0.$$

Therefore, the integral curves in the xv-plane are concentric semicircles in the upper half plane with centers at $(-1, 0)$ and concentric semicircles in the lower half plane with centers at $(1, 0)$ (see Figure 9.28). Since $x' = v$, the flow is left to right for $v > 0$ and right to left for $v < 0$.

To understand the motion of the mass, let's consider the case when the mass is released (with zero velocity) at $x = 7.5$. In the phase plane the mass begins at the point $(7.5, 0)$ and follows the flow along the semicircular trajectory in the lower half plane until it reaches the point $(-5.5, 0)$ (see the colored curve in Figure 9.28). (This corresponds to the mass

---

[†] *Historical Footnote:* The resulting xv-plane is referred to as the **Poincaré phase plane** in honor of **J. Henri Poincaré** (1854–1912), who did fundamental work in celestial mechanics, differential equations, and topology.

moving from $x = 7.5$ to $x = -5.5$.) The flow now takes the mass from $(-5.5, 0)$ along the semicircular trajectory in the upper half plane to the point $(3.5, 0)$. (The mass has moved from $x = -5.5$ to $x = 3.5$.) The mass now follows the trajectory in the lower half plane to $(-1.5, 0)$ and then the trajectory in the upper half plane to $(-0.5, 0)$. (The mass is now at $x = -0.5$.) At the point $(-0.5, 0)$ we have a problem! The trajectory in the lower half plane that touches the point $(-0.5, 0)$ flows *into* that point. In other words, $(-0.5, 0)$ sits at the endpoint of the two trajectories connecting it.[†] Hence the mass is trapped or stopped at $x = -0.5$. To understand what is happening, let's look at the physical situation. When $x = -0.5$, the force on the mass due to the spring is just $kx = -0.5$. But friction can apply a force up to 1. Therefore, friction cancels out the force due to the spring and prevents the mass from moving away from the spot $x = -0.5$. Consequently, the motion is stopped at $x = -0.5$.   ●

For other applications of phase plane analysis, we refer the reader to Chapter 13— "Stability of Autonomous Systems."[††]

EXERCISES 9.6

In Problems 1 and 2 verify that the pair $x(t)$, $y(t)$, is a solution to the given system. Find the trajectory of the given solution and sketch its graph in the phase plane.

1. $\dfrac{dx}{dt} = 3y^3$, $\quad \dfrac{dy}{dt} = y$; $\quad x(t) = e^{3t}$, $\quad y(t) = e^t$.

2. $\dfrac{dx}{dt} = 1$, $\quad \dfrac{dy}{dt} = 3x^2$;

$\qquad x(t) = t + 1$, $\qquad y(t) = t^3 + 3t^2 + 3t$.

In Problems 3 through 8 find the critical point set for the given system.

3. $\dfrac{dx}{dt} = x - y$,

$\quad \dfrac{dy}{dt} = x^2 + y^2 - 1$.

4. $\dfrac{dx}{dt} = y - 1$,

$\quad \dfrac{dy}{dt} = x + y + 5$.

5. $\dfrac{dx}{dt} = x^2 - 2xy$,

$\quad \dfrac{dy}{dt} = 3xy - y^2$.

6. $\dfrac{dx}{dt} = 3x^2 - xy$,

$\quad \dfrac{dy}{dt} = 4xy - 3y^2$.

7. $\dfrac{dx}{dt} = y^2 - 3y + 2$,

$\quad \dfrac{dy}{dt} = (x - 1)(y - 2)$.

8. $\dfrac{dx}{dt} = (x + 1)(y - 2)$,

$\quad \dfrac{dy}{dt} = x^2 - x - 2$.

In Problems 9 through 14 determine the integral curves for the given system.

9. $\dfrac{dx}{dt} = y - 1$,

$\quad \dfrac{dy}{dt} = e^{x+y}$,

10. $\dfrac{dx}{dt} = x^2$,

$\quad \dfrac{dy}{dt} = x^2 + y^2 + xy$.[†††]

11. $\dfrac{dx}{dt} = \dfrac{x^2}{y}$,

$\quad \dfrac{dy}{dt} = x - y$.[†††]

12. $\dfrac{dx}{dt} = x^2 - 2y^{-3}$,

$\quad \dfrac{dy}{dt} = 3x^2 - 2xy$.

13. $\dfrac{dx}{dt} = 2y - x$,

$\quad \dfrac{dy}{dt} = e^x + y$.

14. $\dfrac{dx}{dt} = x + y + 6$,

$\quad \dfrac{dy}{dt} = 3x - y - 6$.[†††]

[†] Note that these trajectories do not end in a critical point. This does not violate Theorem 1 because system (9) fails to satisfy the continuity assumptions in that theorem.

[††] All references to Chapters 12–14 refer to the expanded text *Fundamentals of Differential Equations and Boundary Value Problems*.

[†††] Use techniques from Section 2.6.

In Problems 15 through 20 sketch (by hand or using software) a phase plane diagram for the given system.

15. $\dfrac{dx}{dt} = 2y,$

$\dfrac{dy}{dt} = 2x.$

16. $\dfrac{dx}{dt} = -8y,$

$\dfrac{dy}{dt} = 18x.$

17. $\dfrac{dx}{dt} = (y - x)(y - 1),$

$\dfrac{dy}{dt} = (x - y)(x - 1).$

18. $\dfrac{dx}{dt} = (x - 4)(1 - y),$

$\dfrac{dy}{dt} = (x + 1)(x - 4).$

19. $\dfrac{dx}{dt} = \dfrac{3}{y},$

$\dfrac{dy}{dt} = \dfrac{2}{x}.$

20. $\dfrac{dx}{dt} = xy^2 - x,$

$\dfrac{dy}{dt} = 2y^3 - 4y.$

In Problems 21 through 26 use a software package such as PHASEPLANE to construct a phase plane diagram containing all the critical points for the given system.

21. $\dfrac{dx}{dt} = 2x + y + 3,$

$\dfrac{dy}{dt} = -3x - 2y - 4.$

22. $\dfrac{dx}{dt} = -5x + 2y,$

$\dfrac{dy}{dt} = x - 4y.$

23. $\dfrac{dx}{dt} = 2x + 13y,$

$\dfrac{dy}{dt} = -x - 2y.$

24. $\dfrac{dx}{dt} = -2x + y,$

$\dfrac{dy}{dt} = -5x - 4y.$

25. $\dfrac{dx}{dt} = x(7 - x - 2y),$

$\dfrac{dy}{dt} = y(5 - x - y).$

26. $\dfrac{dx}{dt} = 5x - 3y,$

$\dfrac{dy}{dt} = 4x - 3y.$

In Problems 27 through 30 sketch (by hand or using software) an xv–phase plane diagram for the given equation.

27. $\dfrac{d^2x}{dt^2} - x = 0.$

28. $\dfrac{d^2x}{dt^2} + x = 0.$

29. $\dfrac{d^2x}{dt^2} + x + x^5 = 0.$

30. $\dfrac{d^2x}{dt^2} + x^3 = 0.$

In Problems 31 and 32 use a software package such as PHASEPLANE to construct an xv–phase plane diagram containing all the critical points for the given equation.

31. $x''(t) + x(t) - x^4(t) = 0.$

32. $x''(t) + x(t) - x^3(t) = 0.$

33. Epidemic Model. A model for the spread of a disease through a population, or an **epidemic model,** is given by the system

$$\frac{dS}{dt} = -aSI,$$

$$\frac{dI}{dt} = aSI - bI,$$

where a and b are positive constants. Here $S(t)$ represents the susceptible population and $I(t)$ the infected population. Give a phase plane analysis for this system. (Only the first quadrant is of any interest; why?) Explain why an epidemic occurs (the number of infected persons increases) when $S(0) > b/a$.

34. Simple Pendulum. The motion of a simple pendulum is governed by the equation

$$\frac{d^2\theta}{dt^2} + \sin\theta = 0,$$

where θ represents the angular displacement from the vertical of the pendulum. Give a phase plane analysis for the equation and interpret the results physically.

35. Hard Spring. The motion of a spring-mass system in which the spring is "hard" and does not satisfy Hooke's law is governed by the equation

$$\frac{d^2x}{dt^2} + x + x^3 = 0,$$

where x represents the displacement from equilibrium. Give a phase plane analysis for the equation and interpret the results physically.

36. Falling Object. The motion of an object falling through the air is governed by the equation

$$\frac{d^2x}{dt^2} = g - \frac{g}{V^2}\frac{dx}{dt}\left|\frac{dx}{dt}\right|,$$

where x is the distance fallen and V is a constant called the terminal velocity. Give a phase plane analysis for the equation and interpret the results physically.

37. **Restrained Bar.** The motion of a bar restrained by springs and attracted by a parallel current passing through a wire (see Figure 9.29) is governed by the equation

$$\frac{d^2x}{dt^2} + a\left[x - \frac{b}{C - x}\right] = 0,$$

where a, b, and C are positive constants. Give a phase plane analysis for the equation for $-x_0 < x < C$, where x_0 is the distance from the wall to the equilibrium position. Interpret the results physically.

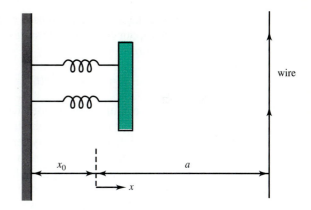

Figure 9.29 Bar restrained by springs and attracted by a parallel current

•9.7• DYNAMICAL SYSTEMS, POINCARÉ MAPS, STRANGE ATTRACTORS, AND CHAOS

In this section we are going to take an excursion through an area of mathematics that is currently receiving a lot of attention both for the interesting mathematical phenomena being observed and for its application to fields such as meteorology, heat conduction, fluid mechanics, lasers, chemical reactions, and nonlinear circuits, among others. The area is that of dynamical systems and, in particular, nonlinear dynamical systems.[†]

A **dynamical system** is any system that allows one to determine (at least theoretically) the future states of the system given its present or past state. For example, the map (recurrence relation)

$$x_{n+1} = (1.05)x_n, \qquad n = 0, 1, 2, \ldots,$$

is a dynamical system since we can determine the next state, x_{n+1}, given the previous state, x_n. If we know x_0, then we can compute any future state (indeed $x_{n+1} = x_0(1.05)^{n+1}$). As this example illustrates, a dynamical system may be defined by a difference equation, such as the ones studied in Section 5.5.

Another example of a dynamical system is provided by the differential equation

$$\frac{dx}{dt} = -2x,$$

[†] For a more detailed study of dynamical systems we refer the reader to *An Introduction to Chaotic Dynamical Systems* by R. L. Devaney, Benjamin/Cummings Publishing Company, Menlo Park, California, 1985, or *Nonlinear Oscillations, Dynamical Systems and Bifurcation of Vector Fields* by J. Guckenheimer and P. J. Holmes, Springer-Verlag, New York, 1983.

where the solution $x(t)$ specifies the state of the system at "time" t. If we know $x(t_0) = x_0$, then we can determine the state of the system at any future time $t > t_0$ by solving the initial value problem

$$\frac{dx}{dt} = -2x, \qquad x(t_0) = x_0.$$

Indeed, a simple calculation yields $x(t) = x_0 e^{-2(t-t_0)}$ for $t \geq t_0$.

For a dynamical system defined by a differential equation, it is often helpful to work with a related dynamical system defined by a difference equation. For example, when we cannot express the solution to a differential equation using elementary functions, we can use a numerical technique such as improved Euler's method or Runge-Kutta to approximate the solution to an initial value problem. This numerical scheme defines a new (but related) dynamical system which is often easier to study.

In the previous section, we used phase plane diagrams to study autonomous systems in the plane. Many important features of the system can be detected just by looking at these diagrams. For example, a closed trajectory corresponds to a periodic solution. The phase plane diagrams for *non*autonomous systems are much more complicated to decipher. One technique that is helpful in this regard is the so-called **Poincaré map.** As we shall see, these maps replace the study of a nonautonomous system with the study of a dynamical system defined by the location in the xv-plane of the solution at regularly spaced moments in time such as $t = 2\pi n$, where $n = 0, 1, 2, \ldots$. The advantage in using the Poincaré map will become clear when the method is applied to a nonlinear problem for which no explicit solution is known. In such a case, the trajectories are computed using a numerical scheme such as Runge-Kutta. Software packages such as PHASEPLANE and PHASER have options that will construct Poincaré maps for a given system.

To illustrate the Poincaré map, let's consider the equation

(1) $\qquad x''(t) + \omega^2 x(t) = F \cos t,$

where F and ω are positive constants. We studied similar equations in Section 5.3 and found that a general solution for $\omega \neq 1$ is given by

(2) $\qquad x(t) = A \sin(\omega t + \phi) + \dfrac{F}{\omega^2 - 1} \cos t,$

where the amplitude A and the phase angle ϕ are arbitrary constants. If we let $v = x'$, then

$$v(t) = \omega A \cos(\omega t + \phi) - \frac{F}{\omega^2 - 1} \sin t.$$

Since the forcing function $F \cos t$ is 2π-periodic, it is natural to seek 2π-periodic solutions to (1). For this purpose, we define the Poincaré map

(3) $\qquad \begin{aligned} x_n &:= x(2\pi n) = A \sin(2\pi \omega n + \phi) + F/(\omega^2 - 1), \\ v_n &:= v(2\pi n) = \omega A \cos(2\pi \omega n + \phi), \end{aligned}$

for $n = 0, 1, 2, \ldots$. In Figure 9.30 we have plotted the first eleven ($n = 0, 1, \ldots, 10$) values of (x_n, v_n) in the xv-plane for different choices of ω. For simplicity, we have taken $A = F = 1$ and $\phi = 0$. These graphs are called **Poincaré sections.**

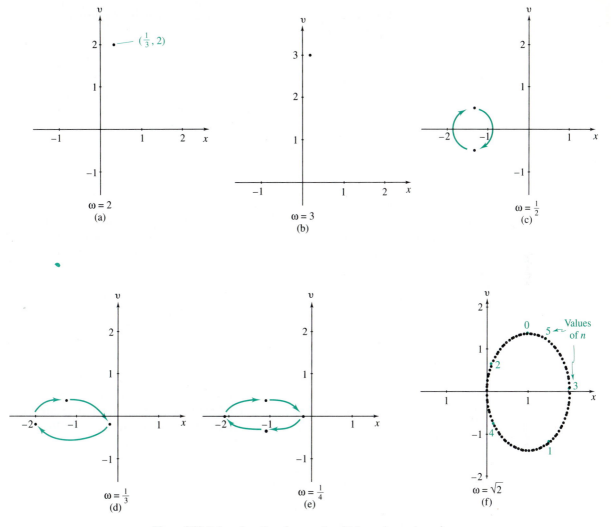

Figure 9.30 Poincaré sections for equation (1) for various values of ω

Now let's play the following game. We agree to ignore the fact that we already know the formula for $x(t)$ for all $t \geq 0$, and see what information about the solution we can glean just from the Poincaré section and the form of the differential equation.

Notice that the first two Poincaré sections in Figure 9.30, corresponding to $\omega = 2$ and 3, consist of a single point. This tells us that, starting with $t = 0$, every increment 2π of t returns us to the same point in the phase plane. This implies that equation (1) has a 2π-periodic solution, which can be seen as follows: For $\omega = 2$, let $x(t)$ be the solution to (1) with $(x(0), v(0)) = (\frac{1}{3}, 2)$ and let $X(t) := x(t + 2\pi)$. Since the Poincaré section is just the point $(\frac{1}{3}, 2)$, we have $X(0) = x(2\pi) = \frac{1}{3}$ and $X'(0) = x'(2\pi) = 2$. Thus, $x(t)$ and $X(t)$ have the same initial values at $t = 0$. Furthermore, since $\cos t$ is 2π-periodic, we also have

$$X''(t) + \omega^2 X(t) = x''(t + 2\pi) + \omega^2 x(t + 2\pi) = \cos(t + 2\pi) = \cos t.$$

Consequently, $x(t)$ and $X(t)$ satisfy the same initial value problem. By the existence and uniqueness theorem (see Theorem 2 on page 136), these functions must agree on the interval $[0, \infty)$. Hence, $x(t) = X(t) = x(t + 2\pi)$ for all $t \geq 0$; that is, $x(t)$ is 2π-periodic. Using a similar argument, it follows from the Poincaré section for $\omega = \frac{1}{2}$ that there is a solution of period 4π that alternates between the two points displayed in Figure 9.30(c) as t is incremented by 2π. For the case $\omega = \frac{1}{3}$, we deduce that there is a solution of period 6π rotating among three points, and for $\omega = \frac{1}{4}$, there is an 8π-periodic solution rotating among four points. We call these last three solution **subharmonics.**

The case $\omega = \sqrt{2}$ is different. So far, in Figure 9.30(f), none of the points has repeated. Did we stop too soon? Will the points ever repeat? Here, the fact that $\sqrt{2}$ is irrational plays a crucial role. It turns out that every integer n yields a distinct point in the Poincaré section (see Problem 8). However, there is a pattern developing. The points all appear to lie on a simple curve, possibly an ellipse. To see that this is indeed the case, notice that when $\omega = \sqrt{2}$, $A = F = 1$, and $\phi = 0$, we have

$$x_n = \sin(2\sqrt{2}\,\pi n) + 1, \qquad v_n = \sqrt{2}\cos(2\sqrt{2}\,\pi n), \qquad n = 0, 1, 2, \ldots.$$

It is then an easy computation to show that each (x_n, v_n) lies on the ellipse

$$(x - 1)^2 + \frac{v^2}{2} = 1.$$

In our investigation of equation (1), we concentrated on 2π-periodic solutions because the forcing term $F \cos t$ has period 2π. (We observed subharmonics when $\omega = \frac{1}{2}, \frac{1}{3}$, and $\frac{1}{4}$, that is, solutions with periods $2(2\pi)$, $3(2\pi)$, and $4(2\pi)$.) When a damping term is introduced into the differential equation, the Poincaré map displays a different behavior. Recall that the solution will now be the sum of a transient and a steady-state term. For example, let's consider the equation

(4) $$x''(t) + bx'(t) + \omega^2 x(t) = F \cos t,$$

where b, F, and ω are positive constants.

When $b^2 < 4\omega^2$, the solution to (4) can be expressed as

(5) $$x(t) = Ae^{-(b/2)t} \sin\left(\frac{\sqrt{4\omega^2 - b^2}}{2} t + \phi\right) + \frac{F}{\sqrt{(\omega^2 - 1)^2 + b^2}} \sin(t + \theta),$$

where $\tan\theta = (\omega^2 - 1)/b$ and A and ϕ are arbitrary constants (cf. equations (7) and (8) in Section 5.3). The first term on the right-hand side of (5) is the transient solution and the second term, the steady-state solution. Let's construct the Poincaré map using $t = 2\pi n$, $n = 0, 1, 2, \ldots$. We will take $b = 0.22$, $\omega = A = F = 1$, and $\phi = 0$ to simplify the computations. Since $\tan\theta = (\omega^2 - 1)/b = 0$, we will take $\theta = 0$ as well. Then, we have

$$x(2\pi n) = x_n = e^{-0.22\pi n} \sin(\sqrt{0.9879}\,2\pi n),$$

$$x'(2\pi n) = v_n = -0.11e^{-0.22\pi n} \sin(\sqrt{0.9879}\,2\pi n)$$

$$+ \sqrt{0.9879}\,e^{-0.22\pi n} \cos(\sqrt{0.9879}\,2\pi n) + \frac{1}{(0.22)}.$$

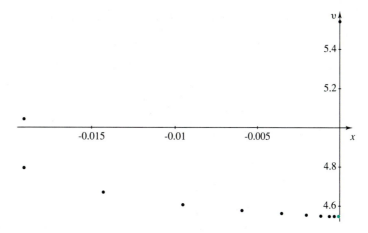

Figure 9.31 Poincaré section for equation (4) with $F = 1$, $b = 0.22$, and $\omega = 1$

The Poincaré maps for $n = 0, 1, 2, \ldots, 10$ are shown in Figure 9.31 (black points). After just a few iterations, we observe that $x_n \approx 0$ and $v_n \approx 1/(0.22) \approx 4.545$; that is, the points of the Poincaré section are approaching a single point in the xv-plane (colored point). Thus, we might expect that there is a 2π-periodic solution corresponding to a particular choice of A and ϕ. (In this example, where we can explicitly represent the solution, we see that indeed a 2π-periodic solution arises when we take $A = 0$ in (5).)

There is an important difference between the Poincaré sections for equation (2) and those for equation (4). In Figure 9.30, the location of all of the points in (a)–(e) depend upon the initial value selected (here $A = 1$ and $\phi = 0$). (See Problem 10.) However, in Figure 9.31, the first few points (black points) depend upon the initial conditions while the limit point (colored point) does not (see Problem 6). The latter behavior is typical for equations that have a "damping" term (i.e., $b > 0$); namely, the Poincaré section has a limit set[†] that is essentially independent of the initial conditions.

For equations with damping, the limit set may be more complicated than just a point. For example, the Poincaré map for the equation

(6) $x''(t) + (0.22)x'(t) + x(t) = \cos t + \cos(\sqrt{2}\,t),$

has a limit set consisting of an ellipse (see Problem 11). This is illustrated in Figure 9.32 on page 464 for the initial values $x_0 = 2$, $v_0 = 4$ and $x_0 = -1$, $v_0 = 1$.

So far we have seen limit sets for the Poincaré map which were either a single point or an ellipse—independent of the initial values. These particular limit sets are **attractors.** In general, an attractor is a set A with the property that there exists an open set[††] B containing A, such that whenever the Poincaré map enters B, its points remain in B *and* the limit set of the Poincaré map is a subset of A. Furthermore, we asssume that A has the *invariant property*: whenever the Poincaré map starts at a point in A, it remains in A.

In the previous examples the attractors of the dynamical system (Poincaré map) were

---

[†] The **limit set** for a map (x_n, v_n), $n = 1, 2, 3, \ldots$, is the set of points (p, q) such that $\lim_{k \to \infty}(x_{n_k}, v_{n_k}) = (p, q)$ where $n_1 < n_2 < n_3 < \cdots$ is some subsequence of the positive integers.

[††] A set $B \subset \mathbf{R}^2$ is an **open set** if for each point $p \in B$ there is an open disk V containing p such that $V \subset B$.

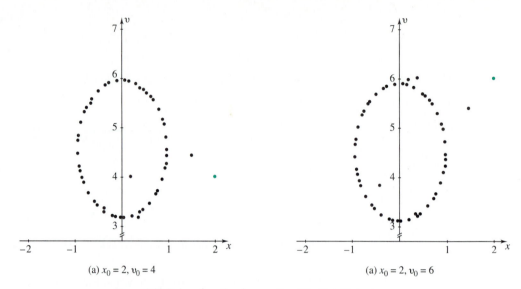

(a) $x_0 = 2$, $v_0 = 4$ (a) $x_0 = 2$, $v_0 = 6$

Figure 9.32 Poincaré section for equation (6) with initial values x_0, v_0

easy to describe. In recent years, however, many investigators, working on a variety of applications, have encountered dynamical systems that do *not* behave as orderly—their attractor sets are very complicated (not just isolated points or familiar geometric objects such as ellipses). The behavior of such systems is called **chaotic** and the corresponding limit sets are referred to as **strange attractors.**

To illustrate chaotic behavior and what is meant by a strange attractor, we will discuss two nonlinear differential equations and a simple difference equation. First, let's consider the **Duffing equation**

$$(7) \qquad x''(t) + bx'(t) - x(t) + x^3(t) = F \sin \gamma t.$$

Since we cannot express the solution to (7) in any explicit form, we must obtain the Poincaré map by numerically approximating the solution to (7) for fixed initial values and then plot the approximations for $x(2\pi n/\gamma)$ and $v(2\pi n/\gamma) = x'(2\pi n/\gamma)$. (Since the forcing term $F \sin \gamma t$ has period $2\pi/\gamma$, we seek $2\pi/\gamma$-periodic solutions and subharmonics.) In Figure 9.33 we display the Poincaré sections when $b = 0.3$ and $\gamma = 1.2$ in the cases (a) $F = 0.2$; (b) $F = 0.28$; (c) $F = 0.29$; and (d) $F = 0.37$.

Notice that as the constant F increases, the Poincaré map changes character. When $F = 0.2$, the Poincaré section tells us that there is a $2\pi/\gamma$-periodic solution. When $F = 0.28$, there is a subharmonic of period $4\pi/\gamma$, and for $F = 0.29$ and 0.37, there are subharmonics with periods $8\pi/\gamma$ and $10\pi/\gamma$, respectively.

Things are dramatically different when $F = 0.5$; the solution is neither $2\pi/\gamma$-periodic nor subharmonic. The Poincaré section for $F = 0.5$ is illustrated in Figure 9.34. This section was generated by numerically approximating the solution to (7) when $\gamma = 1.2$, $b = 0.3$, and

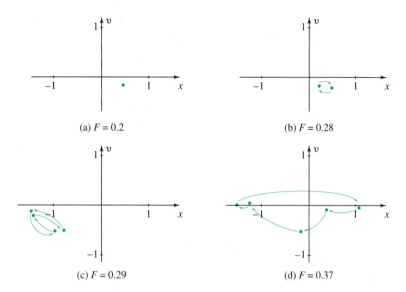

(a) $F = 0.2$ (b) $F = 0.28$

(c) $F = 0.29$ (d) $F = 0.37$

Figure 9.33 Poincaré sections for the Duffing equation (7) with $b = 0.3$ and $\gamma = 1.2$

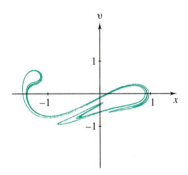

Figure 9.34 Poincaré section for the Duffing equation (7) with $b = 0.3$, $\gamma = 1.2$, and $F = 0.5$

$F = 0.5$, for fixed initial values[†]. Not all of the approximations $x(2\pi n/\gamma)$ and $v(2\pi n/\gamma)$ that were calculated are graphed; because of the presence of a transient solution, the first few points were omitted. It turns out that the plotted set is essentially independent of the initial values and has the property that once a point is in the set, all subsequent points lie in the set. Because of the complicated shape of the set, it is indeed a strange attractor. While the shape of the strange attractor does not depend upon the initial values, the picture does

---

[†] *Historical Footnote*: When researchers first encountered these strange looking Poincaré sections, they would check their computations using different computers and different numerical schemes (see Hénon and Heiles, *The applicability of the third integral of motion: some numerical experiments*, Astronomical Journal, Vol. 69, 1964, page 75). For special types of dynamical systems, such as the Hénon map, it can be shown that there exists a true trajectory that *shadows* the numerical trajectory (see Hammel, Yorke, and Grebogi, *Numerical orbits of chaotic processes represent true orbits*, Bulletin American Mathematical Society, Vol. 19, 1988, pages 465–469).

change if we consider different sections; for example, $t = (2\pi n + \pi/2)/\gamma$, $n = 0, 1, 2, \ldots$, yields a different configuration.

Another example of a strange attractor occurs when we consider the **forced pendulum equation**

$$(8) \qquad x''(t) + bx'(t) + \sin(x(t)) = F \cos t,$$

where the $x(t)$ term in (4) has been replaced by $\sin(x(t))$. Here $x(t)$ is the angle between the pendulum and the vertical rest position, b is related to friction, and F represents the strength of the forcing (see Figure 9.35). For $F = 2.7$ and $b = 0.22$, we have graphed in Figure 9.36 approximately 90,000 points in the Poincaré map. Since we cannot express the solution to (8) in any explicit form, the Poincaré map was obtained by numerically approximating the solution to (8) for fixed initial values and plotting the approximations for $x(2\pi n)$ and $v(2\pi n) = x'(2\pi n)$.

The Poincaré maps for the Duffing equation and for the forced pendulum equation not only illustrate the idea of a strange attractor, but they exhibit another peculiar behavior called **chaos.** Chaos occurs when small changes in the initial conditions lead to major changes in the behavior of the solution. Henri Poincaré described the situation as follows:

> "... it may happen that small differences in the initial conditions will produce very large ones in the final phenomena. A small error in the former produces an enormous error in the latter. Prediction becomes impossible...."

In a physical experiment we can never *exactly* (with infinite accuracy) reproduce the same initial conditions. Consequently, if the behavior is chaotic, even a slight difference in the initial conditions will lead to quite different values for the corresponding Poincaré map when n is large. Such behavior does not occur for solutions to either equation (4) or equation (1) (see Problems 6 and 7). However, two solutions to the Duffing equation (7)

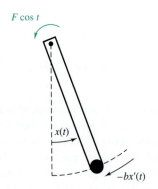

Figure 9.35 Forced damped pendulum

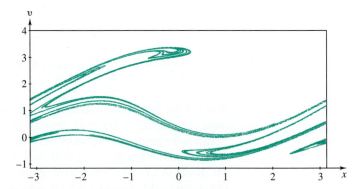

Figure 9.36 Poincaré section for the forced pendulum equation (8) with $b = 0.22$ and $F = 2.7$

with $F = 0.5$ that correspond to two different, but close, initial values have Poincaré maps which do *not* remain close together. While they both are attracted to the same set, their locations with respect to this set may be relatively far apart.

The phenomenon of chaos can also be illustrated by the following simple map. Let x_0 lie in $[0, 1)$ and define

(9) $x_{n+1} = 2x_n \pmod 1$,

where by (mod 1) we mean the decimal part of the number if it is greater than or equal to 1; that is

$$x_{n+1} = \begin{cases} 2x_n, & \text{for } 0 \le x_n < \frac{1}{2}, \\ 2x_n - 1, & \text{for } \frac{1}{2} \le x_n < 1. \end{cases}$$

When $x_0 = 1/3$, we find

$$x_1 = 2(\tfrac{1}{3})(\text{mod } 1) = \tfrac{2}{3},$$
$$x_2 = 2(\tfrac{2}{3})(\text{mod } 1) = \tfrac{4}{3}(\text{mod } 1) = \tfrac{1}{3},$$
$$x_3 = 2(\tfrac{1}{3})(\text{mod } 1) = \tfrac{2}{3},$$
$$x_4 = 2(\tfrac{2}{3})(\text{mod } 1) = \tfrac{1}{3},$$

etc.

Written as a sequence, we get $\{\tfrac{1}{3}, \tfrac{2}{3}, \overline{\tfrac{1}{3}, \tfrac{2}{3}}, \ldots\}$.

What happens when we pick a starting value x_0 near $\frac{1}{3}$? Does the sequence cluster about $\frac{1}{3}$ and $\frac{2}{3}$ as does the mapping when $x_0 = \frac{1}{3}$? For example, when $x_0 = 0.3$, we get the sequence

$$\{0.3, 0.6, 0.2, 0.4, 0.8, \overline{0.6, 0.2, 0.4, 0.8}, \ldots\}.$$

In Figure 9.37 we have plotted the values of x_n for $x_0 = 0.3, 0.33$, and 0.333. We have not plotted the first few terms, but only those that repeat. (This omission of the first few terms parallels the situation depicted in Figure 9.34 where transient solutions arise.)

It is clear from Figure 9.37 that while the values for x_0 are getting closer to $\frac{1}{3}$, the corresponding maps are spreading out over the whole interval $[0, 1]$ and *not* clustering near $\frac{1}{3}$ and $\frac{2}{3}$. This behavior is chaotic, since the Poincaré maps for initial values near $\frac{1}{3}$ behave quite differently from the map for $x_0 = \frac{1}{3}$. If we had selected x_0 to be irrational

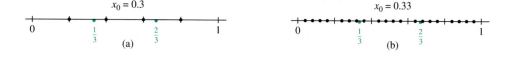

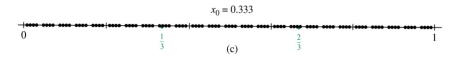

Figure 9.37 Plots of the map $x_{n+1} = 2x_n(\text{mod } 1)$ for $x_0 = 0.3, 0.33$, and 0.333

(which we can't do with a calculator), the sequence would *not* repeat and would be dense in $[0, 1]$.

Systems that exhibit chaotic behavior arise in many applications. The challenge to engineers is to design systems that avoid this chaos and, instead, enjoy the property of **stability.** The topic of stable systems is discussed in Chapter 13[†].

EXERCISES 9.7

The reader will find it helpful to have a microcomputer available or access to a mainframe. Software such as MACMATH, PHASEPLANE, or PHASER may be used.

1. Compute and graph the points of the Poincaré map with $t = 2\pi n$, $n = 0, 1, \ldots, 20$ for equation (1) taking $A = F = 1$, $\phi = 0$, and $\omega = \frac{3}{2}$. Repeat, taking $\omega = \frac{3}{5}$. Do you think the equation has a 2π-periodic solution for either choice of ω? A subharmonic solution?

2. Compute and graph the points of the Poincaré map with $t = 2\pi n$, $n = 0, 1, \ldots, 20$ for equation (1) taking $A = F = 1$, $\phi = 0$, and $\omega = \frac{1}{\sqrt{3}}$. Describe the limit set for this system.

3. Compute and graph the points of the Poincaré map with $t = 2\pi n$, $n = 0, 1, \ldots, 20$ for the equation (5) taking $A = F = 1$, $\phi = 0$, $\omega = 1$ and $b = -0.1$. What is happening to these points as $n \to \infty$?

4. Compute and graph the Poincaré map with $t = 2\pi n$, $n = 0, 1, \ldots, 20$ for equation (5) taking $A = F = 1$, $\phi = 0$, $\omega = 1$, and $b = 0.1$. Describe the attractor for this system.

5. Compute and graph the Poincaré map with $t = 2\pi n$, $n = 0, 1, \ldots, 20$ for equation (5) taking $A = F = 1$, $\phi = 0$, $\omega = \frac{1}{3}$, and $b = 0.22$. Describe the attractor for this system.

6. Show that the Poincaré map for equation (4) is not chaotic by showing that as t gets large

$$x_n = x(2\pi n) \approx \frac{F}{\sqrt{(\omega^2 - 1)^2 + b^2}} \sin(2\pi n + \theta),$$

$$v_n = x'(2\pi n) \approx \frac{F}{\sqrt{(\omega^2 - 1)^2 + b^2}} \cos(2\pi n + \theta),$$

independent of the initial values $x_0 = x(0)$ and $v_0 = x'(0)$.

7. Show that the Poincaré map for equation (1) is not chaotic by showing that if (x_0, v_0) and (x_0^*, v_0^*) are two initial values that define the Poincaré maps $\{(x_n, v_n)\}$ and $\{(x_n^*, v_n^*)\}$, respectively, using the recursive formulas in (3), then one can make the distance between (x_n, v_n) and (x_n^*, v_n^*) small by making the distance between (x_0, v_0) and (x_0^*, v_0^*) small. [Hint: Let (A, ϕ) and (A^*, ϕ^*) be the polar coordinates of two points in the plane. From the law of cosines, it follows that the distance d between them is given by $d^2 = (A - A^*)^2 + 2AA^*[1 - \cos(\phi - \phi^*)]$.]

8. Consider the Poincaré maps defined in (3) with $\omega = \sqrt{2}$, $A = F = 1$, and $\phi = 0$. If this map were ever to repeat, then, for two distinct positive integers n and m, $\sin(2\sqrt{2}\pi n) = \sin(2\sqrt{2}\pi m)$. Using basic properties of the sine function, show that this would imply that $\sqrt{2}$ is rational. It follows from this contradiction that the points of the Poincaré map do not repeat.

9. The doubling modulo 1 map defined by equation (9) exhibits some fascinating behavior. Compute the sequence obtained when

(a) $x_0 = k/7$ for $k = 1, 2, \ldots, 6$.
(b) $x_0 = k/15$ for $k = 1, 2, \ldots, 14$.
(c) $x_0 = k/2^j$, where j is a positive integer and $k = 1, 2, \ldots, 2^j - 1$.

Numbers of the form $k/2^j$ are called **dyadic numbers** and are dense in $[0, 1]$. That is, there is a dyadic number arbitrarily close to any real number (rational or irrational).

[†] All references to Chapters 12–14 refer to the expanded text *Fundamentals of Differential Equations and Boundary Value Problems.*

10. To show that the limit set of the Poincaré map given in (3) depends upon the initial values:

    (a) Show that when $\omega = 2$ or 3, the Poincaré map consists of the single point

$$\left(A \sin\phi + \frac{F}{\omega^2 - 1}, \; \omega A \cos\phi \right).$$

    (b) Show that when $\omega = 1/2$, the Poincaré map alternates between the two points

$$\left(\frac{F}{\omega^2 - 1} \pm A \sin\phi, \; \pm \omega A \cos\phi \right).$$

    (c) Use the results of parts (a) and (b) to show that when $\omega = 2, 3$, or $\frac{1}{2}$, the Poincaré map (3) depends upon the initial values (x_0, v_0).

11. To show that the limit set for the Poincaré map $x_n := x(2\pi n)$, $v_n := x'(2\pi n)$, where $x(t)$ is a solution to equation (6), is an ellipse and that this ellipse is the same for any initial values x_0, v_0:

    (a) Argue that since the initial values affect only the transient solution to (6), the limit set for the Poincaré map is independent of the initial values.

    (b) Now show that for n large,

$$x_n \approx a \sin(2\sqrt{2}\,\pi n + \psi),$$

$$v_n \approx c + \sqrt{2}\,a \cos(2\sqrt{2}\,\pi n + \psi),$$

    where $a = (1 + 2(0.22)^2)^{-1/2}$, $c = (0.22)^{-1}$, and $\psi = \arctan\{-[(0.22)\sqrt{2}]^{-1}\}$.

    (c) Use the result of part (b) to conclude that the ellipse

$$x^2 + \frac{(v - c)^2}{2} = a^2$$

    contains the limit set of the Poincaré map.

12. Using a numerical scheme such as Runge-Kutta or software such as **PHASEPLANE**, calculate the Poincaré map for equation (7) when $b = 0.3$, $\gamma = 1.2$, and $F = 0.2$. (Notice that the closer you start to the limiting point, the sooner the transient part will die out.) Compare your map with Figure 9.33 (a) on page 465. Redo for $F = 0.28$.

13. Redo Problem 12 with $F = 0.31$. What kind of behavior does the solution exhibit?

14. Redo Problem 12 with $F = 0.65$. What kind of behavior does the solution exhibit?

15. **Chaos Machine.** Chaos can be illustrated using a long ruler, a short ruler, a pin, and a tie tack (pivot). Construct the double pendulum as shown in Figure 9.38(a). The pendulum is set in motion by releasing it from a position such as the one shown in Figure 9.38(b). Repeatedly, set the pendulum in motion, each time trying to release it from the same position. Record the number of times the short ruler flips over and the direction in which it was moving. If the pendulum was released in *exactly* the same position each time, then the motion would be the same. However, from your experiments you have observed that even beginning close to the same position leads to very different motions. This double pendulum exhibits chaotic behavior.

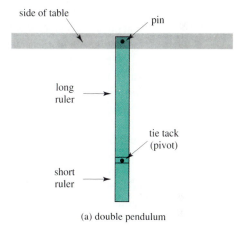

(a) double pendulum

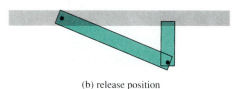

(b) release position

Figure 9.38 Double pendulum as a chaos machine

• CHAPTER SUMMARY •

Systems of differential equations arise in a variety of areas, such as electric networks, coupled spring-mass systems, multiple compartmental analysis models, and predator-prey models. There are two elementary methods for solving a **linear system with constant coefficients.** The first is the **elimination method.** For this method, we begin by expressing the system in operator form. For example, a linear system of two equations and two unknowns has the form

(1) $L_1[x] + L_2[y] = f_1,$

(2) $L_3[x] + L_4[y] = f_2.$

If we formally eliminate y from the above system, we obtain the single equation

(3) $(L_1 L_4 - L_2 L_3)[x] = g_1,$

where $g_1 := L_4[f_1] - L_2[f_2]$. We can solve this linear equation with constant coefficients for x by using the techniques discussed in Chapters 4 and 6. This solution will involve n arbitrary constants where n is the order of the operator $L_1 L_4 - L_2 L_3$. We then solve for y in terms of these same constants.

The second technique is the method of **Laplace transforms** and is applicable to initial value problems. As is the case for a single differential equation, taking the Laplace transform of each of the equations transforms the system of differential equations with initial conditions into a system of algebraic equations in which the unknowns are the Laplace transforms $X_i(s)$ of the unknown solutions $x_i(t)$. Solving for the transforms and then calculating their inverse transforms gives the solution to the original problem. Properties of the Laplace transform are discussed in Chapter 7.

In certain instances it is helpful to express a higher order differential equation as a system of first order equations in **normal form.** This is the case when one wants to use a numerical method such as a fourth order Runge-Kutta algorithm to obtain an approximation for the solution to an initial value problem for a higher order equation, or when one wants to use phase plane analysis on a second order nonlinear differential equation. To reduce the mth order equation $y^{(m)}(t) = f(t, y, y', \ldots, y^{(m-1)})$ to a normal system, set $x_1 = y$, $x_2 = y', \ldots,$ and $x_m = y^{(m-1)}$. This leads to

(4) $x_1' = x_2, \quad x_2' = x_3, \ldots, \quad x_{m-1}' = x_m, \quad x_m' = f(t, x_1, \ldots, x_m).$

The numerical techniques for first order equations discussed in Sections 3.5 and 3.6 each have extensions to first order systems in normal form. These techniques also apply to higher order equations once they have been reduced to a system in normal form. In Section 9.5 we discussed the extension of the fourth order Runge-Kutta algorithm to systems.

A technique for studying the qualitative behavior of solutions to the nonlinear **autonomous system**

(5) $\dfrac{dx}{dt} = f(x, y), \qquad \dfrac{dy}{dt} = g(x, y)$

is the method of **phase plane analysis.** To sketch a phase plane diagram for (5), we begin by finding the **critical points,** which are solutions to the system

$$f(x, y) = 0, \qquad g(x, y) = 0.$$

Next we solve (either analytically or numerically) the first order equation $dy/dx = g(x, y)/f(x, y)$ and use the solutions to sketch some integral curves for (5). (Software such as MACMATH, PHASEPLANE, or PHASER is helpful in constructing a phase plane diagram.) Returning to the original equations in (5), we can determine the **flow** along each integral curve and hence sketch the **trajectories** of the system. This information can be used to describe the behavior of the solutions to the system even though the solutions themselves cannot be found.

Nonautonomous systems can be studied by considering a Poincaré map for the system. In Section 9.7 we used a Poincaré map to detect periodic and subharmonic solutions. We also used it to study systems whose solutions exhibited chaotic behavior.

REVIEW PROBLEMS

In Problems 1 through 6 use the elimination method to find a general solution for the given linear system, where x', y', and D denote differentiation with respect to t.

1. $x' = y,$
   $y' = 3x.$

2. $x' + y' = 2y,$
   $y' = x - y.$

3. $(D + 2)[x] + (D - 1)[y] = t,$
   $(D - 1)[x] + (D + 1)[y] = 0.$

4. $(D - 3)[x] + D[y] = 0,$
   $(D + 1)[x] - y = 0.$

5. $x'' + y' + y = 0,$
   $x' - x + y = 0.$

6. $x'' - x + 5y = 2,$
   $2x + y'' + 2y = 0.$

In Problems 7 through 10 use the method of Laplace transforms to solve the given initial value problem.

7. $x' + y' = x - y,$ $x(0) = 1,$
   $x' - y' = x - y,$ $y(0) = 0.$

8. $x' = x - 2y,$ $x(0) = 0,$
   $y' = 3x + y,$ $y(0) = 1.$

9. $x'' = y,$ $x(0) = 1,$ $x'(0) = 0,$
   $y' = x,$ $y(0) = 0.$

10. $x' = 2x + y,$ $x(0) = 1,$
   $y' = x + y,$ $y(0) = 2.$

In Problems 11 and 12 convert the given initial value problem into an initial value problem for a system in normal form.

11. $y'''(t) + ty''(t) + 3y'(t) + 2y(t) = 2e^{-t};$
   $y(0) = 2,$ $y'(0) = 3,$ $y''(0) = 2.$

12. $y''(t) + y'(t) - \tan y(t) = \cos t;$
   $y(3) = 4,$ $y'(3) = -6.$

In Problems 13 through 18 sketch a phase plane diagram for the given system.

13. $\dfrac{dx}{dt} = 2x,$

   $\dfrac{dy}{dt} = y.$

14. $\dfrac{dx}{dt} = \dfrac{1}{y},$

   $\dfrac{dy}{dt} = \dfrac{3}{x}.$

15. $\dfrac{dx}{dt} = y,$

   $\dfrac{dy}{dt} = 4x.$

16. $\dfrac{dx}{dt} = -y,$

   $\dfrac{dy}{dt} = 9x.$

17. $\dfrac{dx}{dt} = y,$

   $\dfrac{dy}{dt} = -y^2 \tan x.$

18. $\dfrac{dx}{dt} = 1 - y,$

   $\dfrac{dy}{dt} = x + 3.$

19. Arms Race.   A mathematical model for an arms race between two countries whose expenditures for defense are expressed by the variables $x(t)$ and $y(t)$ is given by the linear system

$$\frac{dx}{dt} = 2y - x + a, \qquad x(0) = 1,$$

$$\frac{dy}{dt} = 4x - 3y + b, \qquad y(0) = 4,$$

where a and b are constants that measure the trust (or distrust) each country has for the other. Determine whether there is going to be disarmament (x and y approach 0 as t increases), a stabilized arms race (x and y approach a constant as $t \to \infty$), or a runaway arms race (x and y approach ∞ as $t \to \infty$).

20. Mixing Problem.   Two large tanks, each holding 50 L of liquid, are interconnected by pipes, with the liquid flowing from tank A into tank B at a rate of 4 L/min and from B into A at a rate of 1 L/min (see Figure 9.39). The liquid inside each tank is kept well stirred. A brine solution with a concentration of 5 kg/L of salt flows into tank A at a rate of 3 L/min. The (diluted) solution flows out of the system from tank B at 3 L/min. If, initially, tank A contains 70 kg of salt and tank B contains only water, determine the mass of salt in each tank at time $t \geq 0$.

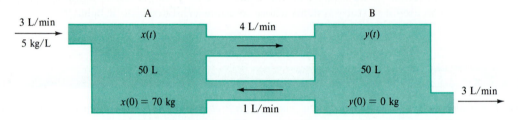

Figure 9.39 Mixing problem for interconnected tanks

TECHNICAL WRITING EXERCISES

1. The elimination method discussed in Section 9.2 for systems of linear differential equations is similar to the elimination method used to solve systems of linear algebraic equations. Describe those features common to both and indicate some of the differences.

2. Describe the similarities and differences between a phase plane diagram for an autonomous first order system in the plane and the sketch of a direction field for a first order differential equation.

3. Discuss how a Poincaré map can be used to determine if a nonautonomous first order system in the plane has a periodic solution. Explain how you detect a subharmonic solution and how your choice of a Poincaré map depends upon the forcing function.

GROUP PROJECTS FOR CHAPTER 9

▶ A. Cleaning Up the Great Lakes

A simple mathematical model that can be used to determine the time it would take to clean up the Great Lakes can be developed using a multiple compartmental analysis approach. In particular, we can view each Great Lake as a tank that contains a liquid in which is dissolved a particular pollutant (DDT, phosphorus, mercury). Schematically, we view the Great Lakes as consisting of five tanks connected as indicated in Figure 9.40.

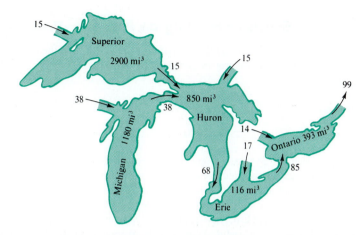

Figure 9.40 Compartmental model of the Great Lakes with flow rates (mi³/year) and volumes (mi³)

For our model we make the following assumptions:

1. The volume of each lake remains constant.
2. The flow rates are constant throughout the year.
3. When a liquid enters the lake, perfect mixing occurs and the pollutants are uniformly distributed.
4. Pollutants are dissolved in the water and enter or leave by inflow or outflow of solution.

Before using this model to obtain estimates on the cleanup times for the lakes, we consider some simpler models.

(a) Use the outflow rates given in Figure 9.40 to determine the time it would take to "drain" each lake. This gives a lower bound on how long it would take to remove all the pollutants.

(b) A better estimate is obtained by assuming that each lake is a separate tank with *only* clean water flowing in. Use this approach to determine how long it would take for the pollution level in each lake to be reduced to 50% of its original level. How long would it take to reduce the pollution to 5% of its original level?

(c) Finally, to take into account the fact that pollution from one lake flows into the next lake in the chain, use the entire multiple compartment model given in Figure 9.40 to determine when the pollution level in each lake has been reduced to 50% of its original level, assuming that pollution has ceased (that is, inflows not from a lake are clean water). Assume that all the lakes initially have the same pollution concentration p. How long would it take for the pollution to be reduced to 5% of its original level?

For a detailed discussion of this model see *An Introduction to Mathematical Modeling*, by Edward A. Bender, John Wiley & Sons, Inc., New York, 1978, Chapter 8.

B. Effects of Hunting on Predator-Prey Systems

Cyclic variations in the populations of predators and their prey have been studied using the Volterra-Lotka predator-prey model given by the system

(1) $$\frac{dx}{dt} = Ax - Bxy,$$

(2) $$\frac{dy}{dt} = -Cy + Dxy,$$

where A, B, C, and D are positive constants, $x(t)$ is the population of prey at time t, and $y(t)$ is the population of predators. It can be shown that such a system has a periodic solution (see Project C). That is, there exists some constant T such that $x(t) = x(t + T)$ and $y(t) = y(t + T)$ for all t. This periodic or cyclic variation in the populations has been observed in various systems such as sharks–food fish, lynx–rabbits, and ladybird beetles–cottony cushion scale. Because of this periodic behavior, it is useful to consider the average populations $\bar{x}$ and $\bar{y}$ defined by

$$\bar{x} := \frac{1}{T} \int_0^T x(t)\, dt, \qquad \bar{y} := \frac{1}{T} \int_0^T y(t)\, dt.$$

(a) Show that $\bar{x} = C/D$ and $\bar{y} = A/B$. [Hint: Use equation (1) and the fact that $x(0) = x(T)$ to show that

$$\int_0^T (A - By(t))\, dt = \int_0^T \frac{x'(t)}{x(t)}\, dt = 0.]$$

(b) To determine the effect of indiscriminate hunting on the populations, assume that hunting reduces the rate of change in a population by a constant times the population. Then, the predator-prey system satisfies the new set of equations

(3) $$\frac{dx}{dt} = Ax - Bxy - \varepsilon x = (A - \varepsilon)x - Bxy,$$

(4) $$\frac{dy}{dt} = -Cy + Dxy - \delta y = -(C + \delta)y + Dxy,$$

where ε and δ are positive constants with $\varepsilon < A$. What effect does this have on the average population of prey? On the average population of predators?

(c) Assume that the hunting is done selectively, as in shooting only rabbits (or shooting only lynx). Then we have $\varepsilon > 0$ and $\delta = 0$ (or $\varepsilon = 0$ and $\delta > 0$) in (3)–(4). What effect does this have on the average populations of predator and prey?

(d) In a rural county, foxes prey mainly on rabbits but occasionally include a chicken in their diet. The farmers decide to put a stop to the chicken killing by hunting the foxes. What do you predict will happen? What happens to the farmers' gardens?

For a discussion of other differential equation models in population biology, see *Differential Equation Models*, by Martin Braun, Courtney S. Coleman, and Donald A. Drew (eds.), Springer-Verlag, New York, 1983, Part V.

C. Periodic Solutions to Volterra-Lotka Systems

As stated in Project B, the Volterra-Lotka predator-prey model is given by the system (1)–(2), where $x(t)$ and $y(t)$ are the populations of the prey and the predators, respectively, and A, B, C, and D are positive constants. For autonomous systems such as (1)–(2), the existence of a closed integral curve that does not contain a critical point implies the existence of a periodic solution (see Section 9.6). To show that (1)–(2) has such an integral curve, proceed as follows:

(a) Determine the critical points of system (1)–(2).

(b) Show that the integral curves of (1)–(2) can be expressed in the form $w(x)z(y) = k$, where $w(x) := x^C e^{-Dx}$ and $z(y) := y^A e^{-By}$.

(c) Show that $w(x) = x^C e^{-Dx}$ has the properties $w(0) = 0$, $\lim_{x\to\infty} w(x) = 0$, and $w(x)$ attains a maximum value of $M_w := (C/D)^C e^{-C}$ for $x > 0$ at $x = C/D$. Sketch the graph of $w(x)$. Similarly, show that $z(y) = y^A e^{-By}$ has the properties $z(0) = 0$, $\lim_{y\to\infty} z(y) = 0$, and $z(y)$ attains a maximum value of $M_z := (A/B)^A e^{-A}$ for $y > 0$ at $y = A/B$. Sketch the graph of $z(y)$.

(d) Show that the product $w(x)z(y)$ is bounded by $M_w M_z$ for x, $y > 0$, and hence there is no solution with x, $y > 0$ for $w(x)z(y) = k$ when $k > M_w M_z$. Moreover, show that the only solution when $k = M_w M_z$ is $x = C/D$, $y = A/B$.

(e) Assume that $0 < k < M_w M_z$. Show that when $y = A/B$, there are two solutions to $w(x)z(y) = k$, one being $x_{\min} < C/D$ and the other $x_{\max} > C/D$.

(f) Again, assume that $0 < k < M_w M_z$. Show that for x fixed, the equation, $w(x)z(y) = k$ has (i) no solution when $x < x_{\min}$ or $x > x_{\max}$, (ii) one solution $y = A/B$ when $x = x_{\min}$ or $x_{\max}$, and (iii) two solutions $y_1(x) < A/B < y_2(x)$ when $x_{\min} < x < x_{\max}$ (see Figure 9.41).

(g) Using the results of parts (e) and (f), argue that system (1)–(2) has a family of closed integral curves in the first quadrant that surround the critical point $(C/D, A/B)$. Hence the system has a periodic solution.

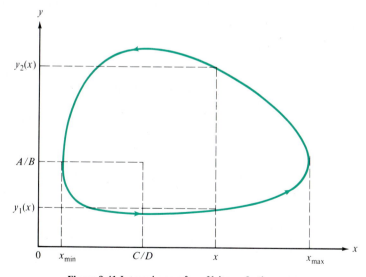

Figure 9.41 Integral curve for a Volterra-Lotka system

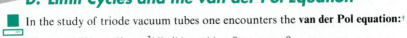

D. Limit Cycles and the van der Pol Equation

In the study of triode vacuum tubes one encounters the **van der Pol equation:**[†]

(5) $x''(t) - \mu(1 - x^2(t))x'(t) + x(t) = 0, \qquad \mu > 0.$

We will study this equation by constructing a phase plane diagram using available software or by using a numerical scheme to generate the trajectories. To obtain better numerical approximations, it is advantageous to work with an equivalent equation called a **Rayleigh equation:**

(6) $y''(t) - \mu[1 - (y'(t))^2]y'(t) + y(t) = 0, \qquad \mu > 0.$

(a) Show that the change of variables $x(t) = \sqrt{3}\,y'(t)$ transforms equation (6) into equation (5).

(b) Show that the Rayleigh equation (6) is equivalent to the system

(7)
$$\frac{dv}{dt} = -w + \mu v(1 - v^2),$$

$$\frac{dw}{dt} = v,$$

where $w(t) = y(t)$. [When you call up the file vdp.ode in PHASEPLANE, you get system (7) with the parameters AL $= \mu$ and EPS $= 1$. When you call up the file vanderpol in PHASER, you get a related system, but with a parameter a unrelated to μ. To do this project in PHASER, you should use the file forcevdp.]

(c) Construct a phase plane diagram for the system (7) with $\mu = 0.1$ about the origin and large enough to include the rectangle $-4 < x < 4$, $-5 < y < 5$. Include trajectories beginning near, but not at, the origin and some beginning outside the circle of radius 3.

Notice that some of the trajectories spiral in while others spiral out. If you continued the approximation long enough in time, then you would find that the curves approach a closed curve that is approximately a circle centered at the origin. This closed curve is called a **limit cycle** for the van der Pol equation, and is the trajectory of a periodic solution to system (7) and hence to van der Pol's equation.

(d) To investigate how the limit cycle for van der Pol's equation depends upon the parameter μ, construct a set of phase plane diagrams corresponding to $\mu = 0.1, 0.2, 1.0, 2.0, 5.0,$ $12.0, 0, -0.5,$ and -2.0. How many limit cycles do you find for each μ? How does the shape change? Do the trajectories spiral toward the limit cycle or away from it?

[†] *Historical Footnote*: Experimental research by **E. V. Appleton** and **B. van der Pol** in 1921 on the oscillations of an electric circuit containing a triode generator (vacuum tube) led to the nonlinear equation now called **van der Pol's equation.** Methods of solution were developed by van der Pol in 1926–1927. **Mary L. Cartwright** continued research into nonlinear oscillation theory and together with **J. E. Littlewood** obtained existence results for forced oscillations in nonlinear systems in 1945.

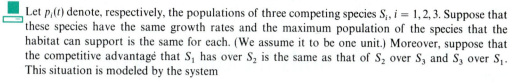

E. Strange Behavior of Competing Species—Part I

Let $p_i(t)$ denote, respectively, the populations of three competing species S_i, $i = 1, 2, 3$. Suppose that these species have the same growth rates and the maximum population of the species that the habitat can support is the same for each. (We assume it to be one unit.) Moreover, suppose that the competitive advantage that S_1 has over S_2 is the same as that of S_2 over S_3 and S_3 over S_1. This situation is modeled by the system

$$
\begin{aligned}
p'_1 &= p_1(1 - p_1 - ap_2 - bp_3), \\
(8) \qquad p'_2 &= p_2(1 - bp_1 - p_2 - ap_3), \\
p'_3 &= p_3(1 - ap_1 - bp_2 - p_3),
\end{aligned}
$$

where a and b are positive constants.

(a) Show that system (8) has four equilibrium populations (solutions) corresponding to the critical points; $(1, 0, 0)$, $(0, 1, 0)$, $(0, 0, 1)$, and (δ, δ, δ) where $\delta := (1 + a + b)^{-1}$.

(b) Let's study the system when $a = b = 0.5$. Since the system is *not* linear, we cannot apply the techniques discussed in Sections 9.2 and 9.3. The system has three unknowns so the phase plane analysis is much more complex than that presented in Section 9.6. Instead, we will perform numerical experiments on the system to see if we can determine the behavior of its solutions. Use a scheme such as fourth order Runge-Kutta method for systems to perform these experiments. Use various initial values, in particular, values near the four equilibrium points. Describe the behavior of the populations as $t \to \infty$.

(c) The system behaves quite differently when $a = 0.5$ and $b = 2.0$. Perform numerical experiments to show that the solution for this system approaches one of the equilibrium solutions—$(1, 0, 0)$, $(0, 1, 0)$, $(0, 0, 1)$—then moves toward another, then moves on toward the third, then back toward the first, and so on. Furthermore, the solutions linger longer and longer about these points before rapidly moving to the next one.

(d) Use the results of part (c) to argue that "most of the time,"

$$
p_1 p_2 \approx 0, \qquad p_2 p_3 \approx 0, \qquad p_1 p_3 \approx 0.
$$

Moreover, $T := p_1 + p_2 + p_3 \approx 1$.

(e) From system (8) derive the equations

$$
\frac{dT}{dt} = T(1 - T) + \eta(p_1 p_2 + p_2 p_3 + p_1 p_3),
$$

$$
\frac{d \ln P}{dt} = -\eta + (3 + \eta)(1 - T),
$$

where $T := p_1 + p_2 + p_3$, $P := p_1 p_2 p_3$, and $\eta := a + b - 2 = 0.5$.

(f) Using the results of parts (d) and (e), show that

$$
\frac{dT}{dt} = T(1 - T), \qquad \frac{d \ln P}{dt} = -\eta
$$

are reasonable simplifications.

(g) Solve the equations in part (f) and verify that as $t \to \infty$, the sum T approaches 1 and the product P approaches zero.

This system is studied in more detail in Project D in Chapter 10.

CHAPTER 10

Matrix Methods for Linear Systems

Remark: Chapter 10 uses matrix-algebra techniques to solve systems of differential equations. After covering the first six sections of this chapter, the reader can return to Chapter 9 to study applications of systems in Section 9.4, numerical methods in Section 9.5, phase plane analysis in Section 9.6, or dynamical systems and chaos in Section 9.7. For those readers who have already studied Chapter 9, several application problems are dispersed throughout Chapter 10.

10.1 INTRODUCTION

In Chapter 9 we presented two elementary techniques for solving linear systems of differential equations and discussed their application to electric networks, mechanical systems with more than one degree of freedom, and multiple compartment systems. These techniques work quite well when the number of variables is small. However, for large electric networks or mechanical systems with many degrees of freedom, these methods are cumbersome and disguise the underlying theory. One way out of the maze is to use matrix methods.

The kinds of systems we shall deal with are those in the **normal form,**

$$
\begin{aligned}
x_1'(t) &= a_{11}(t)x_1(t) + \cdots + a_{1n}(t)x_n(t) + f_1(t),\\
x_2'(t) &= a_{21}(t)x_1(t) + \cdots + a_{2n}(t)x_n(t) + f_2(t),\\
&\ \ \vdots \qquad\qquad \vdots \qquad\qquad\qquad \vdots \qquad\qquad\quad \vdots\\
x_n'(t) &= a_{n1}(t)x_1(t) + \cdots + a_{nn}(t)x_n(t) + f_n(t).
\end{aligned}
\tag{1}
$$

The structure of such a system suggests that there might be a more convenient way of expressing it. This is in fact the case. Using matrices and vectors, system (1) can be written in the compact form

(2) $\mathbf{x}'(t) = \mathbf{A}(t)\mathbf{x}(t) + \mathbf{f}(t),$

where $\mathbf{x}(t) = \text{col}(x_1(t), \ldots, x_n(t))$ and $\mathbf{f}(t) = \text{col}(f_1(t), \ldots, f_n(t))$ are vectors and $\mathbf{A}(t) = [a_{ij}(t)]$ is a matrix. Not only is (2) easier to write and remember, but it also serves as a reminder that first order linear systems behave much like linear first order differential equations.

In this chapter we use matrix methods to study systems of first order differential equations in normal form. We assume that the reader is acquainted with the computational aspects of linear algebra usually taught in a basic linear algebra course[†]—in particular, matrix addition and multiplication, solving systems of linear equations, evaluating determinants, and computing the inverse of a matrix. For the reader's convenience, we give a brief review of matrices in the next section. There we also introduce vector and matrix functions and discuss their calculus.

The advantages of the matrix approach become evident in Section 10.3, where we develop the theory for linear systems in normal form. This development parallels very closely the discussion of linear second order equations in Section 4.3 and the theory of linear higher order equations in Section 6.2. In particular, we discuss linear dependence of vector functions, fundamental solution sets, and the Wronskian of a set of solutions.

In Sections 10.4 and 10.5 we give a procedure for determining a fundamental solution set for a homogeneous normal system with constant coefficients. The approach is analogous to that used for scalar equations. For example, the role previously played by the auxiliary equation and its roots is now taken by the characteristic equation for the matrix and its eigenvalues and eigenvectors. In Section 10.6 we give extensions of the methods of undetermined coefficients and variation of parameters.

The connection between first order normal systems and first order linear equations alluded to earlier is discussed in Section 10.7. There the concept of a matrix exponential function $e^{\mathbf{A}t}$ is discussed. Moreover, we demonstrate how to use the matrix exponential function to obtain a general solution for a homogeneous system.

•10.2• A BRIEF REVIEW OF MATRICES AND VECTORS (Optional)

A **matrix** is a rectangular array of numbers arranged in rows and columns. An $m \times n$ matrix—that is, a matrix with m rows and n columns—is usually denoted by

$$\mathbf{A} := \begin{bmatrix} a_{11} & a_{12} & a_{13} & \cdots & a_{1n} \\ a_{21} & a_{22} & a_{23} & \cdots & a_{2n} \\ \vdots & \vdots & \vdots & & \vdots \\ a_{m1} & a_{m2} & a_{m3} & \cdots & a_{mn} \end{bmatrix},$$

where the element in the ith row and jth column is a_{ij}. The notation $[a_{ij}]$ is also used to

---

[†] A suitable reference is *Elementary Linear Algebra*, Second Edition, by A. Wayne Roberts, Benjamin/Cummings Publishing Co., Menlo Park, California, 1985, or almost any other introductory linear algebra text.

designate $\mathbf{A}$. The matrices we will work with usually consist of real numbers, but in certain instances we allow complex number entries.

The matrix obtained from $\mathbf{A}$ by interchanging its rows and columns is called the **transpose** of $\mathbf{A}$ and is denoted by $\mathbf{A}^T$. For example, if

$$\mathbf{A} = \begin{bmatrix} 1 & 2 & 6 \\ -1 & 2 & -1 \end{bmatrix}, \quad \text{then} \quad \mathbf{A}^T = \begin{bmatrix} 1 & -1 \\ 2 & 2 \\ 6 & -1 \end{bmatrix}.$$

Some matrices of special interest are **square matrices,** which have the same number of rows and columns; **diagonal matrices,** which are square matrices with only zero entries off the main diagonal (that is, $a_{ij} = 0$ if $i \neq j$); and (column) **vectors,** which are $n \times 1$ matrices. For example, if

$$\mathbf{A} = \begin{bmatrix} 3 & 4 & -1 \\ 2 & 6 & 5 \\ 0 & 1 & 4 \end{bmatrix}, \quad \mathbf{B} = \begin{bmatrix} 3 & 0 & 0 \\ 0 & 0 & 0 \\ 0 & 0 & 7 \end{bmatrix}, \quad \mathbf{x} = \begin{bmatrix} 4 \\ 2 \\ 1 \end{bmatrix},$$

then $\mathbf{A}$ is a square matrix, $\mathbf{B}$ is a diagonal matrix, and $\mathbf{x}$ is a vector. An $m \times n$ matrix whose entries are all zero is called a **zero matrix** and is denoted by $\mathbf{0}$. The $n \times n$ diagonal matrix with ones down the main diagonal is called the **identity matrix** and is denoted by $\mathbf{I}$, or by $\mathbf{I}_n$ when the size of the matrix is important.

For consistency, we denote matrices by boldfaced capitals $\mathbf{A}$, $\mathbf{B}$, $\mathbf{C}$, $\mathbf{I}$, $\mathbf{X}$, $\mathbf{Y}$, etc., and reserve boldfaced lowercase letters such as $\mathbf{c}$, $\mathbf{x}$, $\mathbf{y}$, and $\mathbf{z}$ for vectors. We also write $\mathbf{x} = \text{col}(x_1, \dots, x_n)$ to mean the vector whose transpose is $\mathbf{x}^T = [x_1 \, x_2 \dots x_n]$.

Algebra of Matrices

We need the following properties of matrices:

Matrix Addition.   Addition is performed by adding corresponding elements; that is, the *sum* of two $m \times n$ matrices is

$$\mathbf{A} + \mathbf{B} = [a_{ij}] + [b_{ij}] = [a_{ij} + b_{ij}].$$

Scalar Multiplication.   If $\mathbf{A}$ is an $m \times n$ matrix and r is a number (scalar), then the **scalar multiple** of $\mathbf{A}$ by r is the matrix $r\mathbf{A}$ obtained from $\mathbf{A}$ by multiplying each element in $\mathbf{A}$ by the number r. That is, $r\mathbf{A} = r[a_{ij}] = [ra_{ij}]$. The special case of $(-1)\mathbf{A}$ is written $-\mathbf{A}$.

Properties of Matrix Addition and Scalar Multiplication.   If $\mathbf{A}$, $\mathbf{B}$, and $\mathbf{C}$ are $m \times n$ matrices and r, s are scalars, then

$$\mathbf{A} + (\mathbf{B} + \mathbf{C}) = (\mathbf{A} + \mathbf{B}) + \mathbf{C}, \quad \mathbf{A} + \mathbf{B} = \mathbf{B} + \mathbf{A},$$
$$\mathbf{A} + \mathbf{0} = \mathbf{A}, \quad \mathbf{A} + (-\mathbf{A}) = \mathbf{0},$$
$$r(\mathbf{A} + \mathbf{B}) = r\mathbf{A} + r\mathbf{B}, \quad (r + s)\mathbf{A} = r\mathbf{A} + s\mathbf{A},$$
$$r(s\mathbf{A}) = (rs)\mathbf{A} = s(r\mathbf{A}).$$

Matrix Multiplication.   If **A** is an $m \times n$ matrix and **B** is an $n \times p$ matrix, then the **product** of **A** and **B** is the $m \times p$ matrix **AB** defined by

$$\mathbf{AB} := [c_{ij}], \quad \text{where} \quad c_{ij} := \sum_{k=1}^{n} a_{ik} b_{kj}$$

for $i = 1, \ldots, m$ and $j = 1, \ldots, p$. Observe that the ijth entry in **AB** is the dot product of the ith row of **A** and the jth column of **B**. For example,

$$\begin{bmatrix} 1 & 0 & 1 \\ 3 & -1 & 2 \end{bmatrix} \begin{bmatrix} 1 & 2 \\ -1 & -1 \\ 4 & 1 \end{bmatrix} = \begin{bmatrix} 1+0+4 & 2+0+1 \\ 3+1+8 & 6+1+2 \end{bmatrix} = \begin{bmatrix} 5 & 3 \\ 12 & 9 \end{bmatrix}.$$

It is important to keep in mind that matrix multiplication does not commute; that is, **AB** may not equal **BA**. This is certainly true if **A** is 2×3 and **B** is 3×2, since then **AB** is 2×2 while **BA** is 3×3. In some cases **AB** may be defined while **BA** is not. Even for square matrices, **AB** is not necessarily equal to **BA** (see Problem 8).

Properties of Matrix Multiplication.

$(\mathbf{AB})\mathbf{C} = \mathbf{A}(\mathbf{BC})$.	(Associativity)
$(\mathbf{A} + \mathbf{B})\mathbf{C} = \mathbf{AC} + \mathbf{BC}$.	(Distributivity)
$\mathbf{A}(\mathbf{B} + \mathbf{C}) = \mathbf{AB} + \mathbf{AC}$.	(Distributivity)
$(r\mathbf{A})\mathbf{B} = r(\mathbf{AB}) = \mathbf{A}(r\mathbf{B})$.	(Associativity)

Let **A** be an $m \times n$ matrix and let **x** and **y** be $n \times 1$ vectors. Then **Ax** is an $m \times 1$ vector, and so we can think of multiplication by **A** as defining an operator that maps $n \times 1$ vectors into $m \times 1$ vectors. A consequence of the distributivity and associativity properties is that multiplication by **A** defines a **linear operator,** since $\mathbf{A}(\mathbf{x} + \mathbf{y}) = \mathbf{Ax} + \mathbf{Ay}$ and $\mathbf{A}(r\mathbf{x}) = r\mathbf{Ax}$. Moreover, if **A** is an $m \times n$ matrix and **B** is an $n \times p$ matrix, then the $m \times p$ matrix **AB** defines a linear operator that is the composition of the linear operator defined by **B** with the linear operator defined by **A**. That is, $(\mathbf{AB})\mathbf{x} = \mathbf{A}(\mathbf{Bx})$, where **x** is a $p \times 1$ vector.

Inverse of a Matrix.   For square matrices we say that **B** is the **inverse** of **A**, denoted $\mathbf{B} = \mathbf{A}^{-1}$, if $\mathbf{AB} = \mathbf{BA} = \mathbf{I}$. Obviously, if **B** is the inverse of **A**, then **A** is the inverse of **B**. A matrix that has an inverse is called **invertible** or **nonsingular.** If no inverse exists, the matrix is said to be **singular.** For example, the identity matrix is invertible ($\mathbf{I}^{-1} = \mathbf{I}$), but the zero matrix is singular.

Finding the Inverse of a Matrix.   By a **row operation** we mean any one of the following:

(a) Interchanging two rows of the matrix.
(b) Multiplying a row of the matrix by a nonzero scalar.
(c) Adding a scalar multiple of one row of the matrix to another row.

If the $n \times n$ matrix **A** has an inverse, then $\mathbf{A}^{-1}$ can be determined by performing row operations on the $n \times 2n$ matrix $[\mathbf{A} \;\vdots\; \mathbf{I}]$ obtained by writing **A** and **I** side by side. In particular, we perform row operations on the matrix $[\mathbf{A} \;\vdots\; \mathbf{I}]$ until the first n rows and columns form the identity matrix; that is, the new matrix is $[\mathbf{I} \mid \mathbf{B}]$. Then $\mathbf{A}^{-1} = \mathbf{B}$. We remark that if this procedure fails to produce a matrix of the form $[\mathbf{I} \mid \mathbf{B}]$, then $\mathbf{A}$ has no inverse.

● **EXAMPLE 1**   Find the inverse of $\mathbf{A} = \begin{bmatrix} 1 & 2 & 1 \\ 1 & 3 & 2 \\ 1 & 0 & 1 \end{bmatrix}$.

Solution   We first form the matrix $[\mathbf{A} \mid \mathbf{I}]$ and row-reduce the matrix to $[\mathbf{I} \mid \mathbf{A}^{-1}]$. Computing, we find

The matrix $[\mathbf{A} \mid \mathbf{I}]$
$$\left[\begin{array}{ccc|ccc} 1 & 2 & 1 & 1 & 0 & 0 \\ 1 & 3 & 2 & 0 & 1 & 0 \\ 1 & 0 & 1 & 0 & 0 & 1 \end{array}\right].$$

Subtract the first row from the second and third to obtain
$$\left[\begin{array}{ccc|ccc} 1 & 2 & 1 & 1 & 0 & 0 \\ 0 & 1 & 1 & -1 & 1 & 0 \\ 0 & -2 & 0 & -1 & 0 & 1 \end{array}\right].$$

Add 2 times the second row to the third row to obtain
$$\left[\begin{array}{ccc|ccc} 1 & 2 & 1 & 1 & 0 & 0 \\ 0 & 1 & 1 & -1 & 1 & 0 \\ 0 & 0 & 2 & -3 & 2 & 1 \end{array}\right].$$

Subtract 2 times the second row from the first to obtain
$$\left[\begin{array}{ccc|ccc} 1 & 0 & -1 & 3 & -2 & 0 \\ 0 & 1 & 1 & -1 & 1 & 0 \\ 0 & 0 & 2 & -3 & 2 & 1 \end{array}\right].$$

Multiply the third row by $\frac{1}{2}$ to obtain
$$\left[\begin{array}{ccc|ccc} 1 & 0 & -1 & 3 & -2 & 0 \\ 0 & 1 & 1 & -1 & 1 & 0 \\ 0 & 0 & 1 & -\frac{3}{2} & 1 & \frac{1}{2} \end{array}\right].$$

Add the third row to the first, then subtract the third row from the second to obtain
$$\left[\begin{array}{ccc|ccc} 1 & 0 & 0 & \frac{3}{2} & -1 & \frac{1}{2} \\ 0 & 1 & 0 & \frac{1}{2} & 0 & -\frac{1}{2} \\ 0 & 0 & 1 & -\frac{3}{2} & 1 & \frac{1}{2} \end{array}\right].$$

The matrix shown in color is $\mathbf{A}^{-1}$.   ●

Determinants.   For a 2×2 matrix $\mathbf{A}$, the **determinant of** $\mathbf{A}$, denoted $\det \mathbf{A}$ or $|\mathbf{A}|$, is defined by

$$\det \mathbf{A} := \begin{vmatrix} a_{11} & a_{12} \\ a_{21} & a_{22} \end{vmatrix} = a_{11}a_{22} - a_{12}a_{21}.$$

We can define the determinant of a 3×3 matrix $\mathbf{A}$ in terms of its cofactor expansion about the first row; that is,

$$\det \mathbf{A} := \begin{vmatrix} a_{11} & a_{12} & a_{13} \\ a_{21} & a_{22} & a_{23} \\ a_{31} & a_{32} & a_{33} \end{vmatrix} = a_{11} \begin{vmatrix} a_{22} & a_{23} \\ a_{32} & a_{33} \end{vmatrix} - a_{12} \begin{vmatrix} a_{21} & a_{23} \\ a_{31} & a_{33} \end{vmatrix} + a_{13} \begin{vmatrix} a_{21} & a_{22} \\ a_{31} & a_{32} \end{vmatrix}.$$

For example,

$$\begin{vmatrix} 1 & 2 & 1 \\ 0 & 3 & 5 \\ 2 & 1 & -1 \end{vmatrix} = 1 \begin{vmatrix} 3 & 5 \\ 1 & -1 \end{vmatrix} - 2 \begin{vmatrix} 0 & 5 \\ 2 & -1 \end{vmatrix} + 1 \begin{vmatrix} 0 & 3 \\ 2 & 1 \end{vmatrix}$$

$$= 1(-3 - 5) - 2(0 - 10) + 1(0 - 6) = 6.$$

The determinant of an $n \times n$ matrix can be similarly defined by a cofactor expansion involving $(n - 1)$st order determinants; however, a more practical way to evaluate the determinant when n is large is to row-reduce the matrix to upper triangular form. Since here we deal mainly with 2×2 and 3×3 matrices, we refer the reader to an elementary linear algebra text for a further discussion of evaluating determinants.[†]

The next theorem summarizes many of the results from elementary linear algebra that we need.

> ### MATRICES AND SYSTEMS OF EQUATIONS
>
> **Theorem 1.**   Let $\mathbf{A}$ be an $n \times n$ matrix. The following statements are equivalent:
>
> **(a)** $\mathbf{Ax} = \mathbf{0}$ has nontrivial solutions ($\mathbf{x} \neq \mathbf{0}$).
> **(b)** $\mathbf{A}$ is singular.
> **(c)** The determinant of $\mathbf{A}$ is zero.
> **(d)** The columns (rows) of $\mathbf{A}$ form a linearly dependent set.

In part (d) the statement that the n columns of $\mathbf{A}$ are linearly dependent means that there exist scalars $c_1, \dots, c_n$ *not all zero* such that

$$c_1 \mathbf{a}_1 + c_2 \mathbf{a}_2 + \cdots + c_n \mathbf{a}_n = \mathbf{0},$$

where $\mathbf{a}_j$ is the vector forming the jth column of $\mathbf{A}$.

Calculus of Matrices

If we allow the entries $a_{ij}(t)$ in a matrix $\mathbf{A}(t)$ to be functions of the variable t, then $\mathbf{A}(t)$ is a **matrix function of** t. Similarly, if the entries $x_i(t)$ of a vector $\mathbf{x}(t)$ are functions of t, then $\mathbf{x}(t)$ is a **vector function of** t.

These matrix and vector functions have a calculus much like that of real-valued functions. A matrix $\mathbf{A}(t)$ is said to be **continuous at** t_0 if each entry $a_{ij}(t)$ is continuous at t_0.

[†] See *Elementary Linear Algebra*, Second Edition, by A. Wayne Roberts, Benjamin/Cummings Publishing Co., Menlo Park, California, 1985.

Moreover, $\mathbf{A}(t)$ is **differentiable at** t_0 if each entry $a_{ij}(t)$ is differentiable at t_0, and we write

(1) $\dfrac{d\mathbf{A}}{dt}(t_0) = \mathbf{A}'(t_0) := [a'_{ij}(t_0)].$

Similarly, we define

(2) $\displaystyle\int_a^b \mathbf{A}(t)\,dt := \left[\int_a^b a_{ij}(t)\,dt\right].$

● **EXAMPLE 2**   Let $\mathbf{A}(t) = \begin{bmatrix} t^2 + 1 & \cos t \\ e^t & 1 \end{bmatrix}$.

Find: (a) $\mathbf{A}'(t)$.   (b) $\displaystyle\int_0^1 \mathbf{A}(t)\,dt$.

Solution   Using formulas (1) and (2), we compute

(a) $\mathbf{A}'(t) = \begin{bmatrix} 2t & -\sin t \\ e^t & 0 \end{bmatrix}$.   (b) $\displaystyle\int_0^1 \mathbf{A}(t)\,dt = \begin{bmatrix} \frac{4}{3} & \sin 1 \\ e - 1 & 1 \end{bmatrix}$.   ●

The basic properties of differentiation are valid for matrix functions.

Differentiation Formulas for Matrix Functions.

$$\frac{d}{dt}(\mathbf{CA}) = \mathbf{C}\frac{d\mathbf{A}}{dt}, \quad \mathbf{C} \text{ a constant matrix.}$$

$$\frac{d}{dt}(\mathbf{A} + \mathbf{B}) = \frac{d\mathbf{A}}{dt} + \frac{d\mathbf{B}}{dt}.$$

$$\frac{d}{dt}(\mathbf{AB}) = \mathbf{A}\frac{d\mathbf{B}}{dt} + \frac{d\mathbf{A}}{dt}\mathbf{B}.$$

In the last formula the order in which the matrices are written is very important, because matrix multiplication does not always commute.

EXERCISES 10.2

1. Let $\mathbf{A} := \begin{bmatrix} 2 & 1 \\ 3 & 5 \end{bmatrix}$ and $\mathbf{B} := \begin{bmatrix} -1 & 0 \\ 2 & -3 \end{bmatrix}$.

Find: (a) $\mathbf{A}^T$.   (b) $\mathbf{A} + \mathbf{B}$.   (c) $3\mathbf{A} - \mathbf{B}$.

2. Let $\mathbf{A} := \begin{bmatrix} 2 & 0 & 5 \\ 2 & 1 & 1 \end{bmatrix}$ and $\mathbf{B} := \begin{bmatrix} 1 & -1 & 2 \\ 0 & 3 & -2 \end{bmatrix}$.

Find: (a) $\mathbf{B}^T$.   (b) $\mathbf{A} + \mathbf{B}$.   (c) $7\mathbf{A} - 4\mathbf{B}$.

3. Let $\mathbf{A} := \begin{bmatrix} 2 & 4 \\ 1 & 1 \end{bmatrix}$ and $\mathbf{B} := \begin{bmatrix} -1 & 3 \\ 5 & 1 \end{bmatrix}$.

Find: (a) $\mathbf{AB}$.   (b) $\mathbf{A}^2 = \mathbf{AA}$.   (c) $\mathbf{B}^2 = \mathbf{BB}$.

4. Let $\mathbf{A} := \begin{bmatrix} 2 & 1 \\ 0 & 4 \\ -1 & 3 \end{bmatrix}$ and $\mathbf{B} := \begin{bmatrix} 1 & 1 & -1 \\ 0 & 3 & 1 \end{bmatrix}$.

Find: (a) $\mathbf{AB}$.   (b) $\mathbf{BA}$.   (c) $\mathbf{A}^T\mathbf{A}$.

5. Let $\mathbf{A} := \begin{bmatrix} 1 & -2 \\ 2 & -3 \end{bmatrix}$, $\mathbf{B} := \begin{bmatrix} 1 & 0 \\ 1 & 1 \end{bmatrix}$, and

$\mathbf{C} := \begin{bmatrix} -1 & 1 \\ 2 & 1 \end{bmatrix}$.

Find: **(a)** $\mathbf{AB}$. **(b)** $\mathbf{AC}$. **(c)** $\mathbf{A}(\mathbf{B}+\mathbf{C})$.

6. Let $\mathbf{A} := \begin{bmatrix} 1 & 2 \\ 1 & 1 \end{bmatrix}$, $\mathbf{B} := \begin{bmatrix} 0 & 3 \\ 1 & 2 \end{bmatrix}$, and

$\mathbf{C} := \begin{bmatrix} 1 & -4 \\ 1 & 1 \end{bmatrix}$.

Find: **(a)** $\mathbf{AB}$. **(b)** $(\mathbf{AB})\mathbf{C}$. **(c)** $(\mathbf{A}+\mathbf{B})\mathbf{C}$.

7. Let $\mathbf{x} := \begin{bmatrix} 1 \\ 2 \\ 4 \end{bmatrix}$ and $\mathbf{y} := \begin{bmatrix} 2 \\ 4 \\ -1 \end{bmatrix}$.

Find: **(a)** $\mathbf{x}^T\mathbf{x}$. **(b)** $\mathbf{y}^T\mathbf{y}$. **(c)** $\mathbf{x}^T\mathbf{y}$.
(d) $\mathbf{y}^T\mathbf{x}$.

8. Let $\mathbf{A} := \begin{bmatrix} 2 & -1 \\ -3 & 4 \end{bmatrix}$ and $\mathbf{B} := \begin{bmatrix} 1 & 2 \\ 3 & 2 \end{bmatrix}$.

Verify that $\mathbf{AB} \neq \mathbf{BA}$.

In Problems 9 through 14 compute the inverse of the given matrix, if it exists.

9. $\begin{bmatrix} 2 & 1 \\ -1 & 4 \end{bmatrix}$.

10. $\begin{bmatrix} 4 & 1 \\ 5 & 9 \end{bmatrix}$.

11. $\begin{bmatrix} 1 & 1 & 1 \\ 1 & 2 & 1 \\ 2 & 3 & 2 \end{bmatrix}$.

12. $\begin{bmatrix} 1 & 1 & 1 \\ 1 & 2 & 3 \\ 0 & 1 & 1 \end{bmatrix}$.

13. $\begin{bmatrix} -2 & -1 & 1 \\ 2 & 1 & 0 \\ 3 & 1 & -1 \end{bmatrix}$.

14. $\begin{bmatrix} 1 & 1 & 1 \\ 1 & -1 & 2 \\ 1 & 1 & 4 \end{bmatrix}$.

In Problems 15 through 18 find the matrix function $\mathbf{X}^{-1}(t)$ whose value at t is the inverse of the given matrix $\mathbf{X}(t)$.

15. $\mathbf{X}(t) = \begin{bmatrix} e^t & e^{4t} \\ e^t & 4e^{4t} \end{bmatrix}$.

16. $\mathbf{X}(t) = \begin{bmatrix} \sin 2t & \cos 2t \\ 2\cos 2t & -2\sin 2t \end{bmatrix}$.

17. $\mathbf{X}(t) = \begin{bmatrix} e^t & e^{-t} & e^{2t} \\ e^t & -e^{-t} & 2e^{2t} \\ e^t & e^{-t} & 4e^{2t} \end{bmatrix}$.

18. $\mathbf{X}(t) = \begin{bmatrix} e^{3t} & 1 & t \\ 3e^{3t} & 0 & 1 \\ 9e^{3t} & 0 & 0 \end{bmatrix}$.

In Problems 19 through 24 evaluate the given determinant.

19. $\begin{vmatrix} 4 & 3 \\ -1 & 2 \end{vmatrix}$.

20. $\begin{vmatrix} 12 & 8 \\ 3 & 2 \end{vmatrix}$.

21. $\begin{vmatrix} 1 & 0 & 0 \\ 3 & 1 & 2 \\ 1 & 5 & -2 \end{vmatrix}$.

22. $\begin{vmatrix} 1 & 0 & 2 \\ 0 & 3 & -1 \\ -1 & 2 & 1 \end{vmatrix}$.

23. $\begin{vmatrix} 1 & 4 & 3 \\ -1 & -1 & 2 \\ 4 & 5 & 2 \end{vmatrix}$.

24. $\begin{vmatrix} 1 & 4 & 4 \\ 3 & 0 & -3 \\ 1 & 6 & 2 \end{vmatrix}$.

In Problems 25 through 27 determine the values of r for which $det(\mathbf{A} - r\mathbf{I}) = 0$.

25. $\mathbf{A} = \begin{bmatrix} 1 & 1 \\ -2 & 4 \end{bmatrix}$.

26. $\mathbf{A} = \begin{bmatrix} 3 & 3 \\ 2 & 4 \end{bmatrix}$.

27. $\mathbf{A} = \begin{bmatrix} 0 & 0 & 0 \\ 0 & 1 & 0 \\ 1 & 0 & 1 \end{bmatrix}$.

28. Illustrate the equivalence of the assertions (a)–(d) in Theorem 1 for the matrix

$$\mathbf{A} = \begin{bmatrix} 4 & -2 & 2 \\ -2 & 4 & 2 \\ 2 & 2 & 4 \end{bmatrix}$$

as follows.

(a) Determine a nontrivial solution $\mathbf{x}$ to $\mathbf{Ax} = \mathbf{0}$.
(b) Show that the row reduction procedure applied to $[\mathbf{A} \mid \mathbf{I}]$ fails to produce the inverse of $\mathbf{A}$.
(c) Calculate $\det \mathbf{A}$.
(d) Find scalars c_1, c_2, and c_3, *not all zero* so that $c_1\mathbf{a}_1 + c_2\mathbf{a}_2 + c_3\mathbf{a}_3 = \mathbf{0}$, where $\mathbf{a}_1, \mathbf{a}_2$, and $\mathbf{a}_3$ are the columns of $\mathbf{A}$.

In Problems 29 and 30 find $d\mathbf{x}/dt$ for the given vector functions.

29. $\mathbf{x}(t) = \begin{bmatrix} e^{3t} \\ 2e^{3t} \\ -e^{3t} \end{bmatrix}$.

30. $\mathbf{x}(t) = \begin{bmatrix} e^{-t}\sin 3t \\ 0 \\ -e^{-t}\sin 3t \end{bmatrix}$.

In Problems 31 and 32 find $d\mathbf{X}/dt$ for the given matrix functions.

31. $\mathbf{X}(t) = \begin{bmatrix} e^{5t} & 3e^{2t} \\ -2e^{5t} & -e^{2t} \end{bmatrix}$.

32. $\mathbf{X}(t) = \begin{bmatrix} \sin 2t & \cos 2t & e^{-2t} \\ -\sin 2t & 2\cos 2t & 3e^{-2t} \\ 3\sin 2t & \cos 2t & e^{-2t} \end{bmatrix}$.

In Problems 33 and 34 verify that the given vector satisfies the given system.

33. $\mathbf{x}' = \begin{bmatrix} 1 & 1 \\ -2 & 4 \end{bmatrix}\mathbf{x}$, $\mathbf{x} = \begin{bmatrix} e^{3t} \\ 2e^{3t} \end{bmatrix}$.

34. $\mathbf{x}' = \begin{bmatrix} 0 & 0 & 0 \\ 0 & 1 & 0 \\ 1 & 0 & 1 \end{bmatrix}\mathbf{x}$, $\mathbf{x} = \begin{bmatrix} 0 \\ e^t \\ -3e^t \end{bmatrix}$.

In Problems 35 and 36 verify that the given matrix satisfies the given matrix differential equation.

35. $\mathbf{X}' = \begin{bmatrix} 1 & -1 \\ 2 & 4 \end{bmatrix}\mathbf{X}$, $\mathbf{X} = \begin{bmatrix} e^{2t} & e^{3t} \\ -e^{2t} & -2e^{3t} \end{bmatrix}$.

36. $\mathbf{X}' = \begin{bmatrix} 1 & 0 & 0 \\ 0 & 3 & -2 \\ 0 & -2 & 3 \end{bmatrix}\mathbf{X}$, $\mathbf{X} = \begin{bmatrix} e^t & 0 & 0 \\ 0 & e^t & e^{5t} \\ 0 & e^t & -e^{5t} \end{bmatrix}$.

In Problems 37 and 38 the matrices $\mathbf{A}(t)$ and $\mathbf{B}(t)$ are given. Find:

(a) $\displaystyle\int \mathbf{A}(t)\,dt.$ **(b)** $\displaystyle\int_0^1 \mathbf{B}(t)\,dt.$ **(c)** $\dfrac{d}{dt}[\mathbf{A}(t)\mathbf{B}(t)].$

37. $\mathbf{A}(t) = \begin{bmatrix} t & e^t \\ 1 & e^t \end{bmatrix}$, $\mathbf{B}(t) = \begin{bmatrix} \cos t & -\sin t \\ \sin t & \cos t \end{bmatrix}$.

38. $\mathbf{A}(t) = \begin{bmatrix} 1 & e^{-2t} \\ 3 & e^{-2t} \end{bmatrix}$, $\mathbf{B}(t) = \begin{bmatrix} e^{-t} & e^{-t} \\ -e^{-t} & 3e^{-t} \end{bmatrix}$.

39. An $n \times n$ matrix $\mathbf{A}$ is called **symmetric** if $\mathbf{A}^T = \mathbf{A}$; that is, if $a_{ij} = a_{ji}$, for all $i, j = 1, \ldots, n$. Show that if $\mathbf{A}$ is an $n \times n$ matrix, then $\mathbf{A} + \mathbf{A}^T$ is a symmetric matrix.

40. Let $\mathbf{A}$ be an $m \times n$ matrix. Show that $\mathbf{A}^T\mathbf{A}$ is a symmetric $n \times n$ matrix and $\mathbf{A}\mathbf{A}^T$ is a symmetric $m \times m$ matrix (see Problem 39).

•10.3• LINEAR SYSTEMS IN NORMAL FORM

A system of n linear differential equations is in **normal form** if it is expressed as

(1) $\qquad \mathbf{x}'(t) = \mathbf{A}(t)\mathbf{x}(t) + \mathbf{f}(t),$

where $\mathbf{x}(t) = \mathrm{col}(x_1(t), \ldots, x_n(t))$, $\mathbf{f}(t) = \mathrm{col}(f_1(t), \ldots, f_n(t))$, and $\mathbf{A}(t) = [a_{ij}(t)]$ is an $n \times n$ matrix. As with a scalar linear differential equation, a system is called **homogeneous** when $\mathbf{f}(t) \equiv \mathbf{0}$; otherwise, it is called **nonhomogeneous.** When the elements of $\mathbf{A}$ are all constants, the system is said to have **constant coefficients.**

Notice that a linear differential equation

(2) $\qquad y^{(n)}(t) + p_{n-1}(t)y^{(n-1)}(t) + \cdots + p_0(t)y(t) = g(t)$

can be rewritten as a first order system in normal form using the substitution $x_1(t) := y(t)$, $x_2(t) := y'(t), \ldots, x_n(t) := y^{(n-1)}(t)$ (see also Section 9.2). Indeed, equation (2) is equivalent to $\mathbf{x}'(t) = \mathbf{A}(t)\mathbf{x}(t) + \mathbf{f}(t)$, where $\mathbf{x}(t) = \mathrm{col}(x_1(t), \ldots, x_n(t))$, $\mathbf{f}(t) := \mathrm{col}(0, \ldots, 0, g(t))$, and

$$\mathbf{A}(t) := \begin{bmatrix} 0 & 1 & 0 & \cdots & 0 & 0 \\ 0 & 0 & 1 & & 0 & 0 \\ \vdots & \vdots & \vdots & & \vdots & \vdots \\ 0 & 0 & 0 & \cdots & 0 & 1 \\ -p_0(t) & -p_1(t) & -p_2(t) & \cdots & -p_{n-2}(t) & -p_{n-1}(t) \end{bmatrix}.$$

The theory for systems in normal form parallels very closely the theory of linear differential equations presented in Chapters 4 and 6. In many cases the proofs for scalar linear differential equations carry over to normal systems with appropriate modifications. Conversely, results for normal systems apply to scalar linear equations since, as we showed, any scalar linear equation can be expressed as a normal system. This is the case with the existence and uniqueness theorems for linear differential equations.

The **initial value problem** for the normal system (1) is the problem of finding a differentiable vector function $\mathbf{x}(t)$ that satisfies the system on an interval I and also satisfies the **initial condition** $\mathbf{x}(t_0) = \mathbf{x}_0$, where t_0 is a given point of I and $\mathbf{x}_0 = \mathrm{col}(x_{1,0}, \ldots, x_{n,0})$ is a given vector.

EXISTENCE AND UNIQUENESS

Theorem 2.   Suppose that $\mathbf{A}(t)$ and $\mathbf{f}(t)$ are continuous on an open interval I that contains the point t_0. Then, for any choice of the initial vector $\mathbf{x}_0 = \mathrm{col}(x_{1,0}, \ldots, x_{n,0})$, there exists a unique solution $\mathbf{x}(t)$ on the whole interval I to the initial value problem

$$\mathbf{x}'(t) = \mathbf{A}(t)\mathbf{x}(t) + \mathbf{f}(t), \qquad \mathbf{x}(t_0) = \mathbf{x}_0.$$

We give a proof of this result in Chapter 14[†] and obtain as corollaries the existence and uniqueness theorems for second order equations (Theorem 2, Section 4.2) and higher order linear equations (Theorem 1, Section 6.2).

If we rewrite system (1) as $\mathbf{x}' - \mathbf{A}\mathbf{x} = \mathbf{f}$ and define the operator $L[\mathbf{x}] := \mathbf{x}' - \mathbf{A}\mathbf{x}$, then we can express system (1) in the operator form $L[\mathbf{x}] = \mathbf{f}$. Here the operator L maps vector functions into vector functions. Moreover, L is a *linear* operator in the sense that for any scalars a, b and vector functions $\mathbf{x}, \mathbf{y}$ we have

$$L[a\mathbf{x} + b\mathbf{y}] = aL[\mathbf{x}] + bL[\mathbf{y}].$$

The proof of this linearity follows from the properties of matrix multiplication (see Problem 25).

As a consequence of the linearity of L, if $\mathbf{x}_1, \ldots, \mathbf{x}_n$ are solutions to the homogeneous system $\mathbf{x}' = \mathbf{A}\mathbf{x}$, or $L[\mathbf{x}] = \mathbf{0}$ in operator notation, then any linear combination of these vectors, $c_1\mathbf{x}_1 + \cdots + c_n\mathbf{x}_n$, is also a solution. Moreover, we will see that if the solutions $\mathbf{x}_1, \ldots, \mathbf{x}_n$ are linearly independent, then *every* solution to $L[\mathbf{x}] = 0$ can be expressed as $c_1\mathbf{x}_1 + \cdots + c_n\mathbf{x}_n$ for an appropriate choice of the constants $c_1, \ldots, c_n$.

LINEAR DEPENDENCE OF VECTOR FUNCTIONS

Definition 1.   The m vector functions $\mathbf{x}_1, \ldots, \mathbf{x}_m$ are said to be **linearly dependent** on an interval I if there exist constants $c_1, \ldots, c_m$, not all zero, such that

$$c_1\mathbf{x}_1(t) + \cdots + c_m\mathbf{x}_m(t) = \mathbf{0},$$

for all t in I. If the vectors are not linearly dependent, they are said to be **linearly independent on** I.

[†] All references to Chapters 12–14 refer to the expanded text *Fundamentals of Differential Equations and Boundary Value Problems*.

EXAMPLE 1 Show that the vector functions $\mathbf{x}_1(t) = \text{col}(e^t, 0, e^t)$, $\mathbf{x}_2(t) = \text{col}(3e^t, 0, 3e^t)$, and $\mathbf{x}_3(t) = \text{col}(t, 1, 0)$ are linearly dependent on $(-\infty, \infty)$.

Solution Notice that $\mathbf{x}_2$ is just 3 times $\mathbf{x}_1$ and therefore $3\mathbf{x}_1(t) - \mathbf{x}_2(t) + 0 \cdot \mathbf{x}_3(t) = \mathbf{0}$ for all t. Hence, $\mathbf{x}_1$, $\mathbf{x}_2$, and $\mathbf{x}_3$ are linearly dependent on $(-\infty, \infty)$. ●

EXAMPLE 2 Show that the vector functions $\mathbf{x}_1(t) = \text{col}(e^{2t}, 0, e^{2t})$, $\mathbf{x}_2(t) = \text{col}(e^{2t}, e^{2t}, -e^{2t})$, and $\mathbf{x}_3(t) = \text{col}(e^t, 2e^t, e^t)$ are linearly independent on $(-\infty, \infty)$.

Solution To prove independence, we assume that c_1, c_2, and c_3 are constants for which

$$c_1\mathbf{x}_1(t) + c_2\mathbf{x}_2(t) + c_3\mathbf{x}_3(t) = \mathbf{0} \tag{3}$$

holds at every t in $(-\infty, \infty)$ and show that $c_1 = c_2 = c_3 = 0$. Setting $t = 0$ in (3), we have

$$c_1\begin{bmatrix}1\\0\\1\end{bmatrix} + c_2\begin{bmatrix}1\\1\\-1\end{bmatrix} + c_3\begin{bmatrix}1\\2\\1\end{bmatrix} = \mathbf{0},$$

which is equivalent to the system of linear equations

$$\begin{aligned} c_1 + c_2 + c_3 &= 0, \\ c_2 + 2c_3 &= 0, \\ c_1 - c_2 + c_3 &= 0. \end{aligned} \tag{4}$$

Either by solving (4) or by checking that the determinant of its coefficients is nonzero (recall Theorem 1 on page 483) we can verify that (4) has only the trivial solution $c_1 = c_2 = c_3 = 0$. Therefore, the vector functions $\mathbf{x}_1$, $\mathbf{x}_2$, and $\mathbf{x}_3$ are linearly independent on $(-\infty, \infty)$ (in fact, on any interval containing $t = 0$). ●

As the previous example illustrates, if $\mathbf{x}_1(t), \ldots, \mathbf{x}_n(t)$ are n vector functions each having n components, then these vectors will be linearly independent on an interval I if the determinant

$$\det[\mathbf{x}_1(t) \quad \cdots \quad \mathbf{x}_n(t)]$$

is not zero at some point t in I. Because of the analogy with scalar equations, we call this determinant the **Wronskian.**

WRONSKIAN

Definition 2. The **Wronskian** of n vector functions $\mathbf{x}_1(t) = \text{col}(x_{1,1}, \ldots, x_{n,1}), \ldots, \mathbf{x}_n(t) = \text{col}(x_{1,n}, \ldots, x_{n,n})$ is defined to be the real-valued function

$$W[\mathbf{x}_1, \ldots, \mathbf{x}_n](t) := \begin{vmatrix} x_{1,1}(t) & x_{1,2}(t) & \cdots & x_{1,n}(t) \\ x_{2,1}(t) & x_{2,2}(t) & \cdots & x_{2,n}(t) \\ \vdots & \vdots & & \vdots \\ x_{n,1}(t) & x_{n,2}(t) & \cdots & x_{n,n}(t) \end{vmatrix}.$$

If $\mathbf{x}_1, \ldots, \mathbf{x}_n$ are *linearly independent solutions* on I to the same homogeneous system $\mathbf{x}' = \mathbf{A}\mathbf{x}$ where $\mathbf{A}$ is an $n \times n$ matrix, then we can use the existence and uniqueness theorem to prove that the Wronskian $W(t) := W[\mathbf{x}_1, \ldots, \mathbf{x}_n](t)$ is never zero on I. Suppose to the contrary that the determinant $W(t_0) = 0$ at some t_0 in I; then from part (d) of Theorem 1, page 483, there exist scalars $c_1, \ldots, c_n$ *not all zero* such that

$$c_1\mathbf{x}_1(t_0) + \cdots + c_n\mathbf{x}_n(t_0) = \mathbf{0}.$$

But since $c_1\mathbf{x}_1(t) + \cdots + c_n\mathbf{x}_n(t)$ and the vector function $\mathbf{z}(t) \equiv \mathbf{0}$ are both solutions to $\mathbf{x}' = \mathbf{A}\mathbf{x}$ on I, and they agree at the point t_0, these solutions must be identical on I according to Theorem 2. That is,

$$c_1\mathbf{x}_1(t) + \cdots + c_n\mathbf{x}_n(t) = \mathbf{0}$$

for all t in I. But, this contradicts the given information that $\mathbf{x}_1, \ldots, \mathbf{x}_n$ are linearly independent on I. We have shown that $W(t_0) \neq 0$ and since t_0 is an arbitrary point, it follows that $W(t) \neq 0$ for all $t \in I$.

The preceding argument has two important implications that parallel the scalar case. First, *the Wronskian of solutions to* $\mathbf{x}' = \mathbf{A}\mathbf{x}$ *is either identically zero or never zero on* I (see also Problem 31). Second, *a set of n solutions* $\mathbf{x}_1, \ldots, \mathbf{x}_n$ *to* $\mathbf{x}' = \mathbf{A}\mathbf{x}$ *on* I *is linearly independent on* I *if and only if their Wronskian is never zero on* I. With these facts in hand, we can imitate the proof given for the scalar case in Section 6.2 (Theorem 2) to obtain the following representation theorem for the solutions to $\mathbf{x}' = \mathbf{A}\mathbf{x}$.

REPRESENTATION OF SOLUTIONS (HOMOGENEOUS CASE)

Theorem 3.   Let $\mathbf{x}_1, \ldots, \mathbf{x}_n$ be n linearly independent solutions to the homogeneous system

(5) $\mathbf{x}'(t) = \mathbf{A}(t)\mathbf{x}(t)$

on the interval I, where $\mathbf{A}(t)$ is an $n \times n$ matrix function continuous on I. Then every solution to (5) on I can be expressed in the form

(6) $\mathbf{x}(t) = c_1\mathbf{x}_1(t) + \cdots + c_n\mathbf{x}_n(t),$

where $c_1, \ldots, c_n$ are constants.

A set of solutions $\{\mathbf{x}_1, \ldots, \mathbf{x}_n\}$ that are linearly independent on I or, equivalently, whose Wronskian does not vanish on I, is called a **fundamental solution set** for (5) on I. The linear combination in (6), written with arbitrary constants, is referred to as a **general solution** to (5).

If we take the vectors in a fundamental solution set and let them form the columns of a matrix $\mathbf{X}(t)$, that is,

$$\mathbf{X}(t) := \begin{bmatrix} x_{1,1}(t) & x_{1,2}(t) & \cdots & x_{1,n}(t) \\ x_{2,1}(t) & x_{2,2}(t) & \cdots & x_{2,n}(t) \\ \vdots & \vdots & & \vdots \\ x_{n,1}(t) & x_{n,2}(t) & \cdots & x_{n,n}(t) \end{bmatrix},$$

then the matrix $\mathbf{X}(t)$ is called a **fundamental matrix** for (5). We can use it to express the general solution (6) as

$$\mathbf{x}(t) = \mathbf{X}(t)\mathbf{c},$$

where $\mathbf{c} = \text{col}(c_1, \ldots, c_n)$ is an arbitrary constant vector. Since $\det \mathbf{X} = W[\mathbf{x}_1, \ldots, \mathbf{x}_n]$ is never zero on I, it follows from Theorem 1 in Section 10.2 that $\mathbf{X}(t)$ is invertible for every t in I.

● **EXAMPLE 3**   Verify that the set

$$S = \left\{ \begin{bmatrix} e^{2t} \\ e^{2t} \\ e^{2t} \end{bmatrix}, \begin{bmatrix} -e^{-t} \\ 0 \\ e^{-t} \end{bmatrix}, \begin{bmatrix} 0 \\ e^{-t} \\ -e^{-t} \end{bmatrix} \right\}$$

is a fundamental solution set for the system

$$(7) \qquad \mathbf{x}'(t) = \begin{bmatrix} 0 & 1 & 1 \\ 1 & 0 & 1 \\ 1 & 1 & 0 \end{bmatrix} \mathbf{x}(t)$$

on the interval $(-\infty, \infty)$ and find a fundamental matrix for (7). Also determine a general solution for (7).

Solution   Substituting the first vector in the set S into the right-hand side of (7) gives

$$\mathbf{Ax} = \begin{bmatrix} 0 & 1 & 1 \\ 1 & 0 & 1 \\ 1 & 1 & 0 \end{bmatrix} \begin{bmatrix} e^{2t} \\ e^{2t} \\ e^{2t} \end{bmatrix} = \begin{bmatrix} 2e^{2t} \\ 2e^{2t} \\ 2e^{2t} \end{bmatrix} = \mathbf{x}'(t);$$

hence this vector satisfies system (7) for all t. Similar computations verify that the remaining vectors in S are also solutions to (7) on $(-\infty, \infty)$. To show that S is a fundamental solution set, it is enough to observe that the Wronskian

$$W(t) = \begin{vmatrix} e^{2t} & -e^{-t} & 0 \\ e^{2t} & 0 & e^{-t} \\ e^{2t} & e^{-t} & -e^{-t} \end{vmatrix} = e^{2t} \begin{vmatrix} 0 & e^{-t} \\ e^{-t} & -e^{-t} \end{vmatrix} + e^{-t} \begin{vmatrix} e^{2t} & e^{-t} \\ e^{2t} & -e^{-t} \end{vmatrix} = -3$$

is never zero.

A fundamental matrix $\mathbf{X}(t)$ for (7) is just the matrix we used to compute the Wronskian; that is,

$$(8) \qquad \mathbf{X}(t) := \begin{bmatrix} e^{2t} & -e^{-t} & 0 \\ e^{2t} & 0 & e^{-t} \\ e^{2t} & e^{-t} & -e^{-t} \end{bmatrix}.$$

A general solution to (7) can now be expressed as

$$\mathbf{x}(t) = \mathbf{X}(t)\mathbf{c} = c_1 \begin{bmatrix} e^{2t} \\ e^{2t} \\ e^{2t} \end{bmatrix} + c_2 \begin{bmatrix} -e^{-t} \\ 0 \\ e^{-t} \end{bmatrix} + c_3 \begin{bmatrix} 0 \\ e^{-t} \\ -e^{-t} \end{bmatrix}. \quad \bullet$$

It is easy to check that the fundamental matrix in (8) satisfies the equation

$$\mathbf{X}'(t) = \begin{bmatrix} 0 & 1 & 1 \\ 1 & 0 & 1 \\ 1 & 1 & 0 \end{bmatrix} \mathbf{X}(t);$$

indeed, this is equivalent to showing that $\mathbf{x}' = \mathbf{A}\mathbf{x}$ for each $\mathbf{x}$ in S. In general, a fundamental matrix for a system $\mathbf{x}' = \mathbf{A}\mathbf{x}$ satisfies the corresponding **matrix differential equation** $\mathbf{X}' = \mathbf{A}\mathbf{X}$.

Another consequence of the linearity of the operator L defined by $L[\mathbf{x}] := \mathbf{x}' - \mathbf{A}\mathbf{x}$ is the **superposition principle** for linear systems. It states that if $\mathbf{x}_1$ and $\mathbf{x}_2$ are solutions, respectively, to the nonhomogeneous systems

$$L[\mathbf{x}] = \mathbf{g}_1 \quad \text{and} \quad L[\mathbf{x}] = \mathbf{g}_2,$$

then $c_1\mathbf{x}_1 + c_2\mathbf{x}_2$ is a solution to

$$L[\mathbf{x}] = c_1\mathbf{g}_1 + c_2\mathbf{g}_2.$$

Using this superposition principle and the representation theorem for homogeneous systems, we can prove the following theorem.

> ### REPRESENTATION OF SOLUTIONS (NONHOMOGENEOUS CASE)
>
> **Theorem 4.**   Let $\mathbf{x}_p$ be a particular solution to the nonhomogeneous system
>
> **(9)** $\mathbf{x}'(t) = \mathbf{A}(t)\mathbf{x}(t) + \mathbf{f}(t)$
>
> on the interval I, and let $\{\mathbf{x}_1, \dots, \mathbf{x}_n\}$ be a fundamental solution set on I for the corresponding homogeneous system $\mathbf{x}'(t) = \mathbf{A}(t)\mathbf{x}(t)$. Then every solution to (9) on I can be expressed in the form
>
> **(10)** $\mathbf{x}(t) = \mathbf{x}_p(t) + c_1\mathbf{x}_1(t) + \cdots + c_n\mathbf{x}_n(t),$
>
> where $c_1, \dots, c_n$ are constants.

The proof of this theorem is almost identical to the proofs of Theorem 7 in Section 4.7 and Theorem 4 in Section 6.2. We leave it as an exercise.

The linear combination of $\mathbf{x}_p$, $\mathbf{x}_1, \dots, \mathbf{x}_n$ in (10) written with arbitrary constants $c_1, \dots, c_n$ is called a **general solution** of (9). This general solution can also be expressed as $\mathbf{x} = \mathbf{x}_p + \mathbf{X}\mathbf{c}$, where $\mathbf{X}$ is a fundamental matrix for the homogeneous system and $\mathbf{c}$ is an arbitrary constant vector.

Below we summarize the results of this section as they apply to the problem of finding a general solution to a system of n linear first order differential equations in normal form.

▶ **APPROACH TO SOLVING NORMAL SYSTEMS**

1. To obtain a general solution to the $n \times n$ homogeneous system $\mathbf{x}' = \mathbf{Ax}$:

   (a) Find a fundamental solution set $\{\mathbf{x}_1, \ldots, \mathbf{x}_n\}$ that consists of n linearly independent solutions to the homogeneous system.

   (b) A general solution is

   $$\mathbf{x} = \mathbf{Xc} = c_1\mathbf{x}_1 + \cdots + c_n\mathbf{x}_n,$$

   where $\mathbf{c} = \text{col}(c_1, \ldots, c_n)$ is a constant vector and $\mathbf{X} = [\mathbf{x}_1, \ldots, \mathbf{x}_n]$ is the fundamental matrix whose columns are the vectors in the fundamental solution set.

2. To obtain a general solution to the nonhomogeneous system $\mathbf{x}' = \mathbf{Ax} + \mathbf{f}$:

   (a) Find a particular solution $\mathbf{x}_p$ to the nonhomogeneous system.

   (b) A general solution to the nonhomogeneous system is

   $$\mathbf{x} = \mathbf{x}_p + \mathbf{Xc} = \mathbf{x}_p + c_1\mathbf{x}_1 + \cdots + c_n\mathbf{x}_n,$$

   where $\mathbf{Xc} = c_1\mathbf{x}_1 + \cdots + c_n\mathbf{x}_n$ is the general solution to the homogeneous system obtained in Part 1.

The remainder of this chapter is devoted to methods for finding fundamental solution sets for homogeneous systems and particular solutions for nonhomogeneous systems.

EXERCISES 10.3

In Problems 1 through 4 write the given system in the matrix form $\mathbf{x}' = \mathbf{Ax} + \mathbf{f}$.

1. $x'(t) = 3x(t) - y(t) + t^2,$
$\quad y'(t) = -x(t) + 2y(t) + e^t.$

2. $r'(t) = 2r(t) + \sin t,$
$\quad \theta'(t) = r(t) - \theta(t) + 1.$

3. $\dfrac{dx}{dt} = t^2 x - y - z + t,$

$\quad \dfrac{dy}{dt} = e^t z + 5,$

$\quad \dfrac{dz}{dt} = tx - y + 3z - e^t.$

4. $\dfrac{dx}{dt} = x + y + z,$

$\quad \dfrac{dy}{dt} = 2x - y + 3z,$

$\quad \dfrac{dz}{dt} = x + 5z.$

In Problems 5 through 8 rewrite the given scalar equation as a first order system in normal form. Express the system in the matrix form $\mathbf{x}' = \mathbf{Ax} + \mathbf{f}$.

5. $y''(t) - 3y'(t) - 10y(t) = \sin t.$

6. $x''(t) + x(t) = t^2.$

7. $\dfrac{d^4 w}{dt^4} + w = t^2.$

8. $\dfrac{d^3 y}{dt^3} - \dfrac{dy}{dt} + y = \cos t.$

In Problems 9 through 12 write the given system as a set of scalar equations.

9. $\mathbf{x}' = \begin{bmatrix} 5 & 0 \\ -2 & 4 \end{bmatrix} \mathbf{x} + e^{-2t} \begin{bmatrix} 2 \\ -3 \end{bmatrix}.$

10. $\mathbf{x}' = \begin{bmatrix} 2 & 1 \\ -1 & 3 \end{bmatrix} \mathbf{x} + e^t \begin{bmatrix} t \\ 1 \end{bmatrix}.$

11. $\mathbf{x}' = \begin{bmatrix} 1 & 0 & 1 \\ -1 & 2 & 5 \\ 0 & 5 & 1 \end{bmatrix} \mathbf{x} + e^t \begin{bmatrix} 1 \\ 0 \\ 0 \end{bmatrix} + t \begin{bmatrix} 0 \\ 1 \\ 0 \end{bmatrix}.$

12. $\mathbf{x}' = \begin{bmatrix} 0 & 1 & 0 \\ 0 & 0 & 1 \\ -1 & 1 & 2 \end{bmatrix}\mathbf{x} + t\begin{bmatrix} 1 \\ -1 \\ 2 \end{bmatrix} + \begin{bmatrix} 3 \\ 1 \\ 0 \end{bmatrix}.$

In Problems 13 through 18 determine whether the given vector functions are linearly dependent (LD) or linearly independent (LI) on the interval $(-\infty, \infty)$.

13. $\begin{bmatrix} t \\ 3 \end{bmatrix}, \begin{bmatrix} 4 \\ 1 \end{bmatrix}.$

14. $\begin{bmatrix} te^{-t} \\ e^{-t} \end{bmatrix}, \begin{bmatrix} e^{-t} \\ e^{-t} \end{bmatrix}.$

15. $\begin{bmatrix} \sin t \\ \cos t \end{bmatrix}, \begin{bmatrix} \sin 2t \\ \cos 2t \end{bmatrix}.$

16. $e^t\begin{bmatrix} 1 \\ 5 \end{bmatrix}, e^t\begin{bmatrix} -3 \\ -15 \end{bmatrix}.$

17. $e^{2t}\begin{bmatrix} 1 \\ 0 \\ 5 \end{bmatrix}, e^{2t}\begin{bmatrix} 1 \\ 1 \\ -1 \end{bmatrix}, e^{3t}\begin{bmatrix} 0 \\ 1 \\ 0 \end{bmatrix}.$

18. $\begin{bmatrix} 1 \\ 0 \\ 1 \end{bmatrix}, \begin{bmatrix} t \\ 0 \\ t \end{bmatrix}, \begin{bmatrix} t^2 \\ 0 \\ t^2 \end{bmatrix}.$

In Problems 19 through 22 the given vector functions are solutions to the system $\mathbf{x}'(t) = \mathbf{A}\mathbf{x}(t).$ *Determine whether they form a fundamental solution set. If they do, find a fundamental matrix for the system and give a general solution.*

19. $\mathbf{x}_1 = e^{2t}\begin{bmatrix} 1 \\ -2 \end{bmatrix}, \mathbf{x}_2 = e^{2t}\begin{bmatrix} -2 \\ 4 \end{bmatrix}.$

20. $\mathbf{x}_1 = e^{-t}\begin{bmatrix} 3 \\ 2 \end{bmatrix}, \mathbf{x}_2 = e^{4t}\begin{bmatrix} 1 \\ -1 \end{bmatrix}.$

21. $\mathbf{x}_1 = \begin{bmatrix} e^{-t} \\ 2e^{-t} \\ e^{-t} \end{bmatrix}, \mathbf{x}_2 = \begin{bmatrix} e^t \\ 0 \\ e^t \end{bmatrix}, \mathbf{x}_3 = \begin{bmatrix} e^{3t} \\ -e^{3t} \\ 2e^{3t} \end{bmatrix}.$

22. $\mathbf{x}_1 = \begin{bmatrix} e^t \\ e^t \\ e^t \end{bmatrix}, \mathbf{x}_2 = \begin{bmatrix} \sin t \\ \cos t \\ -\sin t \end{bmatrix}, \mathbf{x}_3 = \begin{bmatrix} -\cos t \\ \sin t \\ \cos t \end{bmatrix}.$

23. Verify that the vector functions

$$\mathbf{x}_1 = \begin{bmatrix} e^t \\ e^t \end{bmatrix} \text{ and } \mathbf{x}_2 = \begin{bmatrix} e^{-t} \\ 3e^{-t} \end{bmatrix}$$

are solutions to the homogeneous system

$$\mathbf{x}' = \mathbf{A}\mathbf{x} = \begin{bmatrix} 2 & -1 \\ 3 & -2 \end{bmatrix}\mathbf{x},$$

on $(-\infty, \infty)$, and that

$$\mathbf{x}_p = \frac{3}{2}\begin{bmatrix} te^t \\ te^t \end{bmatrix} - \frac{1}{4}\begin{bmatrix} e^t \\ 3e^t \end{bmatrix} + \begin{bmatrix} t \\ 2t \end{bmatrix} - \begin{bmatrix} 0 \\ 1 \end{bmatrix}$$

is a particular solution to the nonhomogeneous system $\mathbf{x}' = \mathbf{A}\mathbf{x} + \mathbf{f}(t)$, where $\mathbf{f}(t) = \text{col}(e^t, t)$. Find a general solution to $\mathbf{x}' = \mathbf{A}\mathbf{x} + \mathbf{f}(t)$.

24. Verify that the vector functions

$$\mathbf{x}_1 = \begin{bmatrix} e^{3t} \\ 0 \\ e^{3t} \end{bmatrix}, \mathbf{x}_2 = \begin{bmatrix} -e^{3t} \\ e^{3t} \\ 0 \end{bmatrix}, \mathbf{x}_3 = \begin{bmatrix} -e^{-3t} \\ -e^{-3t} \\ e^{-3t} \end{bmatrix}$$

are solutions to the homogeneous system

$$\mathbf{x}' = \mathbf{A}\mathbf{x} = \begin{bmatrix} 1 & -2 & 2 \\ -2 & 1 & 2 \\ 2 & 2 & 1 \end{bmatrix}\mathbf{x},$$

on $(-\infty, \infty)$, and that

$$\mathbf{x}_p = \begin{bmatrix} 5t + 1 \\ 2t \\ 4t + 2 \end{bmatrix}$$

is a particular solution to $\mathbf{x}' = \mathbf{A}\mathbf{x} + \mathbf{f}(t)$, where $\mathbf{f}(t) = \text{col}(-9, 0, -18)$. Find a general solution to $\mathbf{x}' = \mathbf{A}\mathbf{x} + \mathbf{f}(t)$.

25. Prove that the operator defined by $L[\mathbf{x}] := \mathbf{x}' - \mathbf{A}\mathbf{x}$, where $\mathbf{A}$ is an $n \times n$ matrix function and $\mathbf{x}$ is an $n \times 1$ vector function, is a linear operator.

26. Let $\mathbf{X}(t)$ be a fundamental matrix for the system $\mathbf{x}' = \mathbf{A}\mathbf{x}$. Show that $\mathbf{x}(t) = \mathbf{X}(t)\mathbf{X}^{-1}(t_0)\mathbf{x}_0$ is the solution to the initial value problem $\mathbf{x}' = \mathbf{A}\mathbf{x}, \mathbf{x}(t_0) = \mathbf{x}_0$.

In Problems 27 and 28 verify that $\mathbf{X}(t)$ *is a fundamental matrix for the given system and compute* $\mathbf{X}^{-1}(t).$ *Use the result of Problem 26 to find the solution to the given initial value problem.*

27. $\mathbf{x}' = \begin{bmatrix} 0 & 6 & 0 \\ 1 & 0 & 1 \\ 1 & 1 & 0 \end{bmatrix}\mathbf{x}, \quad \mathbf{x}(0) = \begin{bmatrix} -1 \\ 0 \\ 1 \end{bmatrix};$

$$\mathbf{X}(t) = \begin{bmatrix} 6e^{-t} & -3e^{-2t} & 2e^{3t} \\ -e^{-t} & e^{-2t} & e^{3t} \\ -5e^{-t} & e^{-2t} & e^{3t} \end{bmatrix}.$$

28. $\mathbf{x}' = \begin{bmatrix} 2 & 3 \\ 3 & 2 \end{bmatrix}\mathbf{x}, \quad \mathbf{x}(0) = \begin{bmatrix} 3 \\ -1 \end{bmatrix};$

$$\mathbf{X}(t) = \begin{bmatrix} e^{-t} & e^{5t} \\ -e^{-t} & e^{5t} \end{bmatrix}.$$

29. Show that

$$\begin{vmatrix} t^2 & t|t| \\ 2t & 2|t| \end{vmatrix} \equiv 0$$

on $(-\infty, \infty)$, but that the two column vectors

$$\begin{bmatrix} t^2 \\ 2t \end{bmatrix}, \quad \begin{bmatrix} t|t| \\ 2|t| \end{bmatrix}$$

are linearly independent on $(-\infty, \infty)$.

30. Abel's formula. If $\mathbf{x}_1, \ldots, \mathbf{x}_n$ are any n solutions to the $n \times n$ system $\mathbf{x}'(t) = \mathbf{A}(t)\mathbf{x}(t)$, then Abel's formula gives a representation for the Wronskian $W(t) := W[\mathbf{x}_1, \ldots, \mathbf{x}_n](t)$. Namely,

$$W(t) = W(t_0)\exp\left(\int_{t_0}^{t} \{a_{11}(s) + \cdots + a_{nn}(s)\}\, ds\right),$$

where $a_{11}(s), \ldots, a_{nn}(s)$ are the main diagonal elements of $\mathbf{A}(s)$. Prove this formula in the special case when $n = 3$. [Hint: Follow the outline in Problem 31 of Exercises 6.2.]

31. Using Abel's formula, prove that the Wronskian of n solutions to $\mathbf{x}' = \mathbf{A}\mathbf{x}$ on the interval I is either identically zero on I or never zero on I.

32. Prove that a fundamental solution set for the homogeneous system $\mathbf{x}'(t) = \mathbf{A}(t)\mathbf{x}(t)$ always exists on an interval I, provided that $\mathbf{A}(t)$ is continuous on I. [Hint: Use the existence and uniqueness theorem (Theorem 2) and make judicious choices for $\mathbf{x}_0$.]

33. Prove Theorem 3 on the representation of solutions of the homogeneous system.

34. Prove Theorem 4 on the representation of solutions of the nonhomogeneous system.

35. To illustrate the connection between a higher order equation and the equivalent first order system, consider the equation

$$\text{(11)} \qquad y'''(t) - 6y''(t) + 11y'(t) - 6y(t) = 0.$$

(a) Show that $\{e^t, e^{2t}, e^{3t}\}$ is a fundamental solution set for (11).

(b) Using the definition of Section 6.2, compute the Wronskian of $\{e^t, e^{2t}, e^{3t}\}$.

(c) Setting $x_1 = y$, $x_2 = y'$, $x_3 = y''$, show that equation (11) is equivalent to the first order system

$$\text{(12)} \qquad \mathbf{x}' = \mathbf{A}\mathbf{x},$$

where

$$\mathbf{A} := \begin{bmatrix} 0 & 1 & 0 \\ 0 & 0 & 1 \\ 6 & -11 & 6 \end{bmatrix}.$$

(d) The substitution used in part (c) suggests that

$$S := \left\{ \begin{bmatrix} e^t \\ e^t \\ e^t \end{bmatrix}, \begin{bmatrix} e^{2t} \\ 2e^{2t} \\ 4e^{2t} \end{bmatrix}, \begin{bmatrix} e^{3t} \\ 3e^{3t} \\ 9e^{3t} \end{bmatrix} \right\}$$

is a fundamental solution set for system (12). Verify that this is the case.

(e) Compute the Wronskian of S. How does it compare with the Wronskian computed in part (b)?

• 10.4 • HOMOGENEOUS LINEAR SYSTEMS WITH CONSTANT COEFFICIENTS

In this section we discuss a procedure for obtaining a general solution for the homogeneous system

$$\text{(1)} \qquad \mathbf{x}'(t) = \mathbf{A}\mathbf{x}(t),$$

where $\mathbf{A}$ is a (real) *constant* $n \times n$ matrix. The general solution we seek will be defined for all t because the elements of $\mathbf{A}$ are just constant functions, which are continuous on $(-\infty, \infty)$ (recall Theorem 2, page 487). In Section 10.3 we showed that a general solution to (1) can be constructed from a fundamental solution set consisting of n linearly independent solutions to (1). Thus our goal is to find n such vector solutions.

In Chapter 4 we were successful in solving homogeneous linear equations with constant coefficients by guessing that the equation had a solution of the form e^{rt}. Since any scalar linear equation can be expressed as a system, it is reasonable to expect system (1) to have solutions of the form

$$\mathbf{x}(t) = e^{rt}\mathbf{u},$$

where r is a constant and **u** is a constant vector, both of which must be determined. Substituting e^{rt}**u** for **x**(t) in (1) gives

$$re^{rt}\mathbf{u} = \mathbf{A}e^{rt}\mathbf{u} = e^{rt}\mathbf{A}\mathbf{u}.$$

Canceling the factor e^{rt} and rearranging terms, we find that

(2) $(\mathbf{A} - r\mathbf{I})\mathbf{u} = \mathbf{0}$,

where $r\mathbf{I}$ denotes the diagonal matrix with r's along its main diagonal.

The preceding calculation shows that $\mathbf{x}(t) = e^{rt}\mathbf{u}$ is a solution to (1) if and only if r and **u** satisfy equation (2). Since the trivial case, $\mathbf{u} = \mathbf{0}$, is of no help in finding linearly independent solutions to (1), we require that $\mathbf{u} \neq \mathbf{0}$. Such vectors are given a special name.

▶ EIGENVALUES AND EIGENVECTORS

Definition 3.   Let $\mathbf{A} = [a_{ij}]$ be an $n \times n$ constant matrix. The **eigenvalues** of **A** are those (real or complex) numbers r for which $(\mathbf{A} - r\mathbf{I})\mathbf{u} = \mathbf{0}$ has at least one nontrivial solution[†] **u**. The corresponding nontrivial solutions **u** are called the **eigenvectors** of **A** associated with r.

As stated in Theorem 1 of Section 10.2, a linear homogeneous system of n algebraic equations in n unknowns has a nontrivial solution if and only if the determinant of its coefficients is zero. Hence a necessary and sufficient condition for (2) to have a nontrivial solution is that

(3) $|\mathbf{A} - r\mathbf{I}| = 0$.

Expanding this determinant, we find that it is an nth degree polynomial in r; that is,

(4) $|\mathbf{A} - r\mathbf{I}| = p(r)$.

Therefore, *finding the eigenvalues of a matrix* **A** *is equivalent to finding the zeros of the polynomial* $p(r)$. Equation (3) is called the **characteristic equation** of **A**, and $p(r)$ in (4) is the **characteristic polynomial** of **A**. The characteristic equation plays a role for systems similar to the role played by the auxiliary equation for scalar linear equations.

There are several commercially available software packages that can be used to compute the eigenvalues and eigenvectors for a given matrix. Four such packages are: DERIVE available from Soft Warehouse, Inc., 3615 Harding Ave., Suite 505, Honolulu, Hawaii 96816-3735; MATLAB available from The MathWorks, Inc., 21 Eliot Street, South Natick, Massachusetts 01760; MATHEMATICA available from Wolfram Research, Box 6059, Champaign, Illinois 61826; and MAPLE available from the University of Waterloo, Waterloo, Ontario, Canada. Although the reader is encouraged to make use of such packages, the examples and most exercises in this text can be easily carried out without them. Those exercises for which a computer package is desirable are flagged with the icon ▪

[†] We will allow **u** to have complex-number entries.

● **EXAMPLE 1**   Find the eigenvalues and eigenvectors of the matrix

$$\mathbf{A} := \begin{bmatrix} 2 & -3 \\ 1 & -2 \end{bmatrix}.$$

Solution   The characteristic equation for $\mathbf{A}$ is

$$|\mathbf{A} - r\mathbf{I}| = \begin{vmatrix} 2-r & -3 \\ 1 & -2-r \end{vmatrix} = (2-r)(-2-r) + 3 = r^2 - 1 = 0.$$

Hence the eigenvalues of $\mathbf{A}$ are $r_1 = 1$, $r_2 = -1$. To find the eigenvectors corresponding to $r_1 = 1$, we must solve $(\mathbf{A} - r_1\mathbf{I})\mathbf{u} = \mathbf{0}$. Substituting for $\mathbf{A}$ and r_1 gives

$$(5) \qquad \begin{bmatrix} 1 & -3 \\ 1 & -3 \end{bmatrix}\begin{bmatrix} u_1 \\ u_2 \end{bmatrix} = \begin{bmatrix} 0 \\ 0 \end{bmatrix}.$$

Notice that this matrix equation is equivalent to the single scalar equation $u_1 - 3u_2 = 0$. Therefore the solutions to (5) are obtained by assigning an arbitrary value for u_2, say $u_2 = s$, and setting $u_1 = 3u_2 = 3s$. Consequently, the eigenvectors associated with $r_1 = 1$ can be expressed as

$$(6) \qquad \mathbf{u}_1 = s\begin{bmatrix} 3 \\ 1 \end{bmatrix}.$$

For $r_2 = -1$, the equation $(\mathbf{A} - r_2\mathbf{I})\mathbf{u} = \mathbf{0}$ becomes

$$\begin{bmatrix} 3 & -3 \\ 1 & -1 \end{bmatrix}\begin{bmatrix} u_1 \\ u_2 \end{bmatrix} = \begin{bmatrix} 0 \\ 0 \end{bmatrix}.$$

Solving, we obtain $u_1 = s$ and $u_2 = s$. Therefore, the eigenvectors associated with the eigenvalue $r_2 = -1$ are

$$(7) \qquad \mathbf{u}_2 = s\begin{bmatrix} 1 \\ 1 \end{bmatrix}. \quad ●$$

We remark that in the above example the collection (6) of all eigenvectors associated with $r_1 = 1$ forms a one-dimensional subspace when the zero vector is adjoined. The same is true for $r_2 = -1$. These subspaces are called **eigenspaces.**

● **EXAMPLE 2**   Find the eigenvalues and eigenvectors of the matrix

$$\mathbf{A} := \begin{bmatrix} 1 & 2 & -1 \\ 1 & 0 & 1 \\ 4 & -4 & 5 \end{bmatrix}.$$

Solution   The characteristic equation for $\mathbf{A}$ is

$$|\mathbf{A} - r\mathbf{I}| = \begin{vmatrix} 1-r & 2 & -1 \\ 1 & -r & 1 \\ 4 & -4 & 5-r \end{vmatrix} = 0,$$

which simplifies to $(r-1)(r-2)(r-3) = 0$. Hence the eigenvalues of $\mathbf{A}$ are $r_1 = 1, r_2 = 2$, and $r_3 = 3$. To find the eigenvectors corresponding to $r_1 = 1$, we set $r = 1$ in $(\mathbf{A} - r\mathbf{I})\mathbf{u} = \mathbf{0}$. This gives

$$
(8) \qquad \begin{bmatrix} 0 & 2 & -1 \\ 1 & -1 & 1 \\ 4 & -4 & 4 \end{bmatrix} \begin{bmatrix} u_1 \\ u_2 \\ u_3 \end{bmatrix} = \begin{bmatrix} 0 \\ 0 \\ 0 \end{bmatrix}.
$$

Using elementary row operations (Gaussian elimination), we see that (8) is equivalent to the two equations

$$
\begin{aligned}
u_1 - u_2 + u_3 &= 0, \\
2u_2 - u_3 &= 0.
\end{aligned}
$$

Thus we can obtain the solutions to (8) by assigning an arbitrary value to u_2, say $u_2 = s$, solving $2u_2 - u_3 = 0$ for u_3 to get $u_3 = 2s$, and then solving $u_1 - u_2 + u_3 = 0$ for u_1 to get $u_1 = -s$. Hence the eigenvectors associated with $r_1 = 1$ are

$$
(9) \qquad \mathbf{u}_1 = s \begin{bmatrix} -1 \\ 1 \\ 2 \end{bmatrix}.
$$

For $r_2 = 2$, we solve

$$
\begin{bmatrix} -1 & 2 & -1 \\ 1 & -2 & 1 \\ 4 & -4 & 3 \end{bmatrix} \begin{bmatrix} u_1 \\ u_2 \\ u_3 \end{bmatrix} = \begin{bmatrix} 0 \\ 0 \\ 0 \end{bmatrix}
$$

in a similar fashion to obtain the eigenvectors

$$
(10) \qquad \mathbf{u}_2 = s \begin{bmatrix} -2 \\ 1 \\ 4 \end{bmatrix}.
$$

Finally, for $r_3 = 3$, we solve

$$
\begin{bmatrix} -2 & 2 & -1 \\ 1 & -3 & 1 \\ 4 & -4 & 2 \end{bmatrix} \begin{bmatrix} u_1 \\ u_2 \\ u_3 \end{bmatrix} = \begin{bmatrix} 0 \\ 0 \\ 0 \end{bmatrix}
$$

and get the eigenvectors

$$
(11) \qquad \mathbf{u}_3 = s \begin{bmatrix} -1 \\ 1 \\ 4 \end{bmatrix}. \quad \bullet
$$

Let's return to the problem of finding a general solution to a homogeneous system of differential equations. We have already shown that $e^{rt}\mathbf{u}$ is a solution to (1) if r is an eigenvalue and $\mathbf{u}$ a corresponding eigenvector. The question is: Can we obtain n linearly independent solutions to the homogeneous system by finding all the eigenvalues and eigenvectors of $\mathbf{A}$?

---

n LINEARLY INDEPENDENT EIGENVECTORS

Theorem 5.   Suppose that the $n \times n$ constant matrix $\mathbf{A}$ has n linearly independent eigenvectors $\mathbf{u}_1, \mathbf{u}_2, \ldots, \mathbf{u}_n$. Let r_i be the eigenvalue[†] corresponding to $\mathbf{u}_i$. Then

(12) $\{e^{r_1 t}\mathbf{u}_1, e^{r_2 t}\mathbf{u}_2, \ldots, e^{r_n t}\mathbf{u}_n\}$

is a fundamental solution set on $(-\infty, \infty)$ for the homogeneous system $\mathbf{x}' = \mathbf{A}\mathbf{x}$. Consequently, a general solution of $\mathbf{x}' = \mathbf{A}\mathbf{x}$ is

(13) $\mathbf{x}(t) = c_1 e^{r_1 t}\mathbf{u}_1 + c_2 e^{r_2 t}\mathbf{u}_2 + \cdots + c_n e^{r_n t}\mathbf{u}_n,$

where $c_1, \ldots, c_n$ are arbitrary constants.

Proof.   As we have seen, the vector functions listed in (12) are solutions to the homogeneous system. Moreover, their Wronskian is

$$W(t) = \det[e^{r_1 t}\mathbf{u}_1, \quad \ldots, \quad e^{r_n t}\mathbf{u}_n] = e^{(r_1 + \cdots + r_n)t}\det[\mathbf{u}_1, \quad \ldots, \quad \mathbf{u}_n].$$

Since the eigenvectors are assumed to be linearly independent, it follows from Theorem 1 in Section 10.2 that $\det[\mathbf{u}_1, \ldots, \mathbf{u}_n]$ is not zero. Hence the Wronskian $W(t)$ is never zero. This shows that (12) is a fundamental solution set, and consequently a general solution is given by (13).   ◄◄◄

An application of Theorem 5 is given in the next example.

● **EXAMPLE 3**   Find a general solution of

(14) $\mathbf{x}'(t) = \mathbf{A}\mathbf{x}(t), \quad$ where $\quad \mathbf{A} = \begin{bmatrix} 2 & -3 \\ 1 & -2 \end{bmatrix}.$

Solution   In Example 1 we showed that the matrix $\mathbf{A}$ has eigenvalues $r_1 = 1$ and $r_2 = -1$. Taking $s = 1$ in equations (6) and (7), we get the corresponding eigenvectors

$$\mathbf{u}_1 = \begin{bmatrix} 3 \\ 1 \end{bmatrix} \quad \text{and} \quad \mathbf{u}_2 = \begin{bmatrix} 1 \\ 1 \end{bmatrix}.$$

Since $\mathbf{u}_1$ and $\mathbf{u}_2$ are linearly independent, it follows from Theorem 5 that a general solution to (14) is

(15) $\mathbf{x}(t) = c_1 e^t \begin{bmatrix} 3 \\ 1 \end{bmatrix} + c_2 e^{-t} \begin{bmatrix} 1 \\ 1 \end{bmatrix}. \quad ●$

[†] The eigenvalues $r_1, \ldots, r_n$ may be real or complex and need not be distinct. In this section the cases we discuss have real eigenvalues. We consider complex eigenvalues in Section 10.5.

If we sum the vectors on the right-hand side of equation (15) and then write out the expressions for the components of $\mathbf{x}(t) = \mathrm{col}(x_1(t), x_2(t))$, we get

$$x_1(t) = 3c_1 e^t + c_2 e^{-t},$$
$$x_2(t) = c_1 e^t + c_2 e^{-t}.$$

This is the familiar form of a general solution for a system, as discussed in Chapter 9.

A useful property of eigenvectors that concerns their linear independence is stated in the next theorem.

◣ LINEAR INDEPENDENCE OF EIGENVECTORS

Theorem 6.   If $r_1, \ldots, r_m$ are *distinct* eigenvalues for the matrix $\mathbf{A}$ and $\mathbf{u}_i$ is an eigenvector associated with r_i, then $\mathbf{u}_1, \ldots, \mathbf{u}_m$ are linearly independent.

Proof.   Let's first treat the case $m = 2$. Suppose, to the contrary, that $\mathbf{u}_1$ and $\mathbf{u}_2$ are linearly dependent, so that

(16) $\mathbf{u}_1 = c\mathbf{u}_2$

for some constant c. Multiplying both sides of (16) by $\mathbf{A}$ and using the fact that $\mathbf{u}_1$ and $\mathbf{u}_2$ are eigenvectors with corresponding eigenvalues r_1 and r_2, we obtain

(17) $r_1 \mathbf{u}_1 = c r_2 \mathbf{u}_2.$

Next we multiply (16) by r_2 and then subtract from (17) to get

$$(r_1 - r_2)\mathbf{u}_1 = \mathbf{0}.$$

Since $\mathbf{u}_1$ is not the zero vector, we must have $r_1 = r_2$. But this violates the assumption that the eigenvalues are distinct! Hence $\mathbf{u}_1$ and $\mathbf{u}_2$ are linearly independent.

The cases $2 < m \le n$ follow by induction. The details of the proof are left as an exercise. ◄◄◄

Combining Theorems 5 and 6, we get the following corollary.

◣ n DISTINCT EIGENVALUES

Corollary 1.   If the $n \times n$ constant matrix $\mathbf{A}$ has n distinct eigenvalues $r_1, \ldots, r_n$ and $\mathbf{u}_i$ is an eigenvector associated with r_i, then

$$\{e^{r_1 t}\mathbf{u}_1, \ldots, e^{r_n t}\mathbf{u}_n\}$$

is a fundamental solution set for the homogeneous system $\mathbf{x}' = \mathbf{A}\mathbf{x}$.

● **EXAMPLE 4**   Solve the initial value problem

(18) $\mathbf{x}'(t) = \begin{bmatrix} 1 & 2 & -1 \\ 1 & 0 & 1 \\ 4 & -4 & 5 \end{bmatrix} \mathbf{x}(t), \qquad \mathbf{x}(0) = \begin{bmatrix} -1 \\ 0 \\ 0 \end{bmatrix}.$

Solution In Example 2 we showed that the 3×3 coefficient matrix $\mathbf{A}$ has the three distinct eigenvalues $r_1 = 1$, $r_2 = 2$, and $r_3 = 3$. If we set $s = 1$ in equations (9), (10), and (11), we obtain the corresponding eigenvectors

$$\mathbf{u}_1 = \begin{bmatrix} -1 \\ 1 \\ 2 \end{bmatrix}, \qquad \mathbf{u}_2 = \begin{bmatrix} -2 \\ 1 \\ 4 \end{bmatrix}, \qquad \mathbf{u}_3 = \begin{bmatrix} -1 \\ 1 \\ 4 \end{bmatrix},$$

whose linear independence is guaranteed by Theorem 6. Hence a general solution to (18) is

$$(19) \qquad \mathbf{x}(t) = c_1 e^t \begin{bmatrix} -1 \\ 1 \\ 2 \end{bmatrix} + c_2 e^{2t} \begin{bmatrix} -2 \\ 1 \\ 4 \end{bmatrix} + c_3 e^{3t} \begin{bmatrix} -1 \\ 1 \\ 4 \end{bmatrix}$$

$$= \begin{bmatrix} -e^t & -2e^{2t} & -e^{3t} \\ e^t & e^{2t} & e^{3t} \\ 2e^t & 4e^{2t} & 4e^{3t} \end{bmatrix} \begin{bmatrix} c_1 \\ c_2 \\ c_3 \end{bmatrix}.$$

To satisfy the initial condition in (18), we solve

$$\mathbf{x}(0) = \begin{bmatrix} -1 & -2 & -1 \\ 1 & 1 & 1 \\ 2 & 4 & 4 \end{bmatrix} \begin{bmatrix} c_1 \\ c_2 \\ c_3 \end{bmatrix} = \begin{bmatrix} -1 \\ 0 \\ 0 \end{bmatrix}$$

and find that $c_1 = 0$, $c_2 = 1$, and $c_3 = -1$. Inserting these values into (19) gives the desired solution. ●

There is a special class of $n \times n$ matrices that *always* have real eigenvalues and *always* have n linearly independent eigenvectors. These are the real symmetric matrices.

REAL SYMMETRIC MATRICES

Definition 4. A **real symmetric matrix** $\mathbf{A}$ is a matrix with real entries that satisfies $\mathbf{A}^T = \mathbf{A}$.

Taking the transpose of a matrix interchanges its rows and columns, which is equivalent to "flipping" the matrix about its main diagonal. Consequently $\mathbf{A}^T = \mathbf{A}$ if and only if $\mathbf{A}$ is symmetric about its main diagonal.

If $\mathbf{A}$ is an $n \times n$ real symmetric matrix, it is known[†] that there always exist n linearly independent eigenvectors. Thus, Theorem 5 applies, and a general solution to $\mathbf{x}' = \mathbf{A}\mathbf{x}$ is given by (13).

[†] See *Elementary Linear Algebra*, Second Edition, by A. Wayne Roberts, Benjamin/Cummings Publishing Co., Menlo Park, California, 1985, Chapter 6.

● **EXAMPLE 5**   Find a general solution of

(20) $\mathbf{x}'(t) = \mathbf{A}\mathbf{x}(t)$,   where $\mathbf{A} = \begin{bmatrix} 1 & -2 & 2 \\ -2 & 1 & 2 \\ 2 & 2 & 1 \end{bmatrix}$.

Solution   Since $\mathbf{A}$ is symmetric, we are assured that $\mathbf{A}$ has three linearly independent eigenvectors. To find them, we first compute the characteristic equation for $\mathbf{A}$:

$$|\mathbf{A} - r\mathbf{I}| = \begin{vmatrix} 1 - r & -2 & 2 \\ -2 & 1 - r & 2 \\ 2 & 2 & 1 - r \end{vmatrix} = -(r - 3)^2(r + 3) = 0.$$

Thus the eigenvalues of $\mathbf{A}$ are $r_1 = r_2 = 3$ and $r_3 = -3$.

Notice that the eigenvalue $r = 3$ has multiplicity 2 when considered as a root of the characteristic equation. Therefore, we must find *two* linearly independent eigenvectors associated with $r = 3$. Substituting $r = 3$ in $(\mathbf{A} - r\mathbf{I})\mathbf{u} = \mathbf{0}$ gives

$$\begin{bmatrix} -2 & -2 & 2 \\ -2 & -2 & 2 \\ 2 & 2 & -2 \end{bmatrix} \begin{bmatrix} u_1 \\ u_2 \\ u_3 \end{bmatrix} = \begin{bmatrix} 0 \\ 0 \\ 0 \end{bmatrix}.$$

Since this system is equivalent to the single equation $-u_1 - u_2 + u_3 = 0$, we can obtain its solutions by assigning an arbitrary value to u_2, say $u_2 = v$, and an arbitrary value to u_3, say $u_3 = s$. Solving for u_1, we find $u_1 = u_3 - u_2 = s - v$. Therefore, the eigenvectors associated with $r_1 = r_2 = 3$ can be expressed as

$$\mathbf{u} = \begin{bmatrix} s - v \\ v \\ s \end{bmatrix} = s \begin{bmatrix} 1 \\ 0 \\ 1 \end{bmatrix} + v \begin{bmatrix} -1 \\ 1 \\ 0 \end{bmatrix}.$$

By first taking $s = 1, v = 0$ and then taking $s = 0, v = 1$, we get the two linearly independent eigenvectors

(21) $\mathbf{u}_1 = \begin{bmatrix} 1 \\ 0 \\ 1 \end{bmatrix}$, $\mathbf{u}_2 = \begin{bmatrix} -1 \\ 1 \\ 0 \end{bmatrix}$.

For $r_3 = -3$, we solve

$$(\mathbf{A} + 3\mathbf{I})\mathbf{u} = \begin{bmatrix} 4 & -2 & 2 \\ -2 & 4 & 2 \\ 2 & 2 & 4 \end{bmatrix} \begin{bmatrix} u_1 \\ u_2 \\ u_3 \end{bmatrix} = \begin{bmatrix} 0 \\ 0 \\ 0 \end{bmatrix},$$

to obtain the eigenvectors $\mathrm{col}(-s, -s, s)$. Taking $s = 1$ gives

$$\mathbf{u}_3 = \begin{bmatrix} -1 \\ -1 \\ 1 \end{bmatrix}.$$

Since the eigenvectors $\mathbf{u}_1$, $\mathbf{u}_2$, and $\mathbf{u}_3$ are linearly independent, a general solution to (20) is

$$\mathbf{x}(t) = c_1 e^{3t} \begin{bmatrix} 1 \\ 0 \\ 1 \end{bmatrix} + c_2 e^{3t} \begin{bmatrix} -1 \\ 1 \\ 0 \end{bmatrix} + c_3 e^{-3t} \begin{bmatrix} -1 \\ -1 \\ 1 \end{bmatrix}. \quad \bullet$$

It is possible for a matrix to have a repeated eigenvalue but not have two linearly independent corresponding eigenvectors. In particular, the matrix

(22) $\qquad \mathbf{A} = \begin{bmatrix} 1 & -1 \\ 4 & -3 \end{bmatrix}$

has the repeated eigenvalue $r_1 = r_2 = -1$, but all the eigenvectors associated with $r = -1$ are of the form $\mathbf{u} = s\,\text{col}(1, 2)$. Consequently, no two eigenvectors are linearly independent.

Analogous to the situation for a scalar equation, the system $\mathbf{x}' = \mathbf{A}\mathbf{x}$, where $\mathbf{A}$ is given in (22), has two linearly independent solutions of the form

$$\mathbf{x}_1(t) = e^{-t}\mathbf{u}_1, \qquad \mathbf{x}_2(t) = te^{-t}\mathbf{u}_1 + e^{-t}\mathbf{u}_2.$$

Here $\mathbf{u}_1$ is an eigenvector of $\mathbf{A}$, say $\mathbf{u}_1 = \text{col}(1, 2)$, and $\mathbf{u}_2$ is determined by substituting $\mathbf{x}_2(t)$ into $\mathbf{x}' = \mathbf{A}\mathbf{x}$ and solving (see Problem 25).

When a matrix $\mathbf{A}$ has an eigenvalue r of multiplicity 3, then it can be shown that there exist three linearly independent solutions to $\mathbf{x}' = \mathbf{A}\mathbf{x}$ of the form

$$\mathbf{x}_1(t) = e^{rt}\mathbf{k}_1, \qquad \mathbf{x}_2(t) = te^{rt}\mathbf{k}_2 + e^{rt}\mathbf{k}_3,$$

$$\mathbf{x}_3(t) = \frac{t^2}{2}e^{rt}\mathbf{k}_4 + te^{rt}\mathbf{k}_5 + e^{rt}\mathbf{k}_6.$$

The vectors $\mathbf{k}_2$, $\mathbf{k}_4$, or $\mathbf{k}_5$ may be zero, depending on the number of linearly independent eigenvectors associated with the eigenvalue r (see Problems 33–36).

Another approach to handling repeated eigenvalues that involves the concept of the exponential of a matrix is discussed in Section 10.7.

EXERCISES 10.4

In Problems 1 through 8 find the eigenvalues and eigenvectors of the given matrix.

1. $\begin{bmatrix} -4 & 2 \\ 2 & -1 \end{bmatrix}$.

2. $\begin{bmatrix} 6 & -3 \\ 2 & 1 \end{bmatrix}$.

3. $\begin{bmatrix} 1 & -1 \\ 2 & 4 \end{bmatrix}$.

4. $\begin{bmatrix} 1 & 5 \\ 1 & -3 \end{bmatrix}$.

5. $\begin{bmatrix} 1 & 0 & 0 \\ 0 & 0 & 2 \\ 0 & 2 & 0 \end{bmatrix}$.

6. $\begin{bmatrix} 0 & 1 & 1 \\ 1 & 0 & 1 \\ 1 & 1 & 0 \end{bmatrix}$.

7. $\begin{bmatrix} 1 & 0 & 0 \\ 2 & 3 & 1 \\ 0 & 2 & 4 \end{bmatrix}$.

8. $\begin{bmatrix} -3 & 1 & 0 \\ 0 & -3 & 1 \\ 4 & -8 & 2 \end{bmatrix}$.

In Problems 9 through 14 find a general solution of the system $\mathbf{x}'(t) = \mathbf{A}\mathbf{x}(t)$ for the given matrix $\mathbf{A}$.

9. $\mathbf{A} = \begin{bmatrix} -1 & \frac{3}{4} \\ -5 & 3 \end{bmatrix}$.

10. $\mathbf{A} = \begin{bmatrix} 1 & 3 \\ 12 & 1 \end{bmatrix}$.

11. $\mathbf{A} = \begin{bmatrix} 1 & 2 & 2 \\ 2 & 0 & 3 \\ 2 & 3 & 0 \end{bmatrix}$.

12. $\mathbf{A} = \begin{bmatrix} -1 & 1 & 0 \\ 1 & 2 & 1 \\ 0 & 3 & -1 \end{bmatrix}$.

13. $\mathbf{A} = \begin{bmatrix} 1 & 2 & 3 \\ 0 & 1 & 0 \\ 2 & 1 & 2 \end{bmatrix}$.

14. $\mathbf{A} = \begin{bmatrix} -7 & 0 & 6 \\ 0 & 5 & 0 \\ 6 & 0 & 2 \end{bmatrix}$.

In Problems 15 through 20 find a fundamental matrix for the system $\mathbf{x}'(t) = \mathbf{Ax}(t)$ *for the given matrix* **A**.

15. $\mathbf{A} = \begin{bmatrix} -1 & 1 \\ 8 & 1 \end{bmatrix}$.

16. $\mathbf{A} = \begin{bmatrix} 5 & 4 \\ -1 & 0 \end{bmatrix}$.

17. $\mathbf{A} = \begin{bmatrix} 0 & 1 & 0 \\ 0 & 0 & 1 \\ 8 & -14 & 7 \end{bmatrix}$.

18. $\mathbf{A} = \begin{bmatrix} 3 & 1 & -1 \\ 1 & 3 & -1 \\ 3 & 3 & -1 \end{bmatrix}$.

19. $\mathbf{A} = \begin{bmatrix} 2 & 1 & 1 & -1 \\ 0 & -1 & 0 & 1 \\ 0 & 0 & 3 & 1 \\ 0 & 0 & 0 & 7 \end{bmatrix}$.

20. $\mathbf{A} = \begin{bmatrix} 4 & -1 & 0 & 0 \\ 0 & 0 & 0 & 0 \\ 0 & 0 & 2 & -3 \\ 0 & 0 & 1 & -2 \end{bmatrix}$.

21. Using matrix algebra techniques, find a general solution of the system

$$\begin{aligned} x' &= x + 2y - z, \\ y' &= x \quad\;\; + z, \\ z' &= 4x - 4y + 5z. \end{aligned}$$

Compare your solution with the solution to Example 3 in Section 9.2.

22. Using matrix algebra techniques, find a general solution of the system

$$\begin{aligned} x' &= 3x - 4y, \\ y' &= 4x - 7y. \end{aligned}$$

Compare your solution with the solution to Example 1 in Section 9.2.

In Problems 23 through 26 use a linear algebra software package such as DERIVE, MATLAB, MAPLE, or MATHEMATICA to compute the required eigenvalues and eigenvectors and then give a fundamental matrix for the system $\mathbf{x}'(t) = \mathbf{Ax}(t)$ *for the given matrix* **A**.

23. $\mathbf{A} = \begin{bmatrix} 0 & 1.1 & 0 \\ 0 & 0 & 1.3 \\ 0.9 & 1.1 & -6.9 \end{bmatrix}$.

24. $\mathbf{A} = \begin{bmatrix} 2 & 1 & 1 \\ -1 & 1 & 0 \\ 3 & 3 & 3 \end{bmatrix}$.

25. $\mathbf{A} = \begin{bmatrix} 0 & 1 & 0 & 0 \\ 0 & 0 & 1 & 0 \\ 0 & 0 & 0 & 1 \\ 2 & -6 & 3 & 3 \end{bmatrix}$.

26. $\mathbf{A} = \begin{bmatrix} 0 & 1 & 0 & 0 \\ 1 & -1 & 0 & 0 \\ 0 & 0 & 0 & 1 \\ 0 & 0 & -2 & 4 \end{bmatrix}$.

In Problems 27 through 30 solve the given initial value problem.

27. $\mathbf{x}'(t) = \begin{bmatrix} 1 & 3 \\ 3 & 1 \end{bmatrix} \mathbf{x}(t), \qquad \mathbf{x}(0) = \begin{bmatrix} 3 \\ 1 \end{bmatrix}$.

28. $\mathbf{x}'(t) = \begin{bmatrix} 6 & -3 \\ 2 & 1 \end{bmatrix} \mathbf{x}(t), \qquad \mathbf{x}(0) = \begin{bmatrix} -10 \\ -6 \end{bmatrix}$.

29. $\mathbf{x}'(t) = \begin{bmatrix} 1 & -2 & 2 \\ -2 & 1 & -2 \\ 2 & -2 & 1 \end{bmatrix} \mathbf{x}(t), \qquad \mathbf{x}(0) = \begin{bmatrix} -2 \\ -3 \\ 2 \end{bmatrix}$.

30. $\mathbf{x}'(t) = \begin{bmatrix} 0 & 1 & 1 \\ 1 & 0 & 1 \\ 1 & 1 & 0 \end{bmatrix} \mathbf{x}(t), \qquad \mathbf{x}(0) = \begin{bmatrix} -1 \\ 4 \\ 0 \end{bmatrix}$.

31. **(a)** Show that the matrix

$$\mathbf{A} = \begin{bmatrix} 1 & -1 \\ 4 & -3 \end{bmatrix}$$

has the repeated eigenvalue $r = -1$, and all the eigenvectors are of the form $\mathbf{u} = s\,\mathrm{col}(1, 2)$.

(b) Use the result of part (a) to obtain a nontrivial solution $\mathbf{x}_1(t)$ to the system $\mathbf{x}' = \mathbf{Ax}$.

(c) To obtain a second linearly independent solution to $\mathbf{x}' = \mathbf{Ax}$, try $\mathbf{x}_2(t) = te^{-t}\mathbf{u}_1 + e^{-t}\mathbf{u}_2$. [Hint: Substitute $\mathbf{x}_2$ into the system $\mathbf{x}' = \mathbf{Ax}$ and derive the relations

$$(\mathbf{A} + \mathbf{I})\mathbf{u}_1 = \mathbf{0}, \qquad (\mathbf{A} + \mathbf{I})\mathbf{u}_2 = \mathbf{u}_1.$$

Since $\mathbf{u}_1$ must be an eigenvector, set $\mathbf{u}_1 = \mathrm{col}(1, 2)$ and solve for $\mathbf{u}_2$.]

32. Use the method discussed in Problem 31 to find a general solution to the system

$$\mathbf{x}'(t) = \begin{bmatrix} 5 & -3 \\ 3 & -1 \end{bmatrix} \mathbf{x}(t).$$

33. **(a)** Show that the matrix

$$\mathbf{A} = \begin{bmatrix} 2 & 1 & 6 \\ 0 & 2 & 5 \\ 0 & 0 & 2 \end{bmatrix}$$

has the repeated eigenvalue $r = 2$ with multiplicity 3, and that all the eigenvectors of **A** are of the form $\mathbf{u} = s\,\text{col}(1, 0, 0)$.

(b) Use the result of part (a) to obtain a solution to the system $\mathbf{x}' = \mathbf{A}\mathbf{x}$ of the form $\mathbf{x}_1(t) = e^{2t}\mathbf{u}_1$.

(c) To obtain a second linearly independent solution to $\mathbf{x}' = \mathbf{A}\mathbf{x}$, try $\mathbf{x}_2(t) = te^{2t}\mathbf{u}_1 + e^{2t}\mathbf{u}_2$. [Hint: Show that $\mathbf{u}_1$ and $\mathbf{u}_2$ must satisfy

$$(\mathbf{A} - 2\mathbf{I})\mathbf{u}_1 = \mathbf{0}, \qquad (\mathbf{A} - 2\mathbf{I})\mathbf{u}_2 = \mathbf{u}_1.]$$

(d) To obtain a third linearly independent solution to $\mathbf{x}' = \mathbf{A}\mathbf{x}$, try

$$\mathbf{x}_3(t) = \frac{t^2}{2}e^{2t}\mathbf{u}_1 + te^{2t}\mathbf{u}_2 + e^{2t}\mathbf{u}_3.$$

[Hint: Show that $\mathbf{u}_1$, $\mathbf{u}_2$, and $\mathbf{u}_3$ must satisfy

$$(\mathbf{A} - 2\mathbf{I})\mathbf{u}_1 = \mathbf{0}, \qquad (\mathbf{A} - 2\mathbf{I})\mathbf{u}_2 = \mathbf{u}_1,$$

$$(\mathbf{A} - 2\mathbf{I})\mathbf{u}_3 = \mathbf{u}_2.]$$

34. Use the method discussed in Problem 33 to find a general solution to the system

$$\mathbf{x}'(t) = \begin{bmatrix} 3 & -2 & 1 \\ 2 & -1 & 1 \\ -4 & 4 & 1 \end{bmatrix}\mathbf{x}(t).$$

35. **(a)** Show that the matrix

$$\mathbf{A} = \begin{bmatrix} 2 & 1 & 1 \\ 1 & 2 & 1 \\ -2 & -2 & -1 \end{bmatrix}$$

has the repeated eigenvalue $r = 1$ of multiplicity 3 and that all the eigenvectors of **A** are of the form $\mathbf{u} = s\,\text{col}(-1, 1, 0) + v\,\text{col}(-1, 0, 1)$.

(b) Use the result of part (a) to obtain two linearly independent solutions to the system $\mathbf{x}' = \mathbf{A}\mathbf{x}$ of the form

$$\mathbf{x}_1(t) = e^t\mathbf{u}_1 \quad \text{and} \quad \mathbf{x}_2(t) = e^t\mathbf{u}_2.$$

(c) To obtain a third linearly independent solution to $\mathbf{x}' = \mathbf{A}\mathbf{x}$, try

$$\mathbf{x}_3(t) = te^t\mathbf{u}_3 + e^t\mathbf{u}_4.$$

[Hint: Show that $\mathbf{u}_3$ and $\mathbf{u}_4$ must satisfy

$$(\mathbf{A} - \mathbf{I})\mathbf{u}_3 = \mathbf{0}, \qquad (\mathbf{A} - \mathbf{I})\mathbf{u}_4 = \mathbf{u}_3.$$

Choose $\mathbf{u}_3$, an eigenvector of **A**, so that you can solve for $\mathbf{u}_4$.]

36. Use the method discussed in Problem 35 to find a general solution to the system

$$\mathbf{x}'(t) = \begin{bmatrix} 1 & 3 & -2 \\ 0 & 7 & -4 \\ 0 & 9 & -5 \end{bmatrix}\mathbf{x}(t).$$

37. Use the substitution $x_1 = y$, $x_2 = y'$ to convert the linear equation $ay'' + by' + cy = 0$, where a, b, and c are constants, into a normal system. Show that the characteristic equation for this system is the same as the auxiliary equation for the original equation.

38. Show that the **Cauchy-Euler system**

$$t\mathbf{x}'(t) = \mathbf{A}\mathbf{x}(t),$$

where **A** is a constant matrix, has nontrivial solutions of the form $\mathbf{x}(t) = t^r\mathbf{u}$ if and only if r is an eigenvalue of **A** and **u** is a corresponding eigenvector.

In Problems 39 and 40 use the result of Problem 38 to find a general solution of the given system.

39. $t\mathbf{x}'(t) = \begin{bmatrix} 1 & 3 \\ -1 & 5 \end{bmatrix}\mathbf{x}(t), \quad t > 0.$

40. $t\mathbf{x}'(t) = \begin{bmatrix} -4 & 2 \\ 2 & -1 \end{bmatrix}\mathbf{x}(t), \quad t > 0.$

41. **Mixing Between Interconnected Tanks.** Two tanks, each holding 50 L of liquid are interconnected by pipes, with liquid flowing from tank A into tank B at a rate of 4 L/min and from tank B into tank A at 1 L/min (see Figure 10.1). The liquid inside each tank is kept well stirred. Pure water flows into tank A at a rate of 3 L/min, and the solution flows out of tank B at 3 L/min. If, initially, tank A contains 25 kg of salt and tank B contains no salt (only water), determine the mass of salt in each tank at time $t \geq 0$. Graph on the same axes, the two quantities $x_1(t)$ and $x_2(t)$ where $x_1(t)$ is the mass of salt in tank A and $x_2(t)$ the mass in tank B.

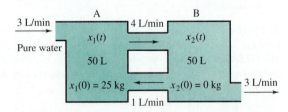

Figure 10.1 Mixing problem for interconnected tanks

42. **Mixing with a Common Drain.** Two tanks, each holding 1 L of liquid, are connected by a pipe through which liquid flows from tank A into tank B at a rate of $3 - \alpha$ L/min. The liquid inside each tank is kept well stirred. Pure water flows into tank A at a rate of 3 L/min. Solution flows out of tank A at α L/min and out of tank B at $3 - \alpha$ L/min. If, initially, tank B contains no salt (only water) and tank A contains 1 kg of salt, determine the mass of salt in each tank at time $t \geq 0$. How does the mass of salt in tank A depend on the choice of α? What is the maximum mass of salt in tank B?

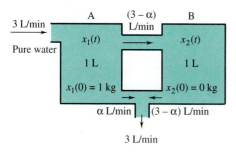

3 L/min Pure water A $x_1(t)$ 1 L $x_1(0) = 1$ kg

$(3 - \alpha)$ L/min

B $x_2(t)$ 1 L $x_2(0) = 0$ kg

α L/min $(3 - \alpha)$ L/min

3 L/min

Figure 10.2 Mixing problem for a common drain, $0 < \alpha < 3$

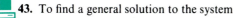

43. To find a general solution to the system

$$\mathbf{x}' = \mathbf{Ax} = \begin{bmatrix} 1 & 3 & -1 \\ 3 & 0 & 1 \\ -1 & 1 & 2 \end{bmatrix} \mathbf{x}$$

proceed as follows:

(a) Use a numerical root finding procedure to approximate the eigenvalues.
(b) If r is an eigenvalue, then let $\mathbf{u} = \text{col}(u_1, u_2, u_3)$ be an eigenvector associated with r. To solve for $\mathbf{u}$, assume $u_1 = 1$. (If not u_1, then either u_2 or u_3 may be chosen to be 1. Why?) Now solve the system

$$(\mathbf{A} - r\mathbf{I}) \begin{bmatrix} 1 \\ u_2 \\ u_3 \end{bmatrix} = \begin{bmatrix} 0 \\ 0 \\ 0 \end{bmatrix}$$

for u_2 and u_3. Use this procedure to find approximations for three linearly independent eigenvectors for $\mathbf{A}$.

(c) Use these approximations to give a general solution to the system.

44. To complete the proof of Theorem 6, assume the induction hypothesis that $\mathbf{u}_1, \ldots, \mathbf{u}_k$, $2 \leq k \leq n - 1$, are linearly independent.

(a) Show that if

$$c_1\mathbf{u}_1 + \cdots + c_k\mathbf{u}_k + c_{k+1}\mathbf{u}_{k+1} = \mathbf{0},$$

then

$$c_1(r_1 - r_{k+1})\mathbf{u}_1 + \cdots + c_k(r_k - r_{k+1})\mathbf{u}_k = \mathbf{0}.$$

(b) Use the result of part (a) and the induction hypothesis to conclude that $\mathbf{u}_1, \ldots, \mathbf{u}_{k+1}$ are linearly independent. The theorem follows by induction.

45. Stability. A homogeneous system $\mathbf{x}' = \mathbf{Ax}$ with constant coefficients is **stable** if it has a fundamental matrix whose entries all remain bounded as $t \to +\infty$. (It will follow from Lemma 1 in Section 10.7 that if one fundamental matrix of the system has this property, then all fundamental matrices for the system do.) Otherwise, the system is **unstable**. A stable system is **asymptotically stable** if all solutions approach the zero solution as $t \to +\infty$. Stability is discussed in more detail in Chapter 13[†].

(a) Show that if $\mathbf{A}$ has all distinct real eigenvalues, then $\mathbf{x}'(t) = \mathbf{Ax}(t)$ is stable if and only if all eigenvalues are nonpositive.
(b) Show that if $\mathbf{A}$ has all distinct real eigenvalues, then $\mathbf{x}'(t) = \mathbf{Ax}(t)$ is asymptotically stable if and only if all eigenvalues are negative.
(c) Show that in parts (a) and (b) we can replace "has distinct real eigenvalues" by "is symmetric" and the statements are still true.

•10.5• COMPLEX EIGENVALUES

In the previous section we showed that the homogeneous system

(1) $\mathbf{x}'(t) = \mathbf{Ax}(t),$

where $\mathbf{A}$ is a constant $n \times n$ matrix, has a solution of the form $\mathbf{x}(t) = e^{rt}\mathbf{u}$ if and only if r is an eigenvalue of $\mathbf{A}$ and $\mathbf{u}$ is a corresponding eigenvector. In this section we show how

---

[†] All references to Chapters 12–14 refer to the expanded text *Fundamentals of Differential Equations and Boundary Value Problems*.

to obtain two real vector solutions to system (1) when $\mathbf{A}$ is real and has a pair[†] of complex conjugate eigenvalues $\alpha + i\beta$ and $\alpha - i\beta$.

Suppose that $r_1 = \alpha + i\beta$ (α and β real numbers) is an eigenvalue of $\mathbf{A}$ with corresponding eigenvector $\mathbf{z} = \mathbf{a} + i\mathbf{b}$, where $\mathbf{a}$ and $\mathbf{b}$ are real constant vectors. We first observe that the complex conjugate of $\mathbf{z}$, namely, $\bar{\mathbf{z}} := \mathbf{a} - i\mathbf{b}$, is an eigenvector associated with the eigenvalue $r_2 = \alpha - i\beta$. To see this, note that taking the complex conjugate of $(\mathbf{A} - r_1\mathbf{I})\mathbf{z} = \mathbf{0}$ yields $(\mathbf{A} - \bar{r}_1\mathbf{I})\bar{\mathbf{z}} = \mathbf{0}$, because the conjugate of the product is the product of the conjugates and $\mathbf{A}$ and $\mathbf{I}$ have real entries ($\bar{\mathbf{A}} = \mathbf{A}$, $\bar{\mathbf{I}} = \mathbf{I}$). Since $r_2 = \bar{r}_1$, we see that $\bar{\mathbf{z}}$ is an eigenvector associated with r_2. Therefore, two linearly independent complex vector solutions to (1) are

(2) $\mathbf{w}_1(t) = e^{r_1 t}\mathbf{z} = e^{(\alpha + i\beta)t}(\mathbf{a} + i\mathbf{b})$,

(3) $\mathbf{w}_2(t) = e^{r_2 t}\bar{\mathbf{z}} = e^{(\alpha - i\beta)t}(\mathbf{a} - i\mathbf{b})$.

As in Section 4.5, where we handled complex roots to the auxiliary equation, let's use one of these complex solutions and Euler's formula to obtain two real vector solutions. With the aid of Euler's formula we rewrite $\mathbf{w}_1(t)$ as

$$\mathbf{w}_1(t) = e^{\alpha t}(\cos \beta t + i \sin \beta t)(\mathbf{a} + i\mathbf{b})$$
$$= e^{\alpha t}\{(\cos \beta t\,\mathbf{a} - \sin \beta t\,\mathbf{b}) + i(\sin \beta t\,\mathbf{a} + \cos \beta t\,\mathbf{b})\}.$$

We have thereby expressed $\mathbf{w}_1(t)$ in the form $\mathbf{w}_1(t) = \mathbf{x}_1(t) + i\mathbf{x}_2(t)$, where $\mathbf{x}_1(t)$ and $\mathbf{x}_2(t)$ are the two real vector functions

(4) $\mathbf{x}_1(t) := e^{\alpha t} \cos \beta t\,\mathbf{a} - e^{\alpha t} \sin \beta t\,\mathbf{b}$,

(5) $\mathbf{x}_2(t) := e^{\alpha t} \sin \beta t\,\mathbf{a} + e^{\alpha t} \cos \beta t\,\mathbf{b}$.

Since $\mathbf{w}_1(t)$ is a solution to (1), then

$$\mathbf{x}_1'(t) + i\mathbf{x}_2'(t) = \mathbf{w}_1'(t) = \mathbf{A}\mathbf{w}_1(t) = \mathbf{A}\mathbf{x}_1(t) + i\mathbf{A}\mathbf{x}_2(t).$$

Equating the real and imaginary parts yields

$$\mathbf{x}_1'(t) = \mathbf{A}\mathbf{x}_1(t) \quad \text{and} \quad \mathbf{x}_2'(t) = \mathbf{A}\mathbf{x}_2(t).$$

Hence $\mathbf{x}_1(t)$ and $\mathbf{x}_2(t)$ are real vector solutions to (1) associated with the complex conjugate eigenvalues $\alpha \pm i\beta$. Since $\mathbf{a}$ and $\mathbf{b}$ are not both the zero vector, it can be shown that $\mathbf{x}_1(t)$ and $\mathbf{x}_2(t)$ are linearly independent vector functions on $(-\infty, \infty)$ (see Problem 15).

Let's summarize our findings.

> ### COMPLEX EIGENVALUES
>
> If the real matrix $\mathbf{A}$ has complex conjugate eigenvalues $\alpha \pm i\beta$ with corresponding eigenvectors $\mathbf{a} \pm i\mathbf{b}$, then two linearly independent real vector solutions to $\mathbf{x}'(t) = \mathbf{A}\mathbf{x}(t)$ are
>
> **(6)** $e^{\alpha t} \cos \beta t\,\mathbf{a} - e^{\alpha t} \sin \beta t\,\mathbf{b}$,
>
> **(7)** $e^{\alpha t} \sin \beta t\,\mathbf{a} + e^{\alpha t} \cos \beta t\,\mathbf{b}$.

[†] Recall that the complex roots of a polynomial equation with real coefficients must occur in complex conjugate pairs.

● **EXAMPLE 1**   Find a general solution of

(8) $$\mathbf{x}'(t) = \begin{bmatrix} -1 & 2 \\ -1 & -3 \end{bmatrix} \mathbf{x}(t).$$

Solution   The characteristic equation for $\mathbf{A}$ is

$$|\mathbf{A} - r\mathbf{I}| = \begin{vmatrix} -1-r & 2 \\ -1 & -3-r \end{vmatrix} = r^2 + 4r + 5 = 0.$$

Hence $\mathbf{A}$ has eigenvalues $r = -2 \pm i$.

To find a general solution we need only find an eigenvector associated with the eigenvalue $r = -2 + i$. Substituting $r = -2 + i$ into $(\mathbf{A} - r\mathbf{I})\mathbf{z} = \mathbf{0}$ gives

$$\begin{bmatrix} 1-i & 2 \\ -1 & -1-i \end{bmatrix} \begin{bmatrix} z_1 \\ z_2 \end{bmatrix} = \begin{bmatrix} 0 \\ 0 \end{bmatrix}.$$

The solutions are $z_1 = 2s$ and $z_2 = (-1 + i)s$, with s arbitrary. Hence, the eigenvectors associated with $r = -2 + i$ are $\mathbf{z} = s\,\text{col}(2, -1 + i)$. Taking $s = 1$ gives the eigenvector

$$\mathbf{z} = \begin{bmatrix} 2 \\ -1+i \end{bmatrix} = \begin{bmatrix} 2 \\ -1 \end{bmatrix} + i \begin{bmatrix} 0 \\ 1 \end{bmatrix}.$$

We have found that $\alpha = -2$, $\beta = 1$, $\mathbf{a} = \text{col}(2, -1)$, and $\mathbf{b} = \text{col}(0, 1)$, and so a general solution to (8) is

$$\mathbf{x}(t) = c_1 \left\{ e^{-2t} \cos t \begin{bmatrix} 2 \\ -1 \end{bmatrix} - e^{-2t} \sin t \begin{bmatrix} 0 \\ 1 \end{bmatrix} \right\}$$

$$+ c_2 \left\{ e^{-2t} \sin t \begin{bmatrix} 2 \\ -1 \end{bmatrix} + e^{-2t} \cos t \begin{bmatrix} 0 \\ 1 \end{bmatrix} \right\}$$

$$= c_1 \begin{bmatrix} 2e^{-2t} \cos t \\ -e^{-2t}(\cos t + \sin t) \end{bmatrix} + c_2 \begin{bmatrix} 2e^{-2t} \sin t \\ e^{-2t}(\cos t - \sin t) \end{bmatrix}. \quad ●$$

Complex eigenvalues occur in modeling coupled spring-mass systems. For example, the motion of the spring-mass system illustrated in Figure 10.3 is governed by the second order system

(9) $$\begin{aligned} m_1 x_1'' &= -k_1 x_1 + k_2(x_2 - x_1), \\ m_2 x_2'' &= -k_2(x_2 - x_1) - k_3 x_2, \end{aligned}$$

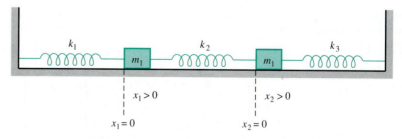

Figure 10.3 Coupled spring-mass system with fixed ends

where x_1 and x_2 represent the displacement of the masses m_1 and m_2 to the right of their equilibrium positions and k_1, k_2, k_3 are the spring constants of the three springs (see also Example 1 in Section 9.4 for a derivation of these equations). If we introduce the new variables $y_1 := x_1, y_2 := x_1', y_3 := x_2, y_4 := x_2'$, then we can rewrite the system in the normal form

$$(10) \qquad \mathbf{y}'(t) = \mathbf{A}\mathbf{y}(t) = \begin{bmatrix} 0 & 1 & 0 & 0 \\ -(k_1 + k_2)/m_1 & 0 & k_2/m_1 & 0 \\ 0 & 0 & 0 & 1 \\ k_2/m_2 & 0 & -(k_2 + k_3)/m_2 & 0 \end{bmatrix} \mathbf{y}(t).$$

For such a system it turns out that $\mathbf{A}$ has only imaginary eigenvalues and they occur in complex conjugate pairs: $\pm\beta_1 i, \pm\beta_2 i$. Hence, any solution will consist of sums of sine and cosine functions. The frequencies of these functions

$$f_1 := \frac{\beta_1}{2\pi} \quad \text{and} \quad f_2 := \frac{\beta_2}{2\pi}$$

are called the **normal** or **natural frequencies** of the system (β_1 and β_2 are the **angular frequencies** of the system).

● **EXAMPLE 2** Determine the normal frequencies for the coupled spring-mass system governed by system (10) when $m_1 = m_2 = 1$ kg, $k_1 = 1$ N/m, $k_2 = 2$ N/m, and $k_3 = 3$ N/m.

Solution To find the eigenvalues of $\mathbf{A}$, we must solve the characteristic equation

$$|\mathbf{A} - r\mathbf{I}| = \begin{vmatrix} -r & 1 & 0 & 0 \\ -3 & -r & 2 & 0 \\ 0 & 0 & -r & 1 \\ 2 & 0 & -5 & -r \end{vmatrix} = r^4 + 8r^2 + 11 = 0.$$

From the quadratic formula we find $r^2 = -4 \pm \sqrt{5}$, and so the four eigenvalues of $\mathbf{A}$ are $\pm i\sqrt{4 - \sqrt{5}}$ and $\pm i\sqrt{4 + \sqrt{5}}$. Hence the two normal frequencies for this system are

$$\frac{\sqrt{4 - \sqrt{5}}}{2\pi} \approx 0.211 \quad \text{and} \quad \frac{\sqrt{4 + \sqrt{5}}}{2\pi} \approx 0.397. \quad ●$$

EXERCISES 10.5

In Problems 1 through 4 find a general solution of the system $\mathbf{x}'(t) = \mathbf{A}\mathbf{x}(t)$ for the given matrix $\mathbf{A}$.

1. $\mathbf{A} = \begin{bmatrix} 2 & -4 \\ 2 & -2 \end{bmatrix}$.

2. $\mathbf{A} = \begin{bmatrix} -2 & -5 \\ 1 & 2 \end{bmatrix}$.

3. $\mathbf{A} = \begin{bmatrix} 1 & 2 & -1 \\ 0 & 1 & 1 \\ 0 & -1 & 1 \end{bmatrix}$.

4. $\mathbf{A} = \begin{bmatrix} 5 & -5 & -5 \\ -1 & 4 & 2 \\ 3 & -5 & -3 \end{bmatrix}$.

In Problems 5 through 8 find a fundamental matrix for the system $\mathbf{x}'(t) = \mathbf{A}\mathbf{x}(t)$ for the given matrix $\mathbf{A}$.

5. $\mathbf{A} = \begin{bmatrix} -1 & -2 \\ 8 & -1 \end{bmatrix}$.

6. $\mathbf{A} = \begin{bmatrix} -2 & -2 \\ 4 & 2 \end{bmatrix}$.

7. $\mathbf{A} = \begin{bmatrix} 0 & 0 & 1 \\ 0 & 0 & -1 \\ 0 & 1 & 0 \end{bmatrix}$.

8. A = $\begin{bmatrix} 0 & 1 & 0 & 0 \\ 1 & 0 & 0 & 0 \\ 0 & 0 & 0 & 1 \\ 0 & 0 & -13 & 4 \end{bmatrix}$.

*In Problems 9 through 12 use a linear algebra software package to compute the required eigenvalues and eigenvectors for the given matrix **A** and then give a fundamental matrix for the system $\mathbf{x}'(t) = \mathbf{A}\mathbf{x}(t)$.*

9. A = $\begin{bmatrix} 0 & 1 & 1 \\ -1 & 0 & 1 \\ -1 & -1 & 0 \end{bmatrix}$.

10. A = $\begin{bmatrix} 0 & 1 & 0 & 0 \\ 0 & 0 & 1 & 0 \\ 0 & 0 & 0 & 1 \\ 13 & -4 & -12 & 4 \end{bmatrix}$.

11. A = $\begin{bmatrix} 0 & 1 & 0 & 0 \\ 0 & 0 & 1 & 0 \\ 0 & 0 & 0 & 1 \\ -2 & 2 & -3 & 2 \end{bmatrix}$.

12. A = $\begin{bmatrix} 1 & 0 & 0 & 0 & 0 \\ 0 & 0 & 1 & 0 & 0 \\ 0 & 1 & 0 & 0 & 0 \\ 0 & 0 & 0 & 0 & 1 \\ 0 & 0 & 0 & -29 & -4 \end{bmatrix}$.

In Problems 13 and 14 find the solution to the given system that satisfies the given initial condition.

13. $\mathbf{x}'(t) = \begin{bmatrix} -3 & -1 \\ 2 & -1 \end{bmatrix} \mathbf{x}(t),$

(a) $\mathbf{x}(0) = \begin{bmatrix} -1 \\ 0 \end{bmatrix}.$ **(b)** $\mathbf{x}(\pi) = \begin{bmatrix} 1 \\ -1 \end{bmatrix}.$

(c) $\mathbf{x}(-2\pi) = \begin{bmatrix} 2 \\ 1 \end{bmatrix}.$ **(d)** $\mathbf{x}(\pi/2) = \begin{bmatrix} 0 \\ 1 \end{bmatrix}.$

14. $\mathbf{x}'(t) = \begin{bmatrix} 1 & 0 & -1 \\ 0 & 2 & 0 \\ 1 & 0 & 1 \end{bmatrix} \mathbf{x}(t),$

(a) $\mathbf{x}(0) = \begin{bmatrix} -2 \\ 2 \\ -1 \end{bmatrix}.$ **(b)** $\mathbf{x}(-\pi) = \begin{bmatrix} 0 \\ 1 \\ 1 \end{bmatrix}.$

15. Show that $\mathbf{x}_1(t)$ and $\mathbf{x}_2(t)$ given by equations (4) and (5) are linearly independent on $(-\infty, \infty)$, provided that $\beta \neq 0$ and **a** and **b** are not both the zero vector.

16. Show that $\mathbf{x}_1(t)$ and $\mathbf{x}_2(t)$ given by equations (4) and (5) can be obtained as linear combinations of $\mathbf{w}_1(t)$ and $\mathbf{w}_2(t)$ given by equations (2) and (3). [Hint: Show that

$$\mathbf{x}_1(t) = \frac{\mathbf{w}_1(t) + \mathbf{w}_2(t)}{2}, \qquad \mathbf{x}_2(t) = \frac{\mathbf{w}_1(t) - \mathbf{w}_2(t)}{2i}.$$]

In Problems 17 and 18 use the results of Problem 38 in Exercises 10.4. to find a general solution to the given Cauchy-Euler system for $t > 0$.

17. $t\mathbf{x}'(t) = \begin{bmatrix} -1 & -1 & 0 \\ 2 & -1 & 1 \\ 0 & 1 & -1 \end{bmatrix} \mathbf{x}(t).$

18. $t\mathbf{x}'(t) = \begin{bmatrix} -1 & -1 \\ 9 & -1 \end{bmatrix} \mathbf{x}(t).$

19. For the coupled spring-mass system governed by system (9), assume that $m_1 = m_2 = 1$ kg, $k_1 = k_2 = 2$ N/m, and $k_3 = 3$ N/m. Determine the normal frequencies for this coupled spring-mass system.

20. For the coupled spring-mass system governed by system (9), assume that $m_1 = m_2 = 1$ kg, $k_1 = k_2 = k_3 = 1$ N/m, and assume initially that $x_1(0) = 0$ m, $x_1'(0) = 0$ m/sec, $x_2(0) = 2$ m, and $x_2'(0) = 0$ m/sec. Using matrix algebra techniques, solve this initial value problem. Compare your solution with Example 1 in Section 9.4.

21. RLC Network. The currents in the RLC network given by the schematic diagram in Figure 10.4 are governed by the following equations:

$$4I_2'(t) + 52q_1(t) = 10,$$
$$13I_3(t) + 52q_1(t) = 10,$$
$$I_1(t) = I_2(t) + I_3(t),$$

where $q_1(t)$ is the charge on the capacitor, $I_1(t) = q_1'(t)$, and initially $q_1(0) = 0$ coulombs and $I_1(0) = 0$ amps. Solve for the currents I_1, I_2, and I_3. [Hint: Express as a normal system with $x_1 = I_2$, $x_2 = I_2'$, and $x_3 = I_3$.]

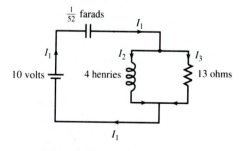

Figure 10.4 RLC network for Problem 21

22. RLC Network. The currents in the RLC network given by the schematic diagram in Figure 10.5 are governed by the following equations:

$$50I'_1(t) + 80I_2(t) = 160,$$
$$50I'_1(t) + 800q_3(t) = 160,$$
$$I_1(t) = I_2(t) + I_3(t),$$

where $q_3(t)$ is the charge on the capacitor, $I_3(t) = q'_3(t)$, and initially $q_3(0) = 0.5$ coulombs and $I_3(0) = 0$ amps. Solve for the currents I_1, I_2, and I_3. [Hint: Express as a normal system with $x_1 = I_1$, $x_2 = I'_1$, and $x_3 = I_2$.]

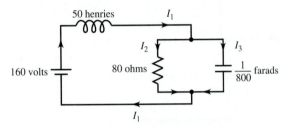

Figure 10.5 RLC network for Problem 22

23. Stability. In Problem 45 of Exercises 10.4 we discussed the idea of stability and asymptotic stability for a linear system of the form $\mathbf{x}'(t) = \mathbf{A}\mathbf{x}(t)$. Assume $\mathbf{A}$ has all distinct eigenvalues (real or complex).

(a) Show that the system is stable if and only if all the eigenvalues of $\mathbf{A}$ have nonpositive real part.

(b) Show that the system is asymptotically stable if and only if all the eigenvalues of $\mathbf{A}$ have negative real part.

• 10.6 • NONHOMOGENEOUS LINEAR SYSTEMS

The techniques discussed in Chapter 4 for finding a particular solution to the nonhomogeneous equation $y'' + p(x)y' + q(x)y = g(x)$ have natural extensions to nonhomogeneous linear systems.

Undetermined Coefficients

The method of undetermined coefficients can be used to find a particular solution to the nonhomogeneous linear system

$$\mathbf{x}'(t) = \mathbf{A}\mathbf{x}(t) + \mathbf{f}(t),$$

when $\mathbf{A}$ is an $n \times n$ *constant* matrix and the entries of $\mathbf{f}(t)$ are polynomials, exponential functions, sines and cosines, or finite sums and products of these functions. We can use Table 4.1 (discussed in Section 4.8 and reproduced inside the back cover) as a *guide* in choosing the form of a particular solution $\mathbf{x}_p(t)$. Some exceptions are discussed in the exercises (see Problems 25–28).

• EXAMPLE 1 Find a general solution of

$$\text{(1)} \qquad \mathbf{x}'(t) = \mathbf{A}\mathbf{x}(t) + t\mathbf{g}, \quad \text{where} \quad \mathbf{A} = \begin{bmatrix} 1 & -2 & 2 \\ -2 & 1 & 2 \\ 2 & 2 & 1 \end{bmatrix} \quad \text{and} \quad \mathbf{g} = \begin{bmatrix} -9 \\ 0 \\ -18 \end{bmatrix}.$$

Solution In Example 5 in Section 10.4 we found that a general solution to the corresponding homogeneous system $\mathbf{x}' = \mathbf{A}\mathbf{x}$ is

$$\text{(2)} \qquad \mathbf{x}_h(t) = c_1 e^{3t}\begin{bmatrix}1\\0\\1\end{bmatrix} + c_2 e^{3t}\begin{bmatrix}-1\\1\\0\end{bmatrix} + c_3 e^{-3t}\begin{bmatrix}-1\\-1\\1\end{bmatrix}.$$

Since the entries in $\mathbf{f}(t) := t\mathbf{g}$ are just linear functions of t, we seek a particular solution of the form[†]

$$\mathbf{x}_p(t) = t\mathbf{a} + \mathbf{b} = t\begin{bmatrix}a_1\\a_2\\a_3\end{bmatrix} + \begin{bmatrix}b_1\\b_2\\b_3\end{bmatrix},$$

where the constant vectors $\mathbf{a}$ and $\mathbf{b}$ are to be determined. Substituting this expression for $\mathbf{x}_p(t)$ into system (1) yields

$$\mathbf{a} = \mathbf{A}(t\mathbf{a} + \mathbf{b}) + t\mathbf{g},$$

which can be written as

$$t(\mathbf{A}\mathbf{a} + \mathbf{g}) + (\mathbf{A}\mathbf{b} - \mathbf{a}) = \mathbf{0}.$$

Setting the "coefficients" of this vector polynomial equal to zero yields the two systems

$$\text{(3)} \qquad \mathbf{A}\mathbf{a} = -\mathbf{g},$$

$$\text{(4)} \qquad \mathbf{A}\mathbf{b} = \mathbf{a}.$$

By Gaussian elimination or by using a linear algebra software package, we can solve (3) for $\mathbf{a}$ and we find $\mathbf{a} = \text{col}(5, 2, 4)$. Next we substitute for $\mathbf{a}$ in (4) and solve for $\mathbf{b}$ to obtain $\mathbf{b} = \text{col}(1, 0, 2)$. Hence, a particular solution for (1) is

$$\text{(5)} \qquad \mathbf{x}_p(t) = t\mathbf{a} + \mathbf{b} = t\begin{bmatrix}5\\2\\4\end{bmatrix} + \begin{bmatrix}1\\0\\2\end{bmatrix} = \begin{bmatrix}5t+1\\2t\\4t+2\end{bmatrix}.$$

A general solution for (1) is $\mathbf{x}(t) = \mathbf{x}_h(t) + \mathbf{x}_p(t)$, where $\mathbf{x}_h(t)$ is given in (2) and $\mathbf{x}_p(t)$ in (5). ●

In the preceding example, the nonhomogeneous term $\mathbf{f}(t)$ was a vector polynomial. If, instead, $\mathbf{f}(t)$ has the form

$$\mathbf{f}(t) = \text{col}(1, t, \sin t),$$

then, using the superposition principle, we would seek a particular solution of the form

$$\mathbf{x}_p(t) = t\mathbf{a} + \mathbf{b} + \sin t\,\mathbf{c} + \cos t\,\mathbf{d}.$$

[†] Notice that none of the terms in $\mathbf{x}_p(t)$ is a solution to the corresponding homogeneous system $\mathbf{x}' = \mathbf{A}\mathbf{x}$.

Similarly, if

$$\mathbf{f}(t) = \text{col}(t, e^t, t^2),$$

we would take

$$\mathbf{x}_p(t) = t^2\mathbf{a} + t\mathbf{b} + \mathbf{c} + e^t\mathbf{d}.$$

Of course we must modify our guess should one of the terms be a solution to the corresponding homogeneous system.

Variation of Parameters

In Section 4.10 we discussed the method of variation of parameters for a second order linear equation. Simply put, the idea is that if a general solution to the homogeneous equation has the form $x_h(t) = c_1 x_1(t) + c_2 x_2(t)$, where $x_1(t)$ and $x_2(t)$ are linearly independent solutions to the homogeneous equation, then a particular solution to the nonhomogeneous equation would have the form $x_p(t) = v_1(t)x_1(t) + v_2(t)x_2(t)$, where $v_1(t)$ and $v_2(t)$ are certain functions of t. A similar idea can be used for systems.

Let $\mathbf{X}(t)$ be a fundamental matrix for the homogeneous system

$$(6) \qquad \mathbf{x}'(t) = \mathbf{A}(t)\mathbf{x}(t),$$

where now *the entries of* $\mathbf{A}$ *may be any continuous functions of* t. Since a general solution to (6) is given by $\mathbf{X}(t)\mathbf{c}$, where $\mathbf{c}$ is a constant $n \times 1$ vector, we seek a particular solution to the nonhomogeneous system

$$(7) \qquad \mathbf{x}'(t) = \mathbf{A}(t)\mathbf{x}(t) + \mathbf{f}(t)$$

of the form

$$(8) \qquad \mathbf{x}_p(t) = \mathbf{X}(t)\mathbf{v}(t),$$

where $\mathbf{v}(t) = \text{col}(v_1(t), \ldots, v_n(t))$ is a vector function of t to be determined.

To derive a formula for $\mathbf{v}(t)$, we first differentiate (8) using the matrix version of the product rule to obtain

$$\mathbf{x}_p'(t) = \mathbf{X}(t)\mathbf{v}'(t) + \mathbf{X}'(t)\mathbf{v}(t).$$

Substituting the expressions for $\mathbf{x}_p(t)$ and $\mathbf{x}_p'(t)$ into (7) yields

$$(9) \qquad \mathbf{X}(t)\mathbf{v}'(t) + \mathbf{X}'(t)\mathbf{v}(t) = \mathbf{A}(t)\mathbf{X}(t)\mathbf{v}(t) + \mathbf{f}(t).$$

Since $\mathbf{X}(t)$ satisfies the matrix equation $\mathbf{X}'(t) = \mathbf{A}(t)\mathbf{X}(t)$, equation (9) becomes

$$\mathbf{X}\mathbf{v}' + \mathbf{A}\mathbf{X}\mathbf{v} = \mathbf{A}\mathbf{X}\mathbf{v} + \mathbf{f},$$
$$\mathbf{X}\mathbf{v}' = \mathbf{f}.$$

Multiplying both sides of the last equation by $\mathbf{X}^{-1}(t)$ (which exists since the columns of $\mathbf{X}(t)$ are linearly independent) gives

$$\mathbf{v}'(t) = \mathbf{X}^{-1}(t)\mathbf{f}(t).$$

Integrating, we obtain

$$\mathbf{v}(t) = \int \mathbf{X}^{-1}(t)\mathbf{f}(t)\,dt.$$

Hence a particular solution to (7) is

$$\textbf{(10)} \qquad \mathbf{x}_p(t) = \mathbf{X}(t)\mathbf{v}(t) = \mathbf{X}(t)\int \mathbf{X}^{-1}(t)\mathbf{f}(t)\,dt.$$

Combining (10) with the solution $\mathbf{X}(t)\mathbf{c}$ to the homogeneous system yields the following general solution to (7):

$$\textbf{(11)} \qquad \mathbf{x}(t) = \mathbf{X}(t)\mathbf{c} + \mathbf{X}(t)\int \mathbf{X}^{-1}(t)\mathbf{f}(t)\,dt.$$

The elegance of the derivation of the variation of parameters formula (10) for systems becomes evident when one compares it with the more lengthy derivations for the scalar case in Sections 4.10 and 6.5.

Given an initial value problem of the form

$$\textbf{(12)} \qquad \mathbf{x}'(t) = \mathbf{A}(t)\mathbf{x}(t) + \mathbf{f}(t), \qquad \mathbf{x}(t_0) = \mathbf{x}_0,$$

we can use the initial condition $\mathbf{x}(t_0) = \mathbf{x}_0$ to solve for $\mathbf{c}$ in (11). Expressing $\mathbf{x}(t)$ using a definite integral, we have

$$\mathbf{x}(t) = \mathbf{X}(t)\mathbf{c} + \mathbf{X}(t)\int_{t_0}^{t} \mathbf{X}^{-1}(s)\mathbf{f}(s)\,ds.$$

Using the initial condition $\mathbf{x}(t_0) = \mathbf{x}_0$, we find

$$\mathbf{x}_0 = \mathbf{x}(t_0) = \mathbf{X}(t_0)\mathbf{c} + \mathbf{X}(t_0)\int_{t_0}^{t_0} \mathbf{X}^{-1}(s)\mathbf{f}(s)\,ds = \mathbf{X}(t_0)\mathbf{c}.$$

Solving for $\mathbf{c}$, we have $\mathbf{c} = \mathbf{X}^{-1}(t_0)\mathbf{x}_0$. Thus, the solution to (12) is given by the formula

$$\textbf{(13)} \qquad \mathbf{x}(t) = \mathbf{X}(t)\mathbf{X}^{-1}(t_0)\mathbf{x}_0 + \mathbf{X}(t)\int_{t_0}^{t} \mathbf{X}^{-1}(s)\mathbf{f}(s)\,ds.$$

To apply the variation of parameters formulas it is necessary first to determine a fundamental matrix $\mathbf{X}(t)$ for the homogeneous system. In the case when the coefficient matrix $\mathbf{A}$ is constant, we have discussed methods for finding $\mathbf{X}(t)$. However, if the entries of $\mathbf{A}$ depend on t, the determination of $\mathbf{X}(t)$ may be extremely difficult.

● EXAMPLE 2 Find the solution to the initial value problem

$$\textbf{(14)} \qquad \mathbf{x}'(t) = \begin{bmatrix} 2 & -3 \\ 1 & -2 \end{bmatrix}\mathbf{x}(t) + \begin{bmatrix} e^{2t} \\ 1 \end{bmatrix}, \quad \mathbf{x}(0) = \begin{bmatrix} -1 \\ 0 \end{bmatrix}.$$

Solution   In Example 3 in Section 10.4 we found two linearly independent solutions to the corresponding homogeneous system; namely,

$$\mathbf{x}_1(t) = \begin{bmatrix} 3e^t \\ e^t \end{bmatrix} \quad \text{and} \quad \mathbf{x}_2(t) = \begin{bmatrix} e^{-t} \\ e^{-t} \end{bmatrix}.$$

Hence a fundamental matrix for the homogeneous system is

$$\mathbf{x}(t) = \begin{bmatrix} 3e^t & e^{-t} \\ e^t & e^{-t} \end{bmatrix}.$$

Although the solution to (14) can be found via the method of undetermined coefficients, we shall find it directly from formula (13). For this purpose we need $\mathbf{X}^{-1}(t)$. One way† to obtain $\mathbf{X}^{-1}(t)$ is to form the augmented matrix

$$\begin{bmatrix} 3e^t & e^{-t} & \vdots & 1 & 0 \\ e^t & e^{-t} & \vdots & 0 & 1 \end{bmatrix}$$

and row-reduce this matrix to the matrix $[\mathbf{I} \ \vdots \ \mathbf{X}^{-1}(t)]$. This gives

$$\mathbf{X}^{-1}(t) = \begin{bmatrix} \frac{1}{2}e^{-t} & -\frac{1}{2}e^{-t} \\ -\frac{1}{2}e^t & \frac{3}{2}e^t \end{bmatrix}.$$

Substituting into formula (13), we obtain the solution

$$\mathbf{x}(t) = \begin{bmatrix} 3e^t & e^{-t} \\ e^t & e^{-t} \end{bmatrix} \begin{bmatrix} \frac{1}{2} & -\frac{1}{2} \\ -\frac{1}{2} & \frac{3}{2} \end{bmatrix} \begin{bmatrix} -1 \\ 0 \end{bmatrix}$$

$$+ \begin{bmatrix} 3e^t & e^{-t} \\ e^t & e^{-t} \end{bmatrix} \int_0^t \begin{bmatrix} \frac{1}{2}e^{-s} & -\frac{1}{2}e^{-s} \\ -\frac{1}{2}e^s & \frac{3}{2}e^s \end{bmatrix} \begin{bmatrix} e^{2s} \\ 1 \end{bmatrix} ds$$

$$= \begin{bmatrix} -\frac{3}{2}e^t + \frac{1}{2}e^{-t} \\ -\frac{1}{2}e^t + \frac{1}{2}e^{-t} \end{bmatrix} + \begin{bmatrix} 3e^t & e^{-t} \\ e^t & e^{-t} \end{bmatrix} \int_0^t \begin{bmatrix} \frac{1}{2}e^s - \frac{1}{2}e^{-s} \\ -\frac{1}{2}e^{3s} + \frac{3}{2}e^s \end{bmatrix} ds$$

$$= \begin{bmatrix} -\frac{3}{2}e^t + \frac{1}{2}e^{-t} \\ -\frac{1}{2}e^t + \frac{1}{2}e^{-t} \end{bmatrix} + \begin{bmatrix} 3e^t & e^{-t} \\ e^t & e^{-t} \end{bmatrix} \begin{bmatrix} \frac{1}{2}e^t + \frac{1}{2}e^{-t} - 1 \\ \frac{3}{2}e^t - \frac{1}{6}e^{3t} - \frac{4}{3} \end{bmatrix}$$

$$= \begin{bmatrix} -\frac{9}{2}e^t - \frac{5}{6}e^{-t} + \frac{4}{3}e^{2t} + 3 \\ -\frac{3}{2}e^t - \frac{5}{6}e^{-t} + \frac{1}{3}e^{2t} + 2 \end{bmatrix}. \quad \bullet$$

---

† This procedure works for an invertible matrix of any dimension. For an arbitrary 2×2 invertible matrix $\mathbf{U}(t)$, a formula for $\mathbf{U}^{-1}(t)$ is derived in Problem 32. Of course, a linear algebra software package can be used to compute the inverse of a matrix.

EXERCISES 10.6

In Problems 1 through 4 use the method of undetermined coefficients to find a general solution to the system $\mathbf{x}'(t) = \mathbf{A}\mathbf{x}(t) + \mathbf{f}(t)$, *where* $\mathbf{A}$ *and* $\mathbf{f}(t)$ *are given.*

1. $\mathbf{A} = \begin{bmatrix} 6 & 1 \\ 4 & 3 \end{bmatrix}$, $\mathbf{f}(t) = \begin{bmatrix} -11 \\ -5 \end{bmatrix}$.

2. $\mathbf{A} = \begin{bmatrix} 1 & 1 \\ 4 & 1 \end{bmatrix}$, $\mathbf{f}(t) = \begin{bmatrix} -t - 1 \\ -4t - 2 \end{bmatrix}$.

3. $\mathbf{A} = \begin{bmatrix} 1 & -2 & 2 \\ -2 & 1 & 2 \\ 2 & 2 & 1 \end{bmatrix}$, $\mathbf{f}(t) = \begin{bmatrix} 2e^t \\ 4e^t \\ -2e^t \end{bmatrix}$.

4. $\mathbf{A} = \begin{bmatrix} 2 & 2 \\ 2 & 2 \end{bmatrix}$, $\mathbf{f}(t) = \begin{bmatrix} -4\cos t \\ -\sin t \end{bmatrix}$.

In Problems 5 through 10 use the method of undetermined coefficients (Table 4.1 on inside back cover) to determine only the form of a particular solution for the system $\mathbf{x}'(t) = \mathbf{A}\mathbf{x}(t) + \mathbf{f}(t)$, *where* $\mathbf{A}$ *and* $\mathbf{f}(t)$ *are given.*

5. $\mathbf{A} = \begin{bmatrix} 1 & 1 \\ 0 & 2 \end{bmatrix}$, $\mathbf{f}(t) = e^{-2t} \begin{bmatrix} t \\ 3 \end{bmatrix}$.

6. $\mathbf{A} = \begin{bmatrix} 1 & -1 \\ 0 & -1 \end{bmatrix}$, $\mathbf{f}(t) = \begin{bmatrix} t \\ e^{3t} \end{bmatrix}$.

7. $\mathbf{A} = \begin{bmatrix} 0 & 1 \\ 2 & 0 \end{bmatrix}$, $\mathbf{f}(t) = \begin{bmatrix} \sin 3t \\ t \end{bmatrix}$.

8. $\mathbf{A} = \begin{bmatrix} -1 & 0 \\ 2 & 2 \end{bmatrix}$, $\mathbf{f}(t) = \begin{bmatrix} t^2 \\ t + 1 \end{bmatrix}$.

9. $\mathbf{A} = \begin{bmatrix} 1 & 1 \\ 0 & 1 \end{bmatrix}$, $\mathbf{f}(t) = \begin{bmatrix} e^{2t} \\ e^{3t} \end{bmatrix}$.

10. $\mathbf{A} = \begin{bmatrix} 2 & -1 \\ 1 & 5 \end{bmatrix}$, $\mathbf{f}(t) = \begin{bmatrix} te^{-t} \\ 3e^{-t} \end{bmatrix}$.

In Problems 11 through 16 use the variation of parameters formula (11) to find a general solution of the system $\mathbf{x}'(t) = \mathbf{A}\mathbf{x}(t) + \mathbf{f}(t)$, *where* $\mathbf{A}$ *and* $\mathbf{f}(t)$ *are given.*

11. $\mathbf{A} = \begin{bmatrix} 0 & 1 \\ -1 & 0 \end{bmatrix}$, $\mathbf{f}(t) = \begin{bmatrix} 1 \\ 0 \end{bmatrix}$.

12. $\mathbf{A} = \begin{bmatrix} 1 & 2 \\ 3 & 2 \end{bmatrix}$, $\mathbf{f}(t) = \begin{bmatrix} 1 \\ -1 \end{bmatrix}$.

13. $\mathbf{A} = \begin{bmatrix} 2 & 1 \\ -3 & -2 \end{bmatrix}$, $\mathbf{f}(t) = \begin{bmatrix} 2e^t \\ 4e^t \end{bmatrix}$.

14. $\mathbf{A} = \begin{bmatrix} 0 & -1 \\ 1 & 0 \end{bmatrix}$, $\mathbf{f}(t) = \begin{bmatrix} t^2 \\ 1 \end{bmatrix}$.

15. $\mathbf{A} = \begin{bmatrix} -4 & 2 \\ 2 & -1 \end{bmatrix}$, $\mathbf{f}(t) = \begin{bmatrix} t^{-1} \\ 4 + 2t^{-1} \end{bmatrix}$.

16. $\mathbf{A} = \begin{bmatrix} 0 & 1 \\ -1 & 0 \end{bmatrix}$, $\mathbf{f}(t) = \begin{bmatrix} 8\sin t \\ 0 \end{bmatrix}$.

In Problems 17 through 20 use the variation of parameters formula (11) and possibly a linear algebra software package to find a general solution of the system $\mathbf{x}'(t) = \mathbf{A}\mathbf{x}(t) + \mathbf{f}(t)$, *where* $\mathbf{A}$ *and* $\mathbf{f}(t)$ *are given.*

17. $\mathbf{A} = \begin{bmatrix} 0 & 1 & 1 \\ 1 & 0 & 1 \\ 1 & 1 & 0 \end{bmatrix}$, $\mathbf{f}(t) = \begin{bmatrix} 3e^t \\ -e^t \\ -e^t \end{bmatrix}$.

18. $\mathbf{A} = \begin{bmatrix} 1 & -1 & 1 \\ 0 & 0 & 1 \\ 0 & -1 & 2 \end{bmatrix}$, $\mathbf{f}(t) = \begin{bmatrix} 0 \\ e^t \\ e^t \end{bmatrix}$.

19. $\mathbf{A} = \begin{bmatrix} 0 & 1 & 0 & 0 \\ -1 & 0 & 0 & 0 \\ 0 & 0 & 0 & 1 \\ 0 & 0 & 1 & 0 \end{bmatrix}$, $\mathbf{f}(t) = \begin{bmatrix} t \\ 0 \\ e^{-t} \\ t \end{bmatrix}$.

20. $\mathbf{A} = \begin{bmatrix} 0 & 1 & 0 & 0 \\ 0 & 0 & 1 & 0 \\ 0 & 0 & 0 & 1 \\ 8 & -4 & -2 & -1 \end{bmatrix}$, $\mathbf{f}(t) = \begin{bmatrix} e^t \\ 0 \\ 1 \\ 0 \end{bmatrix}$.

In Problems 21 and 22 find the solution to the given system that satisfies the given initial condition.

21. $\mathbf{x}'(t) = \begin{bmatrix} 0 & 2 \\ -1 & 3 \end{bmatrix} \mathbf{x}(t) + \begin{bmatrix} e^t \\ -e^t \end{bmatrix}$,

  (a) $\mathbf{x}(0) = \begin{bmatrix} 5 \\ 4 \end{bmatrix}$.   **(b)** $\mathbf{x}(1) = \begin{bmatrix} 0 \\ 1 \end{bmatrix}$.

  (c) $\mathbf{x}(5) = \begin{bmatrix} 1 \\ 0 \end{bmatrix}$.   **(d)** $x(-1) = \begin{bmatrix} -4 \\ 5 \end{bmatrix}$.

22. $\mathbf{x}'(t) = \begin{bmatrix} 0 & 2 \\ 4 & -2 \end{bmatrix} \mathbf{x}(t) + \begin{bmatrix} 4t \\ -4t-2 \end{bmatrix}$,

(a) $\mathbf{x}(0) = \begin{bmatrix} 4 \\ -5 \end{bmatrix}$. **(b)** $\mathbf{x}(2) = \begin{bmatrix} 1 \\ 1 \end{bmatrix}$.

23. Using matrix algebra techniques and the method of undetermined coefficients, find a general solution for

$$x''(t) + y'(t) - x(t) + y(t) = -1,$$
$$x'(t) + y'(t) - x(t) = t^2.$$

Compare your solution with the solution in Example 2 in Section 9.2.

24. Using matrix algebra techniques and the method of undetermined coefficients, solve the initial value problem

$$x'(t) - 2y(t) = 4t, \qquad x(0) = 4,$$
$$y'(t) + 2y(t) - 4x(t) = -4t - 2, \qquad y(0) = -5.$$

Compare your solution with the solution in Example 1 in Section 9.3.

25. To find a general solution to the sytem

$$\mathbf{x}'(t) = \begin{bmatrix} 0 & 1 \\ -2 & 3 \end{bmatrix} \mathbf{x}(t) + \mathbf{f}(t), \quad \text{where} \quad \mathbf{f}(t) = \begin{bmatrix} e^t \\ 0 \end{bmatrix}:$$

(a) Find a fundamental solution set for the corresponding homogeneous system.
(b) The obvious choice for a particular solution would be a vector function of the form $\mathbf{x}_p(t) = e^t \mathbf{a}$; however, the homogeneous system has a solution of this form. The next choice would be $\mathbf{x}_p(t) = te^t \mathbf{a}$. Show that this choice does *not* work.
(c) For systems, multiplying by t is not always sufficient. The proper guess is

$$\mathbf{x}_p(t) = te^t \mathbf{a} + e^t \mathbf{b}.$$

Use this guess to find a particular solution of the given system.
(d) Use the results of parts (a) and (c) to find a general solution of the given system.

26. For the system of Problem 25 we found that a proper guess for a particular solution is $\mathbf{x}_p(t) = te^t \mathbf{a} + e^t \mathbf{b}$. In some cases **a** or **b** may be zero.

(a) Find a particular solution for the system of Problem 25 if $\mathbf{f}(t) = \text{col}(3e^t, 6e^t)$.
(b) Find a particular solution for the system of Problem 25 if $\mathbf{f}(t) = \text{col}(e^t, e^t)$.

27. Find a general solution of the system

$$\mathbf{x}'(t) = \begin{bmatrix} 0 & 1 & 1 \\ 1 & 0 & 1 \\ 1 & 1 & 0 \end{bmatrix} \mathbf{x}(t) + \begin{bmatrix} -1 \\ -1 - e^{-t} \\ -2e^{-t} \end{bmatrix}.$$

[Hint: Use superposition to determine the particular solution.]

28. Find a particular solution for the system

$$\mathbf{x}'(t) = \begin{bmatrix} 1 & -1 \\ -1 & 1 \end{bmatrix} \mathbf{x}(t) + \begin{bmatrix} -3 \\ 1 \end{bmatrix}.$$

[Hint: Try $\mathbf{x}_p(t) = t\mathbf{a} + \mathbf{b}$.]

In Problems 29 and 30 find a general solution to the given Cauchy-Euler system for $t > 0$. Remember to express the system in the form $\mathbf{x}'(t) = \mathbf{A}(t)\mathbf{x}(t) + \mathbf{f}(t)$ before using the variation of parameters formula.

29. $t\mathbf{x}'(t) = \begin{bmatrix} 2 & -1 \\ 3 & -2 \end{bmatrix} \mathbf{x}(t) + \begin{bmatrix} t^{-1} \\ 1 \end{bmatrix}$.

30. $t\mathbf{x}'(t) = \begin{bmatrix} 4 & -3 \\ 8 & -6 \end{bmatrix} \mathbf{x}(t) + \begin{bmatrix} t \\ 2t \end{bmatrix}$.

31. Use the variation of parameters formula (10) to derive a formula for a particular solution y_p to the scalar equation $y'' + p(t)y' + q(t)y = g(t)$ in terms of two linearly independent solutions $y_1(t)$, $y_2(t)$ of the corresponding homogeneous equation. Show that your answer agrees with the formulas derived in Section 4.10. [Hint: First write the scalar equation in system form.]

32. Let $\mathbf{U}(t)$ be the invertible 2×2 matrix

$$\mathbf{U}(t) := \begin{bmatrix} a(t) & b(t) \\ c(t) & d(t) \end{bmatrix}.$$

Show that

$$\mathbf{U}^{-1}(t) = \frac{1}{[a(t)d(t) - b(t)c(t)]} \begin{bmatrix} d(t) & -b(t) \\ -c(t) & a(t) \end{bmatrix}.$$

33. RL Network. The currents in the RL network given by the schematic diagram in Figure 10.6 are governed by the following equations:

$$2I'_1(t) + 90I_2(t) = 9,$$
$$I'_3(t) + 30I_4(t) - 90I_2(t) = 0,$$
$$60I_5(t) - 30I_4(t) = 0,$$
$$I_1(t) = I_2(t) + I_3(t),$$
$$I_3(t) = I_4(t) + I_5(t).$$

Assume the currents are initially zero. Solve for the five currents $I_1, \ldots, I_5$. [Hint: Express as a normal system with $x_1 = I_2$ and $x_2 = I_5$.]

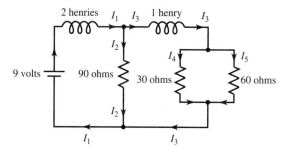

Figure 10.6 RL network for Problem 33

34. **Conventional Combat Model.**   In modeling a pair of conventional forces in combat, the following system arises

$$\mathbf{x}' = \begin{bmatrix} -a & -b \\ -c & -d \end{bmatrix} \mathbf{x} + \begin{bmatrix} p \\ q \end{bmatrix},$$

where $\mathbf{x} = \text{col}(x_1, x_2)$. The variables $x_1(t)$ and $x_2(t)$ represent the strengths of opposing forces at time t. The terms $-ax_1$ and $-dx_2$ represent the *operational loss rates* and the terms $-bx_2$ and $-cx_1$ represent the *combat loss rates* for the troops x_1 and x_2, respectively. The constants p and q represent the respective rates of reinforcement. Let $a = 1$, $b = 4$, $c = 3$, $d = 2$, and $p = q = 5$. By solving the appropriate initial value

problem, determine which forces will win if
(a) $x_1(0) = 20$, $x_2(0) = 20$.
(b) $x_1(0) = 21$, $x_2(0) = 20$.
(c) $x_1(0) = 20$, $x_2(0) = 21$.

35. **Mixing Problem.**   Two tanks A and B, each holding 50 L of liquid, are interconnected by pipes. The liquid flows from tank A into tank B at a rate of 4 L/min and from B into A at a rate of 1 L/min (see Figure 10.7). The liquid inside each tank is kept well stirred. A brine solution that has a concentration of 2 kg/L of salt flows into tank A at a rate of 4 L/min, and a brine solution that has a concentration of 1 kg/L of salt flows into tank B at a rate of 1 L/min. The solutions flow out of the system from both tanks, from tank A at 1 L/min and from tank B at 4 L/min. If, initially, tank A contains only water and tank B contains 5 kg of salt, determine the mass of salt in each tank at time $t \geq 0$. After several minutes have elapsed, which tank has the higher concentration of salt? What is its limiting concentration?

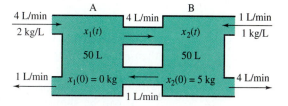

Figure 10.7 Mixing problem for interconnected tanks

•10.7• THE MATRIX EXPONENTIAL FUNCTION

In this chapter we have developed various ways to extend techniques for scalar differential equations to systems. In this section we take a substantial step further by showing that, with the right notation, the formulas for solving normal systems with constant coefficients are identical to the formulas for solving first order equations with constant coefficients. For example, we know that a general solution to the equation $x'(t) = ax(t)$, where a is a constant, is $x(t) = ce^{at}$. Analogously, we shall show that a general solution to the normal system

(1) $\mathbf{x}'(t) = \mathbf{A}\mathbf{x}(t),$

where $\mathbf{A}$ is a constant $n \times n$ matrix, is $\mathbf{x}(t) = e^{\mathbf{A}t}\mathbf{c}$. Our first task is to define the matrix exponential $e^{\mathbf{A}t}$.

If $\mathbf{A}$ is a constant $n \times n$ matrix, we define $e^{\mathbf{A}t}$ by taking the series expansion for e^{at} and replacing a by $\mathbf{A}$; that is,

(2) $$e^{\mathbf{A}t} := \mathbf{I} + \mathbf{A}t + \mathbf{A}^2\frac{t^2}{2!} + \cdots + \mathbf{A}^n\frac{t^n}{n!} + \cdots .$$

By the right-hand side of (2) we mean the $n \times n$ matrix whose elements are power series with coefficients given by the corresponding entries in the matrices $\mathbf{I}, \mathbf{A}, \mathbf{A}^2/2!, \ldots$.
    If $\mathbf{A}$ is a diagonal matrix, then the computation of $e^{\mathbf{A}t}$ is straightforward. For example, if

$$\mathbf{A} = \begin{bmatrix} -1 & 0 \\ 0 & 2 \end{bmatrix},$$

then

$$\mathbf{A}^2 = \mathbf{A}\mathbf{A} = \begin{bmatrix} 1 & 0 \\ 0 & 4 \end{bmatrix}, \quad \mathbf{A}^3 = \begin{bmatrix} -1 & 0 \\ 0 & 8 \end{bmatrix}, \quad \ldots, \quad \mathbf{A}^n = \begin{bmatrix} (-1)^n & 0 \\ 0 & 2^n \end{bmatrix},$$

and so

$$e^{\mathbf{A}t} = \sum_{n=0}^{\infty} \mathbf{A}^n \frac{t^n}{n!} = \begin{bmatrix} \sum_{n=0}^{\infty}(-1)^n\frac{t^n}{n!} & 0 \\ 0 & \sum_{n=0}^{\infty}2^n\frac{t^n}{n!} \end{bmatrix} = \begin{bmatrix} e^{-t} & 0 \\ 0 & e^{2t} \end{bmatrix}.$$

More generally, if $\mathbf{A}$ is an $n \times n$ *diagonal* matrix with $r_1, r_2, \ldots, r_n$ down its main diagonal, then $e^{\mathbf{A}t}$ is the diagonal matrix with $e^{r_1 t}, e^{r_2 t}, \ldots, e^{r_n t}$ down its main diagonal (see Problem 26). If $\mathbf{A}$ is not a diagonal matrix, the computation of $e^{\mathbf{A}t}$ is more involved. We deal with this important problem later in this section.
    It can be shown that the series (2) converges for all t and has many of the same properties[†] as the scalar exponential e^{at}.

▶ **PROPERTIES OF THE MATRIX EXPONENTIAL FUNCTION**

Theorem 7.   Let $\mathbf{A}$ and $\mathbf{B}$ be $n \times n$ constant matrices and r, s, and t be real (or complex) numbers. Then,

(a) $e^{\mathbf{A}0} = e^{\mathbf{0}} = \mathbf{I}$.
(b) $e^{\mathbf{A}(t+s)} = e^{\mathbf{A}t}e^{\mathbf{A}s}$.
(c) $(e^{\mathbf{A}t})^{-1} = e^{-\mathbf{A}t}$.
(d) $e^{(\mathbf{A}+\mathbf{B})t} = e^{\mathbf{A}t}e^{\mathbf{B}t}$, provided that $\mathbf{A}\mathbf{B} = \mathbf{B}\mathbf{A}$.
(e) $e^{r\mathbf{I}t} = e^{rt}\mathbf{I}$.

[†] For proofs of these and other properties of the matrix exponential function, see *Matrices and Linear Transformations*, by Charles G. Cullen, Addison-Wesley Publishing Co., Reading, Massachusetts, 1972, Chapter 8.

Property (c) has profound implications. First, it asserts that for any matrix $\mathbf{A}$, the matrix $e^{\mathbf{A}t}$ *has* an inverse for all t. Moreover, this inverse is obtained by simply replacing t by $-t$. In applying property (d) (the law of exponents), care must be exercised because of the stipulation that the matrices $\mathbf{A}$ and $\mathbf{B}$ commute (see Problem 25).

Another important property of the matrix exponential arises from the fact that we can differentiate the series in (2) term by term. This gives

$$\frac{d}{dt}(e^{\mathbf{A}t}) = \frac{d}{dt}\left(\mathbf{I} + \mathbf{A}t + \mathbf{A}^2\frac{t^2}{2} + \cdots + \mathbf{A}^n\frac{t^n}{n!} + \cdots\right)$$

$$= \mathbf{A} + \mathbf{A}^2 t + \mathbf{A}^3\frac{t^2}{2} + \cdots + \mathbf{A}^n\frac{t^{n-1}}{(n-1)!} + \cdots$$

$$= \mathbf{A}\left[\mathbf{I} + \mathbf{A}t + \mathbf{A}^2\frac{t^2}{2} + \cdots + \mathbf{A}^{n-1}\frac{t^{n-1}}{(n-1)!} + \cdots\right].$$

Hence

$$\frac{d}{dt}(e^{\mathbf{A}t}) = \mathbf{A}e^{\mathbf{A}t},$$

and so $e^{\mathbf{A}t}$ *is a solution to the matrix differential equation* $\mathbf{X}' = \mathbf{A}\mathbf{X}$. Since $e^{\mathbf{A}t}$ is invertible (property (c)), it follows that the columns of $e^{\mathbf{A}t}$ are linearly independent solutions to system (1). Combining these facts we have:

$e^{\mathbf{A}t}$ IS A FUNDAMENTAL MATRIX

Theorem 8.   If $\mathbf{A}$ is an $n \times n$ constant matrix, then the columns of the matrix exponential $e^{\mathbf{A}t}$ form a fundamental solution set for the system $\mathbf{x}'(t) = \mathbf{A}\mathbf{x}(t)$. Therefore, $e^{\mathbf{A}t}$ is a fundamental matrix for the system, and a general solution is $\mathbf{x}(t) = e^{\mathbf{A}t}\mathbf{c}$.

Knowing that $e^{\mathbf{A}t}$ is a fundamental matrix is of practical use provided we can calculate $e^{\mathbf{A}t}$. As we observed, if $\mathbf{A}$ is a diagonal matrix, then we simply exponentiate the diagonal elements (times t) to obtain $e^{\mathbf{A}t}$. Also, if $\mathbf{B}$ is a **nilpotent** matrix, that is, $\mathbf{B}^k = \mathbf{0}$ for some positive integer k, then the series for $e^{\mathbf{B}t}$ has only a finite number of terms, since $\mathbf{B}^k = \mathbf{B}^{k+1} = \cdots = \mathbf{0}$. In such cases $e^{\mathbf{B}t}$ reduces to

$$e^{\mathbf{B}t} = \mathbf{I} + \mathbf{B}t + \cdots + \mathbf{B}^{k-1}\frac{t^{k-1}}{(k-1)!}.$$

Using the law of exponents and this fact about nilpotent matrices, we can determine $e^{\mathbf{A}t}$ for a special class of matrices. Let r be a scalar. Since

$$e^{\mathbf{A}t} = e^{r\mathbf{I}t}e^{(\mathbf{A}-r\mathbf{I})t} = e^{rt}e^{(\mathbf{A}-r\mathbf{I})t},$$

we get a finite representation for $e^{\mathbf{A}t}$ if $\mathbf{B} = \mathbf{A} - r\mathbf{I}$ is nilpotent for some r. In fact, when the characteristic polynomial for $\mathbf{A}$ has the form $p(r) = (r_1 - r)^n$, that is, when $\mathbf{A}$ has one eigenvalue r_1 of multiplicity n, it is a consequence of the Cayley-Hamilton theorem[†] that $(r_1\mathbf{I} - \mathbf{A})^n = \mathbf{0}$. Hence $\mathbf{A} - r_1\mathbf{I}$ is nilpotent and

$$e^{\mathbf{A}t} = e^{r_1 t}\left\{\mathbf{I} + (\mathbf{A} - r_1\mathbf{I})t + \cdots + (\mathbf{A} - r_1\mathbf{I})^{n-1}\frac{t^{n-1}}{(n-1)!}\right\}.$$

● **EXAMPLE 1** Find the fundamental matrix $e^{\mathbf{A}t}$ for the system

$$(3) \qquad \mathbf{x}' = \mathbf{A}\mathbf{x}, \quad \text{where} \quad \mathbf{A} = \begin{bmatrix} 2 & 1 & 1 \\ 1 & 2 & 1 \\ -2 & -2 & -1 \end{bmatrix}.$$

Solution We begin by computing the characteristic polynomial for $\mathbf{A}$:

$$p(r) = |\mathbf{A} - r\mathbf{I}| = \begin{vmatrix} 2-r & 1 & 1 \\ 1 & 2-r & 1 \\ -2 & -2 & -1-r \end{vmatrix} = -r^3 + 3r^2 - 3r + 1 = -(r-1)^3.$$

Thus $r = 1$ is an eigenvalue of $\mathbf{A}$ with multiplicity 3. By the Cayley-Hamilton theorem, $(\mathbf{A} - \mathbf{I})^3 = \mathbf{0}$, and so

$$(4) \qquad e^{\mathbf{A}t} = e^t e^{(\mathbf{A}-\mathbf{I})t} = e^t\left\{\mathbf{I} + (\mathbf{A}-\mathbf{I})t + (\mathbf{A}-\mathbf{I})^2\frac{t^2}{2}\right\}.$$

Computing, we find

$$\mathbf{A} - \mathbf{I} = \begin{bmatrix} 1 & 1 & 1 \\ 1 & 1 & 1 \\ -2 & -2 & -2 \end{bmatrix} \quad \text{and} \quad (\mathbf{A} - \mathbf{I})^2 = \begin{bmatrix} 0 & 0 & 0 \\ 0 & 0 & 0 \\ 0 & 0 & 0 \end{bmatrix}.$$

Substituting into (4) yields

$$(5) \qquad e^{\mathbf{A}t} = e^t\begin{bmatrix} 1 & 0 & 0 \\ 0 & 1 & 0 \\ 0 & 0 & 1 \end{bmatrix} + te^t\begin{bmatrix} 1 & 1 & 1 \\ 1 & 1 & 1 \\ -2 & -2 & -2 \end{bmatrix} = \begin{bmatrix} e^t + te^t & te^t & te^t \\ te^t & e^t + te^t & te^t \\ -2te^t & -2te^t & e^t - 2te^t \end{bmatrix}. \quad ●$$

In the preceding example we used the nilpotency of the matrix $\mathbf{A} - r\mathbf{I}$ to compute $e^{\mathbf{A}t}$ directly. In general, we cannot expect nilpotency to hold, but we can take advantage of the following relationship between fundamental matrices to help compute $e^{\mathbf{A}t}$.

> **RELATIONSHIP BETWEEN FUNDAMENTAL MATRICES**
>
> **Lemma 1.** Let $\mathbf{X}(t)$ and $\mathbf{Y}(t)$ be two fundamental matrices for the same system $\mathbf{x}' = \mathbf{A}\mathbf{x}$. Then, there exists a constant matrix $\mathbf{C}$ such that $\mathbf{X}(t) = \mathbf{Y}(t)\mathbf{C}$.

[†] The Cayley-Hamilton theorem states that a matrix satisfies its own characteristic equation. That is, $p(\mathbf{A}) = \mathbf{0}$. For a discussion of this theorem, see *Matrices and Linear Transformations*, by Charles G. Cullen, Addison-Wesley Publishing Co., Reading, Massachusetts, 1972, Chapter 5.

Proof.   Let $\mathbf{x}_1(t), \ldots, \mathbf{x}_n(t)$ be the columns of $\mathbf{X}(t)$ and let $\mathbf{y}_1(t), \ldots, \mathbf{y}_n(t)$ be the columns of $\mathbf{Y}(t)$. Since $\{\mathbf{y}_1(t), \ldots, \mathbf{y}_n(t)\}$ is a fundamental solution set and $\mathbf{x}_j(t)$, $j = 1, \ldots, n$ are solutions to $\mathbf{x}' = \mathbf{Ax}$, there exist constants $c_{1j}, c_{2j}, \ldots, c_{nj}$ such that

$$\mathbf{x}_j(t) = c_{1j}\mathbf{y}_1(t) + c_{2j}\mathbf{y}_2(t) + \cdots + c_{nj}\mathbf{y}_n(t)$$

for each $j = 1, \ldots, n$. But this is equivalent to writing $\mathbf{X}(t) = \mathbf{Y}(t)\mathbf{C}$, where $\mathbf{C} = [c_{ij}]$.
◄◄◄

We now use Lemma 1 to find a formula for $e^{\mathbf{A}t}$ when a fundamental matrix $\mathbf{X}(t)$ for $\mathbf{x}' = \mathbf{Ax}$ is known. Since $e^{\mathbf{A}t}$ is also a fundamental matrix for the system, Lemma 1 asserts that $e^{\mathbf{A}t} = \mathbf{X}(t)\mathbf{C}$ for some constant matrix $\mathbf{C}$. Setting $t = 0$ yields $\mathbf{I} = \mathbf{X}(0)\mathbf{C}$, and solving for $\mathbf{C}$ we get $\mathbf{C} = \mathbf{X}^{-1}(0)$. Hence

(6) $e^{\mathbf{A}t} = \mathbf{X}(t)\mathbf{X}^{-1}(0)$.

Although this formula is useful, it does place the burden of determining $e^{\mathbf{A}t}$ on finding a fundamental matrix $\mathbf{X}(t)$. Fortunately, we can use the *properties* of the matrix exponential $e^{\mathbf{A}t}$ to help simplify this task. Since the columns of a fundamental matrix must have the form $e^{\mathbf{A}t}\mathbf{u}$, let's try to find n vectors $\mathbf{u}$ for which the computation of $e^{\mathbf{A}t}\mathbf{u}$ is tractable.

We begin by using the relation $e^{\mathbf{A}t} = e^{rt}e^{(\mathbf{A} - r\mathbf{I})t}$ to express $e^{\mathbf{A}t}\mathbf{u}$ as

(7) $e^{\mathbf{A}t}\mathbf{u} = e^{rt}e^{(\mathbf{A} - r\mathbf{I})t}\mathbf{u}$

$$= e^{rt}\left\{\mathbf{u} + t(\mathbf{A} - r\mathbf{I})\mathbf{u} + \cdots + \frac{t^k}{k!}(\mathbf{A} - r\mathbf{I})^k\mathbf{u} + \cdots\right\}.$$

Now we know that if r is an eigenvalue of $\mathbf{A}$ and $\mathbf{u}$ is a corresponding eigenvector, then $e^{rt}\mathbf{u}$ is a solution to (1). Indeed, in this situation $(\mathbf{A} - r\mathbf{I})\mathbf{u} = (\mathbf{A} - r\mathbf{I})^2\mathbf{u} = \cdots = \mathbf{0}$, and so the series in (7) reduces to the first term, $e^{rt}\mathbf{u}$. While it is too much to expect $\mathbf{A} - r\mathbf{I}$ to be nilpotent, it is not too much to ask that $(\mathbf{A} - r\mathbf{I})^k\mathbf{u} = \mathbf{0}$ for some nontrivial vector $\mathbf{u}$ and some positive integer k.

▶ **GENERALIZED EIGENVECTORS**

Definition 5.   Let $\mathbf{A}$ be an $n \times n$ constant matrix and r be an eigenvalue of $\mathbf{A}$. A nontrivial vector $\mathbf{u}$ that satisfies

$(\mathbf{A} - r\mathbf{I})^k\mathbf{u} = \mathbf{0}$

for some positive integer k is called a **generalized eigenvector** associated with r.

A consequence of the primary decomposition theorem in advanced linear algebra[†] is that if the characteristic polynomial for $\mathbf{A}$ is

$$p(r) = (r_1 - r)^{m_1} \cdots (r_k - r)^{m_k},$$

where the r_i's are distinct eigenvalues of $\mathbf{A}$ and m_i is the multiplicity of the eigenvalue r_i,

---

[†] *Matrices and Linear Transformations*, by Charles G. Cullen, Theorem 5.11, page 196.

then for each i there exist m_i linearly independent generalized eigenvectors associated with r_i, and the combined set of $n = m_1 + \cdots + m_k$ generalized eigenvectors is linearly independent. Moreover, if $\mathbf{u}$ is a generalized eigenvector associated with r_i, then $(\mathbf{A} - r_i\mathbf{I})^{m_i}\mathbf{u} = \mathbf{0}$. This leads to the following procedure for finding n linearly independent solutions to system (1).

▶ **SOLVING SYSTEMS USING GENERALIZED EIGENVECTORS**

To obtain a fundamental solution set for $\mathbf{x}' = \mathbf{A}\mathbf{x}$:

(a) Compute the characteristic polynomial $p(t) = |\mathbf{A} - r\mathbf{I}|$ and find the distinct eigenvalues $r_1,\ldots,r_k$.

(b) For each eigenvalue r_i, find m_i linearly independent generalized eigenvectors, where m_i is the multiplicity of the eigenvalue r_i.

(c) Use the n linearly independent generalized eigenvectors obtained in (b) to compute the n linearly independent solutions to $\mathbf{x}' = \mathbf{A}\mathbf{x}$ of the form

$$(8) \qquad e^{\mathbf{A}t}\mathbf{u} = e^{rt}\left\{\mathbf{u} + t(\mathbf{A} - r\mathbf{I})\mathbf{u} + \frac{t^2}{2}(\mathbf{A} - r\mathbf{I})^2\mathbf{u} + \cdots\right\},$$

where r is an eigenvalue and $\mathbf{u}$ is a corresponding generalized eigenvector. If r has multiplicity m_i, then the above series reduces to the first m_i terms.

● **EXAMPLE 2** Find the fundamental matrix $e^{\mathbf{A}t}$ for the system

$$(9) \qquad \mathbf{x}' = \mathbf{A}\mathbf{x}, \quad \text{where} \quad \mathbf{A} = \begin{bmatrix} 1 & 0 & 0 \\ 1 & 3 & 0 \\ 0 & 1 & 1 \end{bmatrix}.$$

Solution We begin by finding the characteristic polynomial for $\mathbf{A}$:

$$p(r) = |\mathbf{A} - r\mathbf{I}| = \begin{vmatrix} 1 - r & 0 & 0 \\ 1 & 3 - r & 0 \\ 0 & 1 & 1 - r \end{vmatrix} = -(r - 1)^2(r - 3).$$

Hence the eigenvalues of $\mathbf{A}$ are $r = 1$ with multiplicity 2 and $r = 3$ with multiplicity 1.

Since $r = 1$ has multiplicity 2, we must determine two linearly independent associated generalized eigenvectors. We begin by solving $(\mathbf{A} - \mathbf{I})\mathbf{u} = \mathbf{0}$; that is,

$$\begin{bmatrix} 0 & 0 & 0 \\ 1 & 2 & 0 \\ 0 & 1 & 0 \end{bmatrix}\begin{bmatrix} u_1 \\ u_2 \\ u_3 \end{bmatrix} = \begin{bmatrix} 0 \\ 0 \\ 0 \end{bmatrix}.$$

Solving, we obtain $u_1 = u_2 = 0$ and $u_3 = s$, where s is arbitrary. Thus there is at most one linearly independent eigenvector corresponding to $r = 1$, and with $s = 1$ we choose $\mathbf{u}_1 = \text{col}(0, 0, 1)$. Hence one solution to (9) is

$$(10) \qquad \mathbf{x}_1(t) = e^t \mathbf{u}_1 = e^t \begin{bmatrix} 0 \\ 0 \\ 1 \end{bmatrix} = \begin{bmatrix} 0 \\ 0 \\ e^t \end{bmatrix}.$$

Next we solve $(\mathbf{A} - \mathbf{I})^2 \mathbf{u} = \mathbf{0}$. From

$$(\mathbf{A} - \mathbf{I})^2 \mathbf{u} = \begin{bmatrix} 0 & 0 & 0 \\ 2 & 4 & 0 \\ 1 & 2 & 0 \end{bmatrix} \begin{bmatrix} u_1 \\ u_2 \\ u_3 \end{bmatrix} = \begin{bmatrix} 0 \\ 0 \\ 0 \end{bmatrix},$$

we find $u_2 = s$, $u_1 = -2u_2 = -2s$, and $u_3 = v$, where s and v are arbitrary. Taking $s = 1$ and $v = 0$, we obtain the generalized eigenvector $\mathbf{u}_2 = \text{col}(-2, 1, 0)$, which is linearly independent of $\mathbf{u}_1$. Now we use $\mathbf{u}_2$ to obtain a second solution to (9). Since $(\mathbf{A} - \mathbf{I})^2 \mathbf{u}_2 = \mathbf{0}$, formula (8) reduces to

$$(11) \qquad \mathbf{x}_2(t) = e^{\mathbf{A}t} \mathbf{u}_2 = e^t \{ \mathbf{u}_2 + t(\mathbf{A} - \mathbf{I}) \mathbf{u}_2 \}$$

$$= e^t \begin{bmatrix} -2 \\ 1 \\ 0 \end{bmatrix} + t e^t \begin{bmatrix} 0 & 0 & 0 \\ 1 & 2 & 0 \\ 0 & 1 & 0 \end{bmatrix} \begin{bmatrix} -2 \\ 1 \\ 0 \end{bmatrix}$$

$$= e^t \begin{bmatrix} -2 \\ 1 \\ 0 \end{bmatrix} + t e^t \begin{bmatrix} 0 \\ 0 \\ 1 \end{bmatrix} = \begin{bmatrix} -2e^t \\ e^t \\ t e^t \end{bmatrix}.$$

For the eigenvalue $r = 3$, we solve $(\mathbf{A} - 3\mathbf{I}) \mathbf{u} = \mathbf{0}$, that is,

$$\begin{bmatrix} -2 & 0 & 0 \\ 1 & 0 & 0 \\ 0 & 1 & -2 \end{bmatrix} \begin{bmatrix} u_1 \\ u_2 \\ u_3 \end{bmatrix} = \begin{bmatrix} 0 \\ 0 \\ 0 \end{bmatrix},$$

to obtain the eigenvector $\mathbf{u}_3 = \text{col}(0, 2, 1)$. Hence a third linearly independent solution to (9) is

$$(12) \qquad \mathbf{x}_3(t) = e^{3t} \mathbf{u}_3 = e^{3t} \begin{bmatrix} 0 \\ 2 \\ 1 \end{bmatrix} = \begin{bmatrix} 0 \\ 2e^{3t} \\ e^{3t} \end{bmatrix}.$$

The matrix $\mathbf{X}(t)$ whose columns are the vectors $\mathbf{x}_1(t)$, $\mathbf{x}_2(t)$, and $\mathbf{x}_3(t)$ given in equations (10),

(11), and (12), respectively, that is,

$$\mathbf{X}(t) = \begin{bmatrix} 0 & -2e^t & 0 \\ 0 & e^t & 2e^{3t} \\ e^t & te^t & e^{3t} \end{bmatrix},$$

is a fundamental matrix for (9). Setting $t = 0$ and then computing $\mathbf{X}^{-1}(0)$, we find

$$\mathbf{X}(0) = \begin{bmatrix} 0 & -2 & 0 \\ 0 & 1 & 2 \\ 1 & 0 & 1 \end{bmatrix} \quad \text{and} \quad \mathbf{X}^{-1}(0) = \begin{bmatrix} -\frac{1}{4} & -\frac{1}{2} & 1 \\ -\frac{1}{2} & 0 & 0 \\ \frac{1}{4} & \frac{1}{2} & 0 \end{bmatrix}.$$

It now follows from formula (6) that

$$e^{\mathbf{A}t} = \mathbf{X}(t)\mathbf{X}^{-1}(0) = \begin{bmatrix} 0 & -2e^t & 0 \\ 0 & e^t & 2e^{3t} \\ e^t & te^t & e^{3t} \end{bmatrix} \begin{bmatrix} -\frac{1}{4} & -\frac{1}{2} & 1 \\ -\frac{1}{2} & 0 & 0 \\ \frac{1}{4} & \frac{1}{2} & 0 \end{bmatrix}$$

$$= \begin{bmatrix} e^t & 0 & 0 \\ -\frac{1}{2}e^t + \frac{1}{2}e^{3t} & e^{3t} & 0 \\ -\frac{1}{4}e^t - \frac{1}{2}te^t + \frac{1}{4}e^{3t} & -\frac{1}{2}e^t + \frac{1}{2}e^{3t} & e^t \end{bmatrix}. \quad \bullet$$

Use of the fundamental matrix $e^{\mathbf{A}t}$ simplifies many computations. For example, the properties $e^{\mathbf{A}t}e^{-\mathbf{A}s} = e^{\mathbf{A}(t-s)}$ and $(e^{\mathbf{A}t_0})^{-1} = e^{-\mathbf{A}t_0}$ enable us to rewrite the variation of parameters formula (13) in Section 10.6 in a simpler form; namely, the solution to the initial value problem $\mathbf{x}' = \mathbf{Ax} + \mathbf{f}(t)$, $\mathbf{x}(t_0) = \mathbf{x}_0$ is given by

$$\textbf{(13)} \qquad \mathbf{x}(t) = e^{\mathbf{A}(t-t_0)}\mathbf{x}_0 + \int_{t_0}^{t} e^{\mathbf{A}(t-s)}\mathbf{f}(s)\,ds,$$

which is a system version of the formula for the solution to the scalar initial value problem $x' = ax + f(t)$, $x(t_0) = x_0$.

EXERCISES 10.7

In Problems 1 through 6: (a) Show that the given matrix $\mathbf{A}$ satisfies $(\mathbf{A} - r\mathbf{I})^k = \mathbf{0}$ for some number r and some positive integer k; (b) use this fact to determine the matrix $e^{\mathbf{A}t}$. [Hint: Compute the characteristic polynomial and use the Cayley-Hamilton theorem.]

1. $\mathbf{A} = \begin{bmatrix} 3 & -2 \\ 0 & 3 \end{bmatrix}.$

2. $\mathbf{A} = \begin{bmatrix} 1 & -1 \\ 1 & 3 \end{bmatrix}.$

3. $\mathbf{A} = \begin{bmatrix} 2 & 1 & -1 \\ -3 & -1 & 1 \\ 9 & 3 & -4 \end{bmatrix}.$

4. $\mathbf{A} = \begin{bmatrix} 2 & 1 & 3 \\ 0 & 2 & -1 \\ 0 & 0 & 2 \end{bmatrix}.$

5. $\mathbf{A} = \begin{bmatrix} -2 & 0 & 0 \\ 4 & -2 & 0 \\ 1 & 0 & -2 \end{bmatrix}.$

6. $\mathbf{A} = \begin{bmatrix} 0 & 1 & 0 \\ 0 & 0 & 1 \\ -1 & -3 & -3 \end{bmatrix}.$

In Problems 7 through 10 determine $e^{\mathbf{A}t}$ by first finding a fundamental matrix $\mathbf{X}(t)$ for $\mathbf{x}' = \mathbf{A}\mathbf{x}$ and then using formula (6).

7. $\mathbf{A} = \begin{bmatrix} 0 & 1 \\ -1 & 0 \end{bmatrix}$.

8. $\mathbf{A} = \begin{bmatrix} 1 & 1 \\ 4 & 1 \end{bmatrix}$.

9. $\mathbf{A} = \begin{bmatrix} 0 & 1 & 0 \\ 0 & 0 & 1 \\ 1 & -1 & 1 \end{bmatrix}$.

10. $\mathbf{A} = \begin{bmatrix} 0 & 2 & 2 \\ 2 & 0 & 2 \\ 2 & 2 & 0 \end{bmatrix}$.

In Problems 11 and 12 determine $e^{\mathbf{A}t}$ by using generalized eigenvectors to find a fundamental matrix and then using formula (6).

11. $\mathbf{A} = \begin{bmatrix} 5 & -4 & 0 \\ 1 & 0 & 2 \\ 0 & 2 & 5 \end{bmatrix}$.

12. $\mathbf{A} = \begin{bmatrix} 1 & 1 & 1 \\ 2 & 1 & -1 \\ 0 & -1 & 1 \end{bmatrix}$.

In Problems 13 through 16 use a linear algebra software package to help in determining $e^{\mathbf{A}t}$.

13. $\mathbf{A} = \begin{bmatrix} 0 & 1 & 0 & 0 & 0 \\ 0 & 0 & 1 & 0 & 0 \\ 1 & -3 & 3 & 0 & 0 \\ 0 & 0 & 0 & 0 & 1 \\ 0 & 0 & 0 & -1 & 0 \end{bmatrix}$.

14. $\mathbf{A} = \begin{bmatrix} 1 & 0 & 0 & 0 & 0 \\ 0 & 0 & 1 & 0 & 0 \\ 0 & -1 & -2 & 0 & 0 \\ 0 & 0 & 0 & 0 & 1 \\ 0 & 0 & 0 & -1 & 0 \end{bmatrix}$.

15. $\mathbf{A} = \begin{bmatrix} 0 & 1 & 0 & 0 & 0 \\ 0 & 0 & 1 & 0 & 0 \\ -1 & -3 & -3 & 0 & 0 \\ 0 & 0 & 0 & 0 & 1 \\ 0 & 0 & 0 & -4 & -4 \end{bmatrix}$.

16. $\mathbf{A} = \begin{bmatrix} -1 & 0 & 0 & 0 & 0 \\ 0 & 0 & 1 & 0 & 0 \\ 0 & -1 & -2 & 0 & 0 \\ 0 & 0 & 0 & 0 & 1 \\ 0 & 0 & 0 & -4 & -4 \end{bmatrix}$.

In Problems 17 through 20 use the generalized eigenvectors of $\mathbf{A}$ to find a general solution to the system $\mathbf{x}'(t) = \mathbf{A}\mathbf{x}(t)$, where $\mathbf{A}$ is given.

17. $\mathbf{A} = \begin{bmatrix} 0 & 1 & 0 \\ 0 & 0 & 1 \\ -2 & -5 & -4 \end{bmatrix}$.

18. $\mathbf{A} = \begin{bmatrix} 0 & 0 & 1 \\ 0 & 1 & 2 \\ 0 & 0 & 1 \end{bmatrix}$.

19. $\mathbf{A} = \begin{bmatrix} 1 & 0 & 1 & 2 \\ 1 & 1 & 2 & 1 \\ 0 & 0 & 2 & 0 \\ 0 & 0 & 1 & 1 \end{bmatrix}$.

20. $\mathbf{A} = \begin{bmatrix} -1 & -8 & 1 \\ -1 & -3 & 2 \\ -4 & -16 & 7 \end{bmatrix}$.

21. Use the results of Problem 5 to find the solution to the initial value problem
$$\mathbf{x}'(t) = \begin{bmatrix} -2 & 0 & 0 \\ 4 & -2 & 0 \\ 1 & 0 & -2 \end{bmatrix}\mathbf{x}(t), \qquad \mathbf{x}(0) = \begin{bmatrix} 1 \\ 1 \\ -1 \end{bmatrix}.$$

22. Use your answer to Problem 12 to find the solution to the initial value problem
$$\mathbf{x}'(t) = \begin{bmatrix} 1 & 1 & 1 \\ 2 & 1 & -1 \\ 0 & -1 & 1 \end{bmatrix}\mathbf{x}(t), \qquad \mathbf{x}(0) = \begin{bmatrix} -1 \\ 0 \\ 3 \end{bmatrix}.$$

23. Use the results of Problem 3 and the variation of parameters formula (13) to find the solution to the initial value problem
$$\mathbf{x}'(t) = \begin{bmatrix} 2 & 1 & -1 \\ -3 & -1 & 1 \\ 9 & 3 & -4 \end{bmatrix}\mathbf{x}(t) + \begin{bmatrix} 0 \\ t \\ 0 \end{bmatrix}, \qquad \mathbf{x}(0) = \begin{bmatrix} 0 \\ 3 \\ 0 \end{bmatrix}.$$

24. Use your answer to Problem 9 and the variation of parameters formula (13) to find the solution to the initial value problem
$$\mathbf{x}'(t) = \begin{bmatrix} 0 & 1 & 0 \\ 0 & 0 & 1 \\ 1 & -1 & 1 \end{bmatrix}\mathbf{x}(t) + \begin{bmatrix} 0 \\ 0 \\ t \end{bmatrix}, \qquad \mathbf{x}(0) = \begin{bmatrix} 1 \\ -1 \\ 0 \end{bmatrix}.$$

25. Let
$$\mathbf{A} = \begin{bmatrix} 1 & 2 \\ -1 & 3 \end{bmatrix} \quad \text{and} \quad \mathbf{B} = \begin{bmatrix} 2 & 1 \\ 0 & 1 \end{bmatrix}.$$

(a) Show that $\mathbf{A}\mathbf{B} \neq \mathbf{B}\mathbf{A}$.

(b) Show that property (d) in Theorem 7 does not hold for these matrices. That is, show that $e^{(\mathbf{A}+\mathbf{B})t} \neq e^{\mathbf{A}t}e^{\mathbf{B}t}$.

26. Let $\mathbf{A}$ be a diagonal $n \times n$ matrix with entries $r_1, \ldots, r_n$ down its main diagonal. To compute $e^{\mathbf{A}t}$:

(a) Show that $\mathbf{A}^k$ is the diagonal matrix with entries $r_1^k, \ldots, r_n^k$ down its main diagonal.

(b) Use the result of part (a) to show that $e^{\mathbf{A}t}$ is the diagonal matrix with entries $e^{r_1 t}, \ldots, e^{r_n t}$ down its main diagonal.

• CHAPTER SUMMARY •

In this chapter we discussed the theory of linear systems in normal form and presented methods for solving such systems. The theory and methods are natural extensions of the development for second order and higher order linear equations. The important properties and techniques are listed below.

Homogeneous Normal Systems

$$\mathbf{x}'(t) = \mathbf{A}(t)\mathbf{x}(t)$$

The $n \times n$ matrix function $\mathbf{A}(t)$ is assumed to be continuous on an interval I.

Fundamental Solution Set: $\{\mathbf{x_1}, \dots, \mathbf{x_n}\}$. The n vector solutions $\mathbf{x}_1(t), \dots, \mathbf{x}_n(t)$ of the homogeneous system on the interval I form a **fundamental solution set,** provided that they are linearly independent on I or, equivalently, that their **Wronskian**

$$W[\mathbf{x}_1, \dots, \mathbf{x}_n](t) := \det[\mathbf{x}_1, \dots, \mathbf{x}_n] = \begin{vmatrix} x_{1,1}(t) & x_{1,2}(t) & \cdots & x_{1,n}(t) \\ x_{2,1}(t) & x_{2,2}(t) & \cdots & x_{2,n}(t) \\ \vdots & \vdots & & \vdots \\ x_{n,1}(t) & x_{n,2}(t) & \cdots & x_{n,n}(t) \end{vmatrix}$$

is never zero on I.

Fundamental Matrix: $\mathbf{X}(t)$. An $n \times n$ matrix function $\mathbf{X}(t)$ whose column vectors form a fundamental solution set for the homogeneous system is called a **fundamental matrix.** The determinant of $\mathbf{X}(t)$ is the Wronskian of the fundamental solution set. Since the Wronskian is never zero on the interval I, then $\mathbf{X}^{-1}(t)$ exists for t in I.

General Solution to Homogeneous System: $\mathbf{Xc} = c_1\mathbf{x_1} + \cdots + c_n\mathbf{x_n}$. If $\mathbf{X}(t)$ is a fundamental matrix whose column vectors are $\mathbf{x}_1, \dots, \mathbf{x}_n$, then a general solution to the homogeneous system is

$$\mathbf{x}(t) = \mathbf{X}(t)\mathbf{c} = c_1\mathbf{x}_1(t) + c_2\mathbf{x}_2(t) + \cdots + c_n\mathbf{x}_n(t),$$

where $\mathbf{c} = \text{col}(c_1, \dots, c_n)$ is an arbitrary constant vector.

Homogeneous Systems with Constant Coefficients. The form of a general solution for a homogeneous system with constant coefficients depends on the eigenvalues and eigenvectors of the $n \times n$ constant matrix $\mathbf{A}$. An **eigenvalue** of $\mathbf{A}$ is a number r such that the system $\mathbf{Au} = r\mathbf{u}$ has a nontrivial solution $\mathbf{u}$ called an **eigenvector** of $\mathbf{A}$ associated with the eigenvalue r. Finding the eigenvalues of $\mathbf{A}$ is equivalent to finding the roots of the **characteristic equation**

$$|\mathbf{A} - r\mathbf{I}| = 0.$$

The corresponding eigenvectors are found by solving the system $(\mathbf{A} - r\mathbf{I})\mathbf{u} = \mathbf{0}$.

If the matrix $\mathbf{A}$ has n linearly independent eigenvectors $\mathbf{u}_1, \ldots, \mathbf{u}_n$ and r_i is the eigenvalue corresponding to the eigenvector $\mathbf{u}_i$, then

$$\{e^{r_1 t}\mathbf{u}_1, e^{r_2 t}\mathbf{u}_2, \ldots, e^{r_n t}\mathbf{u}_n\}$$

is a fundamental solution set for the homogeneous system. A class of matrices that always has n linearly independent eigenvectors is the set of **symmetric** matrices—that is, matrices that satisfy $\mathbf{A} = \mathbf{A}^T$.

If $\mathbf{A}$ has complex conjugate eigenvalues $\alpha \pm i\beta$ and associated eigenvectors $\mathbf{z} = \mathbf{a} \pm i\mathbf{b}$, where $\mathbf{a}$ and $\mathbf{b}$ are real vectors, then two linearly independent real vector solutions to the homogeneous system are

$$e^{\alpha t}\cos\beta t\,\mathbf{a} - e^{\alpha t}\sin\beta t\,\mathbf{b}, \qquad e^{\alpha t}\sin\beta t\,\mathbf{a} + e^{\alpha t}\cos\beta t\,\mathbf{b}.$$

When $\mathbf{A}$ has a repeated eigenvalue r of multiplicity m, then it is possible that $\mathbf{A}$ does *not* have n linearly independent eigenvectors. However, associated with r are m linearly independent solutions to the homogeneous system of the form

$$\mathbf{x}_1(t) = e^{rt}\mathbf{k}_{1,1}, \qquad \mathbf{x}_2(t) = e^{rt}\mathbf{k}_{2,1} + te^{rt}\mathbf{k}_{2,2}, \quad \ldots,$$

$$\mathbf{x}_m(t) = e^{rt}\mathbf{k}_{m,1} + te^{rt}\mathbf{k}_{m,2} + \cdots + \frac{t^{m-1}}{(m-1)!}e^{rt}\mathbf{k}_{m,m}.$$

Nonhomogeneous Normal Systems

$$\mathbf{x}'(t) = \mathbf{A}(t)\mathbf{x}(t) + \mathbf{f}(t)$$

The $n \times n$ matrix function $\mathbf{A}(t)$ and the vector function $\mathbf{f}(t)$ are assumed continuous on an interval I.

General Solution to Nonhomogeneous System: $\mathbf{x}_p + \mathbf{Xc}$. If $\mathbf{x}_p(t)$ is any particular solution for the nonhomogeneous system and $\mathbf{X}(t)$ is a fundamental matrix for the associated homogeneous system, then a general solution for the nonhomogeneous system is

$$\mathbf{x}(t) = \mathbf{x}_p(t) + \mathbf{X}(t)\mathbf{c} = \mathbf{x}_p(t) + c_1\mathbf{x}_1(t) + \cdots + c_n\mathbf{x}_n(t),$$

where $\mathbf{x}_1(t), \ldots, \mathbf{x}_n(t)$ are the column vectors of $\mathbf{X}(t)$ and $\mathbf{c} = \mathrm{col}(c_1, \ldots, c_n)$ is an arbitrary constant vector.

Undetermined Coefficients. If the nonhomogeneous term $\mathbf{f}(t)$ is a vector whose components are polynomials, or exponential or sinusoidal functions, and $\mathbf{A}$ is a constant matrix, then one can use an extension of the method of undetermined coefficients to decide the form of a particular solution to the nonhomogeneous system.

Variation of Parameters: $\mathbf{X}(t)v(t)$. Let $\mathbf{X}(t)$ be a fundamental matrix for the homogeneous system. A particular solution to the nonhomogeneous system is given by the **variation of parameters** formula

$$\mathbf{x}_p(t) = \mathbf{X}(t)v(t) = \mathbf{X}(t)\int \mathbf{X}^{-1}(t)\mathbf{f}(t)\,dt.$$

Matrix Exponential Function

If $\mathbf{A}$ is a constant $n \times n$ matrix, then the matrix exponential function

$$e^{\mathbf{A}t} := \mathbf{I} + \mathbf{A}t + \mathbf{A}^2 \frac{t^2}{2!} + \cdots + \mathbf{A}^n \frac{t^n}{n!} + \cdots$$

is a fundamental matrix for the homogeneous system $\mathbf{x}'(t) = \mathbf{A}\mathbf{x}(t)$. The matrix exponential has some of the same properties as the scalar exponential e^{at}. In particular,

$$e^{\mathbf{0}} = \mathbf{I}, \qquad e^{\mathbf{A}(t+s)} = e^{\mathbf{A}t}e^{\mathbf{A}s}, \qquad (e^{\mathbf{A}t})^{-1} = e^{-\mathbf{A}t}.$$

If $(\mathbf{A} - r\mathbf{I})^k = \mathbf{0}$ for some r and k, then the series for $e^{\mathbf{A}t}$ has only a finite number of terms:

$$e^{\mathbf{A}t} = e^{rt}\left\{\mathbf{I} + (\mathbf{A} - r\mathbf{I})t + \cdots + (\mathbf{A} - r\mathbf{I})^{k-1}\frac{t^{k-1}}{(k-1)!}\right\}.$$

The matrix exponential function $e^{\mathbf{A}t}$ can also be computed from any fundamental matrix $\mathbf{X}(t)$ via the formula

$$e^{\mathbf{A}t} = \mathbf{X}(t)\mathbf{X}^{-1}(0).$$

Generalized Eigenvectors

If r is an eigenvalue of $\mathbf{A}$, then a **generalized eigenvector** associated with r is a nonzero vector $\mathbf{u}$ that satisfies $(\mathbf{A} - r\mathbf{I})^k\mathbf{u} = \mathbf{0}$ for some positive integer k. *Every* matrix $\mathbf{A}$ has a set of n linearly independent generalized eigenvectors that can be used to compute a fundamental solution set. In particular, associated with each generalized eigenvector $\mathbf{u}$ and corresponding eigenvalue r is a solution to the homogeneous system $\mathbf{x}'(t) = \mathbf{A}\mathbf{x}(t)$ of the form

$$e^{\mathbf{A}t}\mathbf{u} = e^{rt}\{\mathbf{u} + t(\mathbf{A} - r\mathbf{I})\mathbf{u} + \frac{t^2}{2!}(\mathbf{A} - r\mathbf{I})^2\mathbf{u} + \cdots\}.$$

If r has multiplicity m, then the above series reduces to the first m terms.

REVIEW PROBLEMS

In Problems 1 through 4 find a general solution for the system $\mathbf{x}'(t) = \mathbf{A}\mathbf{x}(t)$, *where* $\mathbf{A}$ *is given.*

1. $\mathbf{A} = \begin{bmatrix} 6 & -3 \\ 2 & 1 \end{bmatrix}$.

2. $\mathbf{A} = \begin{bmatrix} 3 & 2 \\ -5 & 1 \end{bmatrix}$.

3. $\mathbf{A} = \begin{bmatrix} 1 & 2 & 0 & 0 \\ 2 & 1 & 0 & 0 \\ 0 & 0 & 1 & 2 \\ 0 & 0 & 2 & 1 \end{bmatrix}$.

4. $\mathbf{A} = \begin{bmatrix} 1 & 1 & 0 \\ 0 & 1 & 0 \\ 0 & 0 & 2 \end{bmatrix}$.

In Problems 5 and 6 find a fundamental matrix for the system $\mathbf{x}'(t) = \mathbf{A}\mathbf{x}(t)$, *where* $\mathbf{A}$ *is given.*

5. $\mathbf{A} = \begin{bmatrix} 1 & -1 \\ 2 & 4 \end{bmatrix}$.

6. $\mathbf{A} = \begin{bmatrix} 5 & 0 & 0 \\ 0 & -4 & 3 \\ 0 & 3 & 4 \end{bmatrix}$.

In Problems 7 through 10 find a general solution for the system $\mathbf{x}'(t) = \mathbf{A}\mathbf{x}(t) + \mathbf{f}(t)$, *where* $\mathbf{A}$ *and* $\mathbf{f}(t)$ *are given.*

7. $\mathbf{A} = \begin{bmatrix} 1 & 1 \\ 4 & 1 \end{bmatrix}$, $\quad \mathbf{f}(t) = \begin{bmatrix} 5 \\ 6 \end{bmatrix}$.

8. $\mathbf{A} = \begin{bmatrix} -4 & 2 \\ 2 & -1 \end{bmatrix}$, $\mathbf{f}(t) = \begin{bmatrix} e^{4t} \\ 3e^{4t} \end{bmatrix}$.

9. $\mathbf{A} = \begin{bmatrix} 2 & 1 & -1 \\ -3 & -1 & 1 \\ 9 & 3 & -4 \end{bmatrix}$, $\mathbf{f}(t) = \begin{bmatrix} t \\ 0 \\ 1 \end{bmatrix}$.

10. $\mathbf{A} = \begin{bmatrix} 2 & -2 & 3 \\ 0 & 3 & 2 \\ 0 & -1 & 2 \end{bmatrix}$, $\mathbf{f}(t) = \begin{bmatrix} e^{-t} \\ 2 \\ 1 \end{bmatrix}$.

In Problems 11 and 12 solve the given initial value problem.

11. $\mathbf{x}'(t) = \begin{bmatrix} 0 & 1 \\ -2 & 3 \end{bmatrix} \mathbf{x}(t)$, $\mathbf{x}(0) = \begin{bmatrix} 1 \\ -1 \end{bmatrix}$.

12. $\mathbf{x}'(t) = \begin{bmatrix} 2 & 1 \\ -4 & 2 \end{bmatrix} \mathbf{x}(t) + \begin{bmatrix} te^{2t} \\ e^{2t} \end{bmatrix}$, $\mathbf{x}(0) = \begin{bmatrix} 2 \\ 2 \end{bmatrix}$.

In Problems 13 and 14 find a general solution for the Cauchy-Euler system $t\mathbf{x}'(t) = \mathbf{A}\mathbf{x}(t)$, *where* $\mathbf{A}$ *is given.*

13. $\mathbf{A} = \begin{bmatrix} 0 & 3 & 1 \\ 1 & 2 & 1 \\ 1 & 3 & 0 \end{bmatrix}$.   **14.** $\mathbf{A} = \begin{bmatrix} 1 & 2 & -1 \\ 2 & 1 & 1 \\ -1 & 1 & 0 \end{bmatrix}$.

In Problems 15 and 16 find the fundamental matrix $e^{\mathbf{A}t}$ *for the system* $\mathbf{x}'(t) = \mathbf{A}\mathbf{x}(t)$, *where* $\mathbf{A}$ *is given.*

15. $\mathbf{A} = \begin{bmatrix} 4 & 2 & 3 \\ 2 & 1 & 2 \\ -1 & 2 & 0 \end{bmatrix}$.   **16.** $\mathbf{A} = \begin{bmatrix} 0 & 1 & 4 \\ 0 & 0 & 2 \\ 0 & 0 & 0 \end{bmatrix}$.

TECHNICAL WRITING EXERCISES

1. Explain how the theory of homogeneous linear differential equations (as described in Sections 4.3 and 6.2) follows from the theory of linear systems in normal form (as described in Section 10.3).

2. Discuss the similarities and differences between the method for finding solutions to a linear constant coefficient differential equation (see Sections 4.5, 4.6, and 6.3) and the method for finding solutions to a linear system in normal form that has constant coefficients (see Sections 10.4 and 10.5).

3. Explain how the variation of parameters formulas for linear second order equations derived in Section 4.10 follow from the formulas derived in Section 10.6 for linear systems in normal form.

4. Explain how you would define the matrix functions $\sin \mathbf{A}t$ and $\cos \mathbf{A}t$, where $\mathbf{A}$ is a constant $n \times n$ matrix. How are these functions related to the matrix exponential and how are they connected to the solutions of the system $\mathbf{x}'' + \mathbf{A}^2\mathbf{x} = \mathbf{0}$? The reader will find it informative to consider the cases when $\mathbf{A}$ is

$$\begin{bmatrix} 0 & 0 \\ 0 & 0 \end{bmatrix}, \quad \begin{bmatrix} 0 & 1 \\ 0 & 0 \end{bmatrix}, \quad \begin{bmatrix} 0 & 0 \\ 1 & 0 \end{bmatrix}, \quad \text{and} \quad \begin{bmatrix} 1 & -1 \\ 1 & -1 \end{bmatrix}.$$

GROUP PROJECTS FOR CHAPTER 10

A. Uncoupling Normal Systems

The easiest normal systems to solve are systems of the form

(1) $\mathbf{x}'(t) = \mathbf{D}\mathbf{x}(t)$,

where $\mathbf{D}$ is an $n \times n$ diagonal matrix. Such a system actually consists of n uncoupled equations

(2) $x_i'(t) = d_{ii}x_i(t)$, $i = 1, \ldots, n$, whose solution is

$$x_i(t) = c_i e^{d_{ii} t},$$

where the c_i's are arbitrary constants. This raises the following question: When can we *uncouple* a normal system?

To answer this question, we need the following result from linear algebra. An $n \times n$ matrix $\mathbf{A}$ is diagonalizable if and only if $\mathbf{A}$ has n linearly independent eigenvectors $\mathbf{p}_1, \ldots, \mathbf{p}_n$. Moreover, if $\mathbf{P}$ is the matrix whose columns are $\mathbf{p}_1, \ldots, \mathbf{p}_n$, then

(3) $\mathbf{P}^{-1}\mathbf{A}\mathbf{P} = \mathbf{D}$,

where $\mathbf{D}$ is the diagonal matrix whose entry d_{ii} is the eigenvalue associated with the vector $\mathbf{p}_i$.

   (a) Use the above result to show that the system

(4) $\mathbf{x}'(t) = \mathbf{A}\mathbf{x}(t)$,

   where $\mathbf{A}$ is an $n \times n$ diagonalizable matrix, is equivalent to an uncoupled system

(5) $\mathbf{y}'(t) = \mathbf{D}\mathbf{y}(t)$,

   where $\mathbf{y} = \mathbf{P}^{-1}\mathbf{x}$ and $\mathbf{D} = \mathbf{P}^{-1}\mathbf{A}\mathbf{P}$.
   (b) Solve system (5).
   (c) Use the results of parts (a) and (b) to show that a general solution to (4) is given by
   $\mathbf{x}(t) = c_1 e^{d_{11} t}\mathbf{p}_1 + c_2 e^{d_{22} t}\mathbf{p}_2 + \cdots + c_n e^{d_{nn} t}\mathbf{p}_n.$
   (d) Use the procedure discussed in parts (a)–(c) to obtain a general solution for the system

$$\mathbf{x}'(t) = \begin{bmatrix} 1 & 2 & -1 \\ 1 & 0 & 1 \\ 4 & -4 & 5 \end{bmatrix} \mathbf{x}(t).$$

   Specify $\mathbf{P}$, $\mathbf{D}$, $\mathbf{P}^{-1}$, and $\mathbf{y}$.

B. Matrix Laplace Transform Method

The Laplace transform method for solving linear differential equations with constant coefficients was discussed in Chapter 7. This method can also be used for systems of differential equations as described in Section 9.3. To apply the procedure for equations given in matrix form, we first extend the definition of the Laplace operator $\mathcal{L}$ to a column vector of functions $\mathbf{x} = \mathrm{col}(x_1(t), \ldots, x_n(t))$ by taking the transform of each component:

$$\mathcal{L}\{\mathbf{x}\}(s) := \mathrm{col}(\mathcal{L}\{x_1\}(s), \ldots, \mathcal{L}\{x_n\}(s)).$$

With this notation, the vector analogue of the important property relating the Laplace transform of the derivative of a function (see Theorem 4, Chapter 7, page 287) becomes

(6) $\mathcal{L}\{\mathbf{x}'\}(s) = s\mathcal{L}\{\mathbf{x}\}(s) - \mathbf{x}(0)$.

   Now suppose we are given the initial value problem

(7) $\mathbf{x}' = \mathbf{A}\mathbf{x} + \mathbf{f}(t)$, $\mathbf{x}(0) = \mathbf{x}_0$,

where $\mathbf{A}$ is a constant $n \times n$ matrix. Let $\hat{\mathbf{x}}(s)$ denote the Laplace transform of $\mathbf{x}(t)$ and $\hat{\mathbf{f}}(s)$ denote the transform of $\mathbf{f}(t)$. Then, taking the transform of the system and using the relation (6), we get

$$\mathcal{L}\{\mathbf{x}'\} = \mathcal{L}\{\mathbf{A}\mathbf{x} + \mathbf{f}\},$$
$$s\hat{\mathbf{x}} - \mathbf{x}_0 = \mathbf{A}\hat{\mathbf{x}} + \hat{\mathbf{f}}.$$

Next we collect the $\hat{\mathbf{x}}$ terms and solve for $\hat{\mathbf{x}}$ by premultiplying by $(s\mathbf{I} - \mathbf{A})^{-1}$:

$$(s\mathbf{I} - \mathbf{A})\hat{\mathbf{x}} = \hat{\mathbf{f}} + \mathbf{x}_0$$
$$\hat{\mathbf{x}} = (s\mathbf{I} - \mathbf{A})^{-1}(\hat{\mathbf{f}} + \mathbf{x}_0).$$

Finally, we obtain the solution $\mathbf{x}(t)$ by taking the inverse Laplace transform:

$$\mathbf{x} = \mathscr{L}^{-1}\{\hat{\mathbf{x}}\} = \mathscr{L}^{-1}\{(s\mathbf{I} - \mathbf{A})^{-1}(\hat{\mathbf{f}} + \mathbf{x}_0)\}.$$

In applying the matrix Laplace transform method, it is straightforward (but possibly tedious) to compute $(s\mathbf{I} - \mathbf{A})^{-1}$, but the computation of the inverse transform may require some of the special techniques (such as partial fractions) discussed in Chapter 7.

   (a) In the above procedure we used the property that $\mathscr{L}\{\mathbf{Ax}\} = \mathbf{A}\mathscr{L}\{\mathbf{x}\}$ for any constant $n \times n$ matrix $\mathbf{A}$. Show that this property follows from the linearity of the transform in the scalar case.

   (b) Use the matrix Laplace transform method to solve the following initial value problems:

   (i) $\mathbf{x}'(t) = \begin{bmatrix} 0 & 2 \\ -1 & 3 \end{bmatrix}\mathbf{x}(t), \qquad \mathbf{x}(0) = \begin{bmatrix} -1 \\ 3 \end{bmatrix}.$

   (ii) $\mathbf{x}'(t) = \begin{bmatrix} 3 & -2 \\ 4 & -1 \end{bmatrix}\mathbf{x}(t) + \begin{bmatrix} \sin t \\ -\cos t \end{bmatrix}, \qquad \mathbf{x}(0) = \begin{bmatrix} 0 \\ 0 \end{bmatrix}.$

C. Undamped Second Order Systems

In the study of vibrations of mechanical systems, one encounters systems of the form

(8) $\mathbf{x}''(t) = \mathbf{Bx}(t),$

where $\mathbf{B}$ is an $n \times n$ constant matrix. Experience with studying vibrations for a simple spring-mass system (see Section 5.1) suggests that we seek a solution to (8) of the form

(9) $\mathbf{x} = \cos \omega t \, \mathbf{v} \quad \text{or} \quad \mathbf{x} = \sin \omega t \, \mathbf{v},$

where $\mathbf{v}$ is a constant vector and ω is a positive constant.

   (a) Show that system (8) has a nontrivial solution of the form given in (9) if and only if $-\omega^2$ is an eigenvalue of $\mathbf{B}$.

   (b) If $\mathbf{B}$ is an $n \times n$ constant matrix, then system (8) can be written as a system of $2n$ first order equations in normal form. Thus, a general solution to (8) can be formed from $2n$ linearly independent solutions to (8). Use the result of part (a) to find a general solution to the following second order systems.

   (i) $\mathbf{x}'' = \begin{bmatrix} -5 & -4 \\ 1 & 0 \end{bmatrix}\mathbf{x}.$

   (ii) $\mathbf{x}'' = \begin{bmatrix} -5 & 2 \\ 2 & -2 \end{bmatrix}\mathbf{x}.$

   (iii) $\mathbf{x}'' = \begin{bmatrix} -1 & -2 & 1 \\ -1 & 0 & -1 \\ -4 & 4 & -5 \end{bmatrix}\mathbf{x}.$

   (iv) $\mathbf{x}'' = \begin{bmatrix} 2 & -1 & 0 \\ -1 & 2 & -1 \\ 0 & -1 & 2 \end{bmatrix}\mathbf{x}.$

D. Strange Behavior of Competing Species—Part II

In Project E in Chapter 9 we used numerical experimentation to study the behavior of solutions to the system

$$
\begin{aligned}
p_1' &= p_1(1 - p_1 - ap_2 - bp_3), \\
\textbf{(10)} \qquad p_2' &= p_2(1 - bp_1 - p_2 - ap_3), \\
p_3' &= p_3(1 - ap_1 - bp_2 - p_3),
\end{aligned}
$$

where a and b are positive constants and p_i is the population of species S_i, $i = 1, 2, 3$. We found that when $a = b = 0.5$, the populations approach the equilibrium solution $p_1 = p_2 = p_3 = 1/2$ as $t \to \infty$. However, when $a = 0.5$ and $b = 2$, the populations cycled between the three critical points—$(1, 0, 0)$, $(0, 1, 0)$, $(0, 0, 1)$—getting closer and closer to the points. Moreover, the populations remained near each point for longer and longer periods of time before moving on to the next point. To analyze what is happening near these equilibrium points, we will study the corresponding linearized systems.

(a) Let $f_i(p_1, p_2, p_3)$ denote the function on the right-hand side of the equation for p_i' in (10); in particular, $f_1(p_1, p_2, p_3) = p_1(1 - p_1 - ap_2 - bp_3)$. Now $\mathbf{f} = \text{col}(f_1, f_2, f_3)$ is a mapping from $\mathbf{R}^3$ into $\mathbf{R}^3$. Show that the Jacobian matrix for this mapping is

$$
\mathbf{A} := \begin{bmatrix} 1 - 2p_1 - ap_2 - bp_3 & -ap_1 & -bp_1 \\ -bp_2 & 1 - 2p_2 - ap_3 - bp_1 & -ap_2 \\ -ap_3 & -bp_3 & 1 - 2p_3 - ap_1 - bp_2 \end{bmatrix}.
$$

(b) Show that at the critical point (δ, δ, δ) where $\delta = (1 + a + b)^{-1}$, we have $\mathbf{A} = -\delta \mathbf{B}$, where

$$
\mathbf{B} := \begin{bmatrix} 1 & a & b \\ b & 1 & a \\ a & b & 1 \end{bmatrix}.
$$

(c) Show that the eigenvalues of $\mathbf{B}$ are

$$
1 + a + b, \qquad \left(1 - \frac{a + b}{2}\right) \pm i \frac{\sqrt{3}(a - b)}{2}.
$$

(d) Now, near the critical point (δ, δ, δ), the solutions to (10) should behave like the solutions to $\mathbf{p}' = -\delta \mathbf{B} \mathbf{p}$. Use the results of part (c) to argue that when $a + b < 2$, the solutions to (10) should approach the critical point (δ, δ, δ) and when $a + b > 2$, they should move away.

(e) Using a similar analysis, conclude that when either $a < 1$ or $b < 1$, the solutions to (10) should move away from the critical points $(1, 0, 0)$, $(0, 1, 0)$, and $(0, 0, 1)$ because at least one of the eigenvalues of the linearized system (about the critical point) is positive.

(f) How do the analyses performed in parts (d) and (e) support the conclusions reached from numerical experimentations?

CHAPTER 11

•
•
• Partial Differential Equations
•

•11.1• INTRODUCTION: A MODEL FOR HEAT FLOW

> **Develop a model for the flow of heat through a thin, insulated wire whose ends are kept at a constant temperature of 0°C and whose initial temperature distribution is to be specified.**

Suppose that the wire is placed along the x-axis with $x = 0$ at the left end of the wire and $x = L$ at the right end (see Figure 11.1). If we let u denote the temperature of the wire, then u depends upon the time t and upon the position x within the wire. (We will assume the wire is thin and hence u is constant throughout a cross section of the wire corresponding to a fixed value of x.) Since the wire is insulated, we assume that no heat enters or leaves through the sides of the wire.

To develop a model for heat flow through a thin wire, let's consider the small volume element V of wire between the two cross-sectional planes A and B that are perpendicular to the x-axis, with plane A located at x and plane B located at $x + \Delta x$ (see Figure 11.1).

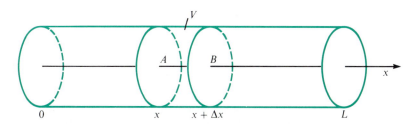

Figure 11.1 Heat flow through a thin piece of wire

The temperature on plane A at time t is $u(x, t)$ and on plane B is $u(x + \Delta x, t)$. We will need the following principles of physics which describe heat flow.[†]

1. **Heat Conduction:** The rate of heat flow (the amount of heat per unit time flowing through a unit of cross-sectional area at A) is proportional to $\partial u/\partial x$, the temperature gradient at A. The proportionality constant k is called the **thermal conductivity** of the material.

2. **Direction of Heat Flow:** The direction of heat flow is always from points of higher temperature to points of lower temperature.

3. **Specific Heat Capacity:** The amount of heat necessary to raise the temperature of an object of mass m by an amount Δu is $cm\,\Delta u$, where the constant c is the **specific heat capacity** of the material.

If we let H represent the amount of heat flowing from left to right through the surface A during an interval of time Δt, then the formula for heat conduction becomes

$$H(x) = -ka\,\Delta t\,\frac{\partial u}{\partial x}(x, t),$$

where a is the cross-sectional area of the wire. The negative sign follows from the second principle—if $\partial u/\partial x$ is positive, then heat flows from right to left (hotter to colder).

Similarly, the amount of heat flowing from left to right across plane B during an interval of time Δt is

$$H(x + \Delta x) = -ka\,\Delta t\,\frac{\partial u}{\partial x}(x + \Delta x, t).$$

The net change in the heat ΔH in volume V is the amount entering at end A minus the amount leaving at end B. That is,

(1) $$\Delta H = H(x) - H(x + \Delta x) = ka\,\Delta t\left[\frac{\partial u}{\partial x}(x + \Delta x, t) - \frac{\partial u}{\partial x}(x, t)\right].$$

Now by the third principle, the net change in heat is given by $\Delta H = cm\,\Delta u$, where Δu is the change in temperature and c is the specific heat capacity. If we assume the change in temperature in the volume V is essentially equal to the change in temperature at x, that is $\Delta u = u(x, t + \Delta t) - u(x, t)$, and that the mass of the volume V of wire is $a\rho\,\Delta x$, where ρ is the density of the wire, a is the cross-sectional area and Δx is the length, then

(2) $$\Delta H = c\rho a\,\Delta x[u(x, t + \Delta t) - u(x, t)].$$

[†] For a discussion of heat transfer see *University Physics*, Sixth Edition, by F. W. Sears, M. W. Zemansky, and H. D. Young, Addison-Wesley Publishing Co., Reading, Massachusetts, 1982.

Equating the change in heat given in equations (1) and (2) yields

$$ka\,\Delta t\left[\frac{\partial u}{\partial x}(x+\Delta x,t)-\frac{\partial u}{\partial x}(x,t)\right]=c\rho a\,\Delta x[u(x,t+\Delta t)-u(x,t)].$$

Now dividing both sides by Δx and Δt and then taking the limits as Δx and Δt approach zero, we obtain

$$k\frac{\partial^2 u}{\partial x^2}(x,t)=c\rho\frac{\partial u}{\partial t}(x,t)$$

or

(3) $\qquad \dfrac{\partial u}{\partial t}(x,t)=\beta\dfrac{\partial^2 u}{\partial x^2}(x,t),$

where the positive constant $\beta:=k/\rho c$ is the **diffusivity** of the material. Equation (3) is the **one-dimensional heat flow equation.**

Equation (3) governs the flow of heat in the wire. We have two other constraints in our original problem. First, we are keeping the ends of the wire at $0°C$. Thus we require that

(4) $\qquad u(0,t)=u(L,t)=0,$

for all t. These are called **boundary conditions.** Second, we must be given the initial temperature distribution $f(x)$. That is, we require

(5) $\qquad u(x,0)=f(x),\qquad 0<x<L.$

Equation (5) is referred to as the **initial condition** on u.

Combining equations (3), (4), and (5), we have the following mathematical model for the heat flow in a wire whose ends are kept at the constant temperature $0°C$:

(6) $\qquad \dfrac{\partial u}{\partial t}(x,t)=\beta\dfrac{\partial^2 u}{\partial x^2}(x,t),\qquad 0<x<L,\qquad t>0,$

(7) $\qquad u(0,t)=u(L,t)=0,\qquad t>0,$

(8) $\qquad u(x,0)=f(x),\qquad 0<x<L.$

This model is an example of an **initial-boundary value problem.**

In higher dimensions the heat flow equation (or just **heat equation**) has the form

$$\frac{\partial u}{\partial t}=\beta\,\Delta u,$$

where Δu is the **Laplacian** of u. In two and three dimensions the Laplacian is defined, respectively, as

$$\Delta u:=\frac{\partial^2 u}{\partial x^2}+\frac{\partial^2 u}{\partial y^2}\quad\text{and}\quad \Delta u:=\frac{\partial^2 u}{\partial x^2}+\frac{\partial^2 u}{\partial y^2}+\frac{\partial^2 u}{\partial z^2}.$$

When the temperature reaches a steady-state, that is, when u does not depend on time, then $\partial u/\partial t = 0$ and the temperature satisfies **Laplace's equation**

$$\Delta u = 0.$$

One classical technique for solving the initial-boundary value problem for the heat equation (6)–(8) is the method of *separation of variables*, which effectively allows us to replace the partial derivatives by ordinary derivatives. This technique is discussed in the next section. In using separation of variables, one is often required to express a given function as a trigonometric series. Such series are called *Fourier series*; their properties are discussed in Sections 11.3 and 11.4. We devote the remaining three sections to the three basic partial differential equations that arise in applications: the heat equation, the wave equation, and Laplace's equation.

•11.2• METHOD OF SEPARATION OF VARIABLES

The method of separation of variables is a classical technique that is effective in solving several types of partial differential equations. The idea is roughly the following. We think of a solution, say $u(x,t)$, to a partial differential equation as being an infinite linear combination of simple component functions $u_n(x,t)$, $n = 0, 1, 2,\ldots$, which also satisfy the equation and certain boundary conditions. (This is a reasonable assumption provided the partial differential equation and the boundary conditions are linear and homogeneous.) To determine a component solution, $u_n(x,t)$, we assume it can be written with its variables separated; that is, as

$$u_n(x,t) = X_n(x)T_n(t).$$

Substituting this form for a solution into the partial differential equation and using the boundary conditions leads to two *ordinary* differential equations for the unknown functions $X_n(x)$ and $T_n(t)$. In this way we have reduced the problem of solving a partial differential equation to the more familiar problem of solving a differential equation that involves only one variable. In this section we will illustrate this technique for the heat equation and the wave equation.

In the previous section we derived the following initial-boundary value problem as a mathematical model for the heat flow in a wire whose ends are kept at the constant temperature zero.

$$(1) \qquad \frac{\partial u}{\partial t}(x,t) = \beta\frac{\partial^2 u}{\partial x^2}(x,t), \qquad 0 < x < L, \qquad t > 0,$$

$$(2) \qquad u(0,t) = u(L,t) = 0, \qquad t > 0,$$

$$(3) \qquad u(x,0) = f(x), \qquad 0 < x < L.$$

To solve this problem by the method of separation of variables, we begin by assuming that equation (1) has a solution of the form

$$u(x,t) = X(x)T(t),$$

where X is a function of x alone and T is a function of t alone. To determine X and T, we first compute the partial derivatives of u to obtain

$$\frac{\partial u}{\partial t} = X(x)T'(t) \quad \text{and} \quad \frac{\partial^2 u}{\partial x^2} = X''(x)T(t).$$

Substituting these expressions into (1) gives

$$X(x)T'(t) = \beta X''(x)T(t),$$

and separating variables yields

(4) $$\frac{T'(t)}{\beta T(t)} = \frac{X''(x)}{X(x)}.$$

We now observe that the functions on the left-hand side of (4) depend only on t, while those on the right-hand side depend only on x. Since x and t are variables that are independent of one another, the two ratios in (4) must equal some *constant* K. Thus,

$$\frac{X''(x)}{X(x)} = K \quad \text{and} \quad \frac{T'(t)}{\beta T(t)} = K,$$

or

(5) $$X''(x) - KX(x) = 0 \quad \text{and} \quad T'(t) - \beta K T(t) = 0.$$

Consequently, for separable solutions, we have reduced the problem of solving the partial differential equation (1) to solving the two *ordinary* differential equations in (5).

Before proceeding, let's consider the boundary conditions in (2). Since $u(x,t) = X(x)T(t)$, these conditions are

$$X(0)T(t) = 0 \quad \text{and} \quad X(L)T(t) = 0, \qquad t > 0.$$

Hence either $T(t) = 0$ for all $t > 0$, which implies that $u(x,t) \equiv 0$, or

(6) $$X(0) = X(L) = 0.$$

Ignoring the trivial solution $u(x,t) \equiv 0$, we combine the boundary conditions in (6) with the differential equation for X in (5) and obtain the *boundary value problem*

(7) $$X''(x) - KX(x) = 0, \qquad X(0) = X(L) = 0,$$

where K can be any constant.

Notice that the function $X(x) \equiv 0$ is a solution for every K and, depending on the choice of K, this may be the *only* solution to the boundary value problem (7). Thus if we seek a nontrivial solution $u(x,t) = X(x)T(t)$ to (1)–(2), we must first determine those values of K for which the boundary value problem (7) has a nontrivial solution. These special values of K are called **eigenvalues,** and the corresponding nontrivial solutions to (7) are the **eigenfunctions.**

To solve (7), we begin with the auxiliary equation $r^2 - K = 0$ and consider three cases.

Case 1. $K > 0$.   In this case, the roots of the auxiliary equation are $\pm\sqrt{K}$, so a general solution to the differential equation in (7) is

$$X(x) = C_1 e^{\sqrt{K}x} + C_2 e^{-\sqrt{K}x}.$$

To determine C_1 and C_2, we appeal to the boundary conditions:

$$X(0) = C_1 + C_2 = 0,$$
$$X(L) = C_1 e^{\sqrt{K}L} + C_2 e^{-\sqrt{K}L} = 0.$$

From the first equation we see that $C_2 = -C_1$. The second equation can then be written as $C_1(e^{\sqrt{K}L} - e^{-\sqrt{K}L}) = 0$ or $C_1(e^{2\sqrt{K}L} - 1) = 0$. Since $K > 0$, it follows that $(e^{2\sqrt{K}L} - 1) > 0$. Therefore, C_1, and hence C_2, is zero. Consequently, there is *no* nontrivial solution to (7) for $K > 0$.

Case 2. $K = 0$. Here $r = 0$ is a repeated root to the auxiliary equation, and a general solution to the differential equation is

$$X(x) = C_1 + C_2 x.$$

The boundary conditions in (7) yield $C_1 = 0$ and $C_1 + C_2 L = 0$, which imply that $C_1 = C_2 = 0$. Thus, for $K = 0$, there is no nontrivial solution to (7).

Case 3. $K < 0$. In this case the roots of the auxiliary equation are $\pm i\sqrt{-K}$. (Note that $-K > 0$ since $K < 0$.) Thus, a general solution to $X'' - KX = 0$ is

(8) $\qquad X(x) = C_1 \cos \sqrt{-K}x + C_2 \sin \sqrt{-K}x.$

This time the boundary conditions $X(0) = X(L) = 0$ give the system

$$C_1 = 0,$$
$$C_1 \cos \sqrt{-K}L + C_2 \sin \sqrt{-K}L = 0.$$

Since $C_1 = 0$, the system reduces to solving $C_2 \sin \sqrt{-K}L = 0$. Hence, either $\sin \sqrt{-K}L = 0$ or $C_2 = 0$. Now $\sin \sqrt{-K}L = 0$ only when $\sqrt{-K}L = n\pi$, where n is an integer. Therefore, (7) has a nontrivial solution ($C_2 \neq 0$) when $\sqrt{-K}L = n\pi$ or $K = -(n\pi/L)^2$, $n = 1, 2, 3, \ldots$ (we exclude $n = 0$, since it makes $K = 0$). Furthermore, the nontrivial solutions (eigenfunctions) X_n corresponding to the eigenvalue $K = -(n\pi/L)^2$ are given by (cf. (8))

(9) $\qquad X_n(x) = a_n \sin\left(\dfrac{n\pi x}{L}\right),$

where the a_n's are arbitrary nonzero constants.

Having determined that $K = -(n\pi/L)^2$ for some positive integer n, let's consider the second equation in (5) with $K = -(n\pi/L)^2$:

$$T'(t) + \beta\left(\frac{n\pi}{L}\right)^2 T(t) = 0.$$

For each $n = 1, 2, 3, \ldots$, a general solution to this linear first order equation is

$$T_n(t) = b_n e^{-\beta(n\pi/L)^2 t}.$$

Combining this with equation (9), we obtain, for each $n = 1, 2, 3, \ldots$, the function

(10) $\quad u_n(x, t) := X_n(x)T_n(t) = a_n \sin(n\pi x/L)b_n e^{-\beta(n\pi/L)^2 t}$

$\qquad\qquad = c_n e^{-\beta(n\pi/L)^2 t} \sin(n\pi x/L),$

where c_n is an arbitrary constant.

We would like to conclude that each $u_n(x, t)$ is a solution to (1)–(2). But, we have shown *only* that *if* (1)–(2) has a solution of the form $u(x, t) = X(x)T(t)$, then u must be one of the functions given in (10). We leave it for the reader to verify, by direct substitution, that the functions in (10) are indeed solutions to (1)–(2).

A simple computation also shows that if u_n and u_m are solutions to (1)–(2), then so is any linear combination $au_n + bu_m$. (This is a consequence of the fact that the operator $\mathcal{L} := \partial/\partial t - \beta \partial^2/\partial x^2$ is a *linear* operator and the boundary conditions in (2) are *homogeneous*.) Furthermore, if we take an *infinite* sum of these functions, that is,

(11) $\quad u(x, t) = \sum_{n=1}^{\infty} u_n(x, t) = \sum_{n=1}^{\infty} c_n e^{-\beta(n\pi/L)^2 t} \sin(n\pi x/L),$

then this formal series will again be a solution to (1)–(2), provided that the infinite series has the proper convergence behavior.

For a solution u of the form (11), we can determine the constants c_n by using the initial condition (3). This gives

$$u(x, 0) = \sum_{n=1}^{\infty} c_n \sin\left(\frac{n\pi x}{L}\right) = f(x), \qquad 0 < x < L.$$

We have thus reduced the problem (1)–(3) of heat flow in a thin wire to the problem of determining an expansion for $f(x)$ of the form

(12) $\quad f(x) = \sum_{n=1}^{\infty} c_n \sin\left(\frac{n\pi x}{L}\right).$

Such an expansion is called a **Fourier sine series** and will be discussed in the next two sections. If we choose the c_n's so that equation (12) holds, then the expansion for $u(x, t)$ in (11) is called a **formal solution** to the heat flow problem (1)–(3). If this expansion converges to a function with continuous second partial derivatives, then the formal solution is an actual (genuine) solution. Moreover, the solution is unique.

● EXAMPLE 1   Find the solution to the heat flow problem

(13) $\qquad \dfrac{\partial u}{\partial t} = 7\dfrac{\partial^2 u}{\partial x^2}; \qquad\qquad 0 < x < \pi, \qquad t > 0,$

(14) $\quad u(0, t) = u(\pi, t) = 0, \qquad\qquad t > 0,$

(15) $\quad u(x, 0) = 3\sin 2x - 6\sin 5x, \qquad 0 < x < \pi.$

Solution   Comparing equation (13) with (1), we see that $\beta = 7$ and $L = \pi$. Hence, we need only determine the values of c_n in formula (12). That is, we must have

$$u(x, 0) = 3\sin 2x - 6\sin 5x = \sum_{n=1}^{\infty} c_n \sin nx.$$

Equating the coefficients of like terms, we find that

$$c_2 = 3 \quad \text{and} \quad c_5 = -6,$$

and the remaining c_n's are zero. Hence, from (11), the solution to the heat flow problem (13)–(15) is

$$u(x,t) = c_2 e^{-\beta(2\pi/L)^2 t} \sin(2\pi x/L) + c_5 e^{-\beta(5\pi/L)^2 t} \sin(5\pi x/L)$$
$$= 3e^{-28t} \sin 2x - 6e^{-175t} \sin 5x. \quad \bullet$$

Another situation in which the separation of variables approach applies nicely occurs in the study of a vibrating string. This concerns the transverse vibrations of a string stretched between two points, such as a guitar string or piano wire. The goal is to find a function $u(x,t)$ that gives the displacement (deflection) of the string at any point x $(0 \leq x \leq L)$ and any time $t \geq 0$ (see Figure 11.2). In developing the mathematical model, it is assumed that the string is perfectly flexible and has constant linear density, the tension on the string is constant, gravity is negligible, and no other forces are acting on the string. Under these conditions and the additional assumption that the displacements $u(x,t)$ are small in comparison to the length of the string, it turns out that the motion of the string is governed by the following initial-boundary value problem.[†]

$$(16) \qquad \frac{\partial^2 u}{\partial t^2} = \alpha^2 \frac{\partial^2 u}{\partial x^2}; \qquad 0 < x < L, \qquad t > 0,$$

$$(17) \qquad u(0,t) = u(L,t) = 0, \qquad t \geq 0,$$

$$(18) \qquad u(x,0) = f(x), \qquad 0 \leq x \leq L,$$

$$(19) \qquad \frac{\partial u}{\partial t}(x,0) = g(x), \qquad 0 \leq x \leq L.$$

The constant α^2 appearing in (16) is strictly positive and depends on the linear density and the tension of the string. The boundary conditions in (17) reflect the fact that the string is held fixed at the two endpoints $x = 0$ and $x = L$. Equations (18) and (19) specify, respectively, the initial displacement of the string and the initial velocity of each point on the string. For the initial and boundary conditions to be consistent, we assume that $f(0) = f(L) = 0$ and $g(0) = g(L) = 0$.

Let apply the method of separation of variables to the initial-boundary value

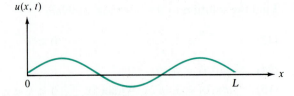

Figure 11.2 Displacement of string at time t

† For a derivation of this mathematical model, see *Applied Differential Equations*, Third Edition, by Murray R. Spiegel, Prentice-Hall, Inc., Englewood Cliffs, New Jersey, 1980.

problem for the vibrating string (16)–(19). Thus, we begin by assuming that equation (16) has a solution of the form

$$u(x, t) = X(x)T(t),$$

where X is a function of x alone and T is a function of t alone. Differentiating u, we obtain

$$\frac{\partial^2 u}{\partial t^2} = X(x)T''(t), \qquad \frac{\partial^2 u}{\partial x^2} = X''(x)T(t).$$

Substituting these expressions into (16), we have

$$X(x)T''(t) = \alpha^2 X''(x)T(t),$$

and separating variables gives

$$\frac{T''(t)}{\alpha^2 T(t)} = \frac{X''(x)}{X(x)}.$$

Just as before, these ratios must equal some constant K:

(20) $\dfrac{X''(x)}{X(x)} = K$ and $\dfrac{T''(t)}{\alpha^2 T(t)} = K.$

Furthermore, with $u(x, t) = X(x)T(t)$, the boundary conditions in (17) give

$$X(0)T(t) = 0, \qquad X(L)T(t) = 0, \qquad t \geq 0.$$

In order for these equations to hold for all $t \geq 0$, either $T(t) \equiv 0$, which implies that $u(x, t) \equiv 0$, or

$$X(0) = X(L) = 0.$$

Ignoring the trivial solution, we combine these boundary conditions with the differential equation for X in (20) and obtain the boundary value problem

(21) $X''(x) - KX(x) = 0, \qquad X(0) = X(L) = 0,$

where K can be any constant.

This is the same boundary value problem that we encountered earlier while solving the heat equation. There we found that the eigenvalues are

$$K = -\left(\frac{n\pi}{L}\right)^2, \qquad n = 1, 2, 3, \ldots,$$

with corresponding eigenfunctions (nontrivial solutions)

(22) $X_n(x) = c_n \sin\left(\dfrac{n\pi x}{L}\right),$

where the c_n's are arbitrary nonzero constants.

Having determined that $K = -(n\pi/L)^2$ for some positive integer n, let's consider the second equation in (20) for such K:

$$T''(t) + \frac{\alpha^2 n^2 \pi^2}{L^2} T(t) = 0.$$

For each $n = 1, 2, 3, \ldots$, a general solution is

$$T_n(t) = c_{n,1} \cos\frac{n\pi\alpha}{L} t + c_{n,2} \sin\frac{n\pi\alpha}{L} t.$$

Combining this with equation (22), we obtain, for each $n = 1, 2, 3, \ldots$, the function

$$u_n(x, t) = X_n(x) T_n(t) = \left(c_n \sin\frac{n\pi x}{L} \right)\left(c_{n,1} \cos\frac{n\pi\alpha}{L} t + c_{n,2} \sin\frac{n\pi\alpha}{L} t \right),$$

(23) $$u_n(x, t) = \left(a_n \cos\frac{n\pi\alpha}{L} t + b_n \sin\frac{n\pi\alpha}{L} t \right) \sin\frac{n\pi x}{L}.$$

Using the fact that linear combinations of solutions to (16)–(19) are again solutions, we consider the infinite sum (superposition) of the functions in (23):

(24) $$u(x, t) = \sum_{n=1}^{\infty} \left[a_n \cos\frac{n\pi\alpha}{L} t + b_n \sin\frac{n\pi\alpha}{L} t \right] \sin\frac{n\pi x}{L}.$$

For a solution of the form (24), substituting it into the initial conditions (18)–(19) gives

(25) $$u(x, 0) = \sum_{n=1}^{\infty} a_n \sin\frac{n\pi x}{L} = f(x), \qquad 0 \le x \le L,$$

(26) $$\frac{\partial u}{\partial t}(x, 0) = \sum_{n=1}^{\infty} \frac{n\pi\alpha}{L} b_n \sin\frac{n\pi x}{L} = g(x), \qquad 0 \le x \le L.$$

We have now reduced the vibrating string problem (16)–(19) to the problem of determining the Fourier sine series expansions for $f(x)$ and $g(x)$:

(27) $$f(x) = \sum_{n=1}^{\infty} a_n \sin\frac{n\pi x}{L}, \qquad g(x) = \sum_{n=1}^{\infty} B_n \sin\frac{n\pi x}{L},$$

where $B_n = (n\pi\alpha/L) b_n$. If we choose the a_n's and b_n's so that the equations in (25) and (26) hold, then the expansion for $u(x, t)$ in (24) is a **formal solution** to the vibrating string problem (16)–(19). If this expansion converges to a function with continuous second partial derivatives, then the formal solution is an actual (genuine) solution.

● **EXAMPLE 2**   Find the solution to the vibrating string problem

(28) $$\frac{\partial^2 u}{\partial t^2} = 4\frac{\partial^2 u}{\partial x^2}; \qquad\qquad 0 < x < \pi, \qquad t > 0,$$

(29) $$u(0, t) = u(\pi, t) = 0, \qquad\qquad t \ge 0,$$

(30) $$u(x, 0) = \sin 3x - 4\sin 10x, \qquad 0 \le x \le \pi,$$

(31) $$\frac{\partial u}{\partial t}(x, 0) = 2\sin 4x + \sin 6x, \qquad 0 \le x \le \pi.$$

Solution   Comparing equation (28) with equation (16), we see that $\alpha = 2$ and $L = \pi$. Hence we need only determine the values of the coefficients a_n and b_n in formula (24). The a_n's are chosen so that equation (25) holds; that is,

$$u(x,0) = \sin 3x - 4\sin 10x = \sum_{n=1}^{\infty} a_n \sin nx.$$

Equating coefficients of like terms, we see that

$$a_3 = 1, \qquad a_{10} = -4,$$

and the remaining a_n's are zero. Similarly, referring to equation (26), we must choose the b_n's so that

$$\frac{\partial u}{\partial t}(x,0) = 2\sin 4x + \sin 6x = \sum_{n=1}^{\infty} n2b_n \sin nx.$$

Comparing coefficients, we find

$$2 = (4)(2)b_4 \quad \text{or} \quad b_4 = \tfrac{1}{4},$$
$$1 = (6)(2)b_6 \quad \text{or} \quad b_6 = \tfrac{1}{12},$$

and the remaining b_n's are zero. Hence, from formula (24), the solution to the vibrating string problem (28)–(31) is

(32) $u(x,t) = \cos 6t \sin 3x + \tfrac{1}{4}\sin 8t \sin 4x$
$\qquad\qquad + \tfrac{1}{12}\sin 12t \sin 6x - 4\cos 20t \sin 10x.$ ●

In later sections the method of separation of variables is used to study a wide variety of problems for the heat, wave, and Laplace's equations. However, to use the method effectively, one must be able to compute trigonometric series (or, more generally, eigenfunction expansions) such as the Fourier sine series that we encountered here. These expansions are discussed in the next two sections.

EXERCISES 11.2

In Problems 1 through 8 determine all the solutions, if any, to the given boundary value problem by first finding a general solution to the differential equation.

1. $y'' - y = 0;$ $0 < x < 1,$ $y(0) = 0,$ $y(1) = -4.$

2. $y'' - 6y' + 5y = 0;$ $0 < x < 2,$
$y(0) = 1,$ $y(2) = 1.$

3. $y'' + 4y = 0;$ $0 < x < \pi,$ $y(0) = 0,$ $y'(\pi) = 0.$

4. $y'' + 9y = 0;$ $0 < x < \pi,$
$y(0) = 0,$ $y'(\pi) = -6.$

5. $y'' - y = 1 - 2x;$ $0 < x < 1,$
$y(0) = 0,$ $y(1) = 1 + e.$

6. $y'' + y = 0;$ $0 < x < 2\pi,$ $y(0) = 0,$ $y(2\pi) = 1.$

7. $y'' + y = 0;$ $0 < x < 2\pi,$ $y(0) = 1,$ $y(2\pi) = 1.$

8. $y'' - 2y' + y = 0;$ $-1 < x < 1,$
$y(-1) = 0,$ $y(1) = 2.$

In Problems 9 through 14 find the values of λ (eigenvalues) for which the given problem has a nontrivial solution. Also determine the corresponding nontrivial solutions (eigenfunctions).

9. $y'' + \lambda y = 0;$ $0 < x < \pi,$ $y(0) = 0,$ $y'(\pi) = 0.$

10. $y'' + \lambda y = 0;$ $0 < x < \pi,$ $y'(0) = 0,$ $y(\pi) = 0.$

11. $y'' + \lambda y = 0;$ $0 < x < 2\pi,$
$y(0) = y(2\pi),$ $y'(0) = y'(2\pi).$

12. $y'' + \lambda y = 0;$ $0 < x < \pi/2,$
$y'(0) = 0,$ $y'(\pi/2) = 0.$

13. $y'' + \lambda y = 0;$ $0 < x < \pi,$
$y(0) - y'(0) = 0,$ $y(\pi) = 0.$

14. $y'' - 2y' + \lambda y = 0;$ $0 < x < \pi,$
$y(0) = 0,$ $y(\pi) = 0.$

In Problems 15 through 18 solve the heat flow problem (1)–(3) with $\beta = 3$, $L = \pi$, and the given initial function $f(x)$.

15. $f(x) = \sin x - 6\sin 4x.$

16. $f(x) = \sin 3x + 5\sin 7x - 2\sin 13x.$

17. $f(x) = \sin x - 7\sin 3x + \sin 5x.$

18. $f(x) = \sin 4x + 3\sin 6x - \sin 10x.$

In Problems 19 through 22 solve the vibrating string problem (16)–(19) with $\alpha = 3$, $L = \pi$ and the given initial functions $f(x)$ and $g(x)$.

19. $f(x) = 3\sin 2x + 12\sin 13x,$ $g(x) \equiv 0.$

20. $f(x) \equiv 0,$ $g(x) = -2\sin 3x + 9\sin 7x - \sin 10x.$

21. $f(x) = 6\sin 2x + 2\sin 6x,$
$g(x) = 11\sin 9x - 14\sin 15x.$

22. $f(x) = \sin x - \sin 2x + \sin 3x,$
$g(x) = 6\sin 3x - 7\sin 5x.$

23. Find the formal solution to the heat flow problem (1)–(3) with $\beta = 2$ and $L = 1$ if
$$f(x) = \sum_{n=1}^{\infty} \frac{1}{n^2} \sin n\pi x.$$

24. Find the formal solution to the vibrating string problem (16)–(19) with $\alpha = 4$, $L = \pi$, and
$$f(x) = \sum_{n=1}^{\infty} \frac{1}{n^2} \sin nx,$$
$$g(x) = \sum_{n=1}^{\infty} \frac{(-1)^{n+1}}{n} \sin nx.$$

25. By considering the behavior of the solutions of the equation
$$T'(t) - \beta KT(t) = 0, \qquad t > 0,$$
give an argument that is based on physical grounds to rule out the case where $K > 0$ in equation (5).

26. Verify that $u_n(x, t)$ given in equation (10) satisfies equa-

tion (1) and the boundary conditions in (2) by substituting $u_n(x, t)$ directly into the equations involved.

In Problems 27 through 30 a partial differential equation (PDE) is given along with the form of a solution having separated variables. Show that such a solution must satisfy the indicated set of ordinary differential equations.

27. $\dfrac{\partial^2 u}{\partial r^2} + \dfrac{1}{r}\dfrac{\partial u}{\partial r} + \dfrac{1}{r^2}\dfrac{\partial^2 u}{\partial \theta^2} = 0,$

with $u(r, \theta) = R(r)T(\theta)$ yields
$r^2 R''(r) + rR'(r) - \lambda R(r) = 0,$
$T''(\theta) + \lambda T(\theta) = 0,$
where λ is a constant.

28. $\dfrac{\partial^2 u}{\partial t^2} + \dfrac{\partial u}{\partial t} + u = \alpha^2 \dfrac{\partial^2 u}{\partial x^2},$

with $u(x, t) = X(x)T(t)$ yields
$X''(x) - \lambda X(x) = 0,$
$T''(t) + T'(t) + (1 - \lambda\alpha^2)T(t) = 0,$
where λ is a constant.

29. $\dfrac{\partial u}{\partial t} = \beta\left\{\dfrac{\partial^2 u}{\partial x^2} + \dfrac{\partial^2 u}{\partial y^2}\right\},$

with $u(x, y, t) = X(x)Y(y)T(t)$ yields
$T'(t) - \beta KT(t) = 0,$
$X''(x) - JX(x) = 0,$
$Y''(y) + (J - K)Y(y) = 0,$
where J, K are constants.

30. $\dfrac{\partial^2 u}{\partial r^2} + \dfrac{1}{r}\dfrac{\partial u}{\partial r} + \dfrac{1}{r^2}\dfrac{\partial^2 u}{\partial \theta^2} + \dfrac{\partial^2 u}{\partial z^2} = 0,$

with $u(r, \theta, z) = R(r)T(\theta)Z(z)$ yields
$T''(\theta) + \mu T(\theta) = 0,$
$Z''(z) + \lambda Z(z) = 0,$
$r^2 R''(r) + rR'(r) - (r^2\lambda + \mu)R(r) = 0,$
where μ, λ are constants.

31. For the PDE in Problem 27 assume that the following boundary conditions are imposed
$$u(r, 0) = u(r, \pi) = 0,$$
$$u(r, \theta) \text{ remains bounded as } r \to 0^+.$$
Show that a nontrivial solution of the form $u(r, \theta) = R(r)T(\theta)$ must satisfy the boundary conditions
$$T(0) = T(\pi) = 0,$$
$$R(r) \text{ remains bounded as } r \to 0^+.$$

32. For the PDE in Problem 29 assume that the following boundary conditions are imposed

$$u(0, y, t) = u(a, y, t) = 0, \qquad 0 \le y \le b, \qquad t \ge 0,$$

$$\frac{\partial u}{\partial y}(x, 0, t) = \frac{\partial u}{\partial y}(x, b, t) = 0, \qquad 0 \le x \le a, \qquad t \ge 0.$$

Show that a nontrivial solution of the form $u(x, y, t) = X(x)Y(y)T(t)$ must satisfy the boundary conditions

$$X(0) = X(a) = 0,$$
$$Y'(0) = Y'(b) = 0.$$

33. When the temperature in a wire reaches a steady state, that is, when u depends only upon x, then $u(x)$ satisfies Laplace's equation $\partial^2 u / \partial x^2 = 0$.

(a) Find the steady-state solution when the ends of the wire are kept at a constant temperature of $50°C$, that is, when $u(0) = u(L) = 50$.

(b) Find the steady-state solution when one end of the wire is kept at $10°C$, while the other is kept at $40°C$, that is, when $u(0) = 10$ and $u(L) = 40$.

•11.3• FOURIER SERIES

While solving the heat flow and vibrating string problems in the previous section, we encountered the problem of expressing a function in a trigonometric series (cf. equations (12) and (25) in Section 11.2). In the next two sections we will study the theory of Fourier series which deals with trigonometric series expansions for periodic functions.

In Section 7.2 we defined a **piecewise continuous** function on $[a, b]$ as a function f that is continuous at every point in $[a, b]$ except possibly for a finite number of points at which f has a *jump discontinuity*. Such functions are necessarily integrable over any finite interval on which they are piecewise continuous. Recall that a function is **periodic of period** T if $f(x + T) = f(x)$ for all x in the domain of f. The smallest positive value of T is called the **fundamental period.** The trigonometric functions $\sin x$ and $\cos x$ are examples of periodic functions with fundamental period 2π and $\tan x$ is periodic with fundamental period π. A constant function is a periodic function with arbitrary period T.

There are two symmetry properties of functions that will be useful in the study of Fourier series. A function f that satisfies $f(-x) = f(x)$ for all x in the domain of f has a graph that is symmetric with respect to the y-axis (see Figure 11.3(a)). We say that

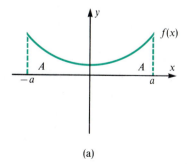

(a)

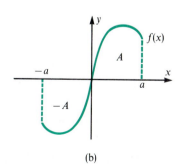

(b)

Figure 11.3 (a) Even function $\int_{-a}^{a} f = A + A = 2\int_{0}^{a} f$
(b) Odd function $\int_{-a}^{a} f = A - A = 0$

such a function is an **even** function. A function f that satisfies $f(-x) = -f(x)$ for all x in the domain of f has a graph that is symmetric with respect to the origin (see Figure 11.3(b)) on page 545. It is said to be an **odd** function. The functions $1, x^2, x^4, \ldots,$ are examples of even functions, while the functions $x, x^3, x^5, \ldots,$ are odd. The trigonometric functions $\sin x$ and $\tan x$ are odd functions and $\cos x$ is an even function.

● **EXAMPLE 1**   Determine whether the given function is even, odd, or neither.

$$\text{(a) } f(x) = \sqrt{1 + x^2}. \qquad \text{(b) } g(x) = x^{1/3} - \sin x. \qquad \text{(c) } h(x) = e^x.$$

Solution   (a) Since $f(-x) = \sqrt{1 + (-x)^2} = \sqrt{1 + x^2} = f(x)$, then $f(x)$ is an even function.
(b) Since $g(-x) = (-x)^{1/3} - \sin(-x) = -x^{1/3} + \sin x = -(x^{1/3} - \sin x) = -g(x)$, then $g(x)$ is an odd function.
(c) Here $h(-x) = e^{-x}$. Since $e^{-x} = e^x$ only when $x = 0$ and e^{-x} is never equal to $-e^x$, then $h(x)$ is neither an even nor an odd function.   ●

Knowing that a function is even or odd can be useful in evaluating definite integrals.

PROPERTIES OF SYMMETRIC FUNCTIONS

Theorem 1.   If f is an even piecewise continuous function on $[-a, a]$, then

(1) $$\int_{-a}^{a} f(x)\,dx = 2\int_{0}^{a} f(x)\,dx.$$

If f is an odd piecewise continuous function on $[-a, a]$, then

(2) $$\int_{-a}^{a} f(x)\,dx = 0.$$

Proof.   If f is an even function, then $f(-x) = f(x)$. Hence,

$$\int_{-a}^{a} f(x)\,dx = \int_{-a}^{0} f(x)\,dx + \int_{0}^{a} f(x)\,dx$$

$$= -\int_{a}^{0} f(u)\,du + \int_{0}^{a} f(x)\,dx = 2\int_{0}^{a} f(x)\,dx,$$

where we used the change of variables $u = -x$. This is illustrated in Figure 11.3(a). Formula (2) can be proved by a similar argument and is illustrated in Figure 11.3(b).   ◄◄◄

● **EXAMPLE 2**   Evaluate the following integrals when m and n are positive integers:

$$\text{(a) } \int_{-T}^{T} \sin\frac{m\pi x}{T} \cos\frac{n\pi x}{T}\,dx. \qquad \text{(b) } \int_{-T}^{T} \sin\frac{m\pi x}{T} \sin\frac{n\pi x}{T}\,dx.$$

Solution (a) Since $\sin\dfrac{m\pi(-x)}{T}\cos\dfrac{n\pi(-x)}{T}=-\sin\dfrac{m\pi x}{T}\cos\dfrac{n\pi x}{T}$, the integrand in part (a) is an odd function. Thus, by Theorem 1

$$(3)\qquad \int_{-T}^{T}\sin\frac{m\pi x}{T}\cos\frac{n\pi x}{T}\,dx=0.$$

(b) Using the fact that the integrand is even (check this!) and the trigonometric identity

$$2\sin A\sin B=\cos(A-B)-\cos(A+B),$$

we find for $m\neq n$

$$\int_{-T}^{T}\sin\frac{m\pi x}{T}\sin\frac{n\pi x}{T}\,dx=2\int_{0}^{T}\sin\frac{m\pi x}{T}\sin\frac{n\pi x}{T}\,dx$$

$$=\int_{0}^{T}\left\{\cos\frac{(m-n)\pi x}{T}-\cos\frac{(m+n)\pi x}{T}\right\}dx$$

$$=\frac{T}{\pi}\left[\frac{\sin\dfrac{(m-n)\pi x}{T}}{m-n}-\frac{\sin\dfrac{(m+n)\pi x}{T}}{m+n}\right]\Bigg|_{0}^{T}=0.$$

When $m=n$, the integral in (b) becomes

$$\int_{-T}^{T}\left(\sin\frac{n\pi x}{T}\right)^{2}dx=2\int_{0}^{T}\left(\sin\frac{n\pi x}{T}\right)^{2}dx=\int_{0}^{T}\left\{1-\cos\frac{2n\pi x}{T}\right\}dx$$

$$=\left[x-\frac{T}{2n\pi}\sin\frac{2n\pi x}{T}\right]\Bigg|_{0}^{T}=T.$$

Hence

$$(4)\qquad \int_{-T}^{T}\sin\frac{m\pi x}{T}\sin\frac{n\pi x}{T}\,dx=\begin{cases}0, & m\neq n,\\ T, & m=n.\end{cases}\quad\bullet$$

We leave it as an exercise for the reader to verify that

$$(5)\qquad \int_{-T}^{T}\cos\frac{m\pi x}{T}\cos\frac{n\pi x}{T}\,dx=\begin{cases}0, & m\neq n,\\ T, & m=n\end{cases}$$

(see Problem 8). Equations (3)–(5) express an **orthogonality condition** satisfied by the set of trigonometric functions $\{\cos x,\sin x,\cos 2x,\sin 2x,\ldots\}$ where $T=\pi$. We will say more about this later in the section.

It is easy to verify that if each of the functions $f_{1},\ldots,f_{n}$ is periodic of period T, then so is any linear combination

$$c_{1}f_{1}(x)+\cdots+c_{n}f_{n}(x).$$

For example, the sum $7+3\cos\pi x-8\sin\pi x+4\cos 2\pi x-6\sin 2\pi x$ has period 2 since each term has period 2. Furthermore, if the infinite series

$$\frac{a_0}{2} + \sum_{n=1}^{\infty} \left(a_n \cos\frac{n\pi x}{T} + b_n \sin\frac{n\pi x}{T} \right)$$

converges for all x, then the function to which it converges will be periodic of period $2T$.

Just as we can associate a Taylor series with a function that has derivatives of all orders at a fixed point, we can identify a particular trigonometric series with a piecewise continuous function. To illustrate this, let's assume that $f(x)$ has the series expansion[†]

$$(6) \qquad f(x) = \frac{a_0}{2} + \sum_{n=1}^{\infty} \left\{ a_n \cos\frac{n\pi x}{T} + b_n \sin\frac{n\pi x}{T} \right\},$$

where the a_n's and b_n's are constants. (Necessarily, f has period $2T$.)

To determine the coefficients $a_0, a_1, b_1, a_2, b_2, \ldots$, we proceed as follows. Let's integrate $f(x)$ from $-T$ to T, assuming we can integrate term-by-term:

$$\int_{-T}^{T} f(x)\,dx = \int_{-T}^{T} \frac{a_0}{2}\,dx + \sum_{n=1}^{\infty} a_n \int_{-T}^{T} \cos\frac{n\pi x}{T}\,dx + \sum_{n=1}^{\infty} b_n \int_{-T}^{T} \sin\frac{n\pi x}{T}\,dx.$$

Since $\sin(n\pi x/T)$ is an odd function, $\int_{-T}^{T} \sin(n\pi x/T)\,dx = 0$. Moreover, for $n = 1, 2, \ldots$, it is easy to check that $\int_{-T}^{T} \cos(n\pi x/T)\,dx = 0$. Hence

$$\int_{-T}^{T} f(x)\,dx = \int_{-T}^{T} \frac{a_0}{2}\,dx = a_0 T,$$

and so

$$a_0 = \frac{1}{T} \int_{-T}^{T} f(x)\,dx.$$

(Notice that $a_0/2$ is the average value of f over one period $2T$.) Next, to find the coefficient a_m when $m \geq 1$, we multiply (6) by $\cos(m\pi x/T)$ and integrate:

$$(7) \qquad \int_{-T}^{T} f(x)\cos\frac{m\pi x}{T}\,dx = \frac{a_0}{2} \int_{-T}^{T} \cos\frac{m\pi x}{T}\,dx + \sum_{n=1}^{\infty} a_n \int_{-T}^{T} \cos\frac{n\pi x}{T} \cos\frac{m\pi x}{T}\,dx$$

$$+ \sum_{n=1}^{\infty} b_n \int_{-T}^{T} \sin\frac{n\pi x}{T} \cos\frac{m\pi x}{T}\,dx.$$

We have already observed that

$$\int_{-T}^{T} \cos\frac{m\pi x}{T}\,dx = 0, \qquad m \geq 1,$$

and, by formula (3),

$$\int_{-T}^{T} \sin\frac{n\pi x}{T} \cos\frac{m\pi x}{T}\,dx = 0.$$

[†] The choice of a constant $a_0/2$ instead of just a_0 will make the formulas easier to remember.

Using the formulas in (5), we also find that

$$\int_{-T}^{T} \cos\frac{n\pi x}{T}\cos\frac{m\pi x}{T}\,dx = \begin{cases} 0, & n \neq m, \\ T, & n = m. \end{cases}$$

Hence, in (7) we see that only one term survives and

$$\int_{-T}^{T} f(x)\cos\frac{m\pi x}{T}\,dx = a_m T.$$

Thus

$$a_m = \frac{1}{T}\int_{-T}^{T} f(x)\cos\frac{m\pi x}{T}\,dx.$$

Similarly, multiplying (6) by $\sin(m\pi x/T)$ and integrating yields

$$\int_{-T}^{T} f(x)\sin\frac{m\pi x}{T}\,dx = b_m T,$$

so that

$$b_m = \frac{1}{T}\int_{-T}^{T} f(x)\sin\frac{m\pi x}{T}\,dx.$$

Motivated by the above computations, we now make the following definition.

FOURIER SERIES

Definition 1.   Let f be a piecewise continuous function on the interval $[-T, T]$. The **Fourier series** of f is the trigonometric series

(8) $$f(x) \sim \frac{a_0}{2} + \sum_{n=1}^{\infty}\left\{a_n\cos\frac{n\pi x}{T} + b_n\sin\frac{n\pi x}{T}\right\},$$

where the a_n's and b_n's are given by the formulas[†]

(9) $$a_n = \frac{1}{T}\int_{-T}^{T} f(x)\cos\frac{n\pi x}{T}\,dx, \qquad n = 0, 1, 2,\dots,$$

(10) $$b_n = \frac{1}{T}\int_{-T}^{T} f(x)\sin\frac{n\pi x}{T}\,dx, \qquad n = 1, 2, 3,\dots.$$

Formulas (9) and (10) are called the **Euler formulas.** We use the symbol $\sim$ in (8) to remind us that this series is associated with $f(x)$, but may not converge to $f(x)$. We will return to the question of convergence later in the section. First let's consider a few examples of Fourier series.

[†] Notice that $f(x)$ need not be defined for every x in $[-T, T]$; we only need that the integrals in (9) and (10) exist.

● **EXAMPLE 3**   Compute the Fourier series for

$$f(x) = \begin{cases} 0, & -\pi < x < 0, \\ x, & 0 < x < \pi. \end{cases}$$

Solution   Here $T = \pi$. Using formulas (9) and (10), we have

$$a_n = \frac{1}{\pi} \int_{-\pi}^{\pi} f(x) \cos nx \, dx = \frac{1}{\pi} \int_{0}^{\pi} x \cos nx \, dx$$

$$= \frac{1}{\pi n^2} \int_{0}^{\pi n} u \cos u \, du = \frac{1}{\pi n^2} [\cos u + u \sin u] \Big|_{0}^{\pi n}$$

$$= \frac{1}{\pi n^2} (\cos n\pi - 1) = \frac{1}{\pi n^2} [(-1)^n - 1], \qquad n = 1, 2, 3, \ldots,$$

$$a_0 = \frac{1}{\pi} \int_{-\pi}^{\pi} f(x) \, dx = \frac{1}{\pi} \int_{0}^{\pi} x \, dx = \frac{x^2}{2\pi} \Big|_{0}^{\pi} = \frac{\pi}{2},$$

$$b_n = \frac{1}{\pi} \int_{-\pi}^{\pi} f(x) \sin nx \, dx = \frac{1}{\pi} \int_{0}^{\pi} x \sin nx \, dx$$

$$= \frac{1}{\pi n^2} \int_{0}^{\pi n} u \sin u \, du = \frac{1}{\pi n^2} [\sin u - u \cos u] \Big|_{0}^{\pi n}$$

$$= \frac{-\cos n\pi}{n} = \frac{(-1)^{n+1}}{n}, \qquad n = 1, 2, 3, \ldots.$$

Therefore,

$$(11) \qquad f(x) \sim \frac{\pi}{4} + \sum_{n=1}^{\infty} \left\{ \frac{1}{\pi n^2} [(-1)^n - 1] \cos nx + \frac{(-1)^{n+1}}{n} \sin nx \right\}$$

$$= \frac{\pi}{4} - \frac{2}{\pi} \left\{ \cos x + \frac{1}{9} \cos 3x + \frac{1}{25} \cos 5x + \cdots \right\}$$

$$+ \left\{ \sin x - \frac{1}{2} \sin 2x + \frac{1}{3} \sin 3x + \cdots \right\}. \qquad ●$$

● **EXAMPLE 4**   Compute the Fourier series for

$$f(x) = \begin{cases} -1 & -\pi < x < 0, \\ 1, & 0 < x < \pi. \end{cases}$$

Solution   Again, $T = \pi$. Notice that f is an odd function. Since the product of an odd function and an even function is odd (see Problem 7), $f(x) \cos nx$ is also an odd function. Thus

$$a_n = \frac{1}{\pi} \int_{-\pi}^{\pi} f(x) \cos nx \, dx = 0, \qquad n = 0, 1, 2, \ldots.$$

Furthermore, $f(x) \sin nx$ is the product of two odd functions and therefore is an even function, so

$$b_n = \frac{1}{\pi} \int_{-\pi}^{\pi} f(x) \sin nx\, dx = \frac{2}{\pi} \int_0^{\pi} \sin nx\, dx$$

$$= \frac{2}{\pi} \left[\frac{-\cos nx}{n} \right]_0^{\pi} = \frac{2}{\pi} \left[\frac{1}{n} - \frac{(-1)^n}{n} \right], \qquad n = 1, 2, 3, \ldots,$$

$$= \begin{cases} 0, & n \text{ even}, \\ \dfrac{4}{\pi n}, & n \text{ odd}. \end{cases}$$

Thus

(12) $$f(x) \sim \frac{2}{\pi} \sum_{n=1}^{\infty} \frac{[1 - (-1)^n]}{n} \sin nx = \frac{4}{\pi} \left[\sin x + \frac{1}{3} \sin 3x + \frac{1}{5} \sin 5x + \cdots \right]. \quad \bullet$$

In Figure 11.4 we have sketched the graph of f along with the graphs of the first two partial sums $(4/\pi) \sin x$ and $(4/\pi)[\sin x + (1/3) \sin 3x]$ of the Fourier expansion in (12).

In Example 4 the odd function f has a Fourier series consisting only of sine functions. It is easy to show that, in general, if f is any odd function, then its Fourier series consists only of sine terms.

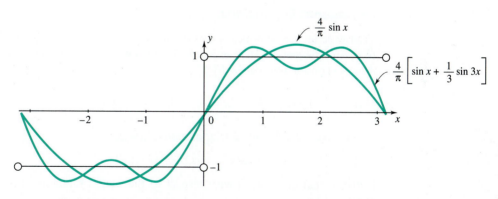

Figure 11.4 Graph of $f(x)$ in Example 4 and the first two partial sums of its Fourier series

● **EXAMPLE 5**   Compute the Fourier series for $f(x) = |x|$, $-1 < x < 1$.

Solution   Here $T = 1$. Since f is an even function, $f(x) \sin n\pi x$ is an odd function. Therefore,

$$b_n = \int_{-1}^{1} f(x) \sin n\pi x\, dx = 0, \qquad n = 1, 2, 3, \ldots.$$

Since $f(x) \cos n\pi x$ is an even function, we have

$$a_0 = \int_{-1}^{1} f(x)\, dx = 2 \int_{0}^{1} x\, dx = x^2 \Big|_{0}^{1} = 1,$$

$$a_n = \int_{-1}^{1} f(x) \cos n\pi x\, dx = 2 \int_{0}^{1} x \cos n\pi x\, dx$$

$$= \frac{2}{\pi^2 n^2} \int_{0}^{\pi n} u \cos u\, du = \frac{2}{\pi^2 n^2} [\cos u + u \sin u] \Big|_{0}^{\pi n} = \frac{2}{\pi^2 n^2} (\cos n\pi - 1)$$

$$= \frac{2}{\pi^2 n^2} [(-1)^n - 1], \qquad n = 1, 2, 3, \dots .$$

Therefore,

(13) $$f(x) \sim \frac{1}{2} + \sum_{n=1}^{\infty} \frac{2}{\pi^2 n^2} [(-1)^n - 1] \cos n\pi x$$

$$= \frac{1}{2} - \frac{4}{\pi^2} \left\{ \cos \pi x + \frac{1}{9} \cos 3\pi x + \frac{1}{25} \cos 5\pi x + \cdots \right\}. \quad \bullet$$

Notice that the even function f of Example 5 has a Fourier series consisting only of cosine functions and the constant function $1 = \cos 0\pi x$. In general, if f is an even function, then its Fourier series consists only of cosine functions (including the constant function).

Orthogonal Expansions

Fourier series are examples of orthogonal expansions.[†] A set of functions $\{f_n(x)\}_{n=1}^{\infty}$ is said to be an **orthogonal system** or just **orthogonal** with respect to the nonnegative weight function $w(x)$ on the interval $[a, b]$ if

(14) $$\int_{a}^{b} f_m(x) f_n(x) w(x)\, dx = 0, \quad \text{whenever} \quad m \neq n.$$

As we have seen, the set of trigonometric functions

(15) $\{1, \cos x, \sin x, \cos 2x, \sin 2x, \dots\}$

is orthogonal on $[-\pi, \pi]$ with respect to the weight function $w(x) \equiv 1$. If we define the **norm** of f as

(16) $$\|f\| := \left[\int_{a}^{b} f^2(x) w(x)\, dx \right]^{1/2},$$

then we say that a set of functions $\{f_n(x)\}_{n=1}^{\infty}$ (or $\{f_n(x)\}_{n=1}^{N}$) is an **orthonormal system with respect to** $w(x)$ if (14) holds and, for each n, $\|f_n\| = 1$. Equivalently, we say the set is

---

[†] Orthogonality is also discussed in Section 8.9 on pages 409–410.

an orthonormal system if

(17)
$$\int_a^b f_m(x)f_n(x)w(x)\,dx = \begin{cases} 0, & m \neq n, \\ 1, & m = n. \end{cases}$$

We can always obtain an orthonormal system from an orthogonal system just by dividing each function by its norm. In particular, since

$$\int_{-\pi}^{\pi} \cos^2 nx\,dx = \int_{-\pi}^{\pi} \sin^2 nx\,dx = \pi, \qquad n = 1, 2, 3, \ldots,$$

and

$$\int_{-\pi}^{\pi} 1\,dx = 2\pi,$$

then the orthogonal system (15) gives rise on $[-\pi, \pi]$ to the orthonormal system

$$\{(2\pi)^{-1/2}, \pi^{-1/2}\cos x, \pi^{-1/2}\sin x, \pi^{-1/2}\cos 2x, \pi^{-1/2}\sin 2x, \ldots\}.$$

If $\{f_n(x)\}_{n=1}^{\infty}$ is an orthogonal system with respect to $w(x)$ on $[a, b]$, we might ask if we can expand a function $f(x)$ in terms of these functions; that is, can we express f in the form

(18) $f(x) = c_1 f_1(x) + c_2 f_2(x) + \cdots$

for a suitable choice of constants $c_1, c_2, \ldots$? Such an expansion is called an **orthogonal expansion,** or a **generalized Fourier series.**

To determine the constants in (18), we can proceed as we did in deriving Euler's formulas for the coefficients of a Fourier series; this time using the orthogonality of the system. Multiply (18) by $f_m(x)w(x)$ and integrate to obtain

(19)
$$\int_a^b f(x)f_m(x)w(x)\,dx = c_1 \int_a^b f_1(x)f_m(x)w(x)\,dx + c_2 \int_a^b f_2(x)f_m(x)w(x)\,dx + \cdots$$

$$= \sum_{n=1}^{\infty} c_n \int_a^b f_n(x)f_m(x)w(x)\,dx.$$

(Here we have again assumed that we can integrate term by term.) Since the system is orthogonal with respect to $w(x)$, every integral on the right-hand side of (19) is zero except when $n = m$. Solving for c_m gives

(20)
$$c_m = \frac{\displaystyle\int_a^b f(x)f_m(x)w(x)\,dx}{\displaystyle\int_a^b f_m^2(x)w(x)\,dx} = \frac{\displaystyle\int_a^b f(x)f_m(x)w(x)\,dx}{\|f_m\|^2}, \qquad n = 1, 2, 3, \ldots.$$

The derivation of the formula for c_m was only *formal* since the question of the convergence of the expansion in (18) was not answered. If the series $\sum_{n=1}^{\infty} c_n f_n(x)$ converges *uniformly* to $f(x)$ on $[a, b]$, then each step can be justified and indeed, the coefficients are given by formula (20). The notion of uniform convergence is discussed in the next subsection and in Section 14.2[†].

Convergence of Fourier Series

Let us now turn to the question of the convergence of a Fourier series. In Example 5 it is possible to use a comparison or limit comparison test to show that the series is absolutely dominated by a *p*-series of the form $\sum_{n=1}^{\infty} 1/n^2$ which converges. However, this is much harder to do in Example 4 since the terms go to zero like $1/n$. Matters can be even worse since there exist Fourier series that *diverge*. Moreover, not every convergent trigonometric series is a Fourier series. For example, although it is not obvious, the series

$$\sum_{n=1}^{\infty} \frac{\sin nx}{\ln(n+1)}$$

converges for all *x*, but is not a Fourier series. We state two theorems that deal with the convergence of a Fourier series and two dealing with the properties of termwise differentiation and integration. For proofs of these results see *Partial Differential Equations of Mathematical Physics*, by Tyn Myint-U, Elsevier North Holland, Inc., New York, 1980, Chapter 5; *Advanced Calculus with Applications*, by N. J. DeLillo, Macmillan, New York, 1982, Chapter 9, or an advanced text on the theory of Fourier series.

Before proceeding we will need the following notation: Let

$$f(x^+) := \lim_{h \to 0^+} f(x+h) \quad \text{and} \quad f(x^-) := \lim_{h \to 0^+} f(x-h).$$

POINTWISE CONVERGENCE OF FOURIER SERIES

Theorem 2.   If f and f' are piecewise continuous on $[-T, T]$, then for any x in $(-T, T)$

(21) $$\frac{a_0}{2} + \sum_{n=1}^{\infty} \left\{ a_n \cos \frac{n\pi x}{T} + b_n \sin \frac{n\pi x}{T} \right\} = \tfrac{1}{2}[f(x^+) + f(x^-)],$$

where the a_n's and b_n's are given by the Euler formulas (9) and (10). For $x = \pm T$, the series converges to $\tfrac{1}{2}[f(-T^+) + f(T^-)]$.

In other words, when f and f' are piecewise continuous on $[-T, T]$, the Fourier series converges to $f(x)$ whenever f is continuous at x and converges to the average of the left- and right-hand limits at points where f is discontinuous.

Observe that the left-hand side of (21) is periodic of period $2T$. This means that if we extend $f(x)$ from the interval $(-T, T)$ to the entire real line using $2T$-periodicity,[††] then equation (21) holds for all x for the $2T$ periodic extension of $f(x)$.

[†] All references to Chapters 12–14 refer to the expanded text *Fundamentals of Differential Equations and Boundary Value Problems*.

[††] From formula (21) we see that it doesn't matter how we define $f(x)$ at $-T$ and T since only the left- and right-hand limits are involved.

● **EXAMPLE 6**   To which function does the Fourier series for

$$f(x) = \begin{cases} -1, & -\pi < x < 0, \\ 1, & 0 < x < \pi, \end{cases}$$

converge?

Solution   In Example 4 we found that the Fourier series for $f(x)$ is given by (12), and in Figure 11.4 we sketched the graphs of two of its partial sums. Now $f(x)$ and $f'(x)$ are piecewise continuous in $[-\pi, \pi]$. Moreover, f is continuous except at $x = 0$. Thus by Theorem 2, the Fourier series of f in (12) converges to the 2π periodic function $g(x)$ where $g(x) = f(x) = -1$ for $-\pi < x < 0$, $g(x) = f(x) = 1$ for $0 < x < \pi$, $g(0) = \frac{1}{2}[f(0^+) + f(0^-)] = 0$, and at $\pm\pi$ we have $g(\pm\pi) = [f(-\pi^+) + f(\pi^-)]/2 = (-1 + 1)/2 = 0$. The graph of $g(x)$ is given in Figure 11.5.   ●

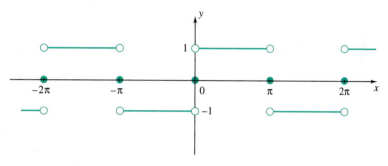

Figure 11.5 The limit function of the Fourier series for $f(x) = \begin{cases} -1, & -\pi < x < 0, \\ 1, & 0 < x < \pi \end{cases}$

When f is a continuous, *piecewise smooth* function on $(-\infty, \infty)$ and is $2T$-periodic, its Fourier series not only converges at each point—it converges *uniformly on* $(-\infty, \infty)$. This means that for any prescribed tolerance $\varepsilon > 0$, the graph of the partial sum

$$s_N(x) := \frac{a_0}{2} + \sum_{n=1}^{N} \left\{ a_n \cos\frac{n\pi x}{T} + b_n \sin\frac{n\pi x}{T} \right\}$$

will, for all N large, lie in an ε-corridor about the graph of f on $(-\infty, \infty)$ (see Figure 11.6).

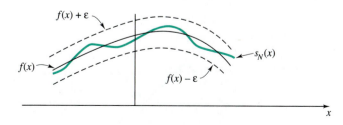

Figure 11.6 An ε-corridor about f

The property of uniform convergence of Fourier series is particularly helpful when one needs to verify that a formal solution to a partial differential equation is an actual (genuine) solution.

UNIFORM CONVERGENCE OF FOURIER SERIES

Theorem 3. Let f be a continuous function on $(-\infty, \infty)$ and periodic of period $2T$. If f' is piecewise continuous on $[-T, T]$, then the Fourier series for f converges uniformly to f on $[-T, T]$ and hence on any interval. That is, for each $\varepsilon > 0$, there exists an integer N_0 (that depends on ε) such that

$$\left| f(x) - \left\{ \frac{a_0}{2} + \sum_{n=1}^{N} \left\{ a_n \cos \frac{n\pi x}{T} + b_n \sin \frac{n\pi x}{T} \right\} \right\} \right| < \varepsilon,$$

for all $N \geq N_0$, and all $x \in (-\infty, \infty)$.

In Example 5 we obtained the Fourier series expansion given in (13) for $f(x) = |x|$, $-1 < x < 1$. Since $g(x)$, the periodic extension of $f(x)$ (see Figure 11.7), is continuous on $(-\infty, \infty)$ and

$$f'(x) = \begin{cases} -1, & -1 < x < 0, \\ 1, & 0 < x < 1, \end{cases}$$

is piecewise continuous on $[-1, 1]$, the Fourier series expansion (13) converges uniformly to $|x|$ on $[-1, 1]$.

The term-by-term differentiation of a Fourier series is not always permissible. For example, the Fourier series for $f(x) = x$, $-\pi < x < \pi$ (see Problem 9) is

(22) $$f(x) \sim 2 \sum_{n=1}^{\infty} (-1)^{n+1} \frac{\sin nx}{n},$$

which converges for all x, whereas its derived series

$$2 \sum_{n=1}^{\infty} (-1)^{n+1} \cos nx$$

diverges for every x. The following theorem gives sufficient conditions for using termwise differentiation.

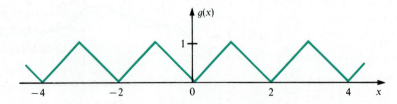

Figure 11.7 Periodic extension of $f(x) = |x|$, $-1 < x < 1$

DIFFERENTIATION OF FOURIER SERIES

Theorem 4. Let $f(x)$ be continuous on $(-\infty, \infty)$ and $2T$ periodic. Let $f'(x)$ and $f''(x)$ be piecewise continuous on $[-T, T]$. Then, the Fourier series of $f'(x)$ can be obtained from the Fourier series for $f(x)$ by termwise differentiation. In particular, if

$$f(x) = \frac{a_0}{2} + \sum_{n=1}^{\infty} \left\{ a_n \cos \frac{n\pi x}{T} + b_n \sin \frac{n\pi x}{T} \right\},$$

then

$$f'(x) \sim \sum_{n=1}^{\infty} \frac{\pi n}{T} \left\{ -a_n \sin \frac{n\pi x}{T} + b_n \cos \frac{n\pi x}{T} \right\}.$$

Notice that Theorem 4 does not apply to the function $f(x)$ and its Fourier series expansion in (22), since the 2π–periodic extension of $f(x)$ fails to be continuous on $(-\infty, \infty)$. Termwise integration of a Fourier series is permissible under much weaker conditions.

INTEGRATION OF FOURIER SERIES

Theorem 5. Let $f(x)$ be piecewise continuous on $[-T, T]$ with Fourier series

$$f(x) \sim \frac{a_0}{2} + \sum_{n=1}^{\infty} \left\{ a_n \cos \frac{n\pi x}{T} + b_n \sin \frac{n\pi x}{T} \right\}.$$

Then, for any x in $[-T, T]$, we have

$$\int_{-T}^{x} f(t)\, dt = \int_{-T}^{x} \frac{a_0}{2}\, dt + \sum_{n=1}^{\infty} \int_{-T}^{x} \left\{ a_n \cos \frac{n\pi t}{T} + b_n \sin \frac{n\pi t}{T} \right\} dt.$$

EXERCISES 11.3

In Problems 1 through 6 determine whether the given function is even, odd, or neither.

1. $f(x) = x^3 + \sin 2x$.

2. $f(x) = \sin^2 x$.

3. $f(x) = (1 - x^2)^{-1/2}$.

4. $f(x) = \sin(x + 1)$.

5. $f(x) = e^{-x} \cos 3x$.

6. $f(x) = x^{1/5} \cos x^2$.

7. Prove the following properties:

    (a) If f and g are even functions, then so is the product fg.

    (b) If f and g are odd functions, then fg is an even function.

    (c) If f is an even function and g is an odd function, then fg is an odd function.

8. Verify formula (5).

In Problems 9 through 16 compute the Fourier series for the given function f on the specified interval.

9. $f(x) = x$, $-\pi < x < \pi$.

10. $f(x) = |x|, \ -\pi < x < \pi.$

11. $f(x) = \begin{cases} 1, & -2 < x < 0, \\ x, & 0 < x < 2. \end{cases}$

12. $f(x) = \begin{cases} 0, & -\pi < x < 0, \\ x^2, & 0 < x < \pi. \end{cases}$

13. $f(x) = x^2, \ -1 < x < 1.$

14. $f(x) = \begin{cases} x, & 0 < x < \pi, \\ x + \pi, & -\pi < x < 0. \end{cases}$

15. $f(x) = e^x, \ -\pi < x < \pi.$

16. $f(x) = \begin{cases} 0, & -\pi < x < -\pi/2, \\ -1, & -\pi/2 < x < 0, \\ 1, & 0 < x < \pi/2, \\ 0, & \pi/2 < x < \pi. \end{cases}$

In Problems 17 through 24 determine the function to which the Fourier series for $f(x)$, given in the indicated problem, converges.

17. Problem 9. **18.** Problem 10.

19. Problem 11. **20.** Problem 12.

21. Problem 13. **22.** Problem 14.

23. Problem 15. **24.** Problem 16.

25. Find the functions represented by the series obtained by the termwise integration of the given series from $-\pi$ to x:

(a) $2 \sum_{n=1}^{\infty} \frac{(-1)^{n+1}}{n} \sin nx = x, \quad -\pi < x < \pi.$

(b) $-\frac{4}{\pi} \sum_{n=0}^{\infty} \frac{\sin(2n+1)x}{(2n+1)}$
$\sim f(x) = \begin{cases} -1, & -\pi < x < 0, \\ 1, & 0 < x < \pi. \end{cases}$

26. Show that the set of functions
$$\left\{ \cos\frac{\pi}{2}x, \sin\frac{\pi}{2}x, \cos\frac{3\pi}{2}x, \sin\frac{3\pi}{2}x, \cdots, \right.$$
$$\left. \cos\frac{(2n-1)\pi}{2}x, \sin\frac{(2n-1)\pi}{2}x, \cdots \right\}$$
is an orthonormal system on $[-1,1]$ with respect to the weight function $w(x) \equiv 1.$

27. Find the orthogonal expansion (generalized Fourier series) for
$$f(x) = \begin{cases} 0, & -1 < x < 0, \\ 1, & 0 < x < 1, \end{cases}$$

in terms of the orthonormal system of Problem 26.

28. (a) Show that the function $f(x) = x^2, \ -\pi < x < \pi$ has the Fourier series
$$f(x) \sim \frac{\pi^2}{3} + 4 \sum_{n=1}^{\infty} \frac{(-1)^n}{n^2} \cos nx.$$

(b) Use the result of part (a) and Theorem 2 to show that
$$\sum_{n=1}^{\infty} \frac{(-1)^{n+1}}{n^2} = \frac{\pi^2}{12}.$$

(c) Use the result of part (a) and Theorem 2 to show that
$$\sum_{n=1}^{\infty} \frac{1}{n^2} = \frac{\pi^2}{6}.$$

29. In Section 8.9 it was shown that the Legendre polynomials $P_n(x)$ are orthogonal on the interval $[-1,1]$ with respect to the weight function $w(x) \equiv 1.$ Using the fact that the first three Legendre polynomials are:
$$P_0(x) \equiv 1, \quad P_1(x) = x, \quad P_2(x) = (3/2)x^2 - (1/2),$$
find the first three coefficients in the expansion
$$f(x) = a_0 P_0(x) + a_1 P_1(x) + a_2 P_2(x) + \cdots,$$
where $f(x)$ is the function
$$f(x) := \begin{cases} -1, & -1 < x < 0, \\ 1, & 0 < x < 1. \end{cases}$$

30. As in Problem 29, find the first three coefficients in the expansion
$$f(x) = a_0 P_0(x) + a_1 P_1(x) + a_2 P_2(x) + \cdots,$$
when $f(x) = |x|, \ -1 < x < 1.$

31. The Hermite polynomials $H_n(x)$ are orthogonal on the interval $(-\infty, \infty)$ with respect to the weight function $W(x) = e^{-x^2}.$ Verify this fact for the first three Hermite polynomials:
$$H_0(x) \equiv 1, \quad H_1(x) = 2x, \quad H_2(x) = 4x^2 - 2.$$

32. The Chebyshev (Tchebichef) polynomials $T_n(x)$ are orthogonal on the interval $[-1,1]$ with respect to the weight function $w(x) = (1 - x^2)^{-1/2}.$ Verify this fact for the first three Chebyshev polynomials:
$$T_0(x) \equiv 1, \quad T_1(x) = x, \quad T_2(x) = 2x^2 - 1.$$

33. Let $\{f_n(x)\}$ be an orthogonal set of functions on the interval $[a,b]$ with respect to the weight function $w(x).$ Show that they satisfy the **Pythagorean Property**
$$\|f_m + f_n\|^2 = \|f_m\|^2 + \|f_n\|^2, \quad \text{if} \quad m \neq n.$$

34. Norm. The norm of a function $\|f\|$ is like the length of a vector in $R^n.$ In particular, show that the norm defined in (16) satisfies the following properties associated with length (assume f and g are continuous and $w(x) > 0$ on $[a, b]$):

(a) $\|f\| \geq 0$, and $\|f\| = 0$ if and only if $f \equiv 0$.
(b) $\|cf\| = |c| \|f\|$, where c is any real number.
(c) $\|f + g\| \leq \|f\| + \|g\|$.

35. Inner Product. The integral in the orthogonality condition (14) is like the dot product of two vectors in R^n. In particular, show that the **inner product** of two functions defined by

(23) $\langle f, g \rangle := \displaystyle\int_a^b f(x)g(x)w(x)\,dx$,

where $w(x)$ is a positive weight function, satisfies the following properties associated with the dot product (assume f, g, and h are continuous on $[a, b]$):

(a) $\langle f + g, h \rangle = \langle f, h \rangle + \langle g, h \rangle$.
(b) $\langle cf, h \rangle = c\langle f, h \rangle$, where c is any real number.
(c) $\langle f, g \rangle = \langle g, f \rangle$.

36. Complex Form of the Fourier Series

(a) Using the Euler formula $e^{i\theta} = \cos\theta + i\sin\theta$, $i = \sqrt{-1}$, prove that

$$\cos nx = \frac{e^{inx} + e^{-inx}}{2} \quad \text{and} \quad \sin nx = \frac{e^{inx} - e^{-inx}}{2i}.$$

(b) Show that the Fourier series

$$f(x) \sim \frac{a_0}{2} + \sum_{n=1}^{\infty} \{a_n \cos nx + b_n \sin nx\}$$

$$= c_0 + \sum_{n=1}^{\infty} \{c_n e^{inx} + c_{-n} e^{-inx}\},$$

where

$$c_0 = \frac{a_0}{2}, \qquad c_n = \frac{a_n - ib_n}{2}, \qquad c_{-n} = \frac{a_n + ib_n}{2}.$$

(c) Finally, use the results of part (b) to show that

$$f(x) \sim \sum_{n=-\infty}^{\infty} c_n e^{inx}, \qquad -\pi < x < \pi,$$

where

$$c_n = \frac{1}{2\pi} \int_{-\pi}^{\pi} f(x) e^{-inx}\,dx.$$

37. Least Squares Approximation Property. The Nth partial sum of the Fourier series gives the best mean square approximation of f by a trigonometric polynomial. To prove this, proceed as follows: Let

$$f(x) \sim \frac{a_0}{2} + \sum_{n=1}^{\infty} \{a_n \cos nx + b_n \sin nx\},$$

and let $F_N(x)$ denote an arbitrary trigonometric polynomial of degree N:

$$F_N(x) = \frac{\alpha_0}{2} + \sum_{n=1}^{N} \{\alpha_n \cos nx + \beta_n \sin nx\}.$$

Define

$$E := \int_{-\pi}^{\pi} [f(x) - F_N(x)]^2\,dx,$$

which is the *total square error*. Expanding the integrand we get

$$E = \int_{-\pi}^{\pi} f^2(x)\,dx - 2\int_{-\pi}^{\pi} f(x)F_N(x)\,dx$$

$$+ \int_{-\pi}^{\pi} F_N^2(x)\,dx.$$

(a) Show that

$$\int_{-\pi}^{\pi} F_N^2(x)\,dx = \pi\left(\frac{\alpha_0^2}{2} + \alpha_1^2 + \cdots + \alpha_N^2\right.$$

$$\left. + \beta_1^2 + \cdots + \beta_N^2\right)$$

and

$$\int_{-\pi}^{\pi} f(x)F_N(x)\,dx = \pi\left(\frac{\alpha_0 a_0}{2} + \alpha_1 a_1 + \cdots + \alpha_N a_N\right.$$

$$\left. + \beta_1 b_1 + \cdots + \beta_N b_N\right).$$

(b) Let E^* be the error when we approximate f by the Nth partial sum of its Fourier series; that is, when we choose $\alpha_n = a_n$ and $\beta_n = b_n$. Show that

$$E^* = \int_{-\pi}^{\pi} f^2(x)\,dx - \pi\left(\frac{a_0^2}{2} + a_1^2 + \cdots + a_N^2\right.$$

$$\left. + b_1^2 + \cdots + b_N^2\right).$$

(c) Using the results of parts (a) and (b), show that $E - E^* \geq 0$, that is, $E \geq E^*$, by proving that

$$E - E^* = \pi\left\{\frac{(\alpha_0 - a_0)^2}{2} + (\alpha_1 - a_1)^2\right.$$

$$+ \cdots + (\alpha_N - a_N)^2 + (\beta_1 - b_1)^2$$

$$\left. + \cdots + (\beta_N - b_N)^2\right\}.$$

Hence the Nth partial sum of the Fourier series gives the least total square error since $E \geq E^*$.

38. Bessel's Inequality. Use the fact that E^*, defined in part (b) of Problem 37, is nonnegative to prove **Bessel's inequality**

(24) $$\frac{a_0^2}{2} + \sum_{n=1}^{\infty} \{a_n^2 + b_n^2\} \le \frac{1}{\pi} \int_{-\pi}^{\pi} f^2(x)\, dx.$$

(If f is piecewise continuous on $[-\pi, \pi]$, then we have equality in (24). This result is called **Parseval's identity**.)

39. Gibbs' Phenomenon. The American mathematician Josiah Willard Gibbs (1839–1903) observed that near points of discontinuity of f the partial sums of the Fourier series for f may overshoot by approximately 9% of the jump. This is illustrated in Figure 11.8 for the function

$$f(x) = \begin{cases} -1, & -\pi < x < 0, \\ 1, & 0 < x < \pi, \end{cases}$$

whose Fourier series has the partial sums

$$f_{2n-1}(x) = \frac{4}{\pi}\left[\sin x + \frac{1}{3}\sin 3x + \cdots + \frac{\sin(2n-1)x}{(2n-1)}\right].$$

To verify this for $f(x)$:

(a) Show that

$$\pi \sin x f'_{2n-1}(x) = 4\sin x[\cos x + \cos 3x$$
$$+ \cdots + \cos(2n-1)x]$$
$$= 2\sin 2nx.$$

(b) Use the result of part (a) to determine that the maximum occurs at $x = \pi/(2n)$ and has the value

$$f_{2n-1}\left(\frac{\pi}{2n}\right) = \frac{4}{\pi}\left[\sin\frac{\pi}{2n} + \frac{1}{3}\sin\frac{3\pi}{2n} + \cdots\right.$$
$$\left. + \frac{1}{2n-1}\sin\frac{(2n-1)\pi}{2n}\right].$$

(c) Show that if one approximates

$$\int_0^\pi \frac{\sin x}{x}\, dx,$$

using the partition $x_k := (2k-1)(\pi/2n)$, $k = 1, 2, \ldots, n$, $\Delta x_k = \pi/n$, and choosing the midpoint of each interval as the place to evaluate the integrand, then

$$\int_0^\pi \frac{\sin x}{x}\, dx \approx \frac{\sin(\pi/2n)}{\pi/2n}\frac{\pi}{n} + \cdots + \frac{\sin[(2n-1)\pi/2n]}{(2n-1)\pi/2n}\frac{\pi}{n}$$

$$= \frac{\pi}{2}f_{2n-1}\left(\frac{\pi}{2n}\right).$$

(d) Use the result of part (c) to show that the overshoot satisfies

$$\lim_{n\to\infty} f_{2n-1}\left(\frac{\pi}{2n}\right) = \frac{2}{\pi}\int_0^\pi \frac{\sin x}{x}\, dx.$$

(e) Using the result of part (d) and a table of values for the **sine integral function**

$$Si(z) := \int_0^z \frac{\sin x}{x}\, dx,$$

show that $\lim_{n\to\infty} f_{2n-1}(\pi/(2n)) \approx 1.18$. Thus, the approximations overshoot the true value of $f(0^+) = 1$ by 0.18 or 9% of the jump from $f(0^-)$ to $f(0^+)$.

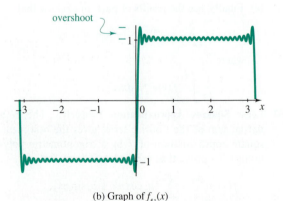

(a) Graph of $f_{11}(x)$

(b) Graph of $f_{51}(x)$

Figure 11.8 Gibbs' phenomenon for partial sums of Fourier series

•11.4• FOURIER COSINE AND SINE SERIES

A typical problem encountered in using separation of variables to solve a partial differential equation is the problem of representing a function defined on some finite interval by a trigonometric series consisting of only sine functions or only cosine functions. For example, in Section 11.2, equation (12), we needed to express the initial values $u(x, 0) = f(x)$, $0 < x < L$, of the solution to the initial-boundary value problem associated with the problem of heat flow as a trigonometric series of the form

$$\sum_{n=1}^{\infty} c_n \sin\left(\frac{n\pi x}{L}\right).$$

Recalling that the Fourier series for an odd function defined on $[-L, L]$ consists entirely of sine terms, we might try to extend the function $f(x), 0 < x < L$, to the interval $(-L, L)$ in such a way that the extended function is odd. This is accomplished by defining the function

$$f_o(x) := \begin{cases} f(x), & 0 < x < L, \\ -f(-x), & -L < x < 0, \end{cases}$$

and extending $f_o(x)$ to all x using $2L$-periodicity. Since $f_o(x)$ is an odd function, it has a Fourier series consisting entirely of sine terms. Moreover, $f_o(x)$ is an extension of $f(x)$, since $f_o(x) = f(x)$ on $(0, L)$. This extension is called the **odd 2L-periodic extension** of $f(x)$. The resulting Fourier series expansion is called a half-range expansion for $f(x)$, since it represents the function $f(x)$ on $(0, L)$, which is half of the interval $(-L, L)$ where it represents $f_o(x)$.

In a similar fashion, we can define the **even 2L-periodic extension** of $f(x)$ as the function

$$f_e(x) := \begin{cases} f(x), & 0 < x < L, \\ f(-x), & -L < x < 0, \end{cases}$$

with $f_e(x + 2L) = f_e(x)$.

To illustrate the various extensions, let's consider the function $f(x) = x, 0 < x < \pi$. If we extend $f(x)$ to the interval $(-\pi, \pi)$ using π periodicity, then the extension $\tilde{f}$ is given by

$$\tilde{f}(x) = \begin{cases} x, & 0 < x < \pi, \\ x + \pi, & -\pi < x < 0, \end{cases}$$

with $\tilde{f}(x + 2\pi) = \tilde{f}(x)$. In Problem 14 in Exercises 11.3 the reader computed the Fourier series for $\tilde{f}(x)$ and found

$$\tilde{f}(x) \sim \frac{\pi}{2} - \sum_{n=1}^{\infty} \frac{1}{n} \sin 2nx,$$

which consists of both odd functions (the sine terms) and even functions (the constant term) since the π-periodic extension $\tilde{f}(x)$ is neither an even nor an odd function. The odd 2π-periodic extension of $f(x)$ is just $f_o(x) = x$, $-\pi < x < \pi$, which has the Fourier series

expansion

(1) $$f_o(x) \sim 2 \sum_{n=1}^{\infty} \frac{(-1)^{n+1}}{n} \sin nx,$$

(see Problem 9 in Exercises 11.3). Since $f_o(x) = f(x)$ on the interval $(0, \pi)$, the expansion in (1) is a half-range expansion for $f(x)$. The even 2π-periodic extension of $f(x)$ is the function $f_e(x) = |x|$, $-\pi < x < \pi$, which has the Fourier series expansion

(2) $$f_e(x) = \frac{\pi}{2} - \frac{4}{\pi} \sum_{n=0}^{\infty} \frac{1}{(2n+1)^2} \cos(2n+1)x$$

(see Problem 10 in Exercises 11.3).

The preceding three extensions, the π-periodic function $\tilde{f}(x)$, the odd 2π-periodic function $f_o(x)$, and the even 2π-periodic function $f_e(x)$, are natural extensions of $f(x)$. There are many other ways of extending $f(x)$. For example, the function

$$g(x) = \begin{cases} x, & 0 < x < \pi, \\ 0, & -\pi < x < 0, \end{cases} \qquad g(x + 2\pi) = g(x),$$

which we studied in Example 3 of Section 11.3 is also an extension of $f(x)$. However, its Fourier series contains both sine and cosine terms and hence is not as useful as previous extensions. The graphs of these extensions of $f(x)$ are given in Figure 11.9.

The Fourier series expansions for $f_o(x)$ and $f_e(x)$ given in (1) and (2) represent $f(x)$ on the interval $(0, \pi)$ (actually, they equal $f(x)$ on $(0, \pi)$). This motivates the following definitions.

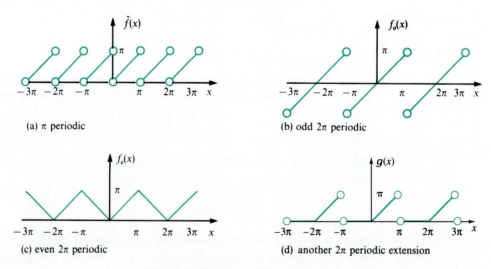

(a) π periodic

(b) odd 2π periodic

(c) even 2π periodic

(d) another 2π periodic extension

Figure 11.9 Extensions of $f(x) = x, 0 < x < \pi$

> **FOURIER COSINE AND SINE SERIES**
>
> **Definition 2.**   Let $f(x)$ be piecewise continuous on the interval $(0, T]$. The
> **Fourier cosine series** of $f(x)$ on $[0, T]$ is
>
> (3) $$\frac{a_0}{2} + \sum_{n=1}^{\infty} a_n \cos \frac{n\pi x}{T},$$
>
> where
>
> (4) $$a_n = \frac{2}{T} \int_0^T f(x) \cos \frac{n\pi x}{T} \, dx, \qquad n = 0, 1, \dots.$$
>
> The **Fourier sine series** of $f(x)$ on $[0, T]$ is
>
> (5) $$\sum_{n=1}^{\infty} b_n \sin \frac{n\pi x}{T},$$
>
> where
>
> (6) $$b_n = \frac{2}{T} \int_0^T f(x) \sin \frac{n\pi x}{T} \, dx, \qquad n = 1, 2, \dots.$$

The trigonometric series in (3) is just the Fourier series for $f_e(x)$, the even $2T$-periodic extension of $f(x)$ and (5) is the Fourier series for $f_o(x)$, the odd $2T$-periodic extension of $f(x)$. These are called **half-range expansions** for $f(x)$.

● **EXAMPLE 1**   Compute the Fourier sine series for

$$f(x) = \begin{cases} x, & 0 < x \le \pi/2, \\ \pi - x, & \pi/2 \le x < \pi. \end{cases}$$

Solution   Using formula (6) with $T = \pi$, we find

$$b_n = \frac{2}{\pi} \int_0^\pi f(x) \sin nx \, dx = \frac{2}{\pi} \int_0^{\pi/2} x \sin nx \, dx + \frac{2}{\pi} \int_{\pi/2}^\pi (\pi - x) \sin nx \, dx$$

$$= \frac{2}{\pi n^2} \int_0^{\pi n/2} u \sin u \, du + 2 \int_{\pi/2}^\pi \sin nx \, dx - \frac{2}{\pi n^2} \int_{\pi n/2}^{\pi n} u \sin u \, du$$

$$= \frac{2}{\pi n^2} \left[\sin u - u \cos u \right] \Big|_0^{\pi n/2} - \frac{2}{n} \left[\cos \pi n - \cos \frac{n\pi}{2} \right]$$

$$\qquad - \frac{2}{\pi n^2} \left[\sin u - u \cos u \right] \Big|_{\pi n/2}^{\pi n}$$

$$= \frac{4}{\pi n^2} \sin \frac{n\pi}{2} = \begin{cases} 0, & n \text{ even}, \\ \dfrac{4(-1)^{(n-1)/2}}{\pi n^2}, & n \text{ odd}. \end{cases}$$

So, on letting $n = 2k + 1$, the Fourier sine series for $f(x)$ is

(7) $\dfrac{4}{\pi} \displaystyle\sum_{k=0}^{\infty} \dfrac{(-1)^k}{(2k+1)^2} \sin(2k+1)x = \dfrac{4}{\pi} \left\{ \sin x - \dfrac{1}{9} \sin 3x + \dfrac{1}{25} \sin 5x + \cdots \right\}.$ ●

Since, in Example 1, the function $f(x)$ is continuous and $f'(x)$ is piecewise continuous on $(0, \pi)$, it follows from Theorem 2 on pointwise convergence of Fourier series that

$$f(x) = \dfrac{4}{\pi} \left\{ \sin x - \dfrac{1}{9} \sin 3x + \dfrac{1}{25} \sin 5x - \dfrac{1}{49} \sin 7x + \cdots \right\},$$

for all x in $(0, \pi)$.

Let's return to the problem of heat flow in one dimension.

● **EXAMPLE 2**   Find the solution to the heat flow problem

(8) $\dfrac{\partial u}{\partial t} = 2 \dfrac{\partial^2 u}{\partial x^2};$ $0 < x < \pi,$ $t > 0,$

(9) $u(0, t) = u(\pi, t) = 0,$ $t > 0,$

(10) $u(x, 0) = \begin{cases} x, & 0 < x \le \pi/2, \\ \pi - x, & \pi/2 \le x < \pi. \end{cases}$

Solution   Comparing equation (8) with equation (1) in Section 11.2, we see that $\beta = 2$ and $L = \pi$. Hence we need only represent $u(x, 0) = f(x)$ in a Fourier sine series (cf. equation (12) on page 539) of the form

$$\sum_{n=1}^{\infty} c_n \sin nx.$$

In Example 1 we obtained such an expansion and showed that

$$c_n = \dfrac{4}{\pi n^2} \sin \dfrac{n\pi}{2} = \begin{cases} 0, & n \text{ even} \\ \dfrac{4(-1)^{(n-1)/2}}{\pi n^2}, & n \text{ odd}. \end{cases}$$

Hence, from equation (11) on page 539, the solution to the heat flow problem (8)–(10) is

(11) $u(x, t) = \displaystyle\sum_{n=1}^{\infty} c_n e^{-2n^2 t} \sin nx$

$$= \dfrac{4}{\pi} \sum_{k=0}^{\infty} \dfrac{(-1)^k}{(2k+1)^2} e^{-2(2k+1)^2 t} \sin(2k+1)x$$

$$= \dfrac{4}{\pi} \left\{ e^{-2t} \sin x - \dfrac{1}{9} e^{-18t} \sin 3x + \dfrac{1}{25} e^{-50t} \sin 5x + \cdots \right\}.$$ ●

EXERCISES 11.4

In Problems 1 through 4 determine (a) the π-periodic extension $\tilde{f}$, (b) the odd 2π-periodic extension f_o, and (c) the even 2π-periodic extension f_e for the given function f and sketch their graphs.

1. $f(x) = x^2, \qquad 0 < x < \pi.$

2. $f(x) = \sin 2x, \qquad 0 < x < \pi.$

3. $f(x) = \begin{cases} 0, & 0 < x < \pi/2, \\ 1, & \pi/2 < x < \pi. \end{cases}$

4. $f(x) = \pi - x, \qquad 0 < x < \pi.$

In Problems 5 through 10 compute the Fourier sine series for the given function.

5. $f(x) = -1, \qquad 0 < x < 1.$

6. $f(x) = \cos x, \qquad 0 < x < \pi.$

7. $f(x) = x^2, \qquad 0 < x < \pi.$

8. $f(x) = \pi - x, \qquad 0 < x < \pi.$

9. $f(x) = x - x^2, \qquad 0 < x < 1.$

10. $f(x) = e^x, \qquad 0 < x < 1.$

In Problems 11 through 16 compute the Fourier cosine series for the given function.

11. $f(x) = \pi - x, \qquad 0 < x < \pi.$

12. $f(x) = 1 + x, \qquad 0 < x < \pi.$

13. $f(x) = e^x, \qquad 0 < x < 1.$

14. $f(x) = e^{-x}, \qquad 0 < x < 1.$

15. $f(x) = \sin x, \qquad 0 < x < \pi.$

16. $f(x) = x - x^2, \qquad 0 < x < 1.$

In Problems 17 and 18 find the solution to the heat flow problem

$$\frac{\partial u}{\partial t} = 5 \frac{\partial^2 u}{\partial x^2}; \qquad 0 < x < \pi, \qquad t > 0,$$

$$u(0, t) = u(\pi, t) = 0, \qquad t > 0,$$

$$u(x, 0) = f(x), \qquad 0 < x < \pi,$$

where $f(x)$ is given.

17. $f(x) = 1 - \cos 2x.$ **18.** $f(x) = x(\pi - x).$

•11.5• THE HEAT EQUATION

In Section 11.1 we developed a model for heat flow in an insulated wire whose ends are kept at the constant temperature $0°C$. In particular, we found that the temperature $u(x, t)$ in the wire is governed by the initial-boundary value problem

(1) $$\frac{\partial u}{\partial t} = \beta \frac{\partial^2 u}{\partial x^2}; \qquad 0 < x < L, \qquad t > 0,$$

(2) $$u(0, t) = u(L, t) = 0, \qquad t > 0,$$

(3) $$u(x, 0) = f(x), \qquad 0 < x < L$$

(see equations (6)–(8) in Section 11.1). Here equation (2) specifies that the temperature at the ends of the wire is zero, whereas equation (3) specifies the initial temperature distribution.

In Section 11.2 we also derived a formal solution to (1)–(3) using separation of variables. There we found the solution to (1)–(3) to have the form

(4) $$u(x, t) = \sum_{n=1}^{\infty} c_n e^{-\beta(n\pi/L)^2 t} \sin \frac{n\pi x}{L},$$

where the c_n's are the coefficients in the Fourier sine series for $f(x)$:

(5) $$f(x) = \sum_{n=1}^{\infty} c_n \sin \frac{n\pi x}{L}.$$

In other words, solving (1)–(3) reduces to computing the Fourier sine series for the initial value function $f(x)$.

In this section we will discuss heat flow problems where the ends of the wire are insulated or kept at a constant, but nonzero, temperature. (The latter involves nonhomogeneous boundary conditions.) We will also discuss the problem in which a heat source is adding heat to the wire. (This results in a nonhomogeneous partial differential equation.) The problem of heat flow in a rectangular plate will also be discussed and leads to the topic of double Fourier series. We will conclude this section with a discussion of the existence and uniqueness of solutions to the heat flow problem.

In the model of heat flow in a wire, let's replace the assumption that the ends of the wire are kept at a constant temperature zero and instead assume that the ends of the wire are *insulated*, that is, no heat flows out (or in) at the ends of the wire. It follows from the principle of heat conduction (see Section 11.1) that the temperature gradient must be zero at these end points, that is,

$$\frac{\partial u}{\partial x}(0, t) = \frac{\partial u}{\partial x}(L, t) = 0, \qquad t > 0.$$

In the next example we obtain the formal solution to the heat flow problem with these boundary conditions.

● **EXAMPLE 1**   Find a formal solution to the heat flow problem governed by the initial-boundary value problem

(6) $$\frac{\partial u}{\partial t} = \beta \frac{\partial^2 u}{\partial x^2}; \qquad 0 < x < L, \qquad t > 0,$$

(7) $$\frac{\partial u}{\partial x}(0, t) = \frac{\partial u}{\partial x}(L, t) = 0, \qquad t > 0,$$

(8) $$u(x, 0) = f(x), \qquad 0 < x < L.$$

Solution   Using the method of separation of variables, we first assume

$$u(x, t) = X(x)T(t).$$

Substituting into equation (6) and separating variables as was done in Section 11.2 (cf. equation (5) on page 537), we get the two equations

(9) $$X''(x) - KX(x) = 0,$$

(10) $$T'(t) - \beta K T(t) = 0,$$

where K is some constant. The boundary conditions in (7) become

$$X'(0)T(t) = 0 \quad \text{and} \quad X'(L)T(t) = 0.$$

In order for these equations to hold for all $t > 0$, either $T(t) \equiv 0$, which implies that $u(x, t) \equiv 0$, or

(11) $$X'(0) = X'(L) = 0.$$

Combining the boundary conditions in (11) with equation (9) gives the boundary value problem

(12) $X''(x) - KX(x) = 0;$ $X'(0) = X'(L) = 0,$

where K can be any constant.

To solve for the nontrivial solutions to (12), we begin with the auxiliary equation $r^2 - K = 0$. When $K > 0$, arguments similar to those used in Section 11.2 show that there are no nontrivial solutions to (12).

When $K = 0$, the auxiliary equation has the repeated root 0 and a general solution to the differential equation is

$$X(x) = A + Bx.$$

The boundary conditions in (12) reduce to $B = 0$ with A arbitrary. Thus, for $K = 0$, the nontrivial solutions to (12) are of the form

$$X(x) = c_0,$$

where c_0 is an arbitrary nonzero constant.

When $K < 0$, the auxiliary equation has the roots $r = \pm i\sqrt{-K}$. Thus, a general solution to the differential equation in (12) is

$$X(x) = C_1 \cos\sqrt{-K}\,x + C_2 \sin\sqrt{-K}\,x.$$

The boundary conditions in (12) lead to the system

$$\sqrt{-K}\,C_2 = 0,$$
$$-\sqrt{-K}\,C_1 \sin\sqrt{-K}\,L + \sqrt{-K}\,C_2 \cos\sqrt{-K}\,L = 0.$$

Hence $C_2 = 0$ and the system reduces to solving $C_1 \sin\sqrt{-K}\,L = 0$. Since $\sin\sqrt{-K}\,L = 0$ only when $\sqrt{-K}\,L = n\pi$, where n is an integer, we obtain a nontrivial solution only when $\sqrt{-K} = n\pi/L$ or $K = -(n\pi/L)^2$, $n = 1, 2, 3, \dots$. Furthermore, the nontrivial solutions (eigenfunctions) X_n corresponding to the eigenvalue $K = -(n\pi/L)^2$ are given by

(13) $X_n(x) = c_n \cos\dfrac{n\pi x}{L},$

where the c_n's are arbitrary nonzero constants. In fact, formula (13) also holds for $n = 0$, since $K = 0$ has the eigenfunctions $X_0(x) = c_0$.

Having determined that $K = -(n\pi/L)^2$, $n = 0, 1, 2, \dots$, let's consider equation (10) for such K:

$$T'(t) + \beta(n\pi/L)^2 T(t) = 0.$$

For $n = 0, 1, 2, \dots$, the general solution is

$$T_n(t) = b_n e^{-\beta(n\pi/L)^2 t},$$

where the b_n's are arbitrary constants. Combining this with equation (13), we obtain the functions

$$u_n(x,t) = X_n(x)T_n(t) = \left[c_n \cos \frac{n\pi x}{L} \right] \left[b_n e^{-\beta(n\pi/L)^2 t} \right],$$

$$u_n(x,t) = a_n e^{-\beta(n\pi/L)^2 t} \cos \frac{n\pi x}{L},$$

where $a_n = b_n c_n$ is an arbitrary constant.

If we take an infinite series of these functions, we obtain

(14) $$u(x,t) = \frac{a_0}{2} + \sum_{n=1}^{\infty} a_n e^{-\beta(n\pi/L)^2 t} \cos \frac{n\pi x}{L},$$

which will be a solution to (6)–(7) provided the series has the proper convergence behavior. Notice that in (14) we have removed the constant term and written it as $a_0/2$, producing the standard form for cosine expansions.

Assuming a solution to (6)–(7) is given by the series in (14) and substituting into the initial condition (8), we get

$$u(x,0) = \frac{a_0}{2} + \sum_{n=1}^{\infty} a_n \cos \frac{n\pi x}{L} = f(x), \qquad 0 < x < L.$$

This means that if we choose the a_n's as the coefficients in the Fourier cosine series for f,

$$a_n = \frac{2}{L} \int_0^L f(x) \cos \frac{n\pi x}{L} dx, \qquad n = 0, 1, 2, \ldots,$$

then $u(x,t)$ given in (14) will be a **formal solution** to the heat flow problem (6)–(8). Again, if this expansion converges to a continuous function with continuous second partial derivatives, then the formal solution is an actual solution. ●

When the ends of the wire are kept at $0°C$ or when the ends are insulated, the boundary conditions are said to be **homogeneous.** But, when the ends of the wire are kept at constant temperatures different from zero, that is,

(15) $$u(0,t) = U_1 \quad \text{and} \quad u(L,t) = U_2, \qquad t > 0,$$

then the boundary conditions are called **nonhomogeneous.**

From our experience with vibration problems in Chapter 5 we expect that the solution to the heat flow problem with nonhomogeneous boundary conditions will consist of a **steady-state solution** $v(x)$ that satisfies the nonhomogeneous boundary conditions in (15) plus a **transient solution** $w(x,t)$. That is,

(16) $$u(x,t) = v(x) + w(x,t),$$

where $w(x,t)$ and its partial derivatives tend to zero as $t \to \infty$. The function $w(x,t)$ will then satisfy homogeneous boundary conditions as illustrated in the next example.

● **EXAMPLE 2**   Find a formal solution to the heat flow problem governed by the initial-boundary value problem

(17) $\dfrac{\partial u}{\partial t} = \beta \dfrac{\partial^2 u}{\partial x^2};$ $0 < x < L,$ $t > 0,$

(18) $u(0, t) = U_1,$ $u(L, t) = U_2,$ $t > 0,$

(19) $u(x, 0) = f(x),$ $0 < x < L.$

Solution   Let's assume the solution $u(x, t)$ consists of a steady-state solution $v(x)$ and a transient solution $w(x, t)$, that is,

(20) $u(x, t) = v(x) + w(x, t).$

Substituting for $u(x, t)$ in equations (17)–(19) leads to

(21) $\dfrac{\partial u}{\partial t} = \dfrac{\partial w}{\partial t} = \beta v''(x) + \beta \dfrac{\partial^2 w}{\partial x^2};$ $0 < x < L,$ $t > 0,$

(22) $v(0) + w(0, t) = U_1,$ $v(L) + w(L, t) = U_2,$ $t > 0,$

(23) $v(x) + w(x, 0) = f(x),$ $0 < x < L.$

If we allow $t \to \infty$ in (21)–(22), assuming $w(x, t)$ is a transient solution, we obtain the steady-state boundary value problem

$v''(x) = 0,$ $0 < x < L,$

$v(0) = U_1,$ $v(L) = U_2.$

Solving for v, we obtain $v(x) = Ax + B$, and choosing A and B so that the boundary conditions are satisifed yields

(24) $v(x) = U_1 + \dfrac{(U_2 - U_1)x}{L},$

as the steady-state solution.

With this choice for $v(x)$, the initial-boundary value problem (21)–(23) becomes the following initial-boundary value problem for $w(x, t)$:

(25) $\dfrac{\partial w}{\partial t} = \beta \dfrac{\partial^2 w}{\partial x^2},$ $0 < x < L,$ $t > 0,$

(26) $w(0, t) = w(L, t) = 0,$ $t > 0,$

(27) $w(x, 0) = f(x) - U_1 - \dfrac{(U_2 - U_1)x}{L},$ $0 < x < L.$

Recall that a formal solution to (25)–(27) is given by equation (4). Hence,

$$w(x, t) = \sum_{n=1}^{\infty} c_n e^{-\beta(n\pi/L)^2 t} \sin \dfrac{n\pi x}{L},$$

where the c_n's are the coefficients of the Fourier sine series expansion

$$f(x) - U_1 - \frac{(U_2 - U_1)x}{L} = \sum_{n=1}^{\infty} c_n \sin\frac{n\pi x}{L}.$$

Therefore, the formal solution to (17)–(19) is

(28) $$u(x,t) = U_1 + \frac{(U_2 - U_1)x}{L} + \sum_{n=1}^{\infty} c_n e^{-\beta(n\pi/L)^2 t} \sin\frac{n\pi x}{L}. \quad \bullet$$

In the next example we consider the heat flow problem when a heat source P is present but is independent of time.

● **EXAMPLE 3**   Find a formal solution to the heat flow problem governed by the initial-boundary value problem

(29) $$\frac{\partial u}{\partial t} = \beta\frac{\partial^2 u}{\partial x^2} + P(x); \qquad 0 < x < L, \qquad t > 0,$$

(30) $$u(0,t) = U_1, \qquad u(L,t) = U_2, \qquad t > 0,$$

(31) $$u(x,0) = f(x), \qquad 0 < x < L.$$

Solution   We begin by assuming that the solution consists of a steady-state solution $v(x)$ and a transient solution $w(x,t)$, namely,

$$u(x,t) = v(x) + w(x,t),$$

where $w(x,t)$ and its partial derivatives tend to zero as $t \to \infty$. Substituting for $u(x,t)$ in (29)–(31) yields

(32) $$\frac{\partial u}{\partial t} = \frac{\partial w}{\partial t} = \beta v''(x) + \beta\frac{\partial^2 w}{\partial x^2} + P(x), \qquad 0 < x < L, \qquad t > 0,$$

(33) $$v(0) + w(0,t) = U_1, \qquad v(L) + w(L,t) = U_2, \qquad t > 0,$$

(34) $$v(x) + w(x,0) = f(x), \qquad 0 < x < L.$$

Letting $t \to \infty$ in (32)–(33), we obtain the steady-state boundary value problem

$$v''(x) = -\frac{1}{\beta}P(x), \qquad 0 < x < L,$$

$$v(0) = U_1, \qquad v(L) = U_2.$$

The solution to this boundary value problem can be obtained by two integrations using the boundary conditions to determine the constants of integration. The reader can verify that the solution is given by the formula

(35) $$v(x) = \left[U_2 - U_1 + \int_0^L \left(\int_0^z \frac{1}{\beta} P(s)\,ds \right) dz \right]\frac{x}{L} + U_1 - \int_0^x \left(\int_0^z \frac{1}{\beta} P(s)\,ds \right) dz.$$

With this choice for $v(x)$, we find that the initial-boundary value problem (32)–(34)

reduces to the following initial-boundary value problem for $w(x, t)$:

(36) $$\frac{\partial w}{\partial t} = \beta \frac{\partial^2 w}{\partial x^2}, \qquad 0 < x < L, \qquad t > 0,$$

(37) $\quad w(0, t) = w(L, t) = 0, \qquad t > 0,$

(38) $\quad w(x, 0) = f(x) - v(x), \qquad 0 < x < L,$

where $v(x)$ is given by formula (35). As before, the solution to this initial-boundary value problem is

(39) $$w(x, t) = \sum_{n=1}^{\infty} c_n e^{-\beta(n\pi/L)^2 t} \sin \frac{n\pi x}{L},$$

where the c_n's are determined from the Fourier sine series expansion of $f(x) - v(x)$:

(40) $$f(x) - v(x) = \sum_{n=1}^{\infty} c_n \sin \frac{n\pi x}{L}.$$

Thus the formal solution to (29)–(31) is given by

$$u(x, t) = v(x) + w(x, t)$$

where $v(x)$ is given in (35) and $w(x, t)$ is prescribed by (39) and (40). $\quad\bullet$

The method of separation of variables is also applicable to problems in higher dimensions. For example, consider the problem of heat flow in a rectangular plate with sides $x = 0$, $x = L$, $y = 0$, and $y = W$. Assuming the two sides $y = 0$, $y = W$ are kept at a constant temperature of $0°C$ and the two sides $x = 0$, $x = L$ are perfectly insulated, then heat flow is governed by the initial-boundary value problem in the following example.

● EXAMPLE 4   Find a formal solution $u(x, y, t)$ to the initial-boundary value problem

(41) $$\frac{\partial u}{\partial t} = \beta \left\{ \frac{\partial^2 u}{\partial x^2} + \frac{\partial^2 u}{\partial y^2} \right\}; \qquad 0 < x < L, \qquad 0 < y < W, \qquad t > 0,$$

(42) $\quad \dfrac{\partial u}{\partial x}(0, y, t) = \dfrac{\partial u}{\partial x}(L, y, t) = 0, \qquad 0 < y < W, \qquad t > 0,$

(43) $\quad u(x, 0, t) = u(x, W, t) = 0, \qquad 0 < x < L, \qquad t > 0,$

(44) $\quad u(x, y, 0) = f(x, y), \qquad 0 < x < L, \qquad 0 < y < W.$

Solution   If we assume a solution of the form $u(x, y, t) = V(x, y)T(t)$, then equation (41) separates into the two equations

(45) $\quad T'(t) - \beta K T(t) = 0,$

(46) $\quad \dfrac{\partial^2 V}{\partial x^2}(x, y) + \dfrac{\partial^2 V}{\partial y^2}(x, y) - K V(x, y) = 0,$

where K can be any constant. To solve equation (46), we again use separation of variables.

Here we assume $V(x, y) = X(x)Y(y)$. This allows us to separate equation (46) into the two equations

(47) $\qquad X''(x) - JX(x) = 0,$

(48) $\quad Y''(y) + (J - K)Y(y) = 0,$

where J can be any constant (see Problem 29 in Exercises 11.2).

To solve for $X(x)$, we observe that the boundary conditions in (42), in terms of the separated variables, become

$$X'(0)Y(y)T(t) = X'(L)Y(y)T(t) = 0, \qquad 0 < y < W, \qquad t > 0.$$

Hence, in order to get a nontrivial solution we must have

(49) $\quad X'(0) = X'(L) = 0.$

The boundary value problem for X given in equations (47) and (49) was solved in Example 1 (cf. equations (12) and (13)). Here $J = -(m\pi/L)^2, m = 0, 1, 2, \ldots,$ and

$$X_m(x) = c_m \cos \frac{m\pi x}{L},$$

where the c_m's are arbitrary.

To solve for $Y(y)$, we first observe that the boundary conditions in (43) become

(50) $\qquad Y(0) = Y(W) = 0.$

Next, substituting $J = -(m\pi/L)^2$ into equation (48) yields

$$Y''(y) - (K + (m\pi/L)^2)Y(y) = 0,$$

which we can rewrite as

(51) $\quad Y''(y) - EY(y) = 0,$

where $E = K + (m\pi/L)^2$. The boundary value problem for Y consisting of (50)–(51) has also been solved before. In Section 11.2 (cf. equations (7) and (9)) we showed that $E = -(n\pi/W)^2, n = 1, 2, 3, \ldots,$ and the nontrivial solutions are given by

$$Y_n(y) = a_n \sin \frac{n\pi y}{W},$$

where the a_n's are arbitrary.

Since $K = E - (m\pi/L)^2$, we have

$$K = -(n\pi/W)^2 - (m\pi/L)^2, \qquad m = 0, 1, 2, \ldots, \qquad n = 1, 2, 3, \ldots.$$

Substituting K into equation (45), we can solve for $T(t)$ and obtain

$$T_{mn}(t) = b_{mn} e^{-(m^2/L^2 + n^2/W^2)\beta\pi^2 t}.$$

Substituting in for X_m, Y_n, and T_{mn}, we get

$$u_{mn}(x, y, t) = \left(c_m \cos \frac{m\pi x}{L} \right) \left(a_n \sin \frac{n\pi y}{W} \right) (b_{mn} e^{-(m^2/L^2 + n^2/W^2)\beta\pi^2 t}),$$

$$u_{mn}(x, y, t) = a_{mn} e^{-(m^2/L^2 + n^2/W^2)\beta\pi^2 t} \cos \frac{m\pi x}{L} \sin \frac{n\pi y}{W},$$

where $a_{mn} := a_n b_{mn} c_m$, $\quad m = 0, 1, 2, \dots$, $\quad n = 1, 2, 3, \dots$, are arbitrary constants.

If we now take a doubly infinite series of such functions, then we obtain the formal series

(52) $$u(x, y, t) = \sum_{m=0}^{\infty} \sum_{n=1}^{\infty} a_{mn} e^{-(m^2/L^2 + n^2/W^2)\beta\pi^2 t} \cos \frac{m\pi x}{L} \sin \frac{n\pi y}{W}.$$

We are now ready to apply the initial conditions (44). Setting $t = 0$, we obtain

(53) $$u(x, y, 0) = f(x, y) = \sum_{m=0}^{\infty} \sum_{n=1}^{\infty} a_{mn} \cos \frac{m\pi x}{L} \sin \frac{n\pi y}{W}.$$

This is a **double Fourier Series.**[†] In fact, it is a double Fourier series for a function $f(x, y)$ that is an even function in the variable x and an odd function in the variable y. The formulas for the coefficients a_{mn} are

(54) $$a_{0n} = \frac{2}{LW} \int_0^L \int_0^W f(x, y) \sin \frac{n\pi y}{W} \, dx \, dy, \qquad n = 1, 2, 3, \dots,$$

and, for $m \geq 1$, $n \geq 1$,

(55) $$a_{mn} = \frac{4}{LW} \int_0^L \int_0^W f(x, y) \cos \frac{m\pi x}{L} \sin \frac{n\pi y}{W} \, dx \, dy.$$

Finally, the solution to the initial-boundary value problem (41)–(44) is given by equation (52) where the coefficients are prescribed by equations (54) and (55).   ●

Existence and Uniqueness of Solutions

In the examples that we have studied in this section and in Section 11.2, we were able to obtain formal solutions in the sense that we could express the solution in terms of a series expansion consisting of exponentials, sines, and cosines. To prove that these series converge to actual solutions requires results on the convergence of Fourier series and results from real analysis on uniform convergence. We will not go into these details here, but refer the reader to Section 6.5 of the text by Tyn Myint-U, (see footnote) for a proof of the existence of a solution to the heat flow problem discussed in Sections 11.1 and 11.2. (A proof of uniqueness is also given there.)

[†] For a discussion of double Fourier series see *Partial Differential Equations of Mathematical Physics*, by Tyn Myint-U, Elsevier North Holland, Inc., New York, 1980, Section 5.14.

As might be expected, using Fourier series and the method of separation of variables one can also obtain "solutions" when the initial data is discontinuous since the formal solutions only require the existence of a convergent Fourier series. This allows one to study idealized problems in which the initial conditions do not agree with the boundary conditions or the initial conditions involve a jump discontinuity. For example, we may assume that initially one half of the wire is at one temperature, whereas the other half is at a different temperature, that is,

$$f(x) = \begin{cases} U_1, & 0 < x < L/2, \\ U_2, & L/2 < x < L. \end{cases}$$

The formal solution that we obtained will make sense for $0 < x < L$, $t > 0$, but we must take care in interpreting the results near the points of discontinuity $x = 0$, $L/2$, and L.

The question of the uniqueness of the solution to the heat flow problem can be answered in various ways. One is tempted to argue that the method of separation of variables yields *formulas* for the solutions and therefore a unique solution. However, this does *not* exclude the possibility of solutions existing that cannot be obtained by the method of separation of variables.

An argument using the physical principles discussed in Section 11.1 can be given which states that the maximum temperature of the wire over all time $t \geq 0$, must occur either at one of the ends or in the initial temperature distribution. These results are called **maximum principles** and exist for the heat equation and Laplace's equation. One such result, which we will not prove, is the following:[†]

MAXIMUM PRINCIPLE FOR THE HEAT EQUATION

Theorem 6.   Let $u(x, t)$ be a continuously differentiable function that satisfies the heat equation

$$\text{(56)} \qquad \frac{\partial u}{\partial t} = \beta \frac{\partial^2 u}{\partial x^2}; \qquad 0 < x < L, \qquad t > 0,$$

and the boundary conditions

$$\text{(57)} \qquad u(0, t) = u(L, t) = 0, \qquad t > 0,$$

Then $u(x, t)$ attains its maximum value at $t = 0$ for some x in $[0, L]$, that is,

$$\max_{\substack{t \geq 0 \\ 0 \leq x \leq L}} u(x, t) = \max_{0 \leq x \leq L} u(x, 0).$$

We can use the maximum principle to show that the heat flow problem has a unique solution.

[†] For a discussion of maximum principles and their applications see *Maximum Principles in Differential Equations*, by M. H. Protter and H. F. Weinberger, Prentice Hall, Inc., Englewood Cliffs, New Jersey, 1967.

> ◢ **UNIQUENESS OF SOLUTION**
>
> **Theorem 7.**   The initial-boundary value problem
>
> (58) $$\frac{\partial u}{\partial t} = \beta \frac{\partial^2 u}{\partial x^2}; \qquad 0 < x < L, \qquad t > 0,$$
>
> (59) $u(0, t) = u(L, t) = 0, \qquad t \ge 0,$
>
> (60) $u(x, 0) = f(x), \qquad 0 \le x \le L,$
>
> has at most one continuously differentiable solution.

Proof.   Assume $u(x, t)$ and $v(x, t)$ are continuously differentiable functions that satisfy the initial-boundary value problem (58)–(60). Let $w = u - v$. Now w is a continuously differentiable solution to the boundary value problem (56)–(57). By the maximum principle, w must attain its maximum at $t = 0$, and since

$$w(x, 0) = u(x, 0) - v(x, 0) = f(x) - f(x) = 0,$$

we have $w(x, t) \le 0$. Hence, $u(x, t) \le v(x, t)$ for all $0 \le x \le L, t \ge 0$. A similar argument using $\hat{w} = v - u$ yields $v(x, t) \le u(x, t)$. Therefore, we have $u(x, t) = v(x, t)$, for all $0 \le x \le L$, $t \ge 0$. Thus, there is at most one continuously differentiable solution to the problem (58)–(60).   ◄◄◄

EXERCISES 11.5

In Problems 1 through 10 find a formal solution to the given initial-boundary value problem.

1. $\dfrac{\partial u}{\partial t} = 5 \dfrac{\partial^2 u}{\partial x^2}; \qquad 0 < x < 1, \qquad t > 0,$

$u(0, t) = u(1, t) = 0, \qquad t > 0,$
$u(x, 0) = (1 - x)x^2, \qquad 0 < x < 1.$

2. $\dfrac{\partial u}{\partial t} = \dfrac{\partial^2 u}{\partial x^2}; \qquad 0 < x < \pi, \qquad t > 0,$

$u(0, t) = u(\pi, t) = 0, \qquad t > 0,$
$u(x, 0) = x^2, \qquad 0 < x < \pi.$

3. $\dfrac{\partial u}{\partial t} = 3 \dfrac{\partial^2 u}{\partial x^2}; \qquad 0 < x < \pi, \qquad t > 0,$

$\dfrac{\partial u}{\partial x}(0, t) = \dfrac{\partial u}{\partial x}(\pi, t) = 0, \qquad t > 0,$
$u(x, 0) = x, \qquad 0 < x < \pi.$

4. $\dfrac{\partial u}{\partial t} = 2 \dfrac{\partial^2 u}{\partial x^2}; \qquad 0 < x < 1, \qquad t > 0,$

$\dfrac{\partial u}{\partial x}(0, t) = \dfrac{\partial u}{\partial x}(1, t) = 0, \qquad t > 0,$
$u(x, 0) = x(1 - x), \qquad 0 < x < 1.$

5. $\dfrac{\partial u}{\partial t} = \dfrac{\partial^2 u}{\partial x^2}; \qquad 0 < x < \pi, \qquad t > 0,$

$\dfrac{\partial u}{\partial x}(0, t) = \dfrac{\partial u}{\partial x}(\pi, t) = 0, \qquad t > 0,$
$u(x, 0) = e^x, \qquad 0 < x < \pi.$

6. $\dfrac{\partial u}{\partial t} = 7 \dfrac{\partial^2 u}{\partial x^2}; \qquad 0 < x < \pi, \qquad t > 0,$

$\dfrac{\partial u}{\partial x}(0, t) = \dfrac{\partial u}{\partial x}(\pi, t) = 0, \qquad t > 0,$
$u(x, 0) = 1 - \sin x, \qquad 0 < x < \pi.$

7. $\dfrac{\partial u}{\partial t} = 2 \dfrac{\partial^2 u}{\partial x^2}; \qquad 0 < x < \pi, \qquad t > 0,$

$u(0, t) = 5, \qquad u(\pi, t) = 10, \qquad t > 0,$
$u(x, 0) = \sin 3x - \sin 5x, \qquad 0 < x < \pi.$

8. $\dfrac{\partial u}{\partial t} = \dfrac{\partial^2 u}{\partial x^2}; \qquad 0 < x < \pi, \qquad t > 0,$

$u(0, t) = 0, \qquad u(\pi, t) = 3\pi, \qquad t > 0,$
$u(x, 0) = 0, \qquad 0 < x < \pi.$

9. $\dfrac{\partial u}{\partial t} = \dfrac{\partial^2 u}{\partial x^2} + e^{-x};$ $0 < x < \pi,$ $t > 0,$

$u(0, t) = u(\pi, t) = 0,$ $t > 0,$
$u(x, 0) = \sin 2x,$ $0 < x < \pi.$

10. $\dfrac{\partial u}{\partial t} = 3\dfrac{\partial^2 u}{\partial x^2} + x;$ $0 < x < \pi,$ $t > 0,$

$u(0, t) = u(\pi, t) = 0,$ $t > 0,$
$u(x, 0) = \sin x,$ $0 < x < \pi.$

11. Find a formal solution to the initial-boundary value problem

$$\dfrac{\partial u}{\partial t} = 4\dfrac{\partial^2 u}{\partial x^2};   0 < x < \pi,   t > 0,$$

$$\dfrac{\partial u}{\partial x}(0, t) = 0,   u(\pi, t) = 0,   t > 0,$$

$$u(x, 0) = f(x),   0 < x < \pi.$$

12. Find a formal solution to the initial-boundary value problem

$$\dfrac{\partial u}{\partial t} = \dfrac{\partial^2 u}{\partial x^2};   0 < x < \pi,   t > 0,$$

$$u(0, t) = 0,   u(\pi, t) + \dfrac{\partial u}{\partial x}(\pi, t) = 0,   t > 0,$$

$$u(x, 0) = f(x),   0 < x < \pi.$$

13. Find a formal solution to the initial-boundary value problem

$$\dfrac{\partial u}{\partial t} = 2\dfrac{\partial^2 u}{\partial x^2} + 4x;   0 < x < \pi,   t > 0,$$

$$u(0, t) = u(\pi, t) = 0,   t > 0,$$
$$u(x, 0) = \sin x,   0 < x < \pi.$$

14. Find a formal solution to the initial-boundary value problem

$$\dfrac{\partial u}{\partial t} = 3\dfrac{\partial^2 u}{\partial x^2} + 5;   0 < x < \pi,   t > 0,$$

$$u(0, t) = u(\pi, t) = 1,   t > 0,$$
$$u(x, 0) = 1,   0 < x < \pi.$$

In Problems 15 through 18 find a formal solution to the initial-boundary value problem

$$\dfrac{\partial u}{\partial t} = \dfrac{\partial^2 u}{\partial x^2} + \dfrac{\partial^2 u}{\partial y^2};$$

$$0 < x < \pi,   0 < y < \pi,   t > 0,$$

$$\dfrac{\partial u}{\partial x}(0, y, t) = \dfrac{\partial u}{\partial x}(\pi, y, t) = 0,   0 < y < \pi,   t > 0,$$

$$u(x, 0, t) = u(x, \pi, t) = 0,   0 < x < \pi,   t > 0,$$

$$u(x, y, 0) = f(x, y),   0 < x < \pi,   0 < y < \pi,$$

for the given function $f(x, y)$.

15. $f(x, y) = \cos 6x \sin 4y - 3\cos x \sin 11y.$

16. $f(x, y) = \cos x \sin y + 4\cos 2x \sin y - 3\cos 3x \sin 4y.$

17. $f(x, y) = y.$

18. $f(x, y) = x \sin y.$

19. Chemical Diffusion.   Chemical diffusion through a thin layer is governed by the equation

$$\dfrac{\partial C}{\partial t} = k\dfrac{\partial^2 C}{\partial x^2} - LC,$$

where $C(x, t)$ is the concentration in moles/cm^3, k is a positive constant with units cm^2/sec, and $L > 0$ is a consumption rate with units sec^{-1}. Assume that the boundary conditions are

$$C(0, t) = C(a, t) = 0,   t > 0,$$

and the initial concentration is given by

$$C(x, 0) = f(x),   0 < x < a.$$

Use the method of separation of variables to formally solve for the concentration $C(x, t)$. What happens to the concentration as $t \to +\infty$?

•11.6• THE WAVE EQUATION

In Section 11.2 we presented a model for the motion of a vibrating string. If $u(x, t)$ represents the displacement (deflection) of the string and the ends of the string are held fixed, then the motion of the string is governed by the initial-boundary value problem

(1) $\dfrac{\partial^2 u}{\partial t^2} = \alpha^2\dfrac{\partial^2 u}{\partial x^2};$ $0 < x < L,$ $t > 0,$

(2) $u(0, t) = u(L, t) = 0,$ $t > 0,$

(3) $u(x, 0) = f(x), \qquad 0 < x < L,$

(4) $\dfrac{\partial u}{\partial t}(x, 0) = g(x), \qquad 0 < x < L.$

Equation (1) is called the **wave equation.**

The constant α^2 appearing in (1) is strictly positive and depends on the linear density and tension of the string. The boundary conditions in (2) reflect the fact that the string is held fixed at the two end points $x = 0$ and $x = L$.

Equations (3) and (4) specify, respectively, the initial displacement and the initial velocity of each point on the string. For the initial and boundary conditions to be consistent, we assume $f(0) = f(L) = 0$ and $g(0) = g(L) = 0$.

Using the method of separation of variables, we found in Section 11.2 that a formal solution to (1)–(4) is given by (cf. equations (24)–(26) on page 542)

(5) $u(x, t) = \displaystyle\sum_{n=1}^{\infty} \left[a_n \cos \frac{n\pi\alpha}{L} t + b_n \sin \frac{n\pi\alpha}{L} t \right] \sin \frac{n\pi x}{L},$

where the a_n's and b_n's are determined from the Fourier sine series

(6) $f(x) = \displaystyle\sum_{n=1}^{\infty} a_n \sin \frac{n\pi x}{L},$

(7) $g(x) = \displaystyle\sum_{n=1}^{\infty} b_n \left(\frac{n\pi\alpha}{L} \right) \sin \frac{n\pi x}{L}.$

Each term in expansion (5) can be viewed as a **standing wave** (a wave that vibrates without lateral motion along the string). For example, the first term,

$$\left(a_1 \cos \frac{\pi\alpha}{L} t + b_1 \sin \frac{\pi\alpha}{L} t \right) \sin \frac{\pi x}{L},$$

consists of a sine wave $\sin(\pi x/L)$ multiplied by a time varying amplitude. The second term is also a sine wave $\sin(2\pi x/L)$ with a time varying amplitude. In the latter case there is a *node* in the middle at $x = L/2$ which never moves. For the nth term, we have a sine wave $\sin(n\pi x/L)$ with a time varying amplitude and the sine wave has $(n - 1)$ nodes. This is illustrated in Figure 11.10. Thus the solution in (5) can be interpreted as the superposition of infinitely many standing waves.

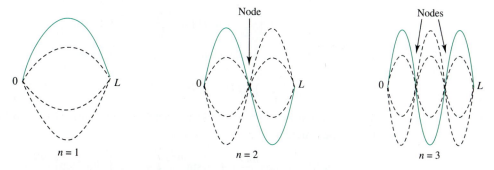

Figure 11.10 Standing waves. Time varying amplitudes are shown by dashed curves.

The method of separation of variables can also be used to solve problems with non-homogeneous boundary conditions and nonhomogeneous equations where the forcing term is time independent, just as was done for the heat equation with nonhomogeneous boundary conditions and with a heat source independent of time. In the next example, we will consider a problem with a time dependent forcing term.

● **EXAMPLE 1**   Find a formal solution to the initial-boundary value problem

$$(8) \qquad \frac{\partial^2 u}{\partial t^2} = \alpha^2 \frac{\partial^2 u}{\partial x^2} + h(x,t); \qquad 0 < x < L, \qquad t > 0,$$

$$(9) \qquad u(0,t) = u(L,t) = 0, \qquad t > 0,$$

$$(10) \qquad u(x,0) = f(x), \qquad 0 < x < L,$$

$$(11) \qquad \frac{\partial u}{\partial t}(x,0) = g(x), \qquad 0 < x < L.$$

Solution   The boundary conditions in (9) certainly require that the solution be zero for $x = 0$ and $x = L$. Motivated by the fact that the solution to the corresponding homogeneous system (1)–(4) consists of a superposition of standing waves, let's try to find a solution to (8)–(11) of the form

$$(12) \qquad u(x,t) = \sum_{n=1}^{\infty} u_n(t) \sin \frac{n\pi x}{L},$$

where the $u_n(t)$'s are functions of t to be determined.

For each fixed t, we can compute a Fourier sine series for $h(x,t)$. If we assume the series is convergent to $h(x,t)$, then

$$(13) \qquad h(x,t) = \sum_{n=1}^{\infty} h_n(t) \sin \frac{n\pi x}{L},$$

where the coefficient $h_n(t)$ is given by (recall equation (6) on page 563)

$$h_n(t) = \frac{2}{L} \int_0^L h(x,t) \sin \frac{n\pi x}{L} dx, \qquad n = 1, 2, \dots .$$

If the series in (13) has the proper convergence properties, then we can substitute (12) and (13) into equation (8) and obtain

$$\sum_{n=1}^{\infty} \left[u_n''(t) + \left(\frac{n\pi \alpha}{L} \right)^2 u_n(t) \right] \sin \frac{n\pi x}{L} = \sum_{n=1}^{\infty} h_n(t) \sin \frac{n\pi x}{L}.$$

Equating the coefficients in each series (why?), we have

$$u_n''(t) + \left(\frac{n\pi \alpha}{L} \right)^2 u_n(t) = h_n(t).$$

This is a nonhomogeneous, constant coefficient equation which can be solved using variation of parameters. The reader should verify that

$$u_n(t) = a_n \cos \frac{n\pi \alpha}{L} t + b_n \sin \frac{n\pi \alpha}{L} t + \frac{L}{n\pi \alpha} \int_0^t h_n(s) \sin \left[\frac{n\pi \alpha}{L} (t - s) \right] ds$$

(cf. Problem 24 in Exercises 4.10). Hence, with this choice of $u_n(t)$, the series in (12) is a formal solution to the partial differential equation (8).

Since[†]

$$u_n(0) = a_n \quad \text{and} \quad u_n'(0) = b_n\left(\frac{n\pi\alpha}{L}\right),$$

substituting (12) into the initial conditions (10)–(11) yields

(14) $$u(x,0) = f(x) = \sum_{n=1}^{\infty} a_n \sin\frac{n\pi x}{L},$$

(15) $$\frac{\partial u}{\partial t}(x,0) = g(x) = \sum_{n=1}^{\infty} b_n\left(\frac{n\pi\alpha}{L}\right)\sin\frac{n\pi x}{L}.$$

Thus, if we choose the a_n's and b_n's so that equations (14) and (15) are satisfied, a formal solution to (8)–(11) is given by

(16) $$u(x,t) = \sum_{n=1}^{\infty} \left\{ a_n \cos\frac{n\pi\alpha}{L}t + b_n \sin\frac{n\pi\alpha}{L}t \right.$$
$$\left. + \frac{L}{n\pi\alpha}\int_0^t h_n(s)\sin\left[\frac{n\pi\alpha}{L}(t-s)\right]ds \right\} \sin\frac{n\pi x}{L}. \quad ●$$

The method of separation of variables can also be used to solve initial-boundary value problems for the wave equation in higher dimensions. For example, a vibrating rectangular membrane of length L and width W (see Figure 11.11) is governed by the following initial-boundary value problem for $u(x, y, t)$:

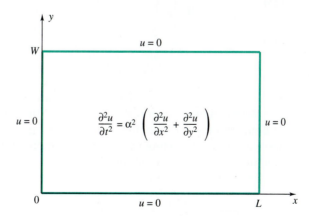

Figure 11.11 Vibrating membrane

[†] To compute $u_n'(0)$, we use the fact that $\dfrac{d}{dt}\displaystyle\int_0^t G(s,t)\,ds = G(t,t) + \int_0^t \dfrac{\partial G}{\partial t}(s,t)\,ds.$

(17)
$$\frac{\partial^2 u}{\partial t^2} = \alpha^2 \left(\frac{\partial^2 u}{\partial x^2} + \frac{\partial^2 u}{\partial y^2} \right); \qquad 0 < x < L, \qquad 0 < y < W, \qquad t > 0,$$

(18) $\quad u(0, y, t) = u(L, y, t) = 0, \qquad 0 < y < W, \qquad t > 0,$

(19) $\quad u(x, 0, t) = u(x, W, t) = 0, \qquad 0 < x < L, \qquad t > 0,$

(20) $\quad u(x, y, 0) = f(x, y), \qquad 0 < x < L, \qquad 0 < y < W,$

(21) $\quad \dfrac{\partial u}{\partial t}(x, y, 0) = g(x, y), \qquad 0 < x < L, \qquad 0 < y < W.$

Using an argument similar to the one given for the problem of heat flow in a rectangular plate (Example 4 in Section 11.5), we find that the initial-boundary value problem (17)–(21) has a formal solution

(22)
$$u(x, y, t) = \sum_{m=1}^{\infty} \sum_{n=1}^{\infty} \left\{ a_{mn} \cos \left(\sqrt{\frac{m^2}{L^2} + \frac{n^2}{W^2}} \, \alpha\pi t \right) \right.$$
$$\left. + b_{mn} \sin \left(\sqrt{\frac{m^2}{L^2} + \frac{n^2}{W^2}} \, \alpha\pi t \right) \right\} \sin \frac{m\pi x}{L} \sin \frac{n\pi y}{W},$$

where the constants a_{mn} and b_{mn} are determined from the double Fourier series

$$f(x, y) = \sum_{m=1}^{\infty} \sum_{n=1}^{\infty} a_{mn} \sin \frac{m\pi x}{L} \sin \frac{n\pi y}{W},$$

$$g(x, y) = \sum_{m=1}^{\infty} \sum_{n=1}^{\infty} \alpha\pi \sqrt{\frac{m^2}{L^2} + \frac{n^2}{W^2}} \, b_{mn} \sin \frac{m\pi x}{L} \sin \frac{n\pi y}{W}.$$

In particular,

(23) $\quad a_{mn} = \dfrac{4}{LW} \displaystyle\int_0^L \int_0^W f(x, y) \sin \frac{m\pi x}{L} \sin \frac{n\pi y}{W} \, dx \, dy,$

(24) $\quad b_{mn} = \dfrac{4}{LW\pi\alpha \sqrt{\dfrac{m^2}{L^2} + \dfrac{n^2}{W^2}}} \displaystyle\int_0^L \int_0^W g(x, y) \sin \frac{m\pi x}{L} \sin \frac{n\pi y}{W} \, dx \, dy.$

We leave the derivation of this solution as an exercise (see Problem 19).

We mentioned earlier that the solution to the vibrating string problem (1)–(4) consisted of a superposition of standing waves. There are also "traveling waves" associated with the wave equation. Traveling waves arise naturally out of d'Alembert's solution to the wave equation for an "infinite" string.

To obtain d'Alembert's solution to the wave equation

$$\frac{\partial^2 u}{\partial t^2} = \alpha^2 \frac{\partial^2 u}{\partial x^2},$$

we use the change of variables

$$\psi = x + \alpha t, \qquad \eta = x - \alpha t.$$

If u has continuous second partial derivatives, then $\partial u/\partial x = \partial u/\partial \psi + \partial u/\partial \eta$ and $\partial u/\partial t = \alpha(\partial u/\partial \psi - \partial u/\partial \eta)$, from which we obtain

$$\frac{\partial^2 u}{\partial x^2} = \frac{\partial^2 u}{\partial \psi^2} + 2\frac{\partial^2 u}{\partial \psi\, \partial \eta} + \frac{\partial^2 u}{\partial \eta^2},$$

$$\frac{\partial^2 u}{\partial t^2} = \alpha^2 \left\{ \frac{\partial^2 u}{\partial \psi^2} - 2\frac{\partial^2 u}{\partial \psi\, \partial \eta} + \frac{\partial^2 u}{\partial \eta^2} \right\}.$$

Substituting these expressions into the wave equation and simplifying yields

$$\frac{\partial^2 u}{\partial \psi\, \partial \eta} = 0.$$

We can solve this equation directly by first integrating with respect to ψ to obtain

$$\frac{\partial u}{\partial \eta} = b(\eta),$$

where $b(\eta)$ is an arbitrary function of η, and then integrating with respect to η to find

$$u(\psi, \eta) = A(\psi) + B(\eta),$$

where $A(\psi)$ and $B(\eta)$ are arbitrary functions. Substituting the original variables x and t gives **d'Alembert's solution**

(25) $u(x, t) = A(x + \alpha t) + B(x - \alpha t).$

It is easy to check by direct substitution that $u(x, t)$, defined by formula (25), is indeed a solution to the wave equation provided A and B are twice differentiable functions.

● **EXAMPLE 2**   Using d'Alembert's formula (25), find a solution to the initial value problem

(26) $$\frac{\partial^2 u}{\partial t^2} = \alpha^2 \frac{\partial^2 u}{\partial x^2}; \qquad -\infty < x < \infty, \qquad t > 0,$$

(27) $u(x, 0) = f(x), \qquad -\infty < x < \infty,$

(28) $\dfrac{\partial u}{\partial t}(x, 0) = g(x), \qquad -\infty < x < \infty.$

Solution   Since a solution to (26) is given by formula (25), we need only choose the functions A and B so that the initial conditions (27)–(28) are satisfied. For this we need

(29) $u(x, 0) = A(x) + B(x) = f(x),$

(30) $\dfrac{\partial u}{\partial t}(x, 0) = \alpha A'(x) - \alpha B'(x) = g(x).$

Integrating equation (30) from x_0 to x (x_0 arbitrary) and dividing by α gives

(31) $A(x) - B(x) = \dfrac{1}{\alpha} \displaystyle\int_{x_0}^{x} g(s)\, ds + C,$

where C is also arbitrary. Solving the system (29) and (31), we obtain

$$A(x) = \frac{1}{2}f(x) + \frac{1}{2\alpha}\int_{x_0}^{x} g(s)\,ds + \frac{C}{2},$$

$$B(x) = \frac{1}{2}f(x) - \frac{1}{2\alpha}\int_{x_0}^{x} g(s)\,ds - \frac{C}{2}.$$

Using these functions in formula (25) gives

$$u(x,t) = \frac{1}{2}\left[f(x+\alpha t) + f(x-\alpha t)\right] + \frac{1}{2\alpha}\left[\int_{x_0}^{x+\alpha t} g(s)\,ds - \int_{x_0}^{x-\alpha t} g(s)\,ds\right],$$

which simplifies to

$$(32) \qquad u(x,t) = \frac{1}{2}\left[f(x+\alpha t) + f(x-\alpha t)\right] + \frac{1}{2\alpha}\int_{x-\alpha t}^{x+\alpha t} g(s)\,ds. \quad \bullet$$

● **EXAMPLE 3**   Find the solution to the initial value problem

$$(33) \qquad \frac{\partial^2 u}{\partial t^2} = 4\frac{\partial^2 u}{\partial x^2}; \qquad -\infty < x < \infty, \qquad t > 0,$$

$$(34) \qquad u(x,0) = \sin x, \qquad -\infty < x < \infty,$$

$$(35) \qquad \frac{\partial u}{\partial t}(x,0) = 1, \qquad -\infty < x < \infty.$$

Solution   This is just a special case of the preceding example where $\alpha = 2$, $f(x) = \sin x$, and $g(x) = 1$. Substituting into (32), we obtain the solution

$$(36) \qquad u(x,t) = \tfrac{1}{2}\left[\sin(x+2t) + \sin(x-2t)\right] + \tfrac{1}{4}\int_{x-2t}^{x+2t} ds$$

$$= \sin x \cos 2t + t. \quad \bullet$$

We now use d'Alembert's formula to show that the solution to the "infinite" string problem consists of traveling waves.

Let $h(x)$ be a function defined on $(-\infty, \infty)$. The function $h(x+a)$, where $a > 0$, is a translation of the function $h(x)$ in the sense that its "shape" is the same as $h(x)$, but the coordinate system has been shifted to the left by an amount a. This is illustrated in Figure 11.12 for a function $h(x)$ whose graph consists of a triangular "bump." If we let $t \geq 0$

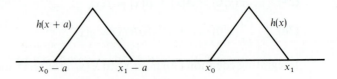

Figure 11.12 Graphs of $h(x)$ and $h(x+a)$

be a parameter (say time), then the functions $h(x + \alpha t)$ represent a family of functions, with the same shape, but shifted further and further to the left as $t \to \infty$. We say that $h(x + \alpha t)$ is a **traveling wave** moving to the left with speed α. In a similar fashion $h(x - \alpha t)$ is a traveling wave moving to the right with speed α.

If we refer to formula (25), we find that the solution to $\partial^2 u / \partial t^2 = \alpha^2 \partial^2 u / \partial x^2$ consists of traveling waves $A(x + \alpha t)$ moving to the left with speed α and $B(x - \alpha t)$ moving to the right at the same speed.

In the special case when the initial velocity $g(x) \equiv 0$, we have

$$u(x, t) = \tfrac{1}{2}[f(x + \alpha t) + f(x - \alpha t)].$$

Hence $u(x, t)$ is the sum of the traveling waves

$$\tfrac{1}{2}f(x + \alpha t) \quad \text{and} \quad \tfrac{1}{2}f(x - \alpha t).$$

These waves are initially superimposed, since

$$u(x, 0) = \tfrac{1}{2}f(x) + \tfrac{1}{2}f(x) = f(x).$$

As t increases, the two waves move away from each other with speed 2α. This is illustrated in Figure 11.13 for a square wave.

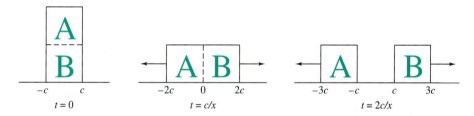

Figure 11.13 Decomposition of initial displacement into traveling waves

Existence and Uniqueness of Solutions

In Example 1, the method of separation of variables was used to derive a formal solution to the given initial-boundary value problem. To show that these series converge to an actual solution requires results from real analysis just as was the case for the formal solutions to the heat equation in Section 11.5. In Examples 2 and 3, the validity of d'Alembert's solution can be established by direct substitution into the initial value problem, assuming sufficient differentiability of the initial functions. We leave it as an exercise for the reader to show that if f has a continuous second derivative and g has a continuous first derivative, then d'Alembert's solution is a true solution (see Problem 12).

The question of the uniqueness of the solution to the initial-boundary value problem (1)–(4) can be answered using an **energy argument.**

> ◢ **UNIQUENESS OF THE SOLUTION TO THE VIBRATING STRING PROBLEM**
>
> **Theorem 8.**   The initial-boundary value problem
>
> (37) $$\frac{\partial^2 u}{\partial t^2} = \alpha^2 \frac{\partial^2 u}{\partial x^2}; \qquad 0 < x < L, \qquad t > 0,$$
>
> (38) $$u(0, t) = u(L, t) = 0, \qquad t \geq 0,$$
>
> (39) $$u(x, 0) = f(x), \qquad 0 \leq x \leq L,$$
>
> (40) $$\frac{\partial u}{\partial t}(x, 0) = g(x), \qquad 0 \leq x \leq L,$$
>
> has at most one twice continuously differentiable solution.

Proof.   Assume that both $u(x, t)$ and $v(x, t)$ are twice continuously differentiable solutions to (37)–(40), and let $w(x, t) := u(x, t) - v(x, t)$. It is easy to check that $w(x, t)$ satisfies the initial-boundary value problem (37)–(40) with zero initial data; that is, for $0 \leq x \leq L$,

(41) $$w(x, 0) = 0 \quad \text{and} \quad \frac{\partial w}{\partial t}(x, 0) = 0.$$

We will now show that $w(x, t) \equiv 0$ for $0 \leq x \leq L, t \geq 0$.

If $w(x, t)$ is the displacement of the vibrating string at location x for time t, then with the appropriate units, the total energy $E(t)$ of the vibrating string at time t is defined by the integral

(42) $$E(t) := \frac{1}{2} \int_0^L \left[\alpha^2 \left(\frac{\partial w}{\partial x} \right)^2 + \left(\frac{\partial w}{\partial t} \right)^2 \right] dx.$$

(The first term in the integrand relates to the stretching of the string at location x and represents the potential energy. The second term is the square of the velocity of the vibrating string at x and represents the kinetic energy.)

We now consider the derivative of $E(t)$:

$$\frac{dE}{dt} = \frac{d}{dt} \left\{ \frac{1}{2} \int_0^L \left[\alpha^2 \left(\frac{\partial w}{\partial x} \right)^2 + \left(\frac{\partial w}{\partial t} \right)^2 \right] dx \right\}.$$

Since w has continuous second partial derivatives (because u and v do), we can interchange the order of integration and differentiation. This gives

(43) $$\frac{dE}{dt} = \int_0^L \left[\alpha^2 \frac{\partial w}{\partial x} \frac{\partial^2 w}{\partial t \, \partial x} + \frac{\partial w}{\partial t} \frac{\partial^2 w}{\partial t^2} \right] dx.$$

Again the continuity of the second partials of w guarantees that the mixed partials are equal; that is,

$$\frac{\partial^2 w}{\partial t \, \partial x} = \frac{\partial^2 w}{\partial x \, \partial t}.$$

Combining this fact with integration by parts, we obtain

(44) $$\int_0^L \alpha^2 \frac{\partial w}{\partial x} \frac{\partial^2 w}{\partial t\, \partial x}\, dx = \int_0^L \alpha^2 \frac{\partial w}{\partial x} \frac{\partial^2 w}{\partial x\, \partial t}\, dx$$

$$= \alpha^2 \frac{\partial w}{\partial x}(L,t) \frac{\partial w}{\partial t}(L,t) - \alpha^2 \frac{\partial w}{\partial x}(0,t) \frac{\partial w}{\partial t}(0,t)$$

$$- \int_0^L \alpha^2 \frac{\partial^2 w}{\partial x^2} \frac{\partial w}{\partial t}\, dx.$$

The boundary conditions $w(0,t) = w(L,t) = 0$, $t \geq 0$, imply that $(\partial w / \partial t)(0,t) = (\partial w / \partial t)(L,t) = 0$, $t \geq 0$. This reduces equation (44) to

$$\int_0^L \alpha^2 \frac{\partial w}{\partial x} \frac{\partial^2 w}{\partial t\, \partial x}\, dx = -\int_0^L \alpha^2 \frac{\partial w}{\partial t} \frac{\partial^2 w}{\partial x^2}\, dx.$$

Substituting this in for the first integrand in (43), we find

$$\frac{dE}{dt} = \int_0^L \frac{\partial w}{\partial t} \left[\frac{\partial^2 w}{\partial t^2} - \alpha^2 \frac{\partial^2 w}{\partial x^2} \right] dx.$$

Since w satisfies equation (37), the integrand is zero for all x. Thus $dE/dt = 0$, and so $E(t) = C$, where C is a constant. This means that the total energy is conserved within the vibrating string.

The first boundary condition in (41) states that $w(x,0) = 0$ for $0 \leq x \leq L$. Hence $(\partial w / \partial x)(x,0) = 0$ for $0 < x < L$. Combining this with the second boundary condition in (41), we find that, when $t = 0$, the integrand in (42) is zero for $0 < x < L$. Therefore $E(0) = 0$. Since $E(t) = C$, we must have $C = 0$, and so

(45) $$E(t) = \frac{1}{2} \int_0^L \left[\alpha^2 \left(\frac{\partial w}{\partial x} \right)^2 + \left(\frac{\partial w}{\partial t} \right)^2 \right] dx \equiv 0.$$

That is, the total energy of the system is zero.

Since the integrand in (45) is nonnegative and continuous and the integral is zero, then the integrand must be zero for $0 \leq x \leq L$. Moreover, the integrand is the sum of two squares and so each term must be zero. Hence

$$\frac{\partial w}{\partial x}(x,t) = 0 \quad \text{and} \quad \frac{\partial w}{\partial t}(x,t) = 0,$$

for all $0 \leq x \leq L$, $t \geq 0$. Thus $w(x,t) = K$, where K is a constant. Physically, this says that there is no motion in the string.

Finally, since w is constant and w is zero when $t = 0$, then $w(x,t) \equiv 0$. Consequently, $u(x,t) = v(x,t)$ and the initial-boundary value problem has at most one solution. ◄◄◄

EXERCISES 11.6

In Problems 1 through 4 find a formal solution to the vibrating string problem governed by the given initial-boundary value problem.

1. $\dfrac{\partial^2 u}{\partial t^2} = \dfrac{\partial^2 u}{\partial x^2}$; $0 < x < 1$, $t > 0$,

$u(0, t) = u(1, t) = 0$, $t > 0$,
$u(x, 0) = x(1 - x)$, $0 < x < 1$,

$\dfrac{\partial u}{\partial t}(x, 0) = \sin 7\pi x$, $0 < x < 1$.

2. $\dfrac{\partial^2 u}{\partial t^2} = 16\dfrac{\partial^2 u}{\partial x^2}$; $0 < x < \pi$, $t > 0$,

$u(0, t) = u(\pi, t) = 0$, $t > 0$,
$u(x, 0) = \sin^2 x$, $0 < x < \pi$,

$\dfrac{\partial u}{\partial t}(x, 0) = 1 - \cos x$, $0 < x < \pi$.

3. $\dfrac{\partial^2 u}{\partial t^2} = 4\dfrac{\partial^2 u}{\partial x^2}$; $0 < x < \pi$, $t > 0$,

$u(0, t) = u(\pi, t) = 0$, $t > 0$,
$u(x, 0) = x^2(\pi - x)$, $0 < x < \pi$,

$\dfrac{\partial u}{\partial t}(x, 0) = 0$, $0 < x < \pi$.

4. $\dfrac{\partial^2 u}{\partial t^2} = 9\dfrac{\partial^2 u}{\partial x^2}$; $0 < x < \pi$, $t > 0$,

$u(0, t) = u(\pi, t) = 0$, $t > 0$,
$u(x, 0) = \sin 4x + 7\sin 5x$, $0 < x < \pi$,

$\dfrac{\partial u}{\partial t}(x, 0) = \begin{cases} x, & 0 < x < \pi/2, \\ \pi - x, & \pi/2 < x < \pi. \end{cases}$

5. The Plucked String.   A vibrating string is governed by the initial-boundary value problem (1)–(4). If the string is lifted to a height h_0 at $x = a$ and released, then the initial conditions are

$$f(x) = \begin{cases} h_0 x/a, & 0 < x \le a, \\ h_0(L - x)/(L - a), & a < x < L, \end{cases}$$

and $g(x) \equiv 0$. Find a formal solution.

6. The Struck String.   A vibrating string is governed by the initial-boundary value problem (1)–(4). If the string is struck at $x = a$, then the initial conditions may be expressed by $f(x) \equiv 0$ and

$$g(x) = \begin{cases} v_0 x/a, & 0 < x \le a, \\ v_0(L - x)/(L - a), & a < x < L, \end{cases}$$

where v_0 is a constant. Find a formal solution.

In Problems 7 and 8 find a formal solution to the vibrating string problem governed by the given nonhomogeneous initial-boundary value problem.

7. $\dfrac{\partial^2 u}{\partial t^2} = \dfrac{\partial^2 u}{\partial x^2} + tx$; $0 < x < \pi$, $t > 0$,

$u(0, t) = u(\pi, t) = 0$, $t > 0$,
$u(x, 0) = \sin x$, $0 < x < \pi$,

$\dfrac{\partial u}{\partial t}(x, 0) = 5\sin 2x - 3\sin 5x$, $0 < x < \pi$.

8. $\dfrac{\partial^2 u}{\partial t^2} = \dfrac{\partial^2 u}{\partial x^2} + x\sin t$; $0 < x < \pi$, $t > 0$,

$u(0, t) = u(\pi, t) = 0$, $t > 0$,
$u(x, 0) = 0$, $0 < x < \pi$,

$\dfrac{\partial u}{\partial t}(x, 0) = 0$, $0 < x < \pi$.

9. If one end of a string is held fixed while the other is free, then the motion of the string is governed by the initial-boundary value problem

$$\dfrac{\partial^2 u}{\partial t^2} = \alpha^2\dfrac{\partial^2 u}{\partial x^2};   0 < x < L,   t > 0,$$

$$u(0, t) = 0   \text{and}   \dfrac{\partial u}{\partial x}(L, t) = 0,   t > 0,$$

$$u(x, 0) = f(x),   0 < x < L,$$

$$\dfrac{\partial u}{\partial t}(x, 0) = g(x),   0 < x < L.$$

Derive a formula for a formal solution.

10. Derive a formula for the solution to the following initial-boundary value problem involving nonhomogeneous boundary conditions

$$\dfrac{\partial^2 u}{\partial t^2} = \alpha^2\dfrac{\partial^2 u}{\partial x^2};   0 < x < L,   t > 0,$$

$$u(0, t) = U_1,   u(L, t) = U_2,   t > 0,$$

$$u(x, 0) = f(x),   0 < x < L,$$

$$\dfrac{\partial u}{\partial t}(x, 0) = g(x),   0 < x < L,$$

where U_1 and U_2 are constants.

11. **The Telegraph Problem.**†   Use the method of separation of variables to derive a formal solution to the telegraph problem

$$\frac{\partial^2 u}{\partial t^2} + \frac{\partial u}{\partial t} + u = \alpha^2 \frac{\partial^2 u}{\partial x^2}; \qquad 0 < x < L, \qquad t > 0,$$

$$u(0,t) = u(L,t) = 0, \qquad t > 0,$$

$$u(x,0) = f(x), \qquad 0 < x < L,$$

$$\frac{\partial u}{\partial t}(x,0) = 0, \qquad 0 < x < L.$$

12. Verify d'Alembert's solution (32) to the initial value problem (26)–(28) when $f(x)$ has a continuous second derivative and $g(x)$ has a continuous first derivative, by substituting it directly into the equations.

In Problems 13 through 18 find the solution to the initial value problem

$$\frac{\partial^2 u}{\partial t^2} = \alpha^2 \frac{\partial^2 u}{\partial x^2}; \qquad -\infty < x < \infty, \qquad t > 0,$$

$$u(x,0) = f(x), \qquad -\infty < x < \infty,$$

$$\frac{\partial u}{\partial t}(x,0) = g(x), \qquad -\infty < x < \infty,$$

for the given functions $f(x)$ and $g(x)$.

13. $f(x) \equiv 0, \qquad g(x) = \cos x.$

14. $f(x) = x^2, \qquad g(x) \equiv 0.$

15. $f(x) = x, \qquad g(x) = x.$

16. $f(x) = \sin 3x, \qquad g(x) \equiv 1.$

17. $f(x) = e^{-x^2}, \qquad g(x) = \sin x.$

18. $f(x) = \cos 2x, \qquad g(x) = 1 - x.$

19. Derive the formal solution given in equations (22)–(24) to the vibrating membrane problem governed by the initial-boundary value problem (17)–(21).

20. **Long Water Waves.**   The motion of long water waves in a channel of constant depth is governed by the **linearized Korteweg and de Vries (KdV) equation**

(46) $\qquad u_t + \alpha u_x + \beta u_{xxx} = 0,$

where $u(x,t)$ is the displacement of the water from its equilibrium depth at location x and at time t, and α and β are positive constants.

(a) Show that equation (46) has a solution of the form

(47) $\qquad \begin{aligned} &u(x,t) = V(z), \\ &z = kx - w(k)t, \end{aligned}$

where k is a fixed constant and $w(k)$ is a function of k, provided V satisfies

(48) $\qquad -w\frac{dV}{dz} + \alpha k \frac{dV}{dz} + \beta k^3 \frac{d^3 V}{dz^3} = 0.$

These solutions, defined by (47), are called **uniform waves.**

(b) Physically, we are interested only in solutions $V(z)$ that are bounded and nonconstant on the infinite interval $(-\infty, \infty)$. Show that such solutions exist only if $\alpha k - w > 0$.

(c) Let $\lambda^2 = (\alpha k - w)/(\beta k^3)$. Show that the solutions from part (b) can be expressed in the form

$$V(x,t) = A \sin[\lambda kx - (\alpha \lambda k - \beta \lambda^3 k^3)t + B],$$

where A and B are arbitrary constants. [Hint: Solve for w in terms of λ and k and use (47).]

(d) Since both λ and k can be chosen arbitrarily and they always appear together as the product λk, we can set $\lambda = 1$ without loss of generality. Hence, we have

$$V(x,t) = A \sin[kx - (\alpha k - \beta k^3)t + B],$$

as a uniform wave solution to (46). The defining relation

$$w(k) = \alpha k - \beta k^3$$

is called the **dispersion relation,** the ratio $w(k)/k = \alpha - \beta k^2$ is called the **phase velocity,** and the derivative $dw/dk = \alpha - 3\beta k^2$ is called the **group velocity.** When the group velocity is not constant, the waves are called **dispersive.** Show that the standard wave equation $u_{tt} = \alpha^2 u_{xx}$ has only nondispersive waves.

21. **Vibrating Drum.**   A vibrating circular membrane of unit radius whose edges are held fixed in a plane and whose displacement $u(r,t)$ depends only on the radial

---

† For a discussion of the telegraph problem see *Methods of Mathematical Physics,* by R. Courant and D. Hibert, Volume II, Wiley-Interscience, New York, 1962.

distance r from the center and on the time t is governed by the initial-boundary value problem

$$\frac{\partial^2 u}{\partial t^2} = \alpha^2 \left(\frac{\partial^2 u}{\partial r^2} + \frac{1}{r} \frac{\partial u}{\partial r} \right); \qquad 0 < r < 1, \qquad t > 0,$$

$$u(1, t) = 0, \qquad t > 0,$$

$u(r, t)$ remains finite as $r \to 0^+$,

$$u(r, 0) = f(r), \qquad 0 < r < 1,$$

$$\frac{\partial u}{\partial t}(r, 0) = g(r), \qquad 0 < r < 1,$$

where f and g are the initial displacements and velocities, respectively. Use the method of separation of variables to derive a formal solution to the vibrating drum problem. [Hint: Show that there is a family of solutions of the form

$$u_n(r, t) = (a_n \cos(k_n \alpha t) + b_n \sin(k_n \alpha t)) J_0(k_n r),$$

where J_0 is the Bessel function of the first kind of order zero and $0 < k_1 < k_2 < \cdots < k_n < \cdots$ are the positive zeros of J_0. Now use superposition.]

•11.7• LAPLACE'S EQUATION

In Section 11.1 we showed how Laplace's equation,

$$\frac{\partial^2 u}{\partial x^2} + \frac{\partial^2 u}{\partial y^2} = 0,$$

arises in the study of steady-state or time-independent solutions to the heat equation. Since these solutions do not depend upon time, initial conditions are irrelevant and only boundary conditions are specified. Other applications include the static displacement $u(x, y)$ of a stretched membrane fastened in space along the boundary of a region (here u must satisfy Laplace's equation inside the region); the electrostatic and gravitational potentials in certain force fields (here u must satisfy Laplace's equation in any region that is free of electrical charges or mass); and, in fluid mechanics for an idealized fluid, the stream function $u(x, y)$ whose level curves (stream lines) $u(x, y) = k$, k constant, represent the path of particles in the fluid (again u satisfies Laplace's equation in the flow region).

There are two basic types of boundary conditions that are usually associated with Laplace's equation: **Dirichlet boundary conditions,** where the solution $u(x, y)$ to Laplace's equation in a domain D is required to satisfy

$$u(x, y) = f(x, y), \qquad \text{on } \partial D,$$

where $f(x, y)$ is a specified function defined on the boundary ∂D of D; and **Neumann boundary conditions,** where the directional derivative $\partial u/\partial n$ along the outward normal to the boundary is required to satisfy

$$\frac{\partial u}{\partial n}(x, y) = g(x, y), \qquad \text{on } \partial D,$$

where $g(x, y)$ is a specified function defined on ∂D. We say that the boundary conditions are **mixed** if the solution is required to satisfy $u(x, y) = f(x, y)$ on part of the boundary and $(\partial u/\partial n)(x, y) = g(x, y)$ on the remaining portion of the boundary.

In this section we will use the method of separation of variables to find solutions to Laplace's equation with various boundary conditions for rectangular, circular, and cylindrical domains. We will also discuss the existence and uniqueness of such solutions.

● **EXAMPLE 1** Find a solution to the following mixed boundary value problem for a rectangle (see Figure 11.14):

(1) $\dfrac{\partial^2 u}{\partial x^2} + \dfrac{\partial^2 u}{\partial y^2} = 0;$ $0 < x < a,$ $0 < y < b,$

(2) $\dfrac{\partial u}{\partial x}(0, y) = \dfrac{\partial u}{\partial x}(a, y) = 0,$ $0 \le y \le b,$

(3) $u(x, b) = 0,$ $0 \le x \le a,$

(4) $u(x, 0) = f(x),$ $0 \le x \le a.$

Solution Separating variables, we first let $u(x, y) = X(x)Y(y)$. Substituting into equation (1), we have

$$X''(x)Y(y) + X(x)Y''(y) = 0,$$

which separates into

$$\frac{X''(x)}{X(x)} = -\frac{Y''(y)}{Y(y)} = K,$$

where K is some constant. This leads to the two ordinary differential equations

(5) $X''(x) - KX(x) = 0,$

(6) $Y''(y) + KY(y) = 0.$

From the boundary condition (2) we observe that

(7) $X'(0) = X'(a) = 0.$

We have encountered the eigenvalue problem in (5) and (7) before (see Example 1 in Section 11.5). The eigenvalues are $K = K_n = -(n\pi/a)^2$, $n = 0, 1, 2, \ldots$, with corresponding solutions

(8) $X_n(x) = a_n \cos\left(\dfrac{n\pi x}{a}\right),$

where the a_n's are arbitrary constants.

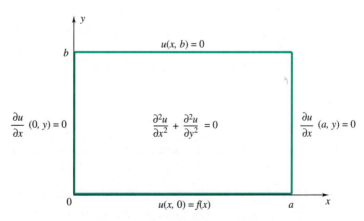

Figure 11.14 Mixed boundary value problem

Setting $K = K_n = -(n\pi/a)^2$ in equation (6) and solving for Y gives[†]

$$Y_0(y) = A_0 + B_0 y,$$

$$Y_n(y) = A_n \cosh\left(\frac{n\pi y}{a}\right) + B_n \sinh\left(\frac{n\pi y}{a}\right), \qquad n = 1, 2, \ldots,$$

which (just as for trigonometric functions) can be written in the form

(9) $$Y_0(y) = A_0 + B_0 y, \qquad Y_n(y) = C_n \sinh\left(\frac{n\pi}{a}(y + D_n)\right), \qquad n = 1, 2, \ldots,$$

where C_n and D_n are arbitrary constants (see Problem 18).

Now the boundary condition $u(x, b) = 0$ in (3) will be satisfied if $Y(b) = 0$. Setting $y = b$ in (9), we want $A_0 + B_0 b = 0$ and

$$C_n \sinh\left(\frac{n\pi}{a}(b + D_n)\right) = 0, \qquad n = 1, 2, \ldots.$$

This is true if we take $A_0 = -bB_0$ and $D_n = -b$. Combining the results in (8) and (9), we find that there are solutions to (1)–(3) of the form

$$u_0(x, y) = X_0(x) Y_0(y) = a_0 B_0(y - b) = E_0(y - b),$$

$$u_n(x, y) = X_n(x) Y_n(y) = a_n \cos\left(\frac{n\pi x}{a}\right) C_n \sinh\left[\frac{n\pi}{a}(y - b)\right]$$

$$= E_n \cos\left(\frac{n\pi x}{a}\right) \sinh\left[\frac{n\pi}{a}(y - b)\right], \qquad n = 1, 2, \ldots,$$

where the E_n's are constants. In fact, by the superposition principle,

(10) $$u(x, y) = E_0(y - b) + \sum_{n=1}^{\infty} E_n \cos\left(\frac{n\pi x}{a}\right) \sinh\left[\frac{n\pi}{a}(y - b)\right]$$

is a formal solution to (1)–(3).

Applying the remaining nonhomogeneous boundary condition in (4), we have

$$u(x, 0) = f(x) = -E_0 b + \sum_{n=1}^{\infty} E_n \sinh\left(-\frac{n\pi b}{a}\right) \cos\left(\frac{n\pi x}{a}\right).$$

This is a Fourier cosine series for $f(x)$ and hence the coefficients are given by the formulas

$$E_0 = \frac{1}{(-ba)} \int_0^a f(x)\, dx,$$

(11) $$E_n = \frac{2}{a \sinh\left(-\dfrac{n\pi b}{a}\right)} \int_0^a f(x) \cos\left(\frac{n\pi x}{a}\right) dx, \qquad n = 1, 2, \ldots.$$

Thus a formal solution is given by (10) with the constants E_n given by (11).   ●

[†] We usually express $Y_n(y) = a_n e^{n\pi y/a} + b_n e^{-n\pi y/a}$. However, computation is simplified in this case by using the hyperbolic functions $\cosh z = (e^z + e^{-z})/2$ and $\sinh z = (e^z - e^{-z})/2$.

In Example 1 the boundary conditions were homogeneous on three sides of the rectangle and nonhomogeneous on the fourth side, $\{(x, y): y = 0, 0 \leq x \leq a\}$. It is important to note that the method used in Example 1 can also be used to solve problems for which the boundary conditions are nonhomogeneous on all sides. This is accomplished by solving four separate boundary value problems in which three sides have homogeneous boundary conditions and only one side is nonhomogeneous. The solution is then obtained by summing these four solutions (see Problem 5).

For problems involving circular domains it is usually more convenient to use polar coordinates. In rectangular coordinates the Laplacian has the form

$$\Delta u = \frac{\partial^2 u}{\partial x^2} + \frac{\partial^2 u}{\partial y^2}.$$

In polar coordinates (r, θ), we let

$$x = r \cos \theta, \qquad y = r \sin \theta,$$

so that

$$r = \sqrt{x^2 + y^2}, \qquad \tan \theta = y/x.$$

With patience and a little care in applying the chain rule, one can show that the Laplacian in polar coordinates is

(12) $\qquad \Delta u = \dfrac{\partial^2 u}{\partial r^2} + \dfrac{1}{r} \dfrac{\partial u}{\partial r} + \dfrac{1}{r^2} \dfrac{\partial^2 u}{\partial \theta^2}$

(see Problem 6). In the next example we obtain a solution to the **Dirichlet problem** in a disk of radius a.

● **EXAMPLE 2**   A circular metal disk of radius a has its top and bottom insulated. The edge ($r = a$) of the disk is kept at a specified temperature which depends upon its location (varies with θ). The steady-state temperature inside the disk satisfies Laplace's equation. Determine the temperature distribution $u(r, \theta)$ inside the disk by finding the solution to the following Dirichlet boundary value problem, depicted in Figure 11.15:

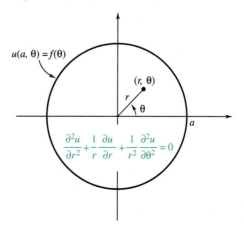

Figure 11.15 Steady-state temperature distribution in a disk

(13) $\quad \dfrac{\partial^2 u}{\partial r^2} + \dfrac{1}{r}\dfrac{\partial u}{\partial r} + \dfrac{1}{r^2}\dfrac{\partial^2 u}{\partial \theta^2} = 0; \qquad 0 \le r < a, \qquad -\pi \le \theta \le \pi,$

(14) $\quad u(a, \theta) = f(\theta), \qquad -\pi \le \theta \le \pi.$

Solution To use the method of separation of variables, we first set

$$u(r, \theta) = R(r)T(\theta),$$

where $0 \le r < a$ and $-\pi \le \theta \le \pi$. Substituting into (13) and separating variables give

$$\frac{r^2 R''(r) + rR'(r)}{R(r)} = -\frac{T''(\theta)}{T(\theta)} = \lambda,$$

where λ is any constant. This leads to the two ordinary differential equations

(15) $\quad r^2 R''(r) + rR'(r) - \lambda R(r) = 0,$

(16) $\quad\quad\quad\quad T''(\theta) + \lambda T(\theta) = 0.$

In order for $u(r, \theta)$ to be continuous in the disk $0 \le r < a$, we need $T(\theta)$ to be 2π-periodic; in particular, we require

(17) $\quad T(-\pi) = T(\pi) \quad$ and $\quad T'(-\pi) = T'(\pi).$

Therefore, we seek nontrivial solutions to the eigenvalue problem (16) and (17).

When $\lambda < 0$, the general solution to (16) is the sum of two exponentials. Hence, we have only trivial 2π-periodic solutions.

When $\lambda = 0$, we find $T(\theta) = A\theta + B$ to be the solution to (16). This linear function is periodic only when $A = 0$, that is, $T_0(\theta) = B$ is the only 2π-periodic solution corresponding to $\lambda = 0$.

When $\lambda > 0$, the general solution to (16) is

$$T(\theta) = A\cos\sqrt{\lambda}\,\theta + B\sin\sqrt{\lambda}\,\theta.$$

Here we get a nontrivial 2π-periodic solution only when $\sqrt{\lambda} = n, n = 1, 2, \ldots$. (The reader can check this using (17).) Hence, we obtain the nontrivial 2π-periodic solutions

(18) $\quad T_n(\theta) = A_n\cos n\theta + B_n\sin n\theta,$

corresponding to $\sqrt{\lambda} = n, n = 1, 2, \ldots$.

Now for $\lambda = n^2, n = 0, 1, 2, \ldots$, equation (15) is the Cauchy-Euler equation

(19) $\quad r^2 R''(r) + rR'(r) - n^2 R(r) = 0$

(see Chapter 4, page 156). When $n = 0$, the general solution is

$$R_0(r) = C + D\ln r.$$

Since $\ln r \to -\infty$ as $r \to 0^+$ this solution is unbounded near $r = 0$ when $D \ne 0$. Therefore, we must choose $D = 0$ if $u(r, \theta)$ is to be continuous at $r = 0$. We now have $R_0(r) = C$ and so $u_0(r, \theta) = R_0(r)T(\theta) = CB$ which, for convenience, we write in the form

(20) $\quad u_0(r, \theta) = \dfrac{A_0}{2},$

where A_0 is an arbitrary constant.

When $\lambda = n^2$, $n = 1, 2, \ldots$, the reader should verify that equation (19) has the general solution

$$R_n(r) = C_n r^n + D_n r^{-n}.$$

Since $r^{-n} \to \infty$ as $r \to 0^+$, we must set $D_n = 0$ in order for $u(r, \theta)$ to be bounded at $r = 0$. Thus

$$R_n(r) = C_n r^n.$$

Now for each $n = 1, 2, \ldots$, we have the solutions

(21) $u_n(r, \theta) = R_n(r) T_n(\theta) = C_n r^n (A_n \cos n\theta + B_n \sin n\theta),$

and by forming an infinite series from the solutions in (20) and (21) we get the following formal solution to (13):

$$u(r, \theta) = \frac{A_0}{2} + \sum_{n=1}^{\infty} C_n r^n (A_n \cos n\theta + B_n \sin n\theta).$$

It is more convenient to write this series in the equivalent form

(22) $u(r, \theta) = \dfrac{a_0}{2} + \displaystyle\sum_{n=1}^{\infty} \left(\frac{r}{a}\right)^n (a_n \cos n\theta + b_n \sin n\theta),$

where the a_n's and b_n's are constants. These constants can be determined from the boundary condition; indeed on setting $r = a$ in (22), condition (14) becomes

$$f(\theta) = \frac{a_0}{2} + \sum_{n=1}^{\infty} (a_n \cos n\theta + b_n \sin n\theta).$$

Hence if $f(\theta)$ is 2π-periodic, we recognize that a_n, b_n are Fourier coefficients. Thus,

(23) $a_n = \dfrac{1}{\pi} \displaystyle\int_{-\pi}^{\pi} f(\theta) \cos n\theta \, d\theta, \qquad n = 0, 1, \ldots,$

(24) $b_n = \dfrac{1}{\pi} \displaystyle\int_{-\pi}^{\pi} f(\theta) \sin n\theta \, d\theta, \qquad n = 1, 2, \ldots.$

To summarize, if a_n and b_n are defined by formulas (23) and (24), then $u(r, \theta)$ given in (22) is a formal solution to the Dirichlet problem (13)–(14).   ●

The procedure in Example 2 can also be used to study the **Neumann problem** in a disk:

(25) $\Delta u = 0, \qquad 0 \le r < a, \qquad -\pi \le \theta \le \pi,$

(26) $\dfrac{\partial u}{\partial r}(a, \theta) = f(\theta), \qquad -\pi \le \theta \le \pi.$

For this problem there is no unique solution since if u is a solution, then the function u plus a constant is also a solution. Moreover f must satisfy the **consistency condition**

$$(27) \qquad \int_{-\pi}^{\pi} f(\theta)\, d\theta = 0.$$

If we interpret the solution $u(r, \theta)$ of equation (25) as the steady-state temperature distribution inside a circular disk that does not contain either a heat source or heat sink, then equation (26) specifies the flow of heat across the boundary of the disk. Here the consistency condition (27) is simply the requirement that the net flow of heat across the boundary is zero. We leave the solution of the Neumann problem and the derivation of the consistency condition for the exercises.

The technique used in Example 2 also applies to annular domains, $\{(r, \theta): 0 < a < r < b\}$, and to exterior domains, $\{(r, \theta): a < r\}$. We leave these applications as exercises.

Laplace's equation in cylindrical coordinates arises in the study of steady-state temperature distributions in a solid cylinder and in determining the electric potential inside a cylinder. In the cylindrical coordinates

$$x = r \cos \theta, \qquad y = r \sin \theta, \qquad z = z,$$

Laplace's equation becomes

$$(28) \qquad \Delta u = \frac{\partial^2 u}{\partial r^2} + \frac{1}{r}\frac{\partial u}{\partial r} + \frac{1}{r^2}\frac{\partial^2 u}{\partial \theta^2} + \frac{\partial^2 u}{\partial z^2} = 0.$$

The Dirichlet problem for the cylinder $\{(r, \theta, z): 0 \le r \le a, 0 \le z \le b\}$ has the boundary conditions

$$(29) \qquad u(a, \theta, z) = f(\theta, z), \qquad -\pi \le \theta \le \pi, \qquad 0 \le z \le b,$$

$$(30) \qquad u(r, \theta, 0) = g(r, \theta), \qquad 0 \le r \le a, \qquad -\pi \le \theta \le \pi,$$

$$(31) \qquad u(r, \theta, b) = h(r, \theta), \qquad 0 \le r \le a, \qquad -\pi \le \theta \le \pi.$$

To solve the Dirichlet boundary value problem (28)–(31), we first solve the three boundary value problems corresponding to: (i) $g \equiv 0$ and $h \equiv 0$; (ii) $f \equiv 0$ and $h \equiv 0$; and (iii) $f \equiv 0$ and $g \equiv 0$. Then by the superposition principle, the solution to (28)–(31) will be the sum of these three solutions. This is the same method that was discussed in dealing with Dirichlet problems on rectangular domains. (See the remarks following Example 1.) In the next example we solve the Dirichlet problem when $g \equiv 0$ and $h \equiv 0$.

● **EXAMPLE 3**   The base ($z = 0$) and the top ($z = b$) of a charge-free cylinder are grounded and therefore are at zero potential. The potential on the lateral surface ($r = a$) of the cylinder is given by $u(a, \theta, z) = f(\theta, z)$, where $f(\theta, 0) = f(\theta, b) = 0$. Inside the cylinder the potential $u(r, \theta, z)$ satisfies Laplace's equation. Determine the potential u inside the cylinder by finding a solution to the Dirichlet boundary value problem

$$(32) \qquad \frac{\partial^2 u}{\partial r^2} + \frac{1}{r}\frac{\partial u}{\partial r} + \frac{1}{r^2}\frac{\partial^2 u}{\partial \theta^2} + \frac{\partial^2 u}{\partial z^2} = 0, \qquad 0 \le r < a, \qquad -\pi \le \theta \le \pi, \qquad 0 < z < b,$$

$$(33) \qquad u(a, \theta, z) = f(\theta, z), \qquad -\pi \le \theta \le \pi, \qquad 0 \le z \le b,$$

$$(34) \qquad u(r, \theta, 0) = u(r, \theta, b) = 0, \qquad 0 \le r < a, \qquad -\pi \le \theta \le \pi.$$

Solution Using the method of separation of variables, we first assume

$$u(r, \theta, z) = R(r)T(\theta)Z(z).$$

Substituting into equation (32) and separating out the Z's, we find

$$\frac{R''(r) + (1/r)R'(r)}{R(r)} + \frac{1}{r^2}\frac{T''(\theta)}{T(\theta)} = -\frac{Z''(z)}{Z(z)} = \lambda,$$

where λ can be any constant. Separating further the R's and T's gives

$$\frac{r^2 R''(r) + rR'(r)}{R(r)} - r^2\lambda = -\frac{T''(\theta)}{T(\theta)} = \mu,$$

where μ can also be any constant. We now have the three ordinary differential equations

(35) $\qquad r^2 R''(r) + rR'(r) - (r^2\lambda + \mu)R(r) = 0,$

(36) $\qquad\qquad\qquad T''(\theta) + \mu T(\theta) = 0,$

(37) $\qquad\qquad\qquad Z''(z) + \lambda Z(z) = 0.$

In order for u to be continuous in the cylinder, $T(\theta)$ must be 2π-periodic. Thus let's begin with the eigenvalue problem

$$T''(\theta) + \mu T(\theta) = 0, \qquad -\pi < \theta < \pi,$$
$$T(-\pi) = T(\pi) \quad \text{and} \quad T'(-\pi) = T'(\pi).$$

In Example 2 we showed that this problem has nontrivial solutions for $\mu = n^2$, $n = 0, 1, 2, \ldots$, that are given by

(38) $\qquad T_n(\theta) = A_n \cos n\theta + B_n \sin n\theta,$

where the A_n's and B_n's are arbitrary constants.

The boundary conditions in (34) imply that $Z(0) = Z(b) = 0$. Therefore, Z must satisfy the eigenvalue problem

$$Z''(z) + \lambda Z(z) = 0, \qquad 0 < z < b,$$
$$Z(0) = Z(b) = 0.$$

We have seen this eigenvalue problem several times before. Nontrivial solutions exist for $\lambda = (m\pi/b)^2$, $m = 1, 2, 3, \ldots$ and are given by

(39) $\qquad Z_m(z) = C_m \sin\left(\frac{m\pi z}{b}\right),$

where the C_m's are arbitrary constants.

Substituting for μ and λ in equation (35) gives

$$r^2 R''(r) + rR'(r) - \left(r^2\left(\frac{m\pi}{b}\right)^2 + n^2\right)R(r) = 0, \qquad 0 \le r < a.$$

The reader should verify that the change of variables $s = (m\pi r/b)$ transforms this equation into the **modified Bessel's equation of order** $n^{\dagger}$

(40) $s^2R''(s) + sR'(s) - (s^2 + n^2)R(s) = 0, \qquad 0 \leq s < \dfrac{m\pi a}{b}.$

The modified Bessel's equation of order n has two linearly independent solutions:

$$I_n(s) = \sum_{k=0}^{\infty} \frac{(s/2)^{2k+n}}{k!\,\Gamma(k+n+1)},$$

the **modified Bessel function of the first kind** which remains bounded near zero, and

$$K_n(s) = \lim_{v \to n} \frac{\pi}{2} \frac{I_{-v}(s) - I_v(s)}{\sin v\pi},$$

the **modified Bessel function of the second kind** which becomes unbounded as $s \to 0$. (Recall that Γ is the gamma function discussed in Section 7.6.) A general solution to (40) has the form $CK_n + DI_n$, where C and D are constants. Since u must remain bounded near $s = 0$, we must take $C = 0$. Thus the desired solutions to (40) have the form

(41) $R_{mn}(r) = D_{mn}I_n\left(\dfrac{m\pi r}{b}\right), \qquad n = 0, 1, \ldots, \qquad m = 1, 2, \ldots,$

where the D_{mn}'s are arbitrary constants.

If we multiply the functions in (38), (39), and (41), and then sum over m and n, we obtain the following series solution to (32) and (34):

(42) $u(r, \theta, z) = \displaystyle\sum_{m=1}^{\infty} a_{m0}I_0\left(\frac{m\pi r}{b}\right)\sin\left(\frac{m\pi z}{b}\right)$

$$+ \sum_{n=1}^{\infty}\sum_{m=1}^{\infty}(a_{mn}\cos n\theta + b_{mn}\sin n\theta)I_n\left(\frac{m\pi r}{b}\right)\sin\left(\frac{m\pi z}{b}\right),$$

where the a_{mn}'s and b_{mn}'s are arbitrary constants.

The constants in (42) can be obtained by imposing boundary condition (33). Setting $r = a$ and rearranging terms, we have

(43) $f(\theta, z) = \displaystyle\sum_{m=1}^{\infty} a_{m0}I_0\left(\frac{m\pi a}{b}\right)\sin\left(\frac{m\pi z}{b}\right)$

$$+ \sum_{n=1}^{\infty}\left[\sum_{m=1}^{\infty} a_{mn}I_n\left(\frac{m\pi a}{b}\right)\sin\left(\frac{m\pi z}{b}\right)\right]\cos n\theta$$

$$+ \sum_{n=1}^{\infty}\left[\sum_{m=1}^{\infty} b_{mn}I_n\left(\frac{m\pi a}{b}\right)\sin\left(\frac{m\pi z}{b}\right)\right]\sin n\theta.$$

---

† The modified Bessel's equation of order n arises in many applications and has been studied extensively. We refer the reader to the text *Special Functions*, by E. D. Raineville, Macmillan Publishing Co., Inc., New York, 1960, for details about its solution.

If we let

(44) $\dfrac{\alpha_0(z)}{2} := \displaystyle\sum_{m=1}^{\infty} a_{m0} I_0\left(\dfrac{m\pi a}{b}\right) \sin\left(\dfrac{m\pi z}{b}\right),$

(45) $\alpha_n(z) := \displaystyle\sum_{m=1}^{\infty} a_{mn} I_n\left(\dfrac{m\pi a}{b}\right) \sin\left(\dfrac{m\pi z}{b}\right), \qquad n = 1, 2, \ldots,$

(46) $\beta_n(z) := \displaystyle\sum_{m=1}^{\infty} b_{mn} I_n\left(\dfrac{m\pi a}{b}\right) \sin\left(\dfrac{m\pi z}{b}\right), \qquad n = 1, 2, \ldots,$

then (43) becomes

(47) $f(\theta, z) = \dfrac{\alpha_0(z)}{2} + \displaystyle\sum_{n=1}^{\infty} [\alpha_n(z) \cos n\theta + \beta_n(z) \sin n\theta].$

For fixed z, equation (47) is a Fourier series for $f(\theta, z)$. Therefore, we can compute the α_n's and β_n's from the formulas

(48) $\alpha_n(z) = \dfrac{1}{\pi} \displaystyle\int_{-\pi}^{\pi} f(\theta, z) \cos n\theta \, d\theta, \qquad n = 0, 1, \ldots,$

(49) $\beta_n(z) = \dfrac{1}{\pi} \displaystyle\int_{-\pi}^{\pi} f(\theta, z) \sin n\theta \, d\theta, \qquad n = 1, 2, \ldots.$

Having determined the functions $\alpha_n(z)$ and $\beta_n(z)$, we can treat the series in (44)–(46) as Fourier sine series. Thus, for $m = 1, 2, \ldots$, we have

(50) $a_{m0} I_0\left(\dfrac{m\pi a}{b}\right) = \dfrac{1}{b} \displaystyle\int_0^b \alpha_0(z) \sin\left(\dfrac{m\pi z}{b}\right) dz,$

(51) $a_{mn} I_n\left(\dfrac{m\pi a}{b}\right) = \dfrac{2}{b} \displaystyle\int_0^b \alpha_n(z) \sin\left(\dfrac{m\pi z}{b}\right) dz,$

(52) $b_{mn} I_n\left(\dfrac{m\pi a}{b}\right) = \dfrac{2}{b} \displaystyle\int_0^b \beta_n(z) \sin\left(\dfrac{m\pi z}{b}\right) dz.$

Consequently, a formal solution to (32)–(34) is given by equation (42) with the constants a_{mn} and b_{mn} determined by equations (48)–(52). ●

Existence and Uniqueness of Solutions

The existence of solutions to the boundary value problems for Laplace's equation can be established by studying the convergence of the formal solutions that we obtained using the method of separation of variables.

To answer the question of the uniqueness of the solution to a Dirichlet boundary value problem for Laplace's equation, recall that Laplace's equation arises in the search for steady-state solutions to the heat equation. Just as there are maximum principles for the heat equation, there are also maximum principles for Laplace's equation. We state one such result here.[†]

[†] A proof can be found in Section 8.2 of the text *Partial Differential Equations of Mathematical Physics*, Second Edition, by Tyn Mýint-U, Elsevier North Holland, Inc., New York, 1980.

> ### MAXIMUM PRINCIPLE FOR LAPLACE'S EQUATION
>
> **Theorem 9.**   Let $u(x, y)$ be a solution to Laplace's equation in a bounded domain D with $u(x, y)$ continuous in $\bar{D}$, the closure of D. ($\bar{D} = D \cup \partial D$, where ∂D denotes the boundary of D.) Then, $u(x, y)$ attains its maximum value on ∂D.

The uniqueness of the solution to the Dirichlet boundary value problem follows from the maximum principle. We state this result in the next theorem, but leave its proof as an exercise (see Problem 19).

> ### UNIQUENESS OF SOLUTION
>
> **Theorem 10.**   Let D be a bounded domain. If there is a continuous solution to the Dirichlet boundary value problem
>
> $$\Delta u(x, y) = 0 \qquad \text{in } D,$$
> $$u(x, y) = f(x, y) \qquad \text{on } \partial D,$$
>
> then the solution is unique.

Solutions to Laplace's equation in two variables are called **harmonic** functions. These functions arise naturally in the study of analytic functions of a single complex variable. Moreover, complex analysis provides many useful results about harmonic functions. For a discussion of this interaction, we refer the reader to an introductory text on complex analysis such as *Fundamentals of Complex Analysis*, Second Edition, by E. B. Saff and A. D. Snider, Prentice-Hall, Inc., Englewood Cliffs, New Jersey, 1993.

EXERCISES 11.7

In Problems 1 through 5 find a formal solution to the given boundary value problem

1. $\dfrac{\partial^2 u}{\partial x^2} + \dfrac{\partial^2 u}{\partial y^2} = 0;$ $0 < x < \pi,$ $0 < y < 1,$

$\dfrac{\partial u}{\partial x}(0, y) = \dfrac{\partial u}{\partial x}(\pi, y) = 0,$ $0 \le y \le 1,$

$u(x, 0) = 4\cos 6x + \cos 7x,$ $0 \le x \le \pi,$

$u(x, 1) = 0,$ $0 \le x \le \pi.$

2. $\dfrac{\partial^2 u}{\partial x^2} + \dfrac{\partial^2 u}{\partial y^2} = 0;$ $0 < x < \pi,$ $0 < y < \pi,$

$\dfrac{\partial u}{\partial x}(0, y) = \dfrac{\partial u}{\partial x}(\pi, y) = 0,$ $0 \le y \le \pi,$

$u(x, 0) = \cos x - 2\cos 4x,$ $0 \le x \le \pi,$

$u(x, \pi) = 0,$ $0 \le x \le \pi.$

3. $\dfrac{\partial^2 u}{\partial x^2} + \dfrac{\partial^2 u}{\partial y^2} = 0;$ $0 < x < \pi,$ $0 < y < \pi,$

$u(0, y) = u(\pi, y) = 0,$ $0 \le y \le \pi,$

$u(x, 0) = f(x),$ $0 \le x \le \pi,$

$u(x, \pi) = 0,$ $0 \le x \le \pi.$

4. $\dfrac{\partial^2 u}{\partial x^2} + \dfrac{\partial^2 u}{\partial y^2} = 0;$ $0 < x < \pi,$ $0 < y < \pi,$

$u(0, y) = u(\pi, y) = 0,$ $0 \le y \le \pi,$

$u(x, 0) = \sin x + \sin 4x,$ $0 \le x \le \pi,$

$u(x, \pi) = 0,$ $0 \le x \le \pi.$

5. $\dfrac{\partial^2 u}{\partial x^2} + \dfrac{\partial^2 u}{\partial y^2} = 0;$ $0 < x < \pi,$ $0 < y < 1,$

$\dfrac{\partial u}{\partial x}(0, y) = \dfrac{\partial u}{\partial x}(\pi, y) = 0,$ $0 \le y \le 1,$

$u(x, 0) = \cos x - \cos 3x,$ $0 \le x \le \pi,$

$u(x, 1) = \cos 2x,$ $0 \le x \le \pi.$

6. Derive the polar coordinate form of the Laplacian given in equation (12).

In Problems 7 and 8 find a solution to the Dirichlet boundary value problem for a disk:

$$\frac{\partial^2 u}{\partial r^2} + \frac{1}{r}\frac{\partial u}{\partial r} + \frac{1}{r^2}\frac{\partial^2 u}{\partial \theta^2} = 0; \quad 0 \le r < 2, \quad -\pi \le \theta \le \pi,$$

$$u(2, \theta) = f(\theta), \quad -\pi \le \theta \le \pi,$$

for the given function $f(\theta)$.

7. $f(\theta) = |\theta|, \quad -\pi \le \theta \le \pi.$

8. $f(\theta) = \cos^2\theta, \quad -\pi \le \theta \le \pi.$

9. Find a solution to the Neumann boundary value problem for a disk:

$$\frac{\partial^2 u}{\partial r^2} + \frac{1}{r}\frac{\partial u}{\partial r} + \frac{1}{r^2}\frac{\partial^2 u}{\partial \theta^2} = 0; \quad 0 \le r < a, \quad -\pi \le \theta \le \pi,$$

$$\frac{\partial u}{\partial r}(a, \theta) = f(\theta), \quad -\pi < \theta < \pi.$$

10. A solution to the Neumann problem (25)–(26) must also satisfy the consistency condition in (27). To show this use **Green's second formula**

$$\iint_D (v\,\Delta u - u\,\Delta v)\,dx\,dy = \int_{\partial D}\left(v\frac{\partial u}{\partial n} - u\frac{\partial v}{\partial n}\right)ds,$$

where $\partial/\partial n$ is the outward normal derivative and ds is the differential of arc length. [Hint: Take $v \equiv 1$ and observe that $\partial u/\partial n = \partial u/\partial r$.]

11. Find a solution to the following Dirichlet problem for an annulus:

$$\frac{\partial^2 u}{\partial r^2} + \frac{1}{r}\frac{\partial u}{\partial r} + \frac{1}{r^2}\frac{\partial^2 u}{\partial \theta^2} = 0; \quad 1 < r < 2, \quad -\pi \le \theta \le \pi,$$

$$u(1, \theta) = \sin 4\theta - \cos\theta, \quad -\pi \le \theta \le \pi,$$

$$u(2, \theta) = \sin\theta, \quad -\pi \le \theta \le \pi.$$

12. Find a solution to the following Dirichlet problem for an annulus:

$$\frac{\partial^2 u}{\partial r^2} + \frac{1}{r}\frac{\partial u}{\partial r} + \frac{1}{r^2}\frac{\partial^2 u}{\partial \theta^2} = 0; \quad 1 < r < 3, \quad -\pi \le \theta \le \pi,$$

$$u(1, \theta) = 0, \quad -\pi \le \theta \le \pi,$$

$$u(3, \theta) = \cos 3\theta + \sin 5\theta, \quad -\pi \le \theta \le \pi.$$

13. Find a solution to the following Dirichlet problem for an exterior domain:

$$\frac{\partial^2 u}{\partial r^2} + \frac{1}{r}\frac{\partial u}{\partial r} + \frac{1}{r^2}\frac{\partial^2 u}{\partial \theta^2} = 0; \quad 1 < r, \quad -\pi \le \theta \le \pi,$$

$$u(1, \theta) = f(\theta), \quad -\pi \le \theta \le \pi,$$

$$u(r, \theta) \text{ remains bounded as } r \to \infty.$$

14. Find a solution to the following Neumann problem for an exterior domain:

$$\frac{\partial^2 u}{\partial r^2} + \frac{1}{r}\frac{\partial u}{\partial r} + \frac{1}{r^2}\frac{\partial^2 u}{\partial \theta^2} = 0; \quad 1 < r, \quad -\pi \le \theta \le \pi,$$

$$\frac{\partial u}{\partial r}(1, \theta) = f(\theta), \quad -\pi \le \theta \le \pi,$$

$$u(r, \theta) \text{ remains bounded as } r \to \infty.$$

15. Find a solution to the following Dirichlet problem for a half disk:

$$\frac{\partial^2 u}{\partial r^2} + \frac{1}{r}\frac{\partial u}{\partial r} + \frac{1}{r^2}\frac{\partial^2 u}{\partial \theta^2} = 0; \quad 0 < r < 1, \quad 0 < \theta < \pi,$$

$$u(r, 0) = 0, \quad 0 \le r \le 1,$$

$$u(r, \pi) = 0, \quad 0 \le r \le 1,$$

$$u(1, \theta) = \sin 3\theta, \quad 0 \le \theta \le \pi,$$

$$u(0, \theta) \text{ bounded.}$$

16. Find a solution to the following Dirichlet problem for a half annulus:

$$\frac{\partial^2 u}{\partial r^2} + \frac{1}{r}\frac{\partial u}{\partial r} + \frac{1}{r^2}\frac{\partial^2 u}{\partial \theta^2} = 0; \quad \pi < r < 2\pi, \quad 0 < \theta < \pi,$$

$$u(r, 0) = \sin r, \quad \pi \le r \le 2\pi,$$

$$u(r, \pi) = 0, \quad \pi \le r \le 2\pi,$$

$$u(\pi, \theta) = u(2\pi, \theta) = 0, \quad 0 \le \theta \le \pi.$$

17. Find a solution to the mixed boundary value problem

$$\frac{\partial^2 u}{\partial r^2} + \frac{1}{r}\frac{\partial u}{\partial r} + \frac{1}{r^2}\frac{\partial^2 u}{\partial \theta^2} = 0; \quad 1 < r < 3, \quad -\pi \le \theta \le \pi,$$

$$u(1, \theta) = f(\theta), \quad -\pi \le \theta \le \pi,$$

$$\frac{\partial u}{\partial r}(3, \theta) = g(\theta), \quad -\pi \le \theta \le \pi.$$

18. Show that

$$A\cosh\theta + B\sinh\theta = C\sinh(\theta + D),$$

where $C^2 = B^2 - A^2$ and $D = \tanh^{-1}(A/B)$.

19. Prove Theorem 10 on the uniqueness of the solution to the Dirichlet problem.

20. **Stability.** Use the maximum principle to prove the following theorem on the continuous dependence of the solution on the boundary conditions:

    Theorem. Let u_1 and u_2 be continuous functions in $\bar{D}$, where D is a bounded domain. For $i = 1$ and 2, let u_i be the solution to the Dirichlet problem

    $$\Delta u = 0, \quad \text{in } D,$$

    $$u = f_i, \quad \text{on } \partial D.$$

    If the boundary values satisfy

    $$|f_1(x, y) - f_2(x, y)| < \varepsilon, \quad \text{for all } (x, y) \text{ on } \partial D,$$

    where $\varepsilon > 0$ is some constant, then

    $$|u_1(x, y) - u_2(x, y)| < \varepsilon, \quad \text{for all } (x, y) \text{ in } D.$$

21. For the Dirichlet problem described in Example 3, let $a = b = \pi$ and assume the potential on the lateral side $(r = \pi)$ of the cylinder is $f(\theta, z) = \sin z$. Use equations (48)–(52) to compute the solution given by equation (42).

22. **Invariance of Laplace's Equation.** A complex-valued function $f(z)$ of the complex variable $z = x + iy$ can be written in the form $f(z) = u(x, y) + iv(x, y)$, where u and v are real-valued functions. If $f(z)$ is analytic in a planar region D, then its real and imaginary parts satisfy the Cauchy-Riemann equations in D; that is, in D

    $$\frac{\partial u}{\partial x} = \frac{\partial v}{\partial y}, \quad \frac{\partial u}{\partial y} = -\frac{\partial v}{\partial x}.$$

    Let f be analytic and one-to-one in D and assume its inverse f^{-1} is analytic in D' where D' is the image of D under f. Then,

    $$\frac{\partial x}{\partial u} = \frac{\partial y}{\partial v}, \quad \frac{\partial x}{\partial v} = -\frac{\partial y}{\partial u},$$

    are just the Cauchy-Riemann equations for f^{-1}. Show that if $\phi(x, y)$ satisfies Laplace's equation for (x, y) in D, then $\psi(u, v) := \phi(x(u, v), y(u, v))$ satisfies Laplace's equation for (u, v) in D'.

23. **Fluid Flow Around a Corner.** The stream lines that describe the fluid flow around a corner (see Figure 11.16) are given by $\phi(x, y) = k$, where k is a constant and ϕ, the stream function, satisfies the boundary value problem

    $$\frac{\partial^2 \phi}{\partial x^2} + \frac{\partial^2 \phi}{\partial y^2} = 0, \quad x > 0, \quad y > 0,$$

    $$\phi(x, 0) = 0, \quad 0 \le x,$$

    $$\phi(0, y) = 0, \quad 0 \le y.$$

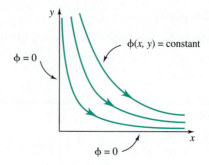

Figure 11.16 Flow around a corner

(a) Using the results of Problem 22, show that this problem can be reduced to finding the flow above a flat plate (see Figure 11.17). That is, show that the problem reduces to finding the solution to

$$\frac{\partial^2 \psi}{\partial u^2} + \frac{\partial^2 \psi}{\partial v^2} = 0, \quad v > 0, \quad -\infty < u < \infty,$$

$$\psi(u, 0) = 0, \quad -\infty < u < \infty,$$

where ψ and ϕ are related as follows: $\psi(u, v) = \phi(x(u, v), y(u, v))$ with the mapping between (u, v) and (x, y) given by the analytic function $f(z) = z^2$.

(b) Verify that a nonconstant solution to the problem in part (a) is given by $\psi(u, v) = v$.

(c) Using the result of part (b), find a stream function $\phi(x, y)$ for the original problem.

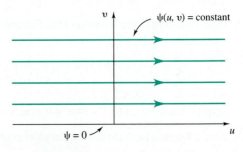

Figure 11.17 Flow above a flat surface

• CHAPTER SUMMARY •

Separation of Variables: A classical technique that is effective in solving boundary value problems for partial differential equations is the **method of separation of variables.** Briefly the idea is to assume first that there exists a solution that can be written with the variables separated; e.g. $u(x, t) = X(x)T(t)$. Substituting $u = XT$ into the partial differential equation and then imposing the boundary conditions leads to the problem of finding the eigenvalues and eigenfunctions for a boundary value problem for an ordinary differential equation. Solving for the eigenvalues and eigenfunctions, one eventually obtains solutions $u_n(x, t) = X_n(x)T_n(t)$, $n = 1, 2, 3, \ldots$, that solve the partial differential equation and the boundary conditions. Taking infinite linear combinations of the $u_n(x, t)$ yields solutions to the partial differential equation and boundary conditions of the form

$$u(x, t) = \sum_{n=1}^{\infty} c_n u_n(x, t).$$

The coefficients c_n are then obtained using the initial conditions or some other boundary conditions.

Fourier Series: Let $f(x)$ be a piecewise continuous function on the interval $[-T, T]$. The **Fourier series** of f is the trigonometric series

$$f(x) \sim \frac{a_0}{2} + \sum_{n=1}^{\infty} \left\{ a_n \cos \frac{n\pi x}{T} + b_n \sin \frac{n\pi x}{T} \right\},$$

where the a_n's and b_n's are given by **Euler's formulas:**

$$a_n = \frac{1}{T} \int_{-T}^{T} f(x) \cos \frac{n\pi x}{T} dx, \qquad n = 0, 1, 2, \ldots,$$

$$b_n = \frac{1}{T} \int_{-T}^{T} f(x) \sin \frac{n\pi x}{T} dx, \qquad n = 1, 2, 3, \ldots.$$

Let f be piecewise continuous on the half-interval $[0, T]$. The **Fourier cosine series** of $f(x)$ on $[0, T]$ is

$$f(x) \sim \frac{a_0}{2} + \sum_{n=1}^{\infty} a_n \cos \frac{n\pi x}{T},$$

where

$$a_n = \frac{2}{T} \int_0^T f(x) \cos \frac{n\pi x}{T} dx.$$

The **Fourier sine series** of $f(x)$ on $[0, T]$ is

$$f(x) \sim \sum_{n=1}^{\infty} b_n \sin \frac{n\pi x}{T},$$

where

$$b_n = \frac{2}{T} \int_0^T f(x) \sin \frac{n\pi x}{T} \, dx.$$

Fourier series and the method of separation of variables are used to solve boundary value problems and initial-boundary value problems for the three classical equations:

Heat equation $\dfrac{\partial u}{\partial t} = \beta \dfrac{\partial^2 u}{\partial x^2}.$

Wave equation $\dfrac{\partial^2 u}{\partial t^2} = \alpha^2 \dfrac{\partial^2 u}{\partial x^2}.$

Laplace's equation $\dfrac{\partial^2 u}{\partial x^2} + \dfrac{\partial^2 u}{\partial y^2} = 0.$

For the heat equation, there is a unique solution $u(x, t)$ when the boundary values at the ends of a conducting wire are specified along with the initial temperature distribution. The wave equation yields the displacement $u(x, t)$ of a vibrating string when the ends are held fixed and the initial displacement and initial velocity at each point of the string are given. In such a case separation of variables yields a solution which is the sum of standing waves. For an infinite string with specified initial displacement and velocity, d'Alembert's solution yields traveling waves. Laplace's equation arises in the study of steady-state solutions to the heat equations where either Dirichlet or Neumann boundary conditions are specified.

TECHNICAL WRITING EXERCISES

1. The method of separation of variables is an important technique in solving initial-boundary value problems and boundary value problems for *linear* partial differential equations. Explain where the linearity of the differential equation plays a critical role in the method of separation of variables.

2. In applying the method of separation of variables, we have encountered a variety of *special functions*, such as sines, cosines, Bessel functions, and modified Bessel functions. Describe three or four examples of partial differential equations that involve other special functions, such as Legendre polynomials, Jacobi functions, Chebyshev polynomials, Hermite polynomials, and Laguerre polynomials. (Some exploring in the library may be needed.)

3. A second order partial differential equation of the form

$$a \frac{\partial^2 u}{\partial x^2} + b \frac{\partial^2 u}{\partial x \, \partial y} + c \frac{\partial^2 u}{\partial y^2} + \text{lower order terms} = 0$$

is classified using the discriminant $D := b^2 - 4ac$. In particular, the equation is called:

parabolic,   if $D = 0,$

hyperbolic,   if $D > 0,$

elliptic,   if $D < 0.$

Verify that the heat equation is parabolic, the wave equation is hyperbolic, and Laplace's equation is elliptic. It can be shown that all parabolic (hyperbolic, elliptic) equations can be transformed by a linear change of variables into the heat (wave, Laplace's) equation. Based upon your knowledge of the heat equation, describe which types of problems (initial value, boundary value, etc.) are appropriate for parabolic equations. Do the same for hyperbolic equations and elliptic equations.

GROUP PROJECTS FOR CHAPTER 11

A. Steady-State Temperature Distribution in a Circular Cylinder

When the temperature u inside a circular cylinder reaches a steady state, it satisfies Laplace's equation $\Delta u = 0$. If the temperature on the lateral surface ($r = a$) is kept at zero, the temperature on the top ($z = b$) is kept at zero, and the temperature on the bottom ($z = 0$) is given by $u(r, \theta, 0) = f(r, \theta)$, then the steady-state temperature satisfies the boundary value problem

$$\frac{\partial^2 u}{\partial r^2} + \frac{1}{r}\frac{\partial u}{\partial r} + \frac{1}{r^2}\frac{\partial^2 u}{\partial \theta^2} + \frac{\partial^2 u}{\partial z^2} = 0, \qquad 0 \le r < a, \qquad -\pi \le \theta \le \pi, \qquad 0 < z < b,$$

$$u(a, \theta, z) = 0, \qquad -\pi \le \theta \le \pi, \qquad 0 \le z \le b,$$

$$u(r, \theta, b) = 0, \qquad 0 \le r < a, \qquad -\pi \le \theta \le \pi,$$

$$u(r, \theta, 0) = f(r, \theta), \qquad 0 \le r < a, \qquad -\pi \le \theta \le \pi,$$

where $f(a, \theta) = 0$ for $-\pi \le \theta \le \pi$ (see Figure 11.18). To find a solution to this boundary value problem, proceed as follows.

(a) Let $u(r, \theta, z) - R(r)T(\theta)Z(z)$. Show that R, T, and Z must satisfy the three ordinary differential equations

$$r^2 R'' + rR' - (r^2\lambda + \mu)R = 0,$$
$$T'' + \mu T = 0,$$
$$Z'' + \lambda Z = 0.$$

(b) Show that $T(\theta)$ has the form

$$T(\theta) = A\cos n\theta + B\sin n\theta$$

for $\mu = n^2$, $n = 0, 1, 2, \ldots$.

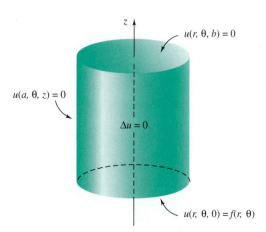

Figure 11.18 Dirichlet problem for cylinder

(c) Show that $Z(z)$ has the form

$$Z(z) = C \sinh \beta(b - z)$$

for $\lambda = -\beta^2$, where $\beta > 0$.

(d) Show that $R(r)$ has the form, for each n,

$$R_n(r) = D J_n(\beta r),$$

where J_n is the Bessel function of the first kind.

(e) Show that the boundary conditions require $R(a) = 0$ and so

$$J_n(\beta a) = 0.$$

Hence, for each n, if $0 < \alpha_{n1} < \alpha_{n2} < \cdots < \alpha_{nm} < \cdots$ are the zeros of J_n, then

$$\beta_{nm} = \alpha_{nm}/a.$$

Moreover,

$$R_n(r) = D J_n(\alpha_{nm} r/a).$$

(f) Use the preceding results to show that $u(r, \theta, z)$ has the form

$$u(r, \theta, z) = \sum_{n=0}^{\infty} \sum_{m=1}^{\infty} J_n\left(\frac{\alpha_{nm} r}{a}\right)(a_{nm} \cos n\theta + b_{nm} \sin n\theta) \sinh\left(\frac{\alpha_{nm}(b - z)}{a}\right),$$

where a_{nm} and b_{nm} are constants.

(g) Use the final boundary condition and a double orthogonal expansion involving Bessel functions and trigonometric functions to derive the formulas

$$a_{0m} = \frac{1}{\pi a^2 \sinh(\alpha_{0m} b/a)[J_1(\alpha_{0m})]^2} \int_0^a \int_0^{2\pi} f(r, \theta) J_0\left(\frac{\alpha_{0m} r}{a}\right) r \, dr \, d\theta,$$

for $m = 1, 2, 3, \ldots$ and for $n, m = 1, 2, 3, \ldots$,

$$a_{nm} = \frac{2}{\pi a^2 \sinh(\alpha_{nm} b/a)[J_{n+1}(\alpha_{nm})]^2} \int_0^a \int_0^{2\pi} f(r, \theta) J_n\left(\frac{\alpha_{nm} r}{a}\right) \cos(n\theta) r \, dr \, d\theta,$$

$$b_{nm} = \frac{2}{\pi a^2 \sinh(\alpha_{nm} b/a)[J_{n+1}(\alpha_{nm})]^2} \int_0^a \int_0^{2\pi} f(r, \theta) J_n\left(\frac{\alpha_{nm} r}{a}\right) \sin(n\theta) r \, dr \, d\theta.$$

B. A Laplace Transform Solution of the Wave Equation

Laplace transforms can be used to solve certain partial differential equations. To illustrate this technique, consider the initial-boundary value problem

(1) $$\frac{\partial^2 u}{\partial t^2} = \alpha^2 \frac{\partial^2 u}{\partial x^2}; \qquad 0 < x < \infty, \qquad t > 0,$$

(2) $$u(0, t) = h(t), \qquad t > 0,$$

(3) $$u(x, 0) = 0, \qquad 0 < x < \infty,$$

(4) $$\frac{\partial u}{\partial t}(x, 0) = 0, \qquad 0 < x < \infty,$$

(5) $$\lim_{x \to \infty} u(x, t) = 0, \qquad t \geq 0.$$

This problem arises in studying a semi-infinite string that is initially horizontal and at rest and where one end is being moved vertically. Let $u(x, t)$ be the solution to (1)–(5). For each x, let

$$U(x, s) := \mathcal{L}\{u(x, t)\}(x, s) = \int_0^\infty e^{-st} u(x, t)\, dt.$$

(a) Using the fact that

$$\mathcal{L}\left\{\frac{\partial^2 u}{\partial x^2}\right\} = \frac{\partial^2}{\partial x^2}\, \mathcal{L}\{u\},$$

show that $U(x, s)$ satisfies the equation

(6) $$s^2 U(x, s) = \alpha^2 \frac{\partial^2 U}{\partial x^2}, \qquad 0 < x < \infty.$$

(b) Show that the general solution to (6) is

$$U(x, s) = A(s)e^{-sx/\alpha} + B(s)e^{sx/\alpha},$$

where $A(s)$ and $B(s)$ are arbitrary functions of s.

(c) Since $u(x, t) \to 0$ as $x \to \infty$ for all $0 \le t < \infty$, we have $U(x, s) \to 0$ as $x \to \infty$. Use this fact to show that the $B(s)$ in part (b) must be zero.

(d) Using equation (2), show that

$$A(s) = F(s) = \mathcal{L}\{f\}(s),$$

where $A(s)$ is given in part (b).

(e) Use the results of parts (b), (c), and (d) to obtain a formal solution to (1)–(5).

C. Green's Function

Let Ω be a region in the xy–plane having a smooth boundary $\partial\Omega$. Associated with Ω is a **Green's function** $G(x, y; \xi, \eta)$ defined for pairs of distinct points (x, y), (ξ, η) in Ω. The function $G(x, y; \xi, \eta)$ has the following property.

Let $\Delta u := \partial^2 u/\partial x^2 + \partial^2 u/\partial y^2$ denote the Laplacian operator on Ω; let $h(x, y)$ be a given continuous function on Ω; and let $f(x, y)$ be a given continuous function on $\partial\Omega$. Then, a continuous solution to the Dirichlet boundary value problem

$$\Delta u(x, y) = h(x, y), \qquad \text{in } \Omega,$$
$$u(x, y) = f(x, y), \qquad \text{on } \partial\Omega,$$

is given by

(7) $$u(x, y) = \iint\limits_\Omega G(x, y; \xi, \eta) h(\xi, \eta)\, d\xi\, d\eta$$

$$+ \int_{\partial\Omega} f(\xi, \eta) \frac{\partial G}{\partial n}(x, y; \xi, \eta)\, d\sigma(\xi, \eta)$$

where n is the outward normal to the boundary $\partial\Omega$ of Ω, and the second integral is the line integral around the boundary of Ω with the interior of Ω on the left as the $\partial\Omega$ is traversed. In (7) we assume that $\partial\Omega$ is sufficiently smooth so that the integrands and integrals exist.

When Ω is the upper half-plane, the Green's function is[†]

$$G(x, y; \xi, \eta) = \frac{1}{4\pi} \ln \left[\frac{(x - \xi)^2 + (y - \eta)^2}{(x - \xi)^2 + (y + \eta)^2} \right].$$

(a) Using (7) show that a solution to

(8) $\Delta u(x, y) = h(x, y), \qquad -\infty < x < \infty, \qquad 0 < y,$

(9) $u(x, 0) = f(x, 0), \qquad -\infty < x < \infty,$

is given by

$$u(x, y) = \frac{y}{\pi} \int_{-\infty}^{\infty} \frac{f(\xi)}{(x - \xi)^2 + y^2} \, d\xi$$

$$+ \frac{1}{4\pi} \int_0^{\infty} \int_{-\infty}^{\infty} \ln \left[\frac{(x - \xi)^2 + (y - \eta)^2}{(x - \xi)^2 + (y + \eta)^2} \right] h(\xi, \eta) \, d\xi \, d\eta.$$

(b) Use the result of part (a) to determine a solution to (8)–(9) when $h \equiv 0$ and $f \equiv 1$.

(c) When Ω is the interior of the unit circle $x^2 + y^2 = 1$, the Green's function, in polar coordinates (r, θ) and (ρ, ϕ), is given by

$$G(r, \theta; \rho, \phi) = \frac{1}{4\pi} \ln[r^2 + \rho^2 - 2r\rho \cos(\phi - \theta)]$$

$$- \frac{1}{4\pi} \ln[r^2 + \rho^{-2} - 2r\rho^{-1} \cos(\phi - \theta)] - \frac{1}{4\pi} \ln \rho.$$

Using (7), show that the solution to

$$\frac{\partial^2 u}{\partial r^2} + \frac{1}{r} \frac{\partial u}{\partial r} + \frac{1}{r^2} \frac{\partial^2 u}{\partial \theta^2} = 0, \qquad 0 \le r < 1, \qquad 0 \le \theta \le 2\pi,$$

$$u(1, \theta) = f(\theta), \qquad 0 \le \theta \le 2\pi,$$

is given by

$$u(r, \theta) = \frac{1}{2\pi} \int_0^{2\pi} \frac{1 - r^2}{1 + r^2 - 2r \cos(\phi - \theta)} f(\phi) \, d\phi.$$

This is known as **Poisson's integral formula.**

(d) Use Poisson's integral formula to derive the following theorem for solutions to Laplace's equation:

▶ **MEAN VALUE PROPERTY**

Theorem 11.   Let $u(x, y)$ satisfy $\Delta u = 0$ in a bounded domain Ω in $\mathbf{R}^2$, and let (x_0, y_0) lie in Ω. Then,

(10) $u(x_0, y_0) = \frac{1}{2\pi} \int_0^{2\pi} u(x_0 + r \cos \theta, y_0 + r \sin \theta) \, d\theta,$

for all $r > 0$ for which the disk $\bar{B}(x_0, y_0; r) = \{(x, y): (x - x_0)^2 + (y - y_0)^2 \le r^2\}$ lies entirely in Ω.

[Hint: Use a change of variables that maps the disk $\bar{B}(x_0, y_0; r)$ to the unit disk $\bar{B}(0, 0; 1)$.]

[†] Techniques for determining Green's functions can be found in texts on partial differential equations such as *Partial Differential Equations of Mathematical Physics*, Second Edition, by Tyn Myint-U, Elsevier North Holland, Inc., 1980, Chapter 10.

▸ D. Numerical Method for Δu = f on a Rectangle

Let R denote the open rectangle

$$R = \{(x, y): a < x < b, c < y < d\}$$

and ∂R be its boundary. Here we describe a numerical technique for solving the generalized Dirichlet problem

(11)
$$\frac{\partial^2 u}{\partial x^2} + \frac{\partial^2 u}{\partial y^2} = f(x, y), \qquad \text{for } (x, y) \text{ in } R,$$

$$u(x, y) = g(x, y), \qquad \text{for } (x, y) \text{ on } \partial R.$$

The method is similar to the finite-difference technique discussed in Chapter 5. We begin by selecting positive integers m and n and step sizes h and k so that $b - a = hm$ and $d - c = kn$. The interval $[a, b]$ is now partitioned into m equal subintervals and $[c, d]$ into n equal subintervals. The partition points are

$$x_i = a + ih, \qquad 0 \le i \le m,$$
$$y_j = c + jk, \qquad 0 \le j \le n,$$

(see Figure 11.19). The (dashed) lines $x = x_i$ and $y = y_j$ are called **grid lines** and their intersections (x_i, y_j) are the **mesh points** of the partition. Our goal is to obtain approximations to the solution $u(x, y)$ of problem (11) at each *interior* mesh point, i.e. at (x_i, y_j) where $1 \le i \le m - 1, 1 \le j \le n - 1$. (Of course, from (11), we are given the values of $u(x, y)$ at the boundary mesh points, e.g. $u(x_0, y_j) = g(x_0, y_j), 0 \le j \le n$.)

The next step is to approximate the partial derivatives $\partial^2 u/\partial x^2$ and $\partial^2 u/\partial y^2$ using the *centered-difference formulas* discussed in Project A, Chapter 5, page 239. That is, on the horizontal grid line $y = y_j$

(12)
$$\frac{\partial^2 u}{\partial x^2}(x_i, y_j) \approx \frac{1}{h^2}[u(x_{i+1}, y_j) - 2u(x_i, y_j) + u(x_{i-1}, y_j)] \quad \text{for } 1 \le i \le m - 1,$$

and on the vertical grid line $x = x_i$

(13)
$$\frac{\partial^2 u}{\partial y^2}(x_i, y_j) \approx \frac{1}{k^2}[u(x_i, y_{j+1}) - 2u(x_i, y_j) + u(x_i, y_{j-1})] \quad \text{for } 1 \le j \le n - 1,$$

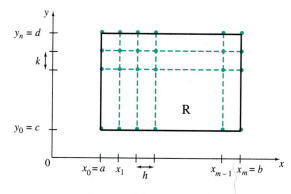

Figure 11.19 Rectangular grid

where the errors in the approximation formulas (12) and (13) are like a constant times h^2 and k^2, respectively.

(a) Show that substituting the approximations (12) and (13) into the Dirichlet problem (11) yields the following system:

(14)
$$2\left[\left(\frac{h}{k}\right)^2 + 1\right]u_{i,j} - (u_{i+1,j} + u_{i-1,j}) - \left(\frac{h}{k}\right)^2 (u_{i,j+1} + u_{i,j-1}) = -h^2 f(x_i, y_j)$$

for $i = 1, 2, \ldots, m-1$, and $j = 1, 2, \ldots, n-1$;

$$u_{0,j} = g(a, y_j), \qquad u_{m,j} = g(b, y_j), \qquad j = 0, 1, \ldots, n,$$
$$u_{i,0} = g(x_i, c), \qquad u_{i,n} = g(x_i, d), \qquad i = 1, 2, \ldots, m-1,$$

where $u_{i,j}$ approximates $u(x_i, y_j)$.

Notice that each equation in (14) involves approximations to the solution that appear in a cross centered at a mesh point (see Figure 11.20).

(b) Show that the system in part (a) is a linear system of $(m-1)(n-1)$ unknowns in $(m-1)(n-1)$ equations.

(c) For Laplace's equation where $f(x, y) \equiv 0$, show that when $h = k$, equation (14) yields

$$u_{i,j} = \frac{1}{4}(u_{i-1,j} + u_{i+1,j} + u_{i,j-1} + u_{i,j+1}).$$

Compare this averaging formula with the mean value property of Theorem 11 (Project C).

(d) For the square plate

$$R = \{(x, y): \quad 0 \le x \le 0.4, \quad 0 \le y \le 0.4\},$$

the boundary is maintained at the following temperatures:

$$u(0, y) = 0, \qquad u(0.4, y) = 150y, \qquad 0 \le y \le 0.4,$$
$$u(x, 0) = 0, \qquad u(x, 0.4) = 200x, \qquad 0 \le x \le 0.4.$$

Using the system in part (a) with $f(x, y) \equiv 0$, $m = n = 4$, and $h = k = 0.1$, find approximations to the steady-state temperatures at the mesh points of the plate. [Hint: It is helpful to label these grid points with a single index, say $p_1, p_2, \ldots, p_q$ choosing the ordering in a book-reading sequence.]

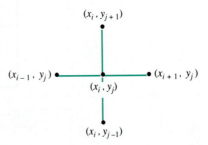

Figure 11.20 Approximate solution $u_{i,j}$ at (x_i, y_j) is obtained from values at the ends of the cross

Appendix

•A• NEWTON'S METHOD

To solve an equation $g(x) = 0$, we must find the point or points where the graph of $y = g(x)$ meets the x-axis. One procedure for approximating a solution is **Newton's method.**

To motivate Newton's method geometrically, let $\tilde{x}$ be a root of $g(x) = 0$ and let x_1 be our guess at the value of $\tilde{x}$. If $g(x_1) = 0$, we are done. If $g(x_1) \neq 0$, then we are off by some amount that we call dy (see Figure A.1). Then

$$\frac{dy}{dx} = g'(x_1),$$

and so

$$(1) \qquad dx = \frac{dy}{g'(x_1)}.$$

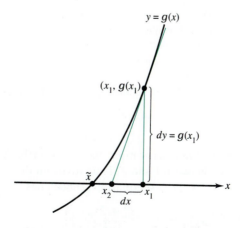

Figure A.1 Tangent line approximation of root

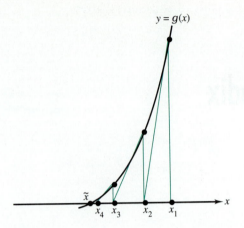

Figure A.2 Sequence of iterations converging to root

Now $dx = x_1 - x_2$ or $x_2 = x_1 - dx$, where x_2 is at the point where the tangent line through $(x_1, g(x_1))$ intersects the x-axis (Figure A.1). Using equation (1) and the fact that $dy = g(x_1)$, we obtain

$$x_2 = x_1 - \frac{dy}{g'(x_1)} = x_1 - \frac{g(x_1)}{g'(x_1)},$$

which we use as the next approximation to the root $\tilde{x}$.

Repeating this process with x_2 in place of x_1, we obtain the next approximation x_3 to the root $\tilde{x}$. In general, we find the next approximation x_{n+1} by the formula

(2) $$x_{n+1} = x_n - \frac{g(x_n)}{g'(x_n)}, \qquad n = 1, 2, \ldots.$$

The process is illustrated in Figure A.2.

If the initial guess x_1 is sufficiently close to a root $\tilde{x}$, then the sequence of iterations $\{x_n\}_{n=1}^{\infty}$ usually converges to the root $\tilde{x}$. However, if we make a bad guess for x_1, then the process may lead away from $\tilde{x}$.

● **EXAMPLE 1**   Find a root to four decimal places of the equation

(3) $$x^3 + 2x - 4 = 0.$$

Solution   Setting $g(x) = x^3 + 2x - 4$, we find that $g'(x) = 3x^2 + 2$ is positive for all x. Hence, g is increasing and has at most one zero. Furthermore, since $g(1) = -1$ and $g(2) = 8$, this zero must lie between 1 and 2. Thus we begin the procedure with the initial guess $x_1 = 1.5$. For $g(x) = x^3 + 2x - 4$, equation (2) becomes

(4) $$x_{n+1} = x_n - \frac{x_n^3 + 2x_n - 4}{3x_n^2 + 2}, \qquad n = 1, 2, \ldots.$$

With $x_1 = 1.5$, equation (4) gives

$$x_2 = 1.5 - \frac{(1.5)^3 + 2(1.5) - 4}{3(1.5)^2 + 2} = 1.5 - \frac{2.375}{8.75} \approx 1.22857.$$

Using x_2 to compute x_3 and so on, we find

$$x_3 = 1.18085,$$
$$x_4 = 1.17951,$$
$$x_5 = 1.17951,$$

where we have rounded off the computations to five decimal places. Since x_4 and x_5 agree to four decimal places and we are uncertain of the fifth decimal place because of roundoff, we surmise that the root $\tilde{x}$ of (3) agrees with 1.1795 to four decimal places. Indeed,

$$g(1.1795) = -0.00005\ldots \quad \text{and} \quad g(1.1796) = 0.00056\ldots,$$

and so $1.1795 < \tilde{x} < 1.1796$. Consequently, $\tilde{x} = 1.1795\ldots$. ●

Observe that Newton's method transforms the problem of finding a root to the equation $g(x) = 0$ into the problem of finding a fixed point for the function $h(x) = x - g(x)/g'(x)$; that is, finding a number x such that $x = h(x)$.

There are several theorems that give conditions that guarantee that the sequence of iterations $\{x_n\}_{n=1}^{\infty}$ defined by (2) will converge to a zero of $g(x)$. We mention one such result.

> **CONVERGENCE OF NEWTON'S METHOD**
>
> **Theorem 1.** Suppose that a zero $\tilde{x}$ of $g(x)$ lies in the interval (a, b), and that in this interval
>
> $$g'(x) > 0 \quad \text{and} \quad g''(x) > 0.$$
>
> If we select x_1 so that $\tilde{x} < x_1 < b$, then the sequence of iterations defined by (2) will decrease to $\tilde{x}$.

We do not give a proof of this theorem, but refer the reader to an introductory numerical analysis text such as *Numerical Analysis*, Third Edition, by R. Burden and J. Faires, Prindle, Weber, and Schmidt, Boston, 1985.

•B• SIMPSON'S RULE

A useful procedure for approximating the value of a definite integral is **Simpson's rule.**

Let the interval $[a, b]$ be divided into $2n$ equal parts and let $x_0, x_1, \ldots, x_{2n}$ be the points of the partition, that is,

$$x_k := a + kh, \qquad k = 0, 1, \ldots, 2n,$$

where $h := (b - a)/(2n)$. If

$$y_k := f(x_k), \qquad k = 0, 1, \ldots, 2n,$$

then the Simpson's rule approximation I_S for the value of the definite integral

$$\int_a^b f(x)\, dx$$

is given by

$$(1) \qquad I_S = \frac{h}{3}[y_0 + 4y_1 + 2y_2 + 4y_3 + \cdots + 2y_{2n-2} + 4y_{2n-1} + y_{2n}]$$

$$= \frac{h}{3}\sum_{k=1}^{n}(y_{2k-2} + 4y_{2k-1} + y_{2k}).$$

If

$$E := \int_a^b f(x)\, dx - I_S$$

is the error that results from using Simpson's rule to approximate the value of the definite integral, then

$$(2) \qquad |E| \leq \frac{(b - a)}{180}h^4 M,$$

where $M := \max|f^{(4)}(x)|$ for all x in $[a, b]$.

● **EXAMPLE 1** Use Simpson's rule with $n = 4$ to approximate the value of the definite integral

$$(3) \qquad \int_0^1 \frac{1}{1 + x^2}\, dx.$$

Solution Here $h = \frac{1}{8}$, $x_k = k/8$, $k = 0, 1, \ldots, 8$, and

$$y_k = (1 + x_k^2)^{-1} = \frac{1}{1 + \dfrac{k^2}{64}} = \frac{64}{64 + k^2}.$$

By Simpson's rule (1), we find

$$I_S = \frac{\left(\frac{1}{8}\right)}{3}\left[1 + 4\left(\frac{64}{64 + 1}\right) + 2\left(\frac{64}{64 + 4}\right) + 4\left(\frac{64}{64 + 9}\right)\right.$$

$$+ 2\left(\frac{64}{64 + 16}\right) + 4\left(\frac{64}{64 + 25}\right) + 2\left(\frac{64}{64 + 36}\right)$$

$$\left. + 4\left(\frac{64}{64 + 49}\right) + \frac{64}{64 + 64}\right] = 0.76456.$$

Hence, the value of the definite integral in (3) is approximately $I_S = 0.76456$. ●

For a more detailed discussion of Simpson's rule, we refer the reader to a numerical analysis book such as *Numerical Analysis*, Third Edition, by R. Burden and J. Faires, Prindle, Weber, and Schmidt, Boston, 1985.

•(• CRAMER'S RULE

When a system of n linear equations in n unknowns has a unique solution, determinants can be used to obtain a formula for the unknowns. This procedure is called **Cramer's rule.** When n is small, these formulas provide a simple procedure for solving the system.

Suppose that, for a system of n linear equations in n unknowns,

(1)
$$
\begin{aligned}
a_{11}x_1 + a_{12}x_2 + \cdots + a_{1n}x_n &= b_1, \\
a_{21}x_1 + a_{22}x_2 + \cdots + a_{2n}x_n &= b_2, \\
\vdots \qquad \vdots \qquad\quad \vdots \qquad\ \ \vdots \\
a_{n1}x_1 + a_{n2}x_2 + \cdots + a_{nn}x_n &= b_n,
\end{aligned}
$$

the coefficient matrix

(2)
$$
\mathbf{A} := \begin{bmatrix}
a_{11} & a_{12} & \cdots & a_{1n} \\
a_{21} & a_{22} & \cdots & a_{2n} \\
\vdots & \vdots & & \vdots \\
a_{n1} & a_{n2} & \cdots & a_{nn}
\end{bmatrix}
$$

has a nonzero determinant. Then Cramer's rule gives the solutions

(3)
$$
x_i = \frac{\det \mathbf{A}_i}{\det \mathbf{A}}, \qquad i = 1, 2, \ldots, n,
$$

where $\mathbf{A}_i$ is the matrix obtained from $\mathbf{A}$ by replacing the ith column of $\mathbf{A}$ by the column vector

$$
\begin{bmatrix}
b_1 \\
b_2 \\
\vdots \\
b_n
\end{bmatrix}
$$

consisting of the constants on the right-hand side of system (1). Again, we assume $\det \mathbf{A} \neq 0$.

● **EXAMPLE 1**   Use Cramer's rule to solve the system

$$
\begin{aligned}
x_1 + 2x_2 - x_3 &= 0, \\
2x_1 + x_2 + x_3 &= 9, \\
x_1 - x_2 - 2x_3 &= 1.
\end{aligned}
$$

Solution We first compute the determinant of the coefficient matrix:

$$\det \begin{bmatrix} 1 & 2 & -1 \\ 2 & 1 & 1 \\ 1 & -1 & -2 \end{bmatrix} = 12.$$

Using formula (3), we find

$$x_1 = \frac{1}{12} \det \begin{bmatrix} 0 & 2 & -1 \\ 9 & 1 & 1 \\ 1 & -1 & -2 \end{bmatrix} = \frac{48}{12} = 4,$$

$$x_2 = \frac{1}{12} \det \begin{bmatrix} 1 & 0 & -1 \\ 2 & 9 & 1 \\ 1 & 1 & -2 \end{bmatrix} = \frac{-12}{12} = -1,$$

$$x_3 = \frac{1}{12} \det \begin{bmatrix} 1 & 2 & 0 \\ 2 & 1 & 9 \\ 1 & -1 & 1 \end{bmatrix} = \frac{24}{12} = 2. \quad \bullet$$

For a more detailed discussion of Cramer's rule, we refer the reader to an introductory linear algebra text such as *Elementary Linear Algebra*, Second Edition, by A. Wayne Roberts, Benjamin/Cummings Publishing Co., Menlo Park, California, 1985.

►Answers to Odd-Numbered Problems

► CHAPTER 1

Exercises 1.1, page 4

1. ODE, 2nd order, ind. var. t, dep. var. x, linear.
3. ODE, 1st order, ind. var. t, dep. var. p, nonlinear.
5. ODE, 1st order, ind. var. x, dep. var. y, nonlinear.
7. ODE, 1st order, ind. var. x, dep. var. y, nonlinear.
9. ODE, 2nd order, ind. var. x, dep. var. y, nonlinear.
11. PDE, 2nd order, ind. var. t, r, dep. var. N.
13. ODE, 2nd order, ind. var. x, dep. var. y, nonlinear.
15. $dA/dt = kA^2$, where k is the proportionality constant.
17. Kevin wins by $6\sqrt{3} - 4\sqrt{6} \approx 0.594$ seconds.

Exercises 1.2, page 12

3. Yes.   **5.** No.   **7.** Yes.   **9.** No.   **11.** Yes.
13. Yes.   **19.** The left-hand side is always ≥ 3.
21. (a) ± 1. (b) $1 \pm \sqrt{6}$.   **23.** Yes.   **25.** Yes.
27. No.   **31.** (a) No.

Exercises 1.3, page 19

1.

Figure B.1

3.
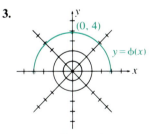
Figure B.2

5.
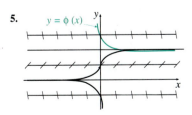
Figure B.3

7. It approaches 2.

9.

x_n	0.1	0.2	0.3	0.4	0.5
y_n	4.000	3.998	3.992	3.985	3.975

(rounded to three decimal places).

11.

x_n	0.1	0.2	0.3	0.4	0.5
y_n	2.700	2.511	2.383	2.291	2.224

(rounded to three decimal places).

13.

x_n	y_n
1.1	0.1
1.2	0.209
1.3	0.32463
1.4	0.44409
1.5	0.56437

15.

n	$\phi(1)$
1	1
2	1.06185
4	1.13920
8	1.19157

17.

x_n	y_n
1.1	-0.9
1.2	-0.81654
1.3	-0.74572
1.4	-0.68480
1.5	-0.63176
1.6	-0.58511
1.7	-0.54371
1.8	-0.50669
1.9	-0.47335
2.0	-0.44314

19. 1.46635.   **23.** $T(1) \approx 82.694$, $T(2) \approx 76.446$.

► CHAPTER 2

Exercises 2.2, page 32

1. Yes.   **3.** Yes.   **5.** No.   **7.** $y = (x^3 - 3x + C)^{1/3}$.

9. $x = Ce^{t^3}$.   **11.** $2y + \sin(2y) = 4\arctan x + C$.

13. $y = \dfrac{Ce^x}{Ce^x - 1}$.   **15.** $y = 1/(C - e^{\cos x})$.

17. $y = 2e^{-x^4/4} + 1$.   **19.** $y = -3e^{-1-\cos\theta}$.

21. $y = \sin^2 x + 2\sin x$.   **23.** $y = \ln\sqrt{4x^4 - 3}$.

25. $y = 4e^{x^3/3} - 1$.

27. (a) $y(x) = \displaystyle\int_0^x e^{t^2}\,dt$.

    (b) $y(x) = \left(1 + 3\displaystyle\int_0^x e^{t^2}\,dt\right)^{1/3}$.

    (c) $y(x) = \tan\left(\displaystyle\int_0^x \sqrt{1 + \sin t}\,dt + \pi/4\right)$.

    (d) $y(0.5) \approx 1.381$.

29. (d) $\partial f/\partial y$ is not continuous at $(0,0)$.

33. 281 kg.   **35. (a)** 82.2 min.   **(b)** 31.8 min.

(c) Never attains desired temperature.

37. (a) \$1105.17.   **(b)** 27.73 years.   **(c)** \$4427.59.

Exercises 2.3, page 41

1. Exact.   **3.** Neither.   **5.** Exact.

7. $y = (C - 3x)/(x^2 - 1)$.   **9.** $e^x \sin y - x^3 + y^{1/3} = C$.

11. $t\ln y + t = C$.   **13.** $r = (C - e^\theta)\sec\theta$.

15. Not exact.   **17.** $x^2 - y^2 + \arctan(xy) = C$.

19. $\ln x + x^2 y^2 - \sin y = \pi^2$.   **21.** $y = -2/(te^t + 2)$.

23. $\sin x - x\cos x = \ln y + 1/y + \pi - 1$ (equation is separable, not exact).

25. (a) $-\ln|y| + f(x)$;

    (b) $\cos x \sin y - y^2/2 + f(x)$; where f is a function of x only.

27. (c) $y = x^2/(C - x)$.   **(d)** Yes, $y \equiv 0$.

33. $\mu = x^2 y^{-2}$; $x^2 + y^2 = Cy$.

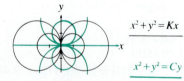

$$x^2 + y^2 = Kx$$
$$x^2 + y^2 = Cy$$

Figure B.4

Exercises 2.4, page 48

1. Linear.   **3.** Exact.   **5.** Exact, linear with x as dep. var.   **7.** Linear.   **9.** $y = 2x^2 + x\ln|x| + Cx$.

11. $y = -t - 2 + Ce^t$.   **13.** $r = \sin\theta + C\cos\theta$.

15. $x = y^3 + Cy^{-2}$.   **17.** $y = 1 + C(x^2 + 1)^{-1/2}$.

19. $y = xe^x - x$.   **21.** $y = x^{-1}/2 - 2x^{-3}$.

23. $y = x^2 \cos x - \pi^2 \cos x$.   **25.** $x = e^{4y}/2 + Ce^{2y}$.

27. (a) $y = e^{-x^2}\displaystyle\int_2^x e^{t^2}\,dt + e^{4-x^2}$.

(b) $y = \left(\displaystyle\int_0^x \sqrt{1 + \sin^2 t}\,dt\right)\Big/\sqrt{1 + \sin^2 x}$.

(c) $y(2.5) \approx 0.297$.

29. (a) $y = x - 1 + Ce^{-x}$.

    (b) $y = x - 1 + 2e^{-x}$.

    (c) $y = x/3 - 1/9 + Ce^{-3x}$.

    (d) $y = \begin{cases} x - 1 + 2e^{-x}, & 0 \le x \le 2, \\ x/3 - 1/9 + (4e^6/9 + 2e^4)e^{-3x}, & 2 < x. \end{cases}$

    (e)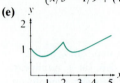

Figure B.5

31. (a) $y = x$ is only solution in a neighborhood of $x = 0$.

    (b) $y = -3x + Cx^2$ satisfies $y(0) = 0$ for any C.

35. $x(t) = \dfrac{1}{2} - \dfrac{2\cos(\pi t/12)}{4 + (\pi/12)^2} - \dfrac{(\pi/12)\sin(\pi t/12)}{4 + (\pi/12)^2}$

$$+ \left(\dfrac{19}{2} + \dfrac{2}{4 + (\pi/12)^2}\right)e^{-2t}.$$

37. $y(t) = (38/3)e^{-5t} - (8/3)e^{-20t}$.

Exercises 2.5, page 54

1. Integrating factor depending on x alone.

3. Integrating factor depending on y alone.

5. Integrating factor depending on y alone.

7. $\mu = x^{-2}$; $y^2/2 - y/x + 3x = C$, or $x \equiv 0$.

9. $\mu = x$; $x^2 y^2 + x^2 y + x^4 = C$.

11. $\mu = y^{-2}$; $x^2 y^{-1} + x = C$ or $y \equiv 0$.

13. $\mu = xy$; $x^2 y^3 - 2x^3 y^2 = C$.

15. $\mu(z) = \exp\left(\displaystyle\int H(z)\,dz\right)$; $z = xy$.

17. (b) $\mu = e^y$; $x = y - 1 + Ce^{-y}$.

Exercises 2.6, page 61

1. $y' = G(ax + by)$.   **3.** Linear coefficients.

5. Bernoulli.   **7.** Homogeneous.

9. $\ln(y^2/x^6) - y^2/x^2 = C$.

11. $y = x/(\ln|x| + C)$ or $y \equiv 0$.

13. $\sqrt{1 + x^2/t^2} = \ln|t| + C$.

15. $(x^2 - 4y^2)^3 x^2 = C$.   **17.** $y = (x + C)^2/4 - x$.

19. $y = x + (6 + 4Ce^{2x})/(1 + Ce^{2x})$.

21. $y = 2/(Cx - x^3)$ or $y \equiv 0$.

23. $y = 5x^2/(x^5 + C)$ or $y \equiv 0$.

25. $x^{-2} = 2t^2 \ln|t| + Ct^2$ or $x \equiv 0$.

27. $r = \theta^2/(C - \theta)$ or $r \equiv 0$.

29. $(y + 2)^2 + 2(x + 1)(y + 2) - 3(x + 1)^2 = C.$
31. $(2x + 2y - 3)^3 = C(2x + y - 2)^2.$
33. $y = 4x + (3 + Ce^{4x})/(1 - Ce^{4x}).$
35. $\ln[(x + 3)^2 + 3(t - 1)^2]$
$\qquad + (2/\sqrt{3})\arctan[(x + 3)/\sqrt{3}(t - 1)] = C.$
37. $y^{-2} = -e^{2x}/2 + Ce^{-2x}$ or $y \equiv 0.$
39. $\theta y^2 = C(\theta + y)^2$ or $y = -\theta.$
41. $(x - y + 2)^2 = Ce^{2x} + 1.$
45. $(y - 4x)^2(y + x)^3 = C.$

Review Problems, page 65

1. $e^x + ye^{-y} = C.$ **3.** $x^2y - x^3 + y^{-2} = C.$
5. $y + x\sin(xy) = C.$ **7.** $y = (7\ln|x| + C)^{-1/7}.$
9. $(x^2 + 4y^2)^3x^2 = C.$ **11.** $\tan(t - x) + t = C.$
13. $y = -(x^2/2)\cos(2x) + (x/4)\sin(2x) + Cx.$
15. $y = 2x + 3 - (x + C)^2/4.$
17. $y = 2/(1 + Ce^{2\theta})$ or $y \equiv 0.$
19. $y^2 = x^2 + Cx^3$ or $x \equiv 0.$
21. $xy - x^2 - x + y^2/2 - 4y = C.$
23. $y^2 + 2xy - x^2 = C.$
25. $x^2y^{-2} - 2xy^{-1} - 4xy^{-2} = C$ or $y \equiv 0.$
27. $[(y - 4)^2 - 3(x - 3)^2]\{[\sqrt{3}(x - 3) + (y - 4)]/$
$\qquad [\sqrt{3}(x - 3) - (y - 4)]\}^{1/\sqrt{3}} = C.$
29. $x^4y^3 - 3x^3y^2 + x^4y^2 = C.$
31. $y = -x^3/2 + 7x/2.$ **33.** $x = -t - 2 + 3e^{-t}.$
35. $y = -2x\sqrt{2x^2 - 1}.$
37. $\ln[(y - 2)^2 + 2(x - 1)^2] + \sqrt{2}\arctan\left[\dfrac{y - 2}{\sqrt{2}(x - 1)}\right] = \ln 2.$
39. $y = \sqrt{(19x^4 - 1)/2}.$

▶ CHAPTER 3

Exercises 3.2, page 81

1. $50 - 45e^{-2t/25}$ kg, 5.07 min.
3. $(0.4)(100 - t) - (39 \times 10^{-8})(100 - t)^4$ L; 19.96 min.
5. 0.0097%; 73.24 hr.
7. $(0.2)(1 - e^{-3t/125})$ g/cm^3; 28.88 sec.
9. 4804; 27,000. **13.** 5769; 6000. **15.** 1527; 1527.
17. $(1/2)\ln 15 \approx 1.354$ years; 14 million tons per year.
19. 1 hr; 2 hr. **21.** 11.7% **23.** 31,606 yr.
25. e^{-2t} kg of Hh, $2e^{-t} - 2e^{-2t}$ kg of It, and
$\qquad 1 - 2e^{-t} + e^{-2t}$ kg of Bu.

Exercises 3.3, page 89

1. 28.3°C; 32.5°C; 1:16 PM.
3. 16.3°C; 19.1°C; 31.7°C; 28.9°C. **5.** 30.4 min.
7. 148.6°F. **9.** 20.7 min. **11.** 22.6 min.

13. $T - M = C(T + M)\exp[2\arctan(T/M) - 4M^3kt]$;
for T near M, $M^4 - T^4 \approx 4M^3(M - T)$, and so
$dT/dt \approx k_1(M - T)$, where $k_1 = 4M^3k.$
15. 9:08 AM.

Exercises 3.4, page 96

1. $(0.981)t + (0.0981)e^{-10t} - 0.0981$ m; 1019 sec.
3. 18.6 sec. **5.** $4.91t + 22.55 - 22.55e^{-2t}$ m; 97.3 sec.
7. 241 sec. **9.** $95.65t + 956.5e^{-t/10} - 956.5$ m; 13.2 sec.
11. $e^{kv}(kv - mg)^{mg} = e^{v_0k}(kv_0 - mg)^{mg}e^{-k^2x/m}.$
13. 2.69 sec.; 101.19 m. **15.** $(\omega_0 - T/k)e^{-kt/I} + T/k.$
17. 1164 sec.
19. $2636e^{-t/20} + 131.8t - 2636$ m; 1.768 sec.
21. $5e^{-2t}/2 + 6t - 5/2$; 6 m/sec. **23.** Sailboat B.
25. (e) 11.18 km/sec. (f) 2.38 km/sec.

Exercises 3.5, page 107

3.

h	"e"
1	3
0.1	2.720551414
0.01	2.718304482
0.001	2.718282082
0.0001	2.718281824

7.

x_n	y_n
1.1	0.10450
1.2	0.21668
1.3	0.33382
1.4	0.45300
1.5	0.57135

9.

x_n	y_n
0.2	0.61784
0.4	1.23864
0.6	1.73653
0.8	1.98111
1.0	1.99705
1.2	1.88461
1.4	1.72447
1.6	1.56184
1.8	1.41732
2.0	1.29779

11. $\phi(1) \approx x(1; 2^{-3}) = 1.25494.$
13. $\phi(1) \approx y(1; 2^{-3}) = 0.71698.$
15. $\phi(1) \approx y(1; 2^{-4}) = 0.71647.$

17.

x_n	$y_n(h = 0.2)$	$y_n(h = 0.1)$	$y_n(h = 0.025)$
0.1		-1	0.06250
0.2	-3	1	0.00391
0.3		-1	0.00024
0.4	9	1	0.00002
0.5		-1	0.00000
0.6	-27	1	0.00000
0.7		-1	0.00000
0.8	81	1	0.00000
0.9		-1	0.00000
1.0	-243	1	0.00000

We conclude that step size can dramatically affect convergence.

19.

x_n	y_n: $r = 1.5$	$r = 2$	$r = 3$
0.25	1.58286	1.53125	1.39063
0.5	2.35144	2.04960	1.55347
0.75	3.26750	2.44003	1.62885
1.0	4.25316	2.68675	1.66999
1.25	5.21675	2.82920	1.69406
1.5	6.08340	2.90804	1.70858
1.75	6.81163	2.95080	1.71748
2.0	7.39215	2.97377	1.72298
2.25	7.83709	2.98604	1.72640
2.5	8.16851	2.99257	1.72852
2.75	8.41036	2.99605	1.72985
3.0	8.58432	2.99790	1.73067
3.25	8.70817	2.99889	1.73119
3.5	8.79571	2.99941	1.73151
3.75	8.85729	2.99969	1.73171
4.0	8.90044	2.99983	1.73184
4.25	8.93062	2.99991	1.73192
4.5	8.95168	2.99995	1.73197
4.75	8.96637	2.99997	1.73200
5.0	8.97660	2.99999	1.73202

21.

Time	T_n: $K = 0.2$	$K = 0.4$	$K = 0.6$
Midnight	65.0000	65.0000	65.0000
4 A.M.	69.1639	68.5644	68.1299
8 A.M.	71.4836	72.6669	73.6678
Noon	72.9089	75.1605	76.9783
4 P.M.	72.0714	73.5977	74.7854
8 P.M.	69.8095	69.5425	69.2832
Midnight	68.3852	67.0500	65.9740

Exercises 3.6, page 116

1. $y_{n+1} = y_n + h\cos(x_n + y_n) - \dfrac{h^2}{2}\sin(x_n + y_n)[1 + \cos(x_n + y_n)]$.

3. $y_{n+1} = y_n + h(x_n - y_n) + \dfrac{h^2}{2}(1 - x_n + y_n) - \dfrac{h^3}{6}(1 - x_n + y_n) + \dfrac{h^4}{24}(1 - x_n + y_n)$.

5. Order 2: $\phi(1) \approx 1.3725$; order 4; $\phi(1) \approx 1.3679$.

7. -11.7679.

9. $h = 0.5$: $\phi(1) \approx 0.75085$; $h = 0.25$: $\phi(1) \approx 0.75007$.

11. 1.36789. **13.** 0.70139 with $h = 0.25$.

15.

x_n	y_n
0.5	0.21462
1.0	0.13890
1.5	−0.02668
2.0	−0.81879
2.5	−1.69491
3.0	−2.99510

19. $v(3) \approx 0.24193$ with $h = 0.0625$.

21. $z(1) \approx 2.87080$ with $h = 0.0625$.

▶ CHAPTER 4

Exercises 4.2, page 137

1. Linear, homogeneous, constant coefficients.

3. Nonlinear.

5. Linear, nonhomogeneous, variable coefficients.

7. Linear, nonhomogeneous, constant coefficients.

9. **(a)** $3x\sin x - (x^2 + 5)\cos x$. **(b)** 0. **(c)** $(r^2 - 4r - 5)x^r$.

13. **(a)** $e^{2x}\cos x - 3e^{2x}\sin x$. **(b)** $-4e^{2x}\cos x + 3e^{2x}\sin x$.

15. Unique solution on $(-\pi/2, \pi/2)$.

17. Unique solution on $(0, 3)$.

19. Does not apply; equation is nonlinear.

21. Unique solution on $(-\infty, 1)$.

23. $D^2 + 7D + 12$. **25.** $D^2 + x$.

27. **(a)** 2. **(b)** $x\sin x + \cos x$. **31.** -2; 1; 5.

33. **(a)** $3x^2 - x^3$. **(b)** $6x + 3x^2 - 2x^3$. **(c)** $3x^2 + 2x^3$. **(d)** $6x + 3x^2 - 2x^3$. **(e)** $D^2 + D - 2$. **(f)** $6x + 3x^2 - 2x^3$.

35. **(a)** $(D + 4)(D - 1)$. **(b)** $(D + 3)(D - 2)$. **(c)** $(2D - 1)(D + 5)$. **(d)** $(D + \sqrt{2})(D - \sqrt{2})$.

Exercises 4.3, page 146

1. Lin. indep.; $2e^{-2x}$. **3.** Lin. indep.; $-e^{4x}$.

5. Lin. dep.; 0.

7. **(b)** $y = c_1e^x\cos 2x + c_2e^x\sin 2x$. **(c)** $y = 2e^x\cos 2x - e^x\sin 2x$.

9. **(b)** $y = c_1x^2 + c_2x^{-1}$. **(c)** $y = -3x^2 + x^{-1}$.

11. **(b)** $y = c_1e^x + c_2(x^2 + 2x + 2)$. **(c)** $y = 5e^{x-1} - x^2 - 2x - 2$.

13. **(c)** $\phi(x) = (1)e^x + (-1)(e^x - e^{-6x})$, $\phi(x) = (-3)e^x + (1)(3e^x + e^{-6x})$.

15. The function in (b) cannot be a Wronskian of solutions.

17. **(a)** True. **(b)** False. **19.** $W = Cxe^{-x}$.

25. **(a)** Lin. indep. **(b)** Lin. dep. **(c)** Lin. indep.

27. **(a)** 2. **(b)** 0. **(c)** $2e^{3x}$.

Exercises 4.4, page 151

1. $y = e^{-5x}$.   **3.** $y = x^{-3}$.   **5.** $x = t + 1$.

7. $xw'' + 2xw' + (x + 1)w = 0$.

9. (a) $y = x \int x^{-2} e^{x^2/2}\, dx$.

(b) $y = \sum_{n=0}^{\infty} \dfrac{x^{2n}}{n!\,2^n(2n-1)} = -1 + \dfrac{x^2}{2} + \dfrac{x^4}{24} + \cdots$.

13. (a) $y = (1 - 2x^2)\int (1 - 2x^2)^{-2} e^{x^2}\, dx$.

(b) $y = (3x - 2x^3)\int (3x - 2x^3)^{-2} e^{x^2}\, dx$.

15. (a) $y = (x - 1)\int e^x x^{-1}(x - 1)^{-2}\, dx$.

(b) $y = (x^2 - 4x + 2)\int e^x x^{-1}(x^2 - 4x + 2)^{-2}\, dx$.

17. (b) $g(2) \approx 10.7983$.

Exercises 4.5, page 158

1. $c_1 e^{-x} + c_2 e^{2x}$.   **3.** $c_1 e^{-4x} + c_2 x e^{-4x}$.

5. $c_1 e^{(-1-\sqrt{5})x/2} + c_2 e^{(-1+\sqrt{5})x/2}$.   **7.** $c_1 e^{x/2} + c_2 e^{-4x}$.

9. $c_1 e^{(1+3\sqrt{5})x/2} + c_2 e^{(1-3\sqrt{5})x/2}$.

11. $c_1 e^{-5x/2} + c_2 x e^{-5x/2}$.   **13.** $3e^{-4x}$.

15. $e^{-x} - 2x e^{-x}$.   **17.** $(\sqrt{3}/2)[e^{(1+\sqrt{3})x} - e^{(1-\sqrt{3})x}]$.

19. $2e^{5(x+1)} + e^{-(x+1)}$.   **21. (a)** $ar + b = 0$.   **(b)** $ce^{-bx/a}$.

23. $ce^{-4x/5}$.   **25.** $ce^{13x/6}$.   **27.** $c_1 e^{-x} + c_2 e^x + c_3 e^{6x}$.

29. $c_1 e^{-x} + c_2 e^{3x} + c_3 e^{5x}$.

31. $c_1 e^x + c_2 e^{(-1+\sqrt{3})x} + c_3 e^{(-1-\sqrt{3})x}$.

33. $c_1 x + c_2 x^{-7}$.   **35.** $c_1 x^{-1} + c_2 x^{-4}$.

37. $c_1 x^{-1/3} + c_2 x^{-1/3} \ln x$.   **39.** $x - 3x^4$.

41. $c_1(x - 2) + c_2(x - 2)^7$.   **43.** No.

45. $c_1 x^{-2} + c_2(-x)^{-1/2}$.   **47.** $c_1 x + c_2 x \ln x + c_3 x^3$.

49. $c_1 e^{r_1} + c_2 e^{r_2} + c_3 e^{r_3}$ where $r_1 = -4.832$,
$r_2 = -1.869$, and $r_3 = 0.701$.

53. $b = 2000\sqrt{3}$, multiply by the factor $\sqrt{2}$.

Exercises 4.6, page 164

1. $c_1 \cos 2x + c_2 \sin 2x$.   **3.** $c_1 e^{3x} \cos x + c_2 e^{3x} \sin x$.

5. $c_1 e^{-2x} \cos \sqrt{2}\,x + c_2 e^{-2x} \sin \sqrt{2}\,x$.

7. $c_1 e^{x/2} \cos(5x/2) + c_2 e^{x/2} \sin(5x/2)$.   **9.** $c_1 e^x + c_2 e^{7x}$.

11. $c_1 e^{-5x} + c_2 x e^{-5x}$.   **13.** $c_1 e^{-x} \cos 2x + c_2 e^{-x} \sin 2x$.

15. $c_1 e^{-5x} \cos 4x + c_2 e^{-5x} \sin 4x$.

17. $c_1 e^{x/2} \cos(3\sqrt{3}\,x/2) + c_2 e^{x/2} \sin(3\sqrt{3}\,x/2)$.

19. $c_1 e^{-2x/3} \cos(\sqrt{23}\,x/3) + c_2 e^{-2x/3} \sin(\sqrt{23}\,x/3)$.

21. $2e^{-x} \cos x + 3e^{-x} \sin x$.

23. $(\sqrt{2}/4)[e^{(2+\sqrt{2})x} - e^{(2-\sqrt{2})x}]$.

25. $e^{2x} \cos x + 4e^{2x} \sin x$.   **27.** $e^x \sin x - e^x \cos x$.

29. (a) $c_1 e^{-x} + c_2 e^x \cos \sqrt{2}\,x + c_3 e^x \sin \sqrt{2}\,x$.

(b) $c_1 e^{2x} + c_2 e^{-2x} \cos 3x + c_3 e^{-2x} \sin 3x$.

33. $c_1 x^2 \cos(\sqrt{2} \ln x) + c_2 x^2 \sin(\sqrt{2} \ln x)$.

35. $c_1 x^{-1} \cos(2 \ln x) + c_2 x^{-1} \sin(2 \ln x)$.

37. (a) $x(t) = 30 e^{-3t} \cos 4t + 20 e^{-3t} \sin 4t$ cm.

(b) $2/\pi$.

(c) Decreases the frequency of oscillation, introduces the factor e^{-3t}, causing the solution to decay to zero.

39. $b \geq 2\sqrt{Ik}$.

Exercises 4.7, page 168

1. $c_1 e^x + c_2 e^{-x} - x$.   **3.** $c_1 e^{2x} + c_2 e^{-x} + x - 1$.

5. $e^{-x}(c_1 \cos \sqrt{3}\,x + c_2 \sin \sqrt{3}\,x) + \sin 2x$.

7. $e^{t/2}[c_1 \cos(\sqrt{3}\,t/2) + c_2 \sin(\sqrt{3}\,t/2)] + \cos t$.

9. $c_1 e^x + c_2 x e^x + x^2 e^x$.

11. (a) $5 \cos x$. **(b)** $\cos x - e^{2x}$. **(c)** $4 \cos x + 6e^{2x}$.

13. (a) $c_1 e^{\sqrt{2}x} + c_2 e^{-\sqrt{2}x} + (1 + \tan x)/2$.

(b) $c_1 e^{\sqrt{2}x} + c_2 e^{-\sqrt{2}x} + 1 - x/2$.

(c) $c_1 e^{\sqrt{2}x} + c_2 e^{-\sqrt{2}x} + \tan x - 3x/2 + 2$.

(d) $c_1 e^{\sqrt{2}x} + c_2 e^{-\sqrt{2}x} - (x + \tan x)/8$.

17. (a) $c_1 e^x + c_2 e^{2x} + c_3 e^{-5x} - x^3 + 2x^2$.

(b) $c_1 e^x + c_2 e^{2x} + c_3 e^{-5x} + 2x^3 + x^2$.

19. (a) $v = e^x/2 + c_1 e^{-x}$.

(b) $y = -e^x/2 + c_1 e^{-x} + c_2 e^{2x}$.

Exercises 4.8, page 176

1. $y_p \equiv -4$.   **3.** e^{2x}.   **5.** $3t^2 - 2t + 4$.   **7.** $\cos 3x$.

9. $re^r/2 + 3e^r/4$.   **11.** $-(t \sin t + \cos t)/2$.

13. $4x^2 e^x$.   **15.** $x^2/9 + 4x/27 + 2/27 + e^x/4$.

17. $c_1 e^x + c_2 e^{-x} + 11x - 1$.

19. $c_1 e^x + c_2 e^{-2x} - x^2/2 + x/2 - 7/4$.

21. $c_1 e^t + c_2 e^{2t} + e^t(\cos t - \sin t)/2$.

23. $e^{-\theta}(c_1 \cos \theta + c_2 \sin \theta) + (\theta e^{-\theta} \sin \theta)/2$.

25. $c_1 e^{2t} + c_2 t e^{2t} + t^3 e^{2t}/6$.

27. $c_1 e^{\sqrt{3}x} + c_2 e^{-\sqrt{3}x} - x^2/3 - 2/9 + e^x/2$.

29. $e^{-x/2}[c_1 \cos(\sqrt{3}\,x/2) + c_2 \sin(\sqrt{3}\,x/2)] + \sin x$
    $+ e^x(-x^2/3 + 2x/3 - 4/9)$.

31. $e^x - 1$.   **33.** $e^{-x} + \sin x - \cos x$.

35. $(3/20) \sin 2x - (1/20) \cos 2x - (3/10) \cos x - (1/10) \sin x$.

37. $x^2/10 + 7x/50 - 161/500 + e^x/4 + 25e^{2x}/4$
    $- 1589 e^{5x}/500$.

39. $-(1/2) \sin \theta - e^{2\theta}/3 + 3e^\theta/4 + 7e^{-\theta}/12$.

41. $(Ax^2 + Bx) \sin x + (Cx^2 + Dx) \cos x + E10^x$.

43. $e^t(A \cos t + B \sin t) + Ct^2 + Dt + E$.

45. $Ae^{5x} + (Bx + C) \cos 3x + (Dx + E) \sin 3x$.

47. $Aue^u + B + C \sin u + D \cos u$.

49. No.   **51.** Yes.   **53.** No.   **55.** Yes.

57. $(1/5)\cos x + (2/5)\sin x$.

59. $c_1 x + c_2 x^{-3} + x^3/12$.

61. $c_1 x^3 + c_2 x^{-2} - (1/5)x^{-2}\ln x + 1$.

63. $c_1 \cos(\ln x) + c_2 \sin(\ln x) - (1/4)(\ln x)^2 \cos(\ln x)$
$\qquad + (1/4)(\ln x)\sin(\ln x)$.

65. (a) $-e^{-x}\sin 2x - 2e^{-x}\cos 2x + 2$.

   (b) $e^{-x}(c_1\sin 2x + c_2\cos 2x)$.

   (c) $y = \begin{cases} -e^{-x}\sin 2x - 2e^{-x}\cos 2x + 2, & 0 \le x \le 3\pi/2, \\ (-1 - e^{3\pi/2})e^{-x}\sin 2x + (-2 - e^{3\pi/2})e^{-x}\cos 2x, \\ \qquad x \ge 3\pi/2. \end{cases}$

69. (a) $x(t) = A\sin\beta t + B\cos\beta t + x_h$, where

$$A = \frac{k - m\beta^2}{(k - m\beta^2)^2 + b^2\beta^2},$$

$$B = \frac{-b\beta}{(k - m\beta^2)^2 + b^2\beta^2} \quad \text{and}$$

$$x_h = e^{-bt/2m}\left[c_1\cos\left(\frac{\sqrt{4mk - b^2}}{2m}t\right)\right.$$
$$\left. + c_2\sin\left(\frac{\sqrt{4mk - b^2}}{2m}t\right)\right].$$

   (b) In each case, $x_h \to 0$ as $t \to \infty$. So as $t \to \infty$, $x(t)$
approaches the function $x_p(t) = A\sin\beta t + B\cos\beta t$.

Exercises 4.9, page 183

1. $[3e^{2x}\cos 3x + e^{2x}(x - 2)\sin 3x]$
$\qquad + i[e^{2x}(x - 2)\cos 3x - 3e^{2x}\sin 3x]$.

3. $[(2x^2 - x)\cos x + (3x^2 + 5x)\sin x]$
$\qquad + i[(2x^2 - x)\sin x - (3x^2 + 5x)\cos x]$.

5. $[\cos 3x - e^{-x}(x^2 - 4)] + i[4xe^{-x} + \sin 3x]$.

7. $e^{(3-i)x}[1 - 5i]$.   **9.** $e^{-2ix}[x^3 - x^2 + 1]$.

11. $e^{(-1-2i)x}[(\sqrt{2}/2) + (\sqrt{2}/2)i]$.

13. $e^{(2-4i)x}[(1 + 3i)x - (1 + 2i)]$.

15. $(1/17)e^x\cos 3x + (4/17)e^x\sin 3x$.

17. $-(1/8)t^2 e^t\cos 2t + (t/16)e^t\sin 2t$.

19. $[(3/40)x + (1/25)]\cos 3x$
$\qquad + [-(1/40)x + (13/400)]\sin 3x$.

21. $-(1/4)x^2 e^{-x}\cos x + (3/4)xe^{-x}\sin x$.

23. $2\cos(x + 3) - 2\sin(x + 3)$.

Exercises 4.10, page 187

1. $c_1\cos 2x + c_2\sin 2x - (1/4)(\cos 2x)\ln|\sec 2x + \tan 2x|$.

3. $c_1 e^{2t} + c_2 e^{-t} + e^{3t}/4$.   **5.** $c_1 e^x + c_2 xe^x + xe^x\ln|x|$.

7. $c_1\cos 4\theta + c_2\sin 4\theta + (\theta/4)\sin 4\theta$
$\qquad + (1/16)(\cos 4\theta)\ln|\cos 4\theta|$.

9. $c_1\cos 2x + c_2\sin 2x$
$\qquad + [(\cos 2x)\ln|\csc 2x + \cot 2x| - 1]/4$.

11. $c_1\cos(3\ln x) + c_2\sin(3\ln x)$
$\qquad + (1/9)\cos(3\ln x)\ln|\sec(3\ln x) + \tan(3\ln x)|$.

13. $c_1\cos x + c_2\sin x + (\sin x)\ln|\sec x + \tan x| - 2$.

15. $c_1\cos 2x + c_2\sin 2x + (1/24)\sec^2 2x - 1/8$
$\qquad + (1/8)(\sin 2x)\ln|\sec 2x + \tan 2x|$.

17. $c_1\cos x + c_2\sin x - x^2 + 3 + 3x\sin x$
$\qquad + 3(\cos x)\ln|\cos x|$.

19. $c_1\cos 2x + c_2\sin 2x - e^x/5$
$\qquad - (1/2)(\cos 2x)\ln|\sec 2x + \tan 2x|$.

21. $c_1 x + c_2 x\ln|x| + (1/2)x(\ln|x|)^2 + 3x(\ln|x|)[\ln(\ln|x|)]$.

23. $c_1 e^x + c_2(x + 1) - x^2$.

25. $c_1(5x - 1) + c_2 e^{-5x} - x^2 e^{-5x}/10$.

27. $y = e^{1-x} - e^{x-1} + \dfrac{e^x}{2}\displaystyle\int_1^x \frac{e^{-t}}{t}dt - \dfrac{e^{-x}}{2}\displaystyle\int_1^x \frac{e^t}{t}dt$,

$\qquad y(2) \approx -1.93$.

Review Problems, page 190

1. $c_1 e^{-9x} + c_2 e^x$.

3. $c_1 e^{x/2}\cos(3x/2) + c_2 e^{x/2}\sin(3x/2)$.

5. $c_1 e^{3x/2} + c_2 e^{x/3}$.

7. $c_1 e^{-x/3}\cos(x/6) + c_2 e^{-x/3}\sin(x/6)$.

9. $c_1 e^{7x/4} + c_2 xe^{7x/4}$.

11. $t^{1/2}\{c_1\cos[(\sqrt{19}/2)\ln t] + c_2\sin[(\sqrt{19}/2)\ln t]\}$.

13. $c_1\cos 4x + c_2\sin 4x + (1/17)xe^x - (2/289)e^x$.

15. $c_1 e^{-2x} + c_2 e^{-x} + c_3 e^{-x/3}$.

17. $c_1 e^x + c_2 e^{-x/2}\cos(\sqrt{43}x/2) + c_3 e^{-x/2}\sin(\sqrt{43}x/2)$.

19. $c_1 e^{-3x} + c_2 e^{x/2} + c_3 xe^{x/2}$.

21. $c_1 e^{3x/2}\cos(\sqrt{19}x/2) + c_2 e^{3x/2}\sin(\sqrt{19}x/2) - e^x/5$
$\qquad + x^2 + 6x/7 + 4/49$.

23. $c_1\cos 4\theta + c_2\sin 4\theta - (1/16)(\cos 4\theta)\ln|\sec 4\theta + \tan 4\theta|$.

25. $c_1 e^{3x/2} + c_2 xe^{3x/2} + e^{3x}/9 + e^{5x}/49$.

27. $c_1 x + c_2 x^{-2} - 2x^{-2}\ln x + x\ln x$.

29. $e^{-2x}\cos(\sqrt{3}x)$.

31. $2e^x\cos 3x - (7/3)e^x\sin 3x - \sin 3x$.

33. $-e^{-x} - 3e^{5x} + e^{8x}$.

35. $\cos\theta + 2\sin\theta + \theta\sin\theta + (\cos\theta)\ln|\cos\theta|$.

37. $c_1 e^{-x} + c_2(x - 1)$.   **39.** $c_1 x + c_2\left[\dfrac{x}{2}\ln\left(\dfrac{1 + x}{1 - x}\right) - 1\right]$.

► CHAPTER 5

Exercises 5.1, page 203

1. $x(t) = (0.05)\cos 7t = (0.05)\sin(7t + \pi/2)$; 0.05;
$\qquad 2\pi/7$; $7/(2\pi)$.

3. $x(t) = (1/4)\sin 4t - (1/2)\cos 4t = (\sqrt{5}/4)\sin(4t + \phi)$,
$\qquad$ where $\phi = -\arctan 2 \approx -1.107$; down at 0.823 ft/sec.

5. $x(t) = (1/10)\cos(7\sqrt{10}\,t/5)$
$\quad -(1/(14\sqrt{10}))\sin(7\sqrt{10}\,t/5)$
$\quad \approx (0.103)\sin(4.427t + 1.793);\ 1.369$ sec.

7. $x(t) = (1/2)\cos 4t - (1/8)\sin 4t = (\sqrt{17}/8)\sin(4t + \phi)$,
where $\quad \phi = \pi - \arctan 4 \approx 1.816;\ 0.331$ sec.

Exercises 5.2, page 210

1. $b = 6$: $x(t) = e^{-3t}\cos\sqrt{7}\,t + (3/\sqrt{7})e^{-3t}\sin\sqrt{7}\,t =$
$(4/\sqrt{7})e^{-3t}\sin(\sqrt{7}\,t + \phi)$, where
$\phi = \arctan\sqrt{7}/3 \approx 0.723$.

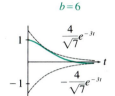

Figure B.6

$b = 8$: $x(t) = (1 + 4t)e^{-4t}$.

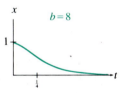

Figure B.7

$b = 10$: $x(t) = (4/3)e^{-2t} - (1/3)e^{-8t}$.

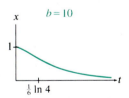

Figure B.8

3. $k = 20$: $x(t) = [(1 + \sqrt{5})/2]e^{(-5+\sqrt{5})t}$
$\quad + [(1 - \sqrt{5})/2]e^{(-5-\sqrt{5})t}$.

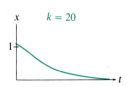

Figure B.9

$k = 25$: $x(t) = (1 + 5t)e^{-5t}$.

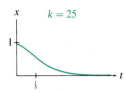

Figure B.10

$k = 30$: $x(t) = e^{-5t}\cos\sqrt{5}\,t + \sqrt{5}e^{-5t}\sin\sqrt{5}\,t =$
$\sqrt{6}e^{-5t}\sin(\sqrt{5}\,t + \phi)$, where
$\phi = \arctan(1/\sqrt{5}) \approx 0.421$.

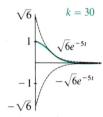

Figure B.11

5. $x(t) = -(1/6)e^{-2t}\cos(2\sqrt{31}\,t)$
$\quad - [5/(12\sqrt{31})]e^{-2t}\sin(2\sqrt{31}\,t)$
$\quad = [\sqrt{149}/(12\sqrt{31})]e^{-2t}\sin(2\sqrt{31}\,t + \phi)$,
where $\quad \phi = \pi + \arctan(2\sqrt{31}/5) \approx 4.290$;
$[\sqrt{149}/(12\sqrt{31})]e^{-2t};\ \pi/\sqrt{31};\ \sqrt{31}/\pi$.

7. 0.242 m.   **9.** $(10/\sqrt{9999})\arctan(\sqrt{9999}) \approx 0.156$ sec.

11. Relative extrema at
$t = [\pi/3 + n\pi - \arctan(\sqrt{3}/2)]/(2\sqrt{3})$
for $n = 0, 1, 2, \ldots$; but touches curves $\pm\sqrt{7/12}\,e^{-2t}$
at $t = [\pi/2 + m\pi - \arctan(\sqrt{3}/2)]/(2\sqrt{3})$
for $m = 0, 1, 2, \ldots$.

Exercises 5.3, page 216

1. $x_p(t) = (0.08)\cos 2t + (0.06)\sin 2t = (0.1)\sin(2t + \theta)$,
where $\theta = \arctan(4/3) \approx 0.927$.

3. $x(t) = -(120/901)e^{-t/2}\cos(\sqrt{255}\,t/2)$
$\quad - [8/(53\sqrt{255})]e^{-t/2}\sin(\sqrt{255}\,t/2)$
$\quad + (4/\sqrt{901})\sin(2t + \theta)$, where
$\theta = \arctan 30 \approx 1.537;\ \sqrt{63.5}/(2\pi) \approx 1.268$.

5. $M(\gamma) = 1/\sqrt{(1-4\gamma^2)^2 + 4\gamma^2}$.

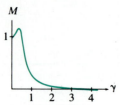

Figure B.12

7. $x(t) = \cos 3t + (1/3)t \sin 3t$.

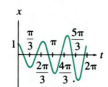

Figure B.13

9. (a) $x(t) = -[F_0/(k - m\gamma^2)]\cos(\sqrt{k/m}t)$
$+ [F_0/(k - m\gamma^2)]\cos \gamma t$
$= (F_0/[m(\omega^2 - \gamma^2)])(\cos \gamma t - \cos \omega t)$.
(c) $x(t) = \sin 8t \sin t$.

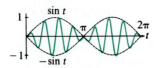

Figure B.14

Exercises 5.4, page 221

1. $I(t) = (19/\sqrt{21})[e^{(-25 - 5\sqrt{21})t/2} - e^{(-25 + 5\sqrt{21})t/2}]$.
3. $I_p(t) = (4/51)\cos 20t - (1/51)\sin 20t$; resonance frequency is $5/\pi$.
5. $M(\gamma) = 1/\sqrt{(100 - 4\gamma^2)^2 + 100\gamma^2}$.

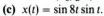

Figure B.15

7. $L = 35$ henrys, $R = 10$ ohms, $C = 1/15$ farads, and $E(t) = 50\cos 10t$ volts.

Exercises 5.5, page 230

1. $c_1(2)^n + c_2(4)^n$.
3. $c_1[-(3/2) + (\sqrt{13}/2)]^n + c_2[-(3/2) - (\sqrt{13}/2)]^n$.
5. $c_1 11^{n/2}\cos n\theta + c_2 11^{n/2}\sin n\theta$, where $\theta = \arctan(\sqrt{2}/3)$.
7. $c_1(-5)^n + c_2 n(-5)^n$.
9. $c_1 2^n \cos(n\theta) + c_2 2^n \sin(n\theta)$,
where $\theta = \arctan\sqrt{15}$.
11. $c_1 \cos(n\pi/2) + c_2 \sin(n\pi/2) + (1/2)n^2 - n$.
13. $c_1 + c_2(-1)^n - (1/2)\cos(3n)$
$+ (1/2)[\sin 6/(1 - \cos 6)]\sin(3n)$.
15. $c_1(-1)^n + c_2(-5)^n + (1/21)2^n$.
17. $c_1 2^n + c_2 n 2^n + (1/8)n^2 2^n$.
19. $c_1(-1)^n + c_2 2^n + c_3(-2)^n$.
21. $c_1 + c_2 2^{n/2}\cos(n\pi/4) + c_3 2^{n/2}\sin(n\pi/4)$.
23. $5^{n/2}\cos n\theta$, where $\theta = \pi - \arctan 2$.
25. $(n/6)3^n - (1/4)3^n + 1/4$.
27. $1 + (7/3)n - (1/2)n^2 + (1/6)n^3$.
31. Unstable for $h \geq 1$.

Exercises 5.6, page 238

1.

i	t_i	$y(t_i)$
1	0.250	0.96924
2	0.500	0.88251
3	0.750	0.75486
4	1.000	0.60656

3.

i	t_i	$y(t_i)$
1	1.250	0.80761
2	1.500	0.71351
3	1.750	0.69724
4	2.000	0.74357

5. $y(8) \approx 24.01540$.   **7.** $y(1) \approx x_1(1, 2^{-3}) = 2.77493$.
9. $t_{n+1} = t_n + h$, $n = 0, 1, 2, \ldots$;

$$x_{n+1,i} = x_{n,i} + \frac{h}{2}[f_i(t_n, x_{n,1}, \ldots, x_{n,m})$$
$$+ f_i(t_n + h, x_{n,1} + hf_1(t_n, x_{n,1}, \ldots, x_{n,m}), \ldots,$$
$$x_{n,m} + hf_m(t_n, x_{n,1}, \ldots, x_{n,m}))],$$
$$i = 1, 2, \ldots, m.$$

11.

i	t_i	$x_1'(t_i) \approx H(t_i)$
1	0.5	0.09573
2	1.0	0.37389
3	1.5	0.81045
4	2.0	1.37361
5	2.5	2.03111
6	3.0	2.75497
7	3.5	3.52322
8	4.0	4.31970
9	4.5	5.13307
10	5.0	5.95554

13. Yes, yes.

▶ CHAPTER 6

Exercises 6.2, page 252

1. $(-\infty, 0)$.   **3.** $(3\pi/2, 5\pi/2)$.   **5.** $(0, \infty)$.
7. Lin. indep.; $48e^{7x}$.   **9.** Lin. dep.; 0.
11. Lin. indep.; $(3/2)x^{-5/2}$.   **13.** Lin. indep.; $12x^4$.
15. $c_1 e^{3x} + c_2 e^{-x} + c_3 e^{-4x}$.   **17.** $c_1 x + c_2 x^2 + c_3 x^3$.
19. (a) $c_1 e^x + c_2 e^{-x} \cos 2x + c_3 e^{-x} \sin 2x + x^2$.
  (b) $-e^x + e^{-x} \sin 2x + x^2$.
21. (a) $c_1 x + c_2 x \ln x + c_3 x (\ln x)^2 + \ln x$.
  (b) $3x - x \ln x + x (\ln x)^2 + \ln x$.
23. (a) $2 \sin x - x$.   (b) $4x - 6 \sin x$.
29. (b) Let $f_1(x) = |x - 1|$ and $f_2(x) = x - 1$.

Exercises 6.3, page 258

1. $c_1 + c_2 e^{2x} + c_3 e^{-4x}$.   **3.** $c_1 e^{-x} + c_2 e^{-2x/3} + c_3 e^{x/2}$.
5. $c_1 e^{-x} + c_2 e^{-x} \cos 5x + c_3 e^{-x} \sin 5x$.
9. $c_1 e^{3x} + c_2 x e^{3x} + c_3 x^2 e^{3x}$.
11. $c_1 e^{-x} + c_2 x e^{-x} + c_3 x^2 e^{-x} + c_4 x^3 e^{-x}$.
13. $c_1 \cos \sqrt{2} x + c_2 x \cos \sqrt{2} x + c_3 \sin \sqrt{2} x + c_4 x \sin \sqrt{2} x$.
15. $c_1 e^x + c_2 x e^x + c_3 e^{-3x} + (c_4 + c_5 x) e^{-x} \cos 2x$
    $+ (c_6 + c_7 x) e^{-x} \sin 2x$.
17. $c_1 e^{-4x} + c_2 e^{3x} + (c_3 + c_4 x + c_5 x^2) e^{-2x}$
    $+ (c_6 + c_7 x) e^{-2x} \cos x + (c_8 + c_9 x) e^{-2x} \sin x$
    $+ c_{10} + c_{11} x + c_{12} x^2 + c_{13} x^3 + c_{14} x^4$.
19. $e^x - 2e^{-2x} - 3e^{2x}$.   **21.** $e^{2x} - \sqrt{2} e^x \sin \sqrt{2} x$.
25. $c_1 e^{1.120x} + c_2 e^{0.296x} + c_3 e^{-0.520x} + c_4 e^{-2.896x}$.
27. $c_1 e^{-0.5x} \cos(0.866x) + c_2 e^{-0.5x} \sin(0.866x)$
    $+ c_3 e^{-0.5x} \cos(1.323x) + c_4 e^{-0.5x} \sin(1.323x)$.
29. (a) $\{x, x^{-1}, x^2\}$ (b) $\{x, x^2, x^{-1}, x^{-2}\}$.
  (c) $\{x, x^2 \cos(3 \ln x), x^2 \sin(3 \ln x)\}$.
31. (b) $x(t) = c_1 \cos t + c_2 \sin t + c_3 \cos \sqrt{6} t + c_4 \sin \sqrt{6} t$.
  (c) $y(t) = 2c_1 \cos t + 2c_2 \sin t - (c_3/2) \cos \sqrt{6} t$
    $- (c_4/2) \sin \sqrt{6} t$.
  (d) $x(t) = (3/5) \cos t + (2/5) \cos \sqrt{6} t$.
    $y(t) = (6/5) \cos t - (1/5) \cos \sqrt{6} t$.
33. $c_1 e^{rx} + c_2 e^{-rx} + c_3 \cos rx + c_4 \sin rx$, where
  $r^4 = k/(EI)$.

Exercises 6.4, page 264

1. $c_1 x e^x + c_2 + c_3 x + c_4 x^2$.   **3.** $c_1 x^2 e^{-2x}$.
5. $c_1 e^x + c_2 e^{3x} + c_3 e^{-2x} - (1/6) x e^x + (1/6) x^2$
  $+ (5/18) x + 37/108$.
7. $c_1 e^x + c_2 e^{-2x} + c_3 x e^{-2x} - (1/6) x^2 e^{-2x}$.
9. $c_1 e^x + c_2 x e^x + c_3 x^2 e^x + (1/6) x^3 e^x$.
11. D^5.   **13.** $D + 7$ **15.** $(D - 2)(D - 1)$.
17. $[(D + 1)^2 + 4]^3$.   **19.** $(D + 2)^2 [(D + 5)^2 + 9]^2$.
21. $c_3 \cos 2x + c_4 \sin 2x + c_5$.

23. $c_3 x e^{3x} + c_4 x^2 + c_5 x + c_6$.
25. $c_3 + c_4 x + c_5 \cos 2x + c_6 \sin 2x$.
27. $c_3 x e^{-x} \cos x + c_4 x e^{-x} \sin x + c_5 x^2 + c_6 x + c_7$.
29. $c_2 x + c_3 x^2 + c_6 x^2 e^x$.
31. $-2e^{3x} + e^{-2x} + x^2 - 1$.   **33.** $x^2 e^{-2x} - x^2 + 3$.
39. $x(t) = (2/5) \cos t + (4/5) \sin t - (2/5) \cos \sqrt{6} t$
    $+ (\sqrt{6}/5) \sin \sqrt{6} t - \sin 2t$;
  $y(t) = (4/5) \cos t + (8/5) \sin t + (1/5) \cos \sqrt{6} t$
    $- (\sqrt{6}/10) \sin \sqrt{6} t - (1/2) \sin 2t$.

Exercises 6.5, page 269

1. $(1/6) x^2 e^{2x}$.   **3.** $e^{2x}/16$.
5. $\ln(\sec x) - (\sin x) \ln(\sec x + \tan x)$.
7. $c_1 x + c_2 x^2 + c_3 x^3 - (1/24) x^{-1}$.
9. $-(1/2) e^x \int e^{-x} g(x) \, dx + (1/6) e^{-x} \int e^x g(x) \, dx$

    $+ (1/3) e^{2x} \int e^{-2x} g(x) \, dx$.
11. $c_1 x + c_2 x^{-1} + c_3 x^3 - x \sin x - 3 \cos x + 3x^{-1} \sin x$.

Review Problems, page 272

1. (a) $(0, \infty)$.   (b) $(-4, -1), (-1, 1), (1, \infty)$.
5. (a) $e^{-5x}(c_1 + c_2 x) + e^{2x}(c_3 + c_4 x + c_5 x^2)$
    $+ (\cos x)(c_6 + c_7 x) + (\sin x)(c_8 + c_9 x)$.
  (b) $c_1 + c_2 x + c_3 x^2 + c_4 x^3 + e^x(c_5 + c_6 x)$
    $+ (e^{-x} \cos \sqrt{3} x)(c_7 + c_8 x)$
    $+ (e^{-x} \sin \sqrt{3} x)(c_9 + c_{10} x)$.
7. (a) D^3.   (b) $D^2(D - 3)$.   (c) $[D^2 + 4]^2$.
  (d) $[(D + 2)^2 + 9]^3$.   (e) $D^3(D + 1)^2(D^2 + 4)(D^2 + 9)$.
9. $c_1 x + c_2 x^5 + c_3 x^{-1} - (1/21) x^{-2}$.

▶ CHAPTER 7

Exercises 7.2, page 285

1. $\dfrac{1}{s^2}$, $s > 0$.   **3.** $\dfrac{1}{s - 6}$, $s > 6$.   **5.** $\dfrac{s}{s^2 + 4}$, $s > 0$.

7. $\dfrac{s - 2}{(s - 2)^2 + 9}$, $s > 2$.   **9.** $e^{-2s}\left(\dfrac{2s + 1}{s^2}\right)$, $s > 0$.

11. $\dfrac{e^{-\pi s} + 1}{s^2 + 1}$, $s > 0$.   **13.** $\dfrac{6}{s + 3} - \dfrac{2}{s^3} + \dfrac{2}{s^2} - \dfrac{8}{s}$, $s > 0$.

15. $\dfrac{6}{s^4} - \dfrac{1}{(s - 1)^2} + \dfrac{s - 4}{(s - 4)^2 + 1}$, $s > 4$.

17. $\dfrac{6}{(s - 3)^2 + 36} - \dfrac{6}{s^4} + \dfrac{1}{s - 1}$, $s > 3$.

19. $\dfrac{24}{(s-5)^5} - \dfrac{s-1}{(s-1)^2+7}$, $s > 5$.

21. Continuous (hence piecewise continuous).

23. Piecewise continuous.

25. Continuous (hence piecewise continuous).

27. Neither.   **29.** All but functions (c), (e), and (h).

Exercises 7.3, page 291

1. $\dfrac{2}{s^3} + \dfrac{2}{(s-1)^2+4}$.   **3.** $\dfrac{s+1}{(s+1)^2+9} + \dfrac{1}{s-6} - \dfrac{1}{s}$.

5. $\dfrac{4}{(s+1)^3} - \dfrac{1}{s^2} + \dfrac{s}{s^2+16}$.   **7.** $\dfrac{24}{s^5} - \dfrac{24}{s^4} + \dfrac{12}{s^3} - \dfrac{4}{s^2} + \dfrac{1}{s}$.

9. $\dfrac{4(s+1)}{[(s+1)^2+4]^2}$.   **11.** $\dfrac{s}{s^2-b^2}$.

13. $\dfrac{1}{2s} - \dfrac{s}{2(s^2+4)}$.   **15.** $\dfrac{3s}{4(s^2+1)} + \dfrac{s}{4(s^2+9)}$.

17. $\dfrac{s}{2(s^2+9)} - \dfrac{s}{2(s^2+49)}$.

19. $\dfrac{n+m}{2[s^2+(n+m)^2]} + \dfrac{m-n}{2[s^2+(m-n)^2]}$.

21. $\dfrac{s-a}{(s-a)^2+b^2}$.   **23.** $\dfrac{s^2+2}{s(s^2+4)}$.

25. (a) $\dfrac{s^2-b^2}{(s^2+b^2)^2}$.   **(b)** $\dfrac{2s^3-6sb^2}{(s^2+b^2)^3}$.

29. $\dfrac{1}{s^2+6s+10}$.

Exercises 7.4, page 300

1. $e^t t^3$.   **3.** $e^{-t} \cos 3t$.   **5.** $(1/2)e^{-2t} \sin 2t$.

7. $2e^{-2t} \cos 3t + 4e^{-2t} \sin 3t$.

9. $(3/2)e^t \cos 2t - 3e^t \sin 2t$.

11. $\dfrac{6}{s+5} - \dfrac{1}{s+2} - \dfrac{4}{s-1}$.   **13.** $\dfrac{1}{(s+1)^2} - \dfrac{2}{s}$.

15. $-\dfrac{3}{s+1} + \dfrac{(s-1)+2}{(s-1)^2+4}$.

17. $-\dfrac{5}{6s} + \dfrac{11}{10(s-2)} - \dfrac{4}{15(s+3)}$.

19. $\dfrac{1}{17}\left[\dfrac{1}{(s-3)} - \dfrac{s+1}{(s+1)^2+1} - \dfrac{4}{(s+1)^2+1}\right]$.

21. $(1/3) + e^t + (14/3)e^{6t}$.   **23.** $-e^{-3t} + 2te^{-3t} + 6e^{-t}$.

25. $8e^{2t} - e^{-t} \cos 2t + 3e^{-t} \sin 2t$.

27. $-(5/3)e^{-t} + (5/12)e^{2t} + (5/4)e^{-2t}$.

29. $3e^{-2t} + 7e^t \cos t + 11e^t \sin t$.

31. $F_1(s) = F_2(s) = F_3(s) = 1/s^2$, $\mathscr{L}^{-1}\{1/s^2\}(t) = f_3(t) = t$.

33. $e^{5t}/t - e^{-2t}/t$.   **35.** $2(\cos t - \cos 3t)/t$.

39. $\dfrac{A}{s} + \dfrac{B}{s-1} + \dfrac{C}{s+2}$, where

$$A = \left.\dfrac{2s+1}{(s-1)(s+2)}\right|_{s=0} = \dfrac{-1}{2},$$

$$B = \left.\dfrac{2s+1}{s(s+2)}\right|_{s=1} = 1,$$

$$C = \left.\dfrac{2s+1}{s(s-1)}\right|_{s=-2} = \dfrac{-1}{2}.$$

41. $2e^{-t} - 4e^{3t} + 5e^{2t}$.   **43.** $\dfrac{4}{s+2} + \dfrac{2(s-1)+2(3)}{(s-1)^2+2^2}$

Exercises 7.5, page 308

1. $2e^t \cos 2t + e^t \sin 2t$.   **3.** $-e^{-3t} + 3te^{-3t}$.

5. $t^2 + \cos t - \sin t$.   **7.** $\cos t - 4e^{5t} + 8e^{2t}$.

9. $3te^t - e^{-6t}$.   **11.** $2 - t + e^{2-t} + 2e^{t-2}$.

13. $(7/5)\sin t + (11/5)\cos t + (3/5)e^{2t-\pi} - e^{-(t-\pi/2)}$.

15. $\dfrac{-s^2+s-1}{(s^2+1)(s-1)(s-2)}$.   **17.** $\dfrac{s^5+s^4+6}{s^4(s^2+s-1)}$.

19. $\dfrac{s^3+5s^2-6s+1}{s(s-1)(s^2+5s-1)}$.   **21.** $\dfrac{s^3+s^2+2s}{(s^2+1)(s-1)^2}$.

23. $\dfrac{-s^3+1+3se^{-2s}-e^{-2s}}{s^2(s^2+4)}$.   **25.** $2e^t - \cos t - \sin t$.

27. $(t^2-4)e^{-t}$.   **29.** $(3a-b)e^t/2 + (b-a)e^{3t}/2$.

31. $5/2 + (a-5/2)e^{-t} \cos t + (a+b-5/2)e^{-t} \sin t$.

35. $t^2/2$.   **37.** $\cos t + t \sin t + c(\sin t - t \cos t)$, ($c$ arbitrary).

39. $e(t) = -a\cos\sqrt{k/I}\,t$.

41. $e(t) = (-2aI/\sqrt{4Ik-\mu^2})e^{-\mu t/2I} \sin(\sqrt{4Ik-\mu^2}\,t/2I)$.

Exercises 7.6, page 319

1. $2e^{-s}/s^3$.   **3.** $e^{-2s}(4s^2+4s+2)/s^3$.

5. $(2e^{-s} - e^{-2s} + 2e^{-3s})/s$.

7. $[(e^{-s} - e^{-2s})(s+1)]/s^2$.   **9.** $e^{t-2}u(t-2)$.

11. $e^{-2(t-2)}u(t-2) - 3e^{-2(t-4)}u(t-4)$.

13. $e^{-2(t-3)}[\cos(t-3) - 2\sin(t-3)]u(t-3)$.

15. $(7e^{6-2t} - 6e^{3-t})u(t-3)$.

17. $10 - 10u(t-3\pi)[1 + e^{-(t-3\pi)}(\cos t + \sin t)]$
    $+ 10u(t-4\pi)[1 - e^{-(t-4\pi)}(\cos t + \sin t)]$.

19. $\dfrac{1-2se^{-2s}-e^{-2s}}{s^2(1-e^{-2s})}$.

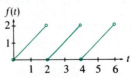

Figure B.16

21. $\dfrac{1}{1-e^{-2s}}\left[\dfrac{1-e^{-s-1}}{s+1}+\dfrac{e^{-s}-e^{-2s}}{s}\right].$

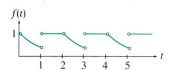

Figure B.17

23. $\dfrac{1}{s(1+e^{-as})}.$ **25.** $\dfrac{1-e^{-as}}{as^2(1+e^{-as})}.$

27. $\sin t+[1-\cos(t-3)]u(t-3).$

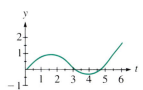

Figure B.18

29. $t+[4-t+\sin(t-2)-2\cos(t-2)]u(t-2).$

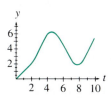

Figure B.19

31. $e^{-t}\cos t+2e^{-t}\sin t$
$+(1/2)[1-e^{2\pi-t}(\cos t+\sin t)]u(t-2\pi)$
$-(1/2)[1-e^{4\pi-t}(\cos t+\sin t)]u(t-4\pi).$

33. $e^{-t}+e^{-2t}+(1/2)[e^{-3t}-2e^{-2(t+1)}+e^{-(t+4)}]u(t-2).$

35. $e^{t}-e^{-t}+3[e^{t-1}/2+e^{-(t-1)}/2-1]u(t-1)$
$-[e^{t-2}+e^{-(t-2)}-2]u(t-2).$

37. $\cos 2t+(1/3)[1-u(t-2\pi)]\sin t$
$+(1/6)[8+u(t-2\pi)]\sin 2t.$

39. $2e^{-2t}-2e^{-3t}$
$+[1/36+(1/6)(t-1)-(1/4)e^{-2(t-1)}$
$+(2/9)e^{-3(t-1)}]u(t-1)-[19/36+(1/6)(t-5)$
$-(7/4)e^{-2(t-5)}+(11/9)e^{-3(t-5)}]u(t-5).$

45. $\dfrac{e^{t-n}}{6}-\dfrac{e^{-t}}{2}\left[\dfrac{e^{n+1}-1-e^{n+2}+e^2}{e-1}\right]$
$+\dfrac{e^{-2t}}{3}\left[\dfrac{e^{2n+2}-1-e^{2n+3}+e^3}{e^2-1}\right],$
for $n<t<n+1.$

47. $\displaystyle\sum_{n=1}^{\infty}\frac{1}{s^n}=\frac{1}{s-1}.$ **49.** $\displaystyle\sum_{n=1}^{\infty}\frac{(-1)^{n+1}}{2ns^{2n}}=\frac{1}{2}\ln(1+1/s^2).$

51. **(a)** $\sqrt{\pi/s}.$ **(b)** $105\sqrt{\pi}/(16s^{9/2}).$

59. $e^{t}-e^{-t}+(1/2)[e^{t-1}+e^{1-t}-2]u(t-1)$
$-(1/2)[e^{t-4}+e^{4-t}-2]u(t-4).$

Exercises 7.7, page 328

1. $2te^{t}-e^{t}+\displaystyle\int_0^t e^{t-v}(t-v)g(v)\,dv.$

3. $\displaystyle\int_0^t g(v)e^{2v-2t}\sin(t-v)\,dv+e^{-2t}\cos t+3e^{-2t}\sin t.$

5. $1-\cos t.$ **7.** $2e^{5t}-2e^{-2t}.$ **9.** $(t/2)\sin t.$

11. $(2/3)e^{-2t}+(1/3)e^{t}.$ **13.** $s^{-2}(s-3)^{-1}.$

15. $t/4+(3/8)\sin 2t.$ **17.** $\cos t.$ **19.** 3.

21. $e^{-t/2}\cos(\sqrt{3}t/2)-(1/\sqrt{3})e^{-t/2}\sin(\sqrt{3}t/2).$

23. $H(s)=(s^2+9)^{-1};\quad h(t)=(1/3)\sin 3t;$
$y_k(t)=2\cos 3t-\sin 3t;$
$y(t)=(1/3)\displaystyle\int_0^t(\sin 3(t-v))g(v)\,dv+2\cos 3t-\sin 3t.$

25. $H(s)=(s^2-s-6)^{-1};\quad h(t)=(e^{3t}-e^{-2t})/5;$
$y_k(t)=2e^{3t}-e^{-2t};$
$y(t)=(1/5)\displaystyle\int_0^t[e^{3(t-v)}-e^{-2(t-v)}]g(v)\,dv+2e^{3t}-e^{-2t}.$

27. $H(s)=(s^2-2s+5)^{-1};\quad h(t)=(1/2)e^{t}\sin 2t;$
$y_k(t)=e^{t}\sin 2t;$
$y(t)=(1/2)\displaystyle\int_0^t e^{(t-v)}(\sin 2(t-v))g(v)\,dv+e^{t}\sin 2t.$

29. $(1/30)\displaystyle\int_0^t e^{-2(t-v)}(\sin 6(t-v))e(v)\,dv-e^{-2t}\cos 6t$
$+e^{-2t}\sin 6t.$

31. $t^2/2.$

Exercises 7.8, page 335

1. -1 **3.** $-1.$ **5.** $e^{-2}.$ **7.** $e^{-s}-e^{-3s}.$

9. $e^{-s}.$ **11.** 0. **13.** $-(\sin t)u(t-\pi).$

15. $e^{t}+e^{-3t}+(1/4)(e^{t-1}-e^{3-3t})u(t-1)$
$-(1/4)(e^{t-2}-e^{6-3t})u(t-2).$

17. $2(e^{t-2}-e^{-(t-2)})u(t-2)+2e^{t}-t^2-2.$

19. $e^{-t}-e^{-5t}+(e/4)(e^{1-t}-e^{5-5t})u(t-1).$

21. $\sin t+(\sin t)u(t-2\pi).$

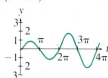

Figure B.20

23. $\sin t + (\sin t)u(t - \pi) + (\sin t)u(t - 2\pi)$.

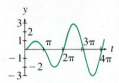

Figure B.21

25. $(1/2)e^{-2t}\sin 2t$.　**27.** $(1/2)e^t\sin 2t$.

29. The mass remains stopped at $x(t) \equiv 0$, $t > \pi/2$.

35. $\dfrac{L}{6EI}[3\lambda x^2 - x^3 + (x - \lambda)^3 u(x - \lambda)]$.

Review Problems, page 339

1. $\dfrac{3}{s} + e^{-2s}\left[\dfrac{1}{s} - \dfrac{1}{s^2}\right]$.　**3.** $\dfrac{2}{(s + 9)^3}$.

5. $\dfrac{1}{s - 2} - \dfrac{6}{s^4} + \dfrac{2}{s^3} - \dfrac{5}{s^2 + 25}$.　**7.** $\dfrac{s^2 - 36}{(s^2 + 36)^2}$.

9. $2e^{-4s}\left[\dfrac{1}{s^3} + \dfrac{4}{s^2} + \dfrac{8}{s}\right]$.　**11.** $(7/2)t^2 e^{-3t}$.

13. $2e^t + 2e^{-2t}\cos 3t + e^{-2t}\sin 3t$.

15. $e^{-2t} + e^{-t} - 2te^{-t}$.　**17.** $[2e^{t-2} + 2e^{4-2t}]u(t - 2)$.

19. $e^{2t} - e^{5t}$.　**21.** $-(3/2) + t + t^2/2 + (3/2)e^{-t}\cos t$.

23. $(2/\sqrt{7})e^{-3t/2}\sin(\sqrt{7}t/2)$
　　$+ \{(1/4) - (3/(4\sqrt{7}))e^{-3(t-1)/2}\sin(\sqrt{7}(t - 1)/2)$
　　$- (1/4)e^{-3(t-1)/2}\cos(\sqrt{7}(t - 1)/2)\}u(t - 1)$.

25. $c[t + te^{-2t} + e^{-2t} - 1]$.

27. $(9/10)e^{-3t} + (1/10)\cos t - (3/10)\sin t$.

29. $(s^2 - 5s + 6)^{-1}$;　$e^{3t} - e^{2t}$.

▶ CHAPTER 8

Exercises 8.2, page 353

1. $[-1, 3)$.　**3.** $(-4, 0)$.　**5.** $[1, 3]$.

7. $\displaystyle\sum_{n=0}^{\infty}\left[\dfrac{1}{n + 1} + 2^{-n-1}\right]x^n$.

9. $x + x^2 + (1/3)x^3 + \cdots$.

11. $1 - 2x + (5/2)x^2 + \cdots$.

13. (c) $1 - (1/2)x + (1/4)x^2 - (1/24)x^3 + \cdots$.

15. $\displaystyle\sum_{n=1}^{\infty}(-1)^n nx^{n-1}$.　**17.** $\displaystyle\sum_{n=1}^{\infty}a_n nx^{n-1}$.

19. $\ln(x + 1) = \displaystyle\sum_{n=0}^{\infty}\dfrac{(-1)^n}{n + 1}x^{n+1}$.　**21.** $\displaystyle\sum_{k=0}^{\infty}(k + 1)a_{k+1}x^k$.

23. $\displaystyle\sum_{k=1}^{\infty}a_{k-1}x^k$.　**25.** $\displaystyle\sum_{k=3}^{\infty}(k - 1)(k - 2)a_{k-1}x^k$.

27. $\displaystyle\sum_{n=0}^{\infty}\dfrac{(-1)^{n+1}(x - \pi)^{2n}}{(2n)!}$.

29. $1 + \displaystyle\sum_{n=1}^{\infty}2x^n$.

31. $6(x - 1) + 3(x - 1)^2 + (x - 1)^3$.

33. (a) $\displaystyle\sum_{n=0}^{\infty}(-1)^n(x - 1)^n$.　**(b)** $\displaystyle\sum_{n=1}^{\infty}\dfrac{(-1)^{n-1}}{n}(x - 1)^n$.

35. $1 + x + x^2 + \cdots$.　**37.** $x + x^2 + (1/2)x^3 + \cdots$.

39. $1 - (1/6)t^3 + (1/180)t^6 + \cdots$.

41. $(1/6)\theta^3 - (1/120)\theta^5 + (1/5040)\theta^7 + \cdots$.

45. $t + (1/2)t^2 - (1/6)t^3 + \cdots$.

Exercises 8.3, page 364

1. -1.　**3.** $\pm\sqrt{2}$.　**5.** $-1, 2$.

7. $x = n\pi$, n an integer.

9. $\theta \leq 0$　and　$\theta = n\pi$, $n = 1, 2, 3, \ldots$.

11. $y = a_0(1 - 2x + (3/2)x^2 - x^3/3 + \cdots)$.

13. $a_0(1 + x^4/12 + \cdots) + a_1(x + x^5/20 + \cdots)$.

15. $a_0(1 - x^2/2 - x^3/6 + \cdots) + a_1(x + x^2/2 - x^3/6 + \cdots)$.

17. $a_0(1 - x^2/2 + \cdots) + a_1(x - x^3/6 + \cdots)$.

19. $a_0 \displaystyle\sum_{n=0}^{\infty}\dfrac{1}{n!}x^{2n}$.

21. $a_0(1 - 2x^2 + x^4/3)$
　　$+ a_1\left(x + \displaystyle\sum_{k=1}^{\infty}\dfrac{(-3)(-1)\cdots(2k - 5)}{(2k + 1)!}x^{2k+1}\right)$.

23. $a_{3k+2} = 0, k = 0, 1, \ldots$.
　　$a_0\left(1 + \displaystyle\sum_{k=1}^{\infty}\dfrac{[1 \cdot 4 \cdot 7 \cdots (3k - 2)]^2}{(3k)!}x^{3k}\right)$
　　$+ a_1\left(x + \displaystyle\sum_{k=1}^{\infty}\dfrac{[2 \cdot 5 \cdot 8 \cdots (3k - 1)]^2}{(3k + 1)!}x^{3k+1}\right)$.

25. $2 + x^2 - (5/12)x^4 + (11/72)x^6 + \cdots$.

27. $x + (1/6)x^3 - (1/12)x^4 + (7/120)x^5 + \cdots$.

29. $1 - 2x + x^2 - (1/6)x^3$.

31. $1 + 2x - (3/4)x^2 - (5/6)x^3$.

35. $10 - 25t^2 + (250/3)t^3 - (775/4)t^4 + \cdots$.

37. (b) $(1/2) - (1/16)t^2 + (3/640)t^3 + (41/38400)t^4 + \cdots$.

Exercises 8.4, page 370

1. 2　**3.** $\sqrt{2}$.　**5.** $\pi/2$.

7. $a_0[1 - (x - 1)^2 + (1/2)(x - 1)^4 - (1/6)(x - 1)^6 + \cdots]$.

9. $a_0[1 + (x - 1)^2 + \cdots]$
　　$+ a_1[(x - 1) + (1/3)(x - 1)^3 + \cdots]$.

11. $a_0[1 - (1/8)(x - 2)^2 + (5/96)(x - 2)^3 + \cdots]$
　　$+ a_1[(x - 2) + (1/8)(x - 2)^2 - (1/96)(x - 2)^2$
　　$+ \cdots]$.

13. $1 - (1/2)t^2 + (1/6)t^4 - (31/720)t^5 + \cdots$.

15. $1 + x + (1/24)x^4 + (1/60)x^5 + \cdots$.

17. $1 - (1/6)(x - \pi)^3 + (1/120)(x - \pi)^5$
$+ (1/180)(x - \pi)^6 + \cdots$.

19. $-1 + x + x^2 + (1/2)x^3 + \cdots$.

21. $a_0[1 + (1/2)x^2 + (1/8)x^4 + (1/48)x^6 + \cdots]$
$+ [(1/2)x^2 + (1/12)x^4 + (11/720)x^6 + \cdots]$.

23. $a_0[1 - (1/2)x^2 + \cdots] + a_1[x - (1/3)x^3 + \cdots]$
$+ [(1/2)x^2 + (1/3)x^3 + \cdots]$.

25. $a_0[1 - (1/2)x^2 + \cdots] + a_1[x + \cdots]$
$+ [(1/2)x^2 - (1/6)x^3 + \cdots]$.

27. $a_0[1 - (1/2)x^2 - (1/6)x^3 + \cdots]$
$+ a_1[x + (1/2)x^2 + \cdots] + [(1/6)x^3 + \cdots]$.

29. (a) $a_0\left[1 + \sum_{k=1}^{\infty}(-1)^k \dfrac{n(n-2)(n-4)\cdots(n-2k+2)(n+1)(n+3)\cdots(n+2k-1)}{(2k)!}x^{2k}\right]$

$+ a_1\left[x + \sum_{k=1}^{\infty}(-1)^k \dfrac{(n-1)(n-3)\cdots(n-2k+1)(n+2)(n+4)\cdots(n+2k)}{(2k+1)!}x^{2k+1}\right]$.

(c) $P_0(x) = 1$, $P_1(x) = x$, $P_2(x) = (1/2)(3x^2 - 1)$.

31. (a) $1 - (1/2)t^2 + (\eta/6)t^3 + [(1 - \eta^2)/24]t^4 + \cdots$.

(b) Yes.

Exercises 8.5, page 374

1. $c_1 x^{-2} + c_2 x^{-3}$.   **3.** $c_1 x \cos(4 \ln x) + c_2 x \sin(4 \ln x)$.

5. $c_1 x^3 \cos(2 \ln x) + c_2 x^3 \sin(2 \ln x)$.

7. $c_1 x + c_2 x^{-1} \cos(3 \ln x) + c_3 x^{-1} \sin(3 \ln x)$.

9. $c_1 x + c_2 x^{-1/2} \cos[(\sqrt{19}/2)\ln x]$
$+ c_3 x^{-1/2} \sin[(\sqrt{19}/2)\ln x]$.

11. $c_1(x - 3)^{-2} + c_2(x - 3)^{1/2}$.

13. $c_1 x + c_2 x^2 + (4/15)x^{-1/2}$.   **15.** $2t^4 + t^{-3}$.

17. $(31/17)x + (3/17)x^{-2} \cos(5 \ln x)$
$- (76/85)x^{-2} \sin(5 \ln x)$.

Exercises 8.6, page 385

1. ± 1 are regular.   **3.** $\pm i$ are regular.

5. 1 is regular, -1 is irregular.

7. 2 is regular, -1 is irregular.

9. -4 is regular and 2 is irregular.

11. $r^2 - 3r - 10 = 0$; $r_1 = 5$, $r_2 = -2$.

13. $r^2 - 5r/9 - 4/3 = 0$; $r_1 = (5 + \sqrt{457})/18$,
$r_2 = (5 - \sqrt{457})/18$.

15. $r^2 - 1 = 0$; $r_1 = 1$, $r_2 = -1$.

17. $r^2 + r - 12 = 0$; $r_1 = 3$, $r_2 = -4$.

19. $a_0[x^{2/3} - (1/2)x^{5/3} + (5/28)x^{8/3} - (1/21)x^{11/3} + \cdots]$.

21. $a_0[1 - (1/4)x^2 + (1/64)x^4 - (1/2304)x^6 + \cdots]$.

23. $a_0[x - (1/3)x^2 + (1/12)x^3 - (1/60)x^4 + \cdots]$.

25. $a_0 \sum_{n=0}^{\infty} \dfrac{(-1)^n x^{n+(3/2)}}{2^{n-1}(n+2)!}$.   **27.** $a_0 \sum_{n=0}^{\infty} \dfrac{x^{2n+2}}{2^{2n}(n+1)!n!}$.

29. $a_0 x^2$.   **31.** $a_0 \sum_{n=0}^{\infty} \dfrac{x^n}{n!} = a_0 e^x$; Yes, $a_0 < 0$.

33. $a_0\left[x^{1/3} + \sum_{n=1}^{\infty} \dfrac{2^{n-2}(3n+4)x^{n+(1/3)}}{3^n n!}\right]$; Yes, $a_0 < 0$.

35. $a_0[x^{5/6} - (1/11)x^{11/6} + (1/374)x^{17/6}$
$- (1/25,806)x^{23/6} + \cdots]$.

37. $a_0 [x^{4/3} + (1/17)x^{7/3} + (1/782)x^{10/3}$
$+ (7/88,442)x^{13/3} + \cdots]$.

39. The expansion $\sum_{n=0}^{\infty} n! x^n$ diverges for $x \neq 0$.

41. The transformed equation is $z\, d^2y/dz^2 +$
$3\, dy/dz - y = 0$. Also $zp(z) = 3$ and $z^2 q(z) = -z$
are analytic at $z = 0$; hence $z = 0$ is a regular singular
point.
$$y_1(x) = a_0[1 + (1/3)x^{-1} + (1/24)x^{-2}$$
$$+ (1/360)x^{-3} + \cdots].$$

45. $a_0[x^{-3} - (1/2)x^{-1} - (1/8)x - (1/144)x^3 + \cdots]$
$+ a_3[1 + (1/10)x^2 + (1/280)x^4$
$+ (1/15,120)x^6 + \cdots]$.

Exercises 8.7, page 394

1. $c_1 y_1(x) + c_2 y_2(x)$, where
$y_1(x) = x^{2/3} - (1/2)x^{5/3} + (5/28)x^{8/3} + \cdots$ and
$y_2(x) = x^{1/3} - (1/2)x^{4/3} + (1/5)x^{7/3} + \cdots$.

3. $c_1 y_1(x) + c_2 y_2(x)$, where
$y_1(x) = 1 - (1/4)x^2 + (1/64)x^4 + \cdots$ and
$y_2(x) = y_1(x)\ln x + (1/4)x^2 - (3/128)x^4$
$+ (11/13,824)x^6 + \cdots$.

5. $c_1[x - (1/3)x^2 + (1/12)x^3 + \cdots] + c_2[x^{-1} - 1]$.

7. $c_1 y_1(x) + c_2 y_2(x)$, where
$y_1(x) = x^{3/2} - (1/6)x^{5/2} + (1/48)x^{7/2} + \cdots$ and
$y_2(x) = x^{-1/2} - (1/2)x^{1/2}$.

9. $c_1 w_1 + c_2 w_2$, where
$w_1(x) = x^2 + (1/8)x^4 + (1/192)x^6 + \cdots$ and
$w_2(x) = w_1(x)\ln x + 2 - (3/32)x^4 - (7/1728)x^6 + \cdots$.

11. $c_1y_1(x) + c_2y_2(x)$, where $y_1(x) = x^2$ and
$y_2(x) = x^2\ln x - 1 + 2x - (1/3)x^3 + (1/24)x^4$

13. $c_1y_1(x) + c_2y_2(x)$, where
$y_1(x) = 1 + x + (1/2)x^2 + \cdots$ and
$y_2(x) = y_1(x)\ln x - [x + (3/4)x^2 + (11/36)x^3 + \cdots]$.

15. $c_1y_1(x) + c_2y_2(x)$, where
$y_1(x) = x^{1/3} + (7/6)x^{4/3} + (5/9)x^{7/3} + \cdots$ and
$y_2(x) = 1 + 2x + (6/5)x^2 + \cdots$; all solutions are
bounded near the origin.

17. $c_1y_1(x) + c_2y_2(x) + c_3y_3(x)$, where
$y_1(x) = x^{5/6} - (1/11)x^{11/6} + (1/374)x^{17/6} + \cdots,$
$y_2(x) = 1 - x + (1/14)x^2 + \cdots,$ and
$y_3(x) = y_2(x)\ln x + 7x - (117/196)x^2$
$\qquad + (4997/298116)x^3 + \cdots.$

19. $c_1y_1(x) + c_2y_2(x) + c_3y_3(x)$, where
$y_1(x) = x^{4/3} + (1/17)x^{7/3} + (1/782)x^{10/3} + \cdots,$
$y_2(x) = 1 - (1/3)x - (1/30)x^2 + \cdots,$ and
$y_3(x) = x^{-1/2} - (1/5)x^{1/2} - (1/10)x^{3/2} + \cdots.$

21. (b) $c_1y_1(t) + c_2y_2(t)$, where
$y_1(t) = 1 - (1/6)(\alpha t)^2 + (1/120)(\alpha t)^4 + \cdots,$ and
$y_2(t) = t^{-1}[1 - (1/2)(\alpha t)^2 + (1/24)(\alpha t)^4 + \cdots].$
(c) $c_1y_1(x) + c_2y_2(x)$, where
$y_1(x) = 1 - (1/6)(\alpha/x)^2 + (1/120)(\alpha/x)^4 + \cdots,$ and
$y_2(x) = x[1 - (1/2)(\alpha/x)^2 + (1/24)(\alpha/x)^4 + \cdots].$

23. $c_1y_1(x) + c_2y_2(x)$, where
$y_1(x) = 1 + 2x + 2x^2$ and
$y_2(x) = -(1/6)x^{-1} - (1/24)x^{-2} - (1/120)x^{-3} + \cdots.$

25. For $n = 0$, $c_1 + c_2[\ln x + x + (1/4)x^2 +$
$(1/18)x^3 + \cdots]$ and for $n = 1$,
$c_1(1 - x) + c_2[(1 - x)\ln x + 3x - (1/4)x^2$
$\qquad\qquad\qquad - (1/36)x^3 + \cdots].$

Exercises 8.8, page 401

1. $y_1(x) = \displaystyle\sum_{n=0}^{\infty} \frac{4^n x^n}{(n!)^2}$ and

$y_2(x) = y_1(x)\ln x$
$\qquad - \displaystyle\sum_{n=1}^{\infty} \frac{2^{2n+1}}{(n!)^2}\left[1 + \frac{1}{2} + \frac{1}{3} + \cdots + \frac{1}{n}\right]x^n.$

3. $y_1(x) = \displaystyle\sum_{n=0}^{\infty} \frac{(-1)^n x^{n+1}}{n!}$ and

$y_2(x) = y_1(x)\ln x$
$\qquad - \displaystyle\sum_{n=1}^{\infty} \frac{(-1)^n}{n!}\left[1 + \frac{1}{2} + \frac{1}{3} + \cdots + \frac{1}{n}\right]x^{n+1}.$

5. $y_1(x) = x^{-1} + 3$ and

$y_2(x) = y_1(x)\ln x - \left[9 + \displaystyle\sum_{n=2}^{\infty} \frac{(-3)^n x^{n-1}}{n(n-1)n!}\right].$

7. $y_1(x) = \displaystyle\sum_{n=0}^{\infty} \frac{3^n x^n}{(n!)^2}$ and

$y_2(x) = y_1(x)\ln x - \displaystyle\sum_{n=1}^{\infty} \frac{2 \cdot 3^n}{(n!)^2}\left[1 + \frac{1}{2} + \frac{1}{3} + \cdots + \frac{1}{n}\right]x^n.$

9. (c) $a_0[x^{-3} - (1/2)x^{-1} - (1/8)x - (1/144)x^3 + \cdots]$
$\qquad + a_3[1 + (1/10)x^2 + (1/280)x^4 + \cdots].$

11. $y_1(x) = 1 + (2/3)x + (1/3)x^2$ and

$y_2(x) = \displaystyle\sum_{n=4}^{\infty} (n - 3)x^n.$

13. $y_1(x) = 2 + x$ and
$y_2(x) = y_1(x)\ln x + x^{-1} - 3 - 4x$

$\qquad + \displaystyle\sum_{n=3}^{\infty} \frac{2(-1)^n x^{n-1}}{n!(n-1)(n-2)}.$

15. $y_1(x) = x^2$ and

$y_2(x) = y_1(x)\ln x - 1 + 2x + 2\displaystyle\sum_{n=3}^{\infty} \frac{(-x)^n}{n!(n-2)}.$

17. $y_1(x) = x^{1/2} + \displaystyle\sum_{n=1}^{\infty} \frac{(4n^2 - 8n + 7)\cdots 39 \cdot 19 \cdot 7 \cdot 3}{[(2n+2)(2n-1)\cdots 5 \cdot 3]^2 2^n n!}x^{n+1/2},$

$y_2(x) = 1 + \displaystyle\sum_{n=1}^{\infty} a_n(0)x^n,$

$y_3(x) = y_2(x)\ln x + \displaystyle\sum_{n=1}^{\infty} a_n(0)\left\{\left[\frac{2n-3}{n^2 - 3n + 3} + \frac{2n-5}{(n-1)^2 - 3(n-1) + 3} + \cdots + \frac{1}{1} + \frac{-1}{1}\right]\right.$

$\qquad\left. -2\left[\frac{1}{n} + \frac{1}{n-1} + \cdots + \frac{1}{1}\right] - 2\left[\frac{1}{2n-1} + \frac{1}{2n-3} + \cdots + \frac{1}{3} + \frac{1}{1}\right]\right\}x^n,$ where

$a_n(0) = \dfrac{(n^2 - 3n + 3)\cdots 7 \cdot 3 \cdot 1 \cdot 1}{[(2n-1)(2n-3)\cdots 3 \cdot 1](n!)^2}.$

19. $y_1(x) = \sum_{n=0}^{\infty} \dfrac{x^n}{(n!)^3},$

$y_2(x) = y_1(x)\ln x - 3\sum_{n=1}^{\infty}\dfrac{1}{(n!)^3}\left[1 + \dfrac{1}{2} + \dfrac{1}{3} + \cdots + \dfrac{1}{n}\right]x^n,$

$y_3(x) = y_1(x)(\ln x)^2 - 6(\ln x)\sum_{n=1}^{\infty}\dfrac{1}{(n!)^3}\left[1 + \dfrac{1}{2} + \dfrac{1}{3} + \cdots + \dfrac{1}{n}\right]x^n$

$+ \sum_{n=1}^{\infty}\dfrac{3}{(n!)^3}\left[1 + \dfrac{1}{2} + \dfrac{1}{3} + \cdots + \dfrac{1}{n}\right]\left[1 + \dfrac{1}{2^2} + \dfrac{1}{3^2} + \cdots + \dfrac{1}{n^2}\right]x^n.$

21. $a_{2n+1} \equiv 0$ and

$a_{2n} = \dfrac{(2n+2+r)(2+r)r}{[(2n+2+r)(2n+r)\cdots(2+r)]^2}.$

Exercises 8.9, page 410

1. $c_1 F\left(1, 2; \dfrac{1}{2}; x\right) + c_2 x^{1/2}F\left(\dfrac{3}{2}, \dfrac{5}{2}; \dfrac{3}{2}; x\right).$

3. $c_1 F\left(1, 1; \dfrac{1}{2}; x\right) + c_2 x^{1/2}F\left(\dfrac{3}{2}, \dfrac{3}{2}; \dfrac{3}{2}; x\right).$

5. $F(1, 1; 2; x) = \sum_{n=0}^{\infty}\dfrac{1}{(n+1)}x^n = -x^{-1}\ln(1-x).$

7. $F\left(\dfrac{1}{2}, 1; \dfrac{3}{2}; x^2\right) = \sum_{n=0}^{\infty}(1/2)_n x^{2n}/(3/2)_n$

$= \sum_{n=0}^{\infty} x^{2n}/(2n+1)$

$= \dfrac{1}{2}x^{-1}\ln\left(\dfrac{1+x}{1-x}\right).$

9. $(1-x)^{-1}$, $(1-x)^{-1}\ln x.$

13. $c_1 J_{1/2}(x) + c_2 J_{-1/2}(x).$ **15.** $c_1 J_1(x) + c_2 Y_1(x).$

17. $c_1 J_{2/3}(x) + c_2 J_{-2/3}(x).$

19. $J_1(x)\ln x - x^{-1} + (3/64)x^3 - (7/2304)x^5 + \cdots.$

21. $x^{3/2}J_{3/2}(x).$

27. $J_0(x)\ln x + \sum_{n=1}^{\infty}\dfrac{(-1)^{n+1}}{2^{2n}(n!)^2}\left[1 + \dfrac{1}{2} + \cdots + \dfrac{1}{n}\right]x^{2n}.$

29. $1,\ x,\ (3x^2 - 1)/2,\ (5x^3 - 3x)/2,$
$(35x^4 - 30x^2 + 3)/8.$

37. $1,\ 2x,\ 4x^2 - 2,\ 8x^3 - 12x.$

39. $1,\ 1 - x,\ (2 - 4x + x^2)/2,$
$(6 - 18x + 9x^2 - x^3)/6.$

Review Problems, page 415

1. (a) $1 - x + (3/2)x^2 - (5/3)x^3 + \cdots.$
 (b) $-1 + x - (1/6)x^3 + (1/6)x^4 + \cdots.$

3. (a) $a_0[1 + x^2 + \cdots] + a_1[x + (1/3)x^3 + \cdots].$
 (b) $a_0[1 + (1/2)x^2 - (1/6)x^3 + \cdots]$
 $+ a_1[x - (1/2)x^2 + (1/2)x^3 + \cdots].$

5. $a_0[1 + (1/2)(x-2)^2 - (1/24)(x-2)^4 + \cdots] + a_1(x-2).$

7. (a) $a_0[x^3 + x^4 + (1/4)x^5 + (1/36)x^6 + \cdots].$
 (b) $a_0[x - (1/4)x^2 + (1/20)x^3 - (1/120)x^4 + \cdots].$

9. (a) $c_1 y_1(x) + c_2 y_2(x),$ where
 $y_1(x) = x + x^2 + (1/2)x^3 + \cdots = xe^x$ and
 $y_2(x) = y_1(x)\ln x - x^2 - (3/4)x^3 - (11/36)x^4 + \cdots.$
 (b) $c_1 y_1(x) + c_2 y_2(x),$ where
 $y_1(x) = 1 + 2x + x^2 + \cdots$ and
 $y_2(x) = y_1(x)\ln x - [4x + 3x^2 + (22/27)x^3 + \cdots].$
 (c) $c_1 y_1(x) + c_2 y_2(x),$ where
 $y_1(x) = 1 - (1/6)x + (1/96)x^2 + \cdots$ and
 $y_2(x) = y_1(x)\ln x - 8x^{-2} - 4x^{-1} + (29/36) + \cdots.$
 (d) $c_1 y_1(x) + c_2 y_2(x),$ where
 $y_1(x) = x^2 - (1/4)x^3 + (1/40)x^4 + \cdots$ and
 $y_2(x) = x^{-1} + (1/2) + (1/4)x + \cdots.$

▶ CHAPTER 9

Exercises 9.2, page 428

1. $x = c_2 e^{-t};\ y = c_1 e^{-2t} + c_2 e^{-t}.$

3. $x = c_1 + c_2 e^{-2t};\ y = c_2 e^{-2t}.$ **5.** $x = -5;\ y = 1.$

7. $u = c_1 + c_2 e^{-t} + (1/2)e^t + (5/3)t;$
 $v = c_1 - 2c_2 e^{-t} + (5/3)t.$

9. $x = c_1 e^t + (1/4)\cos t - (1/4)\sin t;$
 $y = -3c_1 e^t - (3/4)\cos t - (1/4)\sin t.$

11. $u = c_1\cos 2t + c_2\sin 2t + c_3 e^{\sqrt{3}t} + c_4 e^{-\sqrt{3}t} - (3/10)e^t;$
 $v = c_1\cos 2t + c_2\sin 2t - (2/5)c_3 e^{\sqrt{3}t} - (2/5)e^{-\sqrt{3}t}$
 $+ (1/5)e^t.$

13. $x = 2c_2 e^t\cos 2t - 2c_1 e^t\sin 2t;$
 $y = c_1 e^t\cos 2t + c_2 e^t\sin 2t.$

15. $w = -(2/3)c_1 e^{2t} + c_2 e^{7t} + t + 1;$
 $z = c_1 e^{2t} + c_2 e^{7t} - 5t - 2.$

17. $x = c_1\cos t + c_2\sin t - 4c_3\cos(\sqrt{6}t) - 4c_4\sin(\sqrt{6}t);$
 $y = c_1\cos t + c_2\sin t + c_3\cos(\sqrt{6}t) + c_4\sin(\sqrt{6}t).$

19. $x = (1/2)e^t\{(c_1 - c_2)\cos t + (c_1 + c_2)\sin t\} + c_3 e^{2t}$;
$y = e^t(c_1 \cos t + c_2 \sin t)$;
$z = (3/2)e^t\{(c_1 - c_2)\cos t + (c_1 + c_2)\sin t\} + c_3 e^{2t}$.

21. For $\lambda < 1$, solutions go to zero as $t \to +\infty$. For $\lambda = 1$, solutions remain bounded as $t \to +\infty$. For $\lambda > 1$, solutions become unbounded as $t \to +\infty$, that is, they go to either $+\infty$ or $-\infty$.

23. Infinitely many solutions satisfying $x + y = e^t + e^{-2t}$.

25. $x_1' = x_2$, $x_2' = 3x_1 - tx_2 + t^2$; $x_1(0) = 3$, $x_2(0) = -6$.

27. $x_1' = x_2$, $x_2' = x_3$, $x_3' = x_4$, $x_4' = x_4 - 7x_1 + \cos t$; $x_1(0) = x_2(0) = 1$, $x_3(0) = 0$, $x_4(0) = 2$.

29. $x(t) = 1 + t - (1/2)t^2 - t^3 + (1/12)t^4 + \cdots$,
$y(t) = 2 - 3t + t^2 - (1/6)t^3 + (7/24)t^4 + \cdots$.

31. $I_1 = (3/5)e^{-3t/2} - (8/5)e^{-2t/3} + 1$,
$I_2 = (1/5)e^{-3t/2} - (6/5)e^{-2t/3} + 1$,
$I_3 = (2/5)e^{-3t/2} - (2/5)e^{-2t/3}$.

33. $(1/2)I_1' + 2q_3 = \cos 3t$ (where $I_3 = q_3'$),
$(1/2)I_1' + I_2 = 0$, $I_1 = I_2 + I_3$;
$I_1(0) = I_2(0) = I_3(0) = 0$;
$I_1 = -(36/61)e^{-t}\cos\sqrt{3}t$
$\quad - (42\sqrt{3}/61)e^{-t}\sin\sqrt{3}t + (36/61)\cos 3t$
$\quad + (30/61)\sin 3t$,
$I_2 = (45/61)e^{-t}\cos\sqrt{3}t - (39\sqrt{3}/61)e^{-t}\sin\sqrt{3}t$
$\quad - (45/61)\cos 3t + (54/61)\sin 3t$,
$I_3 = -(81/61)e^{-t}\cos\sqrt{3}t$
$\quad - (3\sqrt{3}/61)e^{-t}\sin\sqrt{3}t + (81/61)\cos 3t$
$\quad - (24/61)\sin 3t$.

Exercises 9.3, page 431

1. $x = e^t$; $y = e^t$. **3.** $z = 2e^{-2t} - 3$; $w = 2e^{-2t}$.

5. $x = (7/4)e^t + (7/4)e^{-t} - (3/2)\cos t$;
$y = (7/4)e^t - (7/4)e^{-t} + (1/2)\sin t$.

7. $x = -(150/17)e^{5t/2}\cos(\sqrt{15}t/2)$
$\quad - (334\sqrt{15}/85)e^{5t/2}\sin(\sqrt{15}t/2) - (3/17)e^{-3t}$;
$y = (46/17)e^{5t/2}\cos(\sqrt{15}t/2)$
$\quad - (146\sqrt{15}/85)e^{5t/2}\sin(\sqrt{15}t/2) + (22/17)e^{-3t}$.

9. $x = 4e^{-2t} - e^{-t} - \cos t$; $y = 5e^{-2t} - e^{-t}$.

11. $x = (e^t - e^{-t})/2 - (1/2)[e^{t-2} - e^{-(t-2)}]u(t-2)$;
$y = 1 - (e^t + e^{-t})/2 - [1 - (e^{t-2} + e^{-(t-2)})/2]u(t-2)$.

13. $x = e^{-t} + (1/2)[e^{-(t-\pi)} + \cos t - \sin t]u(t-\pi)$;
$y = e^{-t} - [1 - (1/2)e^{-(t-\pi)}$
$\quad + (1/2)\cos t + (1/2)\sin t]u(t-\pi)$.

15. $x = -7e^{-t} + e^t$; $y = 2e^{-t}$; $z = -13e^{-t} + e^t$.

17. $2I_1 + (0.1)I_3' + (0.2)I_1' = 6$, $(0.1)I_3' - I_2 = 0$,
$I_1 = I_2 + I_3$; $I_1(0) = I_2(0) = I_3(0) = 0$;
$I_1 = -e^{-20t} - 2e^{-5t} + 3$,
$I_2 = -2e^{-20t} + 2e^{-5t}$,
$I_3 = e^{-20t} - 4e^{-5t} + 3$.

Exercises 9.4, page 438

1. $m_1 x_1'' = -k_1 x_1 + k_2(x_2 - x_1)$,
$m_2 x_2'' = -k_2(x_2 - x_1)$;
$x_1 = (2/\sqrt{33})\cos(\sqrt{7 - \sqrt{33}}t/\sqrt{2})$
$\quad - (2/\sqrt{33})\cos(\sqrt{7 + \sqrt{33}}t/\sqrt{2})$,
$x_2 = [(-1 + \sqrt{33})/(2\sqrt{33})]\cos(\sqrt{7 - \sqrt{33}}t/\sqrt{2})$
$\quad + [(1 + \sqrt{33})/(2\sqrt{33})]\cos(\sqrt{7 + \sqrt{33}}t/\sqrt{2})$.

3. $mx_1'' = -kx_1 + k(x_2 - x_1)$,
$mx_2'' = -k(x_2 - x_1) + k(x_3 - x_2)$,
$mx_3'' = -k(x_3 - x_2) - kx_3$;

The normal frequency, $(1/2\pi)\sqrt{(2 + \sqrt{2})(k/m)}$, has the mode $x_1(t) = x_3(t) = -(1/\sqrt{2})x_2(t)$; the normal frequency $(1/2\pi)\sqrt{(2 - \sqrt{2})(k/m)}$ has the mode $x_1(t) = x_3(t) = (1/\sqrt{2})x_2(t)$; and the normal frequency $(1/2\pi)\sqrt{2k/m}$ has the mode $x_1(t) = -x_3(t)$, $x_2(t) \equiv 0$.

5. $x_1 = -e^{-t} - te^{-t} - \cos t$; $x_2 = e^{-t} + te^{-t} - \cos t$.

7. $(1/2\pi)\sqrt{g/l}$; $(1/2\pi)\sqrt{(g/l) + (2k/m)}$.

9. $x = -100e^{-t/10} + 150$;
$y = -50e^{-t/10} - 10te^{-t/10} + 150$.

11. $460/11 \approx 41.8°\text{F}$. **13.** $90.4°\text{F}$.

Exercises 9.5, page 446

1.

i	t_i	$x(t_i)$	$y(t_i)$
1	0.250	-0.59733	1.15365
2	0.500	-1.58112	0.48705
3	0.750	-3.41862	-0.19129
4	1.000	-7.01992	-1.13136

3.

i	t_i	$x(t_i)$	$y(t_i)$
1	0.250	0.98661	0.75492
2	0.500	0.91749	1.04259
3	0.750	0.79823	1.38147
4	1.000	0.63303	1.77948

5.

i	t_i	$y(t_i)$
1	0.25	0.75755
2	0.50	0.55876
3	0.75	0.44552
4	1.00	0.46272

7.

i	t_i	$y(t_i)$
1	0.25	-0.72103
2	0.50	-0.39038
3	0.75	-0.01072
4	1.00	0.42498

9. $x(1) \approx 127.773$; $y(1) \approx -423.476$.

11. $u(1; 2^{-2}) = v(1; 2^{-2}) = 0.36789$.

13. $y(1) \approx x_1(1; 2^{-3}) = 1.25958$.

15. $x(1) \approx 0.80300$; $y(1) \approx 0.59598$; $z(1) \approx 0.82316$.

17. $y(0.1) \approx 0.00647, \ldots, y(2.0) \approx 1.60009$.

19.

		Part (a)		Part (b)		Part (c)	
i	t_i	$x(t_i)$	$y(t_i)$	$x(t_i)$	$y(t_i)$	$x(t_i)$	$y(t_i)$
1	0.5	1.95247	2.25065	1.48118	2.42311	0.91390	2.79704
2	1.0	3.34588	1.83601	2.66294	1.45358	1.63657	1.13415
3	1.5	4.53662	3.36527	5.19629	2.40348	4.49334	1.07811
4	2.0	2.47788	4.32906	3.10706	4.64923	5.96115	5.47788
5	2.5	1.96093	2.71900	1.92574	3.32426	1.51830	5.93110
6	3.0	2.86412	1.96166	2.34143	2.05910	0.95601	2.18079
7	3.5	4.28449	2.77457	3.90106	2.18977	2.06006	0.98131
8	4.0	3.00965	4.11886	3.83241	3.89043	5.62642	1.38072
9	4.5	2.18643	3.14344	2.32171	3.79362	5.10594	5.10462
10	5.0	2.63187	2.25824	2.21926	2.49307	1.74187	5.02491

21. **(a)** $P_1(10) \approx 0.567$, $P_2(10) \approx 0.463$, $P_3(10) \approx 0.463$.
  (b) $P_1(10) \approx 0.463$, $P_2(10) \approx 0.567$, $P_3(10) \approx 0.463$.
  (c) $P_1(10) \approx 0.463$, $P_2(10) \approx 0.463$, $P_3(10) \approx 0.567$.
  All populations approach 0.5.

Exercises 9.6, page 457

1. $x = y^3$, $y > 0$.

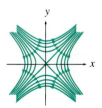

Figure B.22

3. $(1/\sqrt{2}, 1/\sqrt{2})$, $(-1/\sqrt{2}, -1/\sqrt{2})$. **5.** $(0,0)$.
7. $(1,1)$ and line $y = 2$. **9.** $e^x + ye^{-y} = c$.
11. $y = x/(\ln|x| + c)$. **13.** $e^x + xy - y^2 = c$.
15. $y^2 - x^2 = c$.

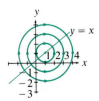

Figure B.23

17. $(x - 1)^2 + (y - 1)^2 = c$.

$y = x$

Figure B.24

19. $y = cx^{2/3}$. **21.**

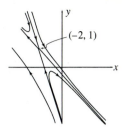

$(-2, 1)$

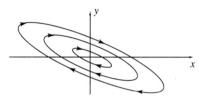

Figure B.25   **Figure B.26**

23.

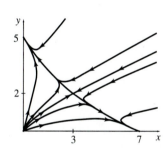

Figure B.27

25.

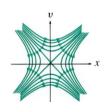

Figure B.28

27. $v^2 - x^2 = c$.

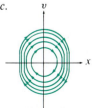

Figure B.29

29. $3v^2 + 3x^2 + x^6 = c$.

Figure B.30

31.

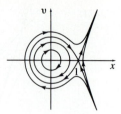

Figure B.31

33. Critical points are the line $I = 0$; integral curves are
$I = -S + (b/a)\ln S + c$; an epidemic occurs because
$I'(t) > 0$ if $I > 0$ and $S > b/a$.

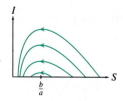

Figure B.32

35. Critical point is $(0,0)$; integral curves are
$2v^2 + 2x^2 + x^4 = c$; motion is periodic about
equilibrium position.

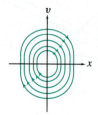

Figure B.33

37. Integral curves are $v^2 + ax^2 + 2ab\ln(C - x) = k$;
When $C^2 - 4b < 0$, the bar is pulled over to the wire
(see Figure B.34). When $C^2 - 4b > 0$, there is a critical
point at $(C - \sqrt{C^2 - 4b}, 0)$, and the bar may oscillate
(see closed trajectories in Figure B.35), or may be
pulled over to the wire.

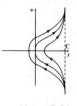

Figure B.34   **Figure B.35**

Exercises 9.7, page 468

1. For $\omega = 3/2$. The Poincaré map alternates between
the points $(0.8, 1.5)$ and $(0.8, -1.5)$. There is a
subharmonic solution of period 4π. For $\omega = 3/5$: The
Poincaré map cycles through the points
$(-1.5625, 0.6)$, $(-2.1503, -0.4854)$, $(-0.6114, 0.1854)$,
$(-2.5136, 0.1854)$, and $(-0.9747, -0.4854)$. There is a
subharmonic solution of period 10π.
3. The Poincaré map converges as $n \to \infty$ to the point
$(-1.060065, 0.262366)$.
5. The attractor is the point $(0, 1.092)$.
9. (a) $\{1/7, 2/7, 4/7, 1/7,...\}$, $\{3/7, 6/7, 5/7, 3/7,...\}$.
   (b) $\{1/15, 2/15, 4/15, 8/15, 1/15,...\}$,
     $\{1/5, 2/5, 4/5, 3/5, 1/5,...\}$, $\{1/3, 2/3, 1/3,...\}$,
     $\{7/15, 14/15, 13/15, 11/15, 7/15,...\}$.
   (c) $x_n = 0$ for $n \ge j$.
13.

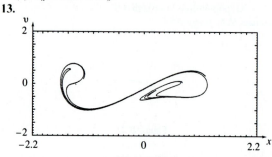

Figure B.36

Review Problems, page 471

1. $x = c_1 e^{\sqrt{3}t} + c_2 e^{-\sqrt{3}t}$; $y = \sqrt{3}c_1 e^{\sqrt{3}t} - \sqrt{3}c_2 e^{-\sqrt{3}t}$.
3. $x = c_1 e^{-t/5} + t - 4$; $y = (3/2)c_1 e^{-t/5} + t - 6$.
5. $x(t) = y(t) \equiv 0$.   **7.** $x = e^t$; $y \equiv 0$.
9. $x = (1/3)e^t + (2/3)e^{-t/2}\cos(\sqrt{3}t/2)$;
   $y = (1/3)e^t - (1/3)e^{-t/2}\cos(\sqrt{3}t/2)$
     $+ (1/\sqrt{3})e^{-t/2}\sin(\sqrt{3}t/2)$.
11. $x_1' = x_2$, $x_2' = x_3$, $x_3' = 2e^{-t} - tx_3 - 3x_2 - 2x_1$;
   $x_1(0) = 2$, $x_2(0) = 3$, $x_3(0) = 2$.
13. Critical point $(0,0)$; integral curves $x = cy^2$.

Figure B.37

15. Critical point $(0,0)$; integral curves $y^2 - 4x^2 = c$.

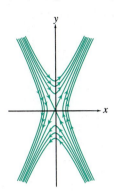

Figure B.38

17. Critical points $y \equiv 0$; integral curves $y = c \cos x$.

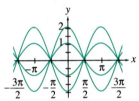

Figure B.39

19. A runaway arms race.

▶ CHAPTER 10

Exercises 10.2, page 484

1. (a) $\begin{bmatrix} 2 & 3 \\ 1 & 5 \end{bmatrix}$. **(b)** $\begin{bmatrix} 1 & 1 \\ 5 & 2 \end{bmatrix}$. **(c)** $\begin{bmatrix} 7 & 3 \\ 7 & 18 \end{bmatrix}$.

3. (a) $\begin{bmatrix} 18 & 10 \\ 4 & 4 \end{bmatrix}$. **(b)** $\begin{bmatrix} 8 & 12 \\ 3 & 5 \end{bmatrix}$. **(c)** $\begin{bmatrix} 16 & 0 \\ 0 & 16 \end{bmatrix}$.

5. (a) $\begin{bmatrix} -1 & -2 \\ -1 & -3 \end{bmatrix}$. **(b)** $\begin{bmatrix} -5 & -1 \\ -8 & -1 \end{bmatrix}$. **(c)** $\begin{bmatrix} -6 & -3 \\ -9 & -4 \end{bmatrix}$.

7. (a) 21. **(b)** 21. **(c)** 6. **(d)** 6.

9. $\begin{bmatrix} 4/9 & -1/9 \\ 1/9 & 2/9 \end{bmatrix}$. **11.** Doesn't exist.

13. $\begin{bmatrix} 1 & 0 & 1 \\ -2 & 1 & -2 \\ 1 & 1 & 0 \end{bmatrix}$. **15.** $\begin{bmatrix} (4/3)e^{-t} & -(1/3)e^{-t} \\ -(1/3)e^{-4t} & (1/3)e^{-4t} \end{bmatrix}$.

17. $\begin{bmatrix} e^{-t} & (1/2)e^{-t} & -(1/2)e^{-t} \\ (1/3)e^{t} & -(1/2)e^{t} & (1/6)e^{t} \\ -(1/3)e^{-2t} & 0 & (1/3)e^{-2t} \end{bmatrix}$.

19. 11. **21.** -12. **23.** 25. **25.** 2, 3.

27. 0, 1, 1. **29.** $\begin{bmatrix} 3e^{3t} \\ 6e^{3t} \\ -3e^{3t} \end{bmatrix}$. **31.** $\begin{bmatrix} 5e^{5t} & 6e^{2t} \\ -10e^{5t} & -2e^{2t} \end{bmatrix}$.

37. (a) $\begin{bmatrix} (1/2)t^2 + c_1 & e^t + c_2 \\ t + c_3 & e^t + c_4 \end{bmatrix}$.

(b) $\begin{bmatrix} \sin 1 & -1 + \cos 1 \\ 1 - \cos 1 & \sin 1 \end{bmatrix}$.

(c) $\begin{bmatrix} (1+e^t)\cos t + (e^t - t)\sin t & (e^t - t)\cos t - (e^t + 1)\sin t \\ (e^t - 1)\sin t + e^t \cos t & (e^t - 1)\cos t - e^t \sin t \end{bmatrix}$.

Exercises 10.3, page 492

1. $\begin{bmatrix} x'(t) \\ y'(t) \end{bmatrix} = \begin{bmatrix} 3 & -1 \\ -1 & 2 \end{bmatrix} \begin{bmatrix} x(t) \\ y(t) \end{bmatrix} + \begin{bmatrix} t^2 \\ e^t \end{bmatrix}$.

3. $\begin{bmatrix} dx/dt \\ dy/dt \\ dz/dt \end{bmatrix} = \begin{bmatrix} t^2 & -1 & -1 \\ 0 & 0 & e^t \\ t & -1 & 3 \end{bmatrix} \begin{bmatrix} x \\ y \\ t \end{bmatrix} + \begin{bmatrix} t \\ 5 \\ -e^t \end{bmatrix}$.

5. $\begin{bmatrix} x_1'(t) \\ x_2'(t) \end{bmatrix} = \begin{bmatrix} 0 & 1 \\ 10 & 3 \end{bmatrix} \begin{bmatrix} x_1(t) \\ x_2(t) \end{bmatrix} + \begin{bmatrix} 0 \\ \sin t \end{bmatrix}$.

7. $\begin{bmatrix} x_1'(t) \\ x_2'(t) \\ x_3'(t) \\ x_4'(t) \end{bmatrix} = \begin{bmatrix} 0 & 1 & 0 & 0 \\ 0 & 0 & 1 & 0 \\ 0 & 0 & 0 & 1 \\ -1 & 0 & 0 & 0 \end{bmatrix} \begin{bmatrix} x_1(t) \\ x_2(t) \\ x_3(t) \\ x_4(t) \end{bmatrix} + \begin{bmatrix} 0 \\ 0 \\ 0 \\ t^2 \end{bmatrix}$.

9. $x_1'(t) = 5x_1(t) + 2e^{-2t}$,
$x_2'(t) = -2x_1(t) + 4x_2(t) - 3e^{-2t}$.

11. $x_1'(t) = x_1(t) + x_3(t) + e^t$;
$x_2'(t) = -x_1(t) + 2x_2(t) + 5x_3(t) + t$;
$x_3'(t) = 5x_2(t) + x_3(t)$.

13. LI. **15.** LI. **17.** LI.

19. Not a fundamental solution set.

21. Yes; $\begin{bmatrix} e^{-t} & e^t & e^{3t} \\ 2e^{-t} & 0 & -e^{3t} \\ e^{-t} & e^t & 2e^{3t} \end{bmatrix}$;

$c_1 \begin{bmatrix} e^{-t} \\ 2e^{-t} \\ e^{-t} \end{bmatrix} + c_2 \begin{bmatrix} e^t \\ 0 \\ e^t \end{bmatrix} + c_3 \begin{bmatrix} e^{3t} \\ -e^{3t} \\ 2e^{3t} \end{bmatrix}$.

23. $\dfrac{3}{2}\begin{bmatrix} te^t \\ te^t \end{bmatrix} - \dfrac{1}{4}\begin{bmatrix} e^t \\ 3e^t \end{bmatrix} + \begin{bmatrix} t \\ 2t \end{bmatrix} - \begin{bmatrix} 0 \\ 1 \end{bmatrix} + c_1 \begin{bmatrix} e^t \\ e^t \end{bmatrix} + c_2 \begin{bmatrix} e^{-t} \\ 3e^{-t} \end{bmatrix}.$

27. $\mathbf{X}^{-1}(t) = \begin{bmatrix} 0 & (1/4)e^t & -(1/4)e^t \\ -(1/5)e^{2t} & (4/5)e^{2t} & -(2/5)e^{2t} \\ (1/5)e^{-3t} & (9/20)e^{-3t} & (3/20)e^{-3t} \end{bmatrix};$

$\mathbf{x}(t) = \begin{bmatrix} -(3/2)e^{-t} + (3/5)e^{-2t} - (1/10)e^{3t} \\ (1/4)e^{-t} - (1/5)e^{-2t} - (1/20)e^{3t} \\ (5/4)e^{-t} - (1/5)e^{-2t} - (1/20)e^{3t} \end{bmatrix}.$

35. (b) $2e^{6t}$. **(e)** $2e^{6t}$; same.

Exercises 10.4, page 502

1. Eigenvalues are $r_1 = 0$ and $r_2 = -5$ with associated eigenvectors $\mathbf{u}_1 = \begin{bmatrix} s \\ 2s \end{bmatrix}$ and $\mathbf{u}_2 = \begin{bmatrix} 2s \\ -s \end{bmatrix}.$

3. Eigenvalues are $r_1 = 2$ and $r_2 = 3$ with associated eigenvectors $\mathbf{u}_1 = s\begin{bmatrix} 1 \\ -1 \end{bmatrix}$ and $\mathbf{u}_2 = s\begin{bmatrix} 1 \\ -2 \end{bmatrix}.$

5. Eigenvalues are $r_1 = 1, r_2 = 2,$ and $r_3 = -2$ with associated eigenvectors $\mathbf{u}_1 = s\begin{bmatrix} 1 \\ 0 \\ 0 \end{bmatrix},$ $\mathbf{u}_2 = s\begin{bmatrix} 0 \\ 1 \\ 1 \end{bmatrix},$ and $\mathbf{u}_3 = s\begin{bmatrix} 0 \\ -1 \\ 1 \end{bmatrix}.$

7. Eigenvalues are $r_1 = 1, r_2 = 2,$ and $r_3 = 5$ with associated eigenvectors $\mathbf{u}_1 = s\begin{bmatrix} 2 \\ -3 \\ 2 \end{bmatrix},$ $\mathbf{u}_2 = s\begin{bmatrix} 0 \\ -1 \\ 1 \end{bmatrix},$ and $\mathbf{u}_3 = s\begin{bmatrix} 0 \\ 1 \\ 2 \end{bmatrix}.$

9. $c_1 e^{3t/2}\begin{bmatrix} 3 \\ 10 \end{bmatrix} + c_2 e^{t/2}\begin{bmatrix} 1 \\ 2 \end{bmatrix}.$

11. $c_1 e^{-t}\begin{bmatrix} -2 \\ 1 \\ 1 \end{bmatrix} + c_2 e^{-3t}\begin{bmatrix} 0 \\ -1 \\ 1 \end{bmatrix} + c_3 e^{5t}\begin{bmatrix} 1 \\ 1 \\ 1 \end{bmatrix}.$

13. $c_1 e^{-t}\begin{bmatrix} 3 \\ 0 \\ -2 \end{bmatrix} + c_2 e^{t}\begin{bmatrix} 1 \\ -6 \\ 4 \end{bmatrix} + c_3 e^{4t}\begin{bmatrix} 1 \\ 0 \\ 1 \end{bmatrix}.$

15. $\begin{bmatrix} e^{3t} & e^{-3t} \\ 4e^{3t} & -2e^{-3t} \end{bmatrix}.$

17. $\begin{bmatrix} e^t & e^{2t} & e^{4t} \\ e^t & 2e^{2t} & 4e^{4t} \\ e^t & 4e^{2t} & 16e^{4t} \end{bmatrix}.$

19. $\begin{bmatrix} e^{2t} & e^{-t} & e^{3t} & -e^{7t} \\ 0 & -3e^{-t} & 0 & e^{7t} \\ 0 & 0 & e^{3t} & 2e^{7t} \\ 0 & 0 & 0 & 8e^{7t} \end{bmatrix}.$

21. $x = -c_1 e^t - 2c_2 e^{2t} - c_3 e^{3t},$
$y = c_1 e^t + c_2 e^{2t} + c_3 e^{3t},$
$z = 2c_1 e^t + 4c_2 e^{2t} + 4c_3 e^{3t}.$

23. $\begin{bmatrix} e^{-0.3473t} & e^{0.5237t} & 0.0286e^{-7.0764t} \\ -0.3157e^{-0.3473t} & 0.4761e^{0.5237t} & -0.1837e^{-7.0764t} \\ 0.0844e^{-0.3473t} & 0.1918e^{0.5237t} & e^{-7.0764t} \end{bmatrix}.$

25. $\begin{bmatrix} 0.0251e^{3.4142t} & -0.2361e^{-1.6180t} & e^{0.6180t} & e^{0.5858t} \\ 0.0858e^{3.4142t} & 0.3820e^{-1.6180t} & 0.6180e^{0.6180t} & 0.5858e^{0.5858t} \\ 0.2929e^{3.4142t} & -0.6180e^{-1.6180t} & 0.3820e^{0.6180t} & 0.3431e^{0.5858t} \\ e^{3.4142t} & e^{-1.6180t} & 0.2361e^{0.6180t} & 0.2010e^{0.5858t} \end{bmatrix}.$

27. $\begin{bmatrix} 2e^{4t} + e^{-2t} \\ 2e^{4t} - e^{-2t} \end{bmatrix}.$

29. $\begin{bmatrix} -3e^{-t} + e^{5t} \\ -2e^{-t} - e^{5t} \\ e^{-t} + e^{5t} \end{bmatrix}.$

31. (b) $\mathbf{x}_1(t) = e^{-t}\begin{bmatrix} 1 \\ 2 \end{bmatrix}.$ **(c)** $\mathbf{x}_2(t) = te^{-t}\begin{bmatrix} 1 \\ 2 \end{bmatrix} + e^{-t}\begin{bmatrix} 1 \\ 1 \end{bmatrix}.$

33. (b) $\mathbf{x}_1(t) = e^{2t}\begin{bmatrix}1\\0\\0\end{bmatrix}$. **(c)** $\mathbf{x}_2(t) = te^{2t}\begin{bmatrix}1\\0\\0\end{bmatrix} + e^{2t}\begin{bmatrix}0\\1\\0\end{bmatrix}$.

(d) $\mathbf{x}_3(t) = \frac{t^2}{2}e^{2t}\begin{bmatrix}1\\0\\0\end{bmatrix} + te^{2t}\begin{bmatrix}0\\1\\0\end{bmatrix} + e^{2t}\begin{bmatrix}0\\-6/5\\1/5\end{bmatrix}$.

35. (b) $\mathbf{x}_1(t) = e^{t}\begin{bmatrix}-1\\1\\0\end{bmatrix}$; $\mathbf{x}_2(t) = e^{t}\begin{bmatrix}-1\\0\\1\end{bmatrix}$.

(c) $\mathbf{x}_3(t) = te^{t}\begin{bmatrix}1\\1\\-2\end{bmatrix} + e^{t}\begin{bmatrix}1\\0\\0\end{bmatrix}$.

39. $c_1\begin{bmatrix}3t^2\\t^2\end{bmatrix} + c_2\begin{bmatrix}t^4\\t^4\end{bmatrix}$.

41. $x_1(t) = (25/2)(e^{-3t/25} + e^{-t/25})$,
$x_2(t) = (25/4)(e^{-t/25} - e^{-3t/25})$.

43. (a) $r_1 = 2.39091, \quad r_2 = -2.94338, \quad r_3 = 3.55247$.

(b) $\mathbf{u}_1 = \begin{bmatrix}1\\-2.64178\\-9.31625\end{bmatrix}$; $\mathbf{u}_2 = \begin{bmatrix}1\\-1.16825\\0.43862\end{bmatrix}$;

$\mathbf{u}_3 = \begin{bmatrix}1\\0.81004\\-0.12236\end{bmatrix}$.

(c) $c_1e^{r_1t}\mathbf{u}_1 + c_2e^{r_2t}\mathbf{u}_2 + c_3e^{r_3t}\mathbf{u}_3$, where the r_i's and the $\mathbf{u}_i$'s are given in parts (a) and (b).

Exercises 10.5, page 508

1. $c_1\begin{bmatrix}2\cos 2t\\\cos 2t + \sin 2t\end{bmatrix} + c_2\begin{bmatrix}2\sin 2t\\\sin 2t - \cos 2t\end{bmatrix}$.

3. $c_1e^t\cos t\begin{bmatrix}-1\\1\\0\end{bmatrix} - c_1e^t\sin t\begin{bmatrix}-2\\0\\1\end{bmatrix} + c_2e^t\sin t\begin{bmatrix}-1\\1\\0\end{bmatrix} + c_2e^t\cos t\begin{bmatrix}-2\\0\\1\end{bmatrix} + c_3e^t\begin{bmatrix}1\\0\\0\end{bmatrix}$.

5. $\begin{bmatrix}e^{-t}\cos 4t & e^{-t}\sin 4t\\2e^{-t}\sin 4t & -2e^{-t}\cos 4t\end{bmatrix}$.

7. $\begin{bmatrix}1 & \cos t & \sin t\\0 & -\cos t & -\sin t\\0 & -\sin t & \cos t\end{bmatrix}$.

9.
$$\begin{bmatrix}1 & \sqrt{3}\sin(\sqrt{3}t) - \cos(\sqrt{3}t) & -\sin(\sqrt{3}t) - \sqrt{3}\cos(\sqrt{3}t)\\-1 & \sqrt{3}\sin(\sqrt{3}t) + \cos(\sqrt{3}t) & \sin(\sqrt{3}t) - \sqrt{3}\cos(\sqrt{3}t)\\1 & 2\cos(\sqrt{3}t) & 2\sin(\sqrt{3}t)\end{bmatrix}.$$

11.
$$\begin{bmatrix}e^t\sin t - e^t\cos t & -e^t\sin t - e^t\cos t & -\cos t & -\sin t\\2e^t\sin t & -2e^t\cos t & \sin t & -\cos t\\2e^t\sin t + 2e^t\cos t & 2e^t\sin t - 2e^t\cos t & \cos t & \sin t\\4e^t\cos t & 4e^t\sin t & -\sin t & \cos t\end{bmatrix}.$$

13. (a) $\begin{bmatrix}e^{-2t}(\sin t - \cos t)\\-2e^{-2t}\sin t\end{bmatrix}$;

(b) $\begin{bmatrix}-e^{-2(t-\pi)}\cos t\\e^{-2(t-\pi)}(\cos t - \sin t)\end{bmatrix}$.

(c) $\begin{bmatrix}e^{-2(t+2\pi)}(2\cos t - 3\sin t)\\e^{-2(t+2\pi)}(\cos t + 5\sin t)\end{bmatrix}$.

(d) $\begin{bmatrix}e^{\pi-2t}\cos t\\e^{\pi-2t}(\sin t - \cos t)\end{bmatrix}$.

17. $c_1\begin{bmatrix}t^{-1}\\0\\-2t^{-1}\end{bmatrix} + c_2\begin{bmatrix}t^{-1}\cos(\ln t)\\t^{-1}\sin(\ln t)\\-t^{-1}\cos(\ln t)\end{bmatrix} + c_3\begin{bmatrix}t^{-1}\sin(\ln t)\\-t^{-1}\cos(\ln t)\\-t^{-1}\sin(\ln t)\end{bmatrix}$.

19. $\frac{\sqrt{9-\sqrt{17}}}{2\sqrt{2\pi}} \approx 0.249$; $\frac{\sqrt{9+\sqrt{17}}}{2\sqrt{2\pi}} \approx 0.408$.

21. $I_1 = (5/6)e^{-2t}\sin 3t$,
$I_2 = -(10/13)e^{-2t}\cos 3t + (25/78)e^{-2t}\sin 3t$,
$I_3 = (10/13)e^{-2t}\cos 3t + (20/39)e^{-2t}\sin 3t$.

Exercises 10.6, page 515

1. $c_1e^{7t}\begin{bmatrix}1\\1\end{bmatrix} + c_2e^{2t}\begin{bmatrix}1\\-4\end{bmatrix} + \begin{bmatrix}2\\-1\end{bmatrix}$.

3. $c_1e^{-3t}\begin{bmatrix}1\\1\\-1\end{bmatrix} + c_2e^{3t}\begin{bmatrix}1\\0\\1\end{bmatrix} + c_3e^{3t}\begin{bmatrix}-1\\1\\0\end{bmatrix} + e^t\begin{bmatrix}1\\0\\-1\end{bmatrix}$.

5. $\boldsymbol{x}_p = e^{-2t}/[\boldsymbol{a} + t\boldsymbol{b}]$. **7.** $\boldsymbol{x}_p = t\boldsymbol{a} + \boldsymbol{b} + \sin 3t\boldsymbol{c} + \cos 3t\boldsymbol{d}$.

9. $\boldsymbol{x}_p = e^{2t}\boldsymbol{a} + e^{3t}\boldsymbol{b}$.

11. $c_1\begin{bmatrix}\sin t\\\cos t\end{bmatrix} + c_2\begin{bmatrix}\cos t\\-\sin t\end{bmatrix} + \begin{bmatrix}0\\-1\end{bmatrix}$.

13. $c_1e^t\begin{bmatrix}1\\-1\end{bmatrix} + c_2e^{-t}\begin{bmatrix}1\\-3\end{bmatrix} + \begin{bmatrix}5te^t + (3/2)e^t\\-5te^t + (9/2)e^t\end{bmatrix}$.

15. $c_1 \begin{bmatrix} 1 \\ 2 \end{bmatrix} + c_2 e^{-5t} \begin{bmatrix} -2 \\ 1 \end{bmatrix} + \begin{bmatrix} \ln|t| + (8/5)t - 8/25 \\ 2\ln|t| + (16/5)t + 4/25 \end{bmatrix}.$

17. $c_1 e^{2t} \begin{bmatrix} 1 \\ 1 \\ 1 \end{bmatrix} + c_2 e^{-t} \begin{bmatrix} -1 \\ 0 \\ 1 \end{bmatrix} + c_3 e^{-t} \begin{bmatrix} -1 \\ 1 \\ 0 \end{bmatrix} + e^t \begin{bmatrix} 1 \\ -1 \\ -1 \end{bmatrix}.$

19. $\begin{bmatrix} c_1 \cos t + c_2 \sin t + 1 \\ -c_1 \sin t + c_2 \cos t - t \\ c_3 e^t + c_4 e^{-t} - (1/4)e^{-t} - t + (1/2)te^{-t} \\ c_3 e^t - c_4 e^{-t} - (1/4)e^{-t} - 1 - (1/2)te^{-t} \end{bmatrix}.$

21. (a) $\begin{bmatrix} 4te^t + 5e^t \\ 2te^t + 4e^t \end{bmatrix}.$

(b) $\begin{bmatrix} -2e^{t-1} + 2e^{2(t-1)} - 3e^{2t-1} + (4t-1)e^t \\ -e^{t-1} + 2e^{2(t-1)} - 3e^{2t-1} + (2t+1)e^t \end{bmatrix}.$

(c) $\begin{bmatrix} (-20+2e^{-5})e^t + (-3e^{-5}-e^{-10})e^{2t} + (4t+3)e^t \\ (-10+e^{-5})e^t + (-3e^{-5}-e^{-10})e^{2t} + (2t+3)e^t \end{bmatrix}.$

(d) $\begin{bmatrix} 2e^{t+1} - 3e^{2t+1} + (4t+7)e^t \\ e^{t+1} - 3e^{2t+1} + (2t+5)e^t \end{bmatrix}.$

23. $x = (c_1 + c_2 t)e^t + c_3 e^{-t} - t^2 - 4t - 6,$
$y = -c_2 e^t - 2c_3 e^{-t} - t^2 - 2t - 3.$

25. (a) $\left\{ \begin{bmatrix} e^t \\ e^t \end{bmatrix}, \begin{bmatrix} e^{2t} \\ 2e^{2t} \end{bmatrix} \right\}.$ **(c)** $te^t \begin{bmatrix} 2 \\ 2 \end{bmatrix} + e^t \begin{bmatrix} 0 \\ 1 \end{bmatrix}.$

(d) $c_1 e^t \begin{bmatrix} 1 \\ 1 \end{bmatrix} + c_2 e^{2t} \begin{bmatrix} 1 \\ 2 \end{bmatrix} + te^t \begin{bmatrix} 2 \\ 2 \end{bmatrix} + e^t \begin{bmatrix} 0 \\ 1 \end{bmatrix}.$

27. $c_1 e^{2t} \begin{bmatrix} 1 \\ 1 \\ 1 \end{bmatrix} + c_2 e^{-t} \begin{bmatrix} -1 \\ 0 \\ 1 \end{bmatrix} + c_3 e^{-t} \begin{bmatrix} -1 \\ 1 \\ 0 \end{bmatrix}$
$+ te^{-t} \begin{bmatrix} 1 \\ 0 \\ -1 \end{bmatrix} + e^{-t} \begin{bmatrix} 1 \\ 0 \\ 0 \end{bmatrix} + \begin{bmatrix} 0 \\ 0 \\ 1 \end{bmatrix}.$

29.
$c_1 t \begin{bmatrix} 1 \\ 1 \end{bmatrix} + c_2 t^{-1} \begin{bmatrix} 1 \\ 3 \end{bmatrix} + \begin{bmatrix} -(3/4)t^{-1} - (1/2)t^{-1}\ln t + 1 \\ -(3/4)t^{-1} - (3/2)t^{-1}\ln t + 2 \end{bmatrix}.$

33. $I_1 = I_2 + 3I_5,$
$I_3 = 3I_5,$
$I_4 = 2I_5,$
where
$I_2 = (1/(20\sqrt{817}))[(13 - \sqrt{817})e^{-(31+\sqrt{817})5t/2}$
$- (13 + \sqrt{817})e^{(-31+\sqrt{817})5t/2}] + 1/10,$
$I_5 = (3/(40\sqrt{817}))[(31 - \sqrt{817})e^{-(31+\sqrt{817})5t/2}$
$- (31 + \sqrt{817})e^{(-31+\sqrt{817})5t/2}] + 3/20.$

35. $x_1 = (2050 - (5845/4)e^{-3t/50} - (2355/4)e^{-7t/50})/21,$
$x_2 = (1850 - (5845/2)e^{-3t/50} + (2355/2)e^{-7t/50})/21.$
Tank A will have the higher concentration. The limiting concentration in tank A is 41/21 kg/L and in tank B is 37/21 kg/L.

Exercises 10.7, page 524

1. (a) $r = 3; \quad k = 2.$ **(b)** $e^{3t} \begin{bmatrix} 1 & -2t \\ 0 & 1 \end{bmatrix}.$

3. (a) $r = -1; \quad k = 3.$

(b) $e^{-t} \begin{bmatrix} 1 + 3t - (3/2)t^2 & t & -t + (1/2)t^2 \\ -3t & 1 & t \\ 9t - (9/2)t^2 & 3t & 1 - 3t + (3/2)t^2 \end{bmatrix}.$

5. (a) $r = -2; \quad k = 2.$ **(b)** $e^{-2t} \begin{bmatrix} 1 & 0 & 0 \\ 4t & 1 & 0 \\ t & 0 & 1 \end{bmatrix}.$

7. $\begin{bmatrix} \cos t & \sin t \\ -\sin t & \cos t \end{bmatrix}.$

9. $\frac{1}{2} \begin{bmatrix} e^t + \cos t - \sin t & 2\sin t & e^t - \cos t - \sin t \\ e^t - \sin t - \cos t & 2\cos t & e^t + \sin t - \cos t \\ e^t - \cos t + \sin t & -2\sin t & e^t + \cos t + \sin t \end{bmatrix}.$

11.
$\frac{1}{25} \begin{bmatrix} -4 + 29e^{5t} - 20te^{5t} & 20 - 20e^{5t} & -8 + 8e^{5t} - 40te^{5t} \\ -5 + 5e^{5t} & 25 & -10 + 10e^{5t} \\ 2 - 2e^{5t} + 10te^{5t} & -10 + 10e^{5t} & 4 + 21e^{5t} + 20te^{5t} \end{bmatrix}.$

13.
$\begin{bmatrix} (1 - t + t^2/2)e^t & (t - t^2)e^t & (t^2/2)e^t & 0 & 0 \\ (t^2/2)e^t & (1 - t + t^2)e^t & (t + t^2/2)e^t & 0 & 0 \\ (t + t^2/2)e^t & (-3t - t^2)e^t & (1 + 2t + t^2/2)e^t & 0 & 0 \\ 0 & 0 & 0 & \cos t & \sin t \\ 0 & 0 & 0 & -\sin t & \cos t \end{bmatrix}.$

15. $\begin{bmatrix} (1 + t + t^2/2)e^{-t} & (t + t^2)e^{-t} & (t^2/2)e^{-t} \\ (-t^2/2)e^{-t} & (1 + t - t^2)e^{-t} & (t - t^2/2)e^{-t} \\ (-t + t^2/2)e^{-t} & (-3t + t^2)e^{-t} & (1 - 2t + t^2/2)e^{-t} \\ 0 & 0 & 0 \\ 0 & 0 & 0 \end{bmatrix}$

$\begin{bmatrix} 0 & 0 \\ 0 & 0 \\ 0 & 0 \\ (1 + 2t)e^{-2t} & te^{-2t} \\ -4te^{-2t} & (1 - 2t)e^{-2t} \end{bmatrix}.$

17. $c_1e^{-t}\begin{bmatrix}1\\-1\\1\end{bmatrix}+c_2e^{-t}\begin{bmatrix}-t\\-1+t\\2-t\end{bmatrix}+c_3e^{-2t}\begin{bmatrix}1\\-2\\4\end{bmatrix}$.

19. $c_1e^{2t}\begin{bmatrix}3\\6\\1\\1\end{bmatrix}+c_2e^{t}\begin{bmatrix}0\\1\\0\\0\end{bmatrix}+c_3e^{t}\begin{bmatrix}1\\t\\0\\0\end{bmatrix}+c_4e^{t}\begin{bmatrix}2t\\t+t^2\\0\\1\end{bmatrix}$.

21. $\begin{bmatrix}e^{-2t}\\4te^{-2t}+e^{-2t}\\te^{-2t}-e^{-2t}\end{bmatrix}$.

23. $e^{-t}\begin{bmatrix}3t\\3\\9t\end{bmatrix}+e^{\mathbf{A}t}\begin{bmatrix}2-e^t(t^2-2t+2)\\1+e^t(t-1)\\6-3e^t(t^2-2t+2)\end{bmatrix}$, where $e^{\mathbf{A}t}$ is the matrix in the answer to Problem 3.

Review Problems, page 528

1. $c_1e^{3t}\begin{bmatrix}1\\1\end{bmatrix}+c_2e^{4t}\begin{bmatrix}3\\2\end{bmatrix}$.

3. $c_1e^{-t}\begin{bmatrix}1\\-1\\0\\0\end{bmatrix}+c_2e^{-t}\begin{bmatrix}0\\0\\1\\-1\end{bmatrix}+c_3e^{3t}\begin{bmatrix}1\\1\\0\\0\end{bmatrix}+c_4e^{3t}\begin{bmatrix}0\\0\\1\\1\end{bmatrix}$.

5. $\begin{bmatrix}e^{2t} & e^{3t}\\-e^{2t} & -2e^{3t}\end{bmatrix}$.

7. $c_1e^{-t}\begin{bmatrix}1\\-2\end{bmatrix}+c_2e^{3t}\begin{bmatrix}1\\2\end{bmatrix}+\begin{bmatrix}-1/3\\-14/3\end{bmatrix}$.

9. $c_1e^{-t}\begin{bmatrix}1\\0\\3\end{bmatrix}+c_2e^{-t}\begin{bmatrix}t\\1\\3t\end{bmatrix}+c_3e^{-t}\begin{bmatrix}-t+(1/2)t^2\\t\\1-3t+(3/2)t^2\end{bmatrix}+\begin{bmatrix}2+t\\7-3t\\10\end{bmatrix}$.

11. $\begin{bmatrix}3e^t-2e^{2t}\\3e^t-4e^{2t}\end{bmatrix}$.

13. $c_1t^{-1}\begin{bmatrix}-3\\1\\0\end{bmatrix}+c_2t^{-1}\begin{bmatrix}-1\\0\\1\end{bmatrix}+c_3t^4\begin{bmatrix}1\\1\\1\end{bmatrix}$.

15. $\begin{bmatrix}(1/2)e^t+(2/3)e^{5t}-(1/6)e^{-t} & -e^t+(2/3)e^{5t}+(1/3)e^{-t} & -(1/2)e^t+(2/3)e^{5t}-(1/6)e^{-t}\\(1/3)e^{5t}-(1/3)e^{-t} & (1/3)e^{5t}+(2/3)e^{-t} & (1/3)e^{5t}-(1/3)e^{-t}\\-(1/2)e^t+(1/2)e^{-t} & e^t-e^{-t} & (1/2)e^t+(1/2)e^{-t}\end{bmatrix}$.

▶ CHAPTER 11

Exercises 11.2, page 543

1. $y=[4/(e-e^{-1})](e^{-x}-e^x)$. **3.** $y\equiv 0$.
5. $y=e^x+2x-1$. **7.** $y=\cos x+c\sin x$; c arbitrary.
9. $\lambda_n=(2n-1)^2/4$ and $y_n=c_n\sin[(2n-1)x/2]$, where $n=1,2,3,\ldots$ and the c_n's are arbitrary.
11. $\lambda_n=n^2$, $n=0,1,2,\ldots$; $y_0=a_0$ and $y_n=a_n\cos nx+b_n\sin nx$, $n=1,2,3,\ldots$, where a_0, a_n, and b_n are arbitrary.
13. The eigenvalues are the roots of $\tan(\sqrt{\lambda_n}\pi)+\sqrt{\lambda_n}=0$, where $\lambda_n>0$. For n large, $\lambda_n\approx(2n-1)^2/4$, n is a positive integer. The eigenfunctions are $y_n=c_n[\sin(\sqrt{\lambda_n}x)+\sqrt{\lambda_n}\cos(\sqrt{\lambda_n}x)]$, where the c_n's are arbitrary.
15. $u(x,t)=e^{-3t}\sin x-6e^{-48t}\sin 4x$.
17. $u(x,t)=e^{-3t}\sin x-7e^{-27t}\sin 3x+e^{-75t}\sin 5x$.
19. $u(x,t)=3\cos 6t\sin 2x+12\cos 39t\sin 13x$.
21. $u(x,t)=6\cos 6t\sin 2x+2\cos 18t\sin 6x+(11/27)\sin 27t\sin 9x-(14/45)\sin 45t\sin 15x$.
23. $u(x,t)=\sum_{n=1}^{\infty}n^{-2}e^{-2\pi^2n^2t}\sin n\pi x$.
25. If $K>0$, then $T(t)$ becomes unbounded as $t\to\infty$, and so the temperature $u(x,t)=X(x)T(t)$ becomes unbounded at each position x. Since the temperature must remain bounded for all time, $K\not>0$.
33. **(a)** $u(x)\equiv 50$. **(b)** $u(x)=30x/L+10$.

Exercises 11.3, page 557

1. Odd. **3.** Even. **5.** Neither.
9. $f(x)\sim\sum_{n=1}^{\infty}\frac{2(-1)^{n+1}}{n}\sin nx$.

11. $f(x) \sim 1 + \sum\limits_{n=1}^{\infty} \left[\dfrac{2}{\pi^2 n^2}(-1+(-1)^n)\cos\dfrac{n\pi x}{2} \right.$

$\left. + \dfrac{1}{\pi n}((-1)^{n+1}-1)\sin\dfrac{n\pi x}{2} \right].$

13. $f(x) \sim \dfrac{1}{3} + \sum\limits_{n=1}^{\infty} \dfrac{4(-1)^n}{n^2\pi^2}\cos n\pi x.$

15. $f(x) \sim [(\sinh \pi)/\pi]\left(1 + \sum\limits_{n=1}^{\infty}\left[\dfrac{(-1)^n}{1+n^2}\cos nx \right.\right.$

$\left.\left. + \dfrac{(-1)^{n+1}n}{1+n^2}\sin nx\right]\right).$

17. The 2π periodic function $g(x)$, where

$$g(x) = \begin{cases} x, & -\pi < x < \pi, \\ 0, & x = \pm\pi. \end{cases}$$

19. The 4 periodic function $g(x)$, where

$$g(x) = \begin{cases} 1, & -2 < x < 0, \\ x, & 0 < x < 2, \\ 1/2, & x = 0, \\ 3/2, & x = \pm 2. \end{cases}$$

21. The 2 periodic function $g(x)$, where

$$g(x) = x^2, \quad -1 \le x \le 1.$$

23. The 2π periodic function $g(x)$, where

$$g(x) = \begin{cases} e^x, & -\pi < x < \pi, \\ (e^\pi + e^{-\pi})/2, & x = \pm\pi. \end{cases}$$

25. (a) $F(x) = (x^2 - \pi^2)/2.$ **(b)** $F(x) = |x| - \pi.$

27. $f(x) \sim \sum\limits_{n=1}^{\infty} \dfrac{2}{(2n-1)\pi}\left[(-1)^{n+1}\cos\left(\dfrac{(2n-1)\pi x}{2}\right)\right.$

$\left. + \sin\left(\dfrac{(2n-1)\pi x}{2}\right)\right].$

29. $a_0 = 0; \quad a_1 = 3/2; \quad a_2 = 0.$

Exercises 11.4, page 565

1. (a) The π periodic function $\tilde{f}(x)$, where

$$\tilde{f}(x) = x^2, \quad 0 < x < \pi.$$

(b) The 2π periodic function $f_o(x)$, where

$$f_o(x) = \begin{cases} x^2, & 0 < x < \pi, \\ -x^2, & -\pi < x < 0. \end{cases}$$

(c) The 2π periodic function $f_e(x)$, where

$$f_e(x) = \begin{cases} x^2, & 0 < x < \pi, \\ x^2, & -\pi < x < 0. \end{cases}$$

3. (a) The π periodic function $\tilde{f}(x)$, where

$$\tilde{f}(x) = \begin{cases} 0, & 0 < x < \pi/2, \\ 1, & \pi/2 < x < \pi. \end{cases}$$

(b) The 2π periodic function $f_o(x)$, where

$$f_o(x) = \begin{cases} -1, & -\pi < x < -\pi/2, \\ 0, & -\pi/2 < x < 0, \\ 0, & 0 < x < \pi/2, \\ 1, & \pi/2 < x < \pi. \end{cases}$$

(c) The 2π periodic function $f_e(x)$, where

$$f_e(x) = \begin{cases} 1, & -\pi < x < -\pi/2, \\ 0, & -\pi/2 < x < 0, \\ 0, & 0 < x < \pi/2, \\ 1, & \pi/2 < x < \pi. \end{cases}$$

5. $f(x) \sim -\dfrac{4}{\pi}\sum\limits_{k=1}^{\infty} \dfrac{1}{2k-1}\sin(2k-1)\pi x.$

7. $f(x) \sim \sum\limits_{n=1}^{\infty}\left[\dfrac{2\pi(-1)^{n+1}}{n} + \dfrac{4}{\pi n^3}((-1)^n - 1)\right]\sin nx.$

9. $f(x) \sim \sum\limits_{k=0}^{\infty} \dfrac{8}{(2k+1)^3\pi^3}\sin(2k+1)\pi x.$

11. $f(x) \sim \dfrac{\pi}{2} + \dfrac{4}{\pi}\sum\limits_{k=1}^{\infty} \dfrac{1}{(2k-1)^2}\cos(2k-1)x.$

13. $f(x) \sim e - 1 + 2\sum\limits_{n=1}^{\infty} \dfrac{(-1)^n e - 1}{1+\pi^2 n^2}\cos n\pi x.$

15. $f(x) \sim \dfrac{2}{\pi} + \dfrac{2}{\pi}\sum\limits_{k=1}^{\infty}\left(\dfrac{1}{2k+1} + \dfrac{1}{2k-1}\right)\cos 2kx.$

17. $u(x,t) = \dfrac{2}{\pi}\sum\limits_{k=1}^{\infty}\left[\dfrac{2}{2k-1} - \dfrac{1}{2k+1} - \dfrac{1}{2k-3}\right]$

$\times e^{-5(2k-1)^2 t}\sin(2k-1)x.$

Exercises 11.5, page 575

1. $u(x,t) = \sum\limits_{n=1}^{\infty} \dfrac{8(-1)^{n+1}-4}{\pi^3 n^3}e^{-5\pi^2 n^2 t}\sin n\pi x.$

3. $u(x,t) = \dfrac{\pi}{2} - \sum\limits_{k=0}^{\infty} \dfrac{4}{\pi(2k+1)^2}$

$\times e^{-3(2k+1)^2 t}\cos(2k+1)x.$

5. $u(x,t) = \dfrac{2(e^\pi - 1)}{\pi} + \sum\limits_{n=1}^{\infty} \dfrac{2e^\pi(-1)^n - 2}{\pi(1+n^2)}e^{-n^2 t}\cos nx.$

7. $u(x,t) = 5 + \dfrac{5}{\pi}x - \dfrac{30}{\pi}e^{-2t}\sin x + \dfrac{5}{\pi}e^{-8t}\sin 2x$

$+ \left(1 - \dfrac{10}{\pi}\right)e^{-18t}\sin 3x$

$+ \dfrac{5}{2\pi}e^{-32t}\sin 4x - \left(1 + \dfrac{6}{\pi}\right)e^{-50t}\sin 5x$

$+ \sum\limits_{n=6}^{\infty} \dfrac{10}{\pi n}[2(-1)^n - 1]e^{-2n^2 t}\sin nx.$

9. $u(x, t) = \dfrac{e^{-\pi} - 1}{\pi} x - e^{-x} + 1 + \sum_{n=1}^{\infty} c_n e^{-n^2 t} \sin nx,$

where

$$c_n = \begin{cases} \dfrac{2e^{-\pi} - 2}{\pi n}(-1)^n + \dfrac{2n}{\pi(1 + n^2)}((-1)^{n+1}e^{-\pi} + 1) \\ \qquad + \dfrac{2}{\pi n}[(-1)^n - 1], & n \neq 2, \\ \dfrac{e^{-\pi} - 1}{\pi} + \dfrac{4}{5\pi}(1 - e^{-\pi}) + 1, & n = 2. \end{cases}$$

11. $u(x,t) = \sum_{n=0}^{\infty} a_n e^{-4(n+1/2)^2 t} \cos(n + 1/2)x$, where

$$f(x) = \sum_{n=0}^{\infty} a_n \cos(n + 1/2)x.$$

13. $u(x,t) = \dfrac{\pi^2}{3}x - \dfrac{1}{3}x^3 - 3e^{-2t}\sin x$

$$+ \sum_{n=2}^{\infty} \frac{4(-1)^n}{n^3} e^{-2n^2 t} \sin nx.$$

15. $u(x, y, t) = e^{-52t}\cos 6x \sin 4y - 3e^{-122t}\cos x \sin 11y.$

17. $u(x, y, t) = 2\sum_{n=1}^{\infty} \dfrac{(-1)^{n+1}}{n} e^{-n^2 t} \sin ny.$

19. $C(x,t) = \sum_{n=1}^{\infty} c_n e^{-[L + kn^2\pi^2/a^2]t} \sin\left(\dfrac{n\pi x}{a}\right),$

where

$$c_n = \frac{2}{a}\int_0^a f(x)\sin\left(\frac{n\pi x}{a}\right)dx.$$

Concentration goes to zero as $t \to +\infty$.

Exercises 11.6, page 586

1. $u(x,t) = \dfrac{1}{7\pi}\sin 7\pi t \sin 7\pi x$

$$+ \sum_{k=0}^{\infty} \frac{8}{((2k+1)\pi)^3}\cos(2k+1)\pi t \sin(2k+1)\pi x.$$

3. $u(x,t) = \sum_{n=1}^{\infty} \dfrac{4}{n^2}[2(-1)^{n+1} - 1]\cos 2nt \sin nx.$

5. $u(x,t) = \dfrac{2h_o L^2}{\pi^2 a(L-a)} \sum_{n=1}^{\infty} \dfrac{1}{n^2}\sin\dfrac{n\pi a}{L}\sin\dfrac{n\pi x}{L}\cos\dfrac{n\pi\alpha t}{L}.$

7. $u(x,t) = \cos t \sin x + \dfrac{5}{2}\sin 2t \sin 2x - \dfrac{3}{5}\sin 5t \sin 5x$

$$+ 2\sum_{n=1}^{\infty} \frac{(-1)^{n+1}}{n^3}\left[t - \frac{\sin nt}{n}\right]\sin nx.$$

9. $u(x,t) = \sum_{n=0}^{\infty}\left(a_n \cos\dfrac{(2n+1)\pi\alpha t}{2L}\right.$

$$\left. + b_n \sin\frac{(2n+1)\pi\alpha t}{2L}\right)\sin\frac{(2n+1)\pi x}{2L}, \text{ where}$$

$$f(x) = \sum_{n=0}^{\infty} a_n \sin\frac{(2n+1)\pi x}{2L} \text{ and}$$

$$g(x) = \sum_{n=0}^{\infty} b_n \frac{(2n+1)\pi\alpha}{2L}\sin\frac{(2n+1)\pi x}{2L}.$$

11. $u(x,t) = \sum_{n=1}^{\infty} a_n T_n(t)\sin\dfrac{n\pi x}{L}$, where

$$a_n = \frac{2}{L}\int_0^L f(x)\sin\frac{n\pi x}{L}dx,$$

and

$$T_n(t) = e^{-t/2}\left(\cos\beta_n t + \frac{1}{2\beta_n}\sin\beta_n t\right),$$

where

$$\beta_n = \frac{1}{2L}\sqrt{3L^2 + 4\alpha^2\pi^2 n^2}.$$

13. $u(x,t) = \dfrac{1}{2\alpha}[\sin(x + \alpha t) - \sin(x - \alpha t)]$

$$= \frac{1}{\alpha}\sin\alpha t\cos x.$$

15. $u(x,t) = x + tx.$

17. $u(x,t) = \dfrac{1}{2}\left[e^{-(x+\alpha t)^2} + e^{-(x-\alpha t)^2}\right.$

$$\left. + \frac{\cos(x - \alpha t) - \cos(x + \alpha t)}{\alpha}\right].$$

21. $u(r,t) = \sum_{n=1}^{\infty}[a_n\cos(k_n\alpha t) + b_n\sin(k_n\alpha t)]J_0(k_n r)$

where

$$a_n = \frac{1}{c_n}\int_0^1 f(r)J_0(k_n r)r\,dr,$$

and

$$b_n = \frac{1}{\alpha k_n c_n}\int_0^1 g(r)J_0(k_n r)r\,dr,$$

with

$$c_n = \int_0^1 J_0^2(k_n r)r\,dr.$$

Exercises 11.7, page 598

1. $u(x,y) = \dfrac{4\cos 6x\sinh(6(y-1))}{\sinh(-6)} + \dfrac{\cos 7x\sinh(7(y-1))}{\sinh(-7)}.$

3. $u(x,y) = \sum_{n=1}^{\infty} A_n \sin nx \sinh(ny - n\pi)$, where

$$A_n = \frac{2}{\pi\sinh(-n\pi)}\int_0^{\pi} f(x)\sin nx\,dx.$$

5. $u(x, y) = \dfrac{\cos x \sinh(y - 1)}{\sinh(-1)} - \dfrac{\cos 3x \sinh(3y - 3)}{\sinh(-3)}$

$\qquad + \dfrac{\cos 2x \sinh 2y}{\sinh(2)}$.

7. $u(r, \theta) = \dfrac{\pi}{2} - \displaystyle\sum_{k=0}^{\infty} \dfrac{r^{2k+1}}{(2k+1)^2 \pi 2^{2k-1}} \cos(2k + 1)\theta$.

9. $u(r, \theta) = \dfrac{a_0}{2} + \displaystyle\sum_{n=1}^{\infty} \left(\dfrac{r}{a}\right)^n (a_n \cos n\theta + b_n \sin n\theta)$, where a_0

   is arbitrary, and for $n = 1, 2, 3, \ldots$

$\qquad a_n = \dfrac{a}{\pi n} \displaystyle\int_{-\pi}^{\pi} f(\theta) \cos n\theta \, d\theta,$

$\qquad b_n = \dfrac{a}{n\pi} \displaystyle\int_{-\pi}^{\pi} f(\theta) \sin n\theta \, d\theta.$

11. $u(r, \theta) = \left(\dfrac{1}{3}r - \dfrac{4}{3}r^{-1}\right)\cos \theta + \left(\dfrac{2}{3}r - \dfrac{2}{3}r^{-1}\right)\sin \theta$

$\qquad + \left(-\dfrac{1}{225}r^4 + \dfrac{256}{255}r^{-4}\right)\sin 4\theta.$

13. $u(r, \theta) = \dfrac{a_0}{2} + \displaystyle\sum_{n=1}^{\infty} r^{-n}(a_n \cos n\theta + b_n \sin n\theta)$, where

$\qquad a_n = \dfrac{1}{\pi} \displaystyle\int_{-\pi}^{\pi} f(\theta) \cos n\theta \, d\theta, \qquad n = 0, 1, 2, \ldots,$

   and

$\qquad b_n = \dfrac{1}{\pi} \displaystyle\int_{-\pi}^{\pi} f(\theta) \sin n\theta \, d\theta, \qquad n = 1, 2, 3, \ldots.$

15. $u(r, \theta) = r^3 \sin 3\theta$.

17. $u(r, \theta) = a_0 + b_0 \ln r + \displaystyle\sum_{n=1}^{\infty} [(a_n r^n + b_n r^{-n}) \cos n\theta$

$\qquad + (c_n r^n + d_n r^{-n}) \sin n\theta], \quad$ where

$\qquad a_0 = \dfrac{1}{2\pi} \displaystyle\int_{-\pi}^{\pi} f(\theta) \, d\theta,$

$\qquad b_0 = \dfrac{3}{2\pi} \displaystyle\int_{-\pi}^{\pi} g(\theta) \, d\theta,$

   and for $n = 1, 2, 3, \ldots$

$\qquad a_n + b_n = \dfrac{1}{\pi} \displaystyle\int_{-\pi}^{\pi} f(\theta) \cos n\theta \, d\theta,$

$\qquad n3^{n-1} a_n - n3^{-n-1} b_n = \dfrac{1}{\pi} \displaystyle\int_{-\pi}^{\pi} g(\theta) \cos n\theta \, d\theta,$

   and

$\qquad c_n + d_n = \dfrac{1}{\pi} \displaystyle\int_{-\pi}^{\pi} f(\theta) \sin n\theta \, d\theta,$

$\qquad n3^{n-1} c_n - n3^{-n-1} d_n = \dfrac{1}{\pi} \displaystyle\int_{-\pi}^{\pi} g(\theta) \sin n\theta \, d\theta.$

21. $u(r, \theta, z) = [I_0(r)/I_0(\pi)] \sin z$.

23. **(c)** $\phi(x, y) = 2xy$.

• Index